$$\int u^\alpha \, du = \frac{u^{\alpha+1}}{\alpha+1} + c, \quad a \neq -1$$

$$\int \frac{du}{u} = $$

$$\int \cos u \, du = \sin u + c$$

$$\int \sin u \, du = -\cos u + c$$

$$\int \tan u \, du = -\ln |\cos u| + c$$

$$\int \cot u \, du = \ln |\sin u| + c$$

$$\int \sec^2 u \, du = \tan u + c$$

$$\int \csc^2 u \, du = -\cot u + c$$

$$\int \sec u \, du = \ln |\sec u + \tan u| + c$$

$$\int \cos^2 u \, du = \frac{u}{2} + \frac{1}{4} \sin 2u + c$$

$$\int \sin^2 u \, du = \frac{u}{2} - \frac{1}{4} \sin 2u + c$$

$$\int \frac{du}{1 + u^2} \, du = \tan^{-1} u + c$$

$$\int \frac{du}{\sqrt{1 - u^2}} \, du = \sin^{-1} u + c$$

$$\int \frac{1}{u^2 - 1} \, du = \frac{1}{2} \ln \left| \frac{u - 1}{u + 1} \right| + c$$

$$\int \cosh u \, du = \sinh u + c$$

$$\int \sinh u \, du = \cosh u + c$$

$$\int u \, dv = uv - \int v \, du$$

$$\int u \cos u \, du = u \sin u + \cos u + c$$

$$\int u \sin u \, du = -u \cos u + \sin u + c$$

$$\int u e^u \, du = u e^u - e^u + c$$

$$\int \ln |u| \, du = u \ln |u| - u + c$$

$$\int u \ln |u| \, du = \frac{u^2 \ln |u|}{2} - \frac{u^2}{4} + c$$

$$\int e^{\lambda u} \cos \omega u \, du = \frac{e^{\lambda u}(\lambda \cos \omega u + \omega \sin \omega u)}{\lambda^2 + \omega^2} + c$$

$$\int e^{\lambda u} \sin \omega u \, du = \frac{e^{\lambda u}(\lambda \sin \omega u - \omega \cos \omega u)}{\lambda^2 + \omega^2} + c$$

$$\int \cos \omega_1 u \cos \omega_2 u \, du = \frac{\sin(\omega_1 + \omega_2)u}{2(\omega_1 + \omega_2)} + \frac{\sin(\omega_1 - \omega_2)u}{2(\omega_1 - \omega_2)} + c \quad (\omega_1 \neq \pm\omega_2)$$

$$\int \sin \omega_1 u \sin \omega_2 u \, du = -\frac{\sin(\omega_1 + \omega_2)u}{2(\omega_1 + \omega_2)} + \frac{\sin(\omega_1 - \omega_2)u}{2(\omega_1 - \omega_2)} + c \quad (\omega_1 \neq \pm\omega_2)$$

$$\int \sin \omega_1 u \cos \omega_2 u \, du = -\frac{\cos(\omega_1 + \omega_2)u}{2(\omega_1 + \omega_2)} - \frac{\cos(\omega_1 - \omega_2)u}{2(\omega_1 - \omega_2)} + c \quad (\omega_1 \neq \pm\omega_2)$$

Elementary
Differential Equations

Elementary Differential Equations

WILLIAM F. TRENCH
Trinity University

Brooks/Cole
Thomson Learning™

Australia • Canada • Denmark • Japan • Mexico • New Zealand • Philippines • Puerto Rico
• Singapore • Spain • United Kingdom • United States

Publisher: *Bob Pirtle*
Marketing Team: *Caroline Croley and Debra Johnston*
Editorial Assistant: *Erin Wickersham*
Production Editor: *Janet Hill*
Production Service: *Martha Emry*
Interior Design: *Vernon T. Boes*

Cover Design: *Roger Knox*
Cover Illustration: *Beverly Trench*
Interior Illustration: *Techsetters, Inc.*
Print Buyer: *John Cronin*
Typesetting: *WestWords, Inc.*
Cover Printing: *The Courier Company, Inc.*
Printing/Binding: *The Courier Company, Inc.*

Printed in United States of America

10   9   8   7   6   5   4   3   2   1

Library of Congress Cataloging-in-Publication Data
Trench, William F. [date]
  Elementary differential equations / William F. Trench.
    p.   cm.
  Includes index.
  ISBN 0-534-36841-7
  1. Differential equations.   I. Title.
QA71.T68 2000
515'.35—dc21   99–34044

To Beverly

Contents

9   Linear Higher Order Equations   *438*

10   Linear Systems of Differential Equations   *476*

Preface

Elementary Differential Equations is a textbook for a first course in differential equations, written for students in science, engineering, and mathematics who have completed calculus through partial differentiation. If the syllabus includes Chapter 10 (Linear Systems of Differential Equations), then the student should have some preparation in linear algebra.

In writing this book I have been guided by the following principles:

- An elementary text should be written so that the student can read it with comprehension, and without too much pain. I have tried to put myself in the student's place, and have chosen to err on the side of too much detail rather than not enough.
- An elementary text cannot be better than its exercises. This text includes 1711 numbered exercises, many with several parts. They range in difficulty from routine to very challenging.
- An elementary text should be written in an informal but mathematically accurate way, illustrated by appropriate graphics. Mathematical concepts have been formulated precisely and accurately, in language that the student can understand. I have minimized the number of explicitly stated theorems and definitions, preferring to deal with concepts in a more conversational way, copiously illustrated by 251 completely worked out examples. Where appropriate, concepts and results are depicted in 158 figures.

Although I believe that the computer is an immensely valuable tool for learning, doing, and writing mathematics, the selection and treatment of topics in this text reflects my pedagogical orientation along traditional lines. However, I have incorporated what I believe to be the best use of modern technology, so you can select the level of technology that you want to include in your course. The text includes 393 exercises—identified by the symbols **C** and **L**—that call for graphics or computation and graphics; 62 of these—identified by **L**—are laboratory exercises that require extensive use of technology. In addition, several sections include informal advice on the use of technology. If you prefer not to emphasize technology, then simply ignore these exercises and the accompanying advice.

There are two schools of thought on whether techniques and applications should be treated together or separately. I have chosen to separate them; thus, Chapter 2 deals with techniques for solving first order equations, and Chapter 4 deals with applications. Similarly, Chapter 5 deals with techniques for solving second order equations, and Chapter 6 deals with applications. However, the exercise sets of the sections dealing with techniques do include some applied problems.

Recently, traditionally oriented texts on elementary differential equations have occasionally been criticized as being collections of unrelated methods for solving miscellaneous problems. To some extent this is true; after all, no single method applies to all situations. Nevertheless, I believe that one idea can go a long way toward unifying some of the techniques for solving diverse problems: variation of parameters. I use variation of parameters at the earliest opportunity in Section 2.1, to solve the nonhomogeneous linear equation, given a nontrivial solution of the complementary equation. You may find this annoying, since most of us learned that one should use integrating factors for this task, while perhaps mentioning the variation of parameters option in an exercise. However, there is little difference between the two approaches, since an integrating factor is nothing more than the reciprocal of a nontrivial solution of the complementary equation. The advantage of using variation of parameters here is that it introduces the concept in its simplest form and focuses the student's attention on the idea of seeking a solution y of a differential equation by writing it as $y = uy_1$, where y_1 is a known solution of a related equation and u is a function to be determined. I use this idea in nonstandard ways, as follows:

- In Section 2.4 to solve nonlinear first order equations, such as Bernoulli's equation and nonlinear homogeneous equations.
- In Chapter 3 for numerical solution of semilinear first order equations.
- In Section 5.2 to avoid the necessity of introducing complex exponentials in solving a second order constant coefficient homogeneous equation whose characteristic polynomial has complex zeros.
- In Sections 5.4, 5.5, and 9.3 for the method of undetermined coefficients. (If the method of annihilators is your preferred approach to this problem, compare the labor involved in solving, for example, $y'' + y' + y = x^4 e^x$ by the method of annihilators and the method used in Section 5.4.)

Introducing variation of parameters as early as possible (Section 2.1) prepares the student for the concept when it appears again in more complex forms in Section 5.6, where reduction of order is used not merely to find a second solution of the complementary equation, but also to find the general solution of the nonhomogeneous equation, and in Sections 5.7, 9.4, and 10.7, which treat the usual variation of parameters problem for second and higher order linear equations and for linear systems.

You may also find the following to be of interest:

- Section 2.6 deals with integrating factors of the form $\mu = p(x)q(y)$, in addition to those of the form $\mu = p(x)$ and $\mu = q(y)$ discussed in most texts.
- Section 4.4 makes phase plane analysis of nonlinear second order autonomous equations accessible to students who have not taken linear algebra, since eigenvalues and eigenvectors do not enter into the treatment. Phase plane analysis of constant coefficient linear systems is included in Sections 10.4–6.
- Section 4.5 presents an extensive discussion of applications of differential equations to curves.
- Section 6.4 studies motion under a central force, which may be useful to students interested in the mathematics of satellite orbits.
- Sections 7.5–7 present the method of Frobenius in more detail than is found in most texts. The approach is to systematize the computations in a way that avoids the necessity of substituting the unknown Frobenius series into each equation. This leads to efficiency in the computation of the coefficients of the Frobenius solution. It also clarifies the case where the roots of the indicial equation differ by an integer (Section 7.7).
- The *Student Solutions Manual* contains solutions of most of the even-numbered exercises.

The text includes more material than can be covered in one semester; I suspect that complete coverage will require two semesters. At Trinity University my colleagues and I have used the following sections in our first course in differential equations, for classes consisting mainly of second semester sophomores or first semester juniors: 1.1–2, 2.1–5, 4.1–3, 5.1–7, 6.1–2, 8.1–7, and 10.1–7. However, many choices can be made. The order of presentation of topics need not be the order in which they appear in the book.

The following observations may be helpful as you choose your own syllabus:

- Section 2.3 is the only specific prerequisite for Chapter 3. To accommodate institutions that offer a separate course in numerical analysis, Chapter 3 is not a prerequisite for any other section in the text.
- The sections in Chapter 4 are independent of each other, and are not prerequisites for any of the later chapters. This is also true of the sections in Chapter 6, except that Section 6.1 is a prerequisite for Section 6.2.
- Chapters 7, 8, and 9 can be covered in any order after the topics selected from Chapter 5. For example, you can proceed directly from Chapter 5 to Chapter 9.
- The second order Euler equation is discussed in Section 7.4, where it sets the stage for the method of Frobenius. As noted at the beginning of Section 7.4, if you wish to include Euler equations in your syllabus while omitting the method of Frobenius, you can skip the introductory paragraphs in Section 7.4 and begin with Definition 7.4.2. You can then cover Section 7.4 immediately after Section 5.2.

I welcome your suggestions for improvements and corrections.

William F. Trench

Acknowledgments

I thank the following reviewers for their comments and useful suggestions for improvement of the manuscript:

Glen Anderson, Michigan State University
David Dudley, Phoenix College
Lynn Erbe, University of Nebraska
Graeme Fairweather, Colorado School of Mines
Charles Friedman, University of Texas at Austin
Kristina Hansen, University of Michigan, Flint
Terry Herdmans, Virginia Polytechnical Institute
Sharon Hill, Rowan University
Charles MacClure, Michigan State University
Chris Pladdy, Nicholls State University
Mohammad Rammaha, University of Nebraska
Joel Smoller, University of Michigan

Special thanks go to my Trinity University colleagues Donald F. Bailey and Roberto Hasfura, who taught from several sections of the manuscript in various stages of development, reviewed other sections in depth, and made numerous suggestions resulting in significant improvements.

I am especially grateful to Howard Anton, who first interested me in writing this book and made many valuable suggestions on style and pedagogy. I have benefited greatly from his experience as a successful author and teacher.

I thank my publisher, Robert Pirtle, for his encouragement and support of this project since its inception. I also thank production editor Janet Hill, designer Vernon Boes, proofreader Marian Selig, and cover designer Roger Knox for their contributions. Especially, many thanks to Martha Emry of Martha Emry Production Services, for her patience, expert guidance, and total competence in converting this work from manuscript to book.

Most of all I am grateful to my wife, Beverly, who endured—with patience, understanding, and support—the ups and downs of the writing and rewriting of this book during the long period from its inception to its completion.

W. F. T.

1 Introduction

IN THIS CHAPTER we begin our study of differential equations.

SECTION 1.1 presents examples of applications that lead to differential equations.

SECTION 1.2 introduces some basic concepts and definitions concerning differential equations.

SECTION 1.3 presents a geometric method of dealing with differential equations that has been known for a very long time, but has become particularly useful and important with the proliferation of readily available differential equations software.

1.1 Some Applications Leading to Differential Equations

In order to apply mathematical methods to a physical or real life problem it is necessary to formulate the problem in mathematical terms; that is, we must construct a **mathematical model** for the problem. Many physical problems concern relationships between changing quantities. Since rates of change are represented mathematically by derivatives, mathematical models often involve equations relating an unknown function and one or more of its derivatives. Such equations are called **differential equations.** They are the subject of this book.

Much of calculus is devoted to learning mathematical techniques that are applied in later courses in mathematics and the sciences; you wouldn't have time to learn much calculus if you insisted upon seeing a specific application of every topic covered in the course. Similarly, much of this book is devoted to methods that can be applied in later courses. Only a relatively small part of the book is devoted to the derivation of specific differential equations from mathematical models, or relating the differential equations that we study to specific applications. In this section we mention a few such applications. Others are given in Chapters 4 and 6 and in problems in the exercise sets of other sections throughout the text.

The mathematical model for an applied problem is almost always simpler than the actual situation being studied, since simplifying assumptions are usually required to obtain a mathematical problem that can be solved. For example, in modeling the motion of a falling object one might neglect air resistance and the gravitational pull of celestial bodies other than Earth, or in modeling population growth one might assume that the population grows continuously rather than in discrete steps.

A good mathematical model has two important properties:

- It is sufficiently simple that the mathematical problem can be solved.
- It represents the actual situation sufficiently well so that the solution to the mathematical problem predicts the behavior of the real problem to within a useful degree of accuracy. If results predicted by the model don't agree with physical observations, then the underlying assumptions of the model must be revised until satisfactory agreement is obtained.

We will now give some examples of mathematical models involving differential equations. We'll return to these problems at the appropriate times, as we learn how to solve the various types of differential equations that occur in the models.

All the examples in this section deal with functions of time, which we will denote by t. If y is a function of t, then y' will denote the derivative of y with respect to t; thus

$$y' = \frac{dy}{dt}.$$

POPULATION GROWTH AND DECAY

Although the number of members of a population (people in a given country, bacteria in a laboratory culture, wildflowers in a forest, etc.) at any given time t is necessarily an integer, models that use differential equations to describe the growth and decay of populations must rest on the simplifying assumption that the number of members of the population can be regarded as a differentiable function $P = P(t)$. In most models it is assumed that the differential equation is of the form

$$P' = a(P)P, \tag{1}$$

where a is a continuous function of P that represents the rate of change of population per unit time per individual.

In the **Malthusian**[1] **model** it is assumed that $a(P)$ is a constant, so (1) becomes

$$P' = aP. \tag{2}$$

This model assumes that the numbers of births and deaths per unit time are both proportional to the population. The constants of proportionality are called the **birth rate** (births per unit time per individual) and the **death rate** (deaths per unit time per individual); a is the birth rate minus the death rate. You learned in calculus that if c is any constant then

$$P = ce^{at} \tag{3}$$

satisfies (2), so (2) has infinitely many solutions. To select the solution of the specific problem that we're considering, we must know the population P_0 at some initial time, say $t = 0$. Setting $t = 0$ in (3) yields $c = P(0) = P_0$, so the applicable solution is

$$P(t) = P_0 e^{at}.$$

From this we see that

$$\lim_{t \to \infty} P(t) = \begin{cases} \infty & \text{if } a > 0, \\ 0 & \text{if } a < 0; \end{cases}$$

that is, the population approaches infinity if the birth rate exceeds the death rate, or zero if the death rate exceeds the birth rate.

To see the limitations of the Malthusian model, suppose we're modeling the population of a country, starting from a time $t = 0$ when the birth rate exceeds the death rate (so $a > 0$) and the country's resources in terms of space, food supply, and other necessities of life can support the existing population. Then the prediction $P = P_0 e^{at}$ may be reasonably accurate as long as it remains within limits that the country's resources can support. However, the model must inevitably lose validity when the prediction exceeds these limits. (If nothing else, eventually there won't be enough space to hold the predicted population!)

This flaw in the Malthusian model suggests the need for a model that accounts for limitations of space and resources that tend to oppose the rate of population growth as the population increases. Perhaps the most famous model of this kind is the **Verhulst**[2] **model,** in which (2) is replaced by

$$P' = aP(1 - \alpha P), \tag{4}$$

where α is a positive constant. As long as P is small compared to $1/\alpha$ the ratio P'/P is approximately equal to a, and therefore the growth is approximately

---

[1] The English mathematician, economist, and ordained minister Thomas Robert Malthus (1766–1834) proposed his model for population growth in a paper published anonymously in 1798.

[2] The Belgian mathematician Pierre Verhulst (1804–1849) modified Malthus's model for population growth in 1845.

exponential; however, as P increases the ratio P'/P decreases, reflecting the fact that phenomena opposing the increase of population become significant as the population increases.

Equation (4) is called the ***logistic equation***. You will learn how to solve it in Section 2.2 (see Exercise 28 of Section 2.2). The solution is

$$P = \frac{P_0}{\alpha P_0 + (1 - \alpha P_0)e^{-at}}$$

where $P_0 = P(0) > 0$. Notice that $\lim_{t\to\infty} P(t) = 1/\alpha$, independent of P_0. Figure 1 shows typical graphs of P versus t for various values of P_0.

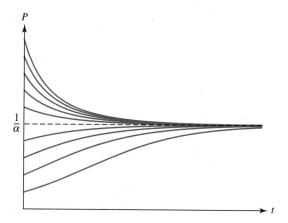

Figure 1   Solutions of the logistic equation

NEWTON'S LAW OF COOLING

Newton's[3] law of cooling states that the temperature of a body changes at a rate proportional to the difference between the temperature of the body and the temperature of the surrounding medium. Thus, if T_m is the temperature of the medium and $T = T(t)$ is the temperature of the body at time t, then T satisfies the differential equation

$$T' = -k(T - T_m) \tag{5}$$

where k is a positive constant and the minus sign occurs because the temperature of the body increases with time if it is less than the temperature of the medium, or decreases if it is greater. We will see in Section 4.2 that if we make the simplifying assumption that T_m is constant, then the solutions of (5) are of the form

$$T = T_m + (T_0 - T_m)e^{-kt},$$

where T_0 is the temperature of the body when $t = 0$. Notice that $\lim_{t\to\infty} T(t) = T_m$, independent of T_0. (Common sense suggests this. Why?) Figure 2 shows typical graphs of T versus t for various values of T_0.

---

[3]Isaac Newton (1642–1727) was a coinventor with Liebniz of calculus. He formulated the laws of mechanics as a body of knowledge that became known as Newtonian physics.

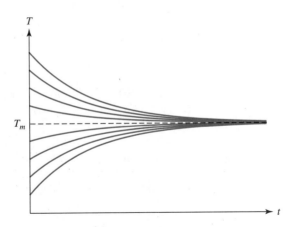

Figure 2 Temperature according to Newton's law of cooling

Assuming that the medium remains at constant temperature seems reasonable if we're considering a cup of coffee cooling in a room, but perhaps not so reasonable if we're cooling a huge cauldron of molten metal in the same room. The difference between the two situations is that the heat lost by the coffee is not likely to raise the temperature of the room appreciably, while the heat lost by the cooling metal is. In this second situation we must use a model that accounts for the heat exchanged between the object and the medium. Let $T = T(t)$ and $T_m = T_m(t)$ be the temperatures of the object and the medium respectively, and let T_0 and T_{m0} be their initial values. Again we assume that T and T_m are related by (5). We also assume that the change in heat of the object as its temperature changes from T_0 to T is $a(T - T_0)$ and that the change in heat of the medium as its temperature changes from T_{m0} to T_m is $a_m(T_m - T_{m0})$, where a and a_m are positive constants depending upon the masses and thermal properties of the object and medium, respectively. If we assume that the total heat of the system consisting of the object and the medium remains constant (that is, energy is conserved), then

$$a(T - T_0) + a_m(T_m - T_{m0}) = 0.$$

Solving this equation for T_m and substituting the result into (5) yields the differential equation

$$T' = -k\left(1 + \frac{a}{a_m}\right)T + k\left(T_{m0} + \frac{a}{a_m}T_0\right)$$

for the temperature of the object. After learning to solve linear first order equations you will be able to show (Exercise 17 of Section 4.2) that

$$T = \frac{aT_0 + a_m T_{m0}}{a + a_m} + \frac{a_m(T_0 - T_{m0})}{a + a_m}e^{-k(1 + a/a_m)t}.$$

GLUCOSE ABSORPTION BY THE BODY

It has been observed that glucose is absorbed by the body at a rate proportional to the amount of glucose present in the bloodstream. Let λ denote the (positive) constant of proportionality. Now suppose that there are G_0 units of glucose in the bloodstream when $t = 0$, and let $G = G(t)$ be the number of units in the bloodstream at time $t > 0$. Then, since the glucose being absorbed by the body is leaving the bloodstream, G must satisfy the equation

$$G' = -\lambda G. \tag{6}$$

From calculus you know that if c is any constant then

$$G = ce^{-\lambda t} \tag{7}$$

satisfies (6), so (6) has infinitely many solutions. Setting $t = 0$ in (7) and requiring that $G(0) = G_0$ yields $c = G_0$, so

$$G(t) = G_0 e^{-\lambda t}.$$

Now let's complicate matters by injecting glucose intravenously at a constant rate of r units of glucose per unit of time. Then the rate of change of the amount of glucose in the bloodstream per unit time is given by

$$G' = -\lambda G + r, \tag{8}$$

where the first term on the right is due to the absorption of the glucose by the body and the second term is due to the injection. After you've studied Section 2.1 you'll be able to show (Exercise 43 of Section 2.1) that the solution of (8) that satisfies $G(0) = G_0$ is

$$G = \frac{r}{\lambda} + \left(G_0 - \frac{r}{\lambda}\right)e^{-\lambda t}.$$

Graphs of this function are similar to those in Figure 2. (Why?)

SPREAD OF EPIDEMICS

One model for the spread of epidemics assumes that the number of people infected changes at a rate proportional to the product of the number of people already infected and the number of people who are susceptible, but not yet infected. It follows that if S denotes the total population of susceptible people and $I = I(t)$ denotes the number of infected people at time t, then $S - I$ is the number of people who are susceptible, but not yet infected. Thus,

$$I' = rI(S - I),$$

where r is a positive constant. Assuming that $I(0) = I_0$, the solution of this equation is

$$I = \frac{SI_0}{I_0 + (S - I_0)e^{-rSt}}$$

(Exercise 29 of Section 2.2). Graphs of this function are similar to those in Figure 1. (Why?) Since $\lim_{t \to \infty} I(t) = S$, this model predicts that all the susceptible people eventually become infected.

NEWTON'S SECOND LAW OF MOTION

Newton's second law of motion says that the instantaneous acceleration a of an object with constant mass m is related to the force F acting on the object by the equation $F = ma$. For simplicity let's assume that $m = 1$ and the motion of the object is along a vertical line. Let y be the displacement of the object from some reference point on Earth's surface, measured positive upward. In many applications there are three kinds of forces that may act on the object:

1. A force such as gravity that depends only on the position y, which we write as $-p(y)$, where $p(y) > 0$ if $y \geq 0$.

2. A force such as atmospheric resistance that depends on the position and velocity of the object, which we write as $-q(y, y')y'$, where q is a nonnegative function and we've put y' "outside" to indicate that the resistive force is always in the direction opposite to the velocity.

3. A force $f = f(t)$, exerted from an external source (such as a towline from a helicopter) that depends only on t.

In this case Newton's second law implies that

$$y'' = -q(y, y')y' - p(y) + f(t),$$

which is usually rewritten as

$$y'' + q(y, y')y' + p(y) = f(t).$$

Since the second (and no higher) order derivative of y occurs in this equation, we say that it is a **second order differential equation.**

INTERACTING SPECIES: COMPETITION

Let $P = P(t)$ and $Q = Q(t)$ be the populations of two species at time t, and assume that each population would grow exponentially if the other didn't exist; that is, in the absence of competition we would have

$$P' = aP \qquad \text{and} \qquad Q' = bQ, \tag{9}$$

where a and b are positive constants. One way to model the effect of competition is to assume that the growth rate of each population is reduced by an amount proportional to the other population, so (9) is replaced by

$$P' = \quad aP - \alpha Q$$
$$Q' = -\beta P + bQ,$$

where α and β are positive constants. (Since negative population doesn't make sense, this system holds only while P and Q are both positive.) Now suppose that $P(0) = P_0 > 0$ and $Q(0) = Q_0 > 0$. It can be shown (Exercise 42 of Section 10.4) that there is a positive constant ρ such that if (P_0, Q_0) is above the line L through the origin with slope ρ then the species with population P becomes extinct in finite time, while if (P_0, Q_0) is below L then the species with population Q becomes extinct in finite time. Figure 3 illustrates this. The curves shown there are given parametrically by $P = P(t), Q = Q(t), t > 0$. The arrows indicate direction along the curves with increasing t.

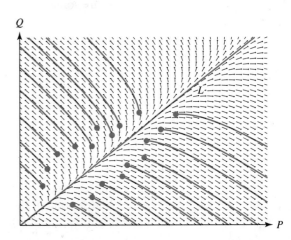

Figure 3 Populations of competing species

1.2   Basic Concepts

A *differential equation* is an equation containing one or more derivatives of an unknown function. The *order* of a differential equation is the order of the highest derivative that it contains. A differential equation is an *ordinary differential equation* if it involves an unknown function of only one variable, or a *partial differential equation* if it involves partial derivatives of a function of more than one variable. For now we will consider only ordinary differential equations, and we'll just call them *differential equations.*

Throughout this text all variables and constants are real unless it is stated otherwise. We'll usually use x for the independent variable unless the independent variable is time; then we'll use t.

The simplest differential equations are first order equations of the form

$$\frac{dy}{dx} = f(x) \qquad \text{or, equivalently,} \qquad y' = f(x),$$

where f is a known function of x. We already know from calculus how to find functions that satisfy this kind of equation. For example, if

$$y' = x^3,$$

then

$$y = \int x^3 \, dx = \frac{x^4}{4} + c,$$

where c is an arbitrary constant. If $n > 1$ we can find functions y that satisfy equations of the form

$$y^{(n)} = f(x) \tag{1}$$

by repeated integration. Again, this is a calculus problem.

Except for illustrative purposes in this section, there is no need to consider differential equations like (1). We will usually consider differential equations that can be written as

$$y^{(n)} = f(x, y, y', \dots, y^{(n-1)}), \tag{2}$$

where at least one of the functions $y, y', \dots, y^{(n-1)}$ actually appears on the right. Here are some examples:

$$\frac{dy}{dx} - x^2 = 0 \qquad \text{(first order)}$$

$$\frac{dy}{dx} + 2xy^2 = -2 \qquad \text{(first order)}$$

$$\frac{d^2y}{dx^2} + 2\frac{dy}{dx} + y = 2x \qquad \text{(second order)}$$

$$xy''' + y^2 = \sin x \qquad \text{(third order)}$$

$$y^{(n)} + xy' + 3y = x \qquad \text{(\textit{n}th order).}$$

Although none of these equations is written as in (2), all of them *can* be written in this form:

$$y' = x^2$$
$$y' = -2 - 2xy^2$$
$$y'' = 2x - 2y' - y$$
$$y''' = \frac{\sin x - y^2}{x}$$
$$y^{(n)} = x - xy' - 3y.$$

SOLUTIONS OF DIFFERENTIAL EQUATIONS

A **solution** of a differential equation is a function that satisfies the differential equation on some open interval; thus, y is a solution of (2) if y is n times differentiable and

$$y^{(n)}(x) = f(x, y(x), y'(x), \ldots, y^{(n-1)}(x))$$

for all x in some open interval (a,b). In this case we also say that y **is a solution of (2) on (a,b).** Functions that satisfy a differential equation at isolated points are of no interest. For example, $y = x^2$ satisfies

$$xy' + x^2 = 3x$$

if and only if $x = 0$ or $x = 1$, but it is not a solution of this differential equation because it does not satisfy the equation on an open interval.

The graph of a solution of a differential equation is called a **solution curve.** More generally, a curve C is said to be an **integral curve** of a differential equation if every function $y = y(x)$ whose graph is a segment of C is a solution of the differential equation. Thus, any solution curve of a differential equation is an integral curve, but an integral curve need not be a solution curve.

EXAMPLE 1 If a is any positive constant then the circle

$$x^2 + y^2 = a^2 \tag{3}$$

is an integral curve of

$$y' = -\frac{x}{y}. \tag{4}$$

To see this, note that the only functions whose graphs are segments of (3) are

$$y_1 = \sqrt{a^2 - x^2} \quad \text{and} \quad y_2 = -\sqrt{a^2 - x^2}.$$

We leave it to you to verify that these functions both satisfy (4) on the open interval $(-a,a)$. However, (3) is not a solution curve of (4), since it is not the graph of a function.

EXAMPLE 2 Verify that

$$y = \frac{x^2}{3} + \frac{1}{x} \tag{5}$$

is a solution of

$$xy' + y = x^2 \tag{6}$$

on $(0,\infty)$ and on $(-\infty,0)$.

Solution Substituting (5) and

$$y' = \frac{2x}{3} - \frac{1}{x^2}$$

into (6) yields

$$xy'(x) + y(x) = x\left(\frac{2x}{3} - \frac{1}{x^2}\right) + \left(\frac{x^2}{3} + \frac{1}{x}\right) = x^2$$

for all $x \neq 0$. Therefore y is a solution of (6) on $(-\infty, 0)$ and $(0, \infty)$. However, y is not a solution of the differential equation on any open interval containing $x = 0$, since y is not defined at $x = 0$. ■

The graph of (5) is shown in Figure 1. The part of the graph of (5) on $(0, \infty)$ is a solution curve of (6), as is the part of the graph on $(-\infty, 0)$.

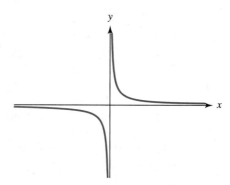

Figure 1 The graph of $y = \dfrac{x^2}{3} + \dfrac{1}{x}$

EXAMPLE 3 Show that if c_1 and c_2 are constants then

$$y = (c_1 + c_2 x)e^{-x} + 2x - 4 \tag{7}$$

is a solution of

$$y'' + 2y' + y = 2x \tag{8}$$

on $(-\infty, \infty)$.

Solution Differentiating (7) twice yields

$$y' = -(c_1 + c_2 x)e^{-x} + c_2 e^{-x} + 2$$

and

$$y'' = (c_1 + c_2 x)e^{-x} - 2c_2 e^{-x},$$

so

$$\begin{aligned}
y'' + 2y' + y &= (c_1 + c_2 x)e^{-x} - 2c_2 e^{-x} \\
&\quad + 2[-(c_1 + c_2 x)e^{-x} + c_2 e^{-x} + 2] \\
&\quad + (c_1 + c_2 x)e^{-x} + 2x - 4 \\
&= (1 - 2 + 1)(c_1 + c_2 x)e^{-x} + (-2 + 2)c_2 e^{-x} \\
&\quad + 4 + 2x - 4 = 2x
\end{aligned}$$

for all values of x. Therefore y is a solution of (8) on $(-\infty, \infty)$.

EXAMPLE 4 Find all solutions of

$$y^{(n)} = e^{2x}. \tag{9}$$

Solution Integrating (9) yields

$$y^{(n-1)} = \frac{e^{2x}}{2} + k_1,$$

where k_1 is a constant. If $n \geq 2$ then integrating again yields

$$y^{(n-2)} = \frac{e^{2x}}{4} + k_1 x + k_2.$$

If $n \geq 3$ then repeatedly integrating yields

$$y = \frac{e^{2x}}{2^n} + k_1 \frac{x^{n-1}}{(n-1)!} + k_2 \frac{x^{n-2}}{(n-2)!} + \cdots + k_n, \tag{10}$$

where $k_1, k_2, \ldots, k_n$ are constants. This shows that every solution of (9) is of the form (10) for some choice of the constants $k_1, k_2, \ldots, k_n$. On the other hand, differentiating (10) n times shows that if $k_1, k_2, \ldots, k_n$ are arbitrary constants, then the function y in (10) satisfies (9). ∎

Since the constants $k_1, k_2, \ldots, k_n$ in (10) are arbitrary, so are the constants

$$\frac{k_1}{(n-1)!}, \frac{k_2}{(n-2)!}, \ldots, k_n.$$

Therefore, Example 4 actually shows that all solutions of (9) can be written as

$$y = \frac{e^{2x}}{2^n} + c_1 + c_2 x + \cdots + c_n x^{n-1},$$

where we have renamed the arbitrary constants in (9) to obtain a simpler formula. As a general rule, arbitrary constants appearing in solutions of differential equations should be simplified if possible. We will see examples of this throughout the text.

INITIAL VALUE PROBLEMS

We saw in Example 4 that the differential equation $y^{(n)} = e^{2x}$ has an infinite family of solutions that depend upon the n arbitrary constants $c_1, c_2, \ldots, c_n$. In the absence of additional conditions, there is no reason to prefer one solution of a differential equation over another. However, we will often be interested in finding a solution of a differential equation that satisfies one or more specific conditions. The following example illustrates this.

EXAMPLE 5 Find a solution of

$$y' = x^3$$

such that $y(1) = 2$.

Solution We saw at the beginning of this section that the solutions of $y' = x^3$ are of form

$$y = \frac{x^4}{4} + c.$$

To determine a value of c for which $y(1) = 2$, we set $x = 1$ and $y = 2$ here to obtain

$$2 = y(1) = \frac{1}{4} + c, \qquad \text{so} \qquad c = \frac{7}{4}.$$

Therefore, the required solution is

$$y = \frac{x^4 + 7}{4}.$$

The graph of this solution is shown in Figure 2. Notice that imposing the condition $y(1) = 2$ is equivalent to requiring that the graph of y pass through the point $(1,2)$. ∎

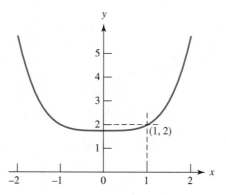

Figure 2   The graph of $y = \dfrac{x^4 + 7}{4}$

We can rewrite the problem considered in Example 5 more briefly as

$$y' = x^3, \qquad y(1) = 2.$$

We call this an ***initial value problem.*** The requirement $y(1) = 2$ is an ***initial condition.*** Initial value problems can also be posed for higher order differential equations. For example,

$$y'' - 2y' + 3y = e^x, \qquad y(0) = 1, \qquad y'(0) = 2 \tag{11}$$

denotes an initial value problem for a second order differential equation in which y and y' are required to have specified values at the point $x = 0$. In general, an initial value problem for an nth order differential equation requires y and its first $n - 1$ derivatives to have specified values at some point x_0. These requirements are the ***initial conditions.***

We will denote an initial value problem for a differential equation by writing the initial conditions after the equation, as in (11). For example, we would write an initial value problem for (2) as

$$y^{(n)} = f(x, y, y', \dots, y^{(n-1)}), \qquad y(x_0) = k_0, \qquad y'(x_0) = k_1, \dots, y^{(n-1)} = k_{n-1}. \tag{12}$$

Consistent with our earlier definition of a solution of the differential equation in (12), we say that y is a solution of the initial value problem (12) if y is n times differentiable and

$$y^{(n)}(x) = f(x, y(x), y'(x), \dots, y^{(n-1)}(x))$$

for all x in some open interval (a, b) *that contains* x_0, and y satisfies the initial conditions in (12). The largest open interval containing x_0 on which y is defined and satisfies the differential equation is called the ***interval of validity*** of y.

EXAMPLE 6   In Example 5 we saw that

$$y = \frac{x^4 + 7}{4} \qquad (13)$$

is a solution of the initial value problem

$$y' = x^3, \qquad y(1) = 2.$$

Since the function in (13) is defined for all x, the interval of validity of this solution is $(-\infty, \infty)$.

EXAMPLE 7   In Example 2 we verified that

$$y = \frac{x^2}{3} + \frac{1}{x} \qquad (14)$$

is a solution of

$$xy' + y = x^2$$

on $(0, \infty)$ and on $(-\infty, 0)$, By evaluating (14) at $x = \pm 1$ you can see that (14) is a solution of the initial value problems

$$xy' + y = x^2, \qquad y(1) = \frac{4}{3} \qquad (15)$$

and

$$xy' + y = x^2, \qquad y(-1) = -\frac{2}{3}. \qquad (16)$$

The interval of validity of (14) as a solution of (15) is $(0, \infty)$, since this is the largest interval containing $x_0 = 1$ on which (14) is defined. Similarly, the interval of validity of (14) as a solution of (16) is $(-\infty, 0)$, since this is the largest interval containing $x_0 = -1$ on which (14) is defined. ■

FREE FALL UNDER CONSTANT GRAVITY

The term ***initial value problem*** originated in problems of motion in which the independent variable is t (representing elapsed time), and the initial conditions are the position and velocity of an object at the initial (starting) time of an experiment.

EXAMPLE 8   An object falls under the influence of gravity near Earth's surface, where it can be assumed that the magnitude of the acceleration due to gravity is a constant, g.

(a) Construct a mathematical model for the motion of the object in the form of an initial value problem for a second order differential equation, assuming that the altitude and velocity of the object at time $t = 0$ are known. Assume that gravity is the only force acting on the object.
(b) Solve the initial value problem derived in **(a)** to obtain the altitude as a function of time.

Solution   **(a)** Let $y(t)$ be the altitude of the object at time t. Since the acceleration of the object has constant magnitude g and is in the downward (negative) direction, y satisfies the second order equation

$$y'' = -g,$$

where the prime now indicates differentiation with respect to t. If y_0 and v_0 denote the altitude and velocity when $t = 0$ then y is a solution of the initial value problem

$$y'' = -g, \qquad y(0) = y_0, \qquad y'(0) = v_0. \tag{17}$$

Solution   **(b)** Integrating (17) twice yields

$$y' = -gt + c_1,$$

$$y = -\frac{gt^2}{2} + c_1 t + c_2.$$

Imposing the initial conditions $y(0) = y_0$ and $y'(0) = v_0$ in these two equations shows that $c_1 = v_0$ and $c_2 = y_0$. Therefore, the solution of the initial value problem (17) is

$$y = -\frac{gt^2}{2} + v_0 t + y_0. \qquad\blacksquare$$

1.2 EXERCISES

1. Find the order of each of the following equations.

   (a) $\dfrac{d^2 y}{dx^2} + 2\dfrac{dy}{dx}\dfrac{d^3 y}{dx^3} + x = 0$

   (b) $y'' - 3y' + 2y = x^7$

   (c) $y' - y^7 = 0$

   (d) $y''y - (y')^2 = 2$

2. Verify that the given function is a solution of the given differential equation on some interval, for any choice of the arbitrary constants appearing in the function.

   (a) $y = ce^{2x}; \quad y' = 2y$

   (b) $y = \dfrac{x^2}{3} + \dfrac{c}{x}; \quad xy' + y = x^2$

   (c) $y = \dfrac{1}{2} + ce^{-x^2}; \quad y' + 2xy = x$

   (d) $y = (1 + ce^{-x^2/2})(1 - ce^{-x^2/2})^{-1}; \quad 2y' + x(y^2 - 1) = 0$

   (e) $y = \tan\left(\dfrac{x^3}{3} + c\right); \quad y' = x^2(1 + y^2)$

   (f) $y = (c_1 + c_2 x)e^x + \sin x + x^2; \quad y'' - 2y' + y = -2\cos x + x^2 - 4x + 2$

   (g) $y = c_1 e^x + c_2 x + \dfrac{2}{x}; \quad (1 - x)y'' + xy' - y = 4(1 - x - x^2)x^{-3}$

   (h) $y = x^{-1/2}(c_1 \sin x + c_2 \cos x) + 4x + 8; \quad x^2 y'' + xy' + \left(x^2 - \dfrac{1}{4}\right)y = 4x^3 + 8x^2 + 3x - 2$

3. Find all solutions of the given equation.

   (a) $y' = -x$

   (b) $y' = -x \sin x$

   (c) $y' = x \ln x$

   (d) $y'' = x \cos x$

   (e) $y'' = 2xe^x$

   (f) $y'' = 2x + \sin x + e^x$

   (g) $y''' = -\cos x$

   (h) $y''' = -x^2 + e^x$

   (i) $y''' = 7e^{4x}$

4. Solve the given initial value problem.

   (a) $y' = -xe^x, \quad y(0) = 1$

   (b) $y' = x \sin x^2, \quad y(\sqrt{\pi/2}) = 1$

   (c) $y' = \tan x, \quad y(\pi/4) = 3$

   (d) $y'' = x^4, \quad y(2) = -1, \quad y'(2) = -1$

   (e) $y'' = xe^{2x}, \quad y(0) = 7, \quad y'(0) = 1$

   (f) $y'' = -x \sin x, \quad y(0) = 1, \quad y'(0) = -3$

   (g) $y''' = x^2 e^x, \quad y(0) = 1, \quad y'(0) = -2, \quad y''(0) = 3$

   (h) $y''' = 2 + \sin 2x, \quad y(0) = 1, \quad y'(0) = -6, \quad y''(0) = 3$

   (i) $y''' = 2x + 1, \quad y(2) = 1, \quad y'(2) = -4, \quad y''(2) = 7$

5. Verify that the given function is a solution of the initial value problem.

   (a) $y = x \cos x; \quad y' = \cos x - y \tan x, \quad y(\pi/4) = \dfrac{\pi}{4\sqrt{2}}$

   (b) $y = \dfrac{1 + 2 \ln x}{x^2} + \dfrac{1}{2}; \quad y' = \dfrac{x^2 - 2x^2 y + 2}{x^3}, \quad y(1) = \dfrac{3}{2}$

   (c) $y = \tan\left(\dfrac{x^2}{2}\right); \quad y' = x(1 + y^2), \quad y(0) = 0$

   (d) $y = \dfrac{2}{x - 2}; \quad y' = \dfrac{-y(y + 1)}{x}, \quad y(1) = -2$

6. Verify that the given function is a solution of the initial value problem.

   (a) $y = x^2(1 + \ln x); \quad y'' = \dfrac{3xy' - 4y}{x^2}, \quad y(e) = 2e^2, \quad y'(e) = 5e$

   (b) $y = \dfrac{x^2}{3} + x - 1; \quad y'' = \dfrac{x^2 - xy' + y + 1}{x^2}, \quad y(1) = \dfrac{1}{3}, \quad y'(1) = \dfrac{5}{3}$

   (c) $y = (1 + x^2)^{-1/2}; \quad y'' = \dfrac{(x^2 - 1)y - x(x^2 + 1)y'}{(x^2 + 1)^2}, \quad y(0) = 1, \quad y'(0) = 0$

   (d) $y = \dfrac{x^2}{1 - x}; \quad y'' = \dfrac{2(x + y)(xy' - y)}{x^3}, \quad y\left(\dfrac{1}{2}\right) = \dfrac{1}{2}, \quad y'\left(\dfrac{1}{2}\right) = 3$

7. Suppose that an object is launched from a point 320 feet above Earth with an initial velocity of 128 ft/s upward, and that the only force acting on it thereafter is gravity. Take $g = 32$ ft/s^2.

   (a) Find the highest altitude attained by the object.

   (b) Determine how long it takes for the object to fall to the ground.

8. Let a be a nonzero real number.

   (a) Verify that if c is an arbitrary constant then

   $$y = (x - c)^a \tag{A}$$

   is a solution of

   $$y' = ay^{(a-1)/a} \tag{B}$$

   on (c, ∞).

   (b) Suppose that $a < 0$ or $a > 1$. Can you think of a solution of (B) that is not of the form (A)?

9. Verify that

   $$y = \begin{cases} e^x - 1, & x \geq 0, \\ 1 - e^{-x}, & x < 0, \end{cases}$$

   is a solution of

$$y' = |y| + 1$$

on $(-\infty, \infty)$.

Hint: *Use the definition of derivative at $x = 0$.*

10. (a) Verify that if c is any real number, then

$$y = c^2 + cx + 2c + 1 \tag{A}$$

satisfies

$$y' = \frac{-(x + 2) + \sqrt{x^2 + 4x + 4y}}{2} \tag{B}$$

on some open interval, and identify the open interval.

(b) Verify that

$$y_1 = \frac{-x(x + 4)}{4}$$

also satisfies (B) on some open interval, and identify the open interval. (Notice that y_1 cannot be obtained by selecting a value of c in (A).)

1.3 Direction Fields for First Order Equations

It is impossible to find explicit formulas for solutions of some differential equations. Even if there are such formulas, they may be so complicated that they're useless. In this case we may resort to graphical or numerical methods to get some idea of how the solutions of the given equation behave.

In Section 2.3 will take up the question of existence of solutions of a first order equation

$$y' = f(x, y). \tag{1}$$

In this section we will simply assume that (1) has solutions and discuss a graphical method for approximating them. In Chapter 3 we discuss numerical methods for obtaining approximate solutions of (1).

Recall that a solution of (1) is a function $y = y(x)$ such that

$$y'(x) = f(x, y(x))$$

for all values of x in some interval, and an integral curve is either the graph of a solution or is made up of segments that are graphs of solutions. Therefore, not being able to solve (1) is equivalent to not knowing the equations of integral curves of (1). However, it is easy to calculate the slopes of these curves. To be specific, the slope of an integral curve of (1) through a given point (x_0, y_0) is given by the number $f(x_0, y_0)$. This is the basis of *the method of direction fields.*

If f is defined on a set R, we can construct a *direction field of* **(1)** *in R* by drawing a short line segment through each point (x, y) in R with slope $f(x, y)$. Of course, as a practical matter we can't actually draw line segments through *every* point in R; rather, we must select a finite set of points in R. For example, suppose that f is defined on the closed rectangular region

$$R : \{a \le x \le b, \ c \le y \le d\}.$$

Let

$$a = x_0 < x_1 < \cdots < x_m = b$$

be equally spaced points in $[a,b]$ and

$$c = y_0 < y_1 < \cdots < y_n = d$$

be equally spaced points in $[c,d]$. We say that the points

$$(x_i, y_j), \qquad 0 \le i \le m, \qquad 0 \le j \le n,$$

form a ***rectangular grid*** (Figure 1). Through each point in the grid we draw a short line segment with slope $f(x_i, y_j)$. The result is an approximation to the direction field of (1) in R. If the grid points are sufficiently numerous and close together we can draw approximate integral curves of (1) by drawing curves through points in the grid tangent to the line segments associated with the points in the grid.

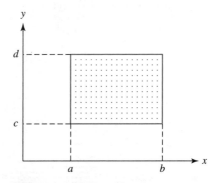

Figure 1   A rectangular grid

Unfortunately, approximating a direction field and graphing integral curves in this way is too tedious to be done effectively by hand. However, there are several software packages for doing this. All of them cheat, in that they don't really generate the integral curves by connecting tangent lines; rather, they use numerical methods to solve the differential equation and plot the results to obtain the integral curves. As you will see, the combination of the direction fields and the integral curves gives very useful insights into the behavior of the solutions of the differential equation, especially in cases where exact solutions can't be obtained.

We will study numerical methods for solving a single first order equation (1) in Chapter 3. These methods can be used to plot solution curves of (1) in a rectangular region R *if* f *is continuous on* R. Figures 2, 3, and 4 show direction fields and solution curves for the differential equations

$$y' = \frac{x^2 - y^2}{1 + x^2 + y^2},$$

$$y' = 1 + xy^2,$$

$$y' = \frac{x - y}{1 + x^2}.$$

These equations are all of the form (1) where f is continuous for all (x, y).

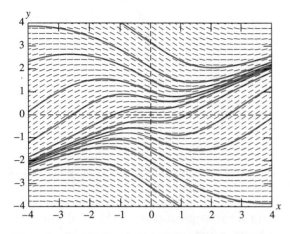

Figure 2   A direction field and solution curves of
$$y' = \frac{x^2 - y^2}{1 + x^2 + y^2}$$

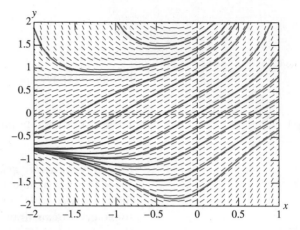

Figure 3   A direction field and solution curves of
$$y' = 1 + xy^2$$

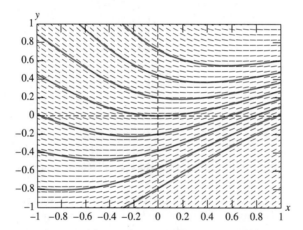

Figure 4   A direction field and solution curves of
$$y' = \frac{x - y}{1 + x^2}$$

The methods of Chapter 3 won't work for the equation

$$y' = -x/y \tag{2}$$

if R contains part of the x-axis, since $f(x, y) = -x/y$ is undefined when $y = 0$. Similarly, they won't work for the equation

$$y' = \frac{x^2}{1 - x^2 - y^2} \tag{3}$$

if R contains any part of the unit circle $x^2 + y^2 = 1$, because the right side of (3) is undefined if $x^2 + y^2 = 1$. However, (2) and (3) can be written in the form

$$y' = \frac{A(x, y)}{B(x, y)} \tag{4}$$

where A and B are continuous on any rectangle R. Because of this, differential equation software packages are based on numerically solving pairs of equations of the form

$$\frac{dx}{dt} = B(x,y), \qquad \frac{dy}{dt} = A(x,y) \qquad (5)$$

where x and y are regarded as functions of a parameter t. If $x = x(t)$ and $y = y(t)$ satisfy these equations, then

$$y' = \frac{dy}{dx} = \frac{dy}{dt} \bigg/ \frac{dx}{dt} = \frac{A(x,y)}{B(x,y)},$$

so $y = y(x)$ satisfies (4).

Equations (2) and (3) can be reformulated as in (4) with

$$\frac{dx}{dt} = -y, \qquad \frac{dy}{dt} = x$$

and

$$\frac{dx}{dt} = 1 - x^2 - y^2, \qquad \frac{dy}{dt} = x^2,$$

respectively. Even if f is continuous and otherwise "nice" throughout R, your software package will probably require you to reformulate the equation $y' = f(x,y)$ as

$$\frac{dx}{dt} = 1, \qquad \frac{dy}{dt} = f(x,y),$$

which is of the form (5) with $A(x,y) = f(x,y)$ and $B(x,y) = 1$.

Figure 5 shows a direction field and some integral curves of (2). As we saw in Example 1 of Section 1.2 and will verify again in Section 2.2, the integral curves of (2) are circles centered at the origin.

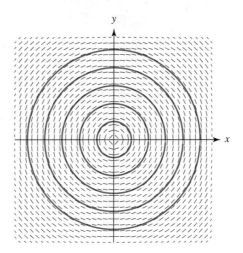

Figure 5 A direction field and integral curves for

$$y' = -\frac{x}{y}$$

Figure 6 shows a direction field and some integral curves of (3). The integral curves near the top and bottom are solution curves. However, the integral curves near the middle are more complicated. For example, Figure 7 shows the integral curve through the origin. The vertices of the dashed rectangle are on the circle $x^2 + y^2 = 1$ ($a \approx .846$, $b \approx .533$), where all integral curves of (3) have infinite slope. There are three solution curves of (3) on the integral curve in the figure: the segment above the level $y = b$ is the graph of a solution on $(-\infty, a)$; the segment below the level $y = -b$ is the graph of a solution on $(-a, \infty)$; and the segment between these two levels is the graph of a solution on $(-a, a)$.

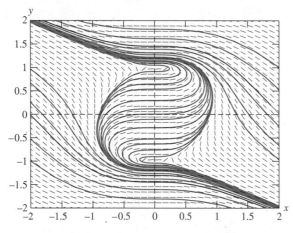

Figure 6   A direction field and integral curves for

$$y' = \frac{x^2}{1 - x^2 - y^2}$$

Figure 7

USING TECHNOLOGY

As you study from this book you will be asked on many occasions to use computer software and graphics. Exercises with this intent are marked as **C** (computer and/or graphics required) or **L** (laboratory work requiring computer and/or graphics).

There are several software packages that include applications to differential equations. Some do symbolic manipulations as well as computation and graphics. Here is a partial list of such packages:

Macsyma®
Maple®
Mathematica™
Derive®
MATLAB®
MathCad®
Phaser™

We urge you to become familiar with one or more such package. Often you may not completely understand how the software does what it does. This is similar to the situation most people are in when they drive automobiles or watch television, and it doesn't decrease the value of using modern technology as an aid to learning. Just be careful that you use the technology as a supplement to thought rather than a substitute for it.

1.3 EXERCISES

In Exercises 1–11 a direction field is drawn for the given equation. Sketch some integral curves.

1. $y' = \dfrac{x}{y}$

2. $y' = \dfrac{2xy^2}{1 + x^2}$

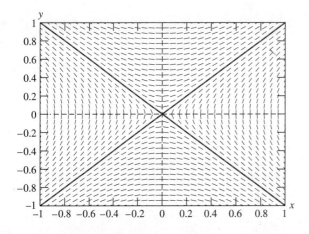
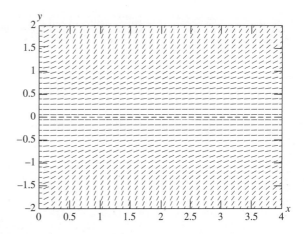

3. $y' = x^2(1 + y^2)$

4. $y' = \dfrac{1}{1 + x^2 + y^2}$

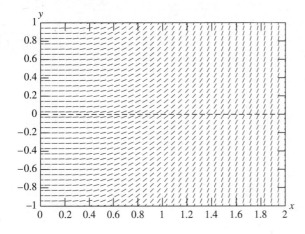
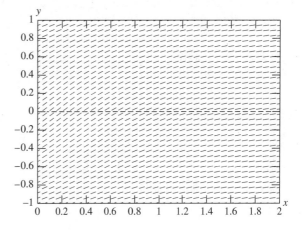

5. $y' = -(2xy^2 + y^3)$

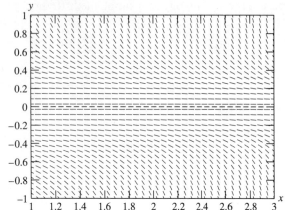

6. $y' = (x^2 + y^2)^{1/2}$

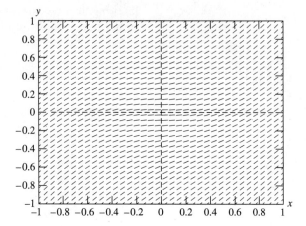

7. $y' = \sin xy$

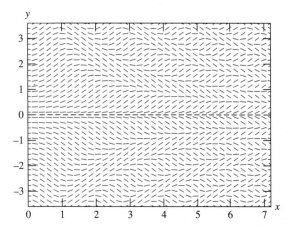

8. $y' = e^{xy}$

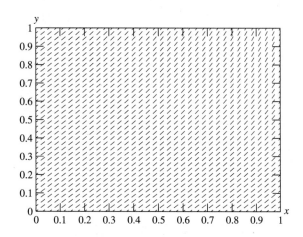

9. $y' = (x - y^2)(x^2 - y)$

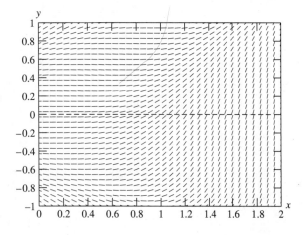

10. $y' = x^3 y^2 + xy^3$

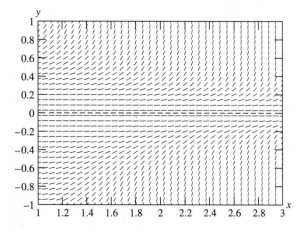

11. $y' = \sin(x - 2y)$

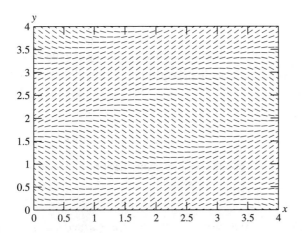

In Exercises 12–22 construct a direction field and plot some integral curves in the indicated rectangular region.

C **12.** $y' = y(y - 1);$ $\{-1 \le x \le 2, -2 \le y \le 2\}$

C **13.** $y' = 2 - 3xy;$ $\{-1 \le x \le 4, -4 \le y \le 4\}$

C **14.** $y' = xy(y - 1);$ $\{-2 \le x \le 2, -4 \le y \le 4\}$

C **15.** $y' = 3x + y;$ $\{-2 \le x \le 2, 0 \le y \le 4\}$

C **16.** $y' = y - x^3;$ $\{-2 \le x \le 2, -2 \le y \le 2\}$

C **17.** $y' = 1 - x^2 - y^2;$ $\{-2 \le x \le 2, -2 \le y \le 2\}$

C **18.** $y' = x(y^2 - 1);$ $\{-3 \le x \le 3, -3 \le y \le 2\}$

C **19.** $y' = \dfrac{x}{y(y^2 - 1)};$ $\{-2 \le x \le 2, -2 \le y \le 2\}$

C **20.** $y' = \dfrac{xy^2}{y - 1};$ $\{-2 \le x \le 2, -1 \le y \le 4\}$

C **21.** $y' = \dfrac{x(y^2 - 1)}{y};$ $\{-1 \le x \le 1, -2 \le y \le 2\}$

C **22.** $y' = -\dfrac{x^2 + y^2}{1 - x^2 - y^2};$ $\{-2 \le x \le 2, -2 \le y \le 2\}$

L **23.** By suitably renaming the constants and dependent variables in the equations

$$T' = -k(T - T_m) \tag{A}$$

and

$$G' = -\lambda G + r \tag{B}$$

discussed in Section 1.1 in connection with Newton's law of cooling and absorption of glucose in the body, we can write both in the form

$$y' = -ay + b, \tag{C}$$

where a is a positive constant and b is an arbitrary constant. Thus, (A) is of the form (C) with $y = T, a = k$, and $b = kT_m$, while (B) is of the form (C) with $y = G, a = \lambda$, and $b = r$. We'll encounter equations of the form (C) in many other applications in Chapter 2.

Choose a positive a and an arbitrary b. Construct a direction field and plot some integral curves of (C) in a rectangular region of the form

$$\{0 \le t \le T, c \le y \le d\}$$

of the ty-plane. Vary $T, c,$ and d until you discover a common property of all the solutions of (C). Repeat this experiment with various choices of a and b until you can state this property precisely in terms of a and b.

L **24.** By suitably renaming the constants and dependent variables in the equations

$$P' = aP(1 - \alpha P) \tag{A}$$

and

$$I' = rI(S - I) \tag{B}$$

discussed in Section 1.1 in connection with Verhulst's population model and the spread of an epidemic, we can write both in the form

$$y' = ay - by^2 \tag{C}$$

where a and b are positive constants. Thus, (A) is of the form (C) with $y = P, a = a,$ and $b = a\alpha,$ while (B) is of the form (C) with $y = I, a = rS,$ and $b = r.$ We'll encounter equations of the form (C) in many other applications in Chapter 2.

(a) Choose positive numbers a and b. Construct a direction field and plot some integral curves of (C) in a rectangular region of the form

$$\{0 \le t \le T, 0 \le y \le d\}$$

of the ty-plane. Vary T and d until you discover a common property of all solutions of (C) with $y(0) > 0$. Repeat this experiment with various choices of a and b until you can state this property precisely in terms of a and b.

(b) Choose positive numbers a and b. Construct a direction field and plot some integral curves of (C) in a rectangular region of the form

$$\{0 \le t \le T, c \le y \le 0\}$$

of the ty-plane. Vary $a, b, T,$ and c until you discover a common property of all solutions of (C) with $y(0) < 0$.

You can verify your results later by doing Exercise 27 of Section 2.2.

2 First Order Equations

IN THIS CHAPTER we study first order equations for which there are general methods of solution.

SECTION 2.1 deals with linear equations, the simplest kind of first order equations. In this section we introduce the method of variation of parameters. The idea underlying this method will be a unifying theme for our approach to solving many different kinds of differential equations throughout the book.

SECTION 2.2 deals with separable equations, the simplest nonlinear equations. In this section we introduce the idea of implicit and constant solutions of differential equations, and we point out some differences between the properties of linear and nonlinear equations.

SECTION 2.3 discusses existence and uniqueness of solutions of nonlinear equations. Although it may seem logical to place this section before Section 2.2, we presented Section 2.2 first so we could have illustrative examples in Section 2.3.

SECTION 2.4 deals with nonlinear equations that are not separable, but can be transformed into separable equations by a procedure similar to variation of parameters.

SECTION 2.5 covers exact differential equations, which are given this name because the method for solving them uses the idea of an exact differential from calculus.

SECTION 2.6 deals with equations that are not exact, but can be made exact by multiplying them by a function known as an *integrating factor.*

2.1   Linear First Order Equations

A first order differential equation is said to be ***linear*** if it can be written as

$$y' + p(x)y = f(x). \tag{1}$$

A first order differential equation that cannot be written like this is ***nonlinear.*** We say that (1) is ***homogeneous*** if $f \equiv 0$; otherwise it is ***nonhomogeneous.*** Since $y \equiv 0$ is obviously a solution of the homgeneous equation

$$y' + p(x)y = 0,$$

we call it the ***trivial solution.*** Any other solution is ***nontrivial.***

EXAMPLE 1   The first order equations

$$x^2 y' + 3y = x^2$$
$$xy' - 8x^2 y = \sin x$$
$$xy' + (\ln x)y = 0$$
$$y' = x^2 y - 2$$

are not in the form (1), but they *are* linear, since they can be rewritten as

$$y' + \frac{3}{x^2} y = 1$$

$$y' - 8xy = \frac{\sin x}{x}$$

$$y' + \frac{\ln x}{x} y = 0$$

$$y' - x^2 y = -2.$$

EXAMPLE 2   The following first order equations are nonlinear:

$$
\begin{array}{ll}
xy' + 3y^2 = 2x & \text{(because } y \text{ is squared),} \\
yy' = 3 & \text{(because of the product } yy'\text{),} \\
y' + xe^y = 12 & \text{(because of } e^y\text{).}
\end{array}
$$

GENERAL SOLUTION OF A LINEAR FIRST ORDER EQUATION

To motivate a definition that we'll need, consider the simple linear first order equation

$$y' = \frac{1}{x^2}. \tag{2}$$

From calculus we know that y satisfies this equation if and only if

$$y = -\frac{1}{x} + c, \tag{3}$$

where c is an arbitrary constant. We call c a ***parameter*** and say that (3) defines a ***one-parameter family*** of functions. For each real number c, the function defined by (3) is a solution of (2) on $(-\infty, 0)$ and $(0, \infty)$; moreover, every solution of (2) on either of these intervals is of the form (3) for some choice of c. We say that (3) is the *general solution* of (2).

We will see that a similar situation occurs in connection with any first order linear equation

$$y' + p(x)y = f(x); \tag{4}$$

that is, if p and f are continuous on some open interval (a,b), then there is a unique formula $y = y(x,c)$ analogous to (3) that involves x and a parameter c, with the following properties:

- For each fixed value of c the resulting function of x is a solution of (4) on (a,b).
- If y is a solution of (4) on (a,b) then y can be obtained from the formula by choosing c appropriately.

We will call $y = y(x,c)$ the *general solution* of (4).

When this has been established it will follow that an equation of the form

$$P_0(x)y' + P_1(x)y = F(x) \tag{5}$$

has a general solution on any open interval (a,b) on which P_0, P_1, and F are all continuous and P_0 has no zeros, since in this case we can rewrite (5) in the form (4) with $p = P_1/P_0$ and $f = F/P_0$, which are both continuous on (a,b).

To avoid awkward wording in examples and exercises we won't specify the interval (a,b) when we ask for the general solution of a specific linear first order equation. Let's agree that this always means that we want the general solution on every open interval on which p and f are continuous if the equation is of the form (4), or on which P_0, P_1, and F are continuous and P_0 has no zeros, if the equation is of the form (5). We leave it to you to identify these intervals in specific examples and exercises.

For completeness we point out that if P_0, P_1, and F are all continuous on an open interval (a,b), but P_0 *does* have a zero in (a,b), then (5) may fail to have a general solution on (a,b) in the sense just defined. Since this is not a major point that needs to be developed in depth, we won't discuss it further; however, see Exercise 44 for an example.

HOMOGENEOUS LINEAR FIRST ORDER EQUATIONS

We begin with the problem of finding the general solution of a homogeneous linear first order equation.

The following example recalls a familiar result from calculus.

EXAMPLE 3   Let a be a constant.

(a) Find the general solution of

$$y' - ay = 0. \tag{6}$$

(b) Solve the initial value problem

$$y' - ay = 0, \qquad y(x_0) = y_0.$$

Solution   **(a)** You already know from calculus that if c is any constant then $y = ce^{ax}$ satisfies (6). However, let's pretend you've forgotten this, and use this problem to illustrate a general method for solving a homogeneous linear first order equation.

We know that (6) has the trivial solution $y \equiv 0$. Now suppose that y is a nontrivial solution of (6). Then, since a differentiable function must be continuous, there must be some open interval I on which y has no zeros. We rewrite (6) as

$$\frac{y'}{y} = a$$

for x in I. Integrating this shows that

$$\ln |y| = ax + k, \qquad \text{so} \qquad |y| = e^k e^{ax}$$

where k is an arbitrary constant. Since e^{ax} can never equal zero, it follows that y has no zeros, so y is either always positive or always negative. Therefore we can rewrite y as

$$y = c e^{ax} \tag{7}$$

where

$$c = \begin{cases} e^k & \text{if } y > 0, \\ -e^k & \text{if } y < 0. \end{cases}$$

This shows that every nontrivial solution of (6) is of the form $y = c e^{ax}$ for some nonzero constant c. Since setting $c = 0$ yields the trivial solution, it follows that *all* solutions of (6) are of the form (7). Conversely, (7) is a solution of (6) for every choice of c, since differentiating (7) yields $y' = ace^{ax} = ay$.

Solution   **(b)** Imposing the initial condition $y(x_0) = y_0$ yields $y_0 = c e^{ax_0}$, so $c = y_0 e^{-ax_0}$ and

$$y = y_0 e^{-ax_0} e^{ax} = y_0 e^{a(x-x_0)}.$$

Graphs of this function with $x_0 = 0$, $y_0 = 1$, and various values of a, are shown in Figure 1.

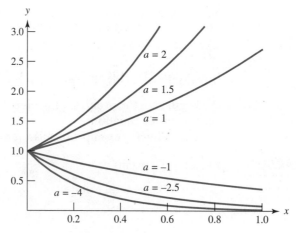

Figure 1   Graphs of solutions of
$$y' - ay = 0, \quad y(0) = 1$$

EXAMPLE 4   **(a)**   Find the general solution of

$$xy' + y = 0. \tag{8}$$

(b) Solve the initial value problem

$$xy' + y = 0, \qquad y(1) = 3. \tag{9}$$

Solution **(a)** We rewrite (8) as

$$y' + \frac{1}{x}y = 0, \tag{10}$$

where x is restricted to either $(-\infty, 0)$ or $(0, \infty)$. If y is a nontrivial solution of (10) then there must be some open interval I on which y has no zeros. We can rewrite (10) as

$$\frac{y'}{y} = -\frac{1}{x}$$

for x in I. Integrating shows that

$$\ln|y| = -\ln|x| + k, \qquad \text{so} \qquad |y| = \frac{e^k}{|x|}.$$

Since a function that satisfies the last equation can't change sign on either $(-\infty, 0)$ or $(0, \infty)$, we can rewrite this result more simply as

$$y = \frac{c}{x} \tag{11}$$

where

$$c = \begin{cases} e^k & \text{if } y > 0, \\ -e^k & \text{if } y < 0. \end{cases}$$

We have now shown that every solution of (10) is given by (11) for some choice of c. (Even though we assumed that y was nontrivial to derive (11), we can get the trivial solution by setting $c = 0$ in (11).) Conversely, any function of the form (11) is a solution of (10), since differentiating (11) yields

$$y' = -\frac{c}{x^2},$$

and substituting this and (11) into (10) yields

$$y' + \frac{1}{x}y = -\frac{c}{x^2} + \frac{1}{x}\frac{c}{x}$$

$$= -\frac{c}{x^2} + \frac{c}{x^2} = 0.$$

Figure 2 show the graphs of some solutions corresponding to various values of c.

Solution **(b)** Imposing the initial condition $y(1) = 3$ in (11) yields $c = 3$. Therefore the solution of (9) is

$$y = \frac{3}{x}.$$

The interval of validity of this solution is $(0, \infty)$. ■

The results in Examples 3**(a)** and 4**(a)** are special cases of the following theorem.

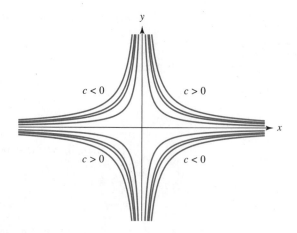

Figure 2   Graphs of solutions
of $xy' + y = 0$ on $(0,\infty)$ and
$(-\infty,0)$

THEOREM 2.1.1

Suppose that p is continuous on (a,b). Then the general solution of the homogeneous equation

$$y' + p(x)y = 0 \tag{12}$$

on (a,b) is

$$y = ce^{-P(x)},$$

where

$$P(x) = \int p(x)\, dx \tag{13}$$

is any antiderivative of p on (a,b); that is,

$$P'(x) = p(x), \qquad a < x < b. \tag{14}$$

PROOF. If $y = ce^{-P(x)}$ then differentiating y and using (14) shows that

$$y' = -P'(x)ce^{-P(x)} = -p(x)ce^{-P(x)} = -p(x)y,$$

so $y' + p(x)y = 0$; that is, y is a solution of (12), for any choice of c.

Now we will show that any solution of (12) can be written as $y = ce^{-P(x)}$ for some constant c. The trivial solution can be written this way, with $c = 0$. Now suppose that y is a nontrivial solution. Then there is an open subinterval I of (a,b) on which y has no zeros. We can rewrite (12) as

$$\frac{y'}{y} = -p(x) \tag{15}$$

for x in I. Integrating (15) and recalling (13) yields

$$\ln|y| = -P(x) + k,$$

where k is a constant. This implies that

$$|y| = e^k e^{-P(x)}.$$

Since P is defined for all x in (a,b) and an exponential function can never equal zero, it follows that we can take $I = (a,b)$ and that y cannot have any zeros on (a,b). Therefore y cannot change sign on (a,b), so we can rewrite the last equation as $y = ce^{-P(x)}$, where

$$c = \begin{cases} e^k & \text{if } y > 0 \text{ on } (a,b), \\ -e^k & \text{if } y < 0 \text{ on } (a,b). \end{cases} \qquad \square$$

REMARK. Rewriting a first order differential equation so that one side depends only on y and y' while the other depends only on x is called ***separation of variables***. We did this in Examples 3 and 4, and in rewriting (12) as (15). We will apply this method to nonlinear equations in Section 2.2.

LINEAR NONHOMOGENEOUS FIRST ORDER EQUATIONS

We will now solve the nonhomogeneous equation

$$y' + p(x)y = f(x). \tag{16}$$

When considering this equation we call

$$y' + p(x)y = 0$$

the ***complementary equation.***

We will find solutions of (16) in the form $y = uy_1$, where y_1 is a nontrivial solution of the complementary equation and u is to be determined. This method of using a solution of the complementary equation to obtain solutions of a nonhomogeneous equation is a special case of a method called ***variation of parameters,*** which you will encounter several times in this book. (Obviously, u can't be constant, since if it were then the left side of (16) would be zero. Recognizing this, the early users of this method viewed u as a "parameter" that varies; hence, the name "variation of parameters.")

If

$$y = uy_1 \qquad \text{then} \qquad y' = u'y_1 + uy_1',$$

Substituting these expressions for y and y' into (16) yields

$$u'y_1 + u(y_1' + p(x)y_1) = f(x),$$

which reduces to

$$u'y_1 = f(x) \tag{17}$$

since y_1 is a solution of the complementary equation; that is,

$$y_1' + p(x)y_1 = 0.$$

In the proof of Theorem 2.1.1 we saw that y_1 has no zeros on an interval where p is continuous. Therefore we can divide (17) through by y_1 to obtain

$$u' = f(x)/y_1(x).$$

We can integrate this (introducing a constant of integration), and multiply the result by y_1 to get the general solution of (16). Before turning to the formal proof of this claim, let's consider some examples.

EXAMPLE 5 Find the general solution of

$$y' + 2y = x^3 e^{-2x}. \tag{18}$$

Solution By applying part **(a)** of Example 3 with $a = -2$ we see that $y_1 = e^{-2x}$ is a solution of the complementary equation $y' + 2y = 0$. Therefore we seek solutions of (18) in the form $y = ue^{-2x}$, so that

$$y' = u'e^{-2x} - 2ue^{-2x} \quad \text{and} \quad y' + 2y = u'e^{-2x} - 2ue^{-2x} + 2ue^{-2x} = u'e^{-2x}. \tag{19}$$

Therefore y is a solution of (18) if and only if

$$u'e^{-2x} = x^3 e^{-2x} \quad \text{or, equivalently,} \quad u' = x^3.$$

Therefore

$$u = \frac{x^4}{4} + c,$$

and

$$y = ue^{-2x} = e^{-2x}\left(\frac{x^4}{4} + c\right)$$

is the general solution of (18).

Figure 3 shows a direction field and some integral curves of (18).

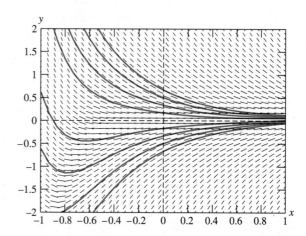

Figure 3 A direction field and integral curves of $y' + 2y = x^3 e^{-2x}$

EXAMPLE 6 **(a)** Find the general solution

$$y' + (\cot x)y = x \csc x. \tag{20}$$

(b) Solve the initial value problem

$$y' + (\cot x)y = x \csc x, \qquad y(\pi/2) = 1. \tag{21}$$

Solution **(a)** Here $p(x) = \cot x$ and $f(x) = x \csc x$ are both continuous except at the points $x = r\pi$, where r is an integer. Therefore we seek solutions of (20) on the intervals $(r\pi, (r + 1)\pi)$. We need a nontrival solution y_1 of the complementary equation; thus, y_1 must satisfy $y_1' + (\cot x)y_1 = 0$, which we rewrite as

$$\frac{y_1'}{y_1} = -\cot x = -\frac{\cos x}{\sin x}. \tag{22}$$

Integrating this yields

$$\ln |y_1| = -\ln |\sin x|,$$

where we take the constant of integration to be zero since we need only *one* function that satisfies (22). Clearly $y_1 = 1/\sin x$ is a suitable choice. Therefore, we seek solutions of (20) in the form

$$y = \frac{u}{\sin x},$$

so that

$$y' = \frac{u'}{\sin x} - \frac{u \cos x}{\sin^2 x} \qquad (23)$$

and

$$y' + (\cot x)y = \frac{u'}{\sin x} - \frac{u \cos x}{\sin^2 x} + \frac{u \cot x}{\sin x}$$

$$= \frac{u'}{\sin x} - \frac{u \cos x}{\sin^2 x} + \frac{u \cos x}{\sin^2 x} \qquad (24)$$

$$= \frac{u'}{\sin x}.$$

Therefore y is a solution of (20) if and only if

$$u'/\sin x = x \csc x = x/\sin x \qquad \text{or, equivalently,} \qquad u' = x.$$

Integrating this yields

$$u = \frac{x^2}{2} + c,$$

and

$$y = \frac{u}{\sin x} = \frac{x^2}{2 \sin x} + \frac{c}{\sin x}. \qquad (25)$$

is the general solution of (20) on every interval $(r\pi, (r+1)\pi)$.

Solution **(b)** Imposing the initial condition $y(\pi/2) = 1$ in (25) yields

$$1 = \frac{\pi^2}{8} + c, \qquad \text{or} \qquad c = 1 - \frac{\pi^2}{8}.$$

Thus,

$$y = \frac{x^2}{2 \sin x} + \frac{(1 - \pi^2/8)}{\sin x}$$

is a solution of (21). The interval of validity of this solution is $(0, \pi)$. Its graph is shown in Figure 4. ■

REMARK. It wasn't necessary to do the computations (23) and (24) in Example 6, since we showed in the discussion preceding Example 5 that if $y = uy_1$ where $y_1' + p(x)y_1 = 0$, then $y' + p(x)y = u'y_1$. We did these computations so you would see this happen in this specific example. We recommend that you

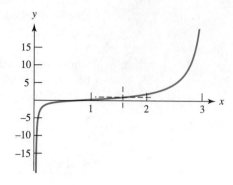

Figure 4   Graph of the solution of
$y' + (\cot x)y = x \csc x,$
$y(\pi/2) = 1$

include these "unnecessary" computations in doing exercises, until you're confident that you really understand the method. After that, omit them.

We summarize the method of variation of parameters for solving

$$y' + p(x)y = f(x) \tag{26}$$

as follows:

1. Find a function y_1 such that

$$\frac{y_1'}{y_1} = -p(x).$$

For convenience, take the constant of integration to be zero.
2. Write

$$y = uy_1 \tag{27}$$

to remind yourself of what you're doing.
3. Write $u'y_1 = f$ and solve for u'; thus $u' = f/y_1$.
4. Integrate u' to obtain u, with an arbitrary constant of integration.
5. Substitute u into (27) to obtain y.

To solve an equation written in the form

$$P_0(x)y' + P_1(x)y = F(x),$$

we recommend that you divide through by $P_0(x)$ to obtain an equation of the form (26) and then follow this procedure.

SOLUTIONS IN INTEGRAL FORM

Sometimes the integrals that arise in solving a linear first order equation cannot be evaluated in terms of elementary functions. In this case the solution must be left in terms of an integral.

EXAMPLE 7   **(a)** Find the general solution of

$$y' - 2xy = 1.$$

(b) Solve the initial value problem

$$y' - 2xy = 1, \qquad y(0) = y_0. \tag{28}$$

Solution (a) To apply variation of parameters we need a nontrivial solution y_1 of the complementary equation; thus, $y_1' - 2xy_1 = 0$, which we rewrite as

$$\frac{y_1'}{y_1} = 2x.$$

Integrating this and taking the constant of integration to be zero yields

$$\ln |y_1| = x^2, \quad \text{so} \quad |y_1| = e^{x^2}.$$

We choose $y_1 = e^{x^2}$ and seek solutions of (28) in the form $y = ue^{x^2}$, where

$$u'e^{x^2} = 1, \quad \text{so} \quad u' = e^{-x^2}.$$

Therefore

$$u = c + \int e^{-x^2}\, dx,$$

but we can't simplify the integral on the right because there is no elementary function with derivative equal to e^{-x^2}. Therefore, the best available form for the general solution of (28) is

$$y = ue^{x^2} = e^{x^2}\left(c + \int e^{-x^2}\, dx\right). \tag{29}$$

Solution (b) Since the initial condition in (28) is imposed at $x_0 = 0$, it is convenient to rewrite (29) as

$$y = e^{x^2}\left(c + \int_0^x e^{-t^2}\, dt\right), \quad \text{since} \quad \int_0^0 e^{-t^2}\, dt = 0.$$

Setting $x = 0$ and $y = y_0$ here shows that $c = y_0$. Therefore, the solution of the initial value problem is

$$y = e^{x^2}\left(y_0 + \int_0^x e^{-t^2}\, dt\right). \tag{30}$$

For a given value of y_0 and each fixed x, the integral on the right can be evaluated by numerical methods. An alternate procedure is to apply the numerical integration procedures discussed in Chapter 3 directly to the initial value problem (28).

The graphs of (30) for several values of y_0 are shown in Figure 5. ■

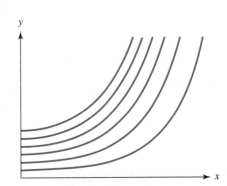

Figure 5 Graphs of solutions of
$y' - 2xy = 1, \quad y(0) = y_0$

EXISTENCE AND UNIQUENESS THEOREMS

The method of variation of parameters leads to the following theorem.

THEOREM 2.1.2

Suppose that p and f are continuous on an open interval (a,b), and let y_1 be any nontrivial solution of the complementary equation

$$y' + p(x)y = 0$$

on (a,b). Then:

(a) *The general solution of the nonhomogeneous equation*

$$y' + p(x)y = f(x) \tag{31}$$

on (a,b) is

$$y = y_1(x)\left(c + \int f(x)/y_1(x)\ dx\right). \tag{32}$$

(b) *If x_0 is an arbitrary point in (a,b) and y_0 is an arbitrary real number, then the initial value problem*

$$y' + p(x)y = f(x), \qquad y(x_0) = y_0$$

has the unique solution

$$y = y_1(x)\left(\frac{y_0}{y_1(x_0)} + \int_{x_0}^{x} \frac{f(t)}{y_1(t)}\ dt\right)$$

on (a,b).

PROOF. **(a)** To show that (32) is the general solution of (31) on (a,b) we must prove the following:

(i) If c is any constant then the function y in (32) is a solution of (31) on (a,b).
(ii) If y is a solution of (31) on (a,b), then y is of the form (32) for some constant c.

To prove **(i)**, we first observe that any function of the form (32) is defined on (a,b), since p and f are continuous on (a,b). Differentiating (32) yields

$$y' = y_1'(x)\left(c + \int f(x)/y_1(x)\ dx\right) + f(x).$$

Since $y_1' = -p(x)y_1$, this and (32) imply that

$$y' = -p(x)y_1(x)\left(c + \int f(x)/y_1(x)\ dx\right) + f(x)$$
$$= -p(x)y(x) + f(x),$$

which implies that y is a solution of (31).

To prove **(ii)**, suppose that y is a solution of (31) on (a,b). From the proof of Theorem 2.1.1 we know that y_1 has no zeros on (a,b), so the function $u = y/y_1$ is defined on (a,b). Moreover, since

$$y' = -py + f \qquad \text{and} \qquad y_1' = -py_1$$

it follows that

$$u' = \frac{y_1 y' - y_1' y}{y_1^2}$$

$$= \frac{y_1(-py + f) - (-py_1)y}{y_1^2} = \frac{f}{y_1}.$$

Integrating $u' = f/y_1$ yields

$$u = \left(c + \int f(x)/y_1(x)\, dx \right),$$

which implies (32), since $y = uy_1$.

   (b) We have proved **(a)** where $\int f(x)/y_1(x)\, dx$ in (32) is an arbitrary anti-derivative of f/y_1. Now it is convenient to choose the antiderivative that equals zero when $x = x_0$, and write the general solution of (31) as

$$y = y_1(x)\left(c + \int_{x_0}^{x} \frac{f(t)}{y_1(t)}\, dt \right).$$

Since

$$y(x_0) = y_1(x_0)\left(c + \int_{x_0}^{x_0} \frac{f(t)}{y_1(t)}\, dt \right) = cy_1(x_0),$$

we see that $y(x_0) = y_0$ if and only if $c = y_0/y_1(x_0)$. □

2.1 EXERCISES

In Exercises 1–5 find the general solution.

1. $y' + ay = 0$ $(a = \text{constant})$

2. $y' + 3x^2 y = 0$

3. $xy' + (\ln x)y = 0$

4. $xy' + 3y = 0$

5. $x^2 y' + y = 0$

In Exercises 6–11 solve the initial value problem.

6. $y' + \left(\dfrac{1 + x}{x} \right)y = 0,$ $y(1) = 1$

7. $xy' + \left(1 + \dfrac{1}{\ln x} \right)y = 0,$ $y(e) = 1$

8. $xy' + (1 + x \cot x)y = 0,$ $y\left(\dfrac{\pi}{2} \right) = 2$

9. $y' - \left(\dfrac{2x}{1 + x^2} \right)y = 0,$ $y(0) = 2$

10. $y' + \dfrac{k}{x}y = 0,$ $y(1) = 3$ $(k = \text{constant})$

11. $y' + (\tan kx)y = 0,$ $y(0) = 2$ $(k = \text{constant})$

In Exercises 12–15 find the general solution. Also, plot a direction field and some integral curves on the rectangular region $\{-2 \le x \le 2, -2 \le y \le 2\}$.

C **12.** $y' + 3y = 1$

C **13.** $y' + \left(\dfrac{1}{x} - 1 \right)y = -\dfrac{2}{x}$

C **14.** $y' + 2xy = xe^{-x^2}$

C **15.** $y' + \dfrac{2x}{1 + x^2}y = \dfrac{e^{-x}}{1 + x^2}$

In Exercises 16–24 find the general solution.

16. $y' + \dfrac{1}{x}y = \dfrac{7}{x^2} + 3$

17. $y' + \dfrac{4}{x - 1}y = \dfrac{1}{(x - 1)^5} + \dfrac{\sin x}{(x - 1)^4}$

18. $xy' + (1 + 2x^2)y = x^3 e^{-x^2}$

19. $xy' + 2y = \dfrac{2}{x^2} + 1$

20. $y' + (\tan x)y = \cos x$

21. $(1 + x)y' + 2y = \dfrac{\sin x}{1 + x}$

22. $(x - 2)(x - 1)y' - (4x - 3)y = (x - 2)^3$

23. $y' + (2\sin x \cos x)y = e^{-\sin^2 x}$

24. $x^2 y' + 3xy = e^x$

In Exercises 25–29 solve the initial value problem and sketch the graph of the solution.

C **25.** $y' + 7y = e^{3x}, \quad y(0) = 0$

C **26.** $(1 + x^2)y' + 4xy = \dfrac{2}{1 + x^2}, \quad y(0) = 1$

C **27.** $xy' + 3y = \dfrac{2}{x(1 + x^2)}, \quad y(-1) = 0$

C **28.** $y' + (\cot x)y = \cos x, \quad y\left(\dfrac{\pi}{2}\right) = 1$

C **29.** $y' + \dfrac{1}{x}y = \dfrac{2}{x^2} + 1, \quad y(-1) = 0$

In Exercises 30–37 solve the initial value problem.

30. $(x - 1)y' + 3y = \dfrac{1}{(x - 1)^3} + \dfrac{\sin x}{(x - 1)^2}, \quad y(0) = 1$

31. $xy' + 2y = 8x^2, \quad y(1) = 3$

32. $xy' - 2y = -x^2, \quad y(1) = 1$

33. $y' + 2xy = x, \quad y(0) = 3$

34. $(x - 1)y' + 3y = \dfrac{1 + (x - 1)\sec^2 x}{(x - 1)^3}, \quad y(0) = -1$

35. $(x + 2)y' + 4y = \dfrac{1 + 2x^2}{x(x + 2)^3}, \quad y(-1) = 2$

36. $(x^2 - 1)y' - 2xy = x(x^2 - 1), \quad y(0) = 4$

37. $(x^2 - 5)y' - 2xy = -2x(x^2 - 5), \quad y(2) = 7$

In Exercises 38–42 solve the initial value problem and leave the answer in a form involving a definite integral. (You can solve these problems numerically by methods discussed in Chapter 3.)

38. $y' + 2xy = x^2, \quad y(0) = 3$

39. $y' + \dfrac{1}{x}y = \dfrac{\sin x}{x^2}, \quad y(1) = 2$

40. $y' + y = \dfrac{e^{-x}\tan x}{x}, \quad y(1) = 0$

41. $y' + \dfrac{2x}{1 + x^2}y = \dfrac{e^x}{(1 + x^2)^2}, \quad y(0) = 1$

42. $xy' + (x + 1)y = e^{x^2}, \quad y(1) = 2$

43. Experiments indicate that glucose is absorbed by the body at a rate proportional to the amount of glucose present in the bloodstream. Let λ denote the (positive) constant of proportionality. Now suppose that glucose is injected into a patient's bloodstream at a constant rate of r units per unit of time. Let $G = G(t)$ be the number of units in the patient's bloodstream at time $t > 0$. Then

$$G' = -\lambda G + r,$$

where the first term on the right is due to the absorption of the glucose by the patient's body and the second term is due to the injection. Determine G for $t > 0$, given that $G(0) = G_0$. Also, find $\lim_{t \to \infty} G(t)$.

L **44.** **(a)** Plot a direction field and some integral curves of

$$xy' - 2y = -1 \qquad\qquad\qquad \text{(A)}$$

on the rectangular region $\{-1 \le x \le 1, \; -0.5 \le y \le 1.5\}$. What do all the integral curves have in common?

(b) Show that the general solution of (A) on $(-\infty, 0)$ and $(0, \infty)$ is

$$y = \dfrac{1}{2} + cx^2.$$

(c) Show that y is a solution of (A) on $(-\infty, \infty)$ if and only if

$$y = \begin{cases} \frac{1}{2} + c_1 x^2, & x \geq 0, \\ \frac{1}{2} + c_2 x^2, & x < 0, \end{cases}$$

where c_1 and c_2 are arbitrary constants.

(d) Conclude from (c) that all solutions of (A) on $(-\infty, \infty)$ are solutions of the initial value problem

$$xy' - 2y = -1, \qquad y(0) = \frac{1}{2}.$$

(e) Use (b) to show that if $x_0 \neq 0$ and y_0 is arbitrary, then the initial value problem

$$xy' - 2y = -1, \qquad y(x_0) = y_0$$

has infinitely many solutions on $(-\infty, \infty)$. Explain why this doesn't contradict Theorem 2.1.1 (b).

45. Suppose that f is continuous on an open interval (a, b) and α is a constant.

(a) Derive a formula for the solution of the initial value problem

$$y' + \alpha y = f(x), \qquad y(x_0) = y_0, \tag{A}$$

where x_0 is in (a, b) and y_0 is an arbitrary real number.

(b) Suppose that $(a, b) = (a, \infty)$, $\alpha > 0$ and $\lim_{x \to \infty} f(x) = L$. Show that if y is the solution of (A), then $\lim_{x \to \infty} y(x) = L/\alpha$.

46. Assume that all functions in this exercise are defined on a common interval (a, b).

(a) Prove: If y_1 and y_2 are solutions of

$$y' + p(x)y = f_1(x)$$

and

$$y' + p(x)y = f_2(x),$$

respectively, and c_1 and c_2 are constants, then $y = c_1 y_1 + c_2 y_2$ is a solution of

$$y' + p(x)y = c_1 f_1(x) + c_2 f_2(x).$$

(This is known as the ***principle of superposition.***)

(b) Use (a) to show that if y_1 and y_2 are solutions of the nonhomogeneous equation

$$y' + p(x)y = f(x), \tag{A}$$

then $y_1 - y_2$ is a solution of the homogeneous equation

$$y' + p(x)y = 0. \tag{B}$$

(c) Use (a) to show that if y_1 is a solution of (A) and y_2 is a solution of (B), then $y_1 + y_2$ is a solution of (A).

47. Some nonlinear equations can be transformed into linear equations by changing the dependent variable. Show that if

$$g'(y)y' + p(x)g(y) = f(x),$$

where y is a function of x and g is a function of y, then the new dependent variable $z = g(y)$ satisfies the linear equation

$$z' + p(x)z = f(x).$$

48. Solve by the method discussed in Exercise 47:

(a) $(\sec^2 y)y' - 3 \tan y = -1$

(b) $e^{y^2}\left(2yy' + \dfrac{2}{x}\right) = \dfrac{1}{x^2}$

(c) $\dfrac{xy'}{y} + 2 \ln y = 4x^2$ **(d)** $\dfrac{y'}{(1+y)^2} - \dfrac{1}{x(1+y)} = -\dfrac{3}{x^2}$

49. We have shown that if p and f are continuous on (a,b) then every solution of

$$y' + p(x)y = f(x) \tag{A}$$

on (a,b) can be written as $y = uy_1$, where y_1 is a nontrivial solution of the complementary equation for (A) and $u' = f/y_1$. Now suppose that $f, f', \ldots, f^{(m)}$ and $p, p', \ldots, p^{(m-1)}$ are continuous on (a,b), where m is a positive integer, and define

$$f_0 = f,$$

$$f_j = f'_{j-1} + pf_{j-1}, \quad 1 \le j \le m.$$

Show that

$$u^{(j+1)} = \frac{f_j}{y_1}, \quad 0 \le j \le m.$$

2.2   Separable Equations

A first order differential equation is said to be **separable** if it can be written in the form

$$h(y)y' = g(x), \tag{1}$$

where the left side is a product of y' and a function of y, while the right side is a function of x. Rewriting a separable differential equation in this form is called **separation of variables.** In Section 2.1 we used separation of variables to solve homogeneous linear equations. In this section we will apply this method to non-linear equations.

To see how to solve (1), let's first assume that y is a solution. Let $G(x)$ and $H(y)$ be antiderivatives of $g(x)$ and $h(y)$; that is,

$$H'(y) = h(y) \quad \text{and} \quad G'(x) = g(x). \tag{2}$$

Then, from the chain rule,

$$\frac{d}{dx} H(y(x)) = H'(y(x))y'(x) = h(y)y'(x).$$

Therefore, (1) is equivalent to

$$\frac{d}{dx} H(y(x)) = \frac{d}{dx} G(x).$$

Integrating both sides of this equation and combining the constants of integration yields

$$H(y(x)) = G(x) + c. \tag{3}$$

Although we derived this equation on the assumption that y is a solution of (1), we can now view it differently: Any differentiable function y that satisfies (3) for some constant c is a solution of (1). To see this we differentiate both sides of (3), using the chain rule on the left, to obtain

$$H'(y(x))y'(x) = G'(x),$$

which is equivalent to

$$h(y(x))y'(x) = g(x)$$

because of (2).

In conclusion, to solve (1) it suffices to find functions $G = G(x)$ and $H = H(y)$ that satisfy (2). Then any differentiable function $y = y(x)$ that satisfies (3) is a solution of (1).

EXAMPLE 1   Solve the equation

$$y' = x(1 + y^2).$$

Solution   Separating variables yields

$$\frac{y'}{1 + y^2} = x.$$

Integrating yields

$$\tan^{-1} y = \frac{x^2}{2} + c.$$

Therefore,

$$y = \tan\left(\frac{x^2}{2} + c\right).$$

EXAMPLE 2   **(a)** Solve the equation

$$y' = -\frac{x}{y}. \tag{4}$$

(b) Solve the initial value problem

$$y' = -\frac{x}{y}, \qquad y(1) = 1. \tag{5}$$

(c) Solve the initial value problem

$$y' = -\frac{x}{y}, \qquad y(1) = -2. \tag{6}$$

Solution   **(a)** Separating variables in (4) yields

$$yy' = -x.$$

Integrating yields

$$\frac{y^2}{2} = -\frac{x^2}{2} + c, \qquad \text{or, equivalently,} \qquad x^2 + y^2 = 2c.$$

The last equation shows that c must be positive if y is to be a solution of (4) on an open interval. Therefore we let $2c = a^2$ (with $a > 0$) and rewrite the last equation as

$$x^2 + y^2 = a^2. \tag{7}$$

This equation has two differentiable solutions for y in terms of x:

$$y = \sqrt{a^2 - x^2}, \qquad -a < x < a, \tag{8}$$

and

$$y = -\sqrt{a^2 - x^2}, \qquad -a < x < a. \tag{9}$$

The solution curves defined by (8) are semicircles above the x-axis, while those defined by (9) are semicircles below the x-axis.

Solution   **(b)** The solution of (5) is positive when $x = 1$; hence, it is of the form (8). Substituting $x = 1$ and $y = 1$ into (7) to satisfy the initial condition yields $a^2 = 2$; hence, the solution of (5) is

$$y = \sqrt{2 - x^2}, \qquad -\sqrt{2} < x < \sqrt{2}.$$

The integral curve for this solution is the semicircle above the x-axis in Figure 1.

Solution   **(c)** The solution of (6) is negative when $x = 1$ and is therefore of the form (9). Substituting $x = 1$ and $y = -2$ into (7) to satisfy the initial condition yields $a^2 = 5$. Hence, the solution of (6) is

$$y = -\sqrt{5 - x^2}, \qquad -\sqrt{5} < x < \sqrt{5}.$$

The integral curve for this solution is the semicircle below the x-axis in Figure 1. ∎

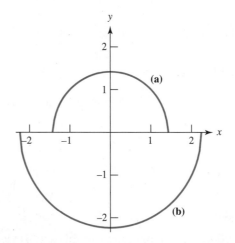

Figure 1   **(a)** $y = \sqrt{2 - x^2}, -\sqrt{2} < x < \sqrt{2}$;
(b) $y = -\sqrt{5 - x^2}, -\sqrt{5} < x < \sqrt{5}$

IMPLICIT SOLUTIONS OF SEPARABLE EQUATIONS

In Examples 1 and 2 we were able to solve the equation $H(y) = G(x) + c$ to obtain explicit formulas for solutions of the given separable differential equations. As we will see in the next example, this is not always possible. In this situation we must broaden our definition of a solution of a separable equation. The following theorem provides the basis for this modification. We omit the proof, which requires a result from advanced calculus known as the ***implicit function theorem.***

THEOREM 2.2.I

Suppose that $g = g(x)$ is continuous on (a,b) and $h = h(y)$ is continuous on (c,d). Let G be an antiderivative of g on (a,b) and H be an antiderivative of h on (c,d). Let x_0 be an arbitrary point in (a,b), let y_0 be a point in (c,d) such that $h(y_0) \neq 0$, and define

$$c = H(y_0) - G(x_0). \tag{10}$$

Then there is a function $y = y(x)$ defined on some open interval (a_1,b_1), where $a \leq a_1 < x_0 < b_1 \leq b$, such that $y(x_0) = y_0$ and

$$H(y) = G(x) + c \tag{11}$$

for $a_1 < x < b_1$. Therefore, y is a solution of the initial value problem

$$h(y)y' = g(x), \qquad y(x_0) = x_0. \tag{12}$$

It is convenient to say that (11) with c arbitrary is an ***implicit solution*** of $h(y)y' = g(x)$. Curves defined by (11) are integral curves of $h(y)y' = g(x)$. If c satisfies (10) then we will say that (11) is an ***implicit solution of the initial value problem*** (12). However, keep the following points in mind:

- For some choices of c there may not be any differentiable functions y that satisfy (11).
- The function y in (11) (not (11) itself) is a solution of $h(y)y' = g(x)$.

EXAMPLE 3   **(a)** Find implicit solutions of

$$y' = \frac{2x + 1}{5y^4 + 1}. \tag{13}$$

(b) Find an implicit solution of

$$y' = \frac{2x + 1}{5y^4 + 1}, \qquad y(2) = 1. \tag{14}$$

Solution   **(a)** Separating variables yields

$$(5y^4 + 1)y' = 2x + 1.$$

Integrating yields the implicit solution

$$y^5 + y = x^2 + x + c. \tag{15}$$

of (13).

Solution   **(b)** Imposing the initial condition $y(2) = 1$ in (15) yields $1 + 1 = 4 + 2 + c$, so $c = -4$. Therefore

$$y^5 + y = x^2 + x - 4$$

is an implicit solution of the initial value problem (14). Although more than one differentiable function $y = y(x)$ satisfies (13) near $x = 1$, it can be shown that there is only one such function that satisfies the initial condition $y(1) = 2$.

Figure 2 shows a direction field and some integral curves of (13). The integral curve through $(2,1)$ is the graph of the solution of the initial value problem (14).

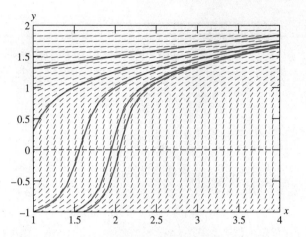

Figure 2 A direction field and integral curves of
$$y' = \frac{2x + 1}{5y^4 + 1}.$$

CONSTANT SOLUTIONS OF SEPARABLE EQUATIONS

An equation of the form

$$y' = g(x)p(y)$$

is separable, since it can be rewritten as

$$\frac{1}{p(y)}y' = g(x).$$

However, the division by $p(y)$ is not legitimate if $p(y) = 0$ for some values of y. The next two examples show how to deal with this problem.

EXAMPLE 4 Find all solutions of

$$y' = 2xy^2. \tag{16}$$

Solution Here we must divide by $p(y) = y^2$ to separate variables. This is not legitimate if y is a solution of (16) that equals zero for some value of x. One such solution can be found by inspection: namely, $y \equiv 0$. Now suppose that y is a solution of (16) that is not identically zero. Since y is continuous there must be an interval on which y is never zero. Since division by y^2 is legitimate for x in this interval, we can separate variables in (16) to obtain

$$\frac{y'}{y^2} = 2x.$$

Integrating this yields

$$-\frac{1}{y} = x^2 + c,$$

which is equivalent to

$$y = -\frac{1}{x^2 + c}. \tag{17}$$

We have now shown that if y is a solution of (16) that is not identically zero, then y must be of the form (17). By substituting (17) into (16) you can show that (17) is in fact a solution of (16). Thus, it follows that the solutions of (16) are $y \equiv 0$ and the functions of the form (17). Note that the solution $y \equiv 0$ is not of the form (17) for any value of c.

Figure 3 shows a direction field and some integral curves of (16).

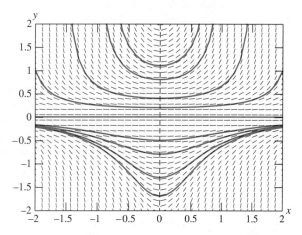

Figure 3   A direction field and integral curves of
$$y' = 2xy^2$$

EXAMPLE 5   Find all solutions of

$$y' = \frac{1}{2}x(1 - y^2). \tag{18}$$

Solution   Here we must divide by $p(y) = 1 - y^2$ to separate variables. This is not legitimate if y is a solution of (18) that equals ± 1 for some value of x. Two such solutions can be found by inspection: namely, $y \equiv 1$ and $y \equiv -1$. Now suppose that y is a solution of (18) such that $1 - y^2$ is not identically zero. Since $1 - y^2$ is continuous there must be an interval on which $1 - y^2$ is never zero. Since division by $1 - y^2$ is legitimate for x in this interval, we can separate variables in (18) to obtain

$$\frac{2y'}{y^2 - 1} = -x.$$

A partial fraction expansion on the left yields

$$\left[\frac{1}{y - 1} - \frac{1}{y + 1} \right] y' = -x,$$

and integrating yields

$$\ln \left| \frac{y - 1}{y + 1} \right| = -\frac{x^2}{2} + k;$$

hence,

$$\left| \frac{y - 1}{y + 1} \right| = e^k e^{-x^2/2}.$$

Since $y(x) \neq \pm 1$ for x on the interval under discussion, $(y - 1)/(y + 1)$ cannot change sign in this interval. Therefore we can rewrite the last equation as

$$\frac{y - 1}{y + 1} = ce^{-x^2/2},$$

where $c = \pm e^k$, depending upon the sign of $(y - 1)/(y + 1)$ on the interval. Solving for y, we obtain

$$y = \frac{1 + ce^{-x^2/2}}{1 - ce^{-x^2/2}}. \tag{19}$$

We have now shown that if y is a solution of (18) that is not identically equal to ± 1, then y must be of the form (19). By substituting (19) into (18) you can show that (19) is in fact a solution of (18). Thus, it follows that the solutions of (18) are $y \equiv 1$, $y \equiv -1$, and the functions of the form (19). Note that the constant solution $y \equiv 1$ can be obtained from this formula by taking $c = 0$; however, the other constant solution, $y \equiv -1$, cannot be obtained in this way.

Figure 4 shows a direction field and some integral curves of (18). ■

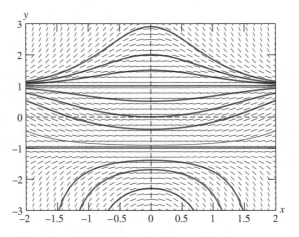

Figure 4 A direction field and integral curves of
$$y' = \frac{1}{2}x(1 - y^2)$$

DIFFERENCES BETWEEN LINEAR AND NONLINEAR EQUATIONS

Theorem 2.1.2 states that if p and f are continuous on (a,b) then every solution of

$$y' + p(x)y = f(x)$$

on (a,b) can be obtained by choosing a value for the constant c in the general solution, and that if x_0 is any point in (a,b) and y_0 is arbitrary, then the initial value problem

$$y' + p(x)y = f(x), \qquad y(x_0) = y_0$$

has a solution on (a,b).

The situation is quite different for nonlinear equations. First, we saw in Examples 4 and 5 that a nonlinear equation may have solutions that cannot be obtained by choosing a specific value of a constant appearing in a one-parameter family of solutions. Second, in general it is impossible to determine the interval of validity of a solution of an initial value problem for a nonlinear equation by simply examining the equation, since the interval of validity may depend on the initial condition. For instance, in Example 2 we saw that the solution of

$$\frac{dy}{dx} = -\frac{x}{y}, \qquad y(x_0) = y_0$$

is valid on $(-a, a)$, where $a = \sqrt{x_0^2 + y_0^2}$.

EXAMPLE 6 Solve the initial value problem

$$y' = 2xy^2, \qquad y(0) = y_0$$

and determine the interval of validity of the solution.

Solution First suppose that $y_0 \neq 0$. From Example 4 we know that y must be of the form

$$y = -\frac{1}{x^2 + c}. \tag{20}$$

Imposing the initial condition shows that $c = -1/y_0$. Substituting this into (20) and rearranging terms yields the solution

$$y = \frac{y_0}{1 - y_0 x^2}$$

This is also the solution if $y_0 = 0$. If $y_0 < 0$ the denominator is not zero for any value of x, so the solution is valid on $(-\infty, \infty)$. If $y_0 > 0$, the solution is valid only on $(-1/\sqrt{y_0}, 1/\sqrt{y_0})$. You can see this graphically in Figure 3, where the integral curves above the line $y \equiv 0$, which correspond to positive values of y_0, are defined only on finite intervals, while those below the line $y \equiv 0$, which correspond to negative values of y_0, are defined for all x. ■

2.2 EXERCISES

In Exercises 1–6 find all solutions.

1. $y' = \dfrac{3x^2 + 2x + 1}{y - 2}$

2. $(\sin x)(\sin y) + (\cos y)y' = 0$

3. $xy' + y^2 + y = 0$

4. $y' \ln |y| + x^2 y = 0$

5. $(3y^3 + 3y \cos y + 1)y' + \dfrac{(2x + 1)y}{1 + x^2} = 0$

6. $x^2 yy' = (y^2 - 1)^{3/2}$

In Exercises 7–10 find all solutions. Also, plot a direction field and some integral curves on the indicated rectangular region.

C **7.** $y' = x^2(1 + y^2); \quad \{-1 \le x \le 1, -1 \le y \le 1\}$

C **8.** $y'(1 + x^2) + xy = 0; \quad \{-2 \le x \le 2, -1 \le y \le 1\}$

C **9.** $y' = (x - 1)(y - 1)(y - 2); \quad \{-2 \le x \le 2, -3 \le y \le 3\}$

C **10.** $(y - 1)^2 y' = 2x + 3; \quad \{-2 \le x \le 2, -2 \le y \le 5\}$

In Exercises 11 and 12 solve the initial value problem.

11. $y' = \dfrac{x^2 + 3x + 2}{y - 2}, \quad y(1) = 4$

12. $y' + x(y^2 + y) = 0, \quad y(2) = 1$

In Exercises 13–16 solve the initial value problem and graph the solution.

C **13.** $(3y^2 + 4y)y' + 2x + \cos x = 0$, $y(0) = 1$

C **14.** $y' + \dfrac{(y + 1)(y - 1)(y - 2)}{x + 1} = 0$, $y(1) = 0$

C **15.** $y' + 2x(y + 1) = 0$, $y(0) = 2$

C **16.** $y' = 2xy(1 + y^2)$, $y(0) = 1$

In Exercises 17–23 solve the initial value problem and find the interval of validity of the solution.

17. $y'(x^2 + 2) + 4x(y^2 + 2y + 1) = 0$, $y(1) = -1$

18. $y' = -2x(y^2 - 3y + 2)$, $y(0) = 3$

19. $y' = \dfrac{2x}{1 + 2y}$, $y(2) = 0$

20. $y' = 2y - y^2$, $y(0) = 1$

21. $x + yy' = 0$, $y(3) = -4$

22. $y' + x^2(y + 1)(y - 2)^2 = 0$, $y(4) = 2$

23. $(x + 1)(x - 2)y' + y = 0$, $y(1) = -3$

24. Solve $y' = (1 + y^2)/(1 + x^2)$ explicitly.

Hint: *Use the identity* $\tan(A + B) = (\tan A + \tan B)/(1 - \tan A \tan B)$.

25. Solve $y'\sqrt{1 - x^2} + \sqrt{1 - y^2} = 0$ explicitly.

Hint: *Use the identity* $\sin(A - B) = \sin A \cos B - \cos A \sin B$.

26. Solve $y' = \cos x/\sin y$, $y(\pi) = \pi/2$ explicitly.

Hint: *Use the identity* $\cos(x + \pi/2) = -\sin x$ *and the periodicity of the cosine.*

27. Solve the initial value problem

$$y' = ay - by^2, \qquad y(0) = y_0.$$

Discuss the behavior of the solution in the following cases: **(a)** $y_0 \geq 0$. **(b)** $y_0 < 0$.

28. The population $P = P(t)$ of a certain species satisfies the logistic equation

$$P' = aP(1 - \alpha P)$$

and $P(0) = P_0 > 0$. Find P for $t > 0$, and find $\lim_{t \to \infty} P(t)$.

29. An epidemic spreads through a population at a rate proportional to the product of the number of people already infected and the number of people susceptible, but not yet infected. It follows that if S denotes the total population of susceptible people and $I = I(t)$ denotes the number of infected people at time t, then

$$I' = rI(S - I)$$

where r is a positive constant. Assuming that $I(0) = I_0$, find $I(t)$ for $t > 0$, and show that $\lim_{t \to \infty} I(t) = S$.

L **30.** The result of Exercise 29 is discouraging: if any susceptible member of the group is initially infected, then in the long run all susceptible members are infected! On a more hopeful note, suppose that the disease spreads according to the model of Exercise 29, but there is a medication that cures the infected population at a rate proportional to the number of infected individuals. Now the equation for the number of infected individuals becomes

$$I' = rI(S - I) - qI \tag{A}$$

where q is a positive constant.

(a) Choose r and S positive. By plotting direction fields and solutions of (A) on suitable rectangular grids

$$R = \{0 \leq t \leq T, 0 \leq I \leq d\}$$

in the tI-plane, verify that if I is any solution of (A) such that $I(0) > 0$ then $\lim_{t \to \infty} I(t) = S - q/r$ if $q < rS$, while $\lim_{t \to \infty} I(t) = 0$ if $q \geq rS$.

(b) To verify the experimental results of **(a)** use separation of variables to solve (A) with initial condition $I(0) = I_0 > 0$, and find $\lim_{t \to \infty} I(t)$.

Hint: *There are three cases to consider:* **(i)** $q < rS$; **(ii)** $q > rS$; **(iii)** $q = rS$.

L **31.** Consider the differential equation

$$y' = ay - by^2 - q, \tag{A}$$

where a, b are positive constants, and q is an arbitrary constant. In the following, y denotes a solution of this equation that satisfies the initial condition $y(0) = y_0$.

(a) Choose a and b positive and $q < a^2/4b$. By plotting direction fields and solutions of (A) on suitable rectangular grids

$$R = \{0 \le t \le T, c \le y \le d\} \tag{B}$$

in the ty-plane, discover that there are numbers y_1 and y_2 with $y_1 < y_2$ such that if $y_0 > y_1$ then $\lim_{t\to\infty} y(t) = y_2$, while if $y_0 < y_1$ then $y(t) = -\infty$ for some finite value of t. (What happens if $y_0 = y_1$?)

(b) Choose a and b positive and $q = a^2/4b$. By plotting direction fields and solutions of (A) on suitable rectangular grids of the form (B), discover that there is a number y_1 such that if $y_0 \ge y_1$ then $\lim_{t\to\infty} y(t) = y_1$, while if $y_0 < y_1$ then $y(t) = -\infty$ for some finite value of t.

(c) Choose a and b positive and $q > a^2/4b$. By plotting direction fields and solutions of (A) on suitable rectangular grids of the form (B), discover that no matter what y_0 is, $y(t) = -\infty$ for some finite value of t.

(d) Verify your experimental results analytically. Start by separating variables in (A) to obtain

$$\frac{y'}{ay - by^2 - q} = 1.$$

To decide what to do next you'll have to use the quadratic formula. This should lead you to see why there are three cases. Take it from there!

Because of its role in the transition between these three cases, $q_0 = a^2/4b$ is called a ***bifurcation value*** of q. In general, if q is a parameter in any differential equation, then q_0 is said to be a bifurcation value of q if the nature of the solutions of the equation with $q < q_0$ is qualitatively different from the nature of the solutions with $q > q_0$.

L **32.** By plotting direction fields and solutions of

$$y' = qy - y^3,$$

convince yourself that $q_0 = 0$ is a bifurcation value (Exercise 31) of q for this equation. Explain what makes you draw this conclusion.

33. Suppose that a disease spreads according to the model of Exercise 29, but there is a medication that cures the infected population at a constant rate of q individuals per unit time, where $q > 0$. Now the equation for the number of infected individuals becomes

$$I' = rI(S - I) - q.$$

Assuming that $I(0) = I_0 > 0$, use the results of Exercise 31 to describe what happens as $t \to \infty$.

34. Assuming that $p \not\equiv 0$, state conditions under which the linear equation

$$y' + p(x)y = f(x)$$

is separable. If the equation satisfies these conditions, solve it by separation of variables and by the method developed in Section 2.1.

Solve the equations in Exercises 35–38 using variation of parameters followed by separation of variables.

35. $y' + y = \dfrac{2xe^{-x}}{1 + ye^x}$

36. $xy' - 2y = \dfrac{x^6}{y + x^2}$

37. $y' - y = \dfrac{(x + 1)e^{4x}}{(y + e^x)^2}$

38. $y' - 2y = \dfrac{xe^{2x}}{1 - ye^{-2x}}$

39. Use variation of parameters to show that the solutions of the following equations are of the form $y = uy_1$, where u satisfies a separable equation $u' = g(x)p(u)$. Find y_1 and g for each equation.

(a) $xy' + y = h(x)p(xy)$

(b) $xy' - y = h(x)p\left(\dfrac{y}{x}\right)$

(c) $y' + y = h(x)p(e^x y)$

(d) $xy' + ry = h(x)p(x^r y)$

(e) $y' + \dfrac{v'(x)}{v(x)} y = h(x)p(v(x)y)$

2.3   Existence and Uniqueness of Solutions of Nonlinear Equations

Although there are methods for solving some nonlinear equations, it is impossible to find useful formulas for the solutions of most. Whether we're looking for exact solutions or numerical approximations, it is useful to know conditions that imply the existence and uniqueness of solutions of initial value problems for nonlinear equations. In this section we state such conditions and illustrate them with examples.

We need the following terminology. An ***open rectangle*** R in the xy-plane is a set of points (x, y) such that

$$a < x < b \quad \text{and} \quad c < y < d$$

(Figure 1). This set is denoted by

$$R : \{a < x < b, c < y < d\}.$$

"Open" means that the boundary rectangle (indicated by the dashed lines in Figure 1) is not included in R.

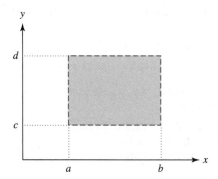

Figure I   An open rectangle

The following theorem gives sufficient conditions for existence and uniqueness of solutions of initial value problems for first order nonlinear differential equations. We omit the proof, which is beyond the scope of this book.

THEOREM 2.3.I

(a) *If f is continuous on an open rectangle*

$$R : \{a < x < b, c < y < d\}$$

containing (x_0, y_0) then the initial value problem

$$y' = f(x, y), \qquad y(x_0) = y_0 \tag{1}$$

has at least one solution on some open subinterval of (a, b) containing x_0.

(b) *If both f and f_y are continuous on R then (1) has a unique solution on some open subinterval of (a, b) containing x_0.*

It is important to understand exactly what Theorem 2.3.1 says.

▪ Part **(a)** is an ***existence theorem.*** It guarantees that a solution exists on some open interval containing x_0, but provides no information on how to find the solution, or to determine the open interval on which it exists. Moreover, part **(a)** provides no information on the number of solutions that (1) may have. It leaves open the possibility that (1) may have two or more solutions that differ for values of x arbitrarily close to x_0. We will see in Example 6 that this can happen.

▪ Part **(b)** is a ***uniqueness theorem.*** It guarantees that (1) has a unique solution on some open interval (a,b) containing x_0. However, if $(a,b) \neq (-\infty,\infty)$ then (1) may have more than one solution on a larger interval containing (a,b). For example, it may happen that $b < \infty$ and all solutions have the same values on (a,b), but two solutions y_1 and y_2 are defined on some interval (a,b_1) with $b_1 > b$, and have different values for $b < x < b_1$; thus, the graphs of y_1 and y_2 "branch off" in different directions at $x = b$. (See Example 7 and Figure 3). In this case continuity implies that $y_1(b) = y_2(b)$ (call their common value $\overline{y}$), and y_1 and y_2 are both solutions of the initial value problem

$$y' = f(x,y), \qquad y(b) = \overline{y} \tag{2}$$

that differ on every open interval containing b. It follows that f or f_y must have a discontinuity at some point in each open rectangle containing $(b,\overline{y})$, since if this were not so, (2) would have a unique solution on some open interval containing b. We leave it to you to give a similar analysis of the case where $a > -\infty$.

EXAMPLE 1 Consider the initial value problem

$$y' = \frac{x^2 - y^2}{1 + x^2 + y^2}, \qquad y(x_0) = y_0. \tag{3}$$

Since

$$f(x,y) = \frac{x^2 - y^2}{1 + x^2 + y^2} \quad \text{and} \quad f_y(x,y) = -\frac{2y(1 + 2x^2)}{(1 + x^2 + y^2)^2}$$

are continuous for all (x,y), Theorem 2.3.1 implies that if (x_0,y_0) is arbitrary then (3) has a unique solution on some open interval containing x_0.

EXAMPLE 2 Consider the initial value problem

$$y' = \frac{x^2 - y^2}{x^2 + y^2}, \qquad y(x_0) = y_0. \tag{4}$$

Here

$$f(x,y) = \frac{x^2 - y^2}{x^2 + y^2} \quad \text{and} \quad f_y(x,y) = -\frac{4x^2 y}{(x^2 + y^2)^2}$$

are continuous everywhere except at $(0,0)$. If $(x_0,y_0) \neq (0,0)$, then there is an open rectangle R containing (x_0,y_0) that does not contain $(0,0)$. Since f and f_y are continuous on R, Theorem 2.3.1 implies that if $(x_0,y_0) \neq (0,0)$, then (4) has a unique solution on some open interval containing x_0.

EXAMPLE 3   Consider the initial value problem

$$y' = \frac{x + y}{x - y}, \qquad y(x_0) = y_0. \tag{5}$$

Here

$$f(x, y) = \frac{x + y}{x - y} \quad \text{and} \quad f_y(x, y) = \frac{2x}{(x - y)^2}$$

are continuous everywhere except on the line $y = x$. If $y_0 \neq x_0$ there is an open rectangle R containing (x_0, y_0) that does not intersect the line $y = x$. Since f and f_y are continuous on R, Theorem 2.3.1 implies that if $y_0 \neq x_0$ then (5) has a unique solution on some open interval containing x_0.

EXAMPLE 4   In Example 4 of Section 2.2 we saw that the solutions of

$$y' = 2xy^2 \tag{6}$$

are

$$y \equiv 0 \quad \text{and} \quad y = -\frac{1}{x^2 + c},$$

where c is an arbitrary constant. In particular, this implies that no solution of (6) other than $y \equiv 0$ can equal zero for any value of x. Show that Theorem 2.3.1**(b)** implies this.

Solution   We will obtain a contradiction by assuming that (6) has a solution y_1 that equals zero for some value of x, but is not identically zero. If y_1 has this property then there is a point x_0 such that $y_1(x_0) = 0$, but $y_1(x) \neq 0$ for some value of x in every open interval containing x_0. This means that the initial value problem

$$y' = 2xy^2, \qquad y(x_0) = 0 \tag{7}$$

has two solutions $y \equiv 0$ and $y = y_1$ that differ for some value of x in every open interval containing x_0. This contradicts Theorem 2.3.1**(b)**, since in (6) the functions

$$f(x, y) = 2xy^2 \quad \text{and} \quad f_y(x, y) = 4xy$$

are both continuous for all (x, y), which implies that (7) has a unique solution on some open interval containing x_0.

EXAMPLE 5   Consider the initial value problem

$$y' = \frac{10}{3} xy^{2/5}, \qquad y(x_0) = y_0. \tag{8}$$

(a) For what points (x_0, y_0) does Theorem 2.3.1**(a)** imply that (8) has a solution?
(b) For what points (x_0, y_0) does Theorem 2.3.1**(b)** imply that (8) has a unique solution on some open interval containing x_0?

Solution   **(a)** Since

$$f(x, y) = \frac{10}{3} xy^{2/5}$$

is continuous for all (x, y), Theorem 2.3.1 implies that (8) has a solution for every (x_0, y_0).

Solution   **(b)** Here

$$f_y(x, y) = \frac{4}{3} xy^{-3/5}$$

is continuous for all (x, y) with $y \neq 0$. Therefore, if $y_0 \neq 0$ there is an open rectangle on which both f and f_y are continuous, and Theorem 2.3.1 implies that (8) has a unique solution on some open interval containing x_0.

If $y = 0$ then $f_y(x, y)$ is undefined, and therefore discontinuous; hence Theorem 2.3.1**(b)** does not apply to (8) if $y_0 = 0$.

EXAMPLE 6   Example 5 leaves open the possibility that the initial value problem

$$y' = \frac{10}{3} xy^{2/5}, \qquad y(0) = 0 \tag{9}$$

has more than one solution on every open interval containing $x_0 = 0$. Show that this is true.

Solution   By inspection, $y \equiv 0$ is a solution of the differential equation

$$y' = \frac{10}{3} xy^{2/5}. \tag{10}$$

Since $y \equiv 0$ satisfies the initial condition $y(0) = 0$, it is a solution of (9).

Now suppose that y is a solution of (10) that is not identically zero. Separating variables in (10) yields

$$y^{-2/5} y' = \frac{10}{3} x$$

on any open interval where y has no zeros. Integrating this and rewriting the arbitrary constant as $5c/3$ yields

$$\frac{5}{3} y^{3/5} = \frac{5}{3}(x^2 + c).$$

Therefore

$$y = (x^2 + c)^{5/3}. \tag{11}$$

Since we divided by y to separate variables in (10), our derivation of (11) is legitimate only on open intervals where y has no zeros. However, (11) actually defines y for all x, and differentiating (11) shows that

$$y' = \frac{10}{3} x(x^2 + c)^{2/3} = \frac{10}{3} xy^{2/5}, \qquad -\infty < x < \infty.$$

Therefore (11) satisfies (10) on $(-\infty, \infty)$ even if $c \leq 0$, so that $y(\sqrt{|c|}) = y(-\sqrt{|c|}) = 0$. In particular, taking $c = 0$ in (11) yields

$$y = x^{10/3}$$

as a second solution of (9). Both solutions are defined on $(-\infty, \infty)$, and they differ on every open interval containing $x_0 = 0$ (see Figure 2). (As a matter of fact, there are *four* distinct solutions of (9) that are defined on $(-\infty, \infty)$ and differ from each other on every open interval containing $x_0 = 0$. Can you identify the other two?)

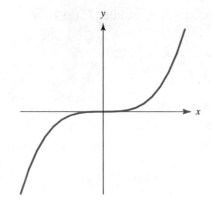

Figure 2 Two solutions ($y \equiv 0$ and $y = x^{10/3}$) of (9) that differ on every open interval containing $x_0 = 0$

EXAMPLE 7 From Example 5 the initial value problem

$$y' = \frac{10}{3} xy^{2/5}, \qquad y(0) = -1 \tag{12}$$

has a unique solution on some open interval containing $x_0 = 0$. Find a solution and determine the largest open interval (a,b) on which it is unique.

Solution Let y be any solution of (12). Because of the initial condition $y(0) = -1$ and the continuity of y, there is some open interval I containing $x_0 = 0$ on which y has no zeros, and is consequently of the form (11). Setting $x = 0$ and $y = -1$ in (11) yields $c = -1$, so

$$y = (x^2 - 1)^{5/3} \tag{13}$$

for x in I. It now follows that every solution of (12) differs from zero and is given by (13) on $(-1,1)$; that is, (13) is the unique solution of (12) on $(-1,1)$. This is the largest open interval on which (12) has a unique solution. To see this, note that (13) is a solution of (12) on $(-\infty,\infty)$. From Exercise 15 there are infinitely many other solutions of (12) that differ from (13) on every open interval larger than $(-1,1)$. One such solution is

$$y = \begin{cases} (x^2 - 1)^{5/3}, & -1 \leq x \leq 1, \\ 0, & |x| > 1. \end{cases}$$

(See Figure 3.)

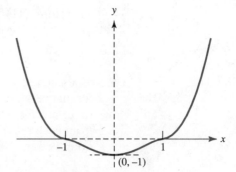

Figure 3 Two solutions of (12) on $(-\infty,\infty)$ that coincide on $(-1,1)$, but on no larger open interval

EXAMPLE 8   From Example 5, the initial value problem

$$y' = \frac{10}{3} xy^{2/5}, \qquad y(0) = 1 \tag{14}$$

has a unique solution on some open interval containing $x_0 = 0$. Find the solution and determine the largest open interval on which it is unique.

Solution   Let y be any solution of (14). Because of the initial condition $y(0) = 1$ and the continuity of y, there is some open interval I containing $x_0 = 0$ on which y has no zeros, and is consequently of the form (11). Setting $x = 0$ and $y = 1$ in (11) yields $c = 1$, so

$$y = (x^2 + 1)^{5/3} \tag{15}$$

for x in I. It now follows that every solution of (14) differs from zero and is given by (15) on $(-\infty, \infty)$; that is, (15) is the unique solution of (14) on $(-\infty, \infty)$. The graph of this solution is shown in Figure 4. ■

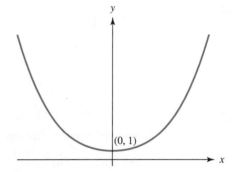

Figure 4   Graph of the unique solution of (14)

2.3 EXERCISES

In Exercises 1–13 find all (x_0, y_0) for which Theorem 2.3.1 implies that the initial value problem $y' = f(x,y)$, $y(x_0) = y_0$ has **(a)** *a solution;* **(b)** *a unique solution on some open interval containing x_0.*

1. $y' = \dfrac{x^2 + y^2}{\sin x}$

2. $y' = \dfrac{e^x + y}{x^2 + y^2}$

3. $y' = \tan xy$

4. $y' = \dfrac{x^2 + y^2}{\ln xy}$

5. $y' = (x^2 + y^2)y^{1/3}$

6. $y' = 2xy$

7. $y' = \ln(1 + x^2 + y^2)$

8. $y' = \dfrac{2x + 3y}{x - 4y}$

9. $y' = (x^2 + y^2)^{1/2}$

10. $y' = x(y^2 - 1)^{2/3}$

11. $y' = (x^2 + y^2)^2$

12. $y' = (x + y)^{1/2}$

13. $y' = \dfrac{\tan y}{x - 1}$

14. Apply Theorem 2.3.1 to the initial value problem

$$y' + p(x)y = q(x), \qquad y(x_0) = y_0$$

for a linear equation, and compare the conclusions that can be drawn from it to those that follow from Theorem 2.1.2.

15. (a) Verify that the function

$$y = \begin{cases} (x^2 - 1)^{5/3}, & -1 < x < 1, \\ 0, & |x| \geq 1, \end{cases}$$

is a solution of the initial value problem

$$y' = \frac{10}{3} xy^{2/5}, \qquad y(0) = -1$$

on $(-\infty, \infty)$.

Hint: *You will need the definition*

$$y'(\bar{x}) = \lim_{x \to \bar{x}} \frac{y(x) - y(\bar{x})}{x - \bar{x}}$$

to verify that y satisfies the differential equation at $\bar{x} = \pm 1$.

(b) Verify that if $\epsilon_i = 0$ or 1 for $i = 1, 2$ and $a, b > 1$, then the function

$$y = \begin{cases} \epsilon_1(x^2 - a^2)^{5/3}, & -\infty < x < -a, \\ 0, & -a \leq x \leq -1, \\ (x^2 - 1)^{5/3}, & -1 < x < 1, \\ 0, & 1 \leq x \leq b, \\ \epsilon_2(x^2 - b^2)^{5/3}, & b < x < \infty \end{cases}$$

is a solution of the initial value problem of **(a)** on $(-\infty, \infty)$.

16. Use the ideas developed in Exercise 15 to find infinitely many solutions on $(-\infty, \infty)$ of the initial value problem

$$y' = y^{2/5}, \qquad y(0) = 1.$$

17. Consider the initial value problem

$$y' = 3x(y - 1)^{1/3}, \qquad y(x_0) = y_0. \tag{A}$$

(a) For what points (x_0, y_0) does Theorem 2.3.1 imply that (A) has a solution?

(b) For what points (x_0, y_0) does Theorem 2.3.1 imply that (A) has a unique solution on some open interval containing x_0?

18. Find nine solutions of the initial value problem

$$y' = 3x(y - 1)^{1/3}, \qquad y(0) = 1$$

that are all defined on $(-\infty, \infty)$ and differ from each other for values of x in every open interval containing $x_0 = 0$.

19. From Theorem 2.3.1 the initial value problem

$$y' = 3x(y - 1)^{1/3}, \qquad y(0) = 9$$

has a unique solution on an open interval containing $x_0 = 0$. Find the solution and determine the largest open interval on which it is unique.

20. (a) From Theorem 2.3.1 the initial value problem

$$y' = 3x(y - 1)^{1/3}, \qquad y(3) = -7 \tag{A}$$

has a unique solution on some open interval containing $x_0 = 3$. Determine the largest such open interval, and find the solution on this interval.

(b) Find infinitely many solutions of (A), all defined on $(-\infty, \infty)$.

21. Prove:

(a) If

$$f(x, y_0) = 0, \qquad a < x < b, \tag{A}$$

and x_0 is in (a, b) then $y \equiv y_0$ is a solution of

$$y' = f(x, y), \qquad y(x_0) = y_0$$

on (a, b).

(b) If f and f_y are continuous on an open rectangle containing (x_0, y_0) and (A) holds, then no solution of $y' = f(x, y)$ other than $y \equiv y_0$ can equal y_0 at any point in (a, b).

2.4 Transformation of Nonlinear Equations into Separable Equations

In Section 2.1 we found that the solutions of a linear nonhomogeneous equation

$$y' + p(x)y = f(x)$$

are of the form $y = uy_1$, where y_1 is a nontrivial solution of the complementary equation

$$y' + p(x)y = 0 \tag{1}$$

and u is a solution of

$$u'y_1(x) = f(x).$$

Notice that this last equation is separable, since it can be rewritten as

$$u' = \frac{f(x)}{y_1(x)}.$$

In this section we will consider nonlinear differential equations that are not separable to begin with, but can be solved in a similar fashion by writing their solutions in the form $y = uy_1$, where y_1 is a suitably chosen known function and u satisfies a separable equation. We will say in this case that we have **transformed** the given equation into a separable equation.

BERNOULLI'S EQUATION

A **Bernoulli**[1] **equation** is an equation of the form

$$y' + p(x)y = f(x)y^r, \tag{2}$$

where r can be any real number other than 0 or 1 . (Note that (2) is linear if and only if $r = 0$ or $r = 1$.) We can transform (2) into a separable equation by variation of parameters: if y_1 is a nontrivial solution of (1) then substituting $y = uy_1$ into (2) yields

$$u'y_1 + u(y_1' + p(x)y_1) = f(x)(uy_1)^r,$$

which is equivalent to the separable equation

$$u'y_1(x) = f(x)(y_1(x))^r u^r \qquad \text{or} \qquad \frac{u'}{u^r} = f(x)(y_1(x))^{r-1},$$

since $y_1' + p(x)y_1 = 0$.

---

[1]Jacob Bernoulli (1654–1705) belonged to a Swiss family that made fundamental contributions to the sciences over a period of centuries.

EXAMPLE 1   Solve the Bernoulli equation

$$y' - y = xy^2. \tag{3}$$

Solution   Since $y_1 = e^x$ is a solution of $y' - y = 0$ we look for solutions of (3) in the form $y = ue^x$, where

$$u'e^x = xu^2e^{2x} \quad \text{or, equivalently,} \quad u' = xu^2e^x.$$

Separating variables yields

$$\frac{u'}{u^2} = xe^x,$$

and integrating yields

$$-\frac{1}{u} = (x - 1)e^x + c.$$

Hence,

$$u = -\frac{1}{(x - 1)e^x + c}$$

and

$$y = -\frac{1}{x - 1 + ce^{-x}}.$$

A direction field and some integral curves of (3) are shown in Figure 1. ■

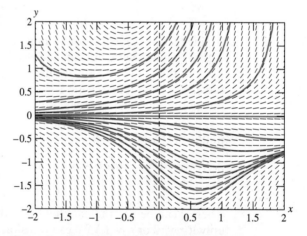

Figure 1   A direction field and integral curves of
$$y' - y = xy^2$$

OTHER NONLINEAR EQUATIONS THAT CAN BE TRANSFORMED INTO SEPARABLE EQUATIONS

We have seen that the nonlinear Bernoulli equation can be transformed into a separable equation by means of the substitution $y = uy_1$ if y_1 is suitably chosen. Now let's discover a sufficient condition for a nonlinear first order differential equation

$$y' = f(x, y) \tag{4}$$

to be transformable into a separable equation involving a new unknown u by means of the substitution $y = uy_1$, where y_1 is some known differentiable function of x. Substituting $y = uy_1$ into (4) yields

$$u'y_1(x) + uy_1'(x) = f(x, uy_1(x)),$$

which is equivalent to

$$u'y_1(x) = f(x, uy_1(x)) - uy_1'(x). \tag{5}$$

If

$$f(x, uy_1(x)) = q(u)y_1'(x)$$

for some function q then (5) becomes

$$u'y_1(x) = (q(u) - u)y_1'(x), \tag{6}$$

which is separable. After checking for constant solutions $u \equiv u_0$ such that $q(u_0) = u_0$, we can separate variables to obtain

$$\frac{u'}{q(u) - u} = \frac{y_1'(x)}{y_1(x)}.$$

HOMOGENEOUS NONLINEAR EQUATIONS

In the text we'll consider only the most widely studied class of equations for which the method of the preceding paragraph works. Other types of equations appear in Exercises 44–51.

The differential equation (4) is said to be **homogeneous** if x and y occur in f in such a way that $f(x, y)$ depends only on the ratio y/x; that is, (4) can be written as

$$y' = q(y/x) \tag{7}$$

where $q = q(u)$ is a function of a single variable. For example,

$$y' = \frac{y + xe^{-y/x}}{x} = \frac{y}{x} + e^{-y/x}$$

and

$$y' = \frac{y^2 + xy - x^2}{x^2} = \left(\frac{y}{x}\right)^2 + \frac{y}{x} - 1$$

are of the form (7), with

$$q(u) = u + e^{-u} \quad \text{and} \quad q(u) = u^2 + u - 1,$$

respectively. The general method discussed above can be applied to (7) with $y_1 = x$ (and, therefore, $y_1' = 1$). Thus, substituting $y = ux$ in (7) yields

$$u'x + u = q(u),$$

and separation of variables (after checking for constant solutions $u \equiv u_0$ such that $q(u_0) = u_0$) yields

$$\frac{u'}{q(u) - u} = \frac{1}{x}.$$

Before turning to examples we point out something that you may have already noticed: the definition of **homogeneous equation** given here is not the same as the definition given in Section 2.1, where we said that a linear equation of the form

$$y' + p(x)y = 0$$

is homogeneous. We make no apology for this inconsistency, since we didn't create it; historically, **homogeneous** has been used in these two inconsistent ways. Of the two meanings, the one having to do with linear equations is the most important. This is the only section of the book in which the meaning defined here will apply.

Since y/x is in general undefined if $x = 0$, we will consider solutions of nonhomogeneous equations only on open intervals that do not contain the point $x = 0$.

EXAMPLE 2   Solve

$$y' = \frac{y + xe^{-y/x}}{x}. \tag{8}$$

Solution   Substituting $y = ux$ into (8) yields

$$u'x + u = \frac{ux + xe^{-ux/x}}{x} = u + e^{-u}.$$

Simplifying and separating variables, we obtain

$$e^u u' = \frac{1}{x}.$$

Integrating yields

$$e^u = \ln|x| + c.$$

Therefore

$$u = \ln(\ln|x| + c),$$

so

$$y = ux = x\ln(\ln|x| + c).$$

A direction field and some integral curves of (8) are shown in Figure 2.

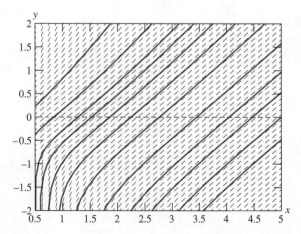

Figure 2   A direction field and integral curves of

$$y' = \frac{y + xe^{-y/x}}{x}$$

<u>**EXAMPLE 3**</u>   **(a)** Solve

$$x^2 y' = y^2 + xy - x^2. \tag{9}$$

(b) Solve the initial value problem

$$x^2 y' = y^2 + xy - x^2, \qquad y(1) = 2. \tag{10}$$

Solution   **(a)** We first find solutions of (9) on open intervals that do not contain $x = 0$. We can rewrite (9) as

$$y' = \frac{y^2 + xy - x^2}{x^2}$$

for x in any such interval. Substituting $y = ux$ yields

$$u'x + u = \frac{(ux)^2 + x(ux) - x^2}{x^2} = u^2 + u - 1,$$

so

$$u'x = u^2 - 1. \tag{11}$$

By inspection this equation has the constant solutions $u \equiv 1$ and $u \equiv -1$. Therefore, $y = x$ and $y = -x$ are solutions of (9). If u is a solution of (11) that does not assume the values ± 1 on some interval, then separating variables yields

$$\frac{u'}{u^2 - 1} = \frac{1}{x},$$

or, after a partial fraction expansion,

$$\frac{1}{2}\left[\frac{1}{u - 1} - \frac{1}{u + 1}\right]u' = \frac{1}{x}.$$

Multiplying by 2 and integrating yields

$$\ln\left|\frac{u - 1}{u + 1}\right| = 2\ln|x| + k,$$

or

$$\left|\frac{u - 1}{u + 1}\right| = e^k x^2,$$

which holds if

$$\frac{u - 1}{u + 1} = cx^2 \tag{12}$$

where c is an arbitrary constant. Solving for u yields

$$u = \frac{1 + cx^2}{1 - cx^2}.$$

Therefore,

$$y = ux = \frac{x(1 + cx^2)}{1 - cx^2} \tag{13}$$

is a solution of (10) for any choice of the constant c. Setting $c = 0$ in (13) yields the solution $y = x$. However, the solution $y = -x$ cannot be obtained from (13). Thus, the solutions of (9) on intervals that do not contain $x = 0$ are $y = -x$ and functions of the form (13).

The situation is more complicated if the open interval of interest contains $x = 0$. First, note that $y = -x$ satisfies (9) on $(-\infty, \infty)$. If c_1 and c_2 are arbitrary constants then the function

$$y = \begin{cases} \dfrac{x(1 + c_1 x^2)}{1 - c_1 x^2}, & a < x < 0, \\ \dfrac{x(1 + c_2 x^2)}{1 - c_2 x^2}, & 0 \le x < b, \end{cases} \tag{14}$$

is a solution of (9) on (a, b), where

$$a = \begin{cases} -\dfrac{1}{\sqrt{c_1}} & \text{if } c_1 > 0, \\ -\infty & \text{if } c_1 \le 0, \end{cases} \quad \text{and} \quad b = \begin{cases} \dfrac{1}{\sqrt{c_2}} & \text{if } c_2 > 0, \\ \infty & \text{if } c_2 \le 0. \end{cases}$$

We leave it to you to verify this. To do so, note that if y is any function of the form (13) then $y(0) = 0$ and $y'(0) = 1$.

Solution   **(b)** We could obtain c by imposing the initial condition $y(1) = 2$ in (13), and then solving for c. However, it is easier to use (12). Since $u = y/x$, the initial condition $y(1) = 2$ implies that $u(1) = 2$. Substituting this into (12) yields $c = 1/3$. Hence, the solution of (10) is

$$y = \frac{x(1 + x^2/3)}{1 - x^2/3}.$$

The interval of validity of this solution is $(-\sqrt{3}, \sqrt{3})$. However, the largest interval on which (10) has a unique solution is $(0, \sqrt{3})$. To see this, note from (14) that any function of the form

$$y = \begin{cases} \dfrac{x(1 + cx^2)}{1 - cx^2}, & a < x \le 0, \\ \dfrac{x(1 + x^2/3)}{1 - x^2/3}, & 0 \le x < \sqrt{3}, \end{cases} \tag{15}$$

is a solution of (10) on $(a, \sqrt{3})$, where $a = -1/\sqrt{c}$ if $c > 0$ or $a = -\infty$ if $c \le 0$. (Why doesn't this contradict Theorem 2.3.1?)

Figure 3 shows a direction field and some integral curves of (9). Figure 4 shows several solutions (15) of the initial value problem (10). Note that these solutions coincide on $(0, \sqrt{3})$. ∎

In the last two examples we were able to solve the given equations explicitly. However, this is not always possible, as you'll see in the exercises.

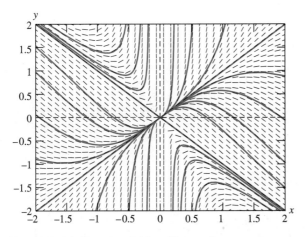

Figure 3 A direction field and integral curves of
$$x^2y' = y^2 + xy - x^2$$

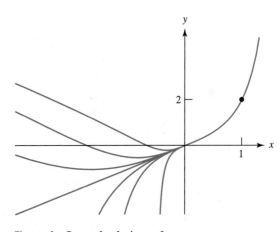

Figure 4 Several solutions of
$$x^2y' = y^2 + xy - x^2, \quad y(1) = 2$$

2.4 EXERCISES

In Exercises 1–4 solve the given Bernoulli equation.

1. $y' + y = y^2$

2. $7xy' - 2y = -\dfrac{x^2}{y^6}$

3. $x^2y' + 2y = 2e^{1/x}y^{1/2}$

4. $(1 + x^2)y' + 2xy = \dfrac{1}{(1 + x^2)y}$

In Exercises 5 and 6 find all solutions. Also, plot a direction field and some integral curves on the indicated rectangular region.

C **5.** $y' - xy = x^3y^3; \quad \{-3 \le x \le 3, 2 \le y \ge 2\}$

C **6.** $y' - \dfrac{1 + x}{3x}y = y^4; \quad \{-2 \le x \le 2, -2 \le y \le 2\}$

In Exercises 7–11 solve the initial value problem.

7. $y' - 2y = xy^3, \quad y(0) = 2\sqrt{2}$

8. $y' - xy = xy^{3/2}, \quad y(1) = 4$

9. $xy' + y = x^4y^4, \quad y(1) = 1/2$

10. $y' - 2y = 2y^{1/2}, \quad y(0) = 1$

11. $y' - 4y = \dfrac{48x}{y^2}, \quad y(0) = 1$

In Exercises 12 and 13 solve the initial value problem and graph the solution.

C **12.** $x^2y' + 2xy = y^3, \quad y(1) = 1/\sqrt{2}$

C **13.** $y' - y = xy^{1/2}, \quad y(0) = 4$

14. You may have noticed that the logistic equation

$$P' = aP(1 - \alpha P)$$

from Verhulst's model for population growth can be written as the Bernoulli equation

$$P' - aP = -a\alpha P^2.$$

This isn't particularly interesting, since the logistic equation is separable, and therefore solvable by the method studied in Section 2.2. So let's consider a more complicated model where a is a positive constant and α is a positive continuous function of t on $[0, \infty)$. The equation for this model is

$$P' - aP = -a\alpha(t)P^2,$$

a non-separable Bernoulli equation.

(a) Assuming that $P(0) = P_0 > 0$, find P for $t > 0$.

Hint: *Express your result in terms of the integral* $\int_0^t \alpha(\tau)e^{a\tau}\,d\tau$.

(b) Verify that your result reduces to the known results for the Malthusian model in which $\alpha = 0$, and for the Verhulst model in which α is a nonzero constant.

(c) Assuming that

$$\lim_{t\to\infty} e^{-at} \int_0^t \alpha(\tau)e^{a\tau}\,d\tau = L$$

exists (finite or infinite), find $\lim_{t\to\infty} P(t)$.

In Exercises 15–18 solve the equation explicitly.

15. $y' = \dfrac{y + x}{x}$

16. $y' = \dfrac{y^2 + 2xy}{x^2}$

17. $xy^3 y' = y^4 + x^4$

18. $y' = \dfrac{y}{x} + \sec\dfrac{y}{x}$

In Exercises 19–21 solve the equation explicitly. Also, plot a direction field and some integral curves on the indicated rectangular region.

C 19. $x^2 y' = xy + x^2 + y^2$; $\{-8 \le x \le 8,\ -8 \le y \le 8\}$

C 20. $xyy' = x^2 + 2y^2$; $\{-4 \le x \le 4,\ -4 \le y \le 4\}$

C 21. $y' = \dfrac{2y^2 + x^2 e^{-(y/x)^2}}{2xy}$; $\{-8 \le x \le 8,\ -8 \le y \le 8\}$

In Exercises 22–27 solve the initial value problem.

22. $y' = \dfrac{xy + y^2}{x^2}$, $y(-1) = 2$

23. $y' = \dfrac{x^3 + y^3}{xy^2}$, $y(1) = 3$

24. $xyy' + x^2 + y^2 = 0$, $y(1) = 2$

25. $y' = \dfrac{y^2 - 3xy - 5x^2}{x^2}$, $y(1) = -1$

26. $x^2 y' = 2x^2 + y^2 + 4xy$, $y(1) = 1$

27. $xyy' = 3x^2 + 4y^2$, $y(1) = \sqrt{3}$

In Exercises 28–34 solve the given homogeneous equation implicitly.

28. $y' = \dfrac{x + y}{x - y}$

29. $(y'x - y)(\ln|y| - \ln|x|) = x$

30. $y' = \dfrac{y^3 + 2xy^2 + x^2 y + x^3}{x(y + x)^2}$

31. $y' = \dfrac{x + 2y}{2x + y}$

32. $y' = \dfrac{y}{y - 2x}$

33. $y' = \dfrac{xy^2 + 2y^3}{x^3 + x^2 y + xy^2}$

34. $y' = \dfrac{x^3 + x^2 y + 3y^3}{x^3 + 3xy^2}$

L 35. (a) Find a solution of the initial value problem

$$x^2 y' = y^2 + xy - 4x^2, \qquad y(-1) = 0 \tag{A}$$

on the interval $(-\infty, 0)$. Verify that this solution is actually valid on $(-\infty, \infty)$.

(b) Use Theorem 2.3.1 to show that (A) has a unique solution on $(-\infty, 0)$.

(c) Plot a direction field for the differential equation in (A) on a square

$$\{-r \le x \le r,\ -r \le y \le r\},$$

where r is any positive number. Graph the solution you obtained in **(a)** on this field.

(d) Graph other solutions of (A) that are defined on $(-\infty, \infty)$.

(e) Graph other solutions of (A) that are defined only on intervals of the form $(-\infty, a)$, where a is a finite positive number.

L **36.** **(a)** Solve the equation

$$xyy' = x^2 - xy + y^2 \tag{A}$$

implicitly.

(b) Plot a direction field for (A) on a square

$$\{0 \le x \le r, 0 \le y \le r\}$$

where r is any positive number.

(c) Let K be a positive integer. (You may have to try several choices for K.) Graph solutions of the initial value problems

$$xyy' = x^2 - xy + y^2, \qquad y(r/2) = \frac{kr}{K},$$

for $k = 1, 2, \ldots, K$. Based on your observations, answer the following question: Under what conditions on the positive numbers x_0 and y_0 does the initial value problem

$$xyy' = x^2 - xy + y^2, \qquad y(x_0) = y_0, \tag{B}$$

have a unique solution **(i)** on $(0, \infty)$; or **(ii)** only on an interval of the form (a, ∞) where $a > 0$?

(d) What can you say about the graph of the solution of (B) as $x \to \infty$? (Again, assume that $x_0 > 0$ and $y_0 > 0$.)

L **37.** **(a)** Solve the equation

$$y' = \frac{2y^2 - xy + 2x^2}{xy + 2x^2} \tag{A}$$

implicitly.

(b) Plot a direction field for (A) on a square

$$\{-r \le x \le r, -r \le y \le r\}$$

where r is any positive number. By graphing solutions of (A), determine necessary and sufficient conditions on (x_0, y_0) such that (A) has a solution on **(i)** $(-\infty, 0)$ or **(ii)** $(0, \infty)$ such that $y(x_0) = y_0$.

L **38.** Follow the instructions of Exercise 37 for the equation

$$y' = \frac{xy + x^2 + y^2}{xy}.$$

L **39.** Pick any nonlinear homogeneous equation $y' = q(y/x)$ you like, and plot direction fields on the square $\{-r \le x \le r, -r \le y \le r\}$, where $r > 0$. What happens to the direction field as you vary r? Why?

40. Prove: If $ad - bc \ne 0$, the equation

$$y' = \frac{ax + by + \alpha}{cx + dy + \beta}$$

can be transformed into the homogeneous nonlinear equation

$$\frac{dY}{dX} = \frac{aX + bY}{cX + dY}$$

by the substitution $x = X - X_0$, $y = Y - Y_0$, where X_0 and Y_0 are suitably chosen constants.

In Exercises 41–43 use a method suggested by Exercise 40 to solve the given equation implicitly.

41. $y' = \dfrac{-6x + y - 3}{2x - y - 1}$

42. $y' = \dfrac{2x + y + 1}{x + 2y - 4}$

43. $y' = \dfrac{-x + 3y - 14}{x + y - 2}$

In Exercises 44–51 find a function y_1 such that the substitution $y = uy_1$ transforms the given equation into a separable equation of the form (6). Then solve the given equation explicitly.

44. $3xy^2y' = y^3 + x$

45. $xyy' = 3x^6 + 6y^2$

46. $x^3y' = 2(y^2 + x^2y - x^4)$

47. $y' = y^2e^{-x} + 4y + 2e^x$

48. $y' = \dfrac{y^2 + y \tan x + \tan^2 x}{\sin^2 x}$

49. $x(\ln x)^2 y' = -4(\ln x)^2 + y \ln x + y^2$

50. $2x(y + 2\sqrt{x})y' = (y + \sqrt{x})^2$

51. $(y + e^{x^2})y' = 2x(y^2 + ye^{x^2} + e^{2x^2})$

52. Solve the initial value problem

$$y' + \frac{2}{x}y = \frac{3x^2y^2 + 6xy + 2}{x^2(2xy + 3)}, \qquad y(2) = 2.$$

53. Solve the initial value problem

$$y' + \frac{3}{x}y = \frac{3x^4y^2 + 10x^2y + 6}{x^3(2x^2y + 5)}, \qquad y(1) = 1.$$

54. Prove: If y is a solution of a homogeneous nonlinear equation $y' = q(y/x)$, then so is $y_1 = y(ax)/a$, where a is any nonzero constant.

55. A **generalized Riccati**[2] **equation** is an equation of the form

$$y' = P(x) + Q(x)y + R(x)y^2. \tag{A}$$

(If $R \equiv -1$ then (A) is a **Riccati equation**.) Let y_1 be a known solution and y an arbitrary solution of (A). Let $z = y - y_1$. Show that z is a solution of a Bernoulli equation with $n = 2$.

In Exercises 56–59, given that y_1 is a solution of the given equation, use the method suggested by Exercise 55 to find other solutions.

56. $y' = 1 + x - (1 + 2x)y + xy^2; \quad y_1 = 1$

57. $y' = e^{2x} + (1 - 2e^x)y + y^2; \quad y_1 = e^x$

58. $xy' = 2 - x + (2x - 2)y - xy^2; \quad y_1 = 1$

59. $xy' = x^3 + (1 - 2x^2)y + xy^2; \quad y_1 = x$

2.5   Exact Equations

In this section it will be convenient to write first order differential equations in the form

$$M(x, y)\, dx + N(x, y)\, dy = 0. \tag{1}$$

This equation can be interpreted as

$$M(x, y) + N(x, y)\frac{dy}{dx} = 0 \tag{2}$$

in which x is the independent variable and y is the dependent variable, or as

---

[2]Count Jacopo Francesco Ricatti (1676–1754) was an Italian mathematician and philosopher who introduced the idea of solving the second order equation bearing his name by reducing it to a first order equation.

$$M(x,y)\,\frac{dx}{dy} + N(x,y) = 0, \tag{3}$$

in which y is the independent variable and x is the dependent variable. Since the solutions of (2) and (3) will often have to be left in implicit form we will say that $F(x,y) = c$ is an implicit solution of (1) if every differentiable function $y = y(x)$ that satisfies $F(x,y) = c$ is a solution of (2), and every differentiable function $x = x(y)$ that satisfies $F(x,y) = c$ is a solution of (3).

The following table shows some examples.

TABLE 1

Equation (1)	Equation (2)	Equation (3)
$3x^2y^2\,dx + 2x^3y\,dy = 0$	$3x^2y^2 + 2x^3y\,\dfrac{dy}{dx} = 0$	$3x^2y^2\,\dfrac{dx}{dy} + 2x^3y = 0$
$(x^2 + y^2)\,dx + 2xy\,dy = 0$	$(x^2 + y^2) + 2xy\,\dfrac{dy}{dx} = 0$	$(x^2 + y^2)\,\dfrac{dx}{dy} + 2xy = 0$
$3y\sin x\,dx - 2xy\cos x\,dy = 0$	$3y\sin x - 2xy\cos x\,\dfrac{dy}{dx} = 0$	$3y\sin x\,\dfrac{dx}{dy} - 2xy\cos x = 0$

Notice that a separable equation can be written in the form (1) as

$$M(x)\,dx + N(y)\,dy = 0.$$

We will develop a method for solving (1) under certain assumptions on M and N. This method is an extension of the method of separation of variables (Exercise 41). Before stating it we consider the following example.

EXAMPLE 1 Show that

$$x^4y^3 + x^2y^5 + 2xy = c \tag{4}$$

is an implicit solution of

$$(4x^3y^3 + 2xy^5 + 2y)\,dx + (3x^4y^2 + 5x^2y^4 + 2x)\,dy = 0. \tag{5}$$

Solution Regarding y as a function of x and differentiating (4) implicitly with respect to x yields

$$(4x^3y^3 + 2xy^5 + 2y) + (3x^4y^2 + 5x^2y^4 + 2x)\,\frac{dy}{dx} = 0.$$

Similarly, regarding x as a function of y and differentiating (4) implicitly with respect to y yields

$$(4x^3y^3 + 2xy^5 + 2y)\,\frac{dx}{dy} + (3x^4y^2 + 5x^2y^4 + 2x) = 0.$$

Therefore (4) is an implicit solution of (5) in either of its possible interpretations. ∎

You may think this example is pointless, since concocting a differential equation that has a given implicit solution is not particularly interesting. However, it illustrates the following important theorem, which we will prove by using implicit differentiation as in the example.

THEOREM 2.5.1

If $F = F(x, y)$ has continuous partial derivatives F_x and F_y, then

$$F(x, y) = c \qquad (c = \text{constant}), \tag{6}$$

is an implicit solution of the differential equation

$$F_x(x, y) \, dx + F_y(x, y) \, dy = 0 \tag{7}$$

PROOF. Regarding y as a function of x and differentiating (6) implicitly with respect to x yields

$$F_x(x, y) + F_y(x, y) \frac{dy}{dx} = 0.$$

On the other hand, regarding x as a function of y and differentiating (6) implicitly with respect to y yields

$$F_x(x, y) \frac{dx}{dy} + F_y(x, y) = 0.$$

Thus, (6) is an implicit solution of (7) in either of its two possible interpretations. $\square$

We will say that the equation

$$M(x, y) \, dx + N(x, y) \, dy = 0 \tag{8}$$

is *exact* on an open rectangle R if there is a function $F = F(x, y)$ such that F_x and F_y are continuous,

$$F_x(x, y) = M(x, y) \qquad \text{and} \qquad F_y(x, y) = N(x, y) \tag{9}$$

for all (x, y) in R. This usage of "exact" is related to its usage in calculus, where the expression

$$F_x(x, y) \, dx + F_y(x, y) \, dy$$

(obtained by substituting (9) into the left side of (8)) is the *exact differential* of F.

Example 1 shows that it is easy to solve (8) if it is exact *and* we know a function F that satisfies (9). The important questions are:

QUESTION 1. Given an equation (8), how can we determine whether it is exact?

QUESTION 2. If (8) is exact, how do we find a function F satisfying (9)?

To discover the answer to Question 1, assume there is a function F that satisfies (9) on some open rectangle R, and in addition that F has continuous mixed partial derivatives F_{xy} and F_{yx}. Then a theorem from calculus implies that

$$F_{xy} = F_{yx}. \tag{10}$$

If $F_x = M$ and $F_y = N$, then differentiating the first of these equations with respect to y and the second with respect to x yields

$$F_{xy} = M_y \quad \text{and} \quad F_{yx} = N_x. \tag{11}$$

From (10) and (11) we conclude that a necessary condition for exactness is that $M_y = N_x$. This motivates the following theorem, which we state without proof.

THEOREM 2.5.2

(The Exactness Condition)

Suppose that M and N are continuous and have continuous partial derivatives M_y and N_x on an open rectangle R. Then

$$M(x,y)\,dx + N(x,y)\,dy = 0$$

is exact on R if and only if

$$M_y(x,y) = N_x(x,y) \tag{12}$$

for all (x,y) in R.

To help you remember the exactness condition, observe that the coefficients of dx and dy are differentiated in (12) with respect to the "opposite" variables; that is, the coefficient of dx is differentiated with respect to y, while the coefficient of dy is differentiated with respect to x.

EXAMPLE 2 Show that the equation

$$3x^2y\,dx + 4x^3\,dy = 0$$

is not exact on any open rectangle.

Solution Here

$$M(x,y) = 3x^2y \quad \text{and} \quad N(x,y) = 4x^3$$

so

$$M_y(x,y) = 3x^2 \quad \text{and} \quad N_x(x,y) = 12x^2.$$

Therefore, $M_y = N_x$ on the line $x = 0$, but not on any open rectangle. Therefore there is no function F such that $F_x(x,y) = M(x,y)$ and $F_y(x,y) = N(x,y)$ for all (x,y) on any open rectangle. ∎

The next example illustrates two possible methods for finding a function F that satisfies the condition $F_x = M$ and $F_y = N$ if $M\,dx + N\,dy = 0$ is exact.

EXAMPLE 3 Solve

$$(4x^3y^3 + 3x^2)\,dx + (3x^4y^2 + 6y^2)\,dy = 0. \tag{13}$$

Solution (Method 1) Here

$$M(x,y) = 4x^3y^3 + 3x^2, \quad N(x,y) = 3x^4y^2 + 6y^2,$$

and

$$M_y(x,y) = N_x(x,y) = 12x^3y^2$$

for all (x,y). Therefore, Theorem 2.5.2 implies that there is a function F such that

$$F_x(x,y) = M(x,y) = 4x^3y^3 + 3x^2 \tag{14}$$

and

$$F_y(x,y) = N(x,y) = 3x^4y^2 + 6y^2 \tag{15}$$

for all (x,y). To find F, we integrate (14) with respect to x to obtain

$$F(x,y) = x^4y^3 + x^3 + \phi(y), \tag{16}$$

where $\phi(y)$ is the "constant" of integration. (Here ϕ is "constant" in that it is independent of x, the variable of integration.) If ϕ is any differentiable function of y then F satisfies (14). To determine ϕ so that F also satisfies (15), assume that ϕ is differentiable and differentiate F with respect to y. This yields

$$F_y(x,y) = 3x^4y^2 + \phi'(y).$$

Comparing this with (15) shows that

$$\phi'(y) = 6y^2.$$

We integrate this with respect to y and take the constant of integration to be zero because we are interested only in finding *some* F that satisfies (14) and (15). This yields

$$\phi(y) = 2y^3.$$

Substituting this into (16) yields

$$F(x,y) = x^4y^3 + x^3 + 2y^3. \tag{17}$$

Now Theorem 2.5.1 implies that

$$x^4y^3 + x^3 + 2y^3 = c$$

is an implicit solution of (13). Solving this for y yields the explicit solution

$$y = \left(\frac{c - x^3}{2 + x^4}\right)^{1/3}.$$

Solution   (Method 2) Instead of first integrating (14) with respect to x, we could begin by integrating (15) with respect to y to obtain

$$F(x,y) = x^4y^3 + 2y^3 + \psi(x), \tag{18}$$

where ψ is an arbitrary function of x. To determine ψ, we assume that ψ is differentiable and differentiate F with respect to x, which yields

$$F_x(x,y) = 4x^3y^3 + \psi'(x).$$

Comparing this with (14) shows that

$$\psi'(x) = 3x^2.$$

Integrating this and again taking the constant of integration to be zero yields

$$\psi(x) = x^3.$$

Substituting this into (18) yields (17).

Figure 1 shows a direction field and some integral curves of (13).   ■

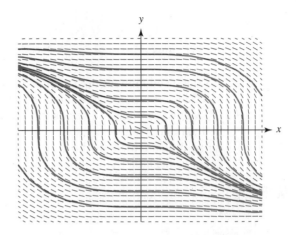

Figure 1   A direction field and integral curves of
$(4x^3y^3 + 3x^2)\,dx + (3x^4y^2 + 6y^2)\,dy = 0$

We summarize the procedure used in Method 1 of this example as follows. The procedure of Method 2 can be summarized similarly.

Procedure for Solving an Exact Equation

Step 1. Check that the equation

$$M(x, y)\,dx + N(x, y)\,dy = 0 \tag{19}$$

satisfies the exactness condition $M_y = N_x$. If not, then don't go further with this procedure.

Step 2. Integrate

$$\frac{\partial F(x, y)}{\partial x} = M(x, y)$$

with respect to x to obtain

$$F(x, y) = G(x, y) + \phi(y), \tag{20}$$

where G is an antiderivative of M with respect to x, and ϕ is an unknown function of y.

Step 3. Differentiate (20) with respect to y to obtain

$$\frac{\partial F(x, y)}{\partial y} = \frac{\partial G(x, y)}{\partial y} + \phi'(y).$$

Step 4. Equate the right side of this equation to N and solve for ϕ'; thus

$$\frac{\partial G(x, y)}{\partial y} + \phi'(y) = N(x, y), \quad \text{so} \quad \phi'(y) = N(x, y) - \frac{\partial G(x, y)}{\partial y}.$$

> **Step 5.** Integrate ϕ' with respect to y, taking the constant of integration to be zero, and substitute the result in (20) to obtain $F(x, y)$.
>
> **Step 6.** Set $F(x, y) = c$ to obtain an implicit solution of (19). If possible, solve for y explicitly as a function of x.

It is a common mistake to omit Step 6. However, it is important to include this step, since F is not itself a solution of (19).

Many equations can be conveniently solved by either of the two methods used in Example 3. However, sometimes the integration required in one approach is more difficult than in the other. In such cases we choose the approach that requires the easier integration.

EXAMPLE 4   Solve the equation

$$(ye^{xy} \tan x + e^{xy} \sec^2 x)\, dx + xe^{xy} \tan x\, dy = 0. \tag{21}$$

Solution   We leave it to you to check that $M_y = N_x$ on any open rectangle where $\tan x$ and $\sec x$ are defined. Here we must find a function F such that

$$F_x(x, y) = ye^{xy} \tan x + e^{xy} \sec^2 x \tag{22}$$

and

$$F_y(x, y) = xe^{xy} \tan x. \tag{23}$$

It is difficult to integrate (22) with respect to x, but easy to integrate (23) with respect to y. This yields

$$F(x, y) = e^{xy} \tan x + \psi(x). \tag{24}$$

Differentiating this with respect to x yields

$$F_x(x, y) = ye^{xy} \tan x + e^{xy} \sec^2 x + \psi'(x).$$

Comparing this with (22) shows that $\psi'(x) = 0$. Hence, ψ is a constant, which we can take to be zero in (24), and

$$e^{xy} \tan x = c$$

is an implicit solution of (21).   ∎

Attempting to apply our procedure to an equation that is not exact will lead to failure in Step 4, since the function

$$N - \frac{\partial G}{\partial y}$$

will not be independent of x if $M_y \neq N_x$ (Exercise 31), and therefore cannot be the derivative of a function of y alone. The following example illustrates this.

EXAMPLE 5   Verify that the equation

$$3x^2y^2\, dx + 6x^3y\, dy = 0 \tag{25}$$

is not exact, and show that the procedure for solving exact equations fails when applied to (25).

Solution   Here

$$M_y(x,y) = 6x^2y \quad \text{and} \quad N_x(x,y) = 18x^2y,$$

so (25) is not exact. Nevertheless, let's try to find a function F such that

$$F_x(x,y) = 3x^2y^2 \tag{26}$$

and

$$F_y(x,y) = 6x^3y. \tag{27}$$

Integrating (26) with respect to x yields

$$F(x,y) = x^3y^2 + \phi(y),$$

and differentiating this with respect to y yields

$$F_y(x,y) = 2x^3y + \phi'(y).$$

For this equation to be consistent with (27) we must have

$$6x^3y = 2x^3y + \phi'(y),$$

or

$$\phi'(y) = 4x^3y.$$

This is a contradiction, since ϕ' must be independent of x. Therefore the procedure fails.   ■

2.5 EXERCISES

In Exercises 1–17 determine which equations are exact, and solve them.

1. $6x^2y^2\,dx + 4x^3y\,dy = 0$

2. $(3y \cos x + 4xe^x + 2x^2e^x)\,dx + (3 \sin x + 3)\,dy = 0$

3. $14x^2y^3\,dx + 21x^2y^2\,dy = 0$ **4.** $(2x - 2y^2)\,dx + (12y^2 - 4xy)\,dy = 0$

5. $(x + y)^2\,dx + (x + y)^2\,dy = 0$ **6.** $(4x + 7y)\,dx + (3x + 4y)\,dy = 0$

7. $(-2y^2 \sin x + 3y^3 - 2x)\,dx + (4y \cos x + 9xy^2)\,dy = 0$

8. $(2x + y)\,dx + (2y + 2x)\,dy = 0$

9. $(3x^2 + 2xy + 4y^2)\,dx + (x^2 + 8xy + 18y)\,dy = 0$

10. $(2x^2 + 8xy + y^2)\,dx + (2x^2 + xy^3/3)\,dy = 0$ **11.** $\left(\dfrac{1}{x} + 2x\right)dx + \left(\dfrac{1}{y} + 2y\right)dy = 0$

12. $(y \sin xy + xy^2 \cos xy)\,dx + (x \sin xy + xy^2 \cos xy)\,dy = 0$

13. $\dfrac{x\,dx}{(x^2 + y^2)^{3/2}} + \dfrac{y\,dy}{(x^2 + y^2)^{3/2}} = 0$

14. $(e^x(x^2y^2 + 2xy^2) + 6x)\,dx + (2x^2ye^x + 2)\,dy = 0$

15. $(x^2e^{x^2+y}(2x^2 + 3) + 4x)\,dx + (x^3e^{x^2+y} - 12y^2)\,dy = 0$

16. $(e^{xy}(x^4y + 4x^3) + 3y)\,dx + (x^5e^{xy} + 3x)\,dy = 0$

17. $(3x^2 \cos xy - x^3y \sin xy + 4x)\,dx + (8y - x^4 \sin xy)\,dy = 0$

In Exercises 18–22 solve the initial value problem.

18. $(4x^3y^2 - 6x^2y - 2x - 3)\,dx + (2x^4y - 2x^3)\,dy = 0, \quad y(1) = 3$

19. $(-4y\cos x + 4\sin x\cos x + \sec^2 x)\,dx + (4y - 4\sin x)\,dy = 0, \quad y(\pi/4) = 0$

20. $(y^3 - 1)e^x\,dx + 3y^2(e^x + 1)\,dy = 0, \quad y(0) = 0$

21. $(\sin x - y\sin x - 2\cos x)\,dx + \cos x\,dy = 0, \quad y(0) = 1$

22. $(2x - 1)(y - 1)\,dx + (x + 2)(x - 3)\,dy = 0, \quad y(1) = -1$

C **23.** Solve the exact equation

$$(7x + 4y)\,dx + (4x + 3y)\,dy = 0.$$

Plot a direction field and some integral curves for this equation on the rectangle $\{-1 \le x \le 1, -1 \le y \le 1\}$.

C **24.** Solve the exact equation

$$e^x(x^4y^2 + 4x^3y^2 + 1)\,dx + (2x^4ye^x + 2y)\,dy = 0.$$

Plot a direction field and some integral curves for this equation on the rectangle $\{-2 \le x \le 2, -1 \le y \le 1\}$.

C **25.** Plot a direction field and some integral curves of the exact equation

$$(x^3y^4 + x)\,dx + (x^4y^3 + y)\,dy = 0$$

on the rectangle $\{-1 \le x \le 1, -1 \le y \le 1\}$. (See Exercise 37**(a)**.)

C **26.** Plot a direction field and some integral curves of the exact equation

$$(3x^2 + 2y)\,dx + (2y + 2x)\,dy = 0$$

on the rectangle $\{-2 \le x \le 2, -2 \le y \le 2\}$. (See Exercise 37**(c)**.)

L **27.** **(a)** Solve the exact equation

$$(x^3y^4 + 2x)\,dx + (x^4y^3 + 3y)\,dy = 0 \tag{A}$$

implicitly.

(b) For what choices of (x_0, y_0) does Theorem 2.3.1 imply that the initial value problem

$$(x^3y^4 + 2x)\,dx + (x^4y^3 + 3y)\,dy = 0, \qquad y(x_0) = y_0, \tag{B}$$

has a unique solution on an open interval (a,b) containing x_0?

(c) Plot a direction field and some integral curves of (A) on a rectangular region centered at the origin. What is the form of the interval of validity of the solution of (B)?

L **28.** **(a)** Solve the exact equation

$$(x^2 + y^2)\,dx + 2xy\,dy = 0 \tag{A}$$

implicitly.

(b) For what choices of (x_0, y_0) does Theorem 2.3.1 imply that the initial value problem

$$(x^2 + y^2)\,dx + 2xy\,dy = 0, \qquad y(x_0) = y_0, \tag{B}$$

has a unique solution $y = y(x)$ on some open interval (a,b) containing x_0?

(c) Plot a direction field and some integral curves of (A). Then, from the plot determine the form of the interval (a,b) of **(b)**, the monotonicity properties (if any) of the solution of (B), and $\lim_{x\to a+} y(x)$ and $\lim_{x\to b-} y(x)$.

Hint: Your answers will depend upon which quadrant contains (x_0, y_0).

29. Find all functions M such that the given equation is exact.

(a) $M(x,y)\,dx + (x^2 - y^2)\,dy = 0$

(b) $M(x,y)\,dx + 2xy\sin x\cos y\,dy = 0$

(c) $M(x,y)\,dx + (e^x - e^y\sin x)\,dy = 0$

30. Find all functions N such that the given equation is exact.

    (a) $(x^3y^2 + 2xy + 3y^2) \, dx + N(x,y) \, dy = 0$ **(b)** $(\ln xy + 2y \sin x) \, dx + N(x,y) \, dy = 0$

    (c) $(x \sin x + y \sin y) \, dx + N(x,y) \, dy = 0$

31. Suppose that M, N, and their partial derivatives are continuous on an open rectangle R, and G is an antiderivative of M with respect to x; that is,

$$\frac{\partial G}{\partial x} = M.$$

Show that if $M_y \neq N_x$ in R then the function

$$N - \frac{\partial G}{\partial y}$$

is not independent of x.

32. Prove: If the equations $M_1 \, dx + N_1 \, dy = 0$ and $M_2 \, dx + N_2 \, dy = 0$ are exact on an open rectangle R, then so is the equation

$$(M_1 + M_2) \, dx + (N_1 + N_2) \, dy = 0.$$

33. Find conditions on the constants A, B, C, and D such that the equation

$$(Ax + By) \, dx + (Cx + Dy) \, dy = 0$$

is exact.

34. Find conditions on the constants A, B, C, D, E, and F such that the equation

$$(Ax^2 + Bxy + Cy^2) \, dx + (Dx^2 + Exy + Fy^2) \, dy = 0$$

is exact.

35. Suppose that M and N are continuous and have continuous partial derivatives M_y and N_x that satisfy the exactness condition $M_y = N_x$ on an open rectangle R. Show that if (x,y) is in R and

$$F(x,y) = \int_{x_0}^{x} M(s, y_0) \, ds + \int_{y_0}^{y} N(x, t) \, dt,$$

then $F_x = M$ and $F_y = N$.

36. Under the assumptions of Exercise 35, show that

$$F(x,y) = \int_{y_0}^{y} N(x_0, s) \, ds + \int_{x_0}^{x} M(t, y) \, dt.$$

37. Use the method suggested by Exercise 35, with $(x_0, y_0) = (0,0)$, to solve the following exact equations:

    (a) $(x^3y^4 + x) \, dx + (x^4y^3 + y) \, dy = 0$ **(b)** $(x^2 + y^2) \, dx + 2xy \, dy = 0$

    (c) $(3x^2 + 2y) \, dx + (2y + 2x) \, dy = 0$

38. Solve the initial value problem

$$y' + \frac{2}{x}y = -\frac{2xy}{x^2 + 2x^2y + 1}, \qquad y(1) = -2.$$

39. Solve the initial value problem

$$y' - \frac{3}{x}y = \frac{2x^4(4x^3 - 3y)}{3x^5 + 3x^3 + 2y}, \qquad y(1) = 1.$$

40. Solve the initial value problem

$$y' + 2xy = -e^{-x^2}\left(\frac{3x + 2ye^{x^2}}{2x + 3ye^{x^2}}\right), \qquad y(0) = -1.$$

41. Rewrite the separable equation

$$h(y)y' = g(x) \tag{A}$$

as an exact equation

$$M(x, y)\,dx + N(x, y)\,dy = 0. \tag{B}$$

Show that applying the method of this section to (B) yields the same solutions that would be obtained by applying the method of separation of variables to (A).

42. Suppose that all second partial derivatives of M and N are continuous, and $M\,dx + N\,dy = 0$ and $-N\,dx + M\,dy = 0$ are exact on an open rectangle R. Show that $M_{xx} + M_{yy} = N_{xx} + N_{yy} = 0$ on R.

43. Suppose that all second partial derivatives of F are continuous, and $F_{xx} + F_{yy} = 0$ on an open rectangle R. (A function with these properties is **harmonic**; see also Exercise 42.) Show that $-F_y\,dx + F_x\,dy = 0$ is exact on R, and therefore there is a function G such that $G_x = -F_y$ and $G_y = F_x$ in R. (A function G with this property is said to be a **harmonic conjugate** of F.)

44. Verify that the following functions are harmonic, and find all their harmonic conjugates. (See Exercise 43.)

(a) $x^2 - y^2$ **(b)** $e^x \cos y$

(c) $x^3 - 3xy^2$ **(d)** $\cos x \cosh y$

(e) $\sin x \cosh y$

2.6   Integrating Factors

We saw in Section 2.5 that if M, N, M_y, and N_x are continuous and $M_y = N_x$ on an open rectangle R then

$$M(x, y)\,dx + N(x, y)\,dy = 0 \tag{1}$$

is exact on R. Sometimes an equation that is not exact can be made exact by multiplying it by an appropriate function. For example,

$$(3x + 2y^2)\,dx + 2xy\,dy = 0 \tag{2}$$

is not exact, since $M_y(x, y) = 4y \neq N_x(x, y) = 2y$ in (2). However, multiplying (2) by x yields

$$(3x^2 + 2xy^2)\,dx + 2x^2y\,dy = 0 \tag{3}$$

which is exact, since $M_y(x, y) = N_x(x, y) = 4xy$ in (3). Solving (3) by the procedure given in Section 2.5 yields the implicit solution

$$x^3 + x^2y^2 = c.$$

A function $\mu = \mu(x, y)$ is an **integrating factor** for (1) if

$$\mu(x, y)M(x, y)\,dx + \mu(x, y)N(x, y)\,dy = 0 \tag{4}$$

is exact. If we know an integrating factor μ for (1), then we can solve the exact equation (4) by the method of Section 2.5. It would be nice if we could say that (1) and (4) always have the same solutions, but this isn't quite accurate. For example, a solution $y = y(x)$ of (4) such that $\mu(x, y(x)) = 0$ on some interval

$a < x < b$ could fail to be a solution of (1) (Exercise 1), while (1) may have a solution $y = y(x)$ such that $\mu(x, y(x))$ isn't even defined (Exercise 2). Similar comments apply if y is the independent variable and x is the dependent variable in (1) and (4). However, if $\mu(x, y)$ is defined and nonzero for all (x, y), then (1) and (4) are equivalent; that is, they have the same solutions.

FINDING INTEGRATING FACTORS

By applying Theorem 2.5.2 (with M and N replaced by μM and μN) we see that (4) is exact on an open rectangle R if μM, μN, $(\mu M)_y$, and $(\mu N)_x$ are continuous and

$$\frac{\partial}{\partial y}(\mu M) = \frac{\partial}{\partial x}(\mu N) \qquad \text{or, equivalently,} \qquad \mu_y M + \mu M_y = \mu_x N + \mu N_x$$

on R. It is better to rewrite the last equation as

$$\mu(M_y - N_x) = \mu_x N - \mu_y M, \tag{5}$$

which reduces to the known result for exact equations; that is, if $M_y = N_x$ then (5) holds with $\mu = 1$, so (1) is exact.

You may think that (5) is of little value, since it involves *partial* derivatives of the unknown integrating factor μ, and we haven't studied methods for solving such equations. However, we will now show that (5) is useful if we restrict our search to integrating factors that are products of a function of x and a function of y; $\mu(x, y) = P(x)Q(y)$. We are not saying that *every* equation $M\,dx + N\,dy = 0$ has an integrating factor of this form; rather, we are saying that *some* equations have such integrating factors. We will now develop a way to determine whether a given equation has such an integrating factor, and a method for finding the integrating factor in this case.

If $\mu(x, y) = P(x)Q(y)$ then $\mu_x(x, y) = P'(x)Q(y)$ and $\mu_y(x, y) = P(x)Q'(y)$, so (5) becomes

$$P(x)Q(y)(M_y - N_x) = P'(x)Q(y)N - P(x)Q'(y)M, \tag{6}$$

or, after dividing through by $P(x)Q(y)$,

$$M_y - N_x = \frac{P'(x)}{P(x)}N - \frac{Q'(y)}{Q(y)}M. \tag{7}$$

Now let

$$p(x) = \frac{P'(x)}{P(x)} \qquad \text{and} \qquad q(y) = \frac{Q'(y)}{Q(y)},$$

so (7) becomes

$$M_y - N_x = p(x)N - q(y)M. \tag{8}$$

We obtained (8) by *assuming* that $M\,dx + N\,dy = 0$ has an integrating factor $\mu(x, y) = P(x)Q(y)$. However, we can now view (7) differently: If there are functions $p = p(x)$ and $q = q(y)$ that satisfy (8) and we define

$$P(x) = \pm e^{\int p(x)\,dx} \qquad \text{and} \qquad Q(y) = \pm e^{\int q(y)\,dy}, \tag{9}$$

then reversing the steps that led from (6) to (8) shows that $\mu(x, y) = P(x)Q(y)$ is an integrating factor for $M\,dx + N\,dy = 0$ In using this result we take the constants of integration in (9) to be zero and choose the signs conveniently so the integrating factor has the simplest form.

Unfortunately, there is no simple general method for ascertaining whether functions $p = p(x)$ and $q = q(y)$ satisfying (8) exist. However, the following theorem gives simple sufficient conditions for the given equation to have an integrating factor that depends on only one of the independent variables x and y, and for finding an integrating factor in this case.

THEOREM 2.6.1

Let M, N, M_y and N_x be continuous on an open rectangle R. Then:

(a) If $(M_y - N_x)/N$ is independent of y on R and we define

$$p(x) = \frac{M_y - N_x}{N}$$

then

$$\mu(x) = \pm e^{\int p(x)\, dx} \tag{10}$$

is an integrating factor for

$$M(x, y)\, dx + N(x, y)\, dy = 0 \tag{11}$$

on R.

(b) If $(N_x - M_y)/M$ is independent of x on R and we define

$$q(y) = \frac{N_x - M_y}{M}$$

then

$$\mu(y) = \pm e^{\int q(y)\, dy} \tag{12}$$

is an integrating factor for (11) on R.

PROOF. (a) If $(M_y - N_x)/N$ is independent of y then (8) holds with $p = (M_y - N_x)/N$ and $q \equiv 0$. Therefore,

$$P(x) = \pm e^{\int p(x)\, dx} \quad \text{and} \quad Q(y) = \pm e^{\int q(y)\, dy} = \pm e^0 = \pm 1,$$

so (10) is an integrating factor for (11) on R.

(b) If $(N_x - M_y)/M$ is independent of x then (8) holds with $p \equiv 0$ and $q = (N_x - M_y)/M$, and a similar argument shows that (12) is an integrating factor for (11) on R. $\square$

The next two examples show how to apply Theorem 2.6.1.

EXAMPLE 1 Find an integrating factor for the equation

$$(2xy^3 - 2x^3y^3 - 4xy^2 + 2x)\, dx + (3x^2y^2 + 4y)\, dy = 0 \tag{13}$$

and solve the equation.

Solution In (13)

$$M = 2xy^3 - 2x^3y^3 - 4xy^2 + 2x, \qquad N = 3x^2y^2 + 4y,$$

and

$$M_y - N_x = (6xy^2 - 6x^3y^2 - 8xy) - 6xy^2 = -6x^3y^2 - 8xy,$$

so (13) is not exact. However,

$$\frac{M_y - N_x}{N} = -\frac{6x^3y^2 + 8xy}{3x^2y^2 + 4y} = -2x$$

is independent of y, so Theorem 2.6.1**(a)** applies with $p(x) = -2x$. Since

$$\int p(x)\, dx = -\int 2x\, dx = -x^2,$$

it follows that $\mu(x) = e^{-x^2}$ is an integrating factor. Multiplying (13) by μ yields the exact equation

$$e^{-x^2}(2xy^3 - 2x^3y^3 - 4xy^2 + 2x)\, dx + e^{-x^2}(3x^2y^2 + 4y)\, dy = 0. \qquad (14)$$

To solve this equation we must find a function F such that

$$F_x(x,y) = e^{-x^2}(2xy^3 - 2x^3y^3 - 4xy^2 + 2x) \qquad (15)$$

and

$$F_y(x,y) = e^{-x^2}(3x^2y^2 + 4y). \qquad (16)$$

Integrating (16) with respect to y yields

$$F(x,y) = e^{-x^2}(x^2y^3 + 2y^2) + \psi(x). \qquad (17)$$

Differentiating this with respect to x yields

$$F_x(x,y) = e^{-x^2}(2xy^3 - 2x^3y^3 - 4xy^2) + \psi'(x).$$

Comparing this with (15) shows that $\psi'(x) = 2xe^{-x^2}$; therefore, we can let $\psi(x) = -e^{-x^2}$ in (17) and conclude that

$$e^{-x^2}(y^2(x^2y + 2) - 1) = c$$

is an implicit solution of (14). It is also an implicit solution of (13).

Figure 1 shows a direction field and some integral curves for (13).

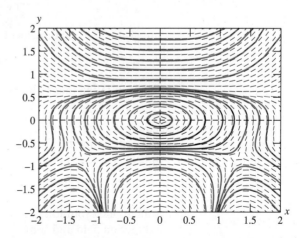

Figure 1   A direction field and integral curves for $(2xy^3 - 2x^3y^3 - 4xy^2 + 2x)\, dx + (3x^2y^2 + 4y)\, dy = 0$

EXAMPLE 2   Find an integrating factor for

$$2xy^3 \, dx + (3x^2y^2 + x^2y^3 + 1) \, dy = 0 \tag{18}$$

and solve the equation.

Solution   In (18)

$$M = 2xy^3, \qquad N = 3x^2y^2 + x^2y^3 + 1,$$

and

$$M_y - N_x = 6xy^2 - (6xy^2 + 2xy^3) = -2xy^3,$$

so (18) is not exact. Moreover,

$$\frac{M_y - N_x}{N} = -\frac{2xy^3}{3x^2y^2 + x^2y^2 + 1}$$

is not independent of y, so Theorem 2.6.1(**a**) does not apply. However, Theorem 2.6.1(**b**) does apply, since

$$\frac{N_x - M_y}{M} = \frac{2xy^3}{2xy^3} = 1$$

is independent of x, so we can take $q(y) = 1$. Since

$$\int q(y) \, dy = \int dy = y$$

it follows that $\mu(y) = e^y$ is an integrating factor. Multiplying (18) by μ yields the exact equation

$$2xy^3e^y \, dx + (3x^2y^2 + x^2y^3 + 1)e^y \, dy = 0. \tag{19}$$

To solve this equation we must find a function F such that

$$F_x(x,y) = 2xy^3e^y \tag{20}$$

and

$$F_y(x,y) = (3x^2y^2 + x^2y^3 + 1)e^y. \tag{21}$$

Integrating (20) with respect to x yields

$$F(x,y) = x^2y^3e^y + \phi(y). \tag{22}$$

Differentiating this with respect to y yields

$$F_y = (3x^2y^2 + x^2y^3)e^y + \phi'(y),$$

and comparing this with (21) shows that $\phi'(y) = e^y$. Therefore, we set $\phi(y) = e^y$ in (22) and conclude that

$$(x^2y^3 + 1)e^y = c$$

is an implicit solution of (19). It is also an implicit solution of (18). Figure 2 shows a direction field and some integral curves for (18).   ∎

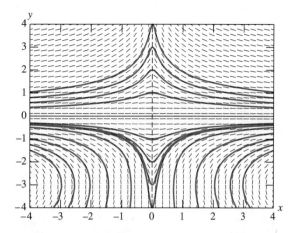

Figure 2   A direction field and integral curves for
$$2xy^3 \, dx + (3x^2y^2 + x^2y^3 + 1) \, dy = 0$$

Theorem 2.6.1 does not apply in the next example, but the more general argument that led to Theorem 2.6.1 provides an integrating factor.

EXAMPLE 3   Find an integrating factor for

$$(3xy + 6y^2) \, dx + (2x^2 + 9xy) \, dy = 0 \tag{23}$$

and solve the equation.

Solution   In (23)

$$M = 3xy + 6y^2, \qquad N = 2x^2 + 9xy,$$

and

$$M_y - N_x = (3x + 12y) - (4x + 9y) = -x + 3y.$$

Therefore

$$\frac{M_y - N_x}{M} = \frac{-x + 3y}{3xy + 6y^2} \quad \text{and} \quad \frac{N_x - M_y}{N} = \frac{x - 3y}{2x^2 + 9xy},$$

so Theorem 2.6.1 does not apply. Following the more general argument that led to Theorem 2.6.1, we look for functions $p = p(x)$ and $q = q(y)$ such that

$$M_y - N_x = p(x)N - q(y)M;$$

that is,

$$-x + 3y = p(x)(2x^2 + 9xy) - q(y)(3xy + 6y^2).$$

Since the left side contains only first degree terms in x and y we rewrite this equation as

$$xp(x)(2x + 9y) - yq(y)(3x + 6y) = -x + 3y.$$

This will be an identity if

$$xp(x) = A \quad \text{and} \quad yq(y) = B, \tag{24}$$

where A and B are constants such that

$$-x + 3y = A(2x + 9y) - B(3x + 6y),$$

or, equivalently,

$$-x + 3y = (2A - 3B)x + (9A - 6B)y.$$

Equating the coefficients of x and y on both sides shows that the last equation holds for all (x, y) if

$$2A - 3B = -1$$
$$9A - 6B = 3,$$

which has the solution $A = 1, B = 1$. Therefore (24) implies that

$$p(x) = \frac{1}{x} \quad \text{and} \quad q(y) = \frac{1}{y}.$$

Since

$$\int p(x)\, dx = \ln|x| \quad \text{and} \quad \int q(y)\, dy = \ln|y|,$$

we can let $P(x) = x$ and $Q(y) = y$; hence, $\mu(x, y) = xy$ is an integrating factor. Multiplying (23) by μ yields the exact equation

$$(3x^2y^2 + 6xy^3)\, dx + (2x^3y + 9x^2y^2)\, dy = 0.$$

We leave it to you to use the method of Section 2.5 to show that this equation has the implicit solution

$$x^3y^2 + 3x^2y^3 = c. \tag{25}$$

This is also an implicit solution of (23). Since $x \equiv 0$ and $y \equiv 0$ satisfy (25), you should check to see that $x \equiv 0$ and $y \equiv 0$ are also solutions of (23). (Why is it necessary to check this?) Figure 3 shows a direction field and some integral curves for (23). ■

See Exercise 28 for a general discussion of equations like (23).

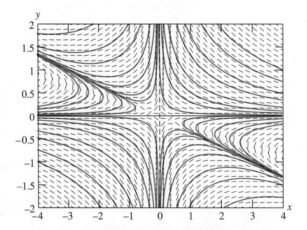

Figure 3   A direction field and integral curves for
$(3xy + 6y^2)\, dx + (2x^2 + 9xy)\, dy = 0$

EXAMPLE 4 The separable equation

$$-y\,dx + (x + x^6)\,dy = 0 \tag{26}$$

can be converted to the exact equation

$$-\frac{dx}{x + x^6} + \frac{dy}{y} = 0 \tag{27}$$

by multiplying through by the integrating factor

$$\mu(x,y) = \frac{1}{y(x + x^6)}.$$

However, to solve (27) by the method of Section 2.5 we would have to evaluate the nasty integral

$$\int \frac{dx}{x + x^6}.$$

Instead, we solve (26) explicitly for y by finding an integrating factor of the form $\mu(x,y) = x^a y^b$.

Solution In (26)

$$M = -y, \qquad N = x + x^6,$$

and

$$M_y - N_x = -1 - (1 + 6x^5) = -2 - 6x^5.$$

We look for functions $p = p(x)$ and $q = q(y)$ such that

$$M_y - N_x = p(x)N - q(y)M;$$

that is,

$$-2 - 6x^5 = p(x)(x + x^6) + q(y)y. \tag{28}$$

The right side will contain the term $-6x^5$ if $p(x) = -6/x$; then (28) becomes

$$-2 - 6x^5 = -6 - 6x^5 + q(y)y,$$

so $q(y) = 4/y$. Since

$$\int p(x)\,dx = -\int \frac{6}{x}\,dx = -6\ln|x| = \ln\frac{1}{x^6},$$

and

$$\int q(y)\,dy = \int \frac{4}{y}\,dy = 4\ln|y| = \ln y^4,$$

we can take $P(x) = x^{-6}$ and $Q(y) = y^4$, which yields the integrating factor $\mu(x,y) = x^{-6}y^4$. Multiplying (26) by μ yields the exact equation

$$-\frac{y^5}{x^6}\,dx + \left(\frac{y^4}{x^5} + y^4\right)dy = 0.$$

We leave it to you to use the method of Section 2.5 to show that this equation has the implicit solution

$$\left(\frac{y}{x}\right)^5 + y^5 = k.$$

Solving for y yields

$$y = k^{1/5}x(1 + x^5)^{-1/5},$$

which we rewrite as

$$y = cx(1 + x^5)^{-1/5}$$

by renaming the arbitrary constant. This is also a solution of (26). Figure 4 shows a direction field and some integral curves for (26). ∎

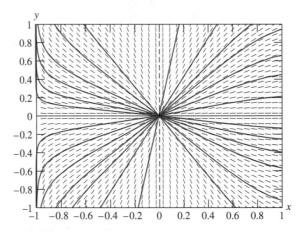

Figure 4   A direction field and integral curves for
$$-y\ dx + (x + x^6)\ dy = 0$$

2.6 EXERCISES

1. **(a)** Verify that $\mu(x,y) = y$ is an integrating factor for

$$y\ dx + \left(2x + \frac{1}{y}\right) dy = 0 \tag{A}$$

on any open rectangle that does not intersect the x-axis or, equivalently, that

$$y^2\ dx + (2xy + 1)\ dy = 0 \tag{B}$$

is exact on any such rectangle.

   (b) Verify that $y \equiv 0$ is a solution of (B), but not of (A).

   (c) Show that

$$y(xy + 1) = c \tag{C}$$

is an implicit solution of (B), and explain why every differentiable function $y = y(x)$ other than $y \equiv 0$ that satisfies (C) is also a solution of (A).

2. **(a)** Verify that $\mu(x,y) = 1/(x - y)^2$ is an integrating factor for

$$-y^2\ dx + x^2\ dy = 0 \tag{A}$$

on any open rectangle that does not intersect the line $y = x$ or, equivalently, that

$$-\frac{y^2}{(x - y)^2}\ dx + \frac{x^2}{(x - y)^2}\ dy = 0 \tag{B}$$

is exact on any such rectangle.

(b) Show that

$$\frac{xy}{(x - y)} = c \qquad\qquad (C)$$

is an implicit solution of (B), and explain why it is also an implicit solution of (A).

(c) Verify that $y = x$ is a solution of (A), even though it cannot be obtained from (C).

In Exercises 3–16 find an integrating factor that is a function of only one variable and solve the given equation.

3. $y\,dx - x\,dy = 0$

4. $3x^2y\,dx + 2x^3\,dy = 0$

5. $2y^3\,dx + 3y^2\,dy = 0$

6. $(5xy + 2y + 5)\,dx + 2x\,dy = 0$

7. $(xy + x + 2y + 1)\,dx + (x + 1)\,dy = 0$

8. $(27xy^2 + 8y^3)\,dx + (18x^2y + 12xy^2)\,dy = 0$

9. $(6xy^2 + 2y)\,dx + (12x^2y + 6x + 3)\,dy = 0$

10. $y^2\,dx + \left(xy^2 + 3xy + \dfrac{1}{y}\right)dy = 0$

11. $(12x^3y + 24x^2y^2)\,dx + (9x^4 + 32x^3y + 4y)\,dy = 0$

12. $(x^2y + 4xy + 2y)\,dx + (x^2 + x)\,dy = 0$

13. $-y\,dx + (x^4 - x)\,dy = 0$

14. $\cos x \cos y\,dx + (\sin x \cos y - \sin x \sin y + y)\,dy = 0$

15. $(2xy + y^2)\,dx + (2xy + x^2 - 2x^2y^2 - 2xy^3)\,dy = 0$

16. $y \sin y\,dx + x(\sin y - y \cos y)\,dy = 0$

In Exercises 17–23 find an integrating factor of the form $\mu(x,y) = P(x)Q(y)$ and solve the given equation.

17. $y(1 + 5 \ln |x|)\,dx + 4x \ln |x|\,dy = 0$

18. $(\alpha y + \gamma xy)\,dx + (\beta x + \delta xy)\,dy = 0$

19. $(3x^2y^3 - y^2 + y)\,dx + (-xy + 2x)\,dy = 0$

20. $2y\,dx + 3(x^2 + x^2y^3)\,dy = 0$

21. $(a \cos xy - y \sin xy)\,dx + (b \cos xy - x \sin xy)\,dy = 0$

22. $x^4y^4\,dx + x^5y^3\,dy = 0$

23. $y(x \cos x + 2 \sin x)\,dx + x(y + 1) \sin x\,dy = 0$

In Exercises 24–27 find an integrating factor and solve the given equation. Plot a direction field and some integral curves of the equation in the indicated rectangular region.

C 24. $(x^4y^3 + y)\,dx + (x^5y^2 - x)\,dy = 0;\quad \{-1 \le x \le 1,\, -1 \le y \le 1\}$

C 25. $(3xy + 2y^2 + y)\,dx + (x^2 + 2xy + x + 2y)\,dy = 0;\quad \{-2 \le x \le 2,\, -2 \le y \le 2\}$

C 26. $(12xy + 6y^3)\,dx + (9x^2 + 10xy^2)\,dy = 0;\quad \{-2 \le x \le 2,\, -2 \le y \le 2\}$

C 27. $(3x^2y^2 + 2y)\,dx + 2x\,dy = 0;\quad \{-4 \le x \le 4,\, -4 \le y \le 4\}$

28. Suppose that $a, b, c,$ and d are constants such that $ad - bc \ne 0$, and let m and n be arbitrary real numbers. Show that

$$(ax^my + by^{n+1})\,dx + (cx^{m+1} + dxy^n)\,dy = 0$$

has an integrating factor $\mu(x,y) = x^\alpha y^\beta$.

29. Suppose that $M, N, M_x,$ and N_y are continuous for all (x,y), and $\mu = \mu(x,y)$ is an integrating factor for

$$M(x,y)\,dx + N(x,y)\,dy = 0. \qquad\qquad (A)$$

Assume that μ_x and μ_y are continuous for all (x,y), and suppose that $y = y(x)$ is a differentiable function such that $\mu(x,y(x)) = 0$ and $\mu_x(x,y(x)) \ne 0$ for all x in some interval I. Show that y is a solution of (A) on I.

30. According to Theorem 2.1.2 the general solution of the linear nonhomogeneous equation

$$y' + p(x)y = f(x) \qquad\qquad (A)$$

is

$$y = y_1(x)\left(c + \int f(x)/y_1(x)\,dx\right), \qquad\qquad (B)$$

where y_1 is any nontrivial solution of the complementary equation $y' + p(x)y = 0$. In this exercise we obtain this conclusion in a different way. You may find it instructive to apply the method suggested here to solve some of the exercises in Section 2.1.

(a) Rewrite (A) as

$$[p(x)y - f(x)] \, dx + dy = 0, \tag{C}$$

and show that $\mu = \pm e^{\int p(x)dx}$ is an integrating factor for (C).

(b) Multiply (A) through by $\mu = \pm e^{\int p(x)dx}$ and verify that the resulting equation can be rewritten as

$$(\mu(x)y)' = \mu(x)f(x).$$

Then integrate both sides of this equation and solve for y to show that the general solution of (A) is

$$y = \frac{1}{\mu(x)}\left(c + \int f(x)\mu(x) \, dx\right).$$

Why is this form of the general solution equivalent to (B)?

3

Numerical Methods

IN THIS CHAPTER we present numerical methods for solving a first order differential equation

$$y' = f(x, y).$$

SECTION 3.1 deals with Euler's[1] method, which is really too crude to be of much use in practical applications. However, its simplicity allows for an introduction to the ideas required to understand the better methods discussed in the other two sections.

SECTION 3.2 discusses improvements on Euler's method.

SECTION 3.3 deals with the Runge[2]–Kutta[3] method, perhaps the most widely used method for numerical solution of differential equations.

---

[1]Leonhard Euler (1707–1783) of Switzerland is regarded as the greatest mathematician of the 18th century. He made fundamental contributions to virtually all the areas of mathematics known in his time. His prodigious mathematical output, unimpeded by his role as father of 13 children and his total blindness for the last 17 years of his life, consisted of over 700 publications.

[2]Carl David Runge (1856–1927) was a German mathematician and physicist who made important contributions to spectroscopy.

[3]Martin Wilhelm Kutta (1867–1944) was a German mathematician and aerodynamicist known for his contributions to airfoil theory.

3.1   Euler's Method

If an initial value problem

$$y' = f(x,y), \qquad y(x_0) = y_0 \tag{1}$$

cannot be solved analytically, then it is necessary to resort to numerical methods to obtain useful approximations to a solution of (1). We will consider such methods in this chapter.

We are interested in computing approximate values of the solution of (1) at equally spaced points $x_0, x_1, \ldots, x_n = b$ in an interval $[x_0, b]$. Thus,

$$x_i = x_0 + ih, \qquad i = 0, 1, \ldots, n,$$

where

$$h = \frac{b - x_0}{n}.$$

We will denote the approximate values of the solution at these points by $y_0, y_1, \ldots, y_n$; thus, y_i is an approximation to $y(x_i)$. We will call

$$e_i = y(x_i) - y_i$$

the ***error at the ith step.*** Because of the initial condition $y(x_0) = y_0$, we will always have $e_0 = 0$. However, in general, $e_i \neq 0$ if $i > 0$.

We encounter two sources of error in applying a numerical method to solve an initial value problem.

- The formulas defining the method are based on some sort of approximation. Errors due to the inaccuracy of the approximation are called ***truncation errors.***
- Computers do arithmetic with a fixed number of digits, and therefore make errors in evaluating the formulas defining the numerical methods. Errors due to the computer's inability to do exact arithmetic are called ***roundoff errors.***

Since a careful analysis of roundoff error is beyond the scope of this book, we will consider only truncation errors.

EULER'S METHOD

The simplest numerical method for solving (1) is ***Euler's method.*** This method is so crude that it is seldom used in practice; however, its simplicity makes it useful for illustrative purposes.

Euler's method is based on the assumption that the tangent line to the integral curve of (1) at $(x_i, y(x_i))$ approximates the integral curve over the interval $[x_i, x_{i+1}]$. Since the slope of the integral curve at $(x_i, y(x_i))$ is $y'(x_i) = f(x_i, y(x_i))$, the equation of the tangent line to the integral curve at $(x_i, y(x_i))$ is

$$y = y(x_i) + f(x_i, y(x_i))(x - x_i). \tag{2}$$

Setting $x = x_{i+1} = x_i + h$ in (2) yields

$$y_{i+1} = y(x_i) + hf(x_i, y(x_i)) \tag{3}$$

as an approximation to $y(x_{i+1})$. Since $y(x_0) = y_0$ is known, we can use (3) with $i = 0$ to compute

$$y_1 = y_0 + hf(x_0, y_0).$$

However, setting $i = 1$ in (3) yields

$$y_2 = y(x_1) + hf(x_1, y(x_1)),$$

which is not useful, since we *don't know* $y(x_1)$. Therefore, we replace $y(x_1)$ by its approximate value y_1 and redefine

$$y_2 = y_1 + hf(x_1, y_1).$$

Having computed y_2, we can then compute

$$y_3 = y_2 + hf(x_2, y_2).$$

In general, Euler's method starts with the known value $y(x_0) = y_0$ and computes $y_1, y_2, \ldots, y_n$ successively by means of the formula

$$y_{i+1} = y_i + hf(x_i, y_i), \qquad 0 \le i \le n - 1. \tag{4}$$

The following example illustrates the computational procedure indicated in Euler's method.

EXAMPLE 1   Use Euler's method with $h = 0.1$ to find approximate values for the solution of the initial value problem

$$y' + 2y = x^3 e^{-2x}, \qquad y(0) = 1 \tag{5}$$

at $x = 0.1, 0.2, 0.3$.

Solution   We rewrite (5) as

$$y' = -2y + x^3 e^{-2x}, \qquad y(0) = 1,$$

which is of the form (1), with

$$f(x, y) = -2y + x^3 e^{-2x}, \qquad x_0 = 0, \qquad \text{and} \qquad y_0 = 1.$$

Euler's method yields

$$y_1 = y_0 + hf(x_0, y_0)$$
$$= 1 + (.1)f(0, 1) = 1 + (.1)(-2) = .8,$$

$$y_2 = y_1 + hf(x_1, y_1)$$
$$= .8 + (.1)f(.1, .8) = .8 + (.1)(-2(.8) + (.1)^3 e^{-.2}) = .640081873,$$

$$y_3 = y_2 + hf(x_2, y_2)$$
$$= .640081873 + (.1)(-2(.640081873) + (.2)^3 e^{-.4}) = .512601754. \quad \blacksquare$$

We've written the details of these computations to ensure that you understand the procedure. However, in the rest of the examples as well as the exercises in this chapter, we will assume that you can use a programmable calculator or a computer to carry out the necessary computations.

EXAMPLES ILLUSTRATING THE ERROR IN EULER'S METHOD

EXAMPLE 2   Use Euler's method with step sizes $h = 0.1$, $h = 0.05$, and $h = 0.025$ to find approximate values of the solution of the initial value problem

$$y' + 2y = x^3 e^{-2x}, \qquad y(0) = 1$$

at $x = 0, 0.1, 0.2, 0.3, \ldots, 1.0$. Compare these approximate values with the values of the exact solution

$$y = \frac{e^{-2x}}{4}(x^4 + 4),\tag{6}$$

which can be obtained by the method of Section 2.1. (Verify.)

Solution   Table 1 shows the values of the exact solution (6) at the specified points, and the approximate values of the solution at these points obtained by Euler's method with step sizes $h = 0.1$, $h = 0.05$, and $h = 0.025$. In examining this table, keep in mind that the approximate values in the column corresponding to $h = .05$ are actually the results of 20 steps with Euler's method. We haven't listed the estimates of the solution obtained for $x = 0.05, 0.15, \ldots$, since there is nothing to compare them with in the column corresponding to $h = 0.1$. Similarly, the approximate values in the column corresponding to $h = 0.025$ are actually the results of 40 steps with Euler's method.   ■

TABLE 1   Numerical solution of $y' + 2y = x^3 e^{-2x}$, $y(0) = 1$, by Euler's method

x	$h = 0.1$	$h = 0.05$	$h = 0.025$	**Exact**
0.0	1.000000000	1.000000000	1.000000000	1.000000000
0.1	0.800000000	0.810005655	0.814518349	0.818751221
0.2	0.640081873	0.656266437	0.663635953	0.670588174
0.3	0.512601754	0.532290981	0.541339495	0.549922980
0.4	0.411563195	0.432887056	0.442774766	0.452204669
0.5	0.332126261	0.353785015	0.363915597	0.373627557
0.6	0.270299502	0.291404256	0.301359885	0.310952904
0.7	0.222745397	0.242707257	0.252202935	0.261398947
0.8	0.186654593	0.205105754	0.213956311	0.222570721
0.9	0.159660776	0.176396883	0.184492463	0.192412038
1.0	0.139778910	0.154715925	0.162003293	0.169169104

You can see from Table 1 that decreasing the step size improves the accuracy of Euler's method. For example,

$$y_{\text{exact}}(1) - y_{\text{approx}}(1) \approx \begin{cases} .0293 \text{ with } h = 0.1, \\ .0144 \text{ with } h = 0.05, \\ .0071 \text{ with } h = 0.025. \end{cases}$$

Based on this scanty evidence, you might conjecture that the error in approximating the exact solution at a *fixed value of x* by Euler's method is roughly halved when the step size is halved. You can find more evidence to support this conjecture by examining Table 2, which lists the approximate values of $y_{\text{exact}} - y_{\text{approx}}$ at $x = 0.1, 0.2, \ldots, 1.0$.

TABLE 2   Errors in approximate solutions of $y' + 2y = x^3 e^{-2x}, y(0) = 1$, obtained by Euler's method

x	$h = 0.1$	$h = 0.05$	$h = 0.025$
0.1	0.0187	0.0087	0.0042
0.2	0.0305	0.0143	0.0069
0.3	0.0373	0.0176	0.0085
0.4	0.0406	0.0193	0.0094
0.5	0.0415	0.0198	0.0097
0.6	0.0406	0.0195	0.0095
0.7	0.0386	0.0186	0.0091
0.8	0.0359	0.0174	0.0086
0.9	0.0327	0.0160	0.0079
1.0	0.0293	0.0144	0.0071

EXAMPLE 3   Tables 3 and 4 show analogous results for the nonlinear initial value problem

$$y' = -2y^2 + xy + x^2, \qquad y(0) = 1, \tag{7}$$

except in this case we can't solve (7) exactly. The results in the "Exact" column here were obtained by using a much more accurate numerical method known as the **_Runge–Kutta_** method, with a small step size. They are exact to eight decimal places.   ∎

TABLE 3   Numerical solution of $y' = -2y^2 + xy + x^2$, $y(0) = 1$, by Euler's method

x	$h = 0.1$	$h = 0.05$	$h = 0.025$	"Exact"
0.0	1.000000000	1.000000000	1.000000000	1.000000000
0.1	0.800000000	0.821375000	0.829977007	0.837584494
0.2	0.681000000	0.707795377	0.719226253	0.729641890
0.3	0.605867800	0.633776590	0.646115227	0.657580377
0.4	0.559628676	0.587454526	0.600045701	0.611901791
0.5	0.535376972	0.562906169	0.575556391	0.587575491
0.6	0.529820120	0.557143535	0.569824171	0.581942225
0.7	0.541467455	0.568716935	0.581435423	0.593629526
0.8	0.569732776	0.596951988	0.609684903	0.621907458
0.9	0.614392311	0.641457729	0.654110862	0.666250842
1.0	0.675192037	0.701764495	0.714151626	0.726015790

TABLE 4 Errors in approximate solutions of $y' = -2y^2 + xy + x^2$, $y(0) = 1$, obtained by Euler's method

x	$h = 0.1$	$h = 0.05$	$h = 0.025$
0.1	0.0376	0.0162	0.0076
0.2	0.0486	0.0218	0.0104
0.3	0.0517	0.0238	0.0115
0.4	0.0523	0.0244	0.0119
0.5	0.0522	0.0247	0.0121
0.6	0.0521	0.0248	0.0121
0.7	0.0522	0.0249	0.0122
0.8	0.0522	0.0250	0.0122
0.9	0.0519	0.0248	0.0121
1.0	0.0508	0.0243	0.0119

Since we think it is important in evaluating the accuracy of the numerical methods that we'll be studying in this chapter, we often include a column listing values of the exact solution of the initial value problem, even if the directions in the example or exercise don't specifically call for it. If quotation marks are included in the heading, then the values were obtained by applying the Runge–Kutta method in a way that is explained in Section 3.3. If quotation marks are not included, then the values were obtained from a known formula for the solution. In either case the values are exact to eight places to the right of the decimal point.

TRUNCATION ERROR IN EULER'S METHOD

Consistent with the results indicated in Tables 1–4, we will now show that under reasonable assumptions on f there is a constant K such that the error in approximating the solution of the initial value problem

$$y' = f(x, y), \qquad y(x_0) = y_0$$

at a given point $b > x_0$ by Euler's method with step size $h = (b - x_0)/n$ satisfies the inequality

$$|y(b) - y_n| \le Kh,$$

where K is a constant independent of n.

There are two sources of error (not counting roundoff) in Euler's method:

1. The error committed in approximating the integral curve by the tangent line (2) over the interval $[x_i, x_{i+1}]$.
2. The error committed in replacing $y(x_i)$ by y_i in (2) and using (4) rather than (2) to compute y_{i+1}.

Euler's method assumes that y_{i+1} defined in (2) is an approximation to $y(x_{i+1})$. We call the error in this approximation the *local truncation error at the ith step,* and denote it by T_i; thus

$$T_i = y(x_{i+1}) - y(x_i) - hf(x_i, y(x_i)). \tag{8}$$

We will now use Taylor's theorem to estimate T_i, assuming for simplicity that f, f_x, and f_y are continuous and bounded for all (x, y). Then y'' exists and is bounded on $[x_0, b]$. To see this we differentiate

$$y'(x) = f(x, y(x))$$

to obtain

$$y''(x) = f_x(x, y(x)) + f_y(x, y(x))y'(x)$$
$$= f_x(x, y(x)) + f_y(x, y(x))f(x, y(x)).$$

Since we have assumed that f, f_x and f_y are bounded, there is a constant M such that

$$|f_x(x, y(x)) + f_y(x, y(x))f(x, y(x))| \le M, \qquad x_0 < x < b,$$

which implies that

$$|y''(x)| \le M, \qquad x_0 < x < b. \tag{9}$$

Since $x_{i+1} = x_i + h$, Taylor's theorem implies that

$$y(x_{i+1}) = y(x_i) + hy'(x_i) + \frac{h^2}{2}y''(\widetilde{x}_i),$$

where $\widetilde{x}_i$ is some number between x_i and x_{i+1}. Since $y'(x_i) = f(x_i, y(x_i))$ this can be written as

$$y(x_{i+1}) = y(x_i) + hf(x_i, y(x_i)) + \frac{h^2}{2}y''(\widetilde{x}_i),$$

or, equivalently,

$$y(x_{i+1}) - y(x_i) - hf(x_i, y(x_i)) = \frac{h^2}{2}y''(\widetilde{x}_i).$$

Comparing this with (8) shows that

$$T_i = \frac{h^2}{2}y''(\widetilde{x}_i).$$

Recalling (9), we can establish the bound

$$|T_i| \le \frac{Mh^2}{2}, \qquad 1 \le i \le n. \tag{10}$$

Although it may be difficult to determine the constant M, what is important is that there is an M such that (10) holds. We say that the local truncation error of Euler's method is *of order h^2*, which we write as $O(h^2)$.

We note that the magnitude of the local truncation error in Euler's method is determined by the second derivative y'' of the solution of the initial value problem. Therefore the local truncation error will be larger where $|y''|$ is large, or smaller where $|y''|$ is small.

Since the local truncation error for Euler's method is $O(h^2)$, it is reasonable to expect that halving h reduces the local truncation error by a factor of 4. This is true, but halving the step size also requires twice as many steps to approximate the solution at a given point. To analyze the overall effect of truncation error in Euler's method it is useful to derive an equation relating the errors

$$e_{i+1} = y(x_{i+1}) - y_{i+1} \quad \text{and} \quad e_i = y(x_i) - y_i.$$

To this end, recall that

$$y(x_{i+1}) = y(x_i) + hf(x_i, y(x_i)) + T_i \tag{11}$$

and

$$y_{i+1} = y_i + hf(x_i, y_i). \tag{12}$$

Subtracting (12) from (11) yields

$$e_{i+1} = e_i + h[f(x_i, y(x_i)) - f(x_i, y_i)] + T_i. \tag{13}$$

The last term on the right is the local truncation error at the ith step. The other terms reflect the way errors made at *previous* steps affect e_{i+1}. Since $|T_i| \le Mh^2/2$ we see from (13) that

$$|e_{i+1}| \le |e_i| + h|f(x_i, y(x_i)) - f(x_i, y_i)| + \frac{Mh^2}{2}. \tag{14}$$

Since we have assumed that f_y is continuous and bounded, the mean value theorem implies that

$$f(x_i, y(x_i)) - f(x_i, y_i) = f_y(x_i, y_i^*)(y(x_i) - y_i) = f_y(x_i, y_i^*)e_i,$$

where y_i^* is between y_i and $y(x_i)$. Therefore,

$$|f(x_i, y(x_i)) - f(x_i, y_i)| \le R|e_i|$$

for some constant R. From this and (14),

$$|e_{i+1}| \le (1 + Rh)|e_i| + \frac{Mh^2}{2}, \quad 0 \le i \le n - 1. \tag{15}$$

For convenience, let $C = 1 + Rh$. Since $e_0 = y(x_0) - y_0 = 0$, applying (15) repeatedly yields

$$|e_1| \le \frac{Mh^2}{2}$$

$$|e_2| \le C|e_1| + \frac{Mh^2}{2} \le (1 + C)\frac{Mh^2}{2}$$

$$|e_3| \le C|e_2| + \frac{Mh^2}{2} \le (1 + C + C^2)\frac{Mh^2}{2}$$

$$\vdots$$

$$|e_n| \le C|e_{n-1}| + \frac{Mh^2}{2} \le (1 + C + \cdots + C^{n-1})\frac{Mh^2}{2}. \tag{16}$$

Recalling the formula for partial sums of a geometric series, we see that

$$1 + C + \cdots + C^{n-1} = \frac{1 - C^n}{1 - C} = \frac{(1 + Rh)^n - 1}{Rh}$$

(since $C = 1 + Rh$). From this and (16)

$$|y(b) - y_n| = |e_n| \le \frac{(1 + Rh)^n - 1}{R}\frac{Mh}{2}. \tag{17}$$

Since Taylor's theorem implies that

$$1 + Rh < e^{Rh}$$

(verify), it follows that

$$(1 + Rh)^n < e^{nRh} = e^{R(b-x_0)} \qquad \text{(since } nh = b - x_0\text{)}.$$

This and (17) imply that

$$|y(b) - y_n| \le Kh, \tag{18}$$

with

$$K = M \frac{e^{R(b-x_0)} - 1}{2R}.$$

Because of (18) we say that the ***global truncation error of Euler's method is of order h,*** which we write as $O(h)$.

SEMILINEAR EQUATIONS AND VARIATION OF PARAMETERS

An equation that can be written in the form

$$y' + p(x)y = h(x,y) \tag{19}$$

with $p \not\equiv 0$ is said to be ***semilinear.*** (Of course, (19) is linear if h is independent of y.) One way to apply Euler's method to an initial value problem

$$y' + p(x)y = h(x,y), \qquad y(x_0) = y_0 \tag{20}$$

for (19) is to think of it as

$$y' = f(x,y), \qquad y(x_0) = y,$$

where

$$f(x,y) = -p(x)y + h(x,y).$$

However, we can also start by applying variation of parameters to (20), as in Sections 2.1 and 2.4; thus, we write the solution of (20) as $y = uy_1$, where y_1 is a nontrivial solution of the complementary equation $y' + p(x)y = 0$. Then $y = uy_1$ is a solution of (20) if and only if u is a solution of the initial value problem

$$u' = h(x, uy_1(x))/y_1(x), \qquad u(x_0) = y(x_0)/y_1(x_0). \tag{21}$$

We can apply Euler's method to obtain approximate values $u_0, u_1, \ldots, u_n$ of this initial value problem, and then take

$$y_i = u_i y_1(x_i)$$

as approximate values of the solution of (20). We will call this procedure the ***Euler semilinear method.***

The next two examples show that the Euler and semilinear Euler methods may yield drastically different results.

EXAMPLE 4 In Example 7 of Section 2.1 we had to leave the solution of the initial value problem

$$y' - 2xy = 1, \qquad y(0) = 3 \tag{22}$$

in the form

$$y = e^{x^2}\left(3 + \int_0^x e^{-t^2}\, dt\right) \qquad (23)$$

because it was impossible to evaluate this integral exactly in terms of elementary functions. Use step sizes $h = 0.2, h = 0.1$, and $h = 0.05$ to find approximate values of the solution of (22) at $x = 0, 0.2, 0.4, 0.6, \ldots, 2.0$ by **(a)** Euler's method; **(b)** the Euler semilinear method.

Solution   **(a)** Rewriting (22) as

$$y' = 1 + 2xy, \qquad y(0) = 3 \qquad (24)$$

and applying Euler's method with $f(x,y) = 1 + 2xy$ yields the results shown in Table 5. Because of the large differences between the estimates obtained for the three values of h, it would be clear that these results are useless even if the "exact" values were not included in the table.

TABLE 5   Numerical solution of $y' - 2xy = 1$, $y(0) = 3$, with Euler's method

x	$h = 0.2$	$h = 0.1$	$h = 0.05$	"Exact"
0.0	3.000000000	3.000000000	3.000000000	3.000000000
0.2	3.200000000	3.262000000	3.294348537	3.327851973
0.4	3.656000000	3.802028800	3.881421103	3.966059348
0.6	4.440960000	4.726810214	4.888870783	5.067039535
0.8	5.706790400	6.249191282	6.570796235	6.936700945
1.0	7.732963328	8.771893026	9.419105620	10.184923955
1.2	11.026148659	13.064051391	14.405772067	16.067111677
1.4	16.518700016	20.637273893	23.522935872	27.289392347
1.6	25.969172024	34.570423758	41.033441257	50.000377775
1.8	42.789442120	61.382165543	76.491018246	98.982969504
2.0	73.797840446	115.440048291	152.363866569	211.954462214

It is easy to see why Euler's method yields such poor results. Recall that the constant M in (10)—which plays an important role in determining the local truncation error in Euler's method—must be an upper bound for the values of the second derivative y'' of the solution of the initial value problem (22) on $(0,2)$. The problem is that y'' assumes very large values on this interval. To see this, we differentiate (24) to obtain

$$y''(x) = 2y(x) + 2xy'(x) = 2y(x) + 2x(1 + 2xy(x)) = 2(1 + 2x^2)y(x) + 2x,$$

where the second equality follows again from (24). Since (23) implies that $y(x) > 3e^{x^2}$ if $x > 0$, it now follows that

$$y''(x) > 6(1 + 2x^2)e^{x^2} + 2x, \qquad x > 0.$$

For example, letting $x = 2$ shows that $y''(2) > 2952$.

Solution **(b)** Since $y_1 = e^{x^2}$ is a solution of the complementary equation $y' - 2xy = 0$, we can apply the Euler semilinear method to (22) with

$$y = ue^{x^2} \quad \text{and} \quad u' = e^{-x^2}, \quad u(0) = 3.$$

The results listed in Table 6 are clearly better than those obtained by Euler's method.

TABLE 6 Numerical solution of $y' - 2xy = 1$, $y(0) = 3$, by the Euler semilinear method

x	$h = 0.2$	$h = 0.1$	$h = 0.05$	"Exact"
0.0	3.000000000	3.000000000	3.000000000	3.000000000
0.2	3.330594477	3.329558853	3.328788889	3.327851973
0.4	3.980734157	3.974067628	3.970230415	3.966059348
0.6	5.106360231	5.087705244	5.077622723	5.067039535
0.8	7.021003417	6.980190891	6.958779586	6.936700945
1.0	10.350076600	10.269170824	10.227464299	10.184923955
1.2	16.381180092	16.226146390	16.147129067	16.067111677
1.4	27.890003380	27.592026085	27.441292235	27.289392347
1.6	51.183323262	50.594503863	50.298106659	50.000377775
1.8	101.424397595	100.206659076	99.595562766	98.982969504
2.0	217.301032800	214.631041938	213.293582978	211.954462214

We can't give a general procedure for determining in advance whether Euler's method or the semilinear Euler method will produce better results for a given semilinear initial value problem (19). As a rule of thumb, the Euler semilinear method will yield better results than Euler's method if $|u''|$ is small on $[x_0, b]$, while Euler's method will yield better results if $|u''|$ is large on $[x_0, b]$. In many cases the results obtained by the two methods don't differ appreciably. However, we propose the following intuitive way to decide which is the better method: Try both methods with multiple step sizes, as we did in Example 4, and accept the results obtained by the method for which the approximations change less as the step size decreases.

EXAMPLE 5 Applying Euler's method with step sizes $h = 0.1$, $h = 0.05$, and $h = 0.025$ to the initial value problem

$$y' - 2y = \frac{x}{1 + y^2}, \quad y(1) = 7, \tag{25}$$

on $[1,2]$ yields the results in Table 7. Applying the Euler semilinear method with

$$y = ue^{2x} \quad \text{and} \quad u' = \frac{xe^{-2x}}{1 + u^2 e^{4x}}, \quad u(1) = 7e^{-2}$$

yields the results in Table 8. Since the latter are clearly less dependent on step size than the former, we conclude that the Euler semilinear method is better than

Euler's method for (25). This conclusion is supported by comparing the approximate results obtained by the two methods with the "exact" values of the solution.

TABLE 7   Numerical solution of $y' - 2y = x/(1 + y^2)$, $y(1) = 7$, by Euler's method

x	h = 0.1	h = 0.05	h = 0.025	"Exact"
1.0	7.000000000	7.000000000	7.000000000	7.000000000
1.1	8.402000000	8.471970569	8.510493955	8.551744786
1.2	10.083936450	10.252570169	10.346014101	10.446546230
1.3	12.101892354	12.406719381	12.576720827	12.760480158
1.4	14.523152445	15.012952416	15.287872104	15.586440425
1.5	17.428443554	18.166277405	18.583079406	19.037865752
1.6	20.914624471	21.981638487	22.588266217	23.253292359
1.7	25.097914310	26.598105180	27.456479695	28.401914416
1.8	30.117766627	32.183941340	33.373738944	34.690375086
1.9	36.141518172	38.942738252	40.566143158	42.371060528
2.0	43.369967155	47.120835251	49.308511126	51.752229656

TABLE 8   Numerical solution of $y' - 2y = x/(1 + y^2)$, $y(1) = 7$, by the Euler semilinear method

x	h = 0.1	h = 0.05	h = 0.025	"Exact"
1.0	7.000000000	7.000000000	7.000000000	7.000000000
1.1	8.552262113	8.551993978	8.551867007	8.551744786
1.2	10.447568674	10.447038547	10.446787646	10.446546230
1.3	12.762019799	12.761221313	12.760843543	12.760480158
1.4	15.588535141	15.587448600	15.586934680	15.586440425
1.5	19.040580614	19.039172241	19.038506211	19.037865752
1.6	23.256721636	23.254942517	23.254101253	23.253292359
1.7	28.406184597	28.403969107	28.402921581	28.401914416
1.8	34.695649222	34.692912768	34.691618979	34.690375086
1.9	42.377544138	42.374180090	42.372589624	42.371060528
2.0	51.760178446	51.756054133	51.754104262	51.752229656

EXAMPLE 6   Applying Euler's method with step sizes $h = 0.1$, $h = 0.05$, and $h = 0.025$ to the initial value problem

$$y' + 3x^2 y = 1 + y^2, \qquad y(2) = 2 \tag{26}$$

on [2,3] yields the results in Table 9. Applying the Euler semilinear method with

$$y = ue^{-x^3} \quad \text{and} \quad u' = e^{x^3}(1 + u^2e^{-2x^3}), \quad u(2) = 2e^8$$

yields the results in Table 10. Noting the close agreement among the three columns of Table 9 (at least for larger values of x), and the lack of any such agreement among the columns of Table 10, we conclude that Euler's method is better than the Euler semilinear method for (26). Comparing the results with the exact values supports this conclusion. ∎

TABLE 9   Numerical solution of $y' + 3x^2y = 1 + y^2$, $y(2) = 2$, by Euler's method

x	$h = 0.1$	$h = 0.05$	$h = 0.025$	"Exact"
2.0	2.000000000	2.000000000	2.000000000	2.000000000
2.1	0.100000000	0.493231250	0.609611171	0.701162906
2.2	0.068700000	0.122879586	0.180113445	0.236986800
2.3	0.069419569	0.070670890	0.083934459	0.103815729
2.4	0.059732621	0.061338956	0.063337561	0.068390786
2.5	0.056871451	0.056002363	0.056249670	0.057281091
2.6	0.050560917	0.051465256	0.051517501	0.051711676
2.7	0.048279018	0.047484716	0.047514202	0.047564141
2.8	0.042925892	0.043967002	0.043989239	0.044014438
2.9	0.042148458	0.040839683	0.040857109	0.040875333
3.0	0.035985548	0.038044692	0.038058536	0.038072838

TABLE 10   Numerical solution of $y' + 3x^2y = 1 + y^2$, $y(2) = 2$, by the Euler semilinear method

x	$h = 0.1$	$h = 0.05$	$h = 0.025$	"Exact"
2.0	2.000000000	2.000000000	2.000000000	2.000000000
2.1	0.708426286	0.702568171	0.701214274	0.701162906
2.2	0.214501852	0.222599468	0.228942240	0.236986800
2.3	0.069861436	0.083620494	0.092852806	0.103815729
2.4	0.032487396	0.047079261	0.056825805	0.068390786
2.5	0.021895559	0.036030018	0.045683801	0.057281091
2.6	0.017332058	0.030750181	0.040189920	0.051711676
2.7	0.014271492	0.026931911	0.036134674	0.047564141
2.8	0.011819555	0.023720670	0.032679767	0.044014438
2.9	0.009776792	0.020925522	0.029636506	0.040875333
3.0	0.008065020	0.018472302	0.026931099	0.038072838

In the next two sections we will study other numerical methods for solving initial value problems, called the *improved Euler method*, the *midpoint method*, *Heun's method*, and the *Runge–Kutta method*. If the initial value problem is semilinear as in (19), then we also have the option of using variation of parameters and then applying the given numerical method to the initial value problem (21) for u. By analogy with the terminology used here, we will call the resulting procedure the *improved Euler semilinear method*, the *midpoint semilinear method, Heun's semilinear method* or the *Runge–Kutta semilinear method*, as the case may be.

3.1 EXERCISES

You may want to save the results of these exercises, since most of them will be revisited in the next two sections.

   In Exercises 1–5 use Euler's method to find approximate values of the solution of the given initial value problem at the points $x_i = x_0 + ih$, where x_0 is the point at which the initial condition is imposed and $i = 1, 2, 3$. The purpose of these exercises is to familiarize you with the computational procedure of Euler's method.

C **1.** $y' = 2x^2 + 3y^2 - 2$, $y(2) = 1$, $h = 0.05$ **C** **2.** $y' = y + \sqrt{x^2 + y^2}$, $y(0) = 1$, $h = 0.1$

C **3.** $y' + 3y = x^2 - 3xy + y^2$, $y(0) = 2$, $h = 0.05$ **C** **4.** $y' = \dfrac{1 + x}{1 - y^2}$, $y(2) = 3$, $h = 0.1$

C **5.** $y' + x^2y = \sin xy$, $y(1) = \pi$, $h = 0.2$

C **6.** Use Euler's method with step sizes $h = 0.1, h = 0.05$, and $h = 0.025$ to find approximate values of the solution of the initial value problem

$$y' + 3y = 7e^{4x}, \qquad y(0) = 2$$

at $x = 0, 0.1, 0.2, 0.3, \ldots, 1.0$. Compare these approximate values with the values of the exact solution $y = e^{4x} + e^{-3x}$, which can be obtained by the method of Section 2.1. Present your results in a table like Table 1.

C **7.** Use Euler's method with step sizes $h = 0.1, h = 0.05$, and $h = 0.025$ to find approximate values of the solution of the initial value problem

$$y' + \frac{2}{x}y = \frac{3}{x^3} + 1, \qquad y(1) = 1$$

at $x = 1.0, 1.1, 1.2, 1.3, \ldots, 2.0$. Compare these approximate values with the values of the exact solution

$$y = \frac{1}{3x^2}(9 \ln x + x^3 + 2),$$

which can be obtained by the method of Section 2.1. Present your results in a table like Table 1.

C **8.** Use Euler's method with step sizes $h = 0.05, h = 0.025$, and $h = 0.0125$ to find approximate values of the solution of the initial value problem

$$y' = \frac{y^2 + xy - x^2}{x^2}, \qquad y(1) = 2$$

at $x = 1.0, 1.05, 1.10, 1.15, \ldots, 1.5$. Compare these approximate values with the values of the exact solution

$$y = \frac{x(1 + x^2/3)}{1 - x^2/3}$$

obtained in Example 3 of Section 2.4. Present your results in a table like Table 1.

C **9.** In Example 3 of Section 2.2 it was shown that

$$y^5 + y = x^2 + x - 4$$

is an implicit solution of the initial value problem

$$y' = \frac{2x + 1}{5y^4 + 1}, \qquad y(2) = 1. \tag{A}$$

Use Euler's method with step sizes $h = 0.1, h = 0.05$, and $h = 0.025$ to find approximate values of the solution of (A) at $x = 2.0, 2.1, 2.2, 2.3, \ldots, 3.0$. Present your results in tabular form. To check the error in these approximate values, construct another table of values of the residual

$$R(x,y) = y^5 + y - x^2 - x + 4$$

for each value of (x, y) appearing in the first table.

C 10. It can be seen from Example 1 of Section 2.5 that

$$x^4 y^3 + x^2 y^5 + 2xy = 4$$

is an implicit solution of the initial value problem

$$y' = -\frac{4x^3 y^3 + 2xy^5 + 2y}{3x^4 y^2 + 5x^2 y^4 + 2x}, \qquad y(1) = 1. \tag{A}$$

Use Euler's method with step sizes $h = 0.1, h = 0.05$, and $h = 0.025$ to find approximate values of the solution of (A) at $x = 1.0, 1.1, 1.2, 1.3, \ldots, 2.0$. Present your results in tabular form. To check the error in these approximate values, construct another table of values of the residual

$$R(x,y) = x^4 y^3 + x^2 y^5 + 2xy - 4$$

for each value of (x, y) appearing in the first table.

C 11. Use Euler's method with step sizes $h = 0.1, h = 0.05$, and $h = 0.025$ to find approximate values of the solution of the initial value problem

$$(3y^2 + 4y)y' + 2x + \cos x = 0, \qquad y(0) = 1 \quad \text{(Exercise 13 of Section 2.2)}$$

at $x = 0, 0.1, 0.2, 0.3, \ldots, 1.0$.

C 12. Use Euler's method with step sizes $h = 0.1, h = 0.05$, and $h = 0.025$ to find approximate values of the solution of the initial value problem

$$y' + \frac{(y + 1)(y - 1)(y - 2)}{x + 1} = 0, \qquad y(1) = 0 \quad \text{(Exercise 14 of Section 2.2)}$$

at $x = 1.0, 1.1, 1.2, 1.3, \ldots, 2.0$.

C 13. Use Euler's method and the Euler semilinear method with step sizes $h = 0.1, h = 0.05$, and $h = 0.025$ to find approximate values of the solution of the initial value problem

$$y' + 3y = 7e^{-3x}, \qquad y(0) = 6$$

at $x = 0, 0.1, 0.2, 0.3, \ldots, 1.0$. Compare these approximate values with the values of the exact solution $y = e^{-3x}(7x + 6)$, which can be obtained by the method of Section 2.1. Do you notice anything special about the results? Explain.

The linear initial value problems in Exercises 14–19 cannot be solved exactly in terms of known elementary functions. In each exercise use Euler's method and the Euler semilinear methods with the indicated step sizes to find approximate values of the solution of the given initial value problem at 11 equally spaced points (including the endpoints) in the interval.

C 14. $y' - 2y = \dfrac{1}{1 + x^2}, \quad y(2) = 2; \quad h = 0.1, 0.05, 0.025$ on $[2,3]$

C 15. $y' + 2xy = x^2, \quad y(0) = 3 \quad$ (Exercise 38 of Section 2.1); $\quad h = 0.2, 0.1, 0.05$ on $[0,2]$

C 16. $y' + \dfrac{1}{x} y = \dfrac{\sin x}{x^2}, \quad y(1) = 2 \quad$ (Exercise 39 of Section 2.1); $\quad h = 0.2, 0.1, 0.05$ on $[1,3]$

C 17. $y' + y = \dfrac{e^{-x} \tan x}{x}, \quad y(1) = 0 \quad$ (Exercise 40 of Section 2.1); $\quad h = 0.05, 0.025, 0.0125$ on $[1,1.5]$

C 18. $y' + \dfrac{2x}{1 + x^2} y = \dfrac{e^x}{(1 + x^2)^2}, \quad y(0) = 1 \quad$ (Exercise 41 of Section 2.1); $\quad h = 0.2, 0.1, 0.05$ on $[0,2]$

C **19.** $xy' + (x + 1)y = e^{x^2}$, $y(1) = 2$ (Exercise 42 of Section 2.1); $h = 0.05, 0.025, 0.0125$ on $[1, 1.5]$

In Exercises 20–22 use Euler's method and the Euler semilinear method with the indicated step sizes to find approximate values of the solution of the given initial value problem at 11 equally spaced points (including the endpoints) in the interval.

C **20.** $y' + 3y = xy^2(y + 1)$, $y(0) = 1$; $h = 0.1, 0.05, 0.025$ on $[0, 1]$

C **21.** $y' - 4y = \dfrac{x}{y^2(y + 1)}$, $y(0) = 1$; $h = 0.1, 0.05, 0.025$ on $[0, 1]$

C **22.** $y' + 2y = \dfrac{x^2}{1 + y^2}$, $y(2) = 1$; $h = 0.1, 0.05, 0.025$ on $[2, 3]$

23. Numerical Quadrature. The fundamental theorem of calculus says that if f is continuous on a closed interval $[a, b]$ then it has an antiderivative F such that $F'(x) = f(x)$ on $[a, b]$ and

$$\int_a^b f(x)\, dx = F(b) - F(a). \tag{A}$$

This solves the problem of evaluating a definite integral if the integrand f has an antiderivative that can be found and evaluated easily. However, if f doesn't have this property then (A) doesn't provide a useful way to evaluate the definite integral. In this case we must resort to approximate methods. There is a class of such methods called *numerical quadrature*, in which the approximation takes the form

$$\int_a^b f(x)\, dx \approx \sum_{i=0}^n c_i f(x_i), \tag{B}$$

where $a = x_0 < x_1 < \cdots < x_n = b$ are suitably chosen points and $c_0, c_1, \ldots, c_n$ are suitably chosen constants. We call (B) a *quadrature formula.*

(a) Derive the quadrature formula

$$\int_a^b f(x)\, dx \approx h \sum_{i=0}^{n-1} f(a + ih) \qquad (\text{where } h = (b - a)/n) \tag{C}$$

by applying Euler's method to the initial value problem

$$y' = f(x), \qquad y(a) = 0.$$

(b) The quadrature formula (C) is sometimes called *the left rectangle rule.* Draw a figure that justifies this terminology.

L **(c)** For several choices of $a, b,$ and A apply (C) to $f(x) = A$, with $n = 10, 20, 40, 80, 160, 320$. Compare your results with the exact answers and explain what you find.

L **(d)** For several choices of $a, b, A,$ and B apply (C) to $f(x) = A + Bx$, with $n = 10, 20, 40, 80, 160, 320$. Compare your results with the exact answers and explain what you find.

3.2   The Improved Euler Method and Related Methods

We saw in Section 3.1 that the global truncation error of Euler's method is $O(h)$, which would seem to imply that we can achieve arbitrarily accurate results with Euler's method by simply choosing the step size sufficiently small. However, this isn't a good idea, for two reasons. First, after a certain point decreasing the step size will increase roundoff errors to the point where the accuracy will deteriorate rather than improve. (However, since most modern software packages carry so many significant figures, this is not usually a severe limitation except in examples specifically contrived to illustrate the point.) The second and more important reason is that in most applications of numerical methods to an initial value problem

$$y' = f(x, y), \qquad y(x_0) = y_0, \tag{1}$$

the expensive part of the computation is the evaluation of f. Therefore we want methods that give good results for a given number of such evaluations. This is what motivates us to look for numerical methods better than Euler's.

To clarify this point, suppose we want to approximate the value of e by applying Euler's method to the initial value problem

$$y' = y, \qquad y(0) = 1 \qquad \text{(with solution } y = e^x\text{)}$$

on $[0,1]$, with $h = 1/12, 1/24$, and $1/48$, respectively. Since each step in Euler's method requires one evaluation of f, the number of evaluations of f in each of these attempts is $n = 12, 24$, and 48, respectively. In each case we accept y_n as an approximation to e. The results are shown in the second column of Table 1. The first column of the table indicates the number of evaluations of f required to obtain the approximation, and the last column contains the value of e rounded to ten significant figures.

In this section we'll study the ***improved Euler method,*** which requires two evaluations of f at each step. We've used this method with $h = 1/6, 1/12$, and $1/24$. The required number of evaluations of f were $12, 24$, and 48, as in the three applications of Euler's method; however, you can see from the third column of Table 1 that the approximation to e obtained by the improved Euler method with only 12 evaluations of f is better than the approximation obtained by Euler's method with 48 evaluations.

In Section 3.3 we'll study the ***Runge–Kutta method,*** which requires four evaluations of f at each step. We've used this method with $h = 1/3, 1/6$, and $1/12$. The required number of evaluations of f were again $12, 24$, and 48, as in the three applications of Euler's method and the improved Euler method; however, you can see from the fourth column of Table 1 that the approximation to e obtained by the Runge–Kutta method with only 12 evaluations of f is better than the approximation obtained by the improved Euler method with 48 evaluations.

TABLE 1 Approximations to e obtained by three numerical methods

n	Euler	Improved Euler	Runge–Kutta	Exact
12	2.613035290	2.707188994	2.718069764	2.718281828
24	2.663731258	2.715327371	2.718266612	2.718281828
48	2.690496599	2.717519565	2.718280809	2.718281828

THE IMPROVED EULER METHOD

The ***improved Euler method*** for solving the initial value problem (1) is based on approximating the integral curve of (1) at $(x_i, y(x_i))$ by the line through $(x_i, y(x_i))$ with slope

$$m_i = \frac{f(x_i, y(x_i)) + f(x_{i+1}, y(x_{i+1}))}{2};$$

that is, m_i is the average of the slopes of the tangents to the integral curve at the endpoints of $[x_i, x_{i+1}]$. The equation of the approximating line is therefore

$$y = y(x_i) + \frac{f(x_i, y(x_i)) + f(x_{i+1}, y(x_{i+1}))}{2}(x - x_i). \tag{2}$$

Setting $x = x_{i+1} = x_i + h$ in (2) yields

$$y_{i+1} = y(x_i) + \frac{h}{2}(f(x_i, y(x_i)) + f(x_{i+1}, y(x_{i+1}))) \tag{3}$$

as an approximation to $y(x_{i+1})$. As in our derivation of Euler's method, we replace $y(x_i)$ (unknown if $i > 0$) by its approximate value y_i; then (3) becomes

$$y_{i+1} = y_i + \frac{h}{2}(f(x_i, y_i) + f(x_{i+1}, y(x_{i+1}))).$$

However, this still won't work, because we don't know $y(x_{i+1})$, which appears on the right. We overcome this by replacing $y(x_{i+1})$ by $y_i + hf(x_i, y_i)$, the value that the Euler method would assign to y_{i+1}. Thus, the improved Euler method starts with the known value $y(x_0) = y_0$ and computes $y_1, y_2, \ldots, y_n$ successively by means of the formula

$$y_{i+1} = y_i + \frac{h}{2}(f(x_i, y_i) + f(x_{i+1}, y_i + hf(x_i, y_i))). \tag{4}$$

The computation indicated here can be conveniently organized as follows: given y_i, compute

$$k_{1i} = f(x_i, y_i),$$
$$k_{2i} = f(x_i + h, y_i + hk_{1i}),$$
$$y_{i+1} = y_i + \frac{h}{2}(k_{1i} + k_{2i}).$$

The improved Euler method requires two evaluations of $f(x, y)$ per step, while Euler's method requires only one. However, we'll see at the end of this section that if f satisfies appropriate assumptions then the local truncation error with the improved Euler method is $O(h^3)$, rather than $O(h^2)$ as with Euler's method. It follows that the global truncation error with the improved Euler method is $O(h^2)$; however, we will not prove this.

We note that the magnitude of the local truncation error in the improved Euler method and other methods discussed in this section is determined by the third derivative y''' of the solution of the initial value problem. Therefore the local truncation error will be larger where $|y'''|$ is large, or smaller where $|y'''|$ is small.

The following example, which deals with the initial value problem considered in Example 1 of Section 3.1, illustrates the computational procedure indicated in the improved Euler method.

EXAMPLE 1   Use the improved Euler method with $h = 0.1$ to find approximate values of the solution of the initial value problem

$$y' + 2y = x^3 e^{-2x}, \qquad y(0) = 1 \tag{5}$$

at $x = 0.1, 0.2, 0.3$.

Solution   As in Example 1 of Section 3.1, we rewrite (5) as

$$y' = -2y + x^3 e^{-2x}, \qquad y(0) = 1,$$

which is of the form (1), with

$$f(x, y) = -2y + x^3 e^{-2x}, \qquad x_0 = 0, \qquad \text{and } y_0 = 1.$$

The improved Euler method yields

$$k_{10} = f(x_0, y_0) = f(0, 1) = -2,$$
$$k_{20} = f(x_1, y_0 + hk_{10}) = f(.1, 1 + (.1)(-2))$$
$$= f(.1, .8) = -2(.8) + (.1)^3 e^{-.2} = -1.599181269,$$
$$y_1 = y_0 + \frac{h}{2}(k_{10} + k_{20}),$$
$$= 1 + (.05)(-2 - 1.599181269) = .820040937,$$

$$k_{11} = f(x_1, y_1) = f(.1, .820040937)$$
$$= -2(.820040937) + (.1)^3 e^{-.2} = -1.639263142,$$
$$k_{21} = f(x_2, y_1 + hk_{11}) = f(.2, .820040937 + .1(-1.639263142)),$$
$$= f(.2, .656114622) = -2(.656114622) + (.2)^3 e^{-.4}$$
$$= -1.306866684,$$

$$y_2 = y_1 + \frac{h}{2}(k_{11} + k_{21}),$$
$$= .820040937 + (.05)(-1.639263142 - 1.306866684) = .672734445,$$

$$k_{12} = f(x_2, y_2) = f(.2, .672734445)$$
$$= -2(.672734445) + (.2)^3 e^{-.4} = -1.340106330,$$
$$k_{22} = f(x_3, y_2 + hk_{12}) = f(.3, .672734445 + .1(-1.340106330)),$$
$$= f(.3, .538723812) = -2(.538723812) + (.3)^3 e^{-.6} = -1.062629710,$$

$$y_3 = y_2 + \frac{h}{2}(k_{12} + k_{22})$$
$$= .672734445 + (.05)(-1.340106330 - 1.062629710) = .552597643.$$

EXAMPLE 2 Table 2 shows results of using the improved Euler method with step sizes $h = 0.1$ and $h = 0.05$ to find approximate values of the solution of the initial value problem

$$y' + 2y = x^3 e^{-2x}, \qquad y(0) = 1$$

at $x = 0, 0.1, 0.2, 0.3, \ldots, 1.0$. For comparison, it also shows the corresponding approximate values obtained with Euler's method in Example 2 of Section 3.1, and the values of the exact solution

$$y = \frac{e^{-2x}}{4}(x^4 + 4).$$

Notice that the results obtained by the improved Euler method with $h = 0.1$ are better than those obtained by Euler's method with $h = 0.05$.

TABLE 2 Numerical solution of $y' + 2y = x^3 e^{-2x}$, $y(0) = 1$, by Euler's method and the improved Euler method.

x	$h = 0.1$	$h = 0.05$	$h = 0.1$	$h = 0.05$	**Exact**
0.0	1.000000000	1.000000000	1.000000000	1.000000000	1.000000000
0.1	0.800000000	0.810005655	0.820040937	0.819050572	0.818751221
	Euler		**Improved Euler**		**Exact**

(continued)

TABLE 2 *(continued)*

x	h = 0.1	h = 0.05	h = 0.1	h = 0.05	Exact
0.2	0.640081873	0.656266437	0.672734445	0.671086455	0.670588174
0.3	0.512601754	0.532290981	0.552597643	0.550543878	0.549922980
0.4	0.411563195	0.432887056	0.455160637	0.452890616	0.452204669
0.5	0.332126261	0.353785015	0.376681251	0.374335747	0.373627557
0.6	0.270299502	0.291404256	0.313970920	0.311652239	0.310952904
0.7	0.222745397	0.242707257	0.264287611	0.262067624	0.261398947
0.8	0.186654593	0.205105754	0.225267702	0.223194281	0.222570721
0.9	0.159660776	0.176396883	0.194879501	0.192981757	0.192412038
1.0	0.139778910	0.154715925	0.171388070	0.169680673	0.169169104
	Euler		**Improved Euler**		**Exact**

EXAMPLE 3 Table 3 shows analogous results for the nonlinear initial value problem

$$y' = -2y^2 + xy + x^2, \qquad y(0) = 1,$$

to which we applied Euler's method in Example 3 of Section 3.1.

TABLE 3 Numerical solution of $y' = -2y^2 + xy + x^2$, $y(0) = 1$, by Euler's method and the improved Euler method

x	h = 0.1	h = 0.05	h = 0.1	h = 0.05	"Exact"
0.0	1.000000000	1.000000000	1.000000000	1.000000000	1.000000000
0.1	0.800000000	0.821375000	0.840500000	0.838288371	0.837584494
0.2	0.681000000	0.707795377	0.733430846	0.730556677	0.729641890
0.3	0.605867800	0.633776590	0.661600806	0.658552190	0.657580377
0.4	0.559628676	0.587454526	0.615961841	0.612884493	0.611901791
0.5	0.535376972	0.562906169	0.591634742	0.588558952	0.587575491
0.6	0.529820120	0.557143535	0.586006935	0.582927224	0.581942225
0.7	0.541467455	0.568716935	0.597712120	0.594618012	0.593629526
0.8	0.569732776	0.596951988	0.626008824	0.622898279	0.621907458
0.9	0.614392311	0.641457729	0.670351225	0.667237617	0.666250842
1.0	0.675192037	0.701764495	0.730069610	0.726985837	0.726015790
	Euler		**Improved Euler**		**"Exact"**

EXAMPLE 4 Use step sizes $h = 0.2$, $h = 0.1$, and $h = 0.05$ to find approximate values of the solution of

$$y' - 2xy = 1, \qquad y(0) = 3 \tag{6}$$

at $x = 0, 0.2, 0.4, 0.6, \ldots, 2.0$ by **(a)** the improved Euler method; **(b)** the improved Euler semilinear method. (We used Euler's method and the Euler semilinear method on this problem in Example 4 of Section 3.1.)

Solution   **(a)** Rewriting (6) as

$$y' = 1 + 2xy, \qquad y(0) = 3$$

and applying the improved Euler method with $f(x, y) = 1 + 2xy$ yields the results shown in Table 4.

TABLE 4   Numerical solution of $y' - 2xy = 1$, $y(0) = 3$, by the improved Euler method

x	$h = 0.2$	$h = 0.1$	$h = 0.05$	"Exact"
0.0	3.000000000	3.000000000	3.000000000	3.000000000
0.2	3.328000000	3.328182400	3.327973600	3.327851973
0.4	3.964659200	3.966340117	3.966216690	3.966059348
0.6	5.057712497	5.065700515	5.066848381	5.067039535
0.8	6.900088156	6.928648973	6.934862367	6.936700945
1.0	10.065725534	10.154872547	10.177430736	10.184923955
1.2	15.708954420	15.970033261	16.041904862	16.067111677
1.4	26.244894192	26.991620960	27.210001715	27.289392347
1.6	46.958915746	49.096125524	49.754131060	50.000377775
1.8	89.982312641	96.200506218	98.210577385	98.982969504
2.0	184.563776288	203.151922739	209.464744495	211.954462214

Solution   **(b)** Since $y_1 = e^{x^2}$ is a solution of the complementary equation $y' - 2xy = 0$, we can apply the improved Euler semilinear method to (6), with

$$y = ue^{x^2} \qquad \text{and} \qquad u' = e^{-x^2}, \qquad u(0) = 3.$$

The results listed in Table 5 are clearly better than those obtained by the improved Euler method.   ■

TABLE 5   Numerical solution of $y' - 2xy = 1$, $y(0) = 3$, by the improved Euler semilinear method

x	$h = 0.2$	$h = 0.1$	$h = 0.05$	"Exact"
0.0	3.000000000	3.000000000	3.000000000	3.000000000
0.2	3.326513400	3.327518315	3.327768620	3.327851973
0.4	3.963383070	3.965392084	3.965892644	3.966059348
0.6	5.063027290	5.066038774	5.066789487	5.067039535
0.8	6.931355329	6.935366847	6.936367564	6.936700945

(continued)

TABLE 5 *(continued)*

x	h = 0.2	h = 0.1	h = 0.05	"Exact"
1.0	10.178248417	10.183256733	10.184507253	10.184923955
1.2	16.059110511	16.065111599	16.066611672	16.067111677
1.4	27.280070674	27.287059732	27.288809058	27.289392347
1.6	49.989741531	49.997712997	49.999711226	50.000377775
1.8	98.971025420	98.979972988	98.982219722	98.982969504
2.0	211.941217796	211.951134436	211.953629228	211.954462214

A FAMILY OF METHODS WITH $O(h^3)$ LOCAL TRUNCATION ERROR

We will now derive a class of methods with $O(h^3)$ local truncation error for solving (1). For simplicity, we assume that $f, f_x, f_y, f_{xx}, f_{yy}$, and f_{xy} are continuous and bounded for all (x, y). This implies that if y is the solution of (1) then y'' and y''' are bounded (Exercise 31).

We begin by approximating the integral curve of (1) at $(x_i, y(x_i))$ by the line through $(x_i, y(x_i))$ with slope

$$m_i = \sigma y'(x_i) + \rho y'(x_i + \theta h),$$

where σ, ρ, and θ are constants that we'll soon specify; however, we insist at the outset that $0 < \theta \leq 1$, so that

$$x_i < x_i + \theta h \leq x_{i+1}.$$

The equation of the approximating line is

$$
\begin{aligned}
y &= y(x_i) + m_i(x - x_i) \\
  &= y(x_i) + [\sigma y'(x_i) + \rho y'(x_i + \theta h)](x - x_i).
\end{aligned}
\tag{7}
$$

Setting $x = x_{i+1} = x_i + h$ in (7) yields

$$\hat{y}_{i+1} = y(x_i) + h[\sigma y'(x_i) + \rho y'(x_i + \theta h)]$$

as an approximation to $y(x_{i+1})$.

To determine σ, ρ, and θ so that the error

$$
\begin{aligned}
E_i &= y(x_{i+1}) - \hat{y}_{i+1} \\
    &= y(x_{i+1}) - y(x_i) - h[\sigma y'(x_i) + \rho y'(x_i + \theta h)]
\end{aligned}
\tag{8}
$$

in this approximation is $O(h^3)$, we begin by recalling from Taylor's theorem that

$$y(x_{i+1}) = y(x_i) + hy'(x_i) + \frac{h^2}{2}y''(x_i) + \frac{h^3}{6}y'''(\hat{x}_i),$$

where $\hat{x}_i$ is in (x_i, x_{i+1}). Since y''' is bounded this implies that

$$y(x_{i+1}) - y(x_i) - hy'(x_i) - \frac{h^2}{2}y''(x_i) = O(h^3).$$

Comparing this with (8) shows that $E_i = O(h^3)$ if

$$\sigma y'(x_i) + \rho y'(x_i + \theta h) = y'(x_i) + \frac{h}{2}y''(x_i) + O(h^2).
\tag{9}$$

However, applying Taylor's theorem to y' shows that

$$y'(x_i + \theta h) = y'(x_i) + \theta h y''(x_i) + \frac{(\theta h)^2}{2} y'''(\bar{x}_i),$$

where $\bar{x}_i$ is in $(x_i, x_i + \theta h)$. Since y''' is bounded, this implies that

$$y'(x_i + \theta h) = y'(x_i) + \theta h y''(x_i) + O(h^2).$$

Substituting this into (9) and noting that the sum of two $O(h^2)$ terms is again $O(h^2)$ shows that $E_i = O(h^3)$ if

$$(\sigma + \rho)y'(x_i) + \rho \theta h y''(x_i) = y'(x_i) + \frac{h}{2}y''(x_i),$$

which will hold if

$$\sigma + \rho = 1 \quad \text{and} \quad \rho\theta = \frac{1}{2}. \tag{10}$$

Since $y' = f(x, y)$, we can now conclude from (8) that

$$y(x_{i+1}) = y(x_i) + h[\sigma f(x_i, y_i) + \rho f(x_i + \theta h, y(x_i + \theta h))] + O(h^3) \tag{11}$$

if σ, ρ, and θ satisfy (10). However, this formula would not be useful even if we knew $y(x_i)$ exactly (as we would for $i = 0$), since we still wouldn't know $y(x_i + \theta h)$ exactly. To overcome this difficulty we again use Taylor's theorem to write

$$y(x_i + \theta h) = y(x_i) + \theta h y'(x_i) + \frac{h^2}{2} y''(\tilde{x}_i),$$

where $\tilde{x}_i$ is in $(x_i, x_i + \theta h)$. Since $y'(x_i) = f(x_i, y(x_i))$ and y'' is bounded, this implies that

$$|y(x_i + \theta h) - y(x_i) - \theta h f(x_i, y(x_i))| \leq K h^2 \tag{12}$$

for some constant K. Since f_y is bounded, the mean value theorem implies that

$$|f(x_i + \theta h, u) - f(x_i + \theta h, v)| \leq M|u - v|$$

for some constant M. Letting

$$u = y(x_i + \theta h) \quad \text{and} \quad v = y(x_i) + \theta h f(x_i, y(x_i))$$

and recalling (12) shows that

$$f(x_i + \theta h, y(x_i + \theta h)) = f(x_i + \theta h, y(x_i) + \theta h f(x_i, y(x_i))) + O(h^2).$$

Substituting this into (11) yields

$$y(x_{i+1}) = y(x_i) + h[\sigma f(x_i, y(x_i)) + \rho f(x_i + \theta h, y(x_i) + \theta h f(x_i, y(x_i)))] + O(h^3).$$

This implies that the formula

$$y_{i+1} = y_i + h[\sigma f(x_i, y_i) + \rho f(x_i + \theta h, y_i + \theta h f(x_i, y_i))]$$

has $O(h^3)$ local truncation error if σ, ρ, and θ satisfy (10). Substituting $\sigma = 1 - \rho$ and $\theta = 1/2\rho$ here yields

$$y_{i+1} = y_i + h\left[(1 - \rho)f(x_i, y_i) + \rho f\left(x_i + \frac{h}{2\rho}, y_i + \frac{h}{2\rho}f(x_i, y_i)\right)\right]. \tag{13}$$

The computation indicated here can be conveniently organized as follows: given y_i, compute

$$k_{1i} = f(x_i, y_i),$$

$$k_{2i} = f\left(x_i + \frac{h}{2\rho}, y_i + \frac{h}{2\rho} k_{1i}\right),$$

$$y_{i+1} = y_i + h[(1 - \rho)k_{1i} + \rho k_{2i}].$$

Consistent with our requirement that $0 < \theta \le 1$, we require that $\rho \ge 1/2$. Letting $\rho = 1/2$ in (13) yields the improved Euler method (4). Letting $\rho = 3/4$ yields **Heun's method,**

$$y_{i+1} = y_i + h\left[\frac{1}{4}f(x_i, y_i) + \frac{3}{4}f\left(x_i + \frac{2}{3}h, y_i + \frac{2}{3}hf(x_i, y_i)\right)\right],$$

which can be organized as

$$k_{1i} = f(x_i, y_i),$$

$$k_{2i} = f\left(x_i + \frac{2h}{3}, y_i + \frac{2h}{3} k_{1i}\right),$$

$$y_{i+1} = y_i + \frac{h}{4}(k_{1i} + 3k_{2i}).$$

Letting $\rho = 1$ yields the **midpoint method,**

$$y_{i+1} = y_i + hf\left(x_i + \frac{h}{2}, y_i + \frac{h}{2}f(x_i, y_i)\right),$$

which can be organized as

$$k_{1i} = f(x_i, y_i),$$

$$k_{2i} = f\left(x_i + \frac{h}{2}, y_i + \frac{h}{2} k_{1i}\right),$$

$$y_{i+1} = y_i + hk_{2i}.$$

Examples involving the midpoint method and Heun's method are given in Exercises 23–30.

3.2 EXERCISES

Most of the following numerical exercises involve initial value problems considered in the exercises in Section 3.1. You will find it instructive to compare the results that you obtain here with the corresponding results that you obtained in Section 3.1.

In Exercises 1–5 use the improved Euler method to find approximate values of the solution of the given initial value problem at the points $x_i = x_0 + ih$, where x_0 is the point at which the initial condition is imposed and $i = 1, 2, 3$.

C **1.** $y' = 2x^2 + 3y^2 - 2, \quad y(2) = 1, \quad h = 0.05$

C **2.** $y' = y + \sqrt{x^2 + y^2}, \quad y(0) = 1, \quad h = 0.1$

C **3.** $y' + 3y = x^2 - 3xy + y^2, \quad y(0) = 2, \quad h = 0.05$

C **4.** $y' = \dfrac{1 + x}{1 - y^2}, \quad y(2) = 3, \quad h = 0.1$

C **5.** $y' + x^2y = \sin xy, \quad y(1) = \pi, \quad h = 0.2$

C **6.** Use the improved Euler method with step sizes $h = 0.1, h = 0.05$, and $h = 0.025$ to find approximate values of the solution of the initial value problem

$$y' + 3y = 7e^{4x}, \qquad y(0) = 2$$

at $x = 0, 0.1, 0.2, 0.3, \ldots, 1.0$. Compare these approximate values with the values of the exact solution $y = e^{4x} + e^{-3x}$, which can be obtained by the method of Section 2.1. Present your results in a table like Table 2.

C **7.** Use the improved Euler method with step sizes $h = 0.1, h = 0.05$, and $h = 0.025$ to find approximate values of the solution of the initial value problem

$$y' + \frac{2}{x}y = \frac{3}{x^3} + 1, \qquad y(1) = 1$$

at $x = 1.0, 1.1, 1.2, 1.3, \ldots, 2.0$. Compare these approximate values with the values of the exact solution

$$y = \frac{1}{3x^2}(9 \ln x + x^3 + 2),$$

which can be obtained by the method of Section 2.1. Present your results in a table like Table 2.

C **8.** Use the improved Euler method with step sizes $h = 0.05, h = 0.025$, and $h = 0.0125$ to find approximate values of the solution of the initial value problem

$$y' = \frac{y^2 + xy - x^2}{x^2}, \qquad y(1) = 2$$

at $x = 1.0, 1.05, 1.10, 1.15, \ldots, 1.5$. Compare these approximate values with the values of the exact solution

$$y = \frac{x(1 + x^2/3)}{1 - x^2/3}$$

obtained in Example 3 of Section 2.4. Present your results in a table like Table 2.

C **9.** In Example 3 of Section 2.2 it was shown that

$$y^5 + y = x^2 + x - 4$$

is an implicit solution of the initial value problem

$$y' = \frac{2x + 1}{5y^4 + 1}, \qquad y(2) = 1. \tag{A}$$

Use the improved Euler method with step sizes $h = 0.1, h = 0.05$, and $h = 0.025$ to find approximate values of the solution of (A) at $x = 2.0, 2.1, 2.2, 2.3, \ldots, 3.0$. Present your results in tabular form. To check the error in these approximate values, construct another table of values of the residual

$$R(x, y) = y^5 + y - x^2 - x + 4$$

for each value of (x, y) appearing in the first table.

C **10.** It can be seen from Example 1 of Section 2.5 that

$$x^4y^3 + x^2y^5 + 2xy = 4$$

is an implicit solution of the initial value problem

$$y' = -\frac{4x^3y^3 + 2xy^5 + 2y}{3x^4y^2 + 5x^2y^4 + 2x}, \qquad y(1) = 1. \tag{A}$$

Use the improved Euler method with step sizes $h = 0.1, h = 0.05$, and $h = 0.025$ to find approximate values of the solution of (A) at $x = 1.0, 1.1, 1.2, 1.3, \ldots, 2.0$. Present your results in tabular form. To check the error in these approximate values, construct another table of values of the residual

$$R(x, y) = x^4y^3 + x^2y^5 + 2xy - 4$$

for each value of (x, y) appearing in the first table.

C **11.** Use the improved Euler method with step sizes $h = 0.1, h = 0.05$, and $h = 0.025$ to find approximate values of the solution of the initial value problem

$$(3y^2 + 4y)y' + 2x + \cos x = 0, \qquad y(0) = 1 \quad \text{(Exercise 13 of Section 2.2)}$$

at $x = 0, 0.1, 0.2, 0.3, \ldots, 1.0$.

C **12.** Use the improved Euler method with step sizes $h = 0.1, h = 0.05$, and $h = 0.025$ to find approximate values of the solution of the initial value problem

$$y' + \frac{(y+1)(y-1)(y-2)}{x+1} = 0, \qquad y(1) = 0 \quad \text{(Exercise 14 of Section 2.2)}$$

at $x = 1.0, 1.1, 1.2, 1.3, \ldots, 2.0$.

C **13.** Use the improved Euler method and the improved Euler semilinear method with step sizes $h = 0.1, h = 0.05$, and $h = 0.025$ to find approximate values of the solution of the initial value problem

$$y' + 3y = e^{-3x}(1 - 2x), \qquad y(0) = 2$$

at $x = 0, 0.1, 0.2, 0.3, \ldots, 1.0$. Compare these approximate values with the values of the exact solution $y = e^{-3x}(2 + x - x^2)$, which can be obtained by the method of Section 2.1. Do you notice anything special about the results? Explain.

The linear initial value problems in Exercises 14–19 cannot be solved exactly in terms of known elementary functions. In each exercise use the improved Euler and improved Euler semilinear methods with the indicated step sizes to find approximate values of the solution of the given initial value problem at 11 equally spaced points (including the endpoints) in the interval.

C **14.** $y' - 2y = \dfrac{1}{1 + x^2}, \quad y(2) = 2; \quad h = 0.1, 0.05, 0.025$ on $[2,3]$

C **15.** $y' + 2xy = x^2, \quad y(0) = 3$ (Exercise 38 of Section 2.1); $\quad h = 0.2, 0.1, 0.05$ on $[0,2]$

C **16.** $y' + \dfrac{1}{x}y = \dfrac{\sin x}{x^2}, \quad y(1) = 2$ (Exercise 39 of Section 2.1); $\quad h = 0.2, 0.1, 0.05$ on $[1,3]$

C **17.** $y' + y = \dfrac{e^{-x} \tan x}{x}, \quad y(1) = 0$ (Exercise 40 of Section 2.1); $\quad h = 0.05, 0.025, 0.0125$ on $[1,1.5]$

C **18.** $y' + \dfrac{2x}{1 + x^2}y = \dfrac{e^x}{(1 + x^2)^2}, \quad y(0) = 1$ (Exercise 41 of Section 2.1); $\quad h = 0.2, 0.1, 0.05$ on $[0,2]$

C **19.** $xy' + (x + 1)y = e^{x^2}, \quad y(1) = 2$ (Exercise 42 of Section 2.1); $\quad h = 0.05, 0.025, 0.0125$ on $[1,1.5]$

In Exercises 20–22 use the improved Euler method and the improved Euler semilinear method with the indicated step sizes to find approximate values of the solution of the given initial value problem at 11 equally spaced points (including the endpoints) in the interval.

C **20.** $y' + 3y = xy^2(y + 1), \quad y(0) = 1; \quad h = 0.1, 0.05, 0.025$ on $[0,1]$

C **21.** $y' - 4y = \dfrac{x}{y^2(y + 1)}, \quad y(0) = 1; \quad h = 0.1, 0.05, 0.025$ on $[0,1]$

C **22.** $y' + 2y = \dfrac{x^2}{1 + y^2}, \quad y(2) = 1; \quad h = 0.1, 0.05, 0.025$ on $[2,3]$

C **23.** Do Exercise 7 with "improved Euler method" replaced by "midpoint method."

C **24.** Do Exercise 7 with "improved Euler method" replaced by "Heun's method."

C **25.** Do Exercise 8 with "improved Euler method" replaced by "midpoint method."

C **26.** Do Exercise 8 with "improved Euler method" replaced by "Heun's method."

C **27.** Do Exercise 11 with "improved Euler method" replaced by "midpoint method."

C **28.** Do Exercise 11 with "improved Euler method" replaced by "Heun's method."

C **29.** Do Exercise 12 with "improved Euler method" replaced by "midpoint method."

C **30.** Do Exercise 12 with "improved Euler method" replaced by "Heun's method."

31. Show that if $f, f_x, f_y, f_{xx}, f_{yy}$, and f_{xy} are continuous and bounded for all (x,y) and y is the solution of the initial value problem

$$y' = f(x,y), \qquad y(x_0) = y_0,$$

then y'' and y''' are bounded.

32. Numerical Quadrature (see Exercise 23 of Section 3.1).

(a) Derive the quadrature formula

$$\int_a^b f(x)\,dx \approx .5h(f(a) + f(b)) + h\sum_{i=1}^{n-1} f(a + ih) \qquad \text{(where } h = (b-a)/n) \tag{A}$$

by applying the improved Euler method to the initial value problem

$$y' = f(x), \qquad y(a) = 0.$$

(b) The quadrature formula (A) is called *the trapezoid rule.* Draw a figure that justifies this terminology.

L (c) For several choices of a, b, A, and B apply (A) to $f(x) = A + Bx$, with $n = 10, 20, 40, 80, 160, 320$. Compare your results with the exact answers and explain what you find.

L (d) For several choices of a, b, A, B, and C apply (A) to $f(x) = A + Bx + Cx^2$, with $n = 10, 20, 40, 80, 160, 320$. Compare your results with the exact answers and explain what you find.

3.3 The Runge–Kutta Method

In general, if k is any positive integer and f satisfies appropriate assumptions, there are numerical methods with local truncation error $O(h^{k+1})$ for solving an initial value problem

$$y' = f(x,y), \qquad y(x_0) = y_0. \tag{1}$$

Moreover, it can be shown that a method with local truncation error $O(h^{k+1})$ has global truncation error $O(h^k)$. In Sections 3.1 and 3.2 we studied numerical methods where $k = 1$ and $k = 2$. We'll skip methods for which $k = 3$ and proceed to the **Runge–Kutta** method, the most widely used method for which $k = 4$. The magnitude of the local truncation error is determined by the fifth derivative $y^{(5)}$ of the solution of the initial value problem. Therefore the local truncation error will be larger where $|y^{(5)}|$ is large, or smaller where $|y^{(5)}|$ is small. The Runge–Kutta method computes approximate values $y_1, y_2, \ldots, y_n$ of the solution of (1) at $x_0, x_0 + h, \ldots, x_0 + nh$ as follows: Given y_i, compute

$$k_{1i} = f(x_i, y_i),$$

$$k_{2i} = f\left(x_i + \frac{h}{2}, y_i + \frac{h}{2}k_{1i}\right),$$

$$k_{3i} = f\left(x_i + \frac{h}{2}, y_i + \frac{h}{2}k_{2i}\right),$$

$$k_{4i} = f(x_i + h, y_i + hk_{3i}),$$

and

$$y_{i+1} = y_i + \frac{h}{6}(k_{1i} + 2k_{2i} + 2k_{3i} + k_{4i}).$$

The following example, which deals with the initial value problem considered in Examples 1 of Section 3.1 and Section 3.2, illustrates the computational procedure indicated in the Runge–Kutta method.

EXAMPLE 1   Use the Runge–Kutta method with $h = 0.1$ to find approximate values for the solution of the initial value problem

$$y' + 2y = x^3 e^{-2x}, \qquad y(0) = 1 \qquad\qquad (2)$$

at $x = 0.1, 0.2$.

Solution   Again we rewrite (2) as

$$y' = -2y + x^3 e^{-2x}, \qquad y(0) = 1,$$

which is of the form (1), with

$$f(x,y) = -2y + x^3 e^{-2x}, \qquad x_0 = 0, \qquad \text{and } y_0 = 1.$$

The Runge–Kutta method yields

$$
\begin{aligned}
k_{10} &= f(x_0, y_0) = f(0,1) = -2, \\
k_{20} &= f(x_0 + h/2, y_0 + hk_{10}/2) = f(.05, 1 + (.05)(-2)) \\
&= f(.05, .9) = -2(.9) + (.05)^3 e^{-.1} = -1.799886895, \\
k_{30} &= f(x_0 + h/2, y_0 + hk_{20}/2) = f(.05, 1 + (.05)(-1.799886895)) \\
&= f(.05, .910005655) = -2(.910005655) + (.05)^3 e^{-.1} = -1.819898206, \\
k_{40} &= f(x_0 + h, y_0 + hk_{30}) = f(.1, 1 + (.1)(-1.819898206)) \\
&= f(.1, .818010179) = -2(.818010179) + (.1)^3 e^{-.2} = -1.635201628,
\end{aligned}
$$

$$y_1 = y_0 + \frac{h}{6}(k_{10} + 2k_{20} + 2k_{30} + k_{40}),$$

$$= 1 + \frac{.1}{6}(-2 + 2(-1.799886895) + 2(-1.819898206) - 1.635201628)$$

$$= .818753803,$$

$$
\begin{aligned}
k_{11} &= f(x_1, y_1) = f(.1, .818753803) \\
&= -2(.818753803)) + (.1)^3 e^{-.2} = -1.636688875, \\
k_{21} &= f(x_1 + h/2, y_1 + hk_{11}/2) = f(.15, .818753803 + (.05)(-1.636688875)) \\
&= f(.15, .736919359) = -2(.736919359) + (.15)^3 e^{-.3} = -1.471338457, \\
k_{31} &= f(x_1 + h/2, y_1 + hk_{21}/2) = f(.15, .818753803 + (.05)(-1.471338457)) \\
&= f(.15, .745186880) = -2(.745186880) + (.15)^3 e^{-.3} = -1.487873498, \\
k_{41} &= f(x_1 + h, y_1 + hk_{31}) = f(.2, .818753803 + (.1)(-1.487873498)) \\
&= f(.2, .669966453) = -2(.669966453) + (.2)^3 e^{-.4} = -1.334570346,
\end{aligned}
$$

$$y_2 = y_1 + \frac{h}{6}(k_{11} + 2k_{21} + 2k_{31} + k_{41}),$$

$$= .818753803 + \frac{.1}{6}(-1.636688875 + 2(-1.471338457)$$

$$+ 2(-1.487873498) - 1.334570346)$$

$$= .670592417.$$

The Runge–Kutta method is sufficiently accurate for most applications.

EXAMPLE 2   Table 1 shows results of using the Runge–Kutta method with step sizes $h = 0.1$ and $h = 0.05$ to find approximate values of the solution of the initial value problem

$$y' + 2y = x^3 e^{-2x}, \qquad y(0) = 1$$

at $x = 0, 0.1, 0.2, 0.3, \ldots, 1.0$. For comparison, it also shows the corresponding approximate values obtained with the improved Euler method in Example 2 of Section 3.2, and the values of the exact solution

$$y = \frac{e^{-2x}}{4}(x^4 + 4).$$

The results obtained by the Runge–Kutta method are clearly better than those obtained by the improved Euler method; in fact, the results obtained by the Runge–Kutta method with $h = 0.1$ are better than those obtained by the improved Euler method with $h = 0.05$.

TABLE I   Numerical solution of $y' + 2y = x^3 e^{-2x}$, $y(0) = 1$, by the Runge–Kutta method and the improved Euler method

x	$h = 0.1$	$h = 0.05$	$h = 0.1$	$h = 0.05$	**Exact**
0.0	1.000000000	1.000000000	1.000000000	1.000000000	1.000000000
0.1	0.820040937	0.819050572	0.818753803	0.818751370	0.818751221
0.2	0.672734445	0.671086455	0.670592417	0.670588418	0.670588174
0.3	0.552597643	0.550543878	0.549928221	0.549923281	0.549922980
0.4	0.455160637	0.452890616	0.452210430	0.452205001	0.452204669
0.5	0.376681251	0.374335747	0.373633492	0.373627899	0.373627557
0.6	0.313970920	0.311652239	0.310958768	0.310953242	0.310952904
0.7	0.264287611	0.262067624	0.261404568	0.261399270	0.261398947
0.8	0.225267702	0.223194281	0.222575989	0.222571024	0.222570721
0.9	0.194879501	0.192981757	0.192416882	0.192412317	0.192412038
1.0	0.171388070	0.169680673	0.169173489	0.169169356	0.169169104
	Improved Euler		**Runge–Kutta**		**Exact**

EXAMPLE 3   Table 2 shows analogous results for the nonlinear initial value problem

$$y' = -2y^2 + xy + x^2, \qquad y(0) = 1,$$

to which we applied the improved Euler method in Example 3 of Section 3.2.

TABLE 2   Numerical solution of $y' = -2y^2 + xy + x^2$, $y(0) = 1$, by the Runge–Kutta method and the improved Euler method

x	$h = 0.1$	$h = 0.05$	$h = 0.1$	$h = 0.05$	**"Exact"**
0.0	1.000000000	1.000000000	1.000000000	1.000000000	1.000000000
0.1	0.840500000	0.838288371	0.837587192	0.837584759	0.837584494
0.2	0.733430846	0.730556677	0.729644487	0.729642155	0.729641890
0.3	0.661600806	0.658552190	0.657582449	0.657580598	0.657580377
	Improved Euler		**Runge–Kutta**		**"Exact"**

(continued)

TABLE 2 *(continued)*

x	h = 0.1	h = 0.05	h = 0.1	h = 0.05	"Exact"
0.4	0.615961841	0.612884493	0.611903380	0.611901969	0.611901791
0.5	0.591634742	0.588558952	0.587576716	0.587575635	0.587575491
0.6	0.586006935	0.582927224	0.581943210	0.581942342	0.581942225
0.7	0.597712120	0.594618012	0.593630403	0.593629627	0.593629526
0.8	0.626008824	0.622898279	0.621908378	0.621907553	0.621907458
0.9	0.670351225	0.667237617	0.666251988	0.666250942	0.666250842
1.0	0.730069610	0.726985837	0.726017378	0.726015908	0.726015790
	Improved Euler		**Runge–Kutta**		**"Exact"**

EXAMPLE 4   Tables 3 and 4 show results obtained by applying the Runge–Kutta and Runge–Kutta semilinear methods to the initial value problem

$$y' - 2xy = 1, \qquad y(0) = 3,$$

which we considered in Examples 4 of Section 3.1 and Section 3.2.   ■

TABLE 3   Numerical solution of $y' - 2xy = 1$, $y(0) = 3$, by the Runge–Kutta method

x	h = 0.2	h = 0.1	h = 0.05	"Exact"
0.0	3.000000000	3.000000000	3.000000000	3.000000000
0.2	3.327846400	3.327851633	3.327851952	3.327851973
0.4	3.966044973	3.966058535	3.966059300	3.966059348
0.6	5.066996754	5.067037123	5.067039396	5.067039535
0.8	6.936534178	6.936690679	6.936700320	6.936700945
1.0	10.184232252	10.184877733	10.184920997	10.184923955
1.2	16.064344805	16.066915583	16.067098699	16.067111677
1.4	27.278771833	27.288605217	27.289338955	27.289392347
1.6	49.960553660	49.997313966	50.000165744	50.000377775
1.8	98.834337815	98.971146146	98.982136702	98.982969504
2.0	211.393800152	211.908445283	211.951167637	211.954462214

TABLE 4   Numerical solution of $y' - 2xy = 1$, $y(0) = 3$, by the Runge–Kutta semilinear method

x	h = 0.2	h = 0.1	h = 0.05	"Exact"
0.0	3.000000000	3.000000000	3.000000000	3.000000000
0.2	3.327853286	3.327852055	3.327851978	3.327851973

(continued)

TABLE 4   *(continued)*

x	$h = 0.2$	$h = 0.1$	$h = 0.05$	"Exact"
0.4	3.966061755	3.966059497	3.966059357	3.966059348
0.6	5.067042602	5.067039725	5.067039547	5.067039535
0.8	6.936704019	6.936701137	6.936700957	6.936700945
1.0	10.184926171	10.184924093	10.184923963	10.184923955
1.2	16.067111961	16.067111696	16.067111678	16.067111677
1.4	27.289389418	27.289392167	27.289392335	27.289392347
1.6	50.000370152	50.000377302	50.000377745	50.000377775
1.8	98.982955511	98.982968633	98.982969450	98.982969504
2.0	211.954439983	211.954460825	211.954462127	211.954462214

THE CASE WHERE x_0 IS NOT THE LEFT ENDPOINT

So far in this chapter we have considered numerical methods for solving an initial value problem

$$y' = f(x,y), \qquad y(x_0) = y_0 \tag{3}$$

on an interval $[x_0, b]$, for which x_0 is the left endpoint. We haven't discussed numerical methods for solving (3) on an interval $[a, x_0]$, for which x_0 is the right endpoint. To be specific, how can we obtain approximate values $y_{-1}, y_{-2}, \ldots, y_{-n}$ of the solution of (3) at $x_0 - h, \ldots, x_0 - nh$, where $h = (x_0 - a)/n$? Here is the answer to this question.

Consider the initial value problem

$$z' = -f(-x,z), \qquad z(-x_0) = y_0 \tag{4}$$

on the interval $[-x_0, -a]$, for which $-x_0$ is the left endpoint. Use a numerical method to obtain approximate values $z_1, z_2, \ldots, z_n$ of the solution of (4) at $-x_0 + h, -x_0 + 2h, \ldots, -x_0 + nh = -a$. Then it follows that $y_{-1} = z_1, y_{-2} = z_2, \ldots, y_{-n} = z_n$ are approximate values of the solution of (3) at $x_0 - h, x_0 - 2h, \ldots, x_0 - nh = a$.

The justification for this answer is sketched in Exercise 23. Notice how easy it is to make the change for the given problem (3) to the modified problem (4): first replace f by $-f$ and then replace $x, x_0,$ and y by $-x, -x_0,$ and z, respectively.

EXAMPLE 5   Use the Runge–Kutta method with step size $h = 0.1$ to find approximate values of the solution of

$$(y - 1)^2 y' = 2x + 3, \qquad y(1) = 4 \tag{5}$$

at $x = 0, 0.1, 0.2, \ldots, 1$.

Solution   We first rewrite (5) in the form (3) as

$$y' = \frac{2x + 3}{(y - 1)^2}, \qquad y(1) = 4. \tag{6}$$

Since the initial condition $y(1) = 4$ is imposed at the right endpoint of the interval $[0,1]$, we apply the Runge–Kutta method to the initial value problem

$$z' = \frac{2x - 3}{(z - 1)^2}, \qquad z(-1) = 4 \tag{7}$$

on the interval $[-1,0]$. (You should verify that (7) is related to (6) as (4) is related to (3).) The results are shown in Table 5. Reversing the order of the rows in Table 5 and changing the signs of the values of x yields the first two columns of Table 6; the last column of Table 6 shows the exact values of y, which are given by

$$y = 1 + (3x^2 + 9x + 15)^{1/3}.$$

(Since the differential equation in (6) is separable, this formula can be obtained by the method of Section 2.2.)   ∎

TABLE 5   Numerical solution of $z' = \dfrac{2x - 3}{(z - 1)^2}$, $z(-1) = 4$ on $[-1,0]$

x	z
−1.0	4.000000000
−0.9	3.944536474
−0.8	3.889298649
−0.7	3.834355648
−0.6	3.779786399
−0.5	3.725680888
−0.4	3.672141529
−0.3	3.619284615
−0.2	3.567241862
−0.1	3.516161955
0.0	3.466212070

TABLE 6   Numerical solution of $(y - 1)^2 y' = 2x + 3$, $y(1) = 4$ on $[0,1]$

x	y	**Exact**
0.00	3.466212070	3.466212074
0.10	3.516161955	3.516161958
0.20	3.567241862	3.567241864
0.30	3.619284615	3.619284617
0.40	3.672141529	3.672141530
0.50	3.725680888	3.725680889
0.60	3.779786399	3.779786399
0.70	3.834355648	3.834355648
0.80	3.889298649	3.889298649
0.90	3.944536474	3.944536474
1.00	4.000000000	4.000000000

We leave it to you to develop a procedure for handling the numerical solution of (3) on an interval $[a,b]$ such that $a < x_0 < b$ (Exercises 26 and 27).

3.3 EXERCISES

Most of the following numerical exercises involve initial value problems considered in the exercises in Sections 3.1 and 3.2. You will find it instructive to compare the results that you obtain here with the corresponding results that you obtained in those sections.

In Exercises 1–5 use the Runge–Kutta method to find approximate values of the solution of the given initial value problem at the points $x_i = x_0 + ih$, where x_0 is the point at which the initial condition is imposed and $i = 1, 2$.

C **1.** $y' = 2x^2 + 3y^2 - 2$, $y(2) = 1$; $h = 0.05$ **C** **2.** $y' = y + \sqrt{x^2 + y^2}$, $y(0) = 1$; $h = 0.1$

C **3.** $y' + 3y = x^2 - 3xy + y^2$, $y(0) = 2$; $h = 0.05$ **C** **4.** $y' = \dfrac{1 + x}{1 - y^2}$, $y(2) = 3$; $h = 0.1$

C **5.** $y' + x^2y = \sin xy$, $y(1) = \pi$; $h = 0.2$

C **6.** Use the Runge–Kutta method with step sizes $h = 0.1$, $h = 0.05$, and $h = 0.025$ to find approximate values of the solution of the initial value problem

$$y' + 3y = 7e^{4x}, \qquad y(0) = 2$$

at $x = 0, 0.1, 0.2, 0.3, \ldots, 1.0$. Compare these approximate values with the values of the exact solution $y = e^{4x} + e^{-3x}$, which can be obtained by the method of Section 2.1. Present your results in a table like Table 1.

C **7.** Use the Runge–Kutta method with step sizes $h = 0.1$, $h = 0.05$, and $h = 0.025$ to find approximate values of the solution of the initial value problem

$$y' + \frac{2}{x}y = \frac{3}{x^3} + 1, \qquad y(1) = 1$$

at $x = 1.0, 1.1, 1.2, 1.3, \ldots, 2.0$. Compare these approximate values with the values of the exact solution

$$y = \frac{1}{3x^2}(9 \ln x + x^3 + 2),$$

which can be obtained by the method of Section 2.1. Present your results in a table like Table 1.

C **8.** Use the Runge–Kutta method with step sizes $h = 0.05$, $h = 0.025$, and $h = 0.0125$ to find approximate values of the solution of the initial value problem

$$y' = \frac{y^2 + xy - x^2}{x^2}, \qquad y(1) = 2$$

at $x = 1.0, 1.05, 1.10, 1.15, \ldots, 1.5$. Compare these approximate values with the values of the exact solution

$$y = \frac{x(1 + x^2/3)}{1 - x^2/3},$$

which was obtained in Example 3 of Section 2.4. Present your results in a table like Table 1.

C **9.** In Example 3 of Section 2.2 it was shown that

$$y^5 + y = x^2 + x - 4$$

is an implicit solution of the initial value problem

$$y' = \frac{2x + 1}{5y^4 + 1}, \qquad y(2) = 1. \tag{A}$$

Use the Runge–Kutta method with step sizes $h = 0.1$, $h = 0.05$, and $h = 0.025$ to find approximate values of the solution of (A) at $x = 2.0, 2.1, 2.2, 2.3, \ldots, 3.0$. Present your results in tabular form. To check the error in these approximate values, construct another table of values of the residual

$$R(x, y) = y^5 + y - x^2 - x + 4$$

for each value of (x, y) appearing in the first table.

C **10.** It can be seen from Example 1 of Section 2.5 that

$$x^4y^3 + x^2y^5 + 2xy = 4$$

is an implicit solution of the initial value problem

$$y' = -\frac{4x^3y^3 + 2xy^5 + 2y}{3x^4y^2 + 5x^2y^4 + 2x}, \qquad y(1) = 1. \tag{A}$$

Use the Runge–Kutta method with step sizes $h = 0.1$, $h = 0.05$, and $h = 0.025$ to find approximate values of the solution of (A) at $x = 1.0, 1.1, 1.2, 1.3, \ldots, 2.0$. Present your results in tabular form. To check the error in these approximate values, construct another table of values of the residual

$$R(x, y) = x^4y^3 + x^2y^5 + 2xy - 4$$

for each value of (x, y) appearing in the first table.

C 11. Use the Runge–Kutta method with step sizes $h = 0.1, h = 0.05$, and $h = 0.025$ to find approximate values of the solution of the initial value problem

$$(3y^2 + 4y)y' + 2x + \cos x = 0, \quad y(0) = 1 \quad \text{(Exercise 13 of Section 2.2)}$$

at $x = 0, 0.1, 0.2, 0.3, \ldots, 1.0$.

C 12. Use the Runge–Kutta method with step sizes $h = 0.1, h = 0.05$, and $h = 0.025$ to find approximate values of the solution of the initial value problem

$$y' + \frac{(y + 1)(y - 1)(y - 2)}{x + 1} = 0, \quad y(1) = 0 \quad \text{(Exercise 14 of Section 2.2)}$$

at $x = 1.0, 1.1, 1.2, 1.3, \ldots, 2.0$.

C 13. Use the Runge–Kutta method and the Runge–Kutta semilinear method with step sizes $h = 0.1, h = 0.05$, and $h = 0.025$ to find approximate values of the solution of the initial value problem

$$y' + 3y = e^{-3x}(1 - 4x + 3x^2 - 4x^3), \quad y(0) = -3$$

at $x = 0, 0.1, 0.2, 0.3, \ldots, 1.0$. Compare these approximate values with the values of the exact solution $y = -e^{-3x}(3 - x + 2x^2 - x^3 + x^4)$, which can be obtained by the method of Section 2.1. Do you notice anything special about the results? Explain.

The linear initial value problems in Exercises 14–19 cannot be solved exactly in terms of known elementary functions. In each exercise use the Runge–Kutta and the Runge–Kutta semilinear methods with the indicated step sizes to find approximate values of the solution of the given initial value problem at 11 equally spaced points (including the endpoints) in the interval.

C 14. $y' - 2y = \dfrac{1}{1 + x^2}$, $y(2) = 2$; $h = 0.1, 0.05, 0.025$ on $[2,3]$

C 15. $y' + 2xy = x^2$, $y(0) = 3$ (Exercise 38 of Section 2.1); $h = 0.2, 0.1, 0.05$ on $[0,2]$

C 16. $y' + \dfrac{1}{x}y = \dfrac{\sin x}{x^2}$, $y(1) = 2$ (Exercise 39 of Section 2.1); $h = 0.2, 0.1, 0.05$ on $[1,3]$

C 17. $y' + y = \dfrac{e^{-x}\tan x}{x}$, $y(1) = 0$ (Exercise 40 of Section 2.1); $h = 0.05, 0.025, 0.0125$ on $[1,1.5]$

C 18. $y' + \dfrac{2x}{1 + x^2}y = \dfrac{e^x}{(1 + x^2)^2}$, $y(0) = 1$ (Exercise 41 of Section 2.1); $h = 0.2, 0.1, 0.05$ on $[0,2]$

C 19. $xy' + (x + 1)y = e^{x^2}$, $y(1) = 2$ (Exercise 42 of Section 2.1); $h = 0.05, 0.025, 0.0125$ on $[1,1.5]$

In Exercises 20–22 use the Runge–Kutta method and the Runge–Kutta semilinear method with the indicated step sizes to find approximate values of the solution of the given initial value problem at 11 equally spaced points (including the endpoints) in the interval.

C 20. $y' + 3y = xy^2(y + 1)$, $y(0) = 1$; $h = 0.1, 0.05, 0.025$ on $[0,1]$

C 21. $y' - 4y = \dfrac{x}{y^2(y + 1)}$, $y(0) = 1$; $h = 0.1, 0.05, 0.025$ on $[0,1]$

C 22. $y' + 2y = \dfrac{x^2}{1 + y^2}$, $y(2) = 1$; $h = 0.1, 0.05, 0.025$ on $[2,3]$

C 23. Suppose that $a < x_0$, so that $-x_0 < -a$. Use the chain rule to show that if z is a solution of

$$z' = -f(-x, z), \quad z(-x_0) = y_0,$$

on $[-x_0, -a]$, then $y = z(-x)$ is a solution of

$$y' = f(x, y), \quad y(x_0) = y_0,$$

on $[a, x_0]$.

C 24. Use the Runge–Kutta method with step sizes $h = 0.1, h = 0.05$, and $h = 0.025$ to find approximate values of the solution of

$$y' = \frac{y^2 + xy - x^2}{x^2}, \qquad y(2) = -1$$

at $x = 1.1, 1.2, 1.3, \ldots, 2.0$. Compare these approximate values with the values of the exact solution

$$y = \frac{x(4 - 3x^2)}{4 + 3x^2},$$

which can be obtained by referring to Example 3 of Section 2.4.

C 25. Use the Runge–Kutta method with step sizes $h = 0.1, h = 0.05$, and $h = 0.025$ to find approximate values of the solution of

$$y' = -x^2 y - xy^2, \qquad y(1) = 1$$

at $x = 0, 0.1, 0.2, 0.3, \ldots, 1$.

C 26. Use the Runge–Kutta method with step sizes $h = 0.1, h = 0.05$, and $h = 0.025$ to find approximate values of the solution of

$$y' + \frac{1}{x} y = \frac{7}{x^2} + 3, \qquad y(1) = \frac{3}{2}$$

at $x = 0.5, 0.6, \ldots, 1.5$. Compare these approximate values with the values of the exact solution

$$y = \frac{7 \ln x}{x} + \frac{3x}{2},$$

which can be obtained by the method discussed in Section 2.1.

C 27. Use the Runge–Kutta method with step sizes $h = 0.1, h = 0.05$, and $h = 0.025$ to find approximate values of the solution of

$$xy' + 2y = 8x^2, \qquad y(2) = 5$$

at $x = 1.0, 1.1, 1.2, \ldots, 3.0$. Compare these approximate values with the values of the exact solution

$$y = 2x^2 - \frac{12}{x^2},$$

which can be obtained by the method discussed in Section 2.1.

28. **Numerical Quadrature** (see Exercise 23 of Section 3.1).

   (a) Derive the quadrature formula

$$\int_a^b f(x)\, dx \approx \frac{h}{6}(f(a) + f(b)) + \frac{h}{3} \sum_{i=1}^{n-1} f(a + ih) + \frac{2h}{3} \sum_{i=1}^{n} f(a + (2i - 1)h/2) \qquad \text{(A)}$$

   (where $h = (b - a)/n$) by applying the Runge–Kutta method to the initial value problem

$$y' = f(x), \qquad y(a) = 0.$$

   This quadrature formula is called *Simpson's Rule.*

   L **(b)** For several choices of a, b, A, B, C, and D apply (A) to $f(x) = A + Bx + Cx + Dx^3$, with $n = 10, 20, 40, 80, 160, 320$. Compare your results with the exact answers and explain what you find.

   L **(c)** For several choices of a, b, A, B, C, D, and E apply (A) to $f(x) = A + Bx + Cx^2 + Dx^3 + Ex^4$, with $n = 10, 20, 40, 80, 160, 320$. Compare your results with the exact answers and explain what you find.

4 Applications of First Order Equations

IN THIS CHAPTER we consider applications of first order differential equations.

SECTION 4.1 begins with a discussion of exponential growth and decay, which you have probably already seen in calculus. We consider applications to radioactive decay, carbon dating, and compound interest. We also consider more complicated problems in which the rate of change of a quantity is in part proportional to the magnitude of the quantity, but is also influenced by other other factors; for example, a radioactive substance is manufactured at a certain rate, but decays at a rate proportional to its mass, or a saver makes regular deposits in a savings account that draws compound interest.

SECTION 4.2 deals with applications of Newton's law of cooling and with mixing problems.

SECTION 4.3 discusses applications to elementary mechanics involving Newton's second law of motion. The problems considered include motion under the influence of gravity in a resistive medium, and determining the initial velocity required to launch a satellite.

SECTION 4.4 discusses methods for dealing with a certain type of second order equation that often arises in applications of Newton's second law of motion, by reformulating it as a first order equation with a different independent variable. Although the method doesn't usually lead to an explicit solution of the given equation, it does provide valuable insights into the behavior of the solutions.

SECTION 4.5 deals with applications of differential equations to curves.

4.1 Growth and Decay

Since the applications in this section deal with functions of time, we will denote the independent variable by t. If Q is a function of t, then Q' will denote the derivative of Q with respect to t; thus,

$$Q' = \frac{dQ}{dt}.$$

EXPONENTIAL GROWTH AND DECAY

One of the most common mathematical models for a physical process is the *exponential model*, in which it is assumed that the rate of change of a quantity Q at time t is proportional to its value at that time; thus,

$$Q' = aQ, \tag{1}$$

where a is the constant of proportionality.

From Example 3 of Section 2.1, the general solution of (1) is

$$Q = ce^{at}$$

and the solution of the initial value problem

$$Q' = aQ, \qquad Q(t_0) = Q_0$$

is

$$Q = Q_0 e^{a(t - t_0)}. \tag{2}$$

Since the solutions of $Q' = aQ$ are exponential functions, we say that a quantity Q that satisfies this equation *grows exponentially* if $a > 0$, or *decays exponentially* if $a < 0$ (Figure 1).

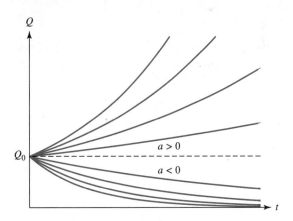

Figure 1 Exponential growth and decay

RADIOACTIVE DECAY

Experimental evidence shows that radioactive material decays at a rate proportional to the mass of the material present. According to this model the mass $Q(t)$ of a radioactive material present at time t satisfies (1), where a is a negative constant whose value for any given material must be determined by experimental observation. It is customary to replace the negative constant a by $-k$, where k is a positive number that we will call the ***decay constant*** of the material. Thus, (1) becomes

$$Q' = -kQ.$$

If the mass of the material present at $t = t_0$ is Q_0, then the mass present at time t is the solution of

$$Q' = -kQ, \qquad Q(t_0) = Q_0.$$

From (2) with $a = -k$, the solution of this initial value problem is

$$Q = Q_0 e^{-k(t-t_0)}. \tag{3}$$

The ***half-life*** τ of a radioactive material is defined to be the time required for half of its mass to decay; that is, if $Q(t_0) = Q_0$, then

$$Q(\tau + t_0) = \frac{Q_0}{2}. \tag{4}$$

From (3) with $t = \tau + t_0$, (4) is equivalent to

$$Q_0 e^{-k\tau} = \frac{Q_0}{2},$$

so

$$e^{-k\tau} = \frac{1}{2}.$$

Taking logarithms yields

$$-k\tau = \ln\frac{1}{2} = -\ln 2,$$

so the half-life is

$$\tau = \frac{1}{k}\ln 2 \tag{5}$$

(Figure 2). The half-life is independent of t_0 and Q_0, since it is determined by the properties of the material, not by the amount of the material present at any particular time.

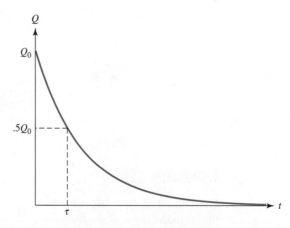

Figure 2 Half-life of a radioactive substance

EXAMPLE 1   A certain radioactive substance has a half-life of 1620 years.

(a) If its mass is now 4 g (grams), how much will be left 810 years from now?
(b) Find the time t_1 when 1.5 g of the substance remain.

Solution   (a) From (3) with $t_0 = 0$ and $Q_0 = 4$,

$$Q = 4e^{-kt}, \tag{6}$$

where we determine k from (5), with $\tau = 1620$ years:

$$k = \frac{\ln 2}{\tau} = \frac{\ln 2}{1620}.$$

Substituting this in (6) yields

$$Q = 4e^{-(t\ln 2)/1620}. \tag{7}$$

Therefore, the mass left after 810 years will be

$$Q(810) = 4e^{-(810\ln 2)/1620} = 4e^{-(\ln 2)/2}$$
$$= 2\sqrt{2}\,\text{g}.$$

Solution   (b) Setting $t = t_1$ in (7) and requiring that $Q(t_1) = 1.5$ yields

$$\frac{3}{2} = 4e^{(-t_1\ln 2)/1620}.$$

Dividing by 4 and taking logarithms yields

$$\ln\frac{3}{8} = -\frac{t_1\ln 2}{1620}.$$

Since $\ln 3/8 = -\ln 8/3$, solving for t_1 yields

$$t_1 = 1620\,\frac{\ln 8/3}{\ln 2} \approx 2292.4 \text{ years}. \qquad \blacksquare$$

INTEREST COMPOUNDED CONTINUOUSLY

Suppose that we deposit an amount of money Q_0 in an interest-bearing account and make no further deposits or withdrawals for t years, during which the account bears interest at a constant annual rate r. In order to calculate the value of the account at the end of t years, we need one more piece of information: how the interest is added to the account, or—as the bankers say—how it is **compounded.** If the interest is compounded annually, then the value of the account is multiplied by $1 + r$ at the end of each year. This means that after t years the value of the account is

$$Q(t) = Q_0(1 + r)^t.$$

If interest is compounded semiannually (every 6 months), then the value of the account is multiplied by $(1 + r/2)$ every 6 months. Since this occurs twice annually, the value of the account after t years is

$$Q(t) = Q_0\left(1 + \frac{r}{2}\right)^{2t}.$$

In general, if interest is compounded n times per year, the value of the account is multiplied n times per year by $(1 + r/n)$; therefore, the value of the account after t years is

$$Q(t) = Q_0 \left(1 + \frac{r}{n} \right)^{nt}. \tag{8}$$

Thus, increasing the frequency of compounding increases the value of the account after a fixed period of time. Table 1 shows the effect of increasing the number of compoundings over $t = 5$ years on an initial deposit of $Q_0 = 100$ (dollars), at an annual interest rate of 6% (that is, $r = .06$).

TABLE 1

n (number of compoundings per year)	$\$100 \left(1 + \dfrac{.06}{n} \right)^{5n}$ (value in dollars after 5 years)
1	$133.82
2	$134.39
4	$134.68
8	$134.83
364	$134.98

It can be seen from Table 1 that the value of the account after 5 years is an increasing function of n. Now suppose that the maximum allowable rate of interest on savings accounts is restricted by law, but the time intervals between successive compoundings is not; then competing banks can attract savers by compounding often. The ultimate step in this direction is to **compound continuously,** by which we mean that $n \to \infty$ in (8). Since we know from calculus that

$$\lim_{n\to\infty} \left(1 + \frac{r}{n} \right)^n = e^r,$$

this yields

$$Q(t) = \lim_{n\to\infty} Q_0 \left(1 + \frac{r}{n} \right)^{nt} = Q_0 \left[\lim_{n\to\infty} \left(1 + \frac{r}{n} \right)^n \right]^t$$
$$= Q_0 e^{rt}.$$

Observe that $Q = Q_0 e^{rt}$ is the solution of the initial value problem

$$Q' = rQ, \qquad Q(0) = Q_0;$$

that is, with continuous compounding the value of the account grows exponentially.

EXAMPLE 2   If $150 is deposited in a bank that pays $5\frac{1}{2}$% annual interest compounded continuously, then the value of the account after t years is

$$Q(t) = 150 e^{.055t}$$

dollars. (Notice that it is necessary to write the interest rate as a decimal; thus, $r = .055$.) Therefore, after $t = 10$ years the value of the account is

$$Q(10) = 150e^{.55} \approx \$259.99.$$

EXAMPLE 3   We wish to accumulate \$10,000 in 10 years by making a single deposit in a savings account bearing $5\frac{1}{2}\%$ annual interest compounded continuously. How much must we deposit in the account?

Solution   The value of the account at time t is

$$Q(t) = Q_0e^{.055t}. \tag{9}$$

Since we want $Q(10)$ to be \$10,000, the initial deposit Q_0 must satisfy the equation

$$10000 = Q_0e^{.55}, \tag{10}$$

obtained by setting $t = 10$ and $Q(10) = 10000$ in (9). Solving (10) for Q_0 yields

$$Q_0 = 10000e^{-.55} \approx \$5769.50. \qquad \blacksquare$$

MIXED GROWTH AND DECAY

EXAMPLE 4   A radioactive substance with decay constant k is produced at a constant rate of a units of mass per unit time.

(a) Assuming that $Q(0) = Q_0$, find the mass $Q(t)$ of the substance present at time t.
(b) Find $\lim_{t\to\infty} Q(t)$.

Solution   **(a)** Here

$$Q' = \text{rate of increase of } Q - \text{rate of decrease of } Q.$$

The rate of increase is the constant a. Since Q is radioactive with decay constant k, the rate of decrease is kQ. Therefore

$$Q' = a - kQ.$$

This is a linear first order differential equation. Rewriting it and imposing the initial condition shows that Q is the solution of the initial value problem

$$Q' + kQ = a, \qquad Q(0) = Q_0. \tag{11}$$

Since e^{-kt} is a solution of the complementary equation, the solutions of (11) are of the form $Q = ue^{-kt}$ where $u'e^{-kt} = a$, so $u' = ae^{kt}$. Hence,

$$u = \frac{a}{k}e^{kt} + c$$

and

$$Q = ue^{-kt} = \frac{a}{k} + ce^{-kt}.$$

Since $Q(0) = Q_0$, setting $t = 0$ here yields

$$Q_0 = \frac{a}{k} + c \qquad \text{or} \qquad c = Q_0 - \frac{a}{k}.$$

Therefore,

$$Q = \frac{a}{k} + \left(Q_0 - \frac{a}{k}\right)e^{-kt}. \tag{12}$$

SOLUTION   **(b)** Since $k > 0$ it follows that $\lim_{t\to\infty} e^{-kt} = 0$, so from (12)

$$\lim_{t\to\infty} Q(t) = \frac{a}{k}.$$

Notice that this limit depends only on a and k, and not on Q_0. We say that a/k is the **steady state** value of Q. From (12) we also see that Q approaches its steady state value from above if $Q_0 > a/k$, or from below if $Q_0 < a/k$. If $Q_0 = a/k$, then Q remains constant (Figure 3). ∎

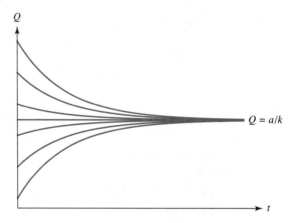

Figure 3 $Q(t)$ approaches the steady state value $\dfrac{a}{k}$ as $t \to \infty$

CARBON DATING

The fact that Q approaches a steady state value in the situation discussed in Example 4 underlies the method of **carbon dating,** devised by W. F. Libby[1].

Carbon-12 is stable, but carbon-14, which is produced by cosmic bombardment of nitrogen in the upper atmosphere, is radioactive with a half-life of approximately 5570 years. Libby assumed that the quantity of carbon-12 in the atmosphere has been constant throughout time, and that the quantity of radioactive carbon-14 achieved its steady state value long ago as a result of its creation and decomposition over millions of years. These assumptions led Libby to conclude that the ratio of carbon-14 to carbon-12 has been nearly constant for a long time. This constant, which we denote by R, has been determined experimentally.

Living cells absorb both carbon-12 and carbon-14 in the proportion in which they are present in the environment. Therefore, the ratio of carbon-14 to carbon-12 in a living cell is always R. However, when the cell dies it ceases to absorb carbon, and the ratio of carbon-14 to carbon-12 decreases exponentially as the

[1]The American chemist Willard F. Libby (1908–1980) won the 1960 Nobel Prize in chemistry for developing the procedure of carbon dating.

radioactive carbon-14 decays. This is the basis for the method of ***carbon dating***, as illustrated in the following example.

EXAMPLE 5

An archaeologist investigating the site of an ancient village finds a burial ground in which the amount of carbon-14 present in individual remains is between 42 and 44 percent of the amount present in live individuals. Estimate the age of the village and the length of time for which it survived.

Solution

Let $Q = Q(t)$ be the quantity of carbon-14 in an individual set of remains t years after death, and let Q_0 be the quantity that would be present in live individuals. Since carbon-14 decays exponentially with half-life 5570 years, its decay constant is

$$k = \frac{\ln 2}{5570}.$$

Therefore

$$Q = Q_0 e^{-t(\ln 2)/5570}$$

if we choose our time scale so that $t_0 = 0$ is the time of death. If we know the present value of Q, then we can solve this equation for t, the number of years since death occurred. This yields

$$t = -5570 \frac{\ln Q/Q_0}{\ln 2}.$$

It is given that $Q = .42Q_0$ in the remains of individuals who died first. Therefore, these deaths occurred approximately

$$t_1 = -5570 \frac{\ln .42}{\ln 2} \approx 6971$$

years ago. For the most recent deaths, $Q = .44Q_0$; hence, these deaths occurred approximately

$$t_2 = -5570 \frac{\ln .44}{\ln 2} \approx 6597$$

years ago. Therefore, it is reasonable to conclude that the village was founded approximately 7000 years ago, and lasted for about 400 years. ∎

A SAVINGS PROGRAM

EXAMPLE 6

A person opens a savings account with an initial deposit of $1000 and subsequently deposits $50 per week. Find the value $Q(t)$ of the account at time $t > 0$, assuming that the bank pays 6% interest compounded continuously.

Solution

Observe that Q is not continuous, since there are 52 discrete deposits per year of $50 each. To construct a mathematical model for this problem in the form of a differential equation, we make the simplifying assumption that the deposits are made continuously at a rate of $2600 per year. This is essential, since solutions of differential equations are continuous functions. With this assumption, Q increases continuously at the rate

$$Q' = 2600 + .06Q$$

and therefore Q satisfies the differential equation

$$Q' - .06Q = 2600. \tag{13}$$

(Of course, we must recognize that the solution of this equation is an approximation to the true value of Q at any given time. We will discuss this further below.) Since $e^{.06t}$ is a solution of the complementary equation, the solutions of (13) are of the form $Q = ue^{.06t}$ where $u'e^{.06t} = 2600$, so $u' = 2600e^{-.06t}$. Hence,

$$u = -\frac{2600}{.06}e^{-.06t} + c$$

and

$$Q = ue^{.06t} = -\frac{2600}{.06} + ce^{.06t}. \tag{14}$$

Setting $t = 0$ and $Q = 1000$ here yields

$$c = 1000 + \frac{2600}{.06},$$

and substituting this into (14) yields

$$Q = 1000e^{.06t} + \frac{2600}{.06}(e^{.06t} - 1), \tag{15}$$

where the first term is the value due to the initial deposit and the second is due to the subsequent weekly deposits. ∎

Mathematical models must be tested for validity by comparing predictions based on them with the actual outcome of experiments. Example 6 is unusual in that we can compute the exact value of the account at any specified time and compare it with the approximate value predicted by (15) (See Exercise 21.). Table 2 gives a comparison for a ten year period. Each exact answer corresponds to the time of the year-end deposit, and each year is assumed to have exactly 52 weeks.

TABLE 2 Value of the Saver's Account

Year	Approximate Value Q (Example 6)	Exact Value P (Exercise 21)	Error $Q - P$	Percentage Error $(Q - P)/P$
1	$ 3741.42	$ 3739.87	$ 1.55	.0413%
2	6652.36	6649.17	3.19	.0479
3	9743.30	9738.37	4.93	.0506
4	13,025.38	13,018.60	6.78	.0521
5	16,510.41	16,501.66	8.75	.0530
6	20,210.94	20,200.11	10.83	.0536
7	24,140.30	24,127.25	13.05	.0541
8	28,312.63	28,297.23	15.40	.0544
9	32,742.97	32,725.07	17.90	.0547
10	37,447.27	37,426.72	20.55	.0549

4.1 EXERCISES

1. The half-life of a certain radioactive substance is 3200 years. Find the quantity $Q(t)$ of the substance left at time $t > 0$ if $Q(0) = 20$ g.

2. The half-life of a radioactive substance is 2 days. Find the time required for a given amount of the material to decay to $1/10$ of its original mass.

3. A certain radioactive material loses 25% of its mass in 10 minutes. What is its half-life?

4. A certain type of tree contains a known percentage p_0 of a radioactive substance with half-life τ. When the tree dies the substance decays and is not replaced. If the percentage of the substance in the fossilized remains of such a tree is found to be p_1, how long has the tree been dead?

5. If t_p and t_q are the times required for a radioactive material to decay to $1/p$ and $1/q$ times its original mass (respectively), how are t_p and t_q related?

6. Find the decay constant k for a radioactive substance, given that the mass of the substance is Q_1 at time t_1 and Q_2 at time t_2.

7. A process creates a radioactive substance at the rate of 2 g/hr and the substance decays at a rate proportional to its mass, with constant of proportionality $k = .1(\text{hr})^{-1}$. If $Q(t)$ is the mass of the substance at time t, find $\lim_{t \to \infty} Q(t)$.

8. A bank pays interest continuously at the rate of 6%. How long does it take for a deposit of Q_0 to grow in value to $2Q_0$?

9. At what rate of interest, compounded continuously, will a bank deposit double in value in 8 years?

10. A savings account pays 5% per annum interest compounded continuously. The initial deposit is Q_0 dollars. Assume that there are no subsequent withdrawals or deposits.

    (a) How long will it take for the value of the account to triple?

    (b) What is Q_0 if the value of the account after 10 years is $100,000 dollars?

11. A candymaker makes 500 pounds of candy per week, while his large family eats the candy at a rate equal to $Q(t)/10$ pounds per week, where $Q(t)$ is the amount of candy present at time t.

    (a) Find $Q(t)$ for $t > 0$ if the candymaker has 250 pounds of candy at $t = 0$.

    (b) Find $\lim_{t \to \infty} Q(t)$.

12. Suppose that a substance decays at a yearly rate numerically equal to half the square of the mass of the substance present. If we start with 50 g of the substance, how long will it be until only 25 g remain?

13. A super bread dough increases in volume at a rate proportional to the volume V present. If V increases by a factor of 10 in 2 hours and $V(0) = V_0$, find V at any time t. How long will it take for V to increase to $100V_0$?

14. A radioactive substance decays at a rate proportional to the amount present, and half the original quantity Q_0 is left after 1500 years. In how many years would the original amount be reduced to $3Q_0/4$? How much will be left after 2000 years?

15. A wizard creates gold continuously at the rate of 1 ounce per hour, but an assistant steals it continuously at the rate of 5% of however much is there per hour. Let $W(t)$ be the number of ounces that the wizard has at time t. Find $W(t)$ and $\lim_{t \to \infty} W(t)$ if $W(0) = 1$.

16. A process creates a radioactive substance at the rate of 1 g/hr, and the substance decays at an hourly rate numerically equal to $1/10$ of the mass present (expressed in grams). Assuming that there are initially 20 g, find the mass $S(t)$ of the substance present at time t, and find $\lim_{t \to \infty} S(t)$.

17. A tank is empty at $t = 0$. Water is added to the tank at the rate of 10 gal/min, but it leaks out at a rate (in gallons per minute) numerically equal to the number of gallons in the tank. What is the smallest capacity the tank can have if this process is to continue forever?

18. A person deposits $25,000 in a bank that pays 5% per year interest, compounded continuously. The person continuously withdraws from the account at the rate of $750 per year. Find $V(t)$, the value of the account at time t after the initial deposit.

19. A person has a fortune that grows at a rate proportional to the square root of its worth. Find the worth W of the fortune as a function of t if it was $1 million 6 months ago and is $4 million today.

20. Let $p = p(t)$ be the quantity of a certain product present at time t. The substance is manufactured continuously at a rate proportional to p, with proportionality constant $1/2$, and it is consumed continuously at a rate proportional to p^2, with proportionality constant $1/8$. Find $p(t)$ if $p(0) = 100$.

21. (a) In the situation of Example 6 find the exact value $P(t)$ of the person's account after t years, where t is an integer. Assume that each year has exactly 52 weeks, and include the year-end deposit in the computation.

Hint: At time t the initial $1000 has been on deposit for t years. There have been $52t$ deposits of $50 each. The first $50 has been on deposit for $t - 1/52$ years, the second for $t - 2/52$ years . . . ; in general, the jth $50 has been on deposit for $t - j/52$ years ($1 \le j \le 52t$). Find the present value of each $50 deposit assuming 6% interest compounded continuously, and use the formula

$$1 + x + x^2 + \cdots + x^n = \frac{1 - x^{n+1}}{1 - x} \qquad (x \ne 1)$$

to find their total value.

(b) Let

$$p(t) = \frac{Q(t) - P(t)}{P(t)}$$

be the relative error after t years. Find

$$p(\infty) = \lim_{t \to \infty} p(t).$$

22. A homebuyer borrows P_0 dollars at an annual interest rate r, agreeing to repay the loan with equal monthly payments of M dollars per month over N years.

(a) Derive a differential equation for the loan principal (amount that the homebuyer owes) $P(t)$ at time $t > 0$, making the simplifying assumption that the homebuyer repays the loan continuously rather than in discrete steps. (See Example 6.)

(b) Solve the equation derived in **(a)**.

(c) Use the result of **(b)** to determine an approximate value for M assuming that each year has exactly 12 months of equal length.

(d) It can be shown that the exact value of M is given by

$$M = \frac{rP_0}{12} \left(1 - (1 + r/12)^{-12N} \right)^{-1}.$$

Compare the value of M obtained from the answer in **(c)** to the exact value if **(i)** $P_0 = \$50,000$, $r = 7\frac{1}{2}\%$, $N = 20$; **(ii)** $P_0 = \$150,000$, $r = 9\%$, $N = 30$.

23. Assume that the homebuyer of Exercise 22 elects to repay the loan continuously at the rate of αM dollars per month, where α is a constant greater than 1. (This is called ***accelerated payment***.)

(a) Determine the time $T(\alpha)$ at which the loan will be paid off and the amount $S(\alpha)$ that the homebuyer will save.

(b) Suppose that $P_0 = \$50,000$, $r = 8\%$, and $N = 15$. Compute the savings realized by accelerated payments with $\alpha = 1.05, 1.10$, and 1.15.

24. A benefactor wishes to establish a trust fund to pay a researcher's salary for T years. The salary is to start at S_0 dollars per year and increase at a fractional rate of a per year. Find the amount of money P_0 that the benefactor must deposit in a trust fund paying interest at a rate r per year. Assume that the researcher's salary is paid continuously, the interest is compounded continuously, and the salary increases are granted continuously.

L 25. A radioactive substance with decay constant k is produced at the rate of

$$\frac{at}{1 + btQ(t)}$$

units of mass per unit time, where a and b are positive constants and $Q(t)$ is the mass of the substance present at time t; thus the rate of production is small at the start and tends to slow when Q is large.

(a) Set up a differential equation for Q.

(b) Choose your own positive values for $a, b, k,$ and $Q_0 = Q(0)$. Use a numerical method to discover what happens to $Q(t)$ as $t \to \infty$. (Be precise, expressing your conclusions in terms of a, b, k. However, no proof is required.)

[L] 26. Follow the instructions of Exercise 25, assuming that the substance is produced at the rate of $at/(1 + bt(Q(t))^2)$ units of mass per unit of time.

[L] 27. Follow the instructions of Exercise 25, assuming that the substance is produced at the rate of $at/(1 + bt)$ units of mass per unit of time.

4.2   Cooling and Mixing

NEWTON'S LAW OF COOLING

Newton's law of cooling states that if an object with temperature $T(t)$ at time t is in a medium with temperature $T_m(t)$, then the rate of change of T at time t is proportional to $T(t) - T_m(t)$; thus, T satisfies a differential equation of the form

$$T' = -k(T - T_m). \tag{1}$$

Here $k > 0$, since the temperature of the object must decrease if $T > T_m$, or increase if $T < T_m$. We will call k the **temperature decay constant of the medium**.

For simplicity, in this section we will assume that the medium is maintained at a constant temperature T_m. This is another example of the way in which we build a simple mathematical model for a physical phenomenon. Like most mathematical models it has its limitations. For example, it seems reasonable to assume that the temperature of a room remains approximately constant if the cooling object is a cup of coffee, but perhaps not if it is a huge cauldron of molten metal. (For more on this see Exercise 17.)

To solve (1) we rewrite it as

$$T' + kT = kT_m.$$

Since e^{-kt} is a solution of the complementary equation, the solutions of this equation are of the form $T = ue^{-kt}$ where $u'e^{-kt} = kT_m$, so $u' = kT_m e^{kt}$. Hence,

$$u = T_m e^{kt} + c,$$

so

$$T = ue^{-kt} = T_m + ce^{-kt}.$$

If $T(0) = T_0$, then setting $t = 0$ here yields $c = T_0 - T_m$, so

$$T = T_m + (T_0 - T_m)e^{-kt}. \tag{2}$$

Notice that $T - T_m$ decays exponentially, with decay constant k.

EXAMPLE 1   A ceramic insulator is baked at 400°C and cooled in a room in which the temperature is 25°C. After 4 minutes the temperature of the insulator is 200°C. What is its temperature after 8 minutes?

Solution   Here $T_0 = 400$ and $T_m = 25$, so (2) becomes

$$T = 25 + 375e^{-kt}. \tag{3}$$

We determine k from the stated condition that $T(4) = 200$; that is,

$$200 = 25 + 375e^{-4k};$$

hence

$$e^{-4k} = \frac{175}{375} = \frac{7}{15}.$$

Taking logarithms and solving for k yields

$$k = -\frac{1}{4} \ln \frac{7}{15} = \frac{1}{4} \ln \frac{15}{7}.$$

Substituting this into (3) yields

$$T = 25 + 375e^{-(t/4)\ln 15/7}$$

(Figure 1), and therefore the temperature of the insulator after 8 minutes is

$$T(8) = 25 + 375e^{-2\ln 15/7}$$

$$= 25 + 375\left(\frac{7}{15}\right)^2 \approx 107°C.$$

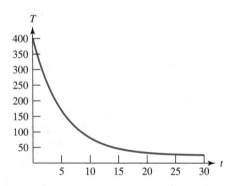

Figure 1 $T = 25 + 375e^{-(t/4)\ln 15/7}$

EXAMPLE 2   An object with temperature 72°F is placed outside where the temperature is −20°F. At 11:05 the temperature of the object is 60°F and at 11:07 its temperature is 50°F. At what time was the object placed outside?

Solution   Let $T(t)$ be the temperature of the object at time t. For convenience, we choose the origin $t_0 = 0$ of the time scale to be 11:05 so that $T_0 = 60$. We must determine the time τ when $T(\tau) = 72$. Substituting $T_0 = 60$ and $T_m = -20$ into (2) yields

$$T = -20 + (60 - (-20))e^{-kt}$$

or

$$T = -20 + 80e^{-kt}. \tag{4}$$

We obtain k from the stated condition that the temperature of the object is 50°F at 11:07. Since 11:07 is $t = 2$ on our time scale, we can determine k by substituting $T = 50$ and $t = 2$ into (4) to obtain

$$50 = -20 + 80e^{-2k};$$

hence

$$e^{-2k} = \frac{70}{80} = \frac{7}{8}.$$

Taking logarithms and solving for k yields

$$k = -\frac{1}{2}\ln\frac{7}{8} = \frac{1}{2}\ln\frac{8}{7}.$$

Substituting this into (4) yields

$$T = -20 + 80e^{-(t/2)\ln 8/7},$$

and the condition $T(\tau) = 72$ implies that

$$72 = -20 + 80e^{-(\tau/2)\ln 8/7};$$

hence

$$e^{-(\tau/2)\ln 8/7} = \frac{92}{80} = \frac{23}{20}.$$

Taking logarithms and solving for τ yields

$$\tau = -\frac{2\ln 23/20}{\ln 8/7} \approx -2.09\ \text{min}.$$

Therefore, the object was placed outside approximately 2 minutes and 5 seconds before 11:05; that is, at 11:02:55. The graph of the solution is shown in Figure 2.

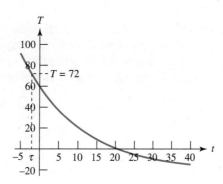

Figure 2 $T = -20 + 80e^{-(t/2)\ln 8/7}$

MIXING PROBLEMS

In the next two examples a saltwater solution with a given concentration (weight of salt per unit volume of solution) is added at a specified rate to a tank that initially contains saltwater with a different concentration. The problem is to determine the quantity of salt in the tank as a function of time. This is an example of a ***mixing problem.*** To construct a tractable mathematical model for mixing problems, we assume in our examples (and most exercises) that the mixture is stirred instantly so that the salt is always uniformly distributed throughout the mixture. Exercises 22 and 23 deal with situations in which this is not so, but the distribution of salt becomes approximately uniform as $t \to \infty$.

EXAMPLE 3   A tank initially contains 40 pounds of salt dissolved in 600 gallons of water. Starting at $t_0 = 0$, water containing 1/2 pound of salt per gallon is poured into the tank at the rate of 4 gal/min and the mixture is drained from the tank at the same rate (Figure 3).

(a) Find a differential equation for the quantity $Q(t)$ of salt in the tank at time $t > 0$, and solve the equation to determine $Q(t)$.

(b) Find $\lim_{t \to \infty} Q(t)$.

4 gal/min; .5 lb/gal

600 gal

4 gal/min

Figure 3   A mixing problem

Solution   **(a)** To find a differential equation for Q, we must use the given information to derive an expression for Q'. But Q' is the rate at which the quantity of salt in the tank changes with time; thus, if *rate in* denotes the rate at which salt enters the tank and *rate out* denotes the rate at which it leaves, then

$$Q' = \text{rate in} - \text{rate out}. \tag{5}$$

The rate in is

$$\left(\frac{1}{2} \text{ lb/gal} \right) \times (4 \text{ gal/min}) = 2 \text{ lb/min}.$$

Determining the rate out requires a little more thought. We are removing 4 gallons of the mixture per minute, and there are always 600 gallons in the tank; that is, we are removing 1/150 of the mixture per minute. Since the salt is evenly distributed in the mixture, we are also removing 1/150 of the salt per minute. Therefore, if there are $Q(t)$ pounds of salt in the tank at time t, then the rate out at any time t is $Q(t)/150$. Alternatively, we can arrive at this conclusion by arguing that

$$
\begin{aligned}
\text{rate out} &= (\text{concentration}) \times (\text{rate of flow out}) \\
&= (\text{lb/gal}) \times (\text{gal/min}) \\
&= \frac{Q(t)}{600} \times 4 \\
&= \frac{Q(t)}{150}.
\end{aligned}
$$

We can now write (5) as

$$Q' = 2 - \frac{Q}{150}.$$

This first order equation can be rewritten as

$$Q' + \frac{Q}{150} = 2.$$

Since $e^{-t/150}$ is a solution of the complementary equation, the solutions of this equation are of the form $Q = ue^{-t/150}$, where $u'e^{-t/150} = 2$, so $u' = 2e^{t/150}$. Hence,

$$u = 300e^{t/150} + c,$$

so

$$Q = ue^{-t/150} = 300 + ce^{-t/150}. \tag{6}$$

Since $Q(0) = 40$, we have $c = -260$; therefore,

$$Q = 300 - 260e^{-t/150}$$

(Figure 4).

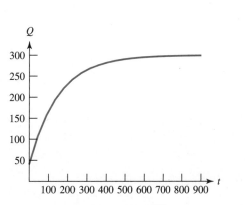

Figure 4 $Q = 300 - 260e^{-t/150}$

4 liters/min; .25 g/liter

$2t + 200$ liters

2 liters/min

Figure 5   Another mixing problem

Solution   **(b)** From (6) we see that $\lim_{t\to\infty} Q(t) = 300$ for any value of $Q(0)$. This is intuitively reasonable, since the incoming solution contains 1/2 pound of salt per gallon and there are always 600 gallons of water in the tank.

EXAMPLE 4   A 500-liter tank initially contains 10 g of salt dissolved in 200 liters of water. Starting at $t_0 = 0$ water containing 1/4 g of salt per liter is poured into the tank at the rate of 4 liters/min and the mixture is drained from the tank at the rate of 2 liters/min (Figure 5). Find a differential equation for the quantity $Q(t)$ of salt in the tank at time t prior to the time when the tank overflows, and find the concentration $K(t)$(g/liter) of salt in the tank at any such time.

Solution   We first determine the amount $W(t)$ of solution in the tank at any time t prior to overflow. Since $W(0) = 200$ and we are adding 4 liters/min while removing only 2 liters/min, there is a net gain of 2 liters/min in the tank; therefore,

$$W(t) = 2t + 200.$$

Since $W(150) = 500$ liters (capacity of the tank), this formula is valid for $0 \leq t \leq 150$.

Now let $Q(t)$ be the number of grams of salt in the tank at time t, where $0 \leq t \leq 150$. As in Example 3,

$$Q' = \text{rate in} - \text{rate out}. \tag{7}$$

The rate in is

$$\left(\frac{1}{4}\text{g/liter}\right) \times (4 \text{ liters/min}) = 1\text{g/min} \tag{8}$$

To determine the rate out, we observe that since the mixture is being removed from the tank at the constant rate of 2 liters/min and there are $2t + 200$ liters in the tank at time t, the fraction of the mixture being removed per minute at time t is

$$\frac{2}{2t + 200} = \frac{1}{t + 100}.$$

We are removing this same fraction of the salt per minute. Therefore, since there are $Q(t)$ grams of salt in the tank at time t,

$$\text{rate out} = \frac{Q(t)}{t + 100}. \tag{9}$$

Alternatively, we can arrive at this conclusion by arguing that

$$\begin{aligned}
\text{rate out} &= (\text{concentration}) \times (\text{rate of flow out}) \\
&= (\text{g/liter}) \times (\text{liters/min}) \\
&= \frac{Q(t)}{2t + 200} \times 2 \\
&= \frac{Q(t)}{t + 100}.
\end{aligned}$$

Substituting (8) and (9) into (7) yields

$$Q' = 1 - \frac{Q}{t + 100},$$

or

$$Q' + \frac{1}{t + 100}Q = 1. \tag{10}$$

By separation of variables, $1/(t + 100)$ is a solution of the complementary equation, so the solutions of (10) are of the form

$$Q = \frac{u}{t + 100} \qquad \text{where} \qquad \frac{u'}{t + 100} = 1, \qquad \text{so} \qquad u' = t + 100.$$

Hence,

$$u = \frac{(t + 100)^2}{2} + c. \tag{11}$$

Since $Q(0) = 10$ and $u = (t + 100)Q$, (11) implies that

$$(100)(10) = \frac{(100)^2}{2} + c,$$

so

$$c = 100(10) - \frac{(100)^2}{2} = -4000$$

and therefore

$$u = \frac{(t + 100)^2}{2} - 4000.$$

Hence,

$$Q = \frac{u}{t + 200} = \frac{t + 100}{2} - \frac{4000}{t + 100}.$$

Now let $K(t)$ be the concentration of salt at time t. Then

$$K(t) = \frac{1}{4} - \frac{2000}{(t + 100)^2}$$

(Figure 6). ◼

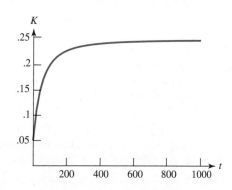

Figure 6 $K(t) = \dfrac{1}{4} - \dfrac{2000}{(t + 100)^2}$

4.2 EXERCISES

1. A thermometer is moved from a room where the temperature is 70°F to a freezer where the temperature is 12°F. After 30 seconds the thermometer reads 40°F. What does it read after 2 minutes?

2. A fluid initially at 100°C is placed outside on a day when the temperature is -10°C, and the temperature of the fluid drops 20°C in 1 minute. Find the temperature $T(t)$ of the fluid for $t > 0$.

3. At 12:00 PM a thermometer reading 10°F is placed in a room where the temperature is 70°F. It reads 56° when it is placed outside, where the temperature is 5°F, at 12:03. What does it read at 12:05 PM?

4. A thermometer initially reading 212°F is placed in a room where the temperature is 70°F. After 2 minutes the thermometer reads 125°F.

(a) What does the thermometer read after 4 minutes?

(b) When will the thermometer read 72°F?

(c) When will the thermometer read 69°F?

5. An object with initial temperature 150°C is placed outside where the temperature is 35°C. Its temperatures at 12:15 and 12:20 are 120°C and 90°C, respectively.

(a) At what time was the object placed outside?

(b) When will its temperature be 40°C?

6. An object is placed in a room where the temperature is 20°C. The temperature of the object drops by 5°C in 4 minutes and by 7°C in 8 minutes. What was the temperature of the object when it was initially placed in the room?

7. A cup of boiling water is placed outside at 1:00 PM. One minute later the temperature of the water is 152°F. After another minute its temperature is 112°F. What is the outside temperature?

8. A tank initially contains 40 gallons of pure water. A solution with 1 gram of salt per gallon of water is added to the tank at 3 gal/min, and the resulting solution leaves at the same rate. Find the quantity $Q(t)$ of salt in the tank at time $t > 0$.

9. A tank initially contains a solution of 10 pounds of salt in 60 gallons of water. Water with 1/2 pound of salt per gallon is added to the tank at 6 gal/min, and the resulting solution leaves at the same rate. Find the quantity $Q(t)$ of salt in the tank at time $t > 0$.

10. A tank initially contains 100 liters of a salt solution with a concentration of .1 g/liter. A solution with a salt concentration of .3 g/liter is added to the tank at 5 liters/min, and the resulting mixture is drained out at the same rate. Find the concentration $K(t)$ of salt in the tank as a function of t.

11. A 200-gallon tank initially contains 100 gallons of water with 20 pounds of salt. A salt solution with 1/4 pound of salt per gallon is added to the tank at 4 gal/min, and the resulting mixture is drained out at 2 gal/min. Find the quantity of salt in the tank as it is about to overflow.

12. Suppose water is added to a tank at 10 gal/min, but leaks out at the rate of 1/5 gal/min for each gallon in the tank. What is the smallest capacity the tank can have if the process is to continue indefinitely?

13. A chemical reaction in a laboratory with volume V (in ft³) produces q_1 ft³/min of a noxious gas as a byproduct. The gas is dangerous at concentrations greater than $\bar{c}$, but harmless at concentrations $\leq \bar{c}$. Intake fans at one end of the laboratory pull in fresh air at the rate of q_2 ft³/min and exhaust fans at the other end exhaust the mixture of gas and air from the laboratory at the same rate. Assuming that the gas is always uniformly distributed in the room and its initial concentration c_0 is at a safe level, find the smallest value of q_2 required to maintain safe conditions in the laboratory for all time.

14. A 1200-gallon tank initially contains 40 pounds of salt dissolved in 600 gallons of water. Starting at $t_0 = 0$, water containing 1/2 pound of salt per gallon is added to the tank at the rate of 6 gal/min and the resulting mixture is drained from the tank at 4 gal/min. Find the quantity $Q(t)$ of salt in the tank at any time $t > 0$ prior to overflow.

15. Tank T_1 initially contains 50 gallons of pure water. Starting at $t_0 = 0$ water containing 1 pound of salt per gallon is poured into T_1 at the rate of 2 gal/min. The mixture is drained from T_1 at the same rate into a second tank T_2, which initially contains 50 gallons of pure water. Also starting at $t_0 = 0$, a mixture from another source containing 2 pounds of salt per gallon is poured into T_2 at the rate of 2 gal/min. The mixture is drained from T_2 at the rate of 4 gal/min.

    (a) Find a differential equation for the quantity $Q(t)$ of salt in tank T_2 at time $t > 0$.

    (b) Solve the equation derived in **(a)** to determine $Q(t)$.

    (c) Find $\lim_{t \to \infty} Q(t)$.

16. Suppose that an object with initial temperature T_0 is placed in a sealed container, which is in turn placed in a medium with temperature T_m. Let the initial temperature of the container be S_0. Assume that the temperature of the object does not affect the temperature of the container, which in turn does not affect the temperature of the medium. (These assumptions are reasonable, for example, if the object is a cup of coffee, the container is a house, and the medium is the atmosphere.)

    (a) Assuming that the container and the medium have distinct temperature decay constants k and k_m respectively, use Newton's law of cooling to find the temperatures $S(t)$ and $T(t)$ of the container and object at time t.

    (b) Assuming that the container and the medium have the same temperature decay constant k, use Newton's law of cooling to find the temperatures $S(t)$ and $T(t)$ of the container and object at time t.

    (c) Find $\lim_{t \to \infty} S(t)$ and $\lim_{t \to \infty} T(t)$.

17. In our previous examples and exercises concerning Newton's law of cooling we assumed that the temperature of the medium remains constant. This model is adequate if the heat lost or gained by the object is insignificant compared to the heat required to cause an appreciable change in the temperature of the medium. If this is not so, then we must use a model that accounts for the heat exchanged between the object and the medium. Let

$T = T(t)$ and $T_m = T_m(t)$ be the temperatures of the object and the medium, respectively, and let T_0 and T_{m0} be their initial values. Again, we assume that T and T_m are related by Newton's law of cooling,

$$T' = -k(T - T_m). \tag{A}$$

We also assume that the change in heat of the object as its temperature changes from T_0 to T is $a(T - T_0)$ and that the change in heat of the medium as its temperature changes from T_{m0} to T_m is $a_m(T_m - T_{m0})$, where a and a_m are positive constants depending upon the masses and thermal properties of the object and medium, respectively. If we assume that the total heat of the system consisting of the object and the medium remains constant (that is, energy is conserved), then

$$a(T - T_0) + a_m(T_m - T_{m0}) = 0. \tag{B}$$

(a) Equation (A) involves two unknown functions T and T_m. Use (A) and (B) to derive a differential equation involving only T.

(b) Find $T(t)$ and $T_m(t)$ for $t > 0$.

(c) Find $\lim_{t \to \infty} T(t)$ and $\lim_{t \to \infty} T_m(t)$.

18. Control mechanisms allow fluid to flow into a tank at a rate proportional to the volume V of fluid in the tank, and to flow out at a rate proportional to V^2. Suppose that $V(0) = V_0$ and the constants of proportionality are a and b, respectively. Find V for $t > 0$ and find $\lim_{t \to \infty} V(t)$.

19. Identical tanks T_1 and T_2 initially contain W gallons each of pure water. Starting at $t_0 = 0$, a salt solution with constant concentration c is pumped into T_1 at r gal/min and drained from T_1 into T_2 at the same rate. The resulting mixture in T_2 is also drained at the same rate. Find the concentrations $c_1(t)$ and $c_2(t)$ in tanks T_1 and T_2 for $t > 0$.

20. An infinite sequence of identical tanks $T_1, T_2, \ldots, T_n, \ldots$, initially contain W gallons each of pure water. They are hooked together so that fluid drains from T_n into T_{n+1} ($n = 1, 2, \ldots$). A salt solution is circulated through the tanks so that it enters and leaves each tank at the constant rate of r gal/min. The solution has a concentration of c pounds of salt per gallon when it enters T_1.

(a) Find the concentration $c_n = c_n(t)$ in tank T_n for $t > 0$.

(b) Find $\lim_{t \to \infty} c_n(t)$ for each n.

21. Tanks T_1 and T_2 have capacities W_1 and W_2 liters, respectively. Initially they are both full of dye solutions with concentrations c_1 and c_2 grams per liter. Starting at $t_0 = 0$ the solution from T_1 is pumped into T_2 at a rate of r liters per minute, and the solution from T_2 is pumped into T_1 at the same rate.

(a) Find the concentrations $k_1(t)$ and $k_2(t)$ of the dye in T_1 and T_2 for $t > 0$.

(b) Find $\lim_{t \to \infty} k_1(t)$ and $\lim_{t \to \infty} k_2(t)$.

L **22.** Consider the mixing problem of Example 3, but without the assumption that the mixture is stirred instantly so that the salt is always uniformly distributed throughout the mixture. Assume instead that the distribution approaches uniformity as $t \to \infty$. In this case the differential equation for Q is of the form

$$Q' + \frac{a(t)}{150} Q = 2$$

where $\lim_{t \to \infty} a(t) = 1$.

(a) Assuming that $Q(0) = Q_0$, can you guess the value of $\lim_{t \to \infty} Q(t)$?

(b) Use numerical methods to confirm your guess in the following cases:

(i) $a(t) = t/(1 + t)$;   **(ii)** $a(t) = 1 - e^{-t^2}$;   **(iii)** $a(t) = 1 - \sin(e^{-t})$.

L **23.** Consider the mixing problem of Example 4 in a tank with infinite capacity, but without the assumption that the mixture is stirred instantly so that the salt is always uniformly distributed throughout the mixture. Assume instead that the distribution approaches uniformity as $t \to \infty$. In this case the differential equation for Q is of the form

$$Q' + \frac{a(t)}{t + 100} Q = 1$$

where $\lim_{t \to \infty} a(t) = 1$.

(a) Let $K(t)$ be the concentration of salt at time t. Assuming that $Q(0) = Q_0$, can you guess the value of $\lim_{t \to \infty} K(t)$?

(b) Use numerical methods to confirm your guess in the following cases:

   (i) $a(t) = t/(1 + t)$;   (ii) $a(t) = 1 - e^{-t^2}$;   (iii) $a(t) = 1 - \sin(e^{-t})$.

4.3   Elementary Mechanics

NEWTON'S SECOND LAW OF MOTION

In this section we consider an object with constant mass m moving along a line under a force F. Let $y = y(t)$ be the displacement of the object from some reference point on the line at time t, and let $v = v(t)$ and $a = a(t)$ be the velocity and acceleration of the object at time t. Thus, $v = y'$ and $a = v' = y''$, where the prime denotes differentiation with respect to t. Newton's second law of motion asserts that the force F and the acceleration a are related by the equation

$$F = ma. \tag{1}$$

UNITS

In applications there are three main sets of units in use for length, mass, force, and time: the cgs, mks, and British systems. All three use the second (s) as the unit of time. The other units are shown in Table 1. Consistent with (1), the unit of force in each system is defined to be the force required to impart an acceleration of (one unit of length)/s^2 to one unit of mass.

TABLE 1

	Length	Force	Mass
cgs	centimeter (cm)	dyne (d)	gram (g)
mks	meter (m)	newton (N)	kilogram (kg)
British	foot (ft)	pound (lb)	slug (sl)

If we assume that Earth is a perfect sphere with constant mass density, then Newton's law of gravitation (discussed later in this section) asserts that the force exerted on an object by Earth's gravitational field is proportional to the mass of the object and inversely proportional to the square of its distance from the center of Earth. However, if the object remains sufficiently close to Earth's surface, we may assume that the gravitational force is constant and equal to its value at the surface. The magnitude of this force is mg, where g is called the ***acceleration due to gravity.*** (To be completely accurate, g should be called the ***magnitude of the acceleration due to gravity at Earth's surface.***) This quantity has been determined experimentally. Approximate values of g are

$$g = 980 \text{ cm/s}^2 \quad \text{(cgs)};$$
$$g = 9.8 \text{ m/s}^2 \quad \text{(mks)};$$
$$g = 32 \text{ ft/s}^2 \quad \text{(British)}.$$

In general, the force F in (1) may depend upon t, y, and y'. Since $a = y''$, it follows that (1) can be written in the form

$$my'' = F(t, y, y'), \tag{2}$$

which is a second order equation. We will consider this equation with restrictions on F later; however, since Chapter 2 dealt only with first order equations, we consider here only problems in which (2) can be recast as a first order equation. This is possible if F does not depend on y, so (2) is of the form

$$my'' = F(t, y').$$

Letting $v = y'$ and $v' = y''$ yields a first order equation for v:

$$mv' = F(t, v). \tag{3}$$

Solving this equation yields v as a function of t. If we know $y(t_0)$ for some time t_0, we can then integrate v to obtain y as a function of t.

Equations of the form (3) occur in problems involving motion through a resisting medium.

MOTION THROUGH A RESISTING MEDIUM
UNDER CONSTANT GRAVITATIONAL FORCE

Now we consider an object moving vertically in some medium. We assume that the only forces acting on the object are gravity and resistance from the medium. We also assume that the motion takes place close to Earth's surface and take the upward direction to be positive, so the gravitational force can be assumed to have the constant value $-mg$. We will see that, under reasonable assumptions on the resisting force, the velocity approaches a limit as $t \to \infty$. We call this limit the ***terminal velocity***.

EXAMPLE 1 An object with mass m moves under constant gravitational force through a medium that exerts a resistance with magnitude proportional to the speed of the object. (Recall that the speed of an object is $|v|$, the absolute value of its velocity v.) Find the velocity of the object as a function of t, and find the terminal velocity. Assume that the initial velocity is v_0.

Solution The total force acting on the object is

$$F = -mg + F_1, \tag{4}$$

where $-mg$ is the force due to gravity and F_1 is the resisting force of the medium, which has magnitude $k|v|$, where k is a positive constant. If the object is moving downward ($v \le 0$), then the resisting force is upward (Figure 1**(a)**), so

$$F_1 = k|v| = k(-v) = -kv.$$

On the other hand, if the object is moving upward ($v \ge 0$), then the resisting force is downward (Figure 1**(b)**), so

$$F_1 = -k|v| = -kv.$$

Thus, (4) can be written as

$$F = -mg - kv, \tag{5}$$

regardless of the sign of the velocity.

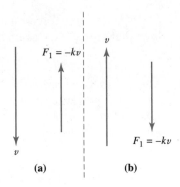

Figure 1

From Newton's second law of motion,

$$F = ma = mv',$$

so (5) yields

$$mv' = -mg - kv,$$

or

$$v' + \frac{k}{m} v = -g. \tag{6}$$

Since $e^{-kt/m}$ is a solution of the complementary equation, the solutions of (6) are of the form $v = ue^{-kt/m}$, where $u'e^{-kt/m} = -g$, so $u' = -ge^{kt/m}$. Hence,

$$u = -\frac{mg}{k} e^{kt/m} + c,$$

so

$$v = ue^{-kt/m} = -\frac{mg}{k} + ce^{-kt/m}. \tag{7}$$

Since $v(0) = v_0$,

$$v_0 = -\frac{mg}{k} + c,$$

so

$$c = v_0 + \frac{mg}{k}$$

and (7) becomes

$$v = -\frac{mg}{k} + \left(v_0 + \frac{mg}{k} \right)e^{-kt/m}.$$

Letting $t \to \infty$ here shows that the terminal velocity is

$$\lim_{t \to \infty} v(t) = -\frac{mg}{k}.$$

Notice that this limit is independent of the initial velocity v_0 (Figure 2).

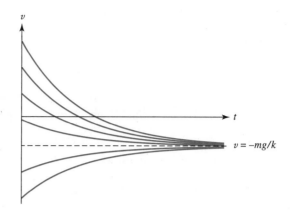

Figure 2   Solutions of $mv' = -mg - kv$

EXAMPLE 2   A 960-lb object is given an initial upward velocity of 60 ft/s near the surface of Earth. The atmosphere resists the motion with a force of 3 lb for each ft/s of speed. Assuming that the only other force acting on the object is constant gravity, find its velocity v as a function of t, and find its terminal velocity.

Solution   Since $mg = 960$ and $g = 32$, we have $m = 960/32 = 30$. The atmospheric resistance is $-3v$ lb if v is expressed in feet per second. Therefore

$$30v' = -960 - 3v,$$

which we rewrite as

$$v' + \frac{1}{10}v = -32.$$

Since $e^{-t/10}$ is a solution of the complementary equation, the solutions of this equation are of the form $v = ue^{-t/10}$, where $u'e^{-t/10} = -32$, so $u' = -32e^{t/10}$. Hence,

$$u = -320e^{t/10} + c,$$

so

$$v = ue^{-t/10} = -320 + ce^{-t/10}. \qquad (8)$$

The initial velocity is 60 ft/s in the upward (positive) direction; hence $v_0 = 60$. Substituting $t = 0$ and $v = 60$ in (8) yields

$$60 = -320 + c,$$

so $c = 380$, and (8) becomes

$$v = -320 + 380e^{-t/10} \text{ ft/s}.$$

The terminal velocity is

$$\lim_{t \to \infty} v(t) = -320 \text{ ft/s}.$$

EXAMPLE 3   A 10-kg mass is given an initial velocity $v_0 \leq 0$ near Earth's surface. The only forces acting on it are gravity and atmospheric resistance proportional to the square of the speed. Assuming that the resistance is 8 N if the speed is 2 m/s, find the velocity of the object as a function of t, and find the terminal velocity.

Solution   Since the object is falling, the resistance is in the upward (positive) direction. Hence,

$$mv' = -mg + kv^2, \tag{9}$$

where k is a constant. Since the magnitude of the resistance is 8 N when $v = 2$ m/s,

$$k(2^2) = 8,$$

so $k = 2$ N-s^2/m^2. Since $m = 10$ and $g = 9.8$, (9) becomes

$$10v' = -98 + 2v^2 = 2(v^2 - 49). \tag{10}$$

If $v_0 = -7$ then $v \equiv -7$ for all $t \geq 0$. If $v_0 \neq -7$ we separate variables to obtain

$$\frac{1}{v^2 - 49} v' = \frac{1}{5}, \tag{11}$$

which is convenient for the required partial fraction expansion

$$\frac{1}{v^2 - 49} = \frac{1}{(v - 7)(v + 7)}$$

$$= \frac{1}{14} \left[\frac{1}{v - 7} - \frac{1}{v + 7} \right]. \tag{12}$$

Substituting (12) into (11) yields

$$\frac{1}{14} \left[\frac{1}{v - 7} - \frac{1}{v + 7} \right] v' = \frac{1}{5},$$

so

$$\left[\frac{1}{v - 7} - \frac{1}{v + 7} \right] v' = \frac{14}{5}.$$

Integrating this yields

$$\ln |v - 7| - \ln |v + 7| = 14t/5 + k.$$

Therefore

$$\left| \frac{v - 7}{v + 7} \right| = e^k e^{14t/5}.$$

Since Theorem 2.3.1 implies that $(v - 7)/(v + 7)$ can't change sign (why?), we can rewrite the last equation as

$$\frac{v - 7}{v + 7} = ce^{14t/5}, \tag{13}$$

which is an implicit solution of (10). Solving this for v yields

$$v = -7 \frac{c + e^{-14t/5}}{c - e^{-14t/5}}. \tag{14}$$

Since $v(0) = v_0$, it follows from (13) that

$$c = \frac{v_0 - 7}{v_0 + 7}.$$

Substituting this in (14) and simplifying yields

$$v = -7 \frac{v_0(1 + e^{-14t/5}) - 7(1 - e^{-14t/5})}{v_0(1 - e^{-14t/5}) - 7(1 + e^{-14t/5})}.$$

Notice that since $v_0 \leq 0$, v is defined and negative for all $t > 0$. The terminal velocity is

$$\lim_{t \to \infty} v(t) = -7 \text{ m/s},$$

independent of v_0. More generally, it can be shown (Exercise 11) that if v is any solution of (9) such that $v_0 \leq 0$, then

$$\lim_{t \to \infty} v(t) = -\sqrt{\frac{mg}{k}}$$

(Figure 3).

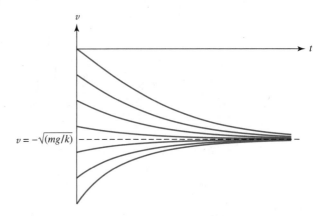

Figure 3 Solutions of $mv' = -mg + kv^2$, $v(0) = v_0 \leq 0$

EXAMPLE 4   A 10-kg mass is launched vertically upward from Earth's surface with an initial velocity of v_0 m/s. The only forces acting on the mass are gravity and atmospheric resistance proportional to the square of the speed. Assuming that the atmospheric resistance is 8 N if the speed is 2 m/s, find the time T required for the mass to reach maximum altitude.

Solution   The mass will climb while $v > 0$ and reach its maximum altitude when $v = 0$. Therefore, $v > 0$ for $0 \leq t < T$ and $v(T) = 0$. Although the mass of the object and our assumptions concerning the forces acting on it are the same as those in Example 3, (10) does not apply here, since the resisting force is negative if $v > 0$; therefore, we replace (10) by

$$10v' = -98 - 2v^2. \tag{15}$$

Separating variables yields

$$\frac{5}{v^2 + 49} v' = -1,$$

and integrating this yields

$$\frac{5}{7} \tan^{-1} \frac{v}{7} = -t + c.$$

(Recall that $\tan^{-1} u$ is the number θ such that $-\pi/2 < \theta < \pi/2$ and $\tan \theta = u$.) Since $v(0) = v_0$,

$$c = \frac{5}{7} \tan^{-1} \frac{v_0}{7},$$

so v is defined implicitly by

$$\frac{5}{7} \tan^{-1} \frac{v}{7} = -t + \frac{5}{7} \tan^{-1} \frac{v_0}{7}, \qquad 0 \le t \le T. \tag{16}$$

Solving this for v yields

$$v = 7 \tan\!\left(-\frac{7t}{5} + \tan^{-1} \frac{v_0}{7}\right). \tag{17}$$

By using the identity

$$\tan(A - B) = \frac{\tan A - \tan B}{1 + \tan A \tan B}$$

with $A = \tan^{-1}(v_0/7)$ and $B = 7t/5$, and noting that $\tan(\tan^{-1} \theta) = \theta$, we can simplify (17) to

$$v = 7\,\frac{v_0 - 7 \tan(7t/5)}{7 + v_0 \tan(7t/5)}.$$

Since $v(T) = 0$ and $\tan^{-1}(0) = 0$, (16) implies that

$$-T + \frac{5}{7} \tan^{-1} \frac{v_0}{7} = 0.$$

Therefore,

$$T = \frac{5}{7} \tan^{-1} \frac{v_0}{7}.$$

Since $\tan^{-1}(v_0/7) < \pi/2$ for all v_0, the time required for the mass to reach its maximum altitude is less than

$$\frac{5\pi}{14} \approx 1.122 \text{ s}$$

regardless of the initial velocity. Figure 4 shows graphs of v over $[0, T]$ for various values of v_0. ■

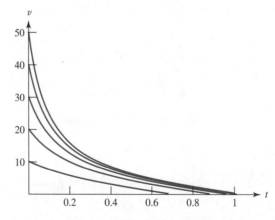

Figure 4   Solutions of (15) for various $v_0 > 0$.

ESCAPE VELOCITY

Suppose that a space vehicle is launched vertically and its fuel is exhausted when the vehicle reaches an altitude h above Earth, where h is sufficiently large that resistance due to Earth's atmosphere can be neglected. Let $t = 0$ be the time at which burnout occurs. Assuming that the gravitational forces of all other celestial bodies can be neglected, the motion of the vehicle for $t > 0$ is that of an object with constant mass m under the influence of Earth's gravitational force, which we now assume to vary inversely with the square of the distance from Earth's center; thus, if we take the upward direction to be positive, the gravitational force on the vehicle at an altitude y above Earth is

$$F = -\frac{K}{(y + R)^2}, \tag{18}$$

where R is Earth's radius (Figure 5).

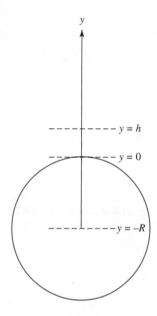

Figure 5

Since $F = -mg$ when $y = 0$, setting $y = 0$ in (18) yields

$$-mg = -\frac{K}{R^2};$$

therefore $K = mgR^2$, and (18) can be written more specifically as

$$F = -\frac{mgR^2}{(y + R)^2}. \tag{19}$$

From Newton's second law of motion,

$$F = m\frac{d^2y}{dt^2},$$

so (19) implies that

$$\frac{d^2y}{dt^2} = -\frac{gR^2}{(y + R)^2}.$$ (20)

We will show that there is a number v_e, called the ***escape velocity***, with the following properties:

1. If $v_0 \geq v_e$ then $v(t) > 0$ for all $t > 0$, and the vehicle continues to climb for all $t > 0$; that is, it "escapes" Earth. (Is it really so obvious that $\lim_{t \to \infty} y(t) = \infty$ in this case? For a proof, see Exercise 20.)
2. If $v_0 < v_e$ then $v(t)$ decreases to zero and becomes negative. Therefore, the vehicle attains a maximum altitude y_m and falls back to Earth.

Since (20) is second order, we can't solve it by methods discussed so far. However, we're concerned with v rather than y, and v is easier to find. Since $v = y'$ the chain rule implies that

$$\frac{d^2y}{dt^2} = \frac{dv}{dt} = \frac{dv}{dy}\frac{dy}{dt} = v\frac{dv}{dy}.$$

Substituting this into (20) yields the first order separable equation

$$v\frac{dv}{dy} = -\frac{gR^2}{(y + R)^2}.$$ (21)

When $t = 0$, the velocity is v_0 and the altitude is h. Therefore, we can obtain v as a function of y by solving the initial value problem

$$v\frac{dv}{dy} = -\frac{gR^2}{(y + R)^2}, \qquad v(h) = v_0.$$

Integrating (21) with respect to y yields

$$\frac{v^2}{2} = \frac{gR^2}{y + R} + c.$$ (22)

Since $v(h) = v_0$,

$$c = \frac{v_0^2}{2} - \frac{gR^2}{h + R},$$

so (22) becomes

$$\frac{v^2}{2} = \frac{gR^2}{y + R} + \left(\frac{v_0^2}{2} - \frac{gR^2}{h + R}\right).$$ (23)

If

$$v_0 \geq \left(\frac{2gR^2}{h + R}\right)^{1/2},$$

then the parenthetical expression in (23) is nonnegative, so $v(y) > 0$ for $y > h$. This proves that there is an escape velocity v_e. We will now prove that

$$v_e = \left(\frac{2gR^2}{h + R}\right)^{1/2}$$

by showing that the vehicle falls back to Earth if

$$v_0 < \left(\frac{2gR^2}{h + R} \right)^{1/2}. \tag{24}$$

If (24) holds then the parenthetical expression in (23) is negative, and the vehicle will attain a maximum altitude $y_m > h$ satisfying the equation

$$0 = \frac{gR^2}{y_m + R} + \left(\frac{v_0^2}{2} - \frac{gR^2}{h + R} \right).$$

The velocity will be zero at the maximum altitude, and the object will then fall to Earth under the influence of gravity.

4.3 EXERCISES

Except where directed otherwise, assume that the magnitude of the gravitational force on an object with mass m is constant and equal to mg. In exercises involving vertical motion take the upward direction to be positive.

1. A firefighter who weighs 192 lb slides down an infinitely long fire pole that exerts a frictional resistive force with magnitude proportional to his speed, with $k = 2.5$ lb-s/ft. Assuming that he starts from rest, find his velocity as a function of time and find his terminal velocity.

2. A firefighter who weighs 192 lb slides down an infinitely long fire pole that exerts a frictional resistive force with magnitude proportional to her speed, with constant of proportionality k. Find k, given that her terminal velocity is -16 ft/s, and then find her velocity v as a function of t. Assume that she starts from rest.

3. A boat weighs 64,000 lb. Its propeller produces a constant thrust of 50,000 lb and the water exerts a resistive force with magnitude proportional to the speed, with $k = 2000$ lb-s/ft. Assuming that the boat starts from rest, find its velocity as a function of time, and find its terminal velocity.

4. A constant horizontal force of 10 N pushes a 20-kg mass through a medium that resists its motion with .5 N for every m/s of speed. The initial velocity of the mass is 7 m/s in the direction opposite to the direction of the applied force. Find the velocity of the mass for $t > 0$.

5. A stone weighing 1/2 lb is thrown upward from an initial height of 5 ft with an initial speed of 32 ft/s. Air resistance is proportional to speed, with $k = 1/128$ lb-s/ft. Find the maximum height attained by the stone.

6. A 3200-lb car is moving at 64 ft/s down a 30-degree grade when it runs out of fuel. Find its velocity after that if friction exerts a resistive force with magnitude proportional to the square of the speed, with $k = 1$ lb-s^2/ft^2. Also find its terminal velocity.

7. A 96-lb weight is dropped from rest in a medium that exerts a resistive force with magnitude proportional to the speed. Find its velocity as a function of time if its terminal velocity is -128 ft/s.

8. An object with mass m moves vertically through a medium that exerts a resistive force with magnitude proportional to the speed. Let $y = y(t)$ be the altitude of the object at time t, with $y(0) = y_0$. Use the results of Example 1 to show that

$$y = y_0 + \frac{m}{k}(v_0 - v - gt).$$

9. An object with mass m is launched vertically upward with initial velocity v_0 from Earth's surface ($y_0 = 0$) in a medium that exerts a resistive force with magnitude proportional to the speed. Find the time T when the object attains its maximum altitude y_m. Then use the result of Exercise 8 to find y_m.

10. An object weighing 256 lb is dropped from rest in a medium that exerts a resistive force with magnitude proportional to the square of the speed. The magnitude of the resisting force is 1 lb when $|v| = 4$ ft/s. Find v for $t > 0$, and find its terminal velocity.

11. An object with mass m is given an initial velocity $v_0 \le 0$ in a medium that exerts a resistive force with magnitude proportional to the square of the speed. Find the velocity of the object for $t > 0$, and find its terminal velocity.

12. An object with mass m is launched vertically upward with initial velocity v_0 in a medium that exerts a resistive force with magnitude proportional to the square of the speed.

    (a) Find the time T at which the object reaches its maximum altitude.

    (b) Use the result of Exercise 11 to find the velocity of the object for $t > T$.

L **13.** An object with mass m is given an initial velocity $v_0 \le 0$ in a medium that exerts a resistive force of the form $a|v|/(1 + |v|)$, where a is positive constant.

    (a) Set up a differential equation for the speed of the object.

    (b) Use your favorite numerical method to solve the equation you found in **(a)**, to convince yourself that there is a unique number a_0 such that $\lim_{t \to \infty} s(t) = \infty$ if $a \le a_0$ and $\lim_{t \to \infty} s(t)$ exists (finite) if $a > a_0$. (We say that a_0 is the **bifurcation value** of a.) Try to find a_0. If $a > a_0$, find $\lim_{t \to \infty} s(t)$.

    Hint: See Exercise 14.

14. An object of mass m falls in a medium that exerts a resistive force $f = f(s)$, where $s = |v|$ is the speed of the object. Assume that $f(0) = 0$ and f is strictly increasing and differentiable on $(0, \infty)$.

    (a) Write a differential equation for the speed $s = s(t)$ of the object. Take it as given that all solutions of this equation with $s(0) \ge 0$ are defined for all $t > 0$ (which makes good sense on physical grounds).

    (b) Show that if $\lim_{t \to \infty} f(s) \le mg$ then $\lim_{t \to \infty} s(t) = \infty$.

    (c) Show that if $\lim_{t \to \infty} f(s) > mg$ then $\lim_{t \to \infty} s(t) = s_T$ (terminal speed), where $f(s_T) = mg$.

    Hint: Use Theorem 2.3.1.

15. A 100-g mass with initial velocity $v_0 \le 0$ falls in a medium that exerts a resistive force proportional to the fourth power of the speed. The resistance is .1 N if the speed is 3 m/s.

    (a) Set up the initial value problem for the velocity v of the mass for $t > 0$.

    (b) Use Exercise 14**(c)** to determine the terminal velocity of the object.

C   **(c)** To confirm your answer to **(b)**, use one of the numerical methods studied in Chapter 3 to compute approximate solutions on $[0, 1]$ (seconds) of the initial value problem of **(a)**, with initial values $v_0 = 0, -2, -4, \ldots, -12$. Present your results in graphical form similar to Figure 3.

16. A 64-lb object with initial velocity $v_0 \le 0$ falls through a dense fluid that exerts a resistive force proportional to the square root of the speed. The resistance is 64 lb if the speed is 16 ft/s.

    (a) Set up the initial value problem for the velocity v of the mass for $t > 0$.

    (b) Use Exercise 14**(c)** to determine the terminal velocity of the object.

C   **(c)** To confirm your answer to **(b)**, use one of the numerical methods studied in Chapter 3 to compute approximate solutions on $[0, 4]$ (seconds) of the initial value problem of **(a)**, with initial values $v_0 = 0, -5, -10, \ldots, -30$. Present your results in graphical form similar to Figure 3.

In Exercises 17–20 assume that the force due to gravity is given by Newton's law of gravitation. Take the upward direction to be positive.

17. A space probe is to be launched from a space station 200 miles above Earth. Determine its escape velocity in miles/s. Take Earth's radius to be 3960 miles.

18. A space vehicle is to be launched from the moon, which has a radius of approximately 1080 miles. The acceleration due to gravity at the surface of the moon is approximately 5.31 ft/s^2. Find the escape velocity in miles/s.

19. (a) Show that Eqn. (23) can be rewritten as

$$v^2 = \frac{h - y}{y + R} v_e^2 + v_0^2.$$

(b) Show that if $v_0 = \rho v_e$ with $0 \le \rho < 1$, then the maximum altitude y_m attained by the space vehicle is

$$y_m = \frac{h + R\rho^2}{1 - \rho^2}.$$

(c) By requiring that $v(y_m) = 0$, use Eqn. (22) to deduce that if $v_0 < v_e$, then

$$|v| = v_e \left[\frac{(1 - \rho^2)(y_m - y)}{y + R} \right]^{1/2},$$

where y_m and ρ are as defined in **(b)** and $y \ge h$.

(d) Deduce from **(c)** that if $v < v_e$, then the vehicle takes equal times to climb from $y = h$ to $y = y_m$ and to fall back from $y = y_m$ to $y = h$.

20. In the situation considered in the discussion of escape velocity, show that $\lim_{t \to \infty} y(t) = \infty$ if $v(t) > 0$ for all $t > 0$.

Hint: Use a proof by contradiction. Assume that there is a number y_m such that $y(t) \le y_m$ for all $t > 0$. Deduce from this that there is positive number α such that $y''(t) \le -\alpha$ for all $t \ge 0$. Show that this contradicts the assumption that $v(t) > 0$ for all $t > 0$.

4.4 Autonomous Second Order Equations

A second order differential equation that can be written as

$$y'' = F(y, y'), \tag{1}$$

where F is independent of t, is said to be ***autonomous.*** An autonomous second order equation can be converted into a first order equation relating $v = y'$ and y. If we let $v = y'$ then (1) becomes

$$v' = F(y, v). \tag{2}$$

Since

$$v' = \frac{dv}{dt} = \frac{dv}{dy} \frac{dy}{dt} = v \frac{dv}{dy}, \tag{3}$$

(2) can be rewritten as

$$v \frac{dv}{dy} = F(y, v). \tag{4}$$

The integral curves of (4) can be plotted in the (y, v) plane, which is called the ***Poincaré*[1] *phase plane*** of (1). If y is a solution of (1) then $y = y(t), v = y'(t)$ is a parametric equation for an integral curve of (4). We'll call these integral curves ***trajectories*** of (1), and we'll call (4) the ***phase plane equivalent*** of (1).

In this section we'll consider autonomous equations that can be written in the form

$$y'' + q(y, y')y' + p(y) = 0. \tag{5}$$

---

[1] Henri Poincaré (1854–1912) of the University of Paris was one of the leading mathematicians of all time. He made fundamental contributions to many areas of mathematics and to celestial mechanics, and initiated the study of the qualitative behavior of solutions of differential equations.

Equations of this form often arise in applications of Newton's second law of motion. For example, suppose y is the displacement of a moving object with mass m. It is quite reasonable to think of two types of time-independent forces acting on the object. One type—such as gravity—depends only on position. We could write such a force as $-mp(y)$. The second type—such as atmospheric resistance or friction—may depend on position and velocity. (Forces that depend on velocity are called *damping forces.*) We write this force as $-mq(y, y')y'$, where $q(y, y')$ is usually a positive function and we've put the factor y' outside to make it explicit that the force is in the direction opposing the motion. In this case Newton's second law of motion leads to (5).

The phase plane equivalent of (5) is

$$v \frac{dv}{dy} + q(y, v)v + p(y) = 0. \tag{6}$$

Some statements that we'll be making about the properties of (5) and (6) are intuitively reasonable, but difficult to prove. Therefore, our presentation in this section will be informal: we'll just say things without proof, all of which are true if we assume that $p = p(y)$ is continuously differentiable for all y and $q = q(y, v)$ is continuously differentiable for all (y, v). We begin with the following statements:

Statement 1. If y_0 and v_0 are arbitrary real numbers then (5) has a unique solution on $(-\infty, \infty)$ such that $y(0) = y_0$ and $y'(0) = v_0$.

Statement 2. If $y = y(t)$ is a solution of (5) and τ is any constant, then $y_1 = y(t - \tau)$ is also a solution of (5), and y and y_1 have the same trajectory.

Statement 3. If two solutions y and y_1 of (5) have the same trajectory then $y_1(t) = y(t - \tau)$ for some constant τ.

Statement 4. Distinct trajectories of (5) cannot intersect; that is, if two trajectories of (5) intersect then they are identical.

Statement 5. If the trajectory of a solution of (5) is a closed curve then the point $(y(t), v(t))$ traverses the trajectory in a finite time T, and the solution is periodic with period T; that is, $y(t + T) = y(t)$ for all t in $(-\infty, \infty)$.

If $\overline{y}$ is a constant such that $p(\overline{y}) = 0$, then $y \equiv \overline{y}$ is a constant solution of (5). We say that $\overline{y}$ is an *equilibrium* of (5) and that $(\overline{y}, 0)$ is a *critical point* of the phase plane equivalent equation (6). We say that the equilibrium and the critical point are *stable* if, for any given $\epsilon > 0$ *no matter how small*, there is a $\delta > 0$, *sufficiently small*, such that if

$$\sqrt{(y_0 - \overline{y})^2 + v_0^2} < \delta,$$

then the solution of the initial value problem

$$y'' + q(y, y')y' + p(y) = 0, \qquad y(0) = y_0, \quad y'(0) = v_0$$

satisfies the inequality

$$\sqrt{(y(t) - \overline{y})^2 + (v(t))^2} < \epsilon$$

for all $t > 0$. Figure 1 illustrates the geometrical interpretation of this definition in the Poincaré phase plane: if (y_0, v_0) is in the smaller shaded circle (with radius δ), then $(y(t), v(t))$ must be in the larger circle (with radius ϵ) for all $t > 0$.

If an equilibrium and the associated critical point are not stable, we say they are *unstable.* To see if you really understand what *stable* means, try to give a

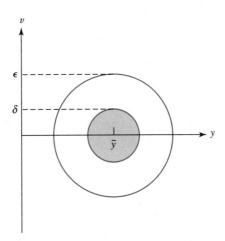

Figure 1 Stability: if (y_0, v_0) is in the smaller circle, then $(y(t), v(t))$ is in the larger circle for all $t > 0$

direct definition of **unstable** (Exercise 22). We'll illustrate both definitions in the following examples.

THE UNDAMPED CASE

We'll begin with the case where $q \equiv 0$, so that (5) reduces to

$$y'' + p(y) = 0. \tag{7}$$

We say that this equation—as well as any physical situation that it may model—is **undamped.** The phase plane equivalent of (7) is the separable equation

$$v \frac{dv}{dy} + p(y) = 0.$$

Integrating this yields

$$\frac{v^2}{2} + P(y) = c, \tag{8}$$

where c is a constant of integration and $P(y) = \int p(y)\, dy$ is an antiderivative of p.

If (7) is the equation of motion of an object of mass m then $mv^2/2$ is the kinetic energy and $mP(y)$ is the potential energy of the object; thus, (8) says that the total energy of the object remains constant, or is **conserved.** In particular, if a trajectory passes through a given point (y_0, v_0) then

$$c = \frac{v_0^2}{2} + P(y_0).$$

EXAMPLE 1

(The Undamped Spring–Mass System)

Consider an object with mass m suspended from a spring and moving vertically. Let y be the displacement of the object from the position it occupies when suspended at rest from the spring (Figure 2).

Assume that if the length of the spring is changed by an amount ΔL (positive or negative), then the spring exerts an opposing force with magnitude $k|\Delta L|$, where k is a positive constant. In Section 6.1 it will be shown that if the mass of

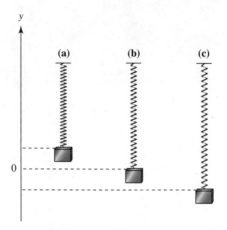

Figure 2 **(a)** $y > 0$; **(b)** $y = 0$; **(c)** $y < 0$

the spring is negligible compared to m and no other forces act on the object, then Newton's second law of motion implies that

$$my'' = -ky, \qquad (9)$$

which can be written in the form (7) with $p(y) = ky/m$. This equation can be solved easily by a method that we'll study in Section 5.2, but that method isn't available here. Instead, we'll consider the phase plane equivalent of (9).

From (3), we can rewrite (9) as the separable equation

$$mv\frac{dv}{dy} = -ky.$$

Integrating this yields

$$\frac{mv^2}{2} = -\frac{ky^2}{2} + c,$$

which implies that

$$mv^2 + ky^2 = \rho \qquad (10)$$

($\rho = 2c$). This defines an ellipse in the Poincaré phase plane (Figure 3).

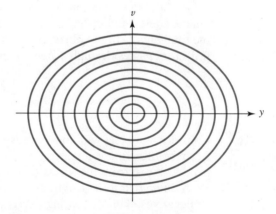

Figure 3 Trajectories of $my'' + ky = 0$

We can identify ρ by setting $t = 0$ in (10); thus, $\rho = mv_0^2 + ky_0^2$, where $y_0 = y(0)$ and $v_0 = v(0)$. To determine the maximum and minimum values of y we set $v = 0$ in (10); thus,

$$y_{max} = R \quad \text{and} \quad y_{min} = -R, \quad \text{with} \quad R = \sqrt{\frac{\rho}{k}}. \tag{11}$$

Equation (9) has exactly one equilibrium, $\bar{y} = 0$, and it is stable. You can see intuitively why this is so: if the object is displaced in either direction from equilibrium, the spring tries to bring it back.

In this case we can find y explicitly as a function of t. (Don't expect this to happen in more complicated problems!) If $v > 0$ on an interval I then (10) implies that

$$\frac{dy}{dt} = v = \sqrt{\frac{\rho - ky^2}{m}}$$

on I. This is equivalent to

$$\frac{\sqrt{k}}{\sqrt{\rho - ky^2}} \frac{dy}{dt} = \omega_0, \quad \text{where} \quad \omega_0 = \sqrt{\frac{k}{m}}. \tag{12}$$

Since

$$\int \frac{\sqrt{k}\, dy}{\sqrt{\rho - ky^2}} = \sin^{-1}\left(\sqrt{\frac{k}{\rho}}\, y\right) + c = \sin^{-1}\left(\frac{y}{R}\right) + c$$

(see (11)), (12) implies that that there is a constant ϕ such that

$$\sin^{-1}\left(\frac{y}{R}\right) = \omega_0 t + \phi$$

or

$$y = R\sin(\omega_0 t + \phi)$$

for all t in I. Although we obtained this function assuming that $v > 0$, you can easily verify that y satisfies (9) for all values of t. Thus, the displacement varies periodically between $-R$ and R, with period $T = 2\pi/\omega_0$ (Figure 4). (If you've taken a course in elementary mechanics you may recognize this as ***simple harmonic motion***.)

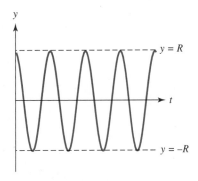

Figure 4 Graph of
$y = R\sin(\omega_0 t + \phi)$

EXAMPLE 2

(The Undamped Pendulum)

Now we consider the motion of a pendulum with mass m, attached to the end of a weightless rod with length L that rotates on a frictionless axle (Figure 5). We assume that there is no air resistance. Let y be the angle measured from the rest position (vertically downward) of the pendulum, as shown in Figure 5. Newton's second law of motion says that the product of m and the tangential acceleration equals the tangential component of the gravitational force; therefore, from Figure 5,

$$mLy'' = -mg \sin y,$$

or

$$y'' = -\frac{g}{L} \sin y. \tag{13}$$

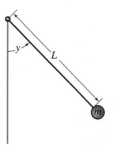

Figure 5   The undamped pendulum

Since $\sin n\pi = 0$ if n is any integer, (13) has infinitely many equilibria $\overline{y}_n = n\pi$. If n is even the mass is directly below the axle (Figure 6(a)), and gravity opposes any deviation from the equilibrium. However, if n is odd the mass is directly above the axle (Figure 6(b)), and gravity increases any deviation from the equilibrium. Therefore, we conclude on physical grounds that $\overline{y}_{2m} = 2m\pi$ is stable and $\overline{y}_{2m+1} = (2m + 1)\pi$ is unstable.

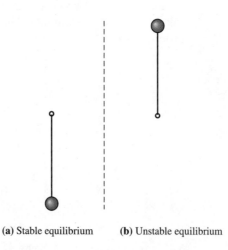

(a) Stable equilibrium   (b) Unstable equilibrium

Figure 6

The phase plane equivalent of (13) is

$$v \frac{dv}{dy} = -\frac{g}{L} \sin y,$$

where $v = y'$ is the angular velocity of the pendulum. Integrating this yields

$$\frac{v^2}{2} = \frac{g}{L} \cos y + c. \tag{14}$$

If $v = v_0$ when $y = 0$ then

$$c = \frac{v_0^2}{2} - \frac{g}{L},$$

so (14) becomes

$$\frac{v^2}{2} = \frac{v_0^2}{2} - \frac{g}{L}(1 - \cos y) = \frac{v_0^2}{2} - \frac{2g}{L} \sin^2 \frac{y}{2},$$

which is equivalent to

$$v^2 = v_0^2 - v_c^2 \sin^2 \frac{y}{2}, \tag{15}$$

where

$$v_c = 2\sqrt{\frac{g}{L}}.$$

The curves defined by (15) are the trajectories of (13). They are periodic with period 2π in y, which is not surprising since if $y = y(t)$ is a solution of (13), then so is $y_n = y(t) + 2n\pi$ for any integer n. Figure 7 shows trajectories over the interval $[-\pi, \pi]$. From (15) you can see that if $|v_0| > v_c$ then v is nonzero for all t, which means that the object whirls in the same direction forever, as in Figure 8. The trajectories associated with this whirling motion are above the upper dashed curve and below the lower dashed curve in Figure 7. You can also see from (15) that if $0 < |v_0| < v_c$ then $v = 0$ when $y = \pm y_{\max}$, where

$$y_{\max} = 2 \sin^{-1}(|v_0|/v_c).$$

In this case the pendulum oscillates periodically between $-y_{\max}$ and $y_{\max}$, as shown in Figure 9. The trajectories associated with this kind of motion are the ovals between the dashed curves in Figure 7. It can be shown (see Exercise 21 for a partial proof) that the period of the oscillation is

$$T = 8 \int_0^{\pi/2} \frac{d\theta}{\sqrt{v_c^2 - v_0^2 \sin^2 \theta}}. \tag{16}$$

Although this integral can't be evaluated in terms of familiar elementary functions, you can see that it is finite if $|v_0| < v_c$.

The dashed curves in Figure 7 contain four trajectories. The critical points $(\pi, 0)$ and $(-\pi, 0)$ are the trajectories of the unstable equilibrium solutions $\bar{y} = \pm\pi$. The upper dashed curve connecting (but not including) them is obtained from initial conditions of the form $y(t_0) = 0$, $v(t_0) = v_c$. If y is any solution with this trajectory, then

$$\lim_{t \to \infty} y(t) = \pi \quad \text{and} \quad \lim_{t \to -\infty} y(t) = -\pi.$$

The lower dashed curve connecting (but not including) them is obtained from initial conditions of the form $y(t_0) = 0$, $v(t_0) = -v_c$. If y is any solution with this trajectory, then

$$\lim_{t \to \infty} y(t) = -\pi \quad \text{and} \quad \lim_{t \to -\infty} y(t) = \pi.$$

Consistent with this, the integral (16) diverges to ∞ if $v_0 = \pm v_c$. (Exercise 21).

Since the dashed curves separate trajectories of whirling solutions from trajectories of oscillating solutions, each of these curves is called a *separatrix*.

■

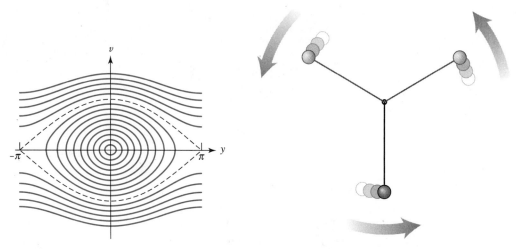

Figure 7 Trajectories of the undamped pendulum

Figure 8 The whirling undamped pendulum

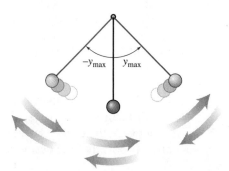

Figure 9 The oscillating undamped pendulum

In general, if (7) has both stable and unstable equilibria, then the separatrices are the curves given by (8) that pass through unstable critical points. Thus, if $(\bar{y}, 0)$ is an unstable critical point then

$$\frac{v^2}{2} + P(y) = P(\bar{y}) \tag{17}$$

defines a separatrix passing through $(\bar{y}, 0)$.

STABILITY AND INSTABILITY CONDITIONS FOR $y'' + p(y) = 0$

It can be shown (Exercise 23) that an equilibrium $\overline{y}$ of an undamped equation

$$y'' + p(y) = 0 \tag{18}$$

is stable if there is an open interval (a, b) containing $\overline{y}$, such that

$$p(y) < 0 \quad \text{if} \quad a < y < \overline{y} \quad \text{and} \quad p(y) > 0 \quad \text{if} \quad \overline{y} < y < b. \tag{19}$$

If we regard $p(y)$ as a force acting on a unit mass, then (19) means that the force resists all sufficiently small displacements from $\overline{y}$.

We've already seen examples illustrating this principle. The equation (9) for the undamped spring-mass system is of the form (18) with $p(y) = ky/m$, which has only the stable equilibrium $\overline{y} = 0$. In this case (19) holds with $a = -\infty$ and $b = \infty$. The equation (13) for the undamped pendulum is of the form (18) with $p(y) = (g/L)\sin y$. We've seen that $\overline{y} = 2m\pi$ is a stable equilibrium if m is an integer. In this case

$$p(y) = \sin y < 0 \quad \text{if} \quad (2m-1)\pi < y < 2m\pi$$

and

$$p(y) > 0 \quad \text{if} \quad 2m\pi < y < (2m+1)\pi.$$

It can also be shown (Exercise 24) that $\overline{y}$ is unstable if there is a $b > \overline{y}$ such that

$$p(y) < 0 \quad \text{if} \quad \overline{y} < y < b \tag{20}$$

or an $a < \overline{y}$ such that

$$p(y) > 0 \quad \text{if} \quad a < y < \overline{y}. \tag{21}$$

If we regard $p(y)$ as a force acting on a unit mass, then (20) means that the force tends to increase all sufficiently small positive displacements from $\overline{y}$, while (21) means that the force tends to increase the magnitude of all sufficiently small negative displacements from $\overline{y}$.

The undamped pendulum also illustrates this principle. We've seen that $\overline{y} = (2m+1)\pi$ is an unstable equilibrium if m is an integer. In this case

$$\sin y < 0 \quad \text{if} \quad (2m+1)\pi < y < (2m+2)\pi$$

so (20) holds with $b = (2m+2)\pi$, and

$$\sin y > 0 \quad \text{if} \quad 2m\pi < y < (2m+1)\pi,$$

so (21) holds with $a = 2m\pi$.

EXAMPLE 3 The equation

$$y'' + y(y-1) = 0 \tag{22}$$

is of the form (18) with $p(y) = y(y-1)$. Therefore $\overline{y} = 0$ and $\overline{y} = 1$ are the equilibria of (22). Since

$$y(y-1) > 0 \quad \text{if} \quad y < 0 \quad \text{or} \quad y > 1,$$
$$< 0 \quad \text{if} \quad 0 < y < 1,$$

it follows that $\overline{y} = 0$ is unstable and $\overline{y} = 1$ is stable.

The phase plane equivalent of (22) is the separable equation

$$v \frac{dv}{dy} + y(y-1) = 0.$$

Integrating yields

$$\frac{v^2}{2} + \frac{y^3}{3} - \frac{y^2}{2} = C,$$

which we rewrite as

$$v^2 + \frac{1}{3}y^2(2y - 3) = c \tag{23}$$

after renaming the constant of integration. These are the trajectories of (22). If y is any solution of (22) then the point $(y(t), v(t))$ moves along the trajectory of y in the direction of increasing y in the upper half plane ($v = y' > 0$), or in the direction of decreasing y in the lower half plane ($v = y' < 0$).

Typical trajectories are shown in Figure 10. The dashed curve through the critical point $(0,0)$, obtained by setting $c = 0$ in (23), separates the yv-plane into regions containing different kinds of trajectories; as in Example 2, we call this curve a **separatrix.** Trajectories in the region bounded by the closed loop **(b)** are closed curves, so solutions associated with them are periodic. Solutions associated with other trajectories are not periodic. If y is any such solution whose trajectory is not on the separatrix then

$$\lim_{t\to\infty} y(t) = -\infty, \qquad \lim_{t\to-\infty} y(t) = -\infty,$$
$$\lim_{t\to\infty} v(t) = -\infty, \qquad \lim_{t\to-\infty} v(t) = \infty.$$

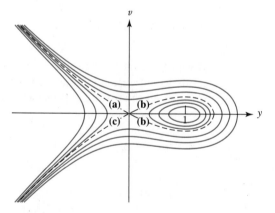

Figure 10   Trajectories of $y'' + y(y - 1) = 0$

The separatrix contains four trajectories of (22). One is the point $(0,0)$, the trajectory of the equilibrium $\bar{y} = 0$. Since distinct trajectories can't intersect, the segments of the separatrix marked **(a)**, **(b)**, and **(c)**—which do not include $(0,0)$—are distinct trajectories, none of which can be traversed in finite time. Solutions with these trajectories have the following asymptotic behavior:

$$\lim_{t\to\infty} y(t) = \quad 0, \qquad \lim_{t\to-\infty} y(t) = -\infty,$$
$$\lim_{t\to\infty} v(t) = \quad 0, \qquad \lim_{t\to-\infty} v(t) = \quad \infty, \qquad \text{(on (a))};$$
$$\lim_{t\to\infty} y(t) = \quad 0, \qquad \lim_{t\to-\infty} y(t) = \quad 0,$$
$$\lim_{t\to\infty} v(t) = \quad 0, \qquad \lim_{t\to-\infty} v(t) = \quad 0, \qquad \text{(on (b))};$$
$$\lim_{t\to\infty} y(t) = -\infty, \qquad \lim_{t\to-\infty} y(t) = \quad 0,$$
$$\lim_{t\to\infty} v(t) = -\infty, \qquad \lim_{t\to-\infty} v(t) = \quad 0, \qquad \text{(on (c))}. \qquad \blacksquare$$

THE DAMPED CASE

The phase plane equivalent of the damped autonomous equation

$$y'' + q(y, y')y' + p(y) = 0 \tag{24}$$

is

$$v \frac{dv}{dy} + q(y, v)v + p(y) = 0.$$

This equation is not separable, so we can't solve it for v in terms of y, as we did in the undamped case, and conservation of energy doesn't hold. (For example, energy expended in overcoming friction is lost.) However, we can study the qualitative behavior of its solutions by rewriting it as

$$\frac{dv}{dy} = -q(y, v) - \frac{p(y)}{v} \tag{25}$$

and considering the direction fields of this equation. In the following examples we'll also be showing computer generated trajectories of this equation, obtained by numerical methods. The exercises call for similar computations. The methods discussed in Sections 3.1, 3.2, and 3.3 are not suitable for this task, since $p(y)/v$ in (25) is undefined on the y-axis of the Poincaré phase plane. Therefore we are forced to apply numerical methods briefly discussed in Section 10.1 to the system

$$y' = v$$
$$v' = -q(y, v)v - p(y),$$

which is equivalent to (24) in the sense defined in Section 10.1. Fortunately, most differential equation software packages enable you to do this painlessly.

In the text we'll confine ourselves to the case where q is constant, so (24) and (25) reduce to

$$y'' + cy' + p(y) = 0 \tag{26}$$

and

$$\frac{dv}{dy} = -c - \frac{p(y)}{v}.$$

(We'll consider more general equations in the exercises.) The constant c is called the ***damping constant.*** In situations where (26) is the equation of motion of an object, c is positive; however, there are other situations in which c may be negative.

EXAMPLE 4

(The Damped Spring–
Mass System)

In Example 1 we considered the spring–mass system under the assumption that the only forces acting on the object were gravity and the spring's resistance to changes in its length. Now we'll assume that some mechanism (for example, friction in the spring or atmospheric resistance) opposes the motion of the object with a force proportional to its velocity. In Section 6.1 it will be shown that in this case Newton's second law of motion implies that

$$my'' + cy' + ky = 0, \tag{27}$$

where $c > 0$ is the ***damping constant.*** Again, this equation can be solved easily by a method that we'll study in Section 5.2, but that method isn't available here. Instead, we'll consider its phase plane equivalent, which can be written in the form (25) as

$$\frac{dv}{dy} = -\frac{c}{m} - \frac{ky}{mv}. \tag{28}$$

(A minor note: the c in (26) actually corresponds to c/m in this equation.) Figure 11 shows a typical direction field for an equation of this form. Recalling that motion along a trajectory must be in the direction of increasing y in the upper half plane ($v > 0$) and in the direction of decreasing y in the lower half plane ($v < 0$), you can infer that all trajectories approach the origin in clockwise fashion. To confirm this, Figure 12 shows the same direction field with some trajectories filled in. All the trajectories shown there correspond to solutions of the initial value problem

$$my'' + cy' + ky = 0, \qquad y(0) = y_0, \qquad y'(0) = v_0,$$

where

$$mv_0^2 + ky_0^2 = \rho \qquad \text{(a positive constant)};$$

thus, if there were no damping ($c = 0$), then all the solutions would have the same dashed elliptic trajectory, shown in Figure 14.

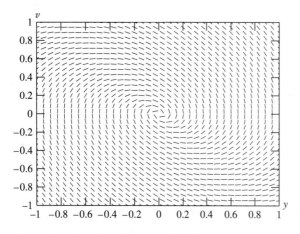

Figure 11 A typical direction field for $my'' + cy' + ky = 0$ with $0 < c < c_1$

Figure 12 Figure 11 with some trajectories added

Solutions corresponding to the trajectories in Figure 12 cross the y-axis infinitely many times. The corresponding solutions are said to be ***oscillatory*** (Figure 13). It is shown in Section 6.2 that there is a number c_1 such that if $0 \leq c < c_1$ then all solutions of (27) are oscillatory, while if $c \geq c_1$ then no solutions of (27) have this property. (In fact, no solution not identically zero can have more than two zeros in this case.) Figure 14 shows a direction field and some integral curves of (28) in this case.

EXAMPLE 5

(The Damped Pendulum)

Now we return to the pendulum of Example 2. If we assume that some mechanism (for example, friction in the axle or atmospheric resistance) opposes the motion of the pendulum with a force proportional to its angular velocity, then Newton's second law of motion implies that

$$mLy'' = -cy' - mg \sin y, \tag{29}$$

where $c > 0$ is the damping constant. (Again, a minor note: the c in (26) actually corresponds to c/mL in this equation.) To plot a direction field for (29) we write its phase plane equivalent as

$$\frac{dv}{dy} = -\frac{c}{mL} - \frac{g}{Lv} \sin y.$$

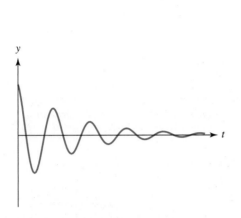

Figure 13 An oscillatory solution of
$my'' + cy' + ky = 0$

Figure 14 A typical direction field for
$my'' + cy' + ky = 0$ with $c > c_1$

Figure 15 shows trajectories of four solutions of (29), all satisfying $y(0) = 0$. For each $m = 0, 1, 2, 3$, imparting the initial velocity $v(0) = v_m$ causes the pendulum to make m complete revolutions and then settle into decaying oscillation about the stable equilibrium $\bar{y} = 2m\pi$. ∎

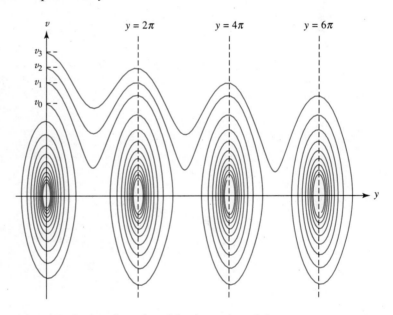

Figure 15 Four trajectories of the damped pendulum

4.4 EXERCISES

In Exercises 1–4 find the equations of the trajectories of the given undamped equation. Identify the equilibrium solutions, determine whether they are stable or unstable, and plot some trajectories.

Hint: *Use Eqn.* (8) *to obtain the equations of the trajectories.*

C **1.** $y'' + y^3 = 0$

C **2.** $y'' + y^2 = 0$

C **3.** $y'' + y|y| = 0$

C **4.** $y'' + ye^{-y} = 0$

In Exercises 5–8 find the equations of the trajectories of the given undamped equation. Identify the equilibrium solutions, determine whether they are stable or unstable, and find the equations of the separatrices (that is, the curves through the unstable equilibria). Plot the separatrices and some trajectories in each of the regions of the Poincaré plane determined by them.

Hint: *Use Eqn. (17) to determine the separatrices.*

C **5.** $y'' - y^3 + 4y = 0$

C **6.** $y'' + y^3 - 4y = 0$

C **7.** $y'' + y(y^2 - 1)(y^2 - 4) = 0$

C **8.** $y'' + y(y - 2)(y - 1)(y + 2) = 0$

*In Exercises 9–12 plot some trajectories of the given equation for various values (positive, negative, zero) of the parameter a. Find the equilibria of the equation and classify them as stable or unstable. Explain why the phase plane plots corresponding to positive and negative values of a differ so markedly. Can you think of a reason why zero deserves to be called the **critical value** of a?*

L **9.** $y'' + y^2 - a = 0$

L **10.** $y'' + y^3 - ay = 0$

L **11.** $y'' - y^3 + ay = 0$

L **12.** $y'' + y - ay^3 = 0$

In Exercises 13–18 plot trajectories of the given equation for c = 0 and small nonzero (positive and negative) values of c to observe the effects of damping.

L **13.** $y'' + cy' + y^3 = 0$

L **14.** $y'' + cy' - y = 0$

L **15.** $y'' + cy' + y^3 = 0$

L **16.** $y'' + cy' + y^2 = 0$

L **17.** $y'' + cy' + y|y| = 0$

L **18.** $y'' + y(y - 1) + cy = 0$

L **19.** The **van der Pol** [2] **equation**

$$y'' - \mu(1 - y^2)y' + y = 0, \tag{A}$$

where μ is a positive constant and y is electrical current (Section 6.3), arises in the study of an electrical circuit whose resistive properties depend upon the current. The damping term $-\mu(1 - y^2)y'$ works to reduce $|y|$ if $|y| < 1$ or to increase $|y|$ if $|y| > 1$. It can be shown that this has the following result: van der Pol's equation has exactly one closed trajectory, which is called a **limit cycle.** Trajectories inside the limit cycle spiral outward to it, while trajectories outside the limit cycle spiral inward to it (Figure 16). Use your favorite differential equations software to verify this for $\mu = 0.5, 1, 1.5, 2$. Use a grid with $-4 \le y \le 4$ and $-4 \le v \le 4$.

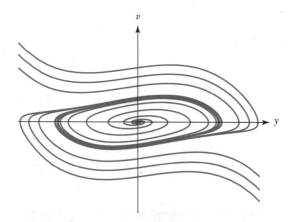

Figure 16   Trajectories of van der Pol's equation

L **20.** **Rayleigh's** [3] **equation**

$$y'' - \mu(1 - (y')^2/3)y' + y = 0$$

also has a limit cycle. Follow the directions of Exercise 19 for this equation.

---

[2] The Dutch mathematician Balthazar van der Pol (1889–1959) formulated the equation named after him in 1924, while studying the behavior of nonlinear electrical circuits.

[3] An English mathematician and physicist, Lord Rayleigh (1842–1919) made fundamental contributions to hydrodynamics, elasticity, and the theory of waves.

21. In Example 2 suppose that $y(0) = 0$ and $y'(0) = v_0$, where $0 < v_0 < v_c$.

    (a) Let T_1 be the time required for y to increase from zero to $y_{max} = 2 \sin^{-1}(v_0/v_c)$. Show that

$$\frac{dy}{dt} = \sqrt{v_0^2 - v_c^2 \, \sin^2 y/2}, \qquad 0 \le t < T_1. \tag{A}$$

    (b) Separate variables in (A) and show that

$$T_1 = \int_0^{y_{max}} \frac{du}{\sqrt{v_0^2 - v_c^2 \, \sin^2 u/2}} \tag{B}$$

    (c) Substitute $\sin u/2 = (v_0/v_c)\sin \theta$ in (B) to obtain

$$T_1 = 2 \int_0^{\pi/2} \frac{d\theta}{\sqrt{v_c^2 - v_0^2 \, \sin^2 \theta}}. \tag{C}$$

    (d) Conclude from symmetry that the time required for the point $(y(t), v(t))$ to traverse the trajectory

$$v^2 = v_0^2 - v_c^2 \, \sin^2 y/2$$

    is $T = 4T_1$, and that consequently $y(t + T) = y(t)$ and $v(t + T) = v(t)$; that is, the oscillation is periodic with period T.

    (e) Show that if $v_0 = v_c$ then the integral in (C) is improper and diverges to ∞. Conclude from this that $y(t) < \pi$ for all t and $\lim_{t\to\infty} y(t) = \pi$.

22. Give a direct definition of an unstable equilibrium of $y'' + p(y) = 0$.

23. Let p be continuous for all y and $p(0) = 0$. Suppose that there is a positive number ρ such that $p(y) > 0$ if $0 < y \le \rho$ and $p(y) < 0$ if $-\rho \le y < 0$. For $0 < r \le \rho$ let

$$\alpha(r) = \min\left\{\int_0^r p(x)\, dx, \ \int_{-r}^0 |p(x)|\, dx\right\} \qquad \text{and} \qquad \beta(r) = \max\left\{\int_0^r p(x)\, dx, \ \int_{-r}^0 |p(x)|\, dx\right\}.$$

Let y be the solution of the initial value problem

$$y'' + p(y) = 0, \qquad y(0) = v_0, \qquad y'(0) = v_0,$$

and define $c(y_0, v_0) = v_0^2 + 2 \int_0^{y_0} p(x)\, dx$.

    (a) Show that

$$0 < c(y_0, v_0) < v_0^2 + 2\beta(|y_0|) \qquad \text{if} \qquad 0 < |y_0| \le \rho.$$

    (b) Show that

$$v^2 + 2 \int_0^y p(x)\, dx = c(y_0, v_0), \qquad t > 0.$$

    (c) Conclude from **(b)** that if $c(y_0, v_0) < 2\alpha(r)$ then $|y| < r$, $t > 0$.

    (d) Given $\epsilon > 0$ let $\delta > 0$ be chosen so that

$$\delta^2 + 2\beta(\delta) < \max\{\epsilon^2/2, 2\alpha(\epsilon/\sqrt{2})\}.$$

    Show that if $\sqrt{y_0^2 + v_0^2} < \delta$ then $\sqrt{y^2 + v^2} < \epsilon$ for $t > 0$, which implies that $\bar{y} = 0$ is a stable equilibrium of $y'' + p(y) = 0$.

    (e) Now let p be continuous for all y and $p(\bar{y}) = 0$, where $\bar{y}$ is not necessarily zero. Suppose that there is a positive number ρ such that $p(y) > 0$ if $\bar{y} < y \le \bar{y} + \rho$ and $p(y) < 0$ if $\bar{y} - \rho \le y < \bar{y}$. Show that $\bar{y}$ is a stable equilibrium of $y'' + p(y) = 0$.

24. Let p be continuous for all y.

    (a) Suppose that $p(0) = 0$ and there is a positive number ρ such that $p(y) < 0$ if $0 < y \leq \rho$. Let ϵ be any number such that $0 < \epsilon < \rho$. Show that if y is the solution of the initial value problem

$$y'' + p(y) = 0, \qquad y(0) = y_0, \qquad y'(0) = 0$$

with $0 < y_0 < \epsilon$, then $y(t) \geq \epsilon$ for some $t > 0$. Conclude that $\overline{y} = 0$ is an unstable equilibrium of $y'' + p(y) = 0$.

    Hint: *Let $k = \min_{y_0 \leq x \leq \epsilon}(-p(x))$, which is positive. Show that if $y(t) < \epsilon$ for $0 \leq t < T$ then $kT^2 < 2(\epsilon - y_0)$.*

    (b) Now let $p(\overline{y}) = 0$, where $\overline{y}$ is not necessarily zero. Suppose that there is a positive number ρ such that $p(y) < 0$ if $\overline{y} < y \leq \overline{y} + \rho$. Show that $\overline{y}$ is an unstable equilibrium of $y'' + p(y) = 0$.

    (c) Modify your proofs of **(a)** and **(b)** to show that if there is a positive number ρ such that $p(y) > 0$ if $\overline{y} - \rho \leq y < \overline{y}$, then $\overline{y}$ is an unstable equilibrium of $y'' + p(y) = 0$.

4.5   Applications to Curves

ONE-PARAMETER FAMILIES OF CURVES

We begin with two examples of families of curves generated by varying a parameter over a set of real numbers.

EXAMPLE 1   For each value of the parameter c the equation

$$y - cx^2 = 0 \tag{1}$$

defines a curve in the xy-plane. If $c \neq 0$, the curve is a parabola through the origin, opening upward if $c > 0$ or downward if $c < 0$; if $c = 0$, the curve is the x-axis (Figure 1).

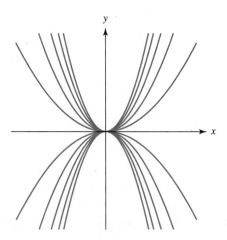

Figure 1

EXAMPLE 2   For each value of the parameter c the equation

$$y = x + c \tag{2}$$

defines a line with slope 1 (Figure 2).

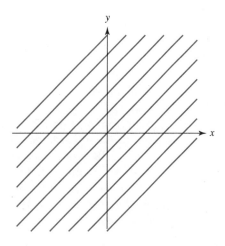

Figure 2

DEFINITION 4.5.1

An equation that can be written in the form

$$H(x, y, c) = 0 \tag{3}$$

is said to define a ***one-parameter family of curves*** if, for each value of c in some nonempty set of real numbers, the set of points (x, y) that satisfy (3) forms a curve in the xy-plane.

Equations (1) and (2) define one-parameter families of curves. (Although (2) is not in the form (3), it can be written in this form as $y - x - c = 0$.)

EXAMPLE 3   If $c > 0$ the graph of the equation

$$x^2 + y^2 - c = 0 \tag{4}$$

is a circle with center at $(0,0)$ and radius $\sqrt{c}$. If $c = 0$ the graph is the single point $(0,0)$. (We do not regard a single point as a curve.) If $c < 0$ the equation has no graph. Hence, (4) defines a one-parameter family of curves for positive values of c. This family consists of all circles centered at $(0,0)$ (Figure 3).

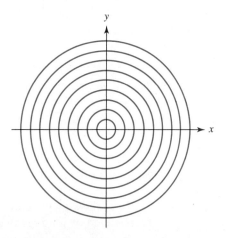

Figure 3

EXAMPLE 4   The equation

$$x^2 + y^2 + c^2 = 0$$

does not define a one-parameter family of curves, since no (x, y) satisfies the equation if $c \neq 0$, and only the single point $(0, 0)$ satisfies it if $c = 0$. ∎

Recall from Section 1.2 that the graphs of solutions of a differential equation are parts of the integral curves of the equation. Solving a first order differential equation usually produces a one-parameter family of integral curves of the equation. Here we are interested in the converse problem:

Given a one-parameter family of curves, is there a first order differential equation for which the members of the family are integral curves?

This suggests the following definition.

DEFINITION 4.5.2

> If every curve in a one-parameter family defined by the equation
>
> $$H(x, y, c) = 0 \tag{5}$$
>
> is an integral curve of the first order differential equation
>
> $$F(x, y, y') = 0, \tag{6}$$
>
> then (6) is said to be a ***differential equation for the family.***

To find a differential equation for a one-parameter family we differentiate its defining equation (5) implicitly with respect to x, to obtain

$$H_x(x, y, c) + H_y(x, y, c)y' = 0. \tag{7}$$

If this equation does not contain c, then it is a differential equation for the family. If it does contain c, then it may be possible to obtain a differential equation for the family by eliminating c between (5) and (7).

EXAMPLE 5   Find a differential equation for the family of curves defined by

$$y = cx^2. \tag{8}$$

Solution   Differentiating (8) with respect to x yields

$$y' = 2cx.$$

Therefore, $c = y'/2x$, and substituting this into (8) yields

$$y = \frac{xy'}{2}$$

as a differential equation for the family of curves defined by (8). The graph of any function of the form $y = cx^2$ is an integral curve of this equation. ∎

The following example shows that members of a given family of curves may be obtained by joining integral curves for more than one differential equation.

EXAMPLE 6 **(a)** Try to find a differential equation for the family of lines tangent to the parabola $y = x^2$.
(b) Find two tangent lines to the parabola $y = x^2$ that pass through $(2, 3)$, and find the points of tangency.

Solution **(a)** The equation of the line through a given point (x_0, y_0) with slope m is

$$y = y_0 + m(x - x_0). \tag{9}$$

If (x_0, y_0) is on the parabola, then $y_0 = x_0^2$ and the slope of the tangent line through (x_0, x_0^2) is $m = 2x_0$; hence, (9) becomes

$$y = x_0^2 + 2x_0(x - x_0),$$

or, equivalently,

$$y = -x_0^2 + 2x_0 x. \tag{10}$$

Here x_0 plays the role of the constant c in Definition 4.5.1; that is, varying x_0 over $(-\infty, \infty)$ produces the family of tangent lines to the parabola $y = x^2$.
Differentiating (10) with respect to x yields

$$y' = 2x_0.$$

We can express x_0 in terms of x and y by rewriting (10) as

$$x_0^2 - 2x_0 x + y = 0$$

and using the quadratic formula to obtain

$$x_0 = x \pm \sqrt{x^2 - y}. \tag{11}$$

We must choose the plus sign in (11) if $x < x_0$ and the minus sign if $x > x_0$; thus,

$$x_0 = \left(x + \sqrt{x^2 - y}\right) \qquad \text{if } x < x_0$$

and

$$x_0 = \left(x - \sqrt{x^2 - y}\right) \qquad \text{if } x > x_0.$$

Since $y' = 2x_0$, this implies that

$$y' = 2\left(x + \sqrt{x^2 - y}\right) \qquad \text{if } x < x_0 \tag{12}$$

and

$$y' = 2\left(x - \sqrt{x^2 - y}\right) \qquad \text{if } x > x_0. \tag{13}$$

Neither (12) nor (13) is a differential equation for the family of tangent lines to the parabola $y = x^2$. However, if each tangent line is regarded as consisting of two *tangent half-lines* joined at the point of tangency, then (12) is a differential equation for the family of tangent half-lines on which x is less than the abscissa of the point of tangency (Figure 4**(a)**), while (13) is a differential equation for the family of tangent half-lines on which x is greater than this abscissa (Figure 4**(b)**). Notice that the parabola $y = x^2$ is also an integral curve of both (12) and (13).

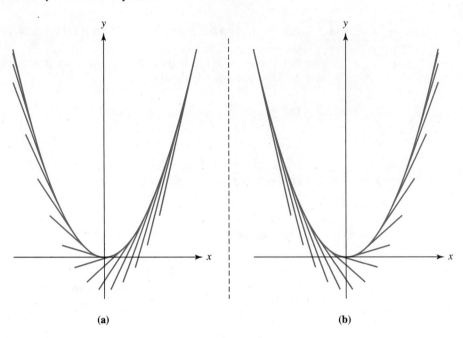

(a)   (b)

Figure 4

Solution   **(b)** From (10) the point $(x, y) = (2, 3)$ is on the tangent line through (x_0, x_0^2) if and only if

$$3 = -x_0^2 + 4x_0,$$

which is equivalent to

$$x_0^2 - 4x_0 + 3 = (x_0 - 3)(x_0 - 1) = 0.$$

Letting $x_0 = 3$ in (10) shows that $(2, 3)$ is on the line

$$y = -9 + 6x,$$

which is tangent to the parabola at $(x_0, x_0^2) = (3, 9)$, as shown in Figure 5.

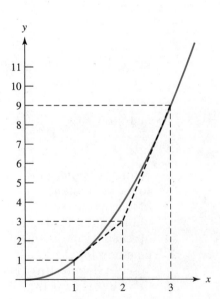

Figure 5

Letting $x_0 = 1$ in (10) shows that $(2,3)$ is on the line

$$y = -1 + 2x,$$

which is tangent to the parabola at $(x_0, x_0^2) = (1,1)$, as shown in Figure 5. ■

GEOMETRIC PROBLEMS

We now consider some geometric problems that can be solved by means of differential equations.

EXAMPLE 7   Find curves $y = y(x)$ such that every point $P = (x_0, y(x_0))$ on the curve is the midpoint of the line segment with endpoints on the coordinate axes and tangent to the curve at $(x_0, y(x_0))$ (Figure 6).

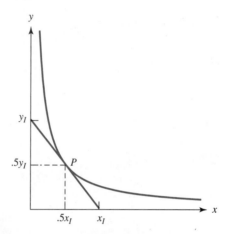

Figure 6

Solution   The equation of the line tangent to the curve at $(x_0, y(x_0))$ is

$$y = y(x_0) + y'(x_0)(x - x_0).$$

If we denote the x- and y-intercepts of the tangent line by x_I and y_I (Figure 6), then

$$0 = y(x_0) + y'(x_0)(x_I - x_0) \tag{14}$$

and

$$y_I = y(x_0) - y'(x_0)x_0. \tag{15}$$

From Figure 6, P is the midpoint of the line segment connecting $(x_I, 0)$ and $(0, y_I)$ if and only if $x_I = 2x_0$ and $y_I = 2y(x_0)$. Substituting the first of these conditions into (14) or the second into (15) yields

$$y(x_0) + y'(x_0)x_0 = 0.$$

Since x_0 is arbitrary we drop the subscript and conclude that $y = y(x)$ satisfies

$$y + xy' = 0,$$

which can be rewritten as

$$(xy)' = 0.$$

Integrating yields $xy = c$, or

$$y = \frac{c}{x}.$$

If $c = 0$ this curve is the line $y = 0$, which does not satisfy the geometric requirements imposed by the problem; thus, $c \neq 0$, and the solutions define a family of hyperbolas (Figure 7).

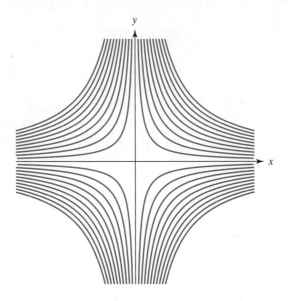

Figure 7

EXAMPLE 8   Find curves $y = y(x)$ such that the tangent line to the curve at any point $(x_0, y(x_0))$ intersects the x-axis at $(x_0^2, 0)$. Figure 8 illustrates the situation in the case where the curve is in the first quadrant and $0 < x < 1$.

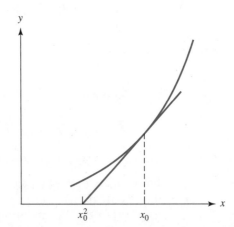

Figure 8

Solution   The equation of the line tangent to the curve at $(x_0, y(x_0))$ is

$$y = y(x_0) + y'(x_0)(x - x_0).$$

Since $(x_0^2, 0)$ is on the tangent line,

$$0 = y(x_0) + y'(x_0)(x_0^2 - x_0).$$

Since x_0 is arbitrary we drop the subscript and conclude that $y = y(x)$ satisfies

$$y + y'(x^2 - x) = 0.$$

Therefore

$$\frac{y'}{y} = -\frac{1}{x^2 - x} = -\frac{1}{x(x-1)} = \frac{1}{x} - \frac{1}{x-1},$$

so

$$\ln |y| = \ln |x| - \ln |x - 1| + k = \ln \left| \frac{x}{x-1} \right| + k,$$

and

$$y = \frac{cx}{x-1}.$$

If $c = 0$ the graph of this function is the x-axis; if $c \neq 0$ it is a hyperbola with vertical asymptote $x = 1$ and horizontal asymptote $y = c$. Figure 9 shows graphs for $c \neq 0$. ∎

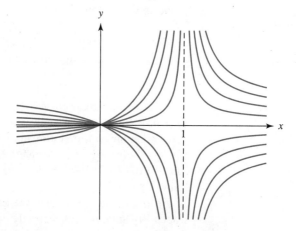

Figure 9

ORTHOGONAL TRAJECTORIES

Two curves C_1 and C_2 are said to be ***orthogonal*** at a point of intersection (x_0, y_0) if they have perpendicular tangents at (x_0, y_0) (Figure 10). A curve is said to be an ***orthogonal trajectory*** of a given family of curves if it is orthogonal to every curve in the family. For example, every line through the origin is an orthogonal trajectory of the family of circles centered at the origin. Conversely, any such circle is an orthogonal trajectory of the family of lines through the origin (Figure 11).

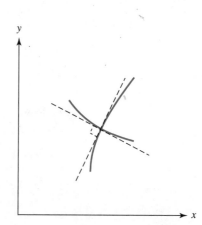

Figure 10 Curves orthogonal at a
point of intersection

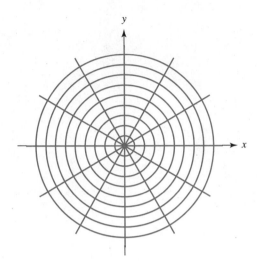

Figure 11   Orthogonal families of circles and lines

Orthogonal trajectories occur in many physical applications. For example, if $u = u(x, y)$ is the temperature at a point (x, y), then the curves defined by

$$u(x, y) = c \tag{16}$$

are called *isothermal* curves. The orthogonal trajectories of this family are called *heat-flow lines,* because at any given point the direction of maximum heat flow is perpendicular to the isothermal through the point. If u represents the potential energy of an object moving under a force that depends upon (x, y) then the curves (16) are called *equipotentials,* and the orthogonal trajectories are called *lines of force.*

From analytic geometry we know that two nonvertical lines L_1 and L_2 with slopes m_1 and m_2, respectively, are perpendicular if and only if $m_2 = -1/m_1$; therefore, the integral curves of the differential equation

$$y' = -\frac{1}{f(x, y)}$$

are orthogonal trajectories of the integral curves of the differential equation

$$y' = f(x, y),$$

because at any point (x_0, y_0) where curves from the two families intersect, the slopes of the respective tangent lines are

$$m_1 = f(x_0, y_0) \qquad \text{and} \qquad m_2 = -\frac{1}{f(x_0, y_0)}.$$

This suggests the following method for finding orthogonal trajectories of a family of integral curves of a first order equation.

Finding Orthogonal Trajectories

Step 1. Find a differential equation

$$y' = f(x, y)$$

for the given family.
Step 2. Solve the differential equation

$$y' = -\frac{1}{f(x, y)}$$

to find the orthogonal trajectories.

EXAMPLE 9 Find the orthogonal trajectories of the family of circles

$$x^2 + y^2 = c^2 \qquad (c > 0). \tag{17}$$

Solution To find a differential equation for the family of circles, we differentiate (17) implicitly with respect to x to obtain

$$2x + 2yy' = 0,$$

or

$$y' = -\frac{x}{y}.$$

Therefore the integral curves of

$$y' = \frac{y}{x}$$

are orthogonal trajectories of the given family. We leave it to you to verify that the general solution of this equation is

$$y = kx,$$

where k is an arbitrary constant. This is the equation of a nonvertical line through $(0,0)$. The y-axis is also an orthogonal trajectory of the given family. Therefore, every line through the origin is an orthogonal trajectory of the given family (17) (Figure 11). This is consistent with the theorem of plane geometry which states that a diameter of a circle and a tangent line to the circle at the end of the diameter are perpendicular.

EXAMPLE 10 Find the orthogonal trajectories of the family of hyperbolas

$$xy = c \qquad (c \neq 0) \tag{18}$$

(Figure 7).

Solution Differentiating (18) implicitly with respect to x yields

$$y + xy' = 0,$$

or

$$y' = -\frac{y}{x};$$

thus, the integral curves of

$$y' = \frac{x}{y}$$

are orthogonal trajectories of the given family. Separating variables, we obtain

$$y'y = x$$

and integrating yields

$$y^2 - x^2 = k,$$

which is the equation of a hyperbola if $k \neq 0$, or of the lines $y = x$ and $y = -x$ if $k = 0$ (Figure 12).

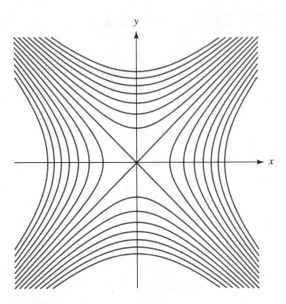

Figure 12 Orthogonal trajectories of the hyperbolas $xy = c$

EXAMPLE 11 Find the orthogonal trajectories of the family of circles defined by

$$(x - c)^2 + y^2 = c^2 \qquad (c \neq 0). \tag{19}$$

These circles are centered on the x-axis and tangent to the y-axis (Figure 13**(a)**).

Solution Expanding the left side of (19) yields

$$x^2 - 2cx + y^2 = 0, \tag{20}$$

and differentiating this implicitly with respect to x yields

$$2(x - c) + 2yy' = 0. \tag{21}$$

From (20),

$$c = \frac{x^2 + y^2}{2x},$$

so

$$x - c = x - \frac{x^2 + y^2}{2x} = \frac{x^2 - y^2}{2x}.$$

Substituting this into (21) and solving for y' yields

$$y' = \frac{y^2 - x^2}{2xy}. \tag{22}$$

The curves defined by (19) are integral curves of (22), and the integral curves of

$$y' = \frac{2xy}{x^2 - y^2}$$

are orthogonal trajectories of the family (19). This is a homogeneous nonlinear equation, which we studied in Section 2.4. Substituting $y = ux$ yields

$$u'x + u = \frac{2x(ux)}{x^2 - (ux)^2} = \frac{2u}{1 - u^2},$$

so

$$u'x = \frac{2u}{1 - u^2} - u = \frac{u(u^2 + 1)}{1 - u^2}.$$

Separating variables yields

$$\frac{1 - u^2}{u(u^2 + 1)} u' = \frac{1}{x},$$

or, equivalently,

$$\left[\frac{1}{u} - \frac{2u}{u^2 + 1}\right] u' = \frac{1}{x}.$$

Therefore

$$\ln |u| - \ln(u^2 + 1) = \ln |x| + k.$$

By substituting $u = y/x$ we see that

$$\ln |y| - \ln |x| - \ln(x^2 + y^2) + \ln(x^2) = \ln |x| + k,$$

which, since $\ln(x^2) = 2 \ln |x|$, is equivalent to

$$\ln |y| - \ln(x^2 + y^2) = k,$$

or

$$|y| = e^k(x^2 + y^2).$$

To see what these curves are we rewrite this equation as

$$x^2 + |y|^2 - e^{-k}|y| = 0$$

and complete the square to obtain

$$x^2 + (|y| - e^{-k}/2)^2 = (e^{-k}/2)^2.$$

This can be rewritten as

$$x^2 + (y - h)^2 = h^2,$$

where

$$h = \begin{cases} \dfrac{e^{-k}}{2} & \text{if } y \geq 0, \\[2mm] -\dfrac{e^{-k}}{2} & \text{if } y \leq 0. \end{cases}$$

Thus, the orthogonal trajectories are circles centered on the y-axis and tangent to the x-axis (Figure 13**(b)**). The circles for which $h > 0$ are above the x-axis, while those for which $h < 0$ are below. ∎

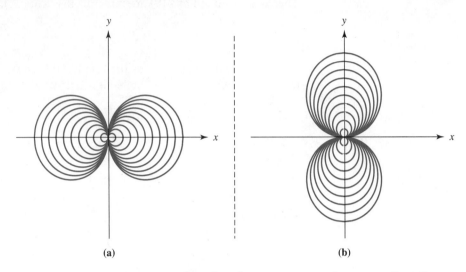

(a)   **(b)**

Figure 13   **(a)** The circles $(x - c)^2 + y^2 = c^2$; **(b)** The circles $x^2 + (y - h)^2 = h^2$

4.5 EXERCISES

In Exercises 1–8 find a first order differential equation for the given family of curves.

1. $y(x^2 + y^2) = c$ **2.** $e^{xy} = cy$

3. $\ln |xy| = c(x^2 + y^2)$ **4.** $y = x^{1/2} + cx$

5. $y = e^{x^2} + ce^{-x^2}$ **6.** $y = x^3 + \dfrac{c}{x}$

7. $y = \sin x + ce^x$ **8.** $y = e^x + c(1 + x^2)$

9. Show that the family of circles

$$(x - x_0)^2 + y^2 = 1, \qquad -\infty < x_0 < \infty,$$

can be obtained by joining integral curves of two first order differential equations. More specifically, find differential equations for the families of semicircles

$$(x - x_0)^2 + y^2 = 1, \qquad x_0 < x < x_0 + 1, \qquad -\infty < x_0 < \infty,$$
$$(x - x_0)^2 + y^2 = 1, \qquad x_0 - 1 < x < x_0, \qquad -\infty < x_0 < \infty.$$

10. Suppose that f and g are differentiable for all x. Find a differential equation for the family of functions $y = f + cg$ (c = constant).

In Exercises 11–13 find a first order differential equation for the given family of curves.

11. Lines through a given point (x_0, y_0).

12. Circles through $(-1, 0)$ and $(1, 0)$.

13. Circles through $(0, 0)$ and $(0, 2)$.

14. Use the method of Example 6**(a)** to find the equations of lines through the given points tangent to the parabola $y = x^2$. Also, find the points of tangency.

    (a) $(5, 9)$ **(b)** $(6, 11)$

    (c) $(-6, 20)$ **(d)** $(-3, 5)$

15. (a) Show that the equation of the line tangent to the circle

$$x^2 + y^2 = 1 \tag{A}$$

at a point (x_0, y_0) on the circle is

$$y = \frac{1 - x_0 x}{y_0} \quad \text{if} \quad x_0 \neq \pm 1. \tag{B}$$

(b) Show that if y' is the slope of a nonvertical tangent line to the circle (A) and (x, y) is a point on the tangent line, then

$$(y')^2(x^2 - 1) - 2xyy' + y^2 - 1 = 0. \tag{C}$$

(c) Show that the segment of the tangent line (B) on which $(x - x_0)/y_0 > 0$ is an integral curve of the differential equation

$$y' = \frac{xy - \sqrt{x^2 + y^2 - 1}}{x^2 - 1}, \tag{D}$$

while the segment on which $(x - x_0)/y_0 < 0$ is an integral curve of the differential equation

$$y' = \frac{xy + \sqrt{x^2 + y^2 - 1}}{x^2 - 1}. \tag{E}$$

Hint: Use the quadratic formula to solve (C) for y'. Then substitute (B) for y and choose the $\pm$ sign in the quadratic formula so that the resulting expression for y' reduces to the known slope $y' = -x_0/y_0$.

(d) Show that the upper and lower semicircles of (A) are also integral curves of (D) and (E).

(e) Find the equations of two lines through $(5, 5)$ tangent to the circle (A), and find the points of tangency.

16. (a) Show that the equation of the line tangent to the parabola

$$x = y^2 \tag{A}$$

at a point $(x_0, y_0) \neq (0,0)$ on the parabola is

$$y = \frac{y_0}{2} + \frac{x}{2y_0}. \tag{B}$$

(b) Show that if y' is the slope of a nonvertical tangent line to the parabola (A) and (x, y) is a point on the tangent line then

$$4x^2(y')^2 - 4xyy' + x = 0. \tag{C}$$

(c) Show that the segment of the tangent line defined in **(a)** on which $x > x_0$ is an integral curve of the differential equation

$$y' = \frac{y + \sqrt{y^2 - x}}{2x}, \tag{D}$$

while the segment on which $x < x_0$ is an integral curve of the differential equation

$$y' = \frac{y - \sqrt{y^2 - x}}{2x}. \tag{E}$$

Hint: Use the quadratic formula to solve (C) for y'. Then substitute (B) for y and choose the $\pm$ sign in the quadratic formula so that the resulting expression for y' reduces to the known slope $y' = 1/2y_0$.

(d) Show that the upper and lower halves of the parabola (A), given by $y = \sqrt{x}$ and $y = -\sqrt{x}$ for $x > 0$, are also integral curves of (D) and (E).

17. Use the results of Exercise 16 to find the equations of two lines tangent to the parabola $x = y^2$ and passing through the given point. Also find the points of tangency.

(a) $(-5, 2)$ **(b)** $(-4, 0)$

(c) $(7, 4)$ **(d)** $(5, -3)$

18. Find a curve $y = y(x)$ through $(1, 2)$ such that the tangent to the curve at any point $(x_0, y(x_0))$ intersects the x-axis at $x_I = x_0/2$.

19. Find all curves $y = y(x)$ such that the tangent to the curve at any point $(x_0, y(x_0))$ intersects the x-axis at $x_I = x_0^3$.

20. Find all curves $y = y(x)$ such that the tangent to the curve at any point passes through a given point (x_1, y_1).

21. Find a curve $y = y(x)$ through $(1, -1)$ such that the tangent to the curve at any point $(x_0, y(x_0))$ intersects the y-axis at $y_I = x_0^3$.

22. Find all curves $y = y(x)$ such that the tangent to the curve at any point $(x_0, y(x_0))$ intersects the y-axis at $y_I = x_0$.

23. Find a curve $y = y(x)$ through $(0, 2)$ such that the normal to the curve at any point $(x_0, y(x_0))$ intersects the x-axis at $x_I = x_0 + 1$.

24. Find a curve $y = y(x)$ through $(2, 1)$ such that the normal to the curve at any point $(x_0, y(x_0))$ intersects the y-axis at $y_I = 2y(x_0)$.

In Exercises 25–29 find the orthogonal trajectories of the given family of curves.

25. $x^2 + 2y^2 = c^2$ **26.** $x^2 + 4xy + y^2 = c$

27. $y = ce^{2x}$ **28.** $xye^{x^2} = c$

29. $y = \dfrac{ce^x}{x}$

30. Find a curve through $(-1, 3)$ orthogonal to every parabola of the form

$$y = 1 + cx^2$$

that it intersects. Which of these parabolas does the desired curve intersect?

31. Show that the orthogonal trajectories of

$$x^2 + 2axy + y^2 = c$$

satisfy

$$|y - x|^{a+1}|y + x|^{a-1} = k.$$

32. If lines L and L_1 intersect at (x_0, y_0) and α is the smallest angle through which L must be rotated counterclockwise about (x_0, y_0) to bring it into coincidence with L_1, we say that **α is the angle from L to L_1**; thus, $0 \le \alpha < \pi$. If L and L_1 are tangents to curves C and C_1, respectively, which intersect at (x_0, y_0), we say that C_1 **intersects C at the angle α.** Use the identity

$$\tan(A + B) = \frac{\tan A + \tan B}{1 - \tan A \tan B}$$

to show that if C and C_1 are intersecting integral curves of

$$y' = f(x, y) \quad \text{and} \quad y' = \frac{f(x, y) + \tan \alpha}{1 - f(x, y)\tan \alpha} \quad \left(\alpha \ne \frac{\pi}{2}\right),$$

respectively, then C_1 intersects C at the angle α.

33. Use the result of Exercise 32 to find a family of curves that intersect every nonvertical line through the origin at the angle $\alpha = \pi/4$.

34. Use the result of Exercise 32 to find a family of curves that intersect every circle centered at the origin at a given angle $\alpha \ne \pi/2$.

5

Linear Second Order Equations

IN THIS CHAPTER we study a particularly important class of second order equations. Because of their many applications in science and engineering, second order differential equation have historically been the most thoroughly studied class of differential equations. Research on the theory of second order differential equations continues to the present day. This chapter is devoted to second order equations that can be written in the form

$$P_0(x)y'' + P_1(x)y' + P_2(x)y = F(x). \tag{1}$$

Such equations are said to be **linear**. As in the case of first order linear equations, (1) is said to be **homogeneous** if $F \equiv 0$, or **nonhomogeneous** if $F \not\equiv 0$.

SECTION 5.1 is devoted to the theory of homogeneous linear equations.

SECTION 5.2 deals with homogeneous equations of the special form

$$ay'' + by' + cy = 0,$$

in which a, b, and c are constant ($a \neq 0$). When you've completed this section you'll know everything there is to know about solving such equations.

SECTION 5.3 presents the theory of nonhomogeneous linear equations.

SECTION 5.4 AND 5.5 present the **method of undetermined coefficients,** which can be used to solve nonhomogeneous equations of the form

$$ay'' + by' + cy = F(x),$$

where a, b, and c are constants and F has a special form that is still sufficiently general to occur in many applications. In this section we make extensive use of the idea of variation of parameters, introduced in Chapter 2.

SECTION 5.6 deals with **reduction of order,** a technique based on the idea of variation of parameters, which enables us to find the general solution of a nonhomogeneous linear second order equation provided that we know one nontrivial (not identically zero) solution of the associated homogeneous equation.

SECTION 5.7 deals with the method traditionally called **variation of parameters,** which enables us to find the general solution of a nonhomogeneous linear second order equation provided that we know two nontrivial solutions (with nonconstant ratio) of the associated homogeneous equation.

5.1   Homogeneous Linear Equations

A second order differential equation is said to be ***linear*** if it can be written as

$$y'' + p(x)y' + q(x)y = f(x). \tag{1}$$

We call the function f on the right a ***forcing function,*** since in physical applications it is often related to a force acting on some system modeled by the differential equation. We say that (1) is ***homogeneous*** if $f \equiv 0$ or ***nonhomogeneous*** if $f \not\equiv 0$. Since these definitions are like the corresponding definitions in Section 2.1 for the linear first order equation

$$y' + p(x)y = f(x), \tag{2}$$

it is natural to expect similarities between methods of solving (1) and (2). However, solving (1) is more difficult than solving (2). For example, while Theorem 2.1.1 gives a formula for the general solution of (2) in the case where $f \equiv 0$ and Theorem 2.1.2 gives a formula for the case where $f \not\equiv 0$, there are no formulas for the general solution of (1) in either case. Therefore we must be content to solve linear second order equations of special forms.

In Section 2.1 we considered the homogeneous equation $y' + p(x)y = 0$ first, and then used a nontrivial solution of this equation to find the general solution of the nonhomogeneous equation $y' + p(x)y = f(x)$. Although the progression from the homogeneous to the nonhomogeneous case is not that simple for the linear second order equation, it is still necessary to solve the homogeneous equation

$$y'' + p(x)y' + q(x)y = 0 \tag{3}$$

in order to solve the nonhomogeneous equation (1). This section is devoted to (3).

The following theorem gives sufficient conditions for existence and uniqueness of solutions of initial value problems for (3). We omit the proof.

THEOREM 5.1.1

> *Suppose that p and q are continuous on an open interval (a,b), let x_0 be any point in (a,b), and let k_0 and k_1 be arbitrary real numbers. Then the initial value problem*
>
> $$y'' + p(x)y' + q(x)y = 0, \qquad y(x_0) = k_0, \qquad y'(x_0) = k_1$$
>
> *has a unique solution on (a,b).*

Since $y \equiv 0$ is obviously a solution of (3) we call it the ***trivial*** solution. Any other solution is ***nontrivial.*** Under the assumptions of Theorem 5.1.1, the only solution of the initial value problem

$$y'' + p(x)y' + q(x)y = 0, \qquad y(x_0) = 0, \qquad y'(x_0) = 0$$

on (a,b) is the trivial solution (Exercise 24).

The next three examples illustrate concepts that we will develop later in this section. You shouldn't be concerned with how to *find* the given solutions of the equations in these examples. This will be explained in later sections.

EXAMPLE 1   The coefficients of y' and y in

$$y'' - y = 0 \tag{4}$$

are the constant functions $p \equiv 0$ and $q \equiv -1$, which are continuous on $(-\infty, \infty)$. Therefore Theorem 5.1.1 implies that every initial value problem for (4) has a unique solution on $(-\infty, \infty)$.

(a) Verify that $y_1 = e^x$ and $y_2 = e^{-x}$ are solutions of (4) on $(-\infty, \infty)$.
(b) Verify that if c_1 and c_2 are arbitrary constants then $y = c_1 e^x + c_2 e^{-x}$ is a solution of (4) on $(-\infty, \infty)$.
(c) Solve the initial value problem

$$y'' - y = 0, \qquad y(0) = 1, \qquad y'(0) = 3. \tag{5}$$

Solution   **(a)** If $y_1 = e^x$ then $y_1' = e^x$ and $y_1'' = e^x = y_1$, so $y_1'' - y_1 = 0$. If $y_2 = e^{-x}$ then $y_2' = -e^{-x}$ and $y_2'' = e^{-x} = y_2$, so $y_2'' - y_2 = 0$.

Solution   **(b)** If

$$y = c_1 e^x + c_2 e^{-x} \tag{6}$$

then

$$y' = c_1 e^x - c_2 e^{-x} \tag{7}$$

and

$$y'' = c_1 e^x + c_2 e^{-x},$$

so

$$y'' - y = (c_1 e^x + c_2 e^{-x}) - (c_1 e^x + c_2 e^{-x})$$
$$= c_1(e^x - e^x) + c_2(e^{-x} - e^{-x}) = 0$$

for all x. Therefore $y = c_1 e^x + c_2 e^{-x}$ is a solution of (4) on $(-\infty, \infty)$.

Solution   **(c)** We can solve (5) by choosing c_1 and c_2 in (6) so that $y(0) = 1$ and $y'(0) = 3$. Setting $x = 0$ in (6) and (7) shows that this is equivalent to

$$c_1 + c_2 = 1$$
$$c_1 - c_2 = 3.$$

Solving these equations yields $c_1 = 2$ and $c_2 = -1$. Therefore $y = 2e^x - e^{-x}$ is the unique solution of (5) on $(-\infty, \infty)$.

EXAMPLE 2   Let ω be a positive constant. The coefficients of y' and y in

$$y'' + \omega^2 y = 0 \tag{8}$$

are the constant functions $p \equiv 0$ and $q \equiv \omega^2$, which are continuous on $(-\infty, \infty)$. Therefore Theorem 5.1.1 implies that every initial value problem for (8) has a unique solution on $(-\infty, \infty)$.

(a) Verify that $y_1 = \cos \omega x$ and $y_2 = \sin \omega x$ are solutions of (8) on $(-\infty, \infty)$.
(b) Verify that if c_1 and c_2 are arbitrary constants then $y = c_1 \cos \omega x + c_2 \sin \omega x$ is a solution of (8) on $(-\infty, \infty)$.
(c) Solve the initial value problem

$$y'' + \omega^2 y = 0, \qquad y(0) = 1, \qquad y'(0) = 3. \tag{9}$$

Solution   **(a)** If $y_1 = \cos \omega x$ then $y_1' = -\omega \sin \omega x$ and $y_1'' = -\omega^2 \cos \omega x = -\omega^2 y_1$, so $y_1'' + \omega^2 y_1 = 0$. If $y_2 = \sin \omega x$ then $y_2' = \omega \cos \omega x$ and $y_2'' = -\omega^2 \sin \omega x = -\omega^2 y_2$, so $y_2'' + \omega^2 y_2 = 0$.

Solution   **(b)** If

$$y = c_1 \cos \omega x + c_2 \sin \omega x \tag{10}$$

then

$$y' = \omega(-c_1 \sin \omega x + c_2 \cos \omega x) \tag{11}$$

and

$$y'' = -\omega^2(c_1 \cos \omega x + c_2 \sin \omega x),$$

so

$$y'' + \omega^2 y = -\omega^2(c_1 \cos \omega x + c_2 \sin \omega x) + \omega^2(c_1 \cos \omega x + c_2 \sin \omega x)$$
$$= c_1\omega^2(-\cos \omega x + \cos \omega x) + c_2\omega^2(-\sin \omega x + \sin \omega x) = 0$$

for all x. Therefore $y = c_1 \cos \omega x + c_2 \sin \omega x$ is a solution of (8) on $(-\infty,\infty)$.

Solution   **(c)** We can solve (9) by choosing c_1 and c_2 in (10) so that $y(0) = 1$ and $y'(0) = 3$. Setting $x = 0$ in (10) and (11) shows that $c_1 = 1$ and $c_2 = 3/\omega$. Therefore

$$y = \cos \omega x + \frac{3}{\omega} \sin \omega x$$

is the unique solution of (9) on $(-\infty,\infty)$.   ∎

Theorem 5.1.1 implies that if k_0 and k_1 are arbitrary real numbers then the initial value problem

$$P_0(x)y'' + P_1(x)y' + P_2(x)y = 0, \qquad y(x_0) = k_0, \qquad y'(x_0) = k_1 \tag{12}$$

has a unique solution on an interval (a,b) containing x_0, provided that P_0, P_1, and P_2 are continuous and P_0 has no zeros on (a,b). To see this, we rewrite the differential equation in (12) as

$$y'' + \frac{P_1(x)}{P_0(x)}y' + \frac{P_2(x)}{P_0(x)}y = 0$$

and apply Theorem 5.1.1 with $p = P_1/P_0$ and $q = P_2/P_0$.

EXAMPLE 3   The equation

$$x^2 y'' + xy' - 4y = 0 \tag{13}$$

has the form of the differential equation in (12), with $P_0(x) = x^2$, $P_1(x) = x$, and $P_2(x) = -4$, which are all continuous on $(-\infty,\infty)$. However, since $P(0) = 0$ we must consider solutions of (13) on $(-\infty,0)$ and $(0,\infty)$. Since P_0 has no zeros on these intervals, Theorem 5.1.1 implies that the initial value problem

$$x^2 y'' + xy' - 4y = 0, \qquad y(x_0) = k_0, \qquad y'(x_0) = k_1$$

has a unique solution on $(0,\infty)$ if $x_0 > 0$, or on $(-\infty,0)$ if $x_0 < 0$.

(a) Verify that $y_1 = x^2$ is a solution of (13) on $(-\infty,\infty)$ and $y_2 = 1/x^2$ is a solution of (13) on $(-\infty,0)$ and $(0,\infty)$.

(b) Verify that if c_1 and c_2 are any constants then $y = c_1 x^2 + c_2/x^2$ is a solution of (13) on $(-\infty, 0)$ and $(0, \infty)$.

(c) Solve the initial value problem

$$x^2 y'' + xy' - 4y = 0, \qquad y(1) = 2, \qquad y'(1) = 0. \tag{14}$$

(d) Solve the initial value problem

$$x^2 y'' + xy' - 4y = 0, \qquad y(-1) = 2, \qquad y'(-1) = 0. \tag{15}$$

Solution **(a)** If $y_1 = x^2$ then $y_1' = 2x$ and $y_1'' = 2$, so

$$x^2 y_1'' + xy_1' - 4y_1 = x^2(2) + x(2x) - 4x^2 = 0$$

for x in $(-\infty, \infty)$. If $y_2 = 1/x^2$ then $y_2' = -2/x^3$ and $y_2'' = 6/x^4$, so

$$x^2 y_2'' + xy_2' - 4y_2 = x^2 \left(\frac{6}{x^4} \right) - x \left(\frac{2}{x^3} \right) - \left(\frac{4}{x^2} \right) = 0$$

for x in $(-\infty, 0)$ or $(0, \infty)$.

Solution **(b)** If

$$y = c_1 x^2 + \frac{c_2}{x^2} \tag{16}$$

then

$$y' = 2c_1 x - \frac{2c_2}{x^3} \tag{17}$$

and

$$y'' = 2c_1 + \frac{6c_2}{x^4},$$

so

$$x^2 y'' + xy' - 4y = x^2 \left(2c_1 + \frac{6c_2}{x^4} \right) + x \left(2c_1 x - \frac{2c_2}{x^3} \right) - 4 \left(c_1 x^2 + \frac{c_2}{x^2} \right)$$

$$= c_1 (2x^2 + 2x^2 - 4x^2) + c_2 \left(\frac{6}{x^2} - \frac{2}{x^2} - \frac{4}{x^2} \right)$$

$$= c_1 \cdot 0 + c_2 \cdot 0 = 0$$

for x in $(-\infty, 0)$ or $(0, \infty)$.

Solution **(c)** We can solve (14) by choosing c_1 and c_2 in (16) so that $y(1) = 2$ and $y'(1) = 0$. Setting $x = 1$ in (16) and (17) shows that this is equivalent to

$$c_1 + c_2 = 2$$
$$2c_1 - 2c_2 = 0.$$

Solving these equations yields $c_1 = 1$ and $c_2 = 1$. Therefore $y = x^2 + 1/x^2$ is the unique solution of (14) on $(0, \infty)$.

Solution **(d)** We can solve (15) by choosing c_1 and c_2 in (16) so that $y(-1) = 2$ and $y'(-1) = 0$. Setting $x = -1$ in (16) and (17) shows that this is equivalent to

$$c_1 + c_2 = 2$$
$$-2c_1 + 2c_2 = 0.$$

Solving these equations yields $c_1 = 1$ and $c_2 = 1$. Therefore $y = x^2 + 1/x^2$ is the unique solution of (15) on $(-\infty, 0)$. ∎

Although the *formulas* for the solutions of (14) and (15) are both $y = x^2 + 1/x^2$, you should not conclude that these two initial value problems have the same solution. Remember that a solution of an initial value problem is defined *on an interval containing the initial point;* therefore, the solution of (14) is $y = x^2 + 1/x^2$ *on the interval* $(0,\infty)$, which contains the initial point $x_0 = 1$, while the solution of (15) is $y = x^2 + 1/x^2$ *on the interval* $(-\infty, 0)$, which contains the initial point $x_0 = -1$.

THE GENERAL SOLUTION OF A HOMOGENEOUS LINEAR SECOND ORDER EQUATION

If y_1 and y_2 are defined on an interval (a,b) and c_1 and c_2 are constants, then

$$y = c_1 y_1 + c_2 y_2$$

is a **linear combination of y_1 and y_2.** For example, $y = 2 \cos x + 7 \sin x$ is a linear combination of $y_1 = \cos x$ and $y_2 = \sin x$, with $c_1 = 2$ and $c_2 = 7$.

The following theorem states a fact that we have already verified in Examples 1, 2, and 3.

THEORM 5.1.2

If y_1 and y_2 are solutions of the homogeneous equation

$$y'' + p(x)y' + q(x)y = 0 \tag{18}$$

on (a,b) then any linear combination

$$y = c_1 y_1 + c_2 y_2 \tag{19}$$

of y_1 and y_2 is also a solution of (18) on (a,b).

PROOF. If

$$y = c_1 y_1 + c_2 y_2$$

then

$$y' = c_1 y_1' + c_2 y_2' \qquad \text{and} \qquad y'' = c_1 y_1'' + c_2 y_2''.$$

Therefore

$$
\begin{aligned}
y'' + p(x)y' + q(x)y &= (c_1 y_1'' + c_2 y_2'') + p(x)(c_1 y_1' + c_2 y_2') + q(x)(c_1 y_1 + c_2 y_2) \\
&= c_1(y_1'' + p(x)y_1' + q(x)y_1) + c_2(y_2'' + p(x)y_2' + q(x)y_2) \\
&= c_1 \cdot 0 + c_2 \cdot 0 = 0,
\end{aligned}
$$

since y_1 and y_2 are solutions of (18). ☐

We say that $\{y_1, y_2\}$ is a **fundamental set of solutions of (18) on (a,b)** if every solution of (18) on (a,b) can be written as a linear combination of y_1 and y_2, as in (19). In this case we say that (19) is **general solution of (18) on (a,b).**

LINEAR INDEPENDENCE

We need a way to determine whether a given set $\{y_1, y_2\}$ of solutions of (18) is a fundamental set. The following definition will enable us to state necessary and sufficient conditions for this.

We say that two functions y_1 and y_2 defined on an interval (a, b) are ***linearly independent on*** (a, b) if neither is a constant multiple of the other on (a, b). (In particular, this means that neither can be the trivial solution of (18), since, for example, if $y_1 \equiv 0$ then we could write $y_1 = 0y_2$.) We will also say that the set $\{y_1, y_2\}$ is ***linearly independent on*** (a, b).

THEOREM 5.1.3

Suppose that p and q are continuous on (a, b). Then a set $\{y_1, y_2\}$ of solutions of

$$y'' + p(x)y' + q(x)y = 0 \tag{20}$$

on (a, b) is a fundamental set if and only if $\{y_1, y_2\}$ is linearly independent on (a, b).

We will present the proof of Theorem 5.1.3 in steps worth regarding as theorems in their own right. However, let's first interpret Theorem 5.1.3 in terms of Examples 1, 2, and 3.

EXAMPLE 4   (a) Since $e^x/e^{-x} = e^{2x}$ is nonconstant, Theorem 5.1.3 implies that $y = c_1e^x + c_2e^{-x}$ is the general solution of $y'' - y = 0$ on $(-\infty, \infty)$.
(b) Since $\cos \omega x/\sin \omega x = \cot \omega x$ is nonconstant, Theorem 5.1.3 implies that $y = c_1 \cos \omega x + c_2 \sin \omega x$ is the general solution of $y'' + \omega^2 y = 0$ on $(-\infty, \infty)$.
(c) Since $x^2/x^{-2} = x^4$ is nonconstant, Theorem 5.1.3 implies that $y = c_1x^2 + c_2/x^2$ is the general solution of $x^2y'' + xy' - 4y = 0$ on $(-\infty, 0)$ and $(0, \infty)$. ∎

THE WRONSKIAN AND ABEL'S FORMULA

To motivate a result that we need in order to prove Theorem 5.1.3, let's see what is required to prove that $\{y_1, y_2\}$ is a fundamental set of solutions of (20) on (a, b). Let x_0 be an arbitrary point in (a, b), and suppose that y is an arbitrary solution of (20) on (a, b). Then y is the unique solution of the initial value problem

$$y'' + p(x)y' + q(x)y = 0, \qquad y(x_0) = k_0, \qquad y'(x_0) = k_1; \tag{21}$$

that is, k_0 and k_1 are the numbers obtained by evaluating y and y' at x_0. Moreover, k_0 and k_1 can be any real numbers, since Theorem 5.1.1 implies that (21) has a solution no matter how k_0 and k_1 are chosen.

It follows that $\{y_1, y_2\}$ is a fundamental set of solutions of (20) on (a, b) if and only if it is possible to write the solution of an arbitrary initial value problem (21) as $y = c_1y_1 + c_2y_2$. This is equivalent to requiring that the system

$$\begin{aligned} c_1y_1(x_0) + c_2y_2(x_0) &= k_0 \\ c_1y_1'(x_0) + c_2y_2'(x_0) &= k_1 \end{aligned} \tag{22}$$

has a solution (c_1, c_2) for every choice of (k_0, k_1). Let's try to solve (22).

Multiplying the first equation in (22) by $y_2'(x_0)$ and the second by $y_2(x_0)$ yields

$$c_1 y_1(x_0) y_2'(x_0) + c_2 y_2(x_0) y_2'(x_0) = y_2'(x_0) k_0$$
$$c_1 y_1'(x_0) y_2(x_0) + c_2 y_2'(x_0) y_2(x_0) = y_2(x_0) k_1,$$

and subtracting the second equation here from the first yields

$$(y_1(x_0) y_2'(x_0) - y_1'(x_0) y_2(x_0)) c_1 = y_2'(x_0) k_0 - y_2(x_0) k_1. \tag{23}$$

Multiplying the first equation in (22) by $y_1'(x_0)$ and the second by $y_1(x_0)$ yields

$$c_1 y_1(x_0) y_1'(x_0) + c_2 y_2(x_0) y_1'(x_0) = y_1'(x_0) k_0$$
$$c_1 y_1'(x_0) y_1(x_0) + c_2 y_2'(x_0) y_1(x_0) = y_1(x_0) k_1,$$

and subtracting the first equation here from the second yields

$$(y_1(x_0) y_2'(x_0) - y_1'(x_0) y_2(x_0)) c_2 = y_1(x_0) k_1 - y_1'(x_0) k_0. \tag{24}$$

If

$$y_1(x_0) y_2'(x_0) - y_1'(x_0) y_2(x_0) = 0$$

then it is impossible to satisfy (23) and (24) (and therefore (22)) unless k_0 and k_1 happen to satisfy

$$y_1(x_0) k_1 - y_1'(x_0) k_0 = 0$$
$$y_2'(x_0) k_0 - y_2(x_0) k_1 = 0.$$

On the other hand, if

$$y_1(x_0) y_2'(x_0) - y_1'(x_0) y_2(x_0) \neq 0 \tag{25}$$

then we can divide (23) and (24) through by the quantity on the left to obtain

$$c_1 = \frac{y_2'(x_0) k_0 - y_2(x_0) k_1}{y_1(x_0) y_2'(x_0) - y_1'(x_0) y_2(x_0)}$$
$$c_2 = \frac{y_1(x_0) k_1 - y_1'(x_0) k_0}{y_1(x_0) y_2'(x_0) - y_1'(x_0) y_2(x_0)}, \tag{26}$$

no matter how k_0 and k_1 are chosen. This motivates us to consider conditions on y_1 and y_2 that imply (25).

THEOREM 5.1.4

Suppose that p and q are continuous on (a, b), let y_1 and y_2 be solutions of

$$y'' + p(x) y' + q(x) y = 0 \tag{27}$$

on (a, b), and define

$$W = y_1 y_2' - y_1' y_2. \tag{28}$$

Let x_0 be any point in (a, b). Then

$$W(x) = W(x_0) \exp\left(-\int_{x_0}^{x} p(t)\, dt\right), \qquad a < x < b. \tag{29}$$

Therefore, either W has no zeros in (a, b) or $W \equiv 0$ on (a, b).

PROOF. Differentiating (28) yields

$$W' = y_1'y_2' + y_1y_2'' - y_1'y_2' - y_1''y_2 = y_1y_2'' - y_1''y_2 \tag{30}$$

Since y_1 and y_2 both satisfy (27),

$$y_1'' = -py_1' - qy_1 \quad \text{and} \quad y_2'' = -py_2' - qy_2.$$

Substituting these into (30) yields

$$\begin{aligned}
W' &= -y_1(py_2' + qy_2) + y_2(py_1' + qy_1) \\
&= -p(y_1y_2' - y_2y_1') - q(y_1y_2 - y_2y_1) \\
&= -p(y_1y_2' - y_2y_1') = -pW.
\end{aligned}$$

Therefore $W' + p(x)W = 0$; that is, W is the solution of the initial value problem

$$y' + p(x)y = 0, \qquad y(x_0) = W(x_0).$$

We leave it to you to verify by separation of variables that this implies (29). If $W(x_0) \neq 0$ then (29) implies that W has no zeros in (a,b), since an exponential is never zero. On the other hand, if $W(x_0) = 0$ then (29) implies that $W(x) = 0$ for all x in (a,b). $\square$

The function W defined in (28) is the ***Wronskian*** of $\{y_1, y_2\}$, after the Polish mathematician Wronski.[1] Formula (29) is ***Abel's***[2] ***formula.***

The Wronskian of $\{y_1, y_2\}$ is usually written as the determinant

$$W = \begin{vmatrix} y_1 & y_2 \\ y_1' & y_2' \end{vmatrix}.$$

The expressions in (26) for c_1 and c_2 can be written in terms of determinants as

$$c_1 = \frac{1}{W(x_0)} \begin{vmatrix} k_0 & y_2(x_0) \\ k_1 & y_2'(x_0) \end{vmatrix} \quad \text{and} \quad c_2 = \frac{1}{W(x_0)} \begin{vmatrix} y_1(x_0) & k_0 \\ y_1'(x_0) & k_1 \end{vmatrix}.$$

If you've taken linear algebra you may recognize this as Cramer's rule.

EXAMPLE 5 Verify Abel's formula for the following differential equations and the corresponding solutions, from Examples 1, 2, and 3:

(a) $y'' - y = 0;$ $\quad y_1 = e^x,$ $\quad y_2 = e^{-x}.$
(b) $y'' + \omega^2 y = 0;$ $\quad y_1 = \cos \omega x,$ $\quad y_2 = \sin \omega x.$
(c) $x^2y'' + xy' - 4y = 0;$ $\quad y_1 = x^2,$ $\quad y_2 = 1/x^2.$

Solution **(a)** Since $p \equiv 0$, we can verify Abel's formula by showing that W is constant, which is true, since

$$W(x) = \begin{vmatrix} e^x & e^{-x} \\ e^x & -e^{-x} \end{vmatrix} = e^x(-e^{-x}) - e^xe^{-x} = -2$$

for all x.

Solution **(b)** Again, since $p \equiv 0$, we can verify Abel's formula by showing that W is constant, which is true, since

---

[1]Wronskian determinants were introduced by Jóseph Maria Hoëné (1778–1853), who changed his name to Wronski. He was born in Poland, educated in Germany, and lived most of his life in France.
[2]Niels Henrik Abel (1802–1829) was a Norwegian mathematician best known for showing that there is no general formula for expressing the zeros of a fifth degree polynomial in terms of its coefficients, and for his contributions to the theory of elliptic functions.

$$W(x) = \begin{vmatrix} \cos \omega x & \sin \omega x \\ -\omega \sin \omega x & \omega \cos \omega x \end{vmatrix}$$

$$= \cos \omega x (\omega \cos \omega x) - (-\omega \sin \omega x) \sin \omega x$$

$$= \omega(\cos^2 \omega x + \sin^2 \omega x) = \omega$$

for all x.

Solution   **(c)** Computing the Wronskian of $y_1 = x^2$ and $y_2 = 1/x^2$ directly yields

$$W = \begin{vmatrix} x^2 & \dfrac{1}{x^2} \\ 2x & -\dfrac{2}{x^3} \end{vmatrix} = x^2\left(-\frac{2}{x^3}\right) - 2x\left(\frac{1}{x^2}\right) = -\frac{4}{x}. \qquad (31)$$

To verify Abel's formula we rewrite the differential equation as

$$y'' + \frac{1}{x}y' - \frac{4}{x^2}y = 0$$

to see that $p(x) = 1/x$. If x_0 and x are either both in $(-\infty, 0)$ or both in $(0, \infty)$ then

$$\int_{x_0}^{x} p(t)\, dt = \int_{x_0}^{x} \frac{dt}{t} = \ln\left(\frac{x}{x_0}\right),$$

so Abel's formula becomes

$$W(x) = W(x_0)e^{-\ln(x/x_0)} = W(x_0)\frac{x_0}{x}$$

$$= -\left(\frac{4}{x_0}\right)\left(\frac{x_0}{x}\right) \qquad \text{from (31)}$$

$$= -\frac{4}{x},$$

which is consistent with (31).   ■

The following theorem will enable us to complete the proof of Theorem 5.1.3.

THEOREM 5.1.5

Suppose that p and q are continuous on an open interval (a,b), let y_1 and y_2 be solutions of

$$y'' + p(x)y' + q(x)y = 0 \qquad (32)$$

on (a,b), and let $W = y_1y_2' - y_1'y_2$. Then y_1 and y_2 are linearly independent on (a,b) if and only if W has no zeros on (a,b).

PROOF. We first show that if $W(x_0) = 0$ for some x_0 in (a,b) then y_1 and y_2 are linearly dependent on (a,b). Let I be a subinterval of (a,b) on which y_1 has no zeros. (If there is no such subinterval, then $y_1 \equiv 0$ on (a,b), so y_1 and y_2 are linearly independent, and we are finished with this part of the proof.) Then y_2/y_1 is defined on I, and

$$\left(\frac{y_2}{y_1}\right)' = \frac{y_1y_2' - y_1'y_2}{y_1^2} = \frac{W}{y_1^2}. \qquad (33)$$

However, if $W(x_0) = 0$ then Theorem 5.1.4 implies that $W \equiv 0$ on (a,b). Therefore (33) implies that $(y_2/y_1)' \equiv 0$, so $y_2/y_1 = c$ (constant) on I. This shows that $y_2(x) = cy_1(x)$ for all x in I. However, we want to show that $y_2 = cy_1(x)$ for all x in (a,b). Let $Y = y_2 - cy_1$. Then Y is a solution of (32) on (a,b) such that $Y \equiv 0$ on I, and therefore $Y' \equiv 0$ on I. Consequently, if x_0 is chosen arbitrarily in I, then Y is a solution of the initial value problem

$$y'' + p(x)y' + q(x)y = 0, \qquad y(x_0) = 0, \qquad y'(x_0) = 0,$$

which implies that $Y \equiv 0$ on (a,b), by the paragraph following Theorem 5.1.1. (See also Exercise 24.) Hence, $y_2 - cy_1 \equiv 0$ on (a,b), which implies that y_1 and y_2 are not linearly independent on (a,b).

Now suppose that W has no zeros on (a,b). Then y_1 cannot be identically zero on (a,b) (why not?), and therefore there is a subinterval I of (a,b) on which y_1 has no zeros. Since (33) implies that y_2/y_1 is nonconstant on I, it follows that y_2 is not a constant multiple of y_1 on (a,b). A similar argument shows that y_1 is not a constant multiple of y_2 on (a,b), since

$$\left(\frac{y_1}{y_2}\right)' = \frac{y_1'y_2 - y_1y_2'}{y_2^2} = -\frac{W}{y_2^2}$$

on any subinterval of (a,b) where y_2 has no zeros. $\qquad\square$

We can now complete the proof of Theorem 5.1.3. From Theorem 5.1.5, two solutions y_1 and y_2 of (32) are linearly independent on (a,b) if and only if W has no zeros on (a,b). From Theorem 5.1.4 and the motivating comments preceding it, $\{y_1, y_2\}$ is a fundamental set of solutions of (32) if and only if W has no zeros on (a,b). It therefore follows that $\{y_1, y_2\}$ is a fundamental set for (32) on (a,b) if and only if $\{y_1, y_2\}$ is linearly independent on (a,b).

The following theorem summarizes the relationships among the concepts discussed in this section.

THEOREM 5.1.6

Suppose that p and q are continuous on an open interval (a,b), and let y_1 and y_2 be solutions of

$$y'' + p(x)y' + q(x)y = 0 \tag{34}$$

on (a,b). Then the following statements are equivalent; that is, they are either all true or all false.

(a) *The general solution of (34) on (a,b) is $y = c_1y_1 + c_2y_2$.*
(b) *$\{y_1, y_2\}$ is a fundamental set of solutions of (34) on (a,b).*
(c) *$\{y_1, y_2\}$ is linearly independent on (a,b).*
(d) *The Wronskian of $\{y_1, y_2\}$ is nonzero at some point in (a,b).*
(e) *The Wronskian of $\{y_1, y_2\}$ is nonzero at all points in (a,b).*

We can apply this theorem to an equation written as

$$P_0(x)y'' + P_1(x)y' + P_2(x)y = 0$$

on an interval (a,b) where P_0, P_1, and P_2 are continuous and P_0 has no zeros.

5.1 EXERCISES

1. **(a)** Verify that $y_1 = e^{2x}$ and $y_2 = e^{5x}$ are solutions of

   $$y'' - 7y' + 10y = 0 \qquad\qquad\text{(A)}$$

   on $(-\infty, \infty)$.

   (b) Verify that if c_1 and c_2 are arbitrary constants then $y = c_1 e^{2x} + c_2 e^{5x}$ is a solution of (A) on $(-\infty, \infty)$.

   (c) Solve the initial value problem

   $$y'' - 7y' + 10y = 0, \qquad y(0) = -1, \qquad y'(0) = 1.$$

   (d) Solve the initial value problem

   $$y'' - 7y' + 10y = 0, \qquad y(0) = k_0, \qquad y'(0) = k_1.$$

2. **(a)** Verify that $y_1 = e^x \cos x$ and $y_2 = e^x \sin x$ are solutions of

   $$y'' - 2y' + 2y = 0 \qquad\qquad\text{(A)}$$

   on $(-\infty, \infty)$.

   (b) Verify that if c_1 and c_2 are arbitrary constants then $y = c_1 e^x \cos x + c_2 e^x \sin x$ is a solution of (A) on $(-\infty, \infty)$.

   (c) Solve the initial value problem

   $$y'' - 2y' + 2y = 0, \qquad y(0) = 3, \qquad y'(0) = -2.$$

   (d) Solve the initial value problem

   $$y'' - 2y' + 2y = 0, \qquad y(0) = k_0, \qquad y'(0) = k_1.$$

3. **(a)** Verify that $y_1 = e^x$ and $y_2 = xe^x$ are solutions of

   $$y'' - 2y' + y = 0 \qquad\qquad\text{(A)}$$

   on $(-\infty, \infty)$.

   (b) Verify that if c_1 and c_2 are arbitrary constants then $y = e^x(c_1 + c_2 x)$ is a solution of (A) on $(-\infty, \infty)$.

   (c) Solve the initial value problem

   $$y'' - 2y' + y = 0, \qquad y(0) = 7, \qquad y'(0) = 4.$$

   (d) Solve the initial value problem

   $$y'' - 2y' + y = 0, \qquad y(0) = k_0, \qquad y'(0) = k_1.$$

4. **(a)** Verify that $y_1 = 1/(x - 1)$ and $y_2 = 1/(x + 1)$ are solutions of

   $$(x^2 - 1)y'' + 4xy' + 2y = 0 \qquad\qquad\text{(A)}$$

   on $(-\infty, -1)$, $(-1, 1)$, and $(1, \infty)$. What is the general solution of (A) on each of these intervals?

   (b) Solve the initial value problem

   $$(x^2 - 1)y'' + 4xy' + 2y = 0, \qquad y(0) = -5, \qquad y'(0) = 1.$$

   What is the interval of validity of the solution?

   C **(c)** Plot the solution of the initial value problem.

   (d) Verify Abel's formula for y_1 and y_2, with $x_0 = 0$.

5. Compute the Wronskians of the given sets of functions.

(a) $\{1, e^x\}$ (b) $\{e^x, e^x \sin x\}$

(c) $\{x + 1, x^2 + 2\}$ (d) $\{x^{1/2}, x^{-1/3}\}$

(e) $\left\{\dfrac{\sin x}{x}, \dfrac{\cos x}{x}\right\}$ (f) $\{x \ln |x|, x^2 \ln |x|\}$

(g) $\{e^x \cos \sqrt{x}, e^x \sin \sqrt{x}\}$

6. Find the Wronskian of a given set $\{y_1, y_2\}$ of solutions of

$$y'' + 3(x^2 + 1)y' - 2y = 0,$$

given that $W(\pi) = 0$.

7. Find the Wronskian of a given set $\{y_1, y_2\}$ of solutions of

$$(1 - x^2)y'' - 2xy' + \alpha(\alpha + 1)y = 0,$$

given that $W(0) = 1$. (This is **Legendre's**[3] **equation.**)

8. Find the Wronskian of a given set $\{y_1, y_2\}$ of solutions of

$$x^2y'' + xy' + (x^2 - \nu^2)y = 0,$$

given that $W(1) = 1$. (This is **Bessel's**[4] **equation.**)

9. (This exercise shows that if you know one nontrivial solution of $y'' + p(x)y' + q(x)y = 0$ then you can use Abel's formula to find another.)

Suppose that p and q are continuous and y_1 is a solution of

$$y'' + p(x)y' + q(x)y = 0 \tag{A}$$

that has no zeros on (a,b). Let $P(x) = \int p(x)\, dx$ be any antiderivative of p on (a,b).

(a) Show that if K is an arbitrary nonzero constant and y_2 satisfies

$$y_1 y_2' - y_1' y_2 = Ke^{-P(x)} \tag{B}$$

on (a,b) then y_2 also satisfies (A) on (a,b), and $\{y_1, y_2\}$ is a fundamental set of solutions on (A) on (a,b).

(b) Conclude from (a) that if $y_2 = uy_1$ where $u' = Ke^{-P(x)}/y_1^2(x)$ then $\{y_1, y_2\}$ is a fundamental set of solutions of (A) on (a,b).

In Exercises 10–23 use the method suggested by Exercise 9 to find a second solution y_2 that is not a constant multiple of the given solution y_1. Choose K conveniently to simplify y_2.

10. $y'' - 2y' - 3y = 0$; $y_1 = e^{3x}$ 11. $y'' - 6y' + 9y = 0$; $y_1 = e^{3x}$

12. $y'' - 2ay' + a^2y = 0$ (a = constant); $y_1 = e^{ax}$ 13. $x^2y'' + xy' - y = 0$; $y_1 = x$

14. $x^2y'' - xy' + y = 0$; $y_1 = x$

15. $x^2y'' - (2a - 1)xy' + a^2y = 0$ (a = nonzero constant), $x > 0$; $y_1 = x^a$

16. $4x^2y'' - 4xy' + (3 - 16x^2)y = 0$; $y_1 = x^{1/2}e^{2x}$ 17. $(x - 1)y'' - xy' + y = 0$; $y_1 = e^x$

18. $x^2y'' - 2xy' + (x^2 + 2)y = 0$; $y_1 = x \cos x$

19. $4x^2(\sin x)y'' - 4x(x \cos x + \sin x)y' + (2x \cos x + 3 \sin x)y = 0$; $y_1 = x^{1/2}$

20. $(3x - 1)y'' - (3x + 2)y' - (6x - 8)y = 0$; $y_1 = e^{2x}$

21. $(x^2 - 4)y'' + 4xy' + 2y = 0$; $y_1 = \dfrac{1}{x - 2}$

[3] The French mathematician Adrien-Marie Legendre (1752–1833) is best known for his work on elliptic integrals. He also contributed to other fields, including number theory and the calculus of variations. Legendre's differential equation arose from his study of gravitation.

[4] Friedrich Wilhelm Bessel (1784–1846) was a German astronomer. He was the first person to measure the distance from Earth to a star. Bessel's equation arose from his study of planetary motion.

22. $(2x + 1)xy'' - 2(2x^2 - 1)y' - 4(x + 1)y = 0; \quad y_1 = \dfrac{1}{x}$

23. $(x^2 - 2x)y'' + (2 - x^2)y' + (2x - 2)y = 0; \quad y_1 = e^x$

24. Suppose that p and q are continuous on an open interval (a,b) and let x_0 be in (a,b). Use Theorem 5.1.1 to show that the only solution of the initial value problem

$$y'' + p(x)y' + q(x)y = 0, \quad y(x_0) = 0, \quad y'(x_0) = 0$$

on (a,b) is the trivial solution $y \equiv 0$.

25. Suppose that P_0, P_1, and P_2 are continuous on (a,b) and let x_0 be in (a,b). Show that if either of the following statements is true then $P_0(x) = 0$ for some x in (a,b).

(a) The initial value problem

$$P_0(x)y'' + P_1(x)y' + P_2(x)y = 0, \quad y(x_0) = k_0, \quad y'(x_0) = k_1$$

has more than one solution on (a,b).

(b) The initial value problem

$$P_0(x)y'' + P_1(x)y' + P_2(x)y = 0, \quad y(x_0) = 0, \quad y'(x_0) = 0$$

has a nontrivial solution on (a,b).

26. Suppose that p and q are continuous on (a,b) and y_1 and y_2 are solutions of

$$y'' + p(x)y' + q(x)y = 0 \tag{A}$$

on (a,b). Let

$$z_1 = \alpha y_1 + \beta y_2$$
$$z_2 = \gamma y_1 + \delta y_2,$$

where $\alpha, \beta, \gamma,$ and δ are constants. Show that if $\{z_1, z_2\}$ is a fundamental set of solutions of (A) on (a,b) then so is $\{y_1, y_2\}$.

27. Suppose that p and q are continuous on (a,b) and $\{y_1, y_2\}$ is a fundamental set of solutions of

$$y'' + p(x)y' + q(x)y = 0 \tag{A}$$

on (a,b). Let

$$z_1 = \alpha y_1 + \beta y_2$$
$$z_2 = \gamma y_1 + \delta y_2,$$

where $\alpha, \beta, \gamma,$ and δ are constants. Show that $\{z_1, z_2\}$ is a fundamental set of solutions of (A) on (a,b) if and only if $\alpha\gamma - \beta\delta \neq 0$.

28. Suppose that y_1 is differentiable on an interval (a,b) and $y_2 = ky_1$, where k is a constant. Show that the Wronskian of $\{y_1, y_2\}$ is identically zero on (a,b).

29. Let

$$y_1 = x^3 \quad \text{and} \quad y_2 = \begin{cases} x^3, & x \geq 0, \\ -x^3, & x < 0. \end{cases}$$

(a) Show that the Wronskian of $\{y_1, y_2\}$ is defined and identically zero on $(-\infty, \infty)$.

(b) Suppose that $a < 0 < b$. Show that $\{y_1, y_2\}$ is linearly independent on (a,b).

(c) Use Exercise 25**(b)** to show that these results do not contradict Theorem 5.1.5, because neither y_1 nor y_2 can be a solution of an equation

$$y'' + p(x)y' + q(x)y = 0$$

on (a,b) if p and q are continuous on (a,b).

30. Suppose that p and q are continuous on (a,b) and $\{y_1, y_2\}$ is a set of solutions of

$$y'' + p(x)y' + q(x)y = 0$$

on (a,b) such that either $y_1(x_0) = y_2(x_0) = 0$ or $y_1'(x_0) = y_2'(x_0) = 0$ for some x_0 in (a,b). Show that $\{y_1, y_2\}$ is linearly dependent on (a,b).

31. Suppose that p and q are continuous on (a,b) and $\{y_1, y_2\}$ is a fundamental set of solutions of

$$y'' + p(x)y' + q(x)y = 0$$

on (a,b). Show that if $y_1(x_1) = y_1(x_2) = 0$ where $a < x_1 < x_2 < b$, then $y_2(x) = 0$ for some x in (x_1, x_2).

Hint: *Show that if y_2 has no zeros in (x_1, x_2) then y_1/y_2 is either strictly increasing or strictly decreasing on (x_1, x_2), and deduce a contradiction.*

32. Suppose that p and q are continuous on (a,b) and every solution of

$$y'' + p(x)y' + q(x)y = 0 \tag{A}$$

on (a,b) can be written as a linear combination of the twice differentiable functions $\{y_1, y_2\}$. Use Theorem 5.1.1 to show that y_1 and y_2 are themselves solutions of (A) on (a,b).

33. Suppose that $p_1, p_2, q_1,$ and q_2 are continuous on (a,b) and the equations

$$y'' + p_1(x)y' + q_1(x)y = 0 \quad \text{and} \quad y'' + p_2(x)y' + q_2(x)y = 0$$

have the same solutions on (a,b). Show that $p_1 = p_2$ and $q_1 = q_2$ on (a,b).

Hint: *Use Abel's formula.*

34. (For this exercise you have to know about 3×3 determinants.) Show that if y_1 and y_2 are twice continuously differentiable on (a,b) and the Wronskian W of $\{y_1, y_2\}$ has no zeros in (a,b), then the equation

$$\frac{1}{W} \begin{vmatrix} y & y_1 & y_2 \\ y' & y_1' & y_2' \\ y'' & y_1'' & y_2'' \end{vmatrix} = 0$$

can be written as

$$y'' + p(x)y' + q(x)y = 0 \tag{A}$$

where p and q are continuous on (a,b) and $\{y_1, y_2\}$ is a fundamental set of solutions of (A) on (a,b).

Hint: *Expand the determinant by cofactors of its first column.*

35. Use the method suggested by Exercise 34 to find a linear homogeneous equation for which the given functions form a fundamental set of solutions on some interval.

(a) $e^x \cos 2x, \quad e^x \sin 2x$ **(b)** $x, \quad e^{2x}$

(c) $x, \quad x \ln x$ **(d)** $\cos(\ln x), \quad \sin(\ln x)$

(e) $\cosh x, \quad \sinh x$ **(f)** $x^2 - 1, \quad x^2 + 1$

36. Suppose that p and q are continuous on (a,b) and $\{y_1, y_2\}$ is a fundamental set of solutions of

$$y'' + p(x)y' + q(x)y = 0 \tag{A}$$

on (a,b). Show that if y is a solution of (A) on (a,b) then there is exactly one way to choose c_1 and c_2 so that $y = c_1 y_1 + c_2 y_2$ on (a,b).

37. Suppose that p and q are continuous on (a,b) and x_0 is in (a,b). Let y_1 and y_2 be the solutions of

$$y'' + p(x)y' + q(x)y = 0 \tag{A}$$

such that

$$y_1(x_0) = 1, \quad y_1'(x_0) = 0 \quad \text{and} \quad y_2(x_0) = 0, \quad y_2'(x_0) = 1.$$

(Theorem 5.1.1 implies that each of these initial value problems has a unique solution on (a,b).)

(a) Show that $\{y_1, y_2\}$ is linearly independent on (a,b).

(b) Show that an arbitrary solution y of (A) on (a,b) can be written as $y = y(x_0)y_1 + y'(x_0)y_2$.

(c) Express the solution of the initial value problem

$$y'' + p(x)y' + q(x)y = 0, \qquad y(x_0) = k_0, \qquad y'(x_0) = k_1$$

as a linear combination of y_1 and y_2.

38. Find solutions y_1 and y_2 of the equation $y'' = 0$ that satisfy the initial conditions

$$y_1(x_0) = 1, \qquad y_1'(x_0) = 0 \qquad \text{and} \qquad y_2(x_0) = 0, \qquad y_2'(x_0) = 1.$$

Then use Exercise 37(c) to write the solution of the initial value problem

$$y'' = 0, \qquad y(0) = k_0, \qquad y'(0) = k_1$$

as a linear combination of y_1 and y_2.

39. Let x_0 be an arbitrary real number. Given (Example 1) that e^x and e^{-x} are solutions of $y'' - y = 0$, find solutions y_1 and y_2 of $y'' - y = 0$ such that

$$y_1(x_0) = 1, \qquad y_1'(x_0) = 0 \qquad \text{and} \qquad y_2(x_0) = 0, \qquad y_2'(x_0) = 1.$$

Then use Exercise 37(c) to write the solution of the initial value problem

$$y'' - y = 0, \qquad y(x_0) = k_0, \qquad y'(x_0) = k_1$$

as a linear combination of y_1 and y_2.

40. Let x_0 be an arbitrary real number. Given (Example 2) that $\cos \omega x$ and $\sin \omega x$ are solutions of $y'' + \omega^2 y = 0$, find solutions of $y'' + \omega^2 y = 0$ such that

$$y_1(x_0) = 1, \qquad y_1'(x_0) = 0 \qquad \text{and} \qquad y_2(x_0) = 0, \qquad y_2'(x_0) = 1.$$

Then use Exercise 37(c) to write the solution of the initial value problem

$$y'' + \omega^2 y = 0, \qquad y(x_0) = k_0, \qquad y'(x_0) = k_1$$

as a linear combination of y_1 and y_2. Use the identities

$$\cos(A + B) = \cos A \cos B - \sin A \sin B$$

$$\sin(A + B) = \sin A \cos B + \cos A \sin B$$

to simplify your expressions for y_1, y_2, and y.

41. Recall from Exercise 4 that $1/(x - 1)$ and $1/(x + 1)$ are solutions of

$$(x^2 - 1)y'' + 4xy' + 2y = 0 \tag{A}$$

on $(-1,1)$. Find solutions of (A) such that

$$y_1(0) = 1, \qquad y_1'(0) = 0 \qquad \text{and} \qquad y_2(0) = 0, \qquad y_2'(0) = 1.$$

Then use Exercise 37(c) to write the solution of the initial value problem

$$(x^2 - 1)y'' + 4xy' + 2y = 0, \qquad y(0) = k_0, \qquad y'(0) = k_1$$

as a linear combination of y_1 and y_2.

42. (a) Verify that $y_1 = x^2$ and $y_2 = x^3$ satisfy

$$x^2y'' - 4xy' + 6y = 0 \tag{A}$$

on $(-\infty,\infty)$, and that $\{y_1, y_2\}$ is a fundamental set of solutions of (A) on $(-\infty,0)$ and $(0,\infty)$.

(b) Let a_1, a_2, b_1, and b_2 be constants. Show that

$$y = \begin{cases} a_1x^2 + a_2x^3, & x \geq 0, \\ b_1x^2 + b_2x^3, & x < 0 \end{cases}$$

is a solution of (A) on $(-\infty, \infty)$ if and only if $a_1 = b_1$. From this, justify the statement that y is a solution of (A) on $(-\infty, \infty)$ if and only if

$$y = \begin{cases} c_1x^2 + c_2x^3 & x \geq 0, \\ c_1x^2 + c_3x^3, & x < 0, \end{cases}$$

where c_1, c_2, and c_3 are arbitrary constants.

(c) For what values of k_0 and k_1 does the initial value problem

$$x^2y'' - 4xy' + 6y = 0, \qquad y(0) = k_0, \qquad y'(0) = k_1$$

have a solution? What are the solutions?

(d) Show that if $x_0 \neq 0$ and k_0, k_1 are arbitrary constants, then the initial value problem

$$x^2y'' - 4xy' + 6y = 0, \qquad y(x_0) = k_0, \qquad y'(x_0) = k_1 \tag{B}$$

has infinitely many solutions on $(-\infty, \infty)$. On what interval does (B) have a unique solution?

43. (a) Verify that $y_1 = x$ and $y_2 = x^2$ satisfy

$$x^2y'' - 2xy' + 2y = 0 \tag{A}$$

on $(-\infty, \infty)$, and that $\{y_1, y_2\}$ is a fundamental set of solutions of (A) on $(-\infty, 0)$ and $(0, \infty)$.

(b) Let a_1, a_2, b_1, and b_2 be constants. Show that

$$y = \begin{cases} a_1x + a_2x^2, & x \geq 0, \\ b_1x + b_2x^2, & x < 0 \end{cases}$$

is a solution of (A) on $(-\infty, \infty)$ if and only if $a_1 = b_1$ and $a_2 = b_2$. From this, justify the statement that the general solution of (A) on $(-\infty, \infty)$ is $y = c_1x + c_2x^2$, where c_1 and c_2 are arbitrary constants.

(c) For what values of k_0 and k_1 does the initial value problem

$$x^2y'' - 2xy' + 2y = 0, \qquad y(0) = k_0, \qquad y'(0) = k_1$$

have a solution? What are the solutions?

(d) Show that if $x_0 \neq 0$ and k_0, k_1 are arbitrary constants, then the initial value problem

$$x^2y'' - 2xy' + 2y = 0, \qquad y(x_0) = k_0, \qquad y'(x_0) = k_1$$

has a unique solution on $(-\infty, \infty)$.

44. (a) Verify that $y_1 = x^3$ and $y_2 = x^4$ satisfy

$$x^2y'' - 6xy' + 12y = 0 \tag{A}$$

on $(-\infty, \infty)$, and that $\{y_1, y_2\}$ is a fundamental set of solutions of (A) on $(-\infty, 0)$ and $(0, \infty)$.

(b) Show that y is a solution of (A) on $(-\infty, \infty)$ if and only if

$$y = \begin{cases} a_1x^3 + a_2x^4 & x \geq 0, \\ b_1x^3 + b_2x^4 & x < 0 \end{cases}$$

where a_1, a_2, b_1, and b_2 are arbitrary constants.

(c) For what values of k_0 and k_1 does the initial value problem

$$x^2y'' - 6xy' + 12y = 0, \qquad y(0) = k_0, \qquad y'(0) = k_1$$

have a solution? What are the solutions?

(d) Show that if $x_0 \neq 0$ and k_0, k_1 are arbitrary constants, then the initial value problem

$$x^2 y'' - 6xy' + 12y = 0, \qquad y(x_0) = k_0, \qquad y'(x_0) = k_1 \qquad \text{(B)}$$

has infinitely many solutions on $(-\infty, \infty)$. On what interval does (B) have a unique solution?

5.2   Constant Coefficient Homogeneous Equations

If a, b, and c are real constants and $a \neq 0$, then

$$ay'' + by' + cy = F(x)$$

is said to be a ***constant coefficient equation.*** In this section we consider the homogeneous constant coefficient equation

$$ay'' + by' + cy = 0. \qquad \text{(1)}$$

As we will see, all solutions of (1) are defined on $(-\infty, \infty)$. This being the case, we'll omit references to the interval on which solutions are defined, or on which a given set of solutions is a fundamental set, etc., since the interval will always be $(-\infty, \infty)$.

The key to solving (1) is that if $y = e^{rx}$ where r is a constant, then the left side of (1) is a multiple of e^{rx}; thus, if $y = e^{rx}$ then $y' = re^{rx}$ and $y'' = r^2 e^{rx}$, so

$$ay'' + by' + cy = ar^2 e^{rx} + bre^{rx} + ce^{rx} = (ar^2 + br + c)e^{rx}. \qquad \text{(2)}$$

The quadratic polynomial

$$p(r) = ar^2 + br + c$$

is the ***characteristic polynomial*** of (1), and $p(r) = 0$ is the ***characteristic equation.*** From (2) we can see that $y = e^{rx}$ is a solution of (1) if and only if $p(r) = 0$.

The roots of the characteristic equation are given by the quadratic formula

$$r = \frac{-b \pm \sqrt{b^2 - 4ac}}{2a}. \qquad \text{(3)}$$

We consider three cases:

Case 1. $b^2 - 4ac > 0$, so the characteristic equation has two distinct real roots.

Case 2. $b^2 - 4ac = 0$, so the characteristic equation has a repeated real root.

Case 3. $b^2 - 4ac < 0$, so the characteristic equation has complex roots.

In each case we will start with an example.

CASE 1: DISTINCT REAL ROOTS

EXAMPLE 1   **(a)** Find the general solution of

$$y'' + 6y' + 5y = 0. \qquad \text{(4)}$$

(b) Solve the initial value problem

$$y'' + 6y' + 5y = 0, \qquad y(0) = 3, \qquad y'(0) = -1. \qquad \text{(5)}$$

Solution   **(a)** The characteristic polynomial of (4) is

$$p(r) = r^2 + 6r + 5 = (r + 1)(r + 5).$$

Since $p(-1) = p(-5) = 0$ it follows that $y_1 = e^{-x}$ and $y_2 = e^{-5x}$ are solutions of (4). Since $y_2/y_1 = e^{-4x}$ is nonconstant, Theorem 5.1.6 implies that the general solution of (4) is

$$y = c_1 e^{-x} + c_2 e^{-5x}. \tag{6}$$

Solution   **(b)** We must determine c_1 and c_2 in (6) so that y satisfies the initial conditions in (5). Differentiating (6) yields

$$y' = -c_1 e^{-x} - 5c_2 e^{-5x}. \tag{7}$$

Imposing the initial conditions $y(0) = 3, y'(0) = -1$ in (6) and (7) yields

$$\begin{aligned} c_1 + c_2 &= 3 \\ -c_1 - 5c_2 &= -1. \end{aligned}$$

The solution of this system is $c_1 = 7/2, c_2 = -1/2$. Therefore the solution of (5) is

$$y = \frac{7}{2} e^{-x} - \frac{1}{2} e^{-5x}.$$

The graph of this solution is shown in Figure 1.   ■

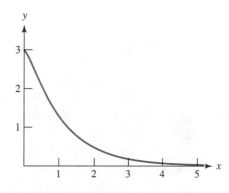

Figure 1   Graph of $y = \dfrac{7}{2} e^{-x} - \dfrac{1}{2} e^{-5x}$

If the characteristic equation has arbitrary distinct real roots r_1 and r_2 then $y_1 = e^{r_1 x}$ and $y_2 = e^{r_2 x}$ are solutions of $ay'' + by' + cy = 0$. Since $y_2/y_1 = e^{(r_2 - r_1)x}$ is nonconstant, Theorem 5.1.6 implies that $\{y_1, y_2\}$ is a fundamental set of solutions of $ay'' + by' + cy = 0$.

CASE 2: A REPEATED REAL ROOT

EXAMPLE 2   **(a)** Find the general solution of

$$y'' + 6y' + 9y = 0. \tag{8}$$

(b) Solve the initial value problem

$$y'' + 6y' + 9y = 0, \qquad y(0) = 3, \qquad y'(0) = -1. \tag{9}$$

Solution **(a)** The characteristic polynomial of (8) is

$$p(r) = r^2 + 6r + 9 = (r + 3)^2,$$

so the characteristic equation has the repeated real root $r_1 = -3$. Therefore $y_1 = e^{-3x}$ is a solution of (8). Since the characteristic equation has no other roots, (8) has no other solutions of the form e^{rx}. We look for solutions of the form $y = uy_1 = ue^{-3x}$, where u is a function that we will now determine. (This should remind you of the method of variation of parameters used in Section 2.1 to solve the nonhomogeneous equation $y' + p(x)y = f(x)$, given a solution y_1 of the complementary equation $y' + p(x)y = 0$. It is also a special case of a method called **reduction of order** that we will study in Section 5.6. For other ways to obtain a second solution of (8) that is not a multiple of e^{-3x}, see Exercise 33 of this section as well as Exercises 9 and 12 of Section 5.1.)

If $y = ue^{-3x}$ then

$$y' = u'e^{-3x} - 3ue^{-3x} \quad \text{and} \quad y'' = u''e^{-3x} - 6u'e^{-3x} + 9ue^{-3x},$$

so

$$y'' + 6y' + 9y = e^{-3x}[(u'' - 6u' + 9u) + 6(u' - 3u) + 9u]$$
$$= e^{-3x}[u'' - (6 - 6)u' + (9 - 18 + 9)u] = u''e^{-3x}.$$

Therefore $y = ue^{-3x}$ is a solution of (8) if and only if $u'' = 0$, which is equivalent to $u = c_1 + c_2x$, where c_1 and c_2 are constants. Therefore any function of the form

$$y = e^{-3x}(c_1 + c_2x) \tag{10}$$

is a solution of (8). Letting $c_1 = 1$ and $c_2 = 0$ yields the solution $y_1 = e^{-3x}$ that we already knew. Letting $c_1 = 0$ and $c_2 = 1$ yields the second solution $y_2 = xe^{-3x}$. Since $y_2/y_1 = x$ is nonconstant, Theorem 5.1.6 implies that $\{y_1, y_2\}$ is a fundamental set of solutions of (8), and (10) is the general solution.

Solution **(b)** Differentiating (10) yields

$$y' = -3e^{-3x}(c_1 + c_2x) + c_2e^{-3x}. \tag{11}$$

Imposing the initial conditions $y(0) = 3, y'(0) = -1$ in (10) and (11) yields $c_1 = 3$ and $-3c_1 + c_2 = -1$, so $c_2 = 8$. Therefore the solution of (9) is

$$y = e^{-3x}(3 + 8x).$$

The graph of this solution is shown in Figure 2. ■

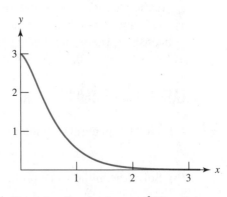

Figure 2 Graph of $y = e^{-3x}(3 + 8x)$

If the characteristic equation of $ay'' + by' + cy = 0$ has an arbitrary repeated root r_1 then the characteristic polynomial must be

$$p(r) = a(r - r_1)^2 = a(r^2 - 2r_1 r + r_1^2).$$

Therefore

$$ar^2 + br + c = ar^2 - (2ar_1)r + ar_1^2,$$

which implies that $b = -2ar_1$ and $c = ar_1^2$. From this it follows that $ay'' + by' + cy = 0$ can be written as $a(y'' - 2r_1 y' + r_1^2 y) = 0$. Since $a \neq 0$ this equation has the same solutions as

$$y'' - 2r_1 y' + r_1^2 y = 0. \tag{12}$$

Since $p(r_1) = 0$ we know that $y_1 = e^{r_1 x}$ is a solution of $ay'' + by' + cy = 0$, and therefore of (12). Proceeding as in Example 2, we look for other solutions of (12) of the form $y_1 = ue^{r_1 x}$; then

$$y' = u'e^{r_1 x} + rue^{r_1 x} \quad \text{and} \quad y'' = u''e^{r_1 x} + 2r_1 u'e^{r_1 x} + r_1^2 ue^{r_1 x},$$

so

$$y'' - 2r_1 y' + r_1^2 y = e^{rx}[(u'' + 2r_1 u' + r_1^2 u) - 2r_1(u' + r_1 u) + r_1^2 u]$$
$$= e^{r_1 x}[u'' + (2r_1 - 2r_1)u' + (r_1^2 - 2r_1^2 + r_1^2)u] = u''e^{r_1 x}.$$

Therefore $y = ue^{r_1 x}$ is a solution of (12) if and only if $u'' = 0$, which is equivalent to $u = c_1 + c_2 x$, where c_1 and c_2 are constants. Hence, any function of the form

$$y = e^{r_1 x}(c_1 + c_2 x) \tag{13}$$

is a solution of (12). Letting $c_1 = 1$ and $c_2 = 0$ here yields the solution $y_1 = e^{r_1 x}$ that we already knew. Letting $c_1 = 0$ and $c_2 = 1$ yields the second solution $y_2 = xe^{r_1 x}$. Since $y_2/y_1 = x$ is nonconstant, Theorem 5.1.6 implies that $\{y_1, y_2\}$ is a fundamental set of solutions of (12), and (13) is the general solution.

CASE 3: COMPLEX CONJUGATE ROOTS

EXAMPLE 3 **(a)** Find the general solution of

$$y'' + 4y' + 13y = 0. \tag{14}$$

(b) Solve the initial value problem

$$y'' + 4y' + 13y = 0, \quad y(0) = 2, \quad y'(0) = -3. \tag{15}$$

Solution **(a)** The characteristic polynomial of (14) is

$$p(r) = r^2 + 4r + 13 = r^2 + 4r + 4 + 9 = (r + 2)^2 + 9.$$

The roots of the characteristic equation are $r_1 = -2 + 3i$ and $r_2 = -2 - 3i$. By analogy with Case 1, it is reasonable to expect that $e^{(-2+3i)x}$ and $e^{(-2-3i)x}$ are solutions of (14). This is true (see Exercise 34); however, there are difficulties here, since you are probably not familiar with exponential functions with complex arguments, and even if you are it is inconvenient to work with them, since they are complex-valued. We will take a simpler approach, which we motivate as follows: The exponential notation suggests that

$$e^{(-2+3i)x} = e^{-2x}e^{3ix} \quad \text{and} \quad e^{(-2-3i)x} = e^{-2x}e^{-3ix},$$

so even though we haven't defined e^{3ix} and e^{-3ix}, it is reasonable to expect that every linear combination of $e^{(-2+3i)x}$ and $e^{(-2-3i)x}$ can be written as $y = ue^{-2x}$, were u depends upon x. To determine u, we note that if $y = ue^{-2x}$ then

$$y' = u'e^{-2x} - 2ue^{-2x} \quad \text{and} \quad y'' = u''e^{-2x} - 4u'e^{-2x} + 4ue^{-2x},$$

so

$$\begin{aligned}
y'' + 4y' + 13y &= e^{-2x}[(u'' - 4u' + 4u) + 4(u' - 2u) + 13u] \\
&= e^{-2x}[u'' - (4-4)u' + (4 - 8 + 13)u] = e^{-2x}(u'' + 9u).
\end{aligned}$$

Therefore $y = ue^{-2x}$ is a solution of (14) if and only if

$$u'' + 9u = 0.$$

From Example 2 of Section 5.1, the general solution of this equation is

$$u = c_1 \cos 3x + c_2 \sin 3x.$$

Therefore any function of the form

$$y = e^{-2x}(c_1 \cos 3x + c_2 \sin 3x) \tag{16}$$

is a solution of (14). Letting $c_1 = 1$ and $c_2 = 0$ yields the solution $y_1 = e^{-2x} \cos 3x$. Letting $c_1 = 0$ and $c_2 = 1$ yields the second solution $y_2 = e^{-2x} \sin 3x$. Since $y_2/y_1 = \tan 3x$ is nonconstant, Theorem 5.1.6 implies that $\{y_1, y_2\}$ is a fundamental set of solutions of (14), and (16) is the general solution.

Solution **(b)** Imposing the condition $y(0) = 2$ in (16) shows that $c_1 = 2$. Differentiating (16) yields

$$y' = -2e^{-2x}(c_1 \cos 3x + c_2 \sin 3x) + 3e^{-2x}(-c_1 \sin 3x + c_2 \cos 3x),$$

and imposing the initial condition $y'(0) = -3$ here yields $-3 = -2c_1 + 3c_2 = -4 + 3c_2$, so $c_2 = 1/3$. Therefore the solution of (15) is

$$y = e^{-2x}\left(2 \cos 3x + \frac{1}{3} \sin 3x\right).$$

The graph of this solution is shown in Figure 3. ∎

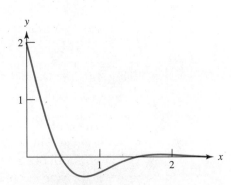

Figure 3 Graph of $y = e^{-2x}\left(2 \cos 3x + \dfrac{1}{3} \sin 3x\right)$

Now suppose that the characteristic equation of $ay'' + by' + cy = 0$ has arbitrary complex roots; thus, $b^2 - 4ac < 0$ and, from (3), the roots are

$$r_1 = \frac{-b + i\sqrt{4ac - b^2}}{2a}, \qquad r_2 = \frac{-b - i\sqrt{4ac - b^2}}{2a},$$

which we rewrite as

$$r_1 = \lambda + i\omega, \qquad r_2 = \lambda - i\omega, \tag{17}$$

with

$$\lambda = -\frac{b}{2a}, \qquad \omega = \frac{\sqrt{4ac - b^2}}{2a}.$$

Don't memorize these formulas. Just remember that r_1 and r_2 are of the form (17), where λ is an arbitrary real number and ω is a positive number; λ and ω are the **real** and **imaginary parts,** respectively, of r_1. Similarly, λ and $-\omega$ are the real and imaginary parts of r_2. We say that r_1 and r_2 are **complex conjugates,** which means that they have the same real part, while their imaginary parts have the same absolute values but opposite signs.

Arguing as in Example 3, it is reasonable to expect that the solutions of $ay'' + by' + cy = 0$ are linear combinations of $e^{(\lambda + i\omega)x}$ and $e^{(\lambda - i\omega)x}$. Again, the exponential notation suggests that

$$e^{(\lambda + i\omega)x} = e^{\lambda x} e^{i\omega x} \qquad \text{and} \qquad e^{(\lambda - i\omega)x} = e^{\lambda x} e^{-i\omega x},$$

so even though we haven't defined $e^{i\omega x}$ and $e^{-i\omega x}$, it is reasonable to expect that every linear combination of $e^{(\lambda + i\omega)x}$ and $e^{(\lambda - i\omega)x}$ can be written as $y = ue^{\lambda x}$ where u depends upon x. To determine u we first observe that since $r_1 = \lambda + i\omega$ and $r_2 = \lambda - i\omega$ are the roots of the characteristic equation, p must be of the form

$$\begin{aligned}
p(r) &= a(r - r_1)(r - r_2) \\
&= a(r - \lambda - i\omega)(r - \lambda + i\omega) \\
&= a\left[(r - \lambda)^2 + \omega^2\right] \\
&= a(r^2 - 2\lambda r + \lambda^2 + \omega^2).
\end{aligned}$$

Now it follows that $ay'' + by' + cy = 0$ can be written as

$$a\left[y'' - 2\lambda y' + (\lambda^2 + \omega^2)y\right] = 0.$$

Since $a \neq 0$ this equation has the same solutions as

$$y'' - 2\lambda y' + (\lambda^2 + \omega^2)y = 0. \tag{18}$$

To determine u we note that if $y = ue^{\lambda x}$ then

$$y' = u'e^{\lambda x} + \lambda ue^{\lambda x} \qquad \text{and} \qquad y'' = u''e^{\lambda x} + 2\lambda u'e^{\lambda x} + \lambda^2 ue^{\lambda x}.$$

Substituting these expressions into (18) and dropping the common factor $e^{\lambda x}$ yields

$$(u'' + 2\lambda u' + \lambda^2 u) - 2\lambda(u' + \lambda u) + (\lambda^2 + \omega^2)u = 0,$$

which simplifies to

$$u'' + \omega^2 u = 0.$$

From Example 2 of Section 5.1, the general solution of this equation is

$$u = c_1 \cos \omega x + c_2 \sin \omega x.$$

Therefore any function of the form

$$y = e^{\lambda x}(c_1 \cos \omega x + c_2 \sin \omega x) \tag{19}$$

satisfies (18). Letting $c_1 = 1$ and $c_2 = 0$ here yields the solution $y_1 = e^{\lambda x} \cos \omega x$. Letting $c_1 = 0$ and $c_2 = 1$ yields a second solution $y_2 = e^{\lambda x} \sin \omega x$. Since $y_2/y_1 = \tan \omega x$ is nonconstant, Theorem 5.1.6 implies that $\{y_1, y_2\}$ is a fundamental set of solutions of (18), and (19) is the general solution.

SUMMARY

The following theorem summarizes the results of this section.

THEOREM 5.2.1

Let $p(r) = ar^2 + br + c$ be the characteristic polynomial of

$$ay'' + by' + cy = 0. \tag{20}$$

Then:

(a) *If $p(r) = 0$ has distinct real roots r_1 and r_2 then the general solution of (20) is*

$$y = c_1 e^{r_1 x} + c_2 e^{r_2 x}.$$

(b) *If $p(r) = 0$ has a repeated root r_1 then the general solution of (20) is*

$$y = e^{r_1 x}(c_1 + c_2 x).$$

(c) *If $p(r) = 0$ has complex conjugate roots $r_1 = \lambda + i\omega$ and $r_2 = \lambda - i\omega$ (where $\omega > 0$) then the general solution of (20) is*

$$y = e^{\lambda x}(c_1 \cos \omega x + c_2 \sin \omega x).$$

5.2 EXERCISES

In Exercises 1–12 find the general solution.

1. $y'' + 5y' - 6y = 0$

2. $y'' - 4y' + 5y = 0$

3. $y'' + 8y' + 7y = 0$

4. $y'' - 4y' + 4y = 0$

5. $y'' + 2y' + 10y = 0$

6. $y'' + 6y' + 10y = 0$

7. $y'' - 8y' + 16y = 0$

8. $y'' + y' = 0$

9. $y'' - 2y' + 3y = 0$

10. $y'' + 6y' + 13y = 0$

11. $4y'' + 4y' + 10y = 0$

12. $10y'' - 3y' - y = 0$

In Exercises 13–17 solve the initial value problem.

13. $y'' + 14y' + 50y = 0$, $y(0) = 2$, $y'(0) = -17$

14. $6y'' - y' - y = 0$, $y(0) = 10$, $y'(0) = 0$

15. $6y'' + y' - y = 0$, $y(0) = -1$, $y'(0) = 3$

16. $4y'' - 4y' - 3y = 0$, $y(0) = \dfrac{13}{12}$, $y'(0) = \dfrac{23}{24}$

17. $4y'' - 12y' + 9y = 0$, $y(0) = 3$, $y'(0) = \dfrac{5}{2}$

In Exercises 18–21 solve the initial value problem and graph the solution.

C **18.** $y'' + 7y' + 12y = 0$, $y(0) = -1$, $y'(0) = 0$ **C** **19.** $y'' - 6y' + 9y = 0$, $y(0) = 0$, $y'(0) = 2$

C **20.** $36y'' - 12y' + y = 0$, $y(0) = 3$, $y'(0) = \dfrac{5}{2}$ **C** **21.** $y'' + 4y' + 10y = 0$, $y(0) = 3$, $y'(0) = -2$

22. (a) Suppose that y is a solution of the constant coefficient homogeneous equation

$$ay'' + by' + cy = 0. \tag{A}$$

Let $z(x) = y(x - x_0)$ where x_0 is an arbitrary real number. Show that

$$az'' + bz' + cz = 0.$$

(b) Let $z_1(x) = y_1(x - x_0)$ and $z_2(x) = y_2(x - x_0)$, where $\{y_1, y_2\}$ is a fundamental set of solutions of (A). Show that $\{z_1, z_2\}$ is also a fundamental set of solutions of (A).

(c) The statement of Theorem 5.2.1 is convenient for solving an initial value problem

$$ay'' + by' + cy = 0, \qquad y(0) = k_0, \qquad y'(0) = k_1,$$

where the initial conditions are imposed at $x_0 = 0$. However, if the initial value problem is

$$ay'' + by' + cy = 0, \qquad y(x_0) = k_0, \qquad y'(x_0) = k_1, \tag{B}$$

where $x_0 \neq 0$, then determining the constants in

$$y = c_1 e^{r_1 x} + c_2 e^{r_2 x}, \qquad y = e^{r_1 x}(c_1 + c_2 x), \qquad \text{or} \qquad y = e^{\lambda x}(c_1 \cos \omega x + c_2 \sin \omega x)$$

(whichever is applicable) is more complicated. Use **(b)** to restate Theorem 5.2.1 in a form more convenient for solving (B).

In Exercises 23–28 use a method suggested by Exercise 22 to solve the initial value problem.

23. $y'' + 3y' + 2y = 0$, $y(1) = -1$, $y'(1) = 4$ **24.** $y'' - 6y' - 7y = 0$, $y(2) = -\dfrac{1}{3}$, $y'(2) = -5$

25. $y'' - 14y' + 49y = 0$, $y(1) = 2$, $y'(1) = 11$ **26.** $9y'' + 6y' + y = 0$, $y(2) = 2$, $y'(2) = -\dfrac{14}{3}$

27. $9y'' + 4y = 0$, $y(\pi/4) = 2$, $y'(\pi/4) = -2$ **28.** $y'' + 3y = 0$, $y(\pi/3) = 2$, $y'(\pi/3) = -1$

29. Prove: If the characteristic equation of

$$ay'' + by' + cy = 0 \tag{A}$$

has a repeated negative root or two roots with negative real parts, then every solution of (A) approaches zero as $x \to \infty$.

30. Suppose that the characteristic polynomial of $ay'' + by' + cy = 0$ has distinct real roots r_1 and r_2. Use a method suggested by Exercise 22 to find a formula for the solution of

$$ay'' + by' + cy = 0, \qquad y(x_0) = k_0, \qquad y'(x_0) = k_1.$$

31. Suppose that the characteristic polynomial of $ay'' + by' + cy = 0$ has a repeated real root r_1. Use a method suggested by Exercise 22 to find a formula for the solution of

$$ay'' + by' + cy = 0, \qquad y(x_0) = k_0, \qquad y'(x_0) = k_1.$$

32. Suppose that the characteristic polynomial of $ay'' + by' + cy = 0$ has complex conjugate roots $\lambda \pm i\omega$. Use a method suggested by Exercise 22 to find a formula for the solution of

$$ay'' + by' + cy = 0, \qquad y(x_0) = k_0, \qquad y'(x_0) = k_1.$$

33. Suppose that the characteristic equation of

$$ay'' + by' + cy = 0 \tag{A}$$

has a repeated real root r_1. Temporarily, think of e^{rx} as a function of two real variables x and r.

(a) Show that

$$a\frac{\partial^2}{\partial x^2}(e^{rx}) + b\frac{\partial}{\partial x}(e^{rx}) + ce^{rx} = a(r - r_1)^2 e^{rx}. \tag{B}$$

(b) Differentiate (B) with respect to r to obtain

$$a\frac{\partial}{\partial r}\left(\frac{\partial^2}{\partial x^2}(e^{rx})\right) + b\frac{\partial}{\partial r}\left(\frac{\partial}{\partial x}(e^{rx})\right) + c(xe^{rx}) = [2 + (r - r_1)x]a(r - r_1)e^{rx}. \tag{C}$$

(c) Reverse the orders of the partial differentiations in the first two terms on the left side of (C) to obtain

$$a\frac{\partial^2}{\partial x^2}(xe^{rx}) + b\frac{\partial}{\partial x}(xe^{rx}) + c(xe^{rx}) = [2 + (r - r_1)x]a(r - r_1)e^{rx}. \tag{D}$$

(d) Set $r = r_1$ in (B) and (D) to see that $y_1 = e^{r_1x}$ and $y_2 = xe^{r_1x}$ are solutions of (A).

34. In calculus you learned that e^u, $\cos u$, and $\sin u$ can be represented by the infinite series

$$e^u = \sum_{n=0}^{\infty} \frac{u^n}{n!} = 1 + \frac{u}{1!} + \frac{u^2}{2!} + \frac{u^3}{3!} + \cdots + \frac{u^n}{n!} + \cdots, \tag{A}$$

$$\cos u = \sum_{n=0}^{\infty} (-1)^n \frac{u^{2n}}{(2n)!} = 1 - \frac{u^2}{2!} + \frac{u^4}{4!} + \cdots + (-1)^n \frac{u^{2n}}{(2n)!} + \cdots, \tag{B}$$

and

$$\sin u = \sum_{n=0}^{\infty} (-1)^n \frac{u^{2n+1}}{(2n + 1)!} = u - \frac{u^3}{3!} + \frac{u^5}{5!} + \cdots + (-1)^n \frac{u^{2n+1}}{(2n + 1)!} + \cdots \tag{C}$$

for all real values of u. Even though you have previously considered (A) only for real values of u, we can set $u = i\theta$, where θ is real, to obtain

$$e^{i\theta} = \sum_{n=0}^{\infty} \frac{(i\theta)^n}{n!}. \tag{D}$$

Given the proper background in the theory of infinite series with complex terms, it can be shown that the series in (D) converges for all real θ.

(a) Recalling that $i^2 = -1$, write enough terms of the sequence $\{i^n\}$ to convince yourself that the sequence is repetitive:

$$1, i, -1, -i, 1, i, -1, -i, 1, i, -1, -i, 1, i, -1, -i, \cdots.$$

Use this to group the terms in (D) as

$$e^{i\theta} = \left(1 - \frac{\theta^2}{2} + \frac{\theta^4}{4} + \cdots\right) + i\left(\theta - \frac{\theta^3}{3!} + \frac{\theta^5}{5!} + \cdots\right)$$

$$= \sum_{n=0}^{\infty} (-1)^n \frac{\theta^{2n}}{(2n)!} + i\sum_{n=0}^{\infty} (-1)^n \frac{\theta^{2n+1}}{(2n + 1)!}.$$

By comparing this result with (B) and (C), conclude that

$$e^{i\theta} = \cos \theta + i \sin \theta. \tag{E}$$

This is known as *Euler's identity.*

(b) Starting from

$$e^{i\theta_1}e^{i\theta_2} = (\cos \theta_1 + i \sin \theta_1)(\cos \theta_2 + i \sin \theta_2),$$

collect the real part (the terms not multiplied by i) and the imaginary part (the terms multiplied by i) on the right, and use the trigonometric identities

$$\cos(\theta_1 + \theta_2) = \cos \theta_1 \cos \theta_2 - \sin \theta_1 \sin \theta_2$$
$$\sin(\theta_1 + \theta_2) = \sin \theta_1 \cos \theta_2 + \cos \theta_1 \sin \theta_2$$

to verify that

$$e^{i(\theta_1+\theta_2)} = e^{i\theta_1}e^{i\theta_2},$$

as you would expect from the use of the exponential notation $e^{i\theta}$.

(c) If α and β are real numbers, define

$$e^{\alpha+i\beta} = e^{\alpha}e^{i\beta} = e^{\alpha}(\cos\beta + i\sin\beta). \tag{F}$$

Show that if $z_1 = \alpha_1 + i\beta_1$ and $z_2 = \alpha_2 + i\beta_2$ then

$$e^{z_1+z_2} = e^{z_1}e^{z_2}.$$

(d) Let a, b, and c be real numbers, with $a \neq 0$. Let $z = u + iv$ where u and v are real-valued functions of x. Then we say that z is a solution of

$$ay'' + by' + cy = 0 \tag{G}$$

if u and v are both solutions of (G). Use Theorem 5.2.1(**c**) to verify that if the characteristic equation of (G) has complex conjugate roots $\lambda \pm i\omega$ then $z_1 = e^{(\lambda+i\omega)x}$ and $z_2 = e^{(\lambda-i\omega)x}$ are both solutions of (G).

5.3   Nonhomogeneous Linear Equations

We will now consider the nonhomogeneous linear second order equation

$$y'' + p(x)y' + q(x)y = f(x), \tag{1}$$

where the forcing function f is not identically zero. The following theorem, an extension of Theorem 5.1.1, gives sufficient conditions for existence and uniqueness of solutions of initial value problems for (1). We omit the proof, which is beyond the scope of this book.

THEOREM 5.3.1

Suppose that p, q, and f are continuous on an open interval (a,b), let x_0 be any point in (a,b), and let k_0 and k_1 be arbitrary real numbers. Then the initial value problem

$$y'' + p(x)y' + q(x)y = f(x), \qquad y(x_0) = k_0, \qquad y'(x_0) = k_1$$

has a unique solution on (a,b).

In order to find the general solution of (1) on an interval (a,b) where p, q, and f are continuous, it is necessary to find the general solution of the associated homogeneous equation

$$y'' + p(x)y' + q(x)y = 0 \tag{2}$$

on (a,b). We call (2) the ***complementary equation*** for (1).

The following theorem shows how to find the general solution of (1) if we know one solution y_p of (1) and a fundamental set of solutions of (2). We call y_p a ***particular solution*** of (1); it can be any solution that we're able to find one way or another.

THEOREM 5.3.2

Suppose that p, q, and f are continuous on (a,b). Let y_p be a particular solution of

$$y'' + p(x)y' + q(x)y = f(x) \tag{3}$$

on (a,b), and let $\{y_1, y_2\}$ be a fundamental set of solutions of the complementary equation

$$y'' + p(x)y' + q(x)y = 0 \tag{4}$$

on (a,b). Then y is a solution of (3) on (a,b) if and only if

$$y = y_p + c_1 y_1 + c_2 y_2, \tag{5}$$

where c_1 and c_2 are constants.

PROOF. We first show that y in (5) is a solution of (3) for any choice of the constants c_1 and c_2. Differentiating (5) twice yields

$$y' = y_p' + c_1 y_1' + c_2 y_2' \quad \text{and} \quad y'' = y_p'' + c_1 y_1'' + c_2 y_2'',$$

so

$$\begin{aligned}
y'' + p(x)y' + q(x)y &= (y_p'' + c_1 y_1'' + c_2 y_2'') + p(x)(y_p' + c_1 y_1' + c_2 y_2') \\
&\quad + q(x)(y_p + c_1 y_1 + c_2 y_2) \\
&= (y_p'' + p(x)y_p' + q(x)y_p) + c_1(y_1'' + p(x)y_1' + q(x)y_1) \\
&\quad + c_2(y_2'' + p(x)y_2' + q(x)y_2) \\
&= f + c_1 \cdot 0 + c_2 \cdot 0 = f,
\end{aligned}$$

since y_p satisfies (3) while y_1 and y_2 satisfy (4).

Now we will show that every solution of (3) is of the form (5) for some choice of the constants c_1 and c_2. Suppose that y is a solution of (3). We will show that $y - y_p$ is a solution of (4), and therefore of the form $y - y_p = c_1 y_1 + c_2 y_2$, which implies (5). To see this we compute

$$\begin{aligned}
(y - y_p)'' + p(x)(y - y_p)' + q(x)(y - y_p) &= (y'' - y_p'') + p(x)(y' - y_p') \\
&\quad + q(x)(y - y_p) \\
&= (y'' + p(x)y' + q(x)y) \\
&\quad - (y_p'' + p(x)y_p' + q(x)y_p) \\
&= f(x) - f(x) = 0,
\end{aligned}$$

since y and y_p both satisfy (3). $\qquad\square$

We say that (5) is the ***general solution of (3) on (a,b).***

If P_0, P_1, and F are continuous and P_0 has no zeros on (a,b), then Theorem 5.3.2 implies that the general solution of

$$P_0(x)y'' + P_1(x)y' + P_2(x)y = F(x) \tag{6}$$

on (a,b) is $y = y_p + c_1 y_1 + c_2 y_2$, where y_p is a particular solution of (6) on (a,b) and $\{y_1, y_2\}$ is a fundamental set of solutions of

$$P_0(x)y'' + P_1(x)y' + P_2(x)y = 0$$

on (a,b). To see this, we rewrite (6) as

$$y'' + \frac{P_1(x)}{P_0(x)} y' + \frac{P_2(x)}{P_0(x)} y = \frac{F(x)}{P_0(x)}$$

and apply Theorem 5.3.2 with $p = P_1/P_0$, $q = P_2/P_0$, and $f = F/P_0$.

To avoid awkward wording in examples and exercises, we won't specify the interval (a,b) when we ask for the general solution of a specific linear second order equation, or for a fundamental set of solutions of a homogeneous linear second order equation. Let's agree that this always means that we want the general solution (or a fundamental set of solutions, as the case may be) on every open interval on which p, q, and f are continuous if the equation is of the form (3), or on which P_0, P_1, P_2, and F are continuous and P_0 has no zeros, if the equation is of the form (6). We leave it to you to identify these intervals in specific examples and exercises.

For completeness we point out that if P_0, P_1, P_2, and F are all continuous on an open interval (a,b), but P_0 *does* have a zero in (a,b), then (6) may fail to have a general solution on (a,b) in the sense just defined. Exercises 42–44 of Section 5.1 illustrate this point for a homogeneous equation.

In this section we have to limit ourselves to applications of Theorem 5.3.2 where we can guess at the form of the particular solution.

EXAMPLE 1   **(a)** Find the general solution of

$$y'' + y = 1. \tag{7}$$

(b) Solve the initial value problem

$$y'' + y = 1, \qquad y(0) = 2, \qquad y'(0) = 7. \tag{8}$$

Solution   **(a)** We can apply Theorem 5.3.2 with $(a,b) = (-\infty,\infty)$, since the functions $p \equiv 0$, $q \equiv 1$, and $f \equiv 1$ in (7) are continuous on $(-\infty,\infty)$. By inspection we see that $y_p \equiv 1$ is a particular solution of (7). Since $y_1 = \cos x$ and $y_2 = \sin x$ form a fundamental set of solutions of the complementary equation $y'' + y = 0$, the general solution of (7) is

$$y = 1 + c_1 \cos x + c_2 \sin x. \tag{9}$$

Solution   **(b)** Imposing the initial condition $y(0) = 2$ in (9) yields $2 = 1 + c_1$, so $c_1 = 1$. Differentiating (9) yields

$$y' = -c_1 \sin x + c_2 \cos x.$$

Imposing the initial condition $y'(0) = 7$ here yields $c_2 = 7$, so the solution of (8) is

$$y = 1 + \cos x + 7 \sin x.$$

The graph of this solution is shown in Figure 1.

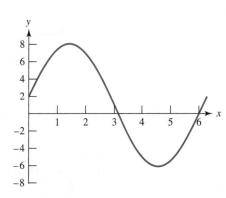

Figure 1   Graph of
$y = 1 + \cos x + 7 \sin x$

EXAMPLE 2   **(a)** Find the general solution of

$$y'' - 2y' + y = -3 - x + x^2. \tag{10}$$

(b) Solve the initial value problem

$$y'' - 2y' + y = -3 - x + x^2, \qquad y(0) = -2, \qquad y'(0) = 1. \tag{11}$$

Solution   **(a)** The characteristic polynomial of the complementary equation

$$y'' - 2y' + y = 0$$

is $r^2 - 2r + 1 = (r - 1)^2$, so $y_1 = e^x$ and $y_2 = xe^x$ form a fundamental set of solutions of the complementary equation. To guess a form for a particular solution of (10), we note that substituting a second degree polynomial $y_p = A + Bx + Cx^2$ into the left side of (10) will produce another second degree polynomial with coefficients that depend upon $A, B,$ and C. The strategy is to choose $A, B,$ and C so the polynomials on the two sides of (10) have the same coefficients; thus, if

$$y_p = A + Bx + Cx^2 \qquad \text{then} \qquad y_p' = B + 2Cx \qquad \text{and} \qquad y_p'' = 2C,$$

so we must have

$$y_p'' - 2y_p' + y_p = 2C - 2(B + 2Cx) + (A + Bx + Cx^2)$$
$$= (2C - 2B + A) + (-4C + B)x + Cx^2 = -3 - x + x^2.$$

Equating coefficients of like powers of x on the two sides of the last equality yields

$$C = 1$$
$$B - 4C = -1$$
$$A - 2B + 2C = -3,$$

so $C = 1$, $B = -1 + 4C = 3$, and $A = -3 - 2C + 2B = 1$. Therefore $y_p = 1 + 3x + x^2$ is a particular solution of (10), and Theorem 5.3.2 implies that

$$y = 1 + 3x + x^2 + e^x(c_1 + c_2 x) \tag{12}$$

is the general solution of (10).

Solution   **(b)** Imposing the initial condition $y(0) = -2$ in (12) yields $-2 = 1 + c_1$, so $c_1 = -3$. Differentiating (12) yields

$$y' = 3 + 2x + e^x(c_1 + c_2 x) + c_2 e^x,$$

and imposing the initial condition $y'(0) = 1$ here yields $1 = 3 + c_1 + c_2$, so $c_2 = 1$. Therefore the solution of (11) is

$$y = 1 + 3x + x^2 - e^x(3 - x).$$

The graph of this solution is shown in Figure 2.

EXAMPLE 3   Find the general solution of

$$x^2 y'' + xy' - 4y = 2x^4 \tag{13}$$

on $(-\infty, 0)$ and $(0, \infty)$.

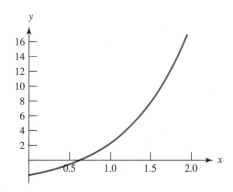

Figure 2 Graph of
$y = 1 + 3x + x^2 - e^x(3 - x)$

Solution   In Example 3 of Section 5.1 we verified that $y_1 = x^2$ and $y_2 = 1/x^2$ form a fundamental set of solutions of the complementary equation

$$x^2 y'' + xy' - 4y = 0$$

on $(-\infty, 0)$ and $(0, \infty)$. To find a particular solution of (13) we note that if we try
a solution of the form $y_p = Ax^4$ where A is a constant, then both sides of (13) will
be constant multiples of x^4, so we may be able to choose A so the two sides are
equal. This is true in this example, since if $y_p = Ax^4$ then

$$x^2 y_p'' + xy_p' - 4y_p = x^2(12Ax^2) + x(4Ax^3) - 4Ax^4 = 12Ax^4 = 2x^4$$

if $A = 1/6$; therefore $y_p = x^4/6$ is a particular solution of (13) on $(-\infty, \infty)$. Theorem 5.3.2 implies that the general solution of (13) on $(-\infty, 0)$ and $(0, \infty)$ is

$$y = \frac{x^4}{6} + c_1 x^2 + \frac{c_2}{x^2}. \qquad \blacksquare$$

THE PRINCIPLE OF SUPERPOSITION

The following theorem enables us to break a nonhomogeous equation into simpler parts, find a particular solution for each part, and then combine their solutions to obtain a particular solution of the original problem.

THEOREM 5.3.3

(The Principle of
Superposition)

Suppose that y_{p_1} is a particular solution of
$$y'' + p(x)y' + q(x)y = f_1(x)$$
on (a, b) and y_{p_2} is a particular solution of
$$y'' + p(x)y' + q(x)y = f_2(x)$$
on (a, b). Then
$$y_p = y_{p_1} + y_{p_2}$$
is a particular solution of
$$y'' + p(x)y' + q(x)y = f_1(x) + f_2(x)$$
on (a, b).

PROOF. If $y_p = y_{p_1} + y_{p_2}$ then

$$
\begin{aligned}
y_p'' + p(x)y_p' + q(x)y_p &= (y_{p_1} + y_{p_2})'' + p(x)(y_{p_1} + y_{p_2})' \\
&\quad + q(x)(y_{p_1} + y_{p_2}) \\
&= (y_{p_1}'' + p(x)y_{p_1}' + q(x)y_{p_1}) \\
&\quad + (y_{p_2}'' + p(x)y_{p_2}' + q(x)y_{p_2}) \\
&= f_1(x) + f_2(x).
\end{aligned}
$$
$\square$

It is straightforward to generalize Theorem 5.3.3 to the equation

$$y'' + p(x)y' + q(x)y = f(x) \tag{14}$$

where

$$f = f_1 + f_2 + \cdots + f_k;$$

thus, if y_{p_i} is a particular solution of

$$y'' + p(x)y' + q(x)y = f_i(x)$$

on (a,b) for $i = 1, 2, \ldots, k$ then $y_{p_1} + y_{p_2} + \cdots + y_{p_k}$ is a particular solution of (14) on (a,b). Moreover, by a proof similar to the proof of Theorem 5.3.3 we can formulate the principle of superposition in terms of a linear equation written in the form

$$P_0(x)y'' + P_1(x)y' + P_2(x)y = F(x)$$

(Exercise 39); that is, if y_{p_1} is a particular solution of

$$P_0(x)y'' + P_1(x)y' + P_2(x)y = F_1(x)$$

on (a,b) and y_{p_2} is a particular solution of

$$P_0(x)y'' + P_1(x)y' + P_2(x)y = F_2(x)$$

on (a,b), then $y_{p_1} + y_{p_2}$ is a solution of

$$P_0(x)y'' + P_1(x)y' + P_2(x)y = F_1(x) + F_2(x)$$

on (a,b).

EXAMPLE 4 The function $y_{p_1} = x^4/15$ is a particular solution of

$$x^2 y'' + 4xy' + 2y = 2x^4 \tag{15}$$

on $(-\infty, \infty)$ and $y_{p_2} = x^2/3$ is a particular solution of

$$x^2 y'' + 4xy' + 2y = 4x^2 \tag{16}$$

on $(-\infty, \infty)$. Use the principle of superposition to find a particular solution of

$$x^2 y'' + 4xy' + 2y = 2x^4 + 4x^2 \tag{17}$$

on $(-\infty, \infty)$.

Solution The right side $F(x) = 2x^4 + 4x^2$ in (17) is the sum of the right sides

$$F_1(x) = 2x^4 \quad \text{and} \quad F_2(x) = 4x^2$$

in (15) and (16). Therefore the principle of superposition implies that

$$y_p = y_{p_1} + y_{p_2} = \frac{x^4}{15} + \frac{x^2}{3}$$

is a particular solution of (17). $\blacksquare$

5.3 EXERCISES

In Exercises 1–6 find a particular solution by the method used in Example 2. Then find the general solution and, where indicated, solve the initial value problem and graph the solution.

1. $y'' + 5y' - 6y = 22 + 18x - 18x^2$

2. $y'' - 4y' + 5y = 1 + 5x$

3. $y'' + 8y' + 7y = -8 - x + 24x^2 + 7x^3$

4. $y'' - 4y' + 4y = 2 + 8x - 4x^2$

C **5.** $y'' + 2y' + 10y = 4 + 26x + 6x^2 + 10x^3$, $y(0) = 2$, $y'(0) = 9$

C **6.** $y'' + 6y' + 10y = 22 + 20x$, $y(0) = 2$, $y'(0) = -2$

7. Show that the method used in Example 2 will not yield a particular solution of

$$y'' + y' = 1 + 2x + x^2; \tag{A}$$

that is, (A) does not have a particular solution of the form $y_p = A + Bx + Cx^2$ where A, B, and C are constants.

In Exercises 8–13 find a particular solution by the method used in Example 3.

8. $x^2 y'' + 7xy' + 8y = \dfrac{6}{x}$

9. $x^2 y'' - 7xy' + 7y = 13x^{1/2}$

10. $x^2 y'' - xy' + y = 2x^3$

11. $x^2 y'' + 5xy' + 4y = \dfrac{1}{x^3}$

12. $x^2 y'' + xy' + y = 10x^{1/3}$

13. $x^2 y'' - 3xy' + 13y = 2x^4$

14. Show that the method suggested for finding a particular solution in Exercises 8–13 will not yield a particular solution of

$$x^2 y'' + 3xy' - 3y = \dfrac{1}{x^3}; \tag{A}$$

that is, (A) does not have a particular solution of the form $y_p = A/x^3$.

15. Prove: If a, b, c, α, and M are constants and $M \neq 0$, then the equation

$$ax^2 y'' + bxy' + cy = Mx^\alpha$$

has a particular solution $y_p = Ax^\alpha$ ($A =$ constant) if and only if $a\alpha(\alpha - 1) + b\alpha + c \neq 0$.

If a, b, c, and α are constants then

$$a(e^{\alpha x})'' + b(e^{\alpha x})' + ce^{\alpha x} = (a\alpha^2 + b\alpha + c)e^{\alpha x}.$$

Use this in Exercises 16–21 to find a particular solution. Then find the general solution and, where indicated, solve the initial value problem and graph the solution.

16. $y'' + 5y' - 6y = 6e^{3x}$

17. $y'' - 4y' + 5y = e^{2x}$

C **18.** $y'' + 8y' + 7y = 10e^{-2x}$, $y(0) = -2$, $y'(0) = 10$

C **19.** $y'' - 4y' + 4y = e^x$, $y(0) = 2$, $y'(0) = 0$

20. $y'' + 2y' + 10y = e^{x/2}$

21. $y'' + 6y' + 10y = e^{-3x}$

22. Show that the method suggested for finding a particular solution in Exercises 16–21 will not yield a particular solution of

$$y'' - 7y' + 12y = 5e^{4x}; \tag{A}$$

that is, (A) does not have a particular solution of the form $y_p = Ae^{4x}$.

23. Prove: If α and M are constants and $M \neq 0$, then the constant coefficient equation

$$ay'' + by' + cy = Me^{\alpha x}$$

has a particular solution $y_p = Ae^{\alpha x}$ ($A =$ constant) if and only if $e^{\alpha x}$ is not a solution of the complementary equation.

If ω is a constant then differentiating a linear combination of cos ωx *and* sin ωx *with respect to x yields another linear combination of* cos ωx *and* sin ωx. *In Exercises 24–29 use this to find a particular solution of the equation. Then find the general solution and, where indicated, solve the initial value problem and graph the solution.*

24. $y'' - 8y' + 16y = 23 \cos x - 7 \sin x$

25. $y'' + y' = -8 \cos 2x + 6 \sin 2x$

26. $y'' - 2y' + 3y = -6 \cos 3x + 6 \sin 3x$

27. $y'' + 6y' + 13y = 18 \cos x + 6 \sin x$

C 28. $y'' + 7y' + 12y = -2 \cos 2x + 36 \sin 2x, \quad y(0) = -3, \quad y'(0) = 3$

C 29. $y'' - 6y' + 9y = 18 \cos 3x + 18 \sin 3x, \quad y(0) = 2, \quad y'(0) = 2$

30. Find the general solution of

$$y'' + \omega_0^2 y = M \cos \omega x + N \sin \omega x,$$

where M and N are constants and ω and ω_0 are distinct positive numbers.

31. Show that the method suggested for finding a particular solution in Exercises 24–29 will not yield a particular solution of

$$y'' + y = \cos x + \sin x; \tag{A}$$

that is, (A) does not have a particular solution of the form $y_p = A \cos x + B \sin x$.

32. Prove: If M and N are constants (not both zero) and $\omega > 0$, then the constant coefficient equation

$$ay'' + by' + cy = M \cos \omega x + N \sin \omega x \tag{A}$$

has a particular solution that is a linear combination of cos ωx and sin ωx if and only if the left side of (A) is not of the form $a(y'' + \omega^2 y)$, so that cos ωx and sin ωx are solutions of the complementary equation.

In Exercises 33–38 refer to the cited exercises and use the principal of superposition to find a particular solution. Then find the general solution.

33. $y'' + 5y' - 6y = 22 + 18x - 18x^2 + 6e^{3x}$ (See Exercises 1 and 16.)

34. $y'' - 4y' + 5y = 1 + 5x + e^{2x}$ (See Exercises 2 and 17.)

35. $y'' + 8y' + 7y = -8 - x + 24x^2 + 7x^3 + 10e^{-2x}$ (See Exercises 3 and 18.)

36. $y'' - 4y' + 4y = 2 + 8x - 4x^2 + e^x$ (See Exercises 4 and 19.)

37. $y'' + 2y' + 10y = 4 + 26x + 6x^2 + 10x^3 + e^{x/2}$ (See Exercises 5 and 20.)

38. $y'' + 6y' + 10y = 22 + 20x + e^{-3x}$ (See Exercises 6 and 21.)

39. Prove: If y_{p_1} is a particular solution of

$$P_0(x)y'' + P_1(x)y' + P_2(x)y = F_1(x)$$

on (a,b) and y_{p_2} is a particular solution of

$$P_0(x)y'' + P_1(x)y' + P_2(x)y = F_2(x)$$

on (a,b), then $y_p = y_{p_1} + y_{p_2}$ is a solution of

$$P_0(x)y'' + P_1(x)y' + P_2(x)y = F_1(x) + F_2(x)$$

on (a,b).

40. Suppose that p, q, and f are continuous on (a,b). Let y_1, y_2, and y_p be twice differentiable on (a,b), such that $y = c_1 y_1 + c_2 y_2 + y_p$ is a solution of

$$y'' + p(x)y' + q(x)y = f$$

on (a,b) for every choice of the constants c_1, c_2. Show that y_1 and y_2 are solutions of the complementary equation on (a,b).

5.4   The Method of Undetermined Coefficients I

In this section we consider the constant coefficient equation

$$ay'' + by' + cy = e^{\alpha x} G(x) \tag{1}$$

where α is a constant and G is a polynomial.

From Theorem 5.3.2 the general solution of (1) is $y = y_p + c_1 y_1 + c_2 y_2$, where y_p is a particular solution of (1) and $\{y_1, y_2\}$ is a fundamental set of solutions of the complementary equation

$$ay'' + by' + cy = 0.$$

In Section 5.2 we showed how to find $\{y_1, y_2\}$. In this section we will show how to find y_p. The procedure that we will use is called **the method of undetermined coefficients.**

Our first example is similar to Exercises 16–21 of Section 5.3.

EXAMPLE 1   Find a particular solution of

$$y'' - 7y' + 12y = 4e^{2x}. \tag{2}$$

Then find the general solution.

Solution   Substituting $y_p = Ae^{2x}$ for y in (2) will produce a constant multiple of Ae^{2x} on the left side of (2), so it may be possible to choose A so that y_p is a solution of (2). Let's try it; if $y_p = Ae^{2x}$ then

$$y_p'' - 7y_p' + 12y_p = 4Ae^{2x} - 14Ae^{2x} + 12Ae^{2x} = 2Ae^{2x} = 4e^{2x}$$

if $A = 2$. Therefore $y_p = 2e^{2x}$ is a particular solution of (2). To find the general solution, we note that the characteristic polynomial of the complementary equation

$$y'' - 7y' + 12y = 0 \tag{3}$$

is $p(r) = r^2 - 7r + 12 = (r - 3)(r - 4)$, so $\{e^{3x}, e^{4x}\}$ is a fundamental set of solutions of (3). Therefore the general solution of (2) is

$$y = 2e^{2x} + c_1 e^{3x} + c_2 e^{4x}.$$

EXAMPLE 2   Find a particular solution of

$$y'' - 7y' + 12y = 5e^{4x}. \tag{4}$$

Then find the general solution.

Solution   Fresh from our success in finding a particular solution of (2)—where we chose $y_p = Ae^{2x}$ because the right side of (2) is a constant multiple of e^{2x}—it may seem reasonable to try $y_p = Ae^{4x}$ as a particular solution of (4). However, this won't work, since we saw in Example 1 that e^{4x} is a solution of the complementary equation (3), so substituting $y_p = Ae^{4x}$ into the left side of (4) produces zero on the left, no matter how A is chosen. To discover a suitable form for y_p, we use the same approach that we used in Section 5.2 to find a second solution of

$$ay'' + by' + cy = 0$$

in the case where the characteristic equation has a repeated real root: we look for solutions of (4) in the form $y = ue^{4x}$, where u is a function to be determined. Substituting

$$y = ue^{4x}, \quad y' = u'e^{4x} + 4ue^{4x}, \quad \text{and} \quad y'' = u''e^{4x} + 8u'e^{4x} + 16ue^{4x} \quad (5)$$

into (4) and canceling the common factor e^{4x} yields

$$(u'' + 8u' + 16u) - 7(u' + 4u) + 12u = 5,$$

or

$$u'' + u' = 5.$$

By inspection we see that $u_p = 5x$ is a particular solution of this equation, so $y_p = 5xe^{4x}$ is a particular solution of (4). Therefore

$$y = 5xe^{4x} + c_1e^{3x} + c_2e^{4x}$$

is the general solution.

EXAMPLE 3 Find a particular solution of

$$y'' - 8y' + 16y = 2e^{4x}. \quad (6)$$

Solution Since the characteristic polynomial of the complementary equation

$$y'' - 8y' + 16y = 0 \quad (7)$$

is $p(r) = r^2 - 8r + 16 = (r - 4)^2$, both $y_1 = e^{4x}$ and $y_2 = xe^{4x}$ are solutions of (7). Therefore (6) does not have a solution of the form $y_p = Ae^{4x}$ or $y_p = Axe^{4x}$. As in Example 2, we look for solutions of (6) in the form $y = ue^{4x}$, where u is a function to be determined. Substituting from (5) into (6) and canceling the common factor e^{4x} yields

$$(u'' + 8u' + 16u) - 8(u' + 4u) + 16u = 2,$$

or

$$u'' = 2.$$

Integrating twice and taking the constants of integration to be zero shows that $u_p = x^2$ is a particular solution of this equation, so $y_p = x^2e^{4x}$ is a particular solution of (4). Therefore

$$y = e^{4x}(x^2 + c_1 + c_2x)$$

is the general solution. ■

The preceding examples illustrate the following facts concerning the form of a particular solution y_p of a constant coefficient equation

$$ay'' + by' + cy = ke^{\alpha x}$$

where k is a nonzero constant:

1. If $e^{\alpha x}$ is not a solution of the complementary equation

$$ay'' + by' + cy = 0, \quad (8)$$

then $y_p = Ae^{\alpha x}$ where A is a constant. (See Example 1.)
2. If $e^{\alpha x}$ is a solution of (8) but $xe^{\alpha x}$ is not, then $y_p = Axe^{\alpha x}$ where A is a constant. (See Example 2.)

3. If both $e^{\alpha x}$ and $xe^{\alpha x}$ are solutions of (8) then $y_p = Ax^2e^{\alpha x}$ where A is a constant. (See Example 3.)

See Exercise 30 for the proofs of these facts.

In all three cases you can just substitute the appropriate form for y_p and its derivatives directly into

$$ay_p'' + by_p' + cy_p = ke^{\alpha x},$$

and solve for the constant A, as we did in Example 1. (See Exercises 31–33.) However, if the equation is

$$ay'' + by' + cy = ke^{\alpha x}G(x)$$

where G is a polynomial of degree greater than zero, we recommend that you use the substitution $y = ue^{\alpha x}$ as we did in Examples 2 and 3. The equation for u will turn out to be

$$au'' + p'(\alpha)u' + p(\alpha)u = G(x), \tag{9}$$

where $p(r) = ar^2 + br + c$ is the characteristic polynomial of the complementary equation and $p'(r) = 2ar + b$ (Exercise 30); however, you shouldn't memorize this since it is easy to derive the equation for u in any particular case. Note, however, that if $e^{\alpha x}$ is a solution of the complementary equation then $p(\alpha) = 0$, so (9) reduces to

$$au'' + p'(\alpha)u' = G(x),$$

while if both $e^{\alpha x}$ and $xe^{\alpha x}$ are solutions of the complementary equation, then $p(r) = a(r - \alpha)^2$ and $p'(r) = 2a(r - \alpha)$, so $p(\alpha) = p'(\alpha) = 0$, and (9) reduces to

$$au'' = G(x).$$

EXAMPLE 4   Find a particular solution of

$$y'' - 3y' + 2y = e^{3x}(-1 + 2x + x^2). \tag{10}$$

Solution   Substituting

$$y = ue^{3x}, \qquad y' = u'e^{3x} + 3ue^{3x}, \qquad \text{and} \qquad y'' = u''e^{3x} + 6u'e^{3x} + 9ue^{3x}$$

into (10) and canceling e^{3x} yields

$$(u'' + 6u' + 9u) - 3(u' + 3u) + 2u = -1 + 2x + x^2,$$

or

$$u'' + 3u' + 2u = -1 + 2x + x^2. \tag{11}$$

As in Example 2 of Section 5.3, in order to guess a form for a particular solution of (11) we note that substituting a second degree polynomial $u_p = A + Bx + Cx^2$ for u in the left side of (11) produces another second degree polynomial with coefficients that depend upon A, B, and C; thus,

if $u_p = A + Bx + Cx^2$ then $u_p' = B + 2Cx$ and $u_p'' = 2C.$

If u_p is to satisfy (11) we must have

$$u_p'' + 3u_p' + 2u_p = 2C + 3(B + 2Cx) + 2(A + Bx + Cx^2)$$
$$= (2C + 3B + 2A) + (6C + 2B)x + 2Cx^2 = -1 + 2x + x^2.$$

Equating coefficients of like powers of x on the two sides of the last equality yields

$$
\begin{aligned}
2C &= 1 \\
2B + 6C &= 2 \\
2A + 3B + 2C &= -1.
\end{aligned}
$$

Solving these equations for C, B, and A (in that order) yields $C = 1/2, B = -1/2$, $A = -1/4$. Therefore

$$u_p = -\frac{1}{4}(1 + 2x - 2x^2)$$

is a particular solution of (11), and

$$y_p = u_p e^{3x} = -\frac{e^{3x}}{4}(1 + 2x - 2x^2)$$

is a particular solution of (10).

EXAMPLE 5 Find a particular solution of

$$y'' - 4y' + 3y = e^{3x}(6 + 8x + 12x^2). \tag{12}$$

Solution Substituting

$$y = ue^{3x}, \qquad y' = u'e^{3x} + 3ue^{3x}, \qquad \text{and} \qquad y'' = u''e^{3x} + 6u'e^{3x} + 9ue^{3x}$$

into (12) and canceling e^{3x} yields

$$(u'' + 6u' + 9u) - 4(u' + 3u) + 3u = 6 + 8x + 12x^2,$$

or

$$u'' + 2u' = 6 + 8x + 12x^2. \tag{13}$$

There is no u term in this equation, since e^{3x} is a solution of the complementary equation for (12). (See Exercise 30.) Therefore (13) does not have a particular solution of the form $u_p = A + Bx + Cx^2$ that we used successfully in Example 4, since with this choice of u_p,

$$u_p'' + 2u_p' = 2C + (B + 2Cx)$$

cannot contain the last term $(12x^2)$ on the right side of (13). Instead, let's try $u_p = Ax + Bx^2 + Cx^3$ on the grounds that

$$u_p' = A + 2Bx + 3Cx^2 \qquad \text{and} \qquad u_p'' = 2B + 6Cx$$

together contain all the powers of x that appear on the right side of (13).

Substituting these expressions in place of u' and u'' in (13) yields

$$(2B + 6Cx) + 2(A + 2Bx + 3Cx^2) = (2B + 2A) + (6C + 4B)x + 6Cx^2$$
$$= 6 + 8x + 12x^2.$$

Comparing coefficients of like powers of x on the two sides of the last equality shows that u_p satisfies (13) if

$$
\begin{aligned}
6C &= 12 \\
4B + 6C &= 8 \\
2A + 2B &= 6.
\end{aligned}
$$

Solving these equations successively yields $C = 2, B = -1$, and $A = 4$. Therefore

$$u_p = x(4 - x + 2x^2)$$

is a particular solution of (13), and

$$y_p = u_p e^{3x} = xe^{3x}(4 - x + 2x^2)$$

is a particular solution of (12).

EXAMPLE 6   Find a particular solution of

$$4y'' + 4y' + y = e^{-x/2}(-8 + 48x + 144x^2). \tag{14}$$

Solution   Substituting

$$y = ue^{-x/2}, \quad y' = u'e^{-x/2} - \frac{1}{2}ue^{-x/2}, \quad \text{and} \quad y'' = u''e^{-x/2} - u'e^{-x/2} + \frac{1}{4}ue^{-x/2}$$

into (14) and canceling $e^{-x/2}$ yields

$$4\left(u'' - u' + \frac{u}{4}\right) + 4\left(u' - \frac{u}{2}\right) + u = 4u'' = -8 + 48x + 144x^2,$$

or

$$u'' = -2 + 12x + 36x^2, \tag{15}$$

which does not contain u or u' because $e^{-x/2}$ and $xe^{-x/2}$ are both solutions of the complementary equation. (See Exercise 30.) To obtain a particular solution of (15) we integrate twice, taking the constants of integration to be zero; thus,

$$u'_p = -2x + 6x^2 + 12x^3 \quad \text{and} \quad u_p = -x^2 + 2x^3 + 3x^4 = x^2(-1 + 2x + 3x^2).$$

Therefore

$$y_p = u_p e^{-x/2} = x^2 e^{-x/2}(-1 + 2x + 3x^2)$$

is a particular solution of (14).   ∎

SUMMARY

The preceding examples illustrate the following facts concerning particular solutions of a constant coefficient equation of the form

$$ay'' + by' + cy = e^{\alpha x}G(x),$$

where G is a polynomial (see Exercise 30):

1. If $e^{\alpha x}$ is not a solution of the complementary equation

$$ay'' + by' + cy = 0, \tag{16}$$

then $y_p = e^{\alpha x}Q(x)$ where Q is a polynomial of the same degree as G. (See Example 4.)
2. If $e^{\alpha x}$ is a solution of (16) but $xe^{\alpha x}$ is not, then $y_p = xe^{\alpha x}Q(x)$ where Q is a polynomial of the same degree as G. (See Example 5.)
3. If both $e^{\alpha x}$ and $xe^{\alpha x}$ are solutions of (16), then $y_p = x^2 e^{\alpha x}Q(x)$ where Q is a polynomial of the same degree as G. (See Example 6.)

In all three cases, you can just substitute the appropriate form for y_p and its derivatives directly into

$$ay_p'' + by_p' + cy_p = e^{\alpha x}G(x),$$

and solve for the coefficients of the polynomial Q. However, if you try this you will see that the computations are more tedious than those that you encounter by making the substitution $y = ue^{\alpha x}$ and finding a particular solution of the resulting equation for u. (See Exercises 34–36.) In item 1 (above) the equation for u will be of the form

$$au'' + p'(\alpha)u' + p(\alpha)u = G(x),$$

with a particular solution of the form $u_p = Q(x)$, a polynomial of the same degree as G, whose coefficients can be found by the method used in Example 4. In item 2 (above) the equation for u will be of the form

$$au'' + p'(\alpha)u' = G(x)$$

(no u term on the left), with a particular solution of the form $u_p = xQ(x)$, where Q is a polynomial of the same degree as G whose coefficients can be found by the method used in Example 5. In item 3 (above) the equation for u will be of the form

$$au'' = G(x)$$

with a particular solution of the form $u_p = x^2Q(x)$ that can be obtained by integrating $G(x)/a$ twice and taking the constants of integration to be zero, as in Example 6.

USING THE PRINCIPLE OF SUPERPOSITION

The next example shows how to combine the method of undetermined coefficients and Theorem 5.3.3, the principle of superposition.

EXAMPLE 7   Find a particular solution of

$$y'' - 7y' + 12y = 4e^{2x} + 5e^{4x}. \tag{17}$$

Solution   In Example 1 we found that $y_{p_1} = 2e^{2x}$ is a particular solution of

$$y'' - 7y' + 12y = 4e^{2x},$$

and in Example 2 we found that $y_{p_2} = 5xe^{4x}$ is a particular solution of

$$y'' - 7y' + 12y = 5e^{4x}.$$

Therefore the principle of superposition implies that $y_p = 2e^{2x} + 5xe^{4x}$ is a particular solution of (17).   ∎

5.4 EXERCISES

In Exercises 1–14 find a particular solution.

1. $y'' - 3y' + 2y = e^{3x}(1 + x)$

2. $y'' - 6y' + 5y = e^{-3x}(35 - 8x)$

3. $y'' - 2y' - 3y = e^x(-8 + 3x)$

4. $y'' + 2y' + y = e^{2x}(-7 - 15x + 9x^2)$

5. $y'' + 4y = e^{-x}(7 - 4x + 5x^2)$

6. $y'' - y' - 2y = e^x(9 + 2x - 4x^2)$

7. $y'' - 4y' - 5y = -6xe^{-x}$

8. $y'' - 3y' + 2y = e^x(3 - 4x)$

9. $y'' + y' - 12y = e^{3x}(-6 + 7x)$

10. $2y'' - 3y' - 2y = e^{2x}(-6 + 10x)$

11. $y'' + 2y' + y = e^{-x}(2 + 3x)$

12. $y'' - 2y' + y = e^x(1 - 6x)$

13. $y'' - 4y' + 4y = e^{2x}(1 - 3x + 6x^2)$

14. $9y'' + 6y' + y = e^{-x/3}(2 - 4x + 4x^2)$

In Exercises 15–19 find the general solution.

15. $y'' - 3y' + 2y = e^{3x}(1 + x)$

16. $y'' - 6y' + 8y = e^x(11 - 6x)$

17. $y'' + 6y' + 9y = e^{2x}(3 - 5x)$

18. $y'' + 2y' - 3y = -16xe^x$

19. $y'' - 2y' + y = e^x(2 - 12x)$

In Exercises 20–23 solve the initial value problem and plot the solution.

C **20.** $y'' - 4y' - 5y = 9e^{2x}(1 + x)$, $\quad y(0) = 0$, $\quad y'(0) = -10$

C **21.** $y'' + 3y' - 4y = e^{2x}(7 + 6x)$, $\quad y(0) = 2$, $\quad y'(0) = 8$

C **22.** $y'' + 4y' + 3y = -e^{-x}(2 + 8x)$, $\quad y(0) = 1$, $\quad y'(0) = 2$

C **23.** $y'' - 3y' - 10y = 7e^{-2x}$, $\quad y(0) = 1$, $\quad y'(0) = -17$

In Exercises 24–29 use the principle of superposition to find a particular solution.

24. $y'' + y' + y = xe^x + e^{-x}(1 + 2x)$

25. $y'' - 7y' + 12y = -e^x(17 - 42x) - e^{3x}$

26. $y'' - 8y' + 16y = 6xe^{4x} + 2 + 16x + 16x^2$

27. $y'' - 3y' + 2y = -e^{2x}(3 + 4x) - e^x$

28. $y'' - 2y' + 2y = e^x(1 + x) + e^{-x}(2 - 8x + 5x^2)$

29. $y'' + y = e^{-x}(2 - 4x + 2x^2) + e^{3x}(8 - 12x - 10x^2)$

30. (a) Prove that y is a solution of the constant coefficient equation

$$ay'' + by' + cy = e^{\alpha x}G(x) \tag{A}$$

if and only if $y = ue^{\alpha x}$ where u satisfies

$$au'' + p'(\alpha)u' + p(\alpha)u = G(x) \tag{B}$$

and $p(r) = ar^2 + br + c$ is the characteristic polynomial of the complementary equation

$$ay'' + by' + cy = 0.$$

For the remainder of this exercise let G be a polynomial. Give the requested proofs for the case where G is of degree three; that is,

$$G(x) = g_0 + g_1x + g_2x^2 + g_3x^3.$$

(b) Prove that if $e^{\alpha x}$ is not a solution of the complementary equation then (B) has a particular solution of the form $u_p = A(x)$, where A is a polynomial of the same degree as G, as in Example 4. Conclude that (A) has a particular solution of the form $y_p = e^{\alpha x}A(x)$.

(c) Show that if $e^{\alpha x}$ is a solution of the complementary equation and $xe^{\alpha x}$ is not, then (B) has a particular solution of the form $u_p = xA(x)$, where A is a polynomial of the same degree as G, as in Example 5. Conclude that (A) has a particular solution of the form $y_p = xe^{\alpha x}A(x)$.

(d) Show that if $e^{\alpha x}$ and $xe^{\alpha x}$ are both solutions of the complementary equation, then (B) has a particular solution of the form $u_p = x^2A(x)$, where A is a polynomial of the same degree as G, and $x^2A(x)$ can be obtained by integrating G/a twice, taking the constants of integration to be zero, as in Example 6. Conclude that (A) has a particular solution of the form $y_p = x^2e^{\alpha x}A(x)$.

Exercises 31–36 treat the equations considered in Examples 1–6. Substitute the suggested form of y_p into the equation and equate the resulting coefficients of like functions on the two sides of the resulting equation to derive a set of simultaneous equations for the coefficients in y_p. Then solve for the coefficients to obtain y_p. Compare the work you've done with the work required to obtain the same results in Examples 1–6.

31. Compare with Example 1:

$$y'' - 7y' + 12y = 4e^{2x}, \qquad y_p = Ae^{2x}$$

32. Compare with Example 2:
$$y'' - 7y' + 12y = 5e^{4x}, \qquad y_p = Axe^{4x}$$

33. Compare with Example 3:
$$y'' - 8y' + 16y = 2e^{4x}, \qquad y_p = Ax^2e^{4x}$$

34. Compare with Example 4:
$$y'' - 3y' + 2y = e^{3x}(-1 + 2x + x^2), \qquad y_p = e^{3x}(A + Bx + Cx^2)$$

35. Compare with Example 5:
$$y'' - 4y' + 3y = e^{3x}(6 + 8x + 12x^2), \qquad y_p = e^{3x}(Ax + Bx^2 + Cx^3)$$

36. Compare with Example 6:
$$4y'' + 4y' + y = e^{-x/2}(-8 + 48x + 144x^2), \qquad y_p = e^{-x/2}(Ax^2 + Bx^3 + Cx^4)$$

37. Make a substitution of the form $y = ue^{\alpha x}$ to find the general solution.

(a) $y'' + 2y' + y = \dfrac{e^{-x}}{\sqrt{x}}$

(b) $y'' + 6y' + 9y = e^{-3x} \ln x$

(c) $y'' - 4y' + 4y = \dfrac{e^{2x}}{1 + x}$

(d) $4y'' + 4y' + y = 4e^{-x/2}\left(\dfrac{1}{x} + x\right)$

38. Suppose that $\alpha \neq 0$ and k is a positive integer. In most calculus books integrals like $\int x^k e^{\alpha x}\, dx$ are evaluated by integrating by parts k times. This exercise presents another method. Let

$$y = \int e^{\alpha x}P(x)\, dx$$

with

$$P(x) = p_0 + p_1 x + \cdots + p_k x^k \quad \text{(where } p_k \neq 0\text{)}.$$

(a) Show that $y = e^{\alpha x}u$ where
$$u' + \alpha u = P(x). \tag{A}$$

(b) Show that (A) has a particular solution of the form
$$u_p = A_0 + A_1 x + \cdots + A_k x^k$$

where $A_k, A_{k-1}, \ldots, A_0$ can be computed successively by equating coefficients of $x^k, x^{k-1}, \ldots, 1$ on both sides of the equation
$$u'_p + \alpha u_p = P(x).$$

(c) Conclude that
$$\int e^{\alpha x}P(x)\, dx = (A_0 + A_1 x + \cdots + A_k x^k)e^{\alpha x} + c,$$

where c is a constant of integration.

39. Use the method of Exercise 38 to evaluate the integral.

(a) $\displaystyle\int e^x(4 + x)\, dx$

(b) $\displaystyle\int e^{-x}(-1 + x^2)\, dx$

(c) $\displaystyle\int x^3 e^{-2x}\, dx$

(d) $\displaystyle\int e^x(1 + x)^2\, dx$

(e) $\displaystyle\int e^{3x}(-14 + 30x + 27x^2)\, dx$

(f) $\displaystyle\int e^{-x}(1 + 6x^2 - 14x^3 + 3x^4)\, dx$

40. Use the method suggested in Exercise 38 to evaluate $\int x^k e^{\alpha x}\, dx$, where k is an arbitrary positive integer and $\alpha \neq 0$.

5.5 The Method of Undetermined Coefficients II

In this section we consider the constant coefficient equation

$$ay'' + by' + cy = e^{\lambda x}(P(x) \cos \omega x + Q(x) \sin \omega x) \tag{1}$$

where λ and ω are real numbers, $\omega \neq 0$, and P and Q are polynomials. We want to find a particular solution of (1). As in Section 5.4, the procedure that we will use is called *the method of undetermined coefficients.*

FORCING FUNCTIONS WITHOUT EXPONENTIAL FACTORS

We begin with the case where $\lambda = 0$ in (1); thus, we want to find a particular solution of

$$ay'' + by' + cy = P(x) \cos \omega x + Q(x) \sin \omega x, \tag{2}$$

where P and Q are polynomials.

Differentiating $x^r \cos \omega x$ and $x^r \sin \omega x$ yields

$$\frac{d}{dx} x^r \cos \omega x = -\omega x^r \sin \omega x + rx^{r-1} \cos \omega x$$

and

$$\frac{d}{dx} x^r \sin \omega x = \omega x^r \cos \omega x + rx^{r-1} \sin \omega x.$$

This implies that if

$$y_p = A(x) \cos \omega x + B(x) \sin \omega x$$

where A and B are polynomials, then

$$ay_p'' + by_p' + cy_p = F(x) \cos \omega x + G(x) \sin \omega x,$$

where F and G are polynomials with coefficients that can be expressed in terms of the coefficients of A and B. This suggests that we try to choose A and B so that $F = P$ and $G = Q$, respectively. Then y_p will be a particular solution of (2). The following theorem tells us how to choose the proper form for y_p. For the proof see Exercise 37.

THEOREM 5.5.1

Suppose that ω is a positive number and P and Q are polynomials. Let k be the larger of the degrees of P and Q. Then the equation

$$ay'' + by' + cy = P(x) \cos \omega x + Q(x) \sin \omega x$$

has a particular solution

$$y_p = A(x) \cos \omega x + B(x) \sin \omega x \tag{3}$$

where

$$A(x) = A_0 + A_1 x + \cdots + A_k x^k$$

and

$$B(x) = B_0 + B_1x + \cdots + B_kx^k,$$

provided that $\cos \omega x$ *and* $\sin \omega x$ *are not solutions of the complementary equation. The solutions of*

$$a(y'' + \omega^2 y) = P(x) \cos \omega x + Q(x) \sin \omega x$$

(for which $\cos \omega x$ *and* $\sin \omega x$ *are solutions of the complementary equation) are of the form* (3), *where*

$$A(x) = A_0x + A_1x^2 + \cdots + A_kx^{k+1}$$

and

$$B(x) = B_0x + B_1x^2 + \cdots + B_kx^{k+1}.$$

For an analog of this theorem that is applicable to (1), see Exercise 38.

EXAMPLE 1   Find a particular solution of

$$y'' - 2y' + y = 5 \cos 2x + 10 \sin 2x. \tag{4}$$

Solution   In (4) the coefficients of $\cos 2x$ and $\sin 2x$ are both zero degree polynomials (constants). Therefore Theorem 5.5.1 implies that (4) has a particular solution

$$y_p = A \cos 2x + B \sin 2x.$$

Since

$$y_p' = -2A \sin 2x + 2B \cos 2x \quad \text{and} \quad y_p'' = -4(A \cos 2x + B \sin 2x),$$

replacing y by y_p in (4) yields

$$
\begin{aligned}
y_p'' - 2y_p' + y_p &= -4(A \cos 2x + B \sin 2x) - 4(-A \sin 2x + B \cos 2x) \\
&\quad + (A \cos 2x + B \sin 2x) \\
&= (-3A - 4B) \cos 2x + (4A - 3B) \sin 2x.
\end{aligned}
$$

Equating the coefficients of $\cos 2x$ and $\sin 2x$ here with the corresponding coefficients on the right side of (4) shows that y_p is a solution of (4) if

$$
\begin{aligned}
-3A - 4B &= 5 \\
4A - 3B &= 10.
\end{aligned}
$$

Solving these equation yields $A = 1, B = -2$. Therefore

$$y_p = \cos 2x - 2 \sin 2x$$

is a particular solution of (4).

EXAMPLE 2   Find a particular solution of

$$y'' + 4y = 8 \cos 2x + 12 \sin 2x. \tag{5}$$

Solution   The procedure used in Example 1 won't work here; substituting $y_p = A \cos 2x + B \sin 2x$ for y in (5) yields

$$y_p'' + 4y_p = -4(A \cos 2x + B \sin 2x) + 4(A \cos 2x + B \sin 2x) = 0$$

for any choice of A and B, since $\cos 2x$ and $\sin 2x$ are both solutions of the complementary equation for (5). We are dealing with the second case mentioned in Theorem 5.5.1, and should therefore try a particular solution of the form

$$y_p = x(A \cos 2x + B \sin 2x). \tag{6}$$

Then

$$y_p' = A \cos 2x + B \sin 2x + 2x(-A \sin 2x + B \cos 2x)$$

and

$$
\begin{aligned}
y_p'' &= -4A \sin 2x + 4B \cos 2x - 4x(A \cos 2x + B \sin 2x) \\
&= -4A \sin 2x + 4B \cos 2x - 4y_p \quad \text{(see (6))},
\end{aligned}
$$

so

$$y_p'' + 4y_p = -4A \sin 2x + 4B \cos 2x.$$

Therefore y_p is a solution of (5) if

$$-4A \sin 2x + 4B \cos 2x = 8 \cos 2x + 12 \sin 2x,$$

which holds if $A = -3$ and $B = 2$. Therefore

$$y_p = -x(3 \cos 2x - 2 \sin 2x)$$

is a particular solution of (5).

EXAMPLE 3 Find a particular solution of

$$y'' + 3y' + 2y = (16 + 20x) \cos x + 10 \sin x. \tag{7}$$

Solution The coefficients of $\cos x$ and $\sin x$ in (7) are polynomials of degree one and zero, respectively. Therefore Theorem 5.5.1 tells us to look for a particular solution of (7) of the form

$$y_p = (A_0 + A_1 x) \cos x + (B_0 + B_1 x) \sin x. \tag{8}$$

Then

$$y_p' = (A_1 + B_0 + B_1 x) \cos x + (B_1 - A_0 - A_1 x) \sin x \tag{9}$$

and

$$y_p'' = (2B_1 - A_0 - A_1 x) \cos x - (2A_1 + B_0 + B_1 x) \sin x, \tag{10}$$

so

$$
\begin{aligned}
y_p'' + 3y_p' + 2y_p = &[A_0 + 3A_1 + 3B_0 + 2B_1 + (A_1 + 3B_1)x] \cos x \\
&+ [B_0 + 3B_1 - 3A_0 - 2A_1 + (B_1 - 3A_1)x] \sin x.
\end{aligned} \tag{11}
$$

Comparing the coefficients of $x \cos x, x \sin x, \cos x$, and $\sin x$ here with the corresponding coefficients in (7) shows that y_p is a solution of (7) if

$$
\begin{aligned}
A_1 + 3B_1 &= 20 \\
-3A_1 + B_1 &= 0 \\
A_0 + 3B_0 + 3A_1 + 2B_1 &= 16 \\
-3A_0 + B_0 - 2A_1 + 3B_1 &= 10.
\end{aligned}
$$

Solving the first two equations yields $A_1 = 2$, $B_1 = 6$. Substituting these into the last two equations yields

$$A_0 + 3B_0 = 16 - 3A_1 - 2B_1 = -2$$
$$-3A_0 + B_0 = 10 + 2A_1 - 3B_1 = -4.$$

Solving these equations yields $A_0 = 1$, $B_0 = -1$. Substituting $A_0 = 1$, $A_1 = 2$, $B_0 = -1$, $B_1 = 6$ into (8) shows that

$$y_p = (1 + 2x) \cos x - (1 - 6x) \sin x$$

is a particular solution of (7). Figure 1 shows a graph of y_p.   ◼

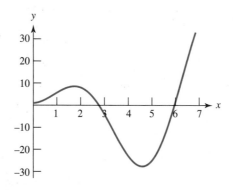

Figure 1   Graph of
$y_p = (1 + 2x) \cos x - (1 - 6x) \sin x$

A USEFUL OBSERVATION

In (9), (10), and (11) the polynomials multiplying $\sin x$ can be obtained by replacing A_0, A_1, B_0, and B_1 by B_0, B_1, $-A_0$, and $-A_1$, respectively, in the polynomials mutiplying $\cos x$. An analogous result applies in general, as follows (Exercise 36).

THEOREM 5.5.2

If

$$y_p = A(x)\cos \omega x + B(x)\sin \omega x,$$

where $A(x)$ and $B(x)$ are polynomials with coefficients $A_0, \dots, A_k$ and $B_0, \dots, B_k$, then the polynomials multiplying $\sin \omega x$ in

$$y_p', \quad y_p'', \quad ay_p'' + by_p' + cy_p \quad \text{and} \quad y_p'' + \omega^2 y_p$$

can be obtained by replacing $A_0, \dots, A_k$ by $B_0, \dots, B_k$ and $B_0, \dots, B_k$ by $-A_0, \dots, -A_k$ in the corresponding polynomials multiplying $\cos \omega x$.

We won't use this theorem in our examples, but we recommend that you use it to check your manipulations when you work the exercises.

EXAMPLE 4   Find a particular solution of

$$y'' + y = (8 - 4x) \cos x - (8 + 8x) \sin x. \tag{12}$$

Solution According to Theorem 5.5.1 we should look for a particular solution of the form

$$y_p = (A_0 x + A_1 x^2) \cos x + (B_0 x + B_1 x^2) \sin x, \tag{13}$$

since $\cos x$ and $\sin x$ are solutions of the complementary equation. However, let's try

$$y_p = (A_0 + A_1 x) \cos x + (B_0 + B_1 x) \sin x \tag{14}$$

first, so you can see why it won't work. From (10),

$$y_p'' = (2B_1 - A_0 - A_1 x) \cos x - (2A_1 + B_0 + B_1 x) \sin x,$$

which together with (14) implies that

$$y_p'' + y_p = 2B_1 \cos x - 2A_1 \sin x.$$

Since the right side of this equation does not contain $x \cos x$ or $x \sin x$, it follows that (14) cannot satisfy (12) no matter how we choose A_0, A_1, B_0, and B_1.

Now let y_p be as in (13). Then

$$\begin{aligned} y_p' &= \left[A_0 + (2A_1 + B_0)x + B_1 x^2 \right] \cos x \\ &\quad + \left[B_0 + (2B_1 - A_0)x - A_1 x^2 \right] \sin x \end{aligned}$$

and

$$\begin{aligned} y_p'' &= \left[2A_1 + 2B_0 - (A_0 - 4B_1)x - A_1 x^2 \right] \cos x \\ &\quad + \left[2B_1 - 2A_0 - (B_0 + 4A_1)x - B_1 x^2 \right] \sin x, \end{aligned}$$

so

$$y_p'' + y_p = (2A_1 + 2B_0 + 4B_1 x) \cos x + (2B_1 - 2A_0 - 4A_1 x) \sin x.$$

Comparing the coefficients of $\cos x$ and $\sin x$ here with the corresponding coefficients in (12) shows that y_p is a solution of (12) if

$$\begin{aligned} 4B_1 &= -4 \\ -4A_1 &= -8 \\ 2B_0 + 2A_1 &= 8 \\ -2A_0 + 2B_1 &= -8. \end{aligned}$$

The solution of this system is $A_1 = 2$, $B_1 = -1$, $A_0 = 3$, $B_0 = 2$. Therefore

$$y_p = x[(3 + 2x) \cos x + (2 - x) \sin x]$$

is a particular solution of (12). Figure 2 shows a graph of y_p.

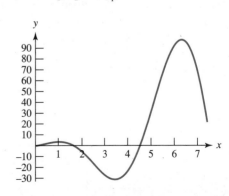

Figure 2 Graph of
$y_p = x[(3 + 2x) \cos x + (2 - x) \sin x]$

FORCING FUNCTIONS WITH EXPONENTIAL FACTORS

To find a particular solution of

$$ay'' + by' + cy = e^{\lambda x}(P(x) \cos \omega x + Q(x) \sin \omega x) \qquad (15)$$

when $\lambda \neq 0$, we recall from Section 5.4 that substituting $y = ue^{\lambda x}$ into (15) will produce a constant coefficient equation for u in which the forcing function is $P(x) \cos \omega x + Q(x) \sin \omega x$. We can find a particular solution u_p of this equation by the procedure that we used in Examples 1–4. Then $y_p = u_p e^{\lambda x}$ is a particular solution of (15).

EXAMPLE 5   Find a particular solution of

$$y'' - 3y' + 2y = e^{-2x}[2 \cos 3x - (34 - 150x) \sin 3x] \qquad (16)$$

Solution   Let $y = ue^{-2x}$. Then

$$y'' - 3y' + 2y = e^{-2x}[(u'' - 4u' + 4u) - 3(u' - 2u) + 2u]$$
$$= e^{-2x}(u'' - 7u' + 12u)$$
$$= e^{-2x}[2 \cos 3x - (34 - 150x) \sin 3x]$$

if

$$u'' - 7u' + 12u = 2 \cos 3x - (34 - 150x) \sin 3x. \qquad (17)$$

Since $\cos 3x$ and $\sin 3x$ are not solutions of the complementary equation

$$u'' - 7u' + 12u = 0,$$

Theorem 5.5.1 tells us to look for a particular solution of (17) of the form

$$u_p = (A_0 + A_1 x) \cos 3x + (B_0 + B_1 x) \sin 3x. \qquad (18)$$

Then

$$u_p' = (A_1 + 3B_0 + 3B_1 x) \cos 3x + (B_1 - 3A_0 - 3A_1 x) \sin 3x$$

and

$$u_p'' = (-9A_0 + 6B_1 - 9A_1 x) \cos 3x - (9B_0 + 6A_1 + 9B_1 x) \sin 3x,$$

so

$$u_p'' - 7u_p' + 12u_p = [3A_0 - 21B_0 - 7A_1 + 6B_1 + (3A_1 - 21B_1)x] \cos 3x$$
$$+ [21A_0 + 3B_0 - 6A_1 - 7B_1 + (21A_1 + 3B_1)x] \sin 3x.$$

Comparing the coefficients of $x \cos 3x$, $x \sin 3x$, $\cos 3x$, and $\sin 3x$ here with the corresponding coefficients on the right side of (17) shows that u_p is a solution of (17) if

$$
\begin{aligned}
3A_1 - 21B_1 &= 0 \\
21A_1 + 3B_1 &= 150 \\
3A_0 - 21B_0 - 7A_1 + 6B_1 &= 2 \\
21A_0 + 3B_0 - 6A_1 - 7B_1 &= -34.
\end{aligned}
\qquad (19)
$$

Solving the first two equations yields $A_1 = 7$, $B_1 = 1$. Substituting these values into the last two equations of (19) yields

$$3A_0 - 21B_0 = 2 + 7A_1 - 6B_1 = 45$$
$$21A_0 + 3B_0 = -34 + 6A_1 + 7B_1 = 15.$$

Solving this system yields $A_0 = 1$, $B_0 = -2$. Substituting $A_0 = 1$, $A_1 = 7$, $B_0 = -2$, and $B_1 = 1$ into (18) shows that

$$u_p = (1 + 7x) \cos 3x - (2 - x) \sin 3x$$

is a particular solution of (17). Therefore

$$y_p = e^{-2x}[(1 + 7x) \cos 3x - (2 - x) \sin 3x]$$

is a particular solution of (16).

EXAMPLE 6   Find a particular solution of

$$y'' + 2y' + 5y = e^{-x}[(6 - 16x) \cos 2x - (8 + 8x) \sin 2x]. \tag{20}$$

Solution   Let $y = ue^{-x}$. Then

$$\begin{aligned} y'' + 2y' + 5y &= e^{-x}[(u'' - 2u' + u) + 2(u' - u) + 5u] \\ &= e^{-x}(u'' + 4u) \\ &= e^{-x}[(6 - 16x) \cos 2x - (8 + 8x) \sin 2x] \end{aligned}$$

if

$$u'' + 4u = (6 - 16x) \cos 2x - (8 + 8x) \sin 2x. \tag{21}$$

Since $\cos 2x$ and $\sin 2x$ are solutions of the complementary equation

$$u'' + 4u = 0,$$

Theorem 5.5.1 tells us to look for a particular solution of (21) of the form

$$u_p = (A_0 x + A_1 x^2) \cos 2x + (B_0 x + B_1 x^2) \sin 2x.$$

Then

$$\begin{aligned} u_p' &= [A_0 + (2A_1 + 2B_0)x + 2B_1 x^2] \cos 2x \\ &\quad + [B_0 + (2B_1 - 2A_0)x - 2A_1 x^2] \sin 2x \end{aligned}$$

and

$$\begin{aligned} u_p'' &= [2A_1 + 4B_0 - (4A_0 - 8B_1)x - 4A_1 x^2] \cos 2x \\ &\quad + [2B_1 - 4A_0 - (4B_0 + 8A_1)x - 4B_1 x^2] \sin 2x, \end{aligned}$$

so

$$u_p'' + 4u_p = (2A_1 + 4B_0 + 8B_1 x) \cos 2x + (2B_1 - 4A_0 - 8A_1 x) \sin 2x.$$

Equating the coefficients of $x \cos 2x$, $x \sin 2x$, $\cos 2x$, and $\sin 2x$ here with the corresponding coefficients on the right side of (21) shows that u_p is a solution of (21) if

$$\begin{aligned} 8B_1 &= -16 \\ -8A_1 &= -8 \\ 4B_0 + 2A_1 &= 6 \\ -4A_0 + 2B_1 &= -8. \end{aligned} \tag{22}$$

The solution of this system is $A_1 = 1$, $B_1 = -2$, $B_0 = 1$, $A_0 = 1$. Therefore

$$u_p = x[(1 + x) \cos 2x + (1 - 2x) \sin 2x]$$

is a particular solution of (21), and

$$y_p = xe^{-x}[(1 + x) \cos 2x + (1 - 2x) \sin 2x]$$

is a particular solution of (20).   ■

It is also possible to find a particular solution of (20) by substituting

$$y_p = xe^{-x}[(A_0 + A_1x)\cos 2x + (B_0 + B_1x)\sin 2x]$$

for y in (20) and equating the coefficients of $xe^{-x}\cos 2x$, $xe^{-x}\sin 2x$, $e^{-x}\cos 2x$, and $e^{-x}\sin 2x$ in the resulting expression for

$$y_p'' + 2y_p' + 5y_p$$

with the corresponding coefficients on the right side of (20). (See Exercise 38.) This leads to the same system (22) of equations for A_0, A_1, B_0, and B_1 that we obtained in Example 6. However, if you try this approach you will see that deriving (22) this way is much more tedious than the way we did it in Example 6.

5.5 EXERCISES

In Exercises 1–17 find a particular solution.

1. $y'' + 3y' + 2y = 7\cos x - \sin x$

2. $y'' + 3y' + y = (2 - 6x)\cos x - 9\sin x$

3. $y'' + 2y' + y = e^x(6\cos x + 17\sin x)$

4. $y'' + 3y' - 2y = -e^{2x}(5\cos 2x + 9\sin 2x)$

5. $y'' - y' + y = e^x(2 + x)\sin x$

6. $y'' + 3y' - 2y = e^{-2x}[(4 + 20x)\cos 3x + (26 - 32x)\sin 3x]$

7. $y'' + 4y = -12\cos 2x - 4\sin 2x$

8. $y'' + y = (-4 + 8x)\cos x + (8 - 4x)\sin x$

9. $4y'' + y = -4\cos x/2 - 8x\sin x/2$

10. $y'' + 2y' + 2y = e^{-x}(8\cos x - 6\sin x)$

11. $y'' - 2y' + 5y = e^x[(6 + 8x)\cos 2x + (6 - 8x)\sin 2x]$

12. $y'' + 2y' + y = 8x^2\cos x - 4x\sin x$

13. $y'' + 3y' + 2y = (12 + 20x + 10x^2)\cos x + 8x\sin x$

14. $y'' + 3y' + 2y = (1 - x - 4x^2)\cos 2x - (1 + 7x + 2x^2)\sin 2x$

15. $y'' - 5y' + 6y = -e^x[(4 + 6x - x^2)\cos x - (2 - 4x + 3x^2)\sin x]$

16. $y'' - 2y' + y = -e^x[(3 + 4x - x^2)\cos x + (3 - 4x - x^2)\sin x]$

17. $y'' - 2y' + 2y = e^x[(2 - 2x - 6x^2)\cos x + (2 - 10x + 6x^2)\sin x]$

In Exercises 18–21 find a particular solution and graph it.

C **18.** $y'' + 2y' + y = e^{-x}[(5 - 2x)\cos x - (3 + 3x)\sin x]$

C **19.** $y'' + 9y = -6\cos 3x - 12\sin 3x$

C **20.** $y'' + 3y' + 2y = (1 - x - 4x^2)\cos 2x - (1 + 7x + 2x^2)\sin 2x$

C **21.** $y'' + 4y' + 3y = e^{-x}[(2 + x + x^2)\cos x + (5 + 4x + 2x^2)\sin x]$

In Exercises 22–26 solve the initial value problem.

22. $y'' - 7y' + 6y = -e^x(17\cos x - 7\sin x)$, $y(0) = 4$, $y'(0) = 2$

23. $y'' - 2y' + 2y = -e^x(6\cos x + 4\sin x)$, $y(0) = 1$, $y'(0) = 4$

24. $y'' + 6y' + 10y = -40e^x\sin x$, $y(0) = 2$, $y'(0) = -3$

25. $y'' - 6y' + 10y = -e^{3x}(6\cos x + 4\sin x)$, $y(0) = 2$, $y'(0) = 7$

26. $y'' - 3y' + 2y = e^{3x}[21\cos x - (11 + 10x)\sin x]$, $y(0) = 0$, $y'(0) = 6$

In Exercises 27–32 use the principle of superposition to find a particular solution. Where indicated, solve the initial value problem.

27. $y'' - 2y' - 3y = 4e^{3x} + e^x(\cos x - 2\sin x)$

28. $y'' + y = 4\cos x - 2\sin x + xe^x + e^{-x}$

29. $y'' - 3y' + 2y = xe^x + 2e^{2x} + \sin x$

30. $y'' - 2y' + 2y = 4xe^x\cos x + xe^{-x} + 1 + x^2$

31. $y'' - 4y' + 4y = e^{2x}(1 + x) + e^{2x}(\cos x - \sin x) + 3e^{3x} + 1 + x$

32. $y'' - 4y' + 4y = 6e^{2x} + 25 \sin x, \quad y(0) = 5, \quad y'(0) = 3$

In Exercises 33–35 solve the initial value problem and graph the solution.

[C] **33.** $y'' + 4y = -e^{-2x}[(4 - 7x) \cos x + (2 - 4x) \sin x], \quad y(0) = 3, \quad y'(0) = 1$

[C] **34.** $y'' + 4y' + 4y = 2 \cos 2x + 3 \sin 2x + e^{-x}, \quad y(0) = -1, \quad y'(0) = 2$

[C] **35.** $y'' + 4y = e^x(11 + 15x) + 8 \cos 2x - 12 \sin 2x, \quad y(0) = 3, \quad y'(0) = 5$

36. (a) Verify that if

$$y_p = A(x) \cos \omega x + B(x) \sin \omega x$$

where A and B are twice differentiable, then

$$y_p' = (A' + \omega B) \cos \omega x + (B' - \omega A) \sin \omega x$$

and

$$y_p'' = (A'' + 2\omega B' - \omega^2 A) \cos \omega x + (B'' - 2\omega A' - \omega^2 B) \sin \omega x.$$

(b) Use the results of **(a)** to verify that

$$ay_p'' + by_p' + cy_p = [(c - a\omega^2)A + b\omega B + 2a\omega B' + bA' + aA''] \cos \omega x$$
$$+ [-b\omega A + (c - a\omega^2)B - 2a\omega A' + bB' + aB''] \sin \omega x.$$

(c) Use the results of **(a)** to verify that

$$y_p'' + \omega^2 y_p = (A'' + 2\omega B') \cos \omega x + (B'' - 2\omega A') \sin \omega x.$$

(d) Prove Theorem 5.5.2.

37. Let a, b, c, and ω be constants, with $a \neq 0$ and $\omega > 0$, and let

$$P(x) = p_0 + p_1 x + \cdots + p_k x^k \quad \text{and} \quad Q(x) = q_0 + q_1 x + \cdots + q_k x^k,$$

where at least one of the coefficients p_k, q_k is nonzero, so k is the larger of the degrees of P and Q.

(a) Show that if $\cos \omega x$ and $\sin \omega x$ are not solutions of the complementary equation

$$ay'' + by' + cy = 0$$

then there are polynomials

$$A(x) = A_0 + A_1 x + \cdots + A_k x^k \quad \text{and} \quad B(x) = B_0 + B_1 x + \cdots + B_k x^k \qquad \text{(A)}$$

such that

$$(c - a\omega^2)A + b\omega B + 2a\omega B' + bA' + aA'' = P$$
$$-b\omega A + (c - a\omega^2)B - 2a\omega A' + bB' + aB'' = Q,$$

where $(A_k, B_k), (A_{k-1}, B_{k-1}), \ldots, (A_0, B_0)$ can be computed successively by solving the systems

$$(c - a\omega^2)A_k + b\omega B_k = p_k$$
$$-b\omega A_k + (c - a\omega^2)B_k = q_k,$$

and, if $1 \leq r \leq k$,

$$(c - a\omega^2)A_{k-r} + b\omega B_{k-r} = p_{k-r} + \cdots$$
$$-b\omega A_{k-r} + (c - a\omega^2)B_{k-r} = q_{k-r} + \cdots,$$

where the terms indicated by " $\cdots$ " depend upon the previously computed coefficients with subscripts greater than $k - r$. Conclude from this and Exercise 36**(b)** that

$$y_p = A(x) \cos \omega x + B(x) \sin \omega x \qquad \text{(B)}$$

is a particular solution of

$$ay'' + by' + cy = P(x) \cos \omega x + Q(x) \sin \omega x.$$

(b) Conclude from Exercise 36**(c)** that the equation

$$a(y'' + \omega^2 y) = P(x) \cos \omega x + Q(x) \sin \omega x \tag{C}$$

does not have a solution of the form (B) with A and B as in (A). Then show that there are polynomials

$$A(x) = A_0 x + A_1 x^2 + \cdots + A_k x^{k+1} \quad \text{and} \quad B(x) = B_0 x + B_1 x^2 + \cdots + B_k x^{k+1}$$

such that

$$a(A'' + 2\omega B') = P$$
$$a(B'' - 2\omega A') = Q,$$

where the pairs $(A_k, B_k), (A_{k-1}, B_{k-1}), \ldots, (A_0, B_0)$ can be computed successively as follows:

$$A_k = -\frac{q_k}{2a\omega(k+1)}$$

$$B_k = \frac{p_k}{2a\omega(k+1)},$$

and, if $k \geq 1$,

$$A_{k-j} = -\frac{1}{2\omega}\left[\frac{q_{k-j}}{a(k-j+1)} - (k-j+2)B_{k-j+1} \right]$$

$$B_{k-j} = \frac{1}{2\omega}\left[\frac{p_{k-j}}{a(k-j+1)} - (k-j+2)A_{k-j+1} \right]$$

for $1 \leq j \leq k$. Conclude that (B) with this choice of the polynomials A and B is a particular solution of (C).

38. Show that Theorem 5.5.1 implies the following theorem: *Suppose that ω is a positive number and P and Q are polynomials. Let k be the larger of the degrees of P and Q. Then the equation*

$$ay'' + by' + cy = e^{\lambda x}(P(x) \cos \omega x + Q(x) \sin \omega x)$$

has a particular solution

$$y_p = e^{\lambda x}(A(x) \cos \omega x + B(x) \sin \omega x), \tag{A}$$

where

$$A(x) = A_0 + A_1 x + \cdots + A_k x^k \quad \text{and} \quad B(x) = B_0 + B_1 x + \cdots + B_k x^k,$$

provided that $e^{\lambda x} \cos \omega x$ and $e^{\lambda x} \sin \omega x$ are not solutions of the complementary equation. The equation

$$a[y'' - 2\lambda y' + (\lambda^2 + \omega^2)y] = e^{\lambda x}(P(x) \cos \omega x + Q(x) \sin \omega x)$$

(for which $e^{\lambda x} \cos \omega x$ and $e^{\lambda x} \sin \omega x$ are solutions of the complementary equation) has a particular solution of the form (A) *where*

$$A(x) = A_0 x + A_1 x^2 + \cdots + A_k x^{k+1} \quad \text{and} \quad B(x) = B_0 x + B_1 x^2 + \cdots + B_k x^{k+1}.$$

39. This exercise presents a method for evaluating the integral

$$y = \int e^{\lambda x}(P(x) \cos \omega x + Q(x) \sin \omega x)\, dx$$

where $\omega \neq 0$ and

$$P(x) = p_0 + p_1 x + \cdots + p_k x^k, \qquad Q(x) = q_0 + q_1 x + \cdots + q_k x^k.$$

(a) Show that $y = e^{\lambda x} u$ where

$$u' + \lambda u = P(x) \cos \omega x + Q(x) \sin \omega x. \tag{A}$$

(b) Show that (A) has a particular solution of the form

$$u_p = A(x) \cos \omega x + B(x) \sin \omega x$$

where

$$A(x) = A_0 + A_1 x + \cdots + A_k x^k, \qquad B(x) = B_0 + B_1 x + \cdots + B_k x^k,$$

and the pairs of coefficients $(A_k, B_k), (A_{k-1}, B_{k-1}), \ldots, (A_0, B_0)$ can be computed successively as the solutions of pairs of equations obtained by equating the coefficients of $x^r \cos \omega x$ and $x^r \sin \omega x$ for $r = k$, $k - 1, \ldots, 0$.

(c) Conclude that

$$\int e^{\lambda x}(P(x) \cos \omega x + Q(x) \sin \omega x) \, dx = e^{\lambda x}(A(x) \cos \omega x + B(x) \sin \omega x) + c$$

where c is a constant of integration.

40. Use the method of Exercise 39 to evaluate the integral.

(a) $\displaystyle\int x^2 \cos x \, dx$

(b) $\displaystyle\int x^2 e^x \cos x \, dx$

(c) $\displaystyle\int xe^{-x} \sin 2x \, dx$

(d) $\displaystyle\int x^2 e^{-x} \sin x \, dx$

(e) $\displaystyle\int x^3 e^x \sin x \, dx$

(f) $\displaystyle\int e^x[x \cos x - (1 + 3x) \sin x] \, dx$

(g) $\displaystyle\int e^{-x}[(1 + x^2) \cos x + (1 - x^2) \sin x] \, dx$

5.6   Reduction of Order

In this section we give a method for finding the general solution of

$$P_0(x)y'' + P_1(x)y' + P_2(x)y = F(x) \tag{1}$$

if we know a nontrivial solution y_1 of the complementary equation

$$P_0(x)y'' + P_1(x)y' + P_2(x)y = 0. \tag{2}$$

The method is called **reduction of order** because it reduces the task of solving (1) to solving a first order equation. Unlike the method of undetermined coefficients, it does not require P_0, P_1, and P_2 to be constants, or F to be of any special form.

By now you shoudn't be surprised that we look for solutions of (1) in the form

$$y = uy_1 \tag{3}$$

where u is to be determined so that y satisfies (1). Substituting (3) and

$$y' = u'y_1 + uy_1'$$
$$y'' = u''y_1 + 2u'y_1' + uy_1''$$

into (1) yields

$$P_0(x)(u''y_1 + 2u'y_1' + uy_1'') + P_1(x)(u'y_1 + uy_1') + P_2(x)uy_1 = F(x).$$

Collecting the coefficients of u, u', and u'' yields

$$(P_0y_1)u'' + (2P_0y_1' + P_1y_1)u' + (P_0y_1'' + P_1y_1' + P_2y_1)u = F. \tag{4}$$

However, the coefficient of u is zero, since y_1 satisfies (2). Therefore (4) reduces to

$$Q_0(x)u'' + Q_1(x)u' = F, \tag{5}$$

with

$$Q_0 = P_0 y_1 \quad \text{and} \quad Q_1 = 2P_0 y_1' + P_1 y_1.$$

(It isn't worthwhile to memorize the formulas for Q_0 and Q_1!) Since (5) is a linear first order equation in u', we can solve it for u' by variation of parameters as in Section 2.1, integrate the solution to obtain u, and then obtain y from (3).

EXAMPLE 1   **(a)** Find the general solution of

$$xy'' - (2x + 1)y' + (x + 1)y = x^2, \tag{6}$$

given that $y_1 = e^x$ is a solution of the complementary equation

$$xy'' - (2x + 1)y' + (x + 1)y = 0. \tag{7}$$

(b) As a byproduct of **(a)**, find a fundamental set of solutions of (7).

Solution   **(a)** If $y = ue^x$ then $y' = u'e^x + ue^x$ and $y'' = u''e^x + 2u'e^x + ue^x$, so

$$
\begin{aligned}
xy'' - (2x + 1)y' + (x + 1)y &= x(u''e^x + 2u'e^x + ue^x) \\
&\quad - (2x + 1)(u'e^x + ue^x) + (x + 1)ue^x \\
&= (xu'' - u')e^x.
\end{aligned}
$$

Therefore $y = ue^x$ is a solution of (6) if and only if

$$(xu'' - u')e^x = x^2,$$

which is a first order equation in u'. We rewrite it as

$$u'' - \frac{u'}{x} = xe^{-x}. \tag{8}$$

To focus on how we apply variation of parameters to this equation, we temporarily write $z = u'$, so that (8) becomes

$$z' - \frac{z}{x} = xe^{-x}. \tag{9}$$

We leave it to you to show (by separation of variables) that $z_1 = x$ is a solution of the complementary equation

$$z' - \frac{z}{x} = 0$$

for (9). By applying variation of parameters as in Section 2.1, we can now see that every solution of (9) is of the form

$$z = vx \quad \text{where} \quad v'x = xe^{-x}, \quad \text{so} \quad v' = e^{-x} \quad \text{and} \quad v = -e^{-x} + C_1.$$

Since $u' = z = vx$, u is a solution of (8) if and only if

$$u' = vx = -xe^{-x} + C_1 x.$$

Integrating this yields

$$u = (x + 1)e^{-x} + \frac{C_1}{2}x^2 + C_2.$$

Therefore the general solution of (6) is

$$y = ue^x = x + 1 + \frac{C_1}{2}x^2 e^x + C_2 e^x. \tag{10}$$

Solution **(b)** By letting $C_1 = C_2 = 0$ in (10), we see that $y_{p_1} = x + 1$ is a solution of (6). By letting $C_1 = 2$ and $C_2 = 0$, we see that $y_{p_2} = x + 1 + x^2e^x$ is also a solution of (6). Since the difference of two solutions of (6) is a solution of (7), it follows that $y_2 = y_{p_1} - y_{p_2} = x^2e^x$ is also a solution of (7). Since y_2/y_1 is nonconstant and we already know that $y_1 = e^x$ is a solution of (6), Theorem 5.1.6 implies that $\{e^x, x^2e^x\}$ is a fundamental set of solutions of (7). ∎

Although (10) is a correct form for the general solution of (6), it is silly to leave the arbitrary coefficient of x^2e^x as $C_1/2$ where C_1 is an arbitrary constant. Moreover, it is sensible to make the subscripts of the coefficients of $y_1 = e^x$ and $y_2 = x^2e^x$ consistent with the subscripts of the functions themselves. Therefore we rewrite (10) as

$$y = x + 1 + c_1e^x + c_2x^2e^x$$

by simply renaming the arbitrary constants. We will also do this in the next two examples, and in the answers to the exercises.

EXAMPLE 2 **(a)** Find the general solution of

$$x^2y'' + xy' - y = x^2 + 1,$$

given that $y_1 = x$ is a solution of the complementary equation

$$x^2y'' + xy' - y = 0. \tag{11}$$

As a byproduct of this result, find a fundamental set of solutions of (11).
(b) Solve the initial value problem

$$x^2y'' + xy' - y = x^2 + 1, \quad y(1) = 2, \quad y'(1) = -3. \tag{12}$$

Solution **(a)** If $y = ux$ then $y' = u'x + u$ and $y'' = u''x + 2u'$, so

$$x^2y'' + xy' - y = x^2(u''x + 2u') + x(u'x + u) - ux$$
$$= x^3u'' + 3x^2u'.$$

Therefore $y = ux$ is a solution of (12) if and only if

$$x^3u'' + 3x^2u' = x^2 + 1,$$

which is a first order equation in u'. We rewrite it as

$$u'' + \frac{3}{x}u' = \frac{1}{x} + \frac{1}{x^3}. \tag{13}$$

To focus on how we apply variation of parameters to this equation, we temporarily write $z = u'$, so that (13) becomes

$$z' + \frac{3}{x}z = \frac{1}{x} + \frac{1}{x^3}. \tag{14}$$

We leave it to you to show by separation of variables that $z_1 = 1/x^3$ is a solution of the complementary equation

$$z' + \frac{3}{x}z = 0$$

for (14). By variation of parameters, every solution of (14) is of the form

$$z = \frac{v}{x^3} \quad \text{where} \quad \frac{v'}{x^3} = \frac{1}{x} + \frac{1}{x^3}, \quad \text{so} \quad v' = x^2 + 1 \quad \text{and} \quad v = \frac{x^3}{3} + x + C_1.$$

Since $u' = z = v/x^3$, u is a solution of (14) if and only if

$$u' = \frac{v}{x^3} = \frac{1}{3} + \frac{1}{x^2} + \frac{C_1}{x^3}.$$

Integrating this yields

$$u = \frac{x}{3} - \frac{1}{x} - \frac{C_1}{2x^2} + C_2.$$

Therefore the general solution of (12) is

$$y = ux = \frac{x^2}{3} - 1 - \frac{C_1}{2x} + C_2 x. \qquad (15)$$

Reasoning as in the solution of Example 1**(a)**, we conclude that $y_1 = x$ and $y_2 = 1/x$ form a fundamental set of solutions for (11).

As we explained above, we rename the constants in (15) and rewrite it as

$$y = \frac{x^2}{3} - 1 + c_1 x + \frac{c_2}{x}. \qquad (16)$$

Solution   **(b)** Differentiating (16) yields

$$y' = \frac{2x}{3} + c_1 - \frac{c_2}{x^2}. \qquad (17)$$

Setting $x = 1$ in (16) and (17) and imposing the initial conditions $y(1) = 2$ and $y'(1) = -3$ yields

$$c_1 + c_2 = \frac{8}{3}$$

$$c_1 - c_2 = -\frac{11}{3}.$$

Solving these equations yields $c_1 = -1/2$, $c_2 = 19/6$. Therefore the solution of (12) is

$$y = \frac{x^2}{3} - 1 - \frac{x}{2} + \frac{19}{6x}.$$

The graph of the solution is shown in Figure 1.   ∎

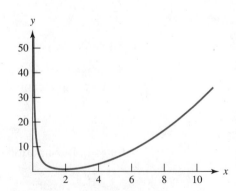

Figure 1   Graph of
$$y = \frac{x^2}{3} - 1 - \frac{x}{2} + \frac{19}{6x}$$

Using reduction of order to find the general solution of a homogeneous linear second order equation leads to a homogeneous linear first order equation in u' that can be solved by separation of variables. The next example illustrates this.

EXAMPLE 3 Find the general solution and a fundamental set of solutions of

$$x^2y'' - 3xy' + 3y = 0, \tag{18}$$

given that $y_1 = x$ is a solution.

Solution If $y = ux$ then $y' = u'x + u$ and $y'' = u''x + 2u'$, so

$$x^2y'' - 3xy' + 3y = x^2(u''x + 2u') - 3x(u'x + u) + 3ux$$
$$= x^3u'' - x^2u'.$$

Therefore $y = ux$ is a solution of (18) if and only if

$$x^3u'' - x^2u' = 0.$$

Separating the variables u' and x yields

$$\frac{u''}{u'} = \frac{1}{x},$$

so

$$\ln|u'| = \ln|x| + k, \qquad \text{or, equivalently,} \qquad u' = C_1x.$$

Therefore

$$u = \frac{C_1}{2}x^2 + C_2,$$

so the general solution of (18) is

$$y = ux = \frac{C_1}{2}x^3 + C_2x,$$

which we rewrite as

$$y = c_1x + c_2x^3.$$

Therefore $\{x, x^3\}$ is a fundamental set of solutions of (18). ∎

5.6 EXERCISES

In Exercises 1–17 find the general solution, given that y_1 satisfies the complementary equation. As a byproduct, find a fundamental set of solutions of the complementary equation.

1. $(2x + 1)y'' - 2y' - (2x + 3)y = (2x + 1)^2; \quad y_1 = e^{-x}$

2. $x^2y'' + xy' - y = \dfrac{4}{x^2}; \quad y_1 = x$

3. $x^2y'' - xy' + y = x; \quad y_1 = x$

4. $y'' - 3y' + 2y = \dfrac{1}{1 + e^{-x}}; \quad y_1 = e^{2x}$

5. $y'' - 2y' + y = 7x^{3/2}e^x; \quad y_1 = e^x$

6. $4x^2y'' + (4x - 8x^2)y' + (4x^2 - 4x - 1)y = 4x^{1/2}e^x(1 + 4x); \quad y_1 = x^{1/2}e^x$

7. $y'' - 2y' + 2y = e^x \sec x;$ $y_1 = e^x \cos x$

8. $y'' + 4xy' + (4x^2 + 2)y = 8e^{-x(x+2)};$ $y_1 = e^{-x^2}$

9. $x^2 y'' + xy' - 4y = -6x - 4;$ $y_1 = x^2$

10. $x^2 y'' + 2x(x - 1)y' + (x^2 - 2x + 2)y = x^3 e^{2x};$ $y_1 = xe^{-x}$

11. $x^2 y'' - x(2x - 1)y' + (x^2 - x - 1)y = x^2 e^x;$ $y_1 = xe^x$

12. $(1 - 2x)y'' + 2y' + (2x - 3)y = (1 - 4x + 4x^2)e^x;$ $y_1 = e^x$

13. $x^2 y'' - 3xy' + 4y = 4x^4;$ $y_1 = x^2$

14. $2xy'' + (4x + 1)y' + (2x + 1)y = 3x^{1/2}e^{-x};$ $y_1 = e^{-x}$

15. $xy'' - (2x + 1)y' + (x + 1)y = -e^x;$ $y_1 = e^x$

16. $4x^2 y'' - 4x(x + 1)y' + (2x + 3)y = 4x^{5/2}e^{2x};$ $y_1 = x^{1/2}$

17. $x^2 y'' - 5xy' + 8y = 4x^2;$ $y_1 = x^2$

In Exercises 18–30 find a fundamental set of solutions, given that y_1 is a solution.

18. $xy'' + (2 - 2x)y' + (x - 2)y = 0;$ $y_1 = e^x$ **19.** $x^2 y'' - 4xy' + 6y = 0;$ $y_1 = x^2$

20. $x^2(\ln |x|)^2 y'' - (2x \ln |x|)y' + (2 + \ln |x|)y = 0;$ $y_1 = \ln |x|$

21. $4xy'' + 2y' + y = 0;$ $y_1 = \sin \sqrt{x}$ **22.** $xy'' - (2x + 2)y' + (x + 2)y = 0;$ $y_1 = e^x$

23. $x^2 y'' - (2a - 1)xy' + a^2 y = 0;$ $y_1 = x^a$ **24.** $x^2 y'' - 2xy' + (x^2 + 2)y = 0;$ $y_1 = x \sin x$

25. $xy'' - (4x + 1)y' + (4x + 2)y = 0;$ $y_1 = e^{2x}$

26. $4x^2(\sin x)y'' - 4x(x \cos x + \sin x)y' + (2x \cos x + 3 \sin x)y = 0;$ $y_1 = x^{1/2}$

27. $4x^2 y'' - 4xy' + (3 - 16x^2)y = 0;$ $y_1 = x^{1/2}e^{2x}$

28. $(2x + 1)xy'' - 2(2x^2 - 1)y' - 4(x + 1)y = 0;$ $y_1 = 1/x$

29. $(x^2 - 2x)y'' + (2 - x^2)y' + (2x - 2)y = 0;$ $y_1 = e^x$

30. $xy'' - (4x + 1)y' + (4x + 2)y = 0;$ $y_1 = e^{2x}$

In Exercises 31–33 solve the initial value problem, given that y_1 satisfies the complementary equation.

31. $x^2 y'' - 3xy' + 4y = 4x^4,$ $y(-1) = 7,$ $y'(-1) = -8;$ $y_1 = x^2$

32. $(3x - 1)y'' - (3x + 2)y' - (6x - 8)y = 0,$ $y(0) = 2,$ $y'(0) = 3;$ $y_1 = e^{2x}$

33. $(x + 1)^2 y'' - 2(x + 1)y' - (x^2 + 2x - 1)y = (x + 1)^3 e^x,$ $y(0) = 1,$ $y'(0) = -1;$ $y_1 = (x + 1)e^x$

In Exercises 34 and 35 solve the initial value problem and graph the solution, given that y_1 satisfies the complementary equation.

C **34.** $x^2 y'' + 2xy' - 2y = x^2,$ $y(1) = \dfrac{5}{4},$ $y'(1) = \dfrac{3}{2};$ $y_1 = x$

C **35.** $(x^2 - 4)y'' + 4xy' + 2y = x + 2,$ $y(0) = -\dfrac{1}{3},$ $y'(0) = -1;$ $y_1 = \dfrac{1}{x - 2}$

36. Suppose that p_1 and p_2 are continuous on (a,b). Let y_1 be a solution of

$$y'' + p_1(x)y' + p_2(x)y = 0 \tag{A}$$

that has no zeros on (a,b), and let x_0 be in (a,b). Use reduction of order to show that y_1 and

$$y_2(x) = y_1(x) \int_{x_0}^x \frac{1}{y_1^2(t)} \exp\left(-\int_{x_0}^t p_1(s) \, ds \right) dt$$

form a fundamental set of solutions of (A) on (a,b). (Note: This exercise is related to Exercise 9 of Section 5.1.)

37. The nonlinear first order equation

$$y' + y^2 + p(x)y + q(x) = 0 \tag{A}$$

is known as a **_Riccati equation._** (See Exercise 55 of Section 2.4.) Assume that p and q are continuous.

(a) Show that y is a solution of (A) if and only if $y = z'/z$, where

$$z'' + p(x)z' + q(x)z = 0. \tag{B}$$

(b) Show that the general solution of (A) is

$$y = \frac{c_1 z_1' + c_2 z_2'}{c_1 z_1 + c_2 z_2} \tag{C}$$

where $\{z_1, z_2\}$ is a fundamental set of solutions of (B) and c_1 and c_2 are arbitrary constants.

(c) Does the formula (C) imply that the first order equation (A) has a two-parameter family of solutions? Explain your answer.

38. Use a method suggested by Exercise 37 to find all solutions of the given equation.

(a) $y' + y^2 + k^2 = 0$ **(b)** $y' + y^2 - 3y + 2 = 0$

(c) $y' + y^2 + 5y - 6 = 0$ **(d)** $y' + y^2 + 8y + 7 = 0$

(e) $y' + y^2 + 14y + 50 = 0$ **(f)** $6y' + 6y^2 - y - 1 = 0$

(g) $36y' + 36y^2 - 12y + 1 = 0$

39. Use a method suggested by Exercise 37 and reduction of order to find all solutions of the given equation, given that y_1 is a solution.

(a) $x^2(y' + y^2) - x(x + 2)y + x + 2 = 0; \quad y_1 = 1/x$

(b) $y' + y^2 + 4xy + 4x^2 + 2 = 0; \quad y_1 = -2x$

(c) $(2x + 1)(y' + y^2) - 2y - (2x + 3) = 0; \quad y_1 = -1$

(d) $(3x - 1)(y' + y^2) - (3x + 2)y - 6x + 8 = 0; \quad y_1 = 2$

(e) $x^2(y' + y^2) + xy + x^2 - \dfrac{1}{4} = 0; \quad y_1 = -\tan x - \dfrac{1}{2x}$

(f) $x^2(y' + y^2) - 7xy + 7 = 0; \quad y_1 = 1/x$

40. The nonlinear first order equation

$$y' + r(x)y^2 + p(x)y + q(x) = 0 \tag{A}$$

is known as the **_generalized Riccati equation._** (See Exercise 55 of Section 2.4.) Assume that p and q are continuous and r is differentiable.

(a) Show that y is a solution of (A) if and only if $y = z'/rz$, where

$$z'' + \left[p(x) - \frac{r'(x)}{r(x)} \right] z' + r(x)q(x)z = 0. \tag{B}$$

(b) Show that the general solution of (A) is

$$y = \frac{c_1 z_1' + c_2 z_2'}{r(c_1 z_1 + c_2 z_2)},$$

where $\{z_1, z_2\}$ is a fundamental set of solutions of (B) and c_1 and c_2 are arbitrary constants.

5.7   Variation of Parameters

In this section we give a method called *variation of parameters* for finding a particular solution of

$$P_0(x)y'' + P_1(x)y' + P_2(x)y = F(x) \tag{1}$$

if we know a fundamental set $\{y_1, y_2\}$ of solutions of the complementary equation

$$P_0(x)y'' + P_1(x)y' + P_2(x)y = 0. \tag{2}$$

Having found a particular solution y_p by this method, we can write the general solution of (1) as

$$y = y_p + c_1 y_1 + c_2 y_2.$$

Since we need only one nontrivial solution of (2) to find the general solution of (1) by reduction of order, it is natural to ask why we're interested in variation of parameters, which requires two linearly independent solutions of (2) to achieve the same goal. We answer this question as follows:

- If we already know two linearly independent solutions of (2) then variation of parameters will probably be simpler than reduction of order.
- Variation of parameters generalizes naturally to a method for finding particular solutions of higher order linear equations (Section 9.4) and linear systems of equations (Section 10.7), while reduction of order doesn't.
- Variation of parameters is a powerful theoretical tool used by researchers in differential equations. Although a detailed discussion of this is beyond the scope of this book, you can get an idea of what it means from Exercises 37–39.

We will now derive the method. As usual, we consider solutions of (1) and (2) on an interval (a, b) where P_0, P_1, P_2, and F are continuous, and P_0 has no zeros. Suppose that $\{y_1, y_2\}$ is a fundamental set of solutions of the complementary equation (2). We look for a particular solution of (1) in the form

$$y_p = u_1 y_1 + u_2 y_2 \tag{3}$$

where u_1 and u_2 are functions to be determined so that y_p satisfies (1). You may not think this is a good idea, since there are now two unknown functions to be determined, rather than one. However, since u_1 and u_2 have to satisfy only one condition (that y_p is a solution of (1)), we can impose a second condition that produces a convenient simplification, as follows.

Differentiating (3) yields

$$y_p' = u_1 y_1' + u_2 y_2' + u_1' y_1 + u_2' y_2. \tag{4}$$

As our second condition on u_1 and u_2 we require that

$$u_1' y_1 + u_2' y_2 = 0. \tag{5}$$

Then (4) becomes

$$y_p' = u_1 y_1' + u_2 y_2'; \tag{6}$$

that is, (5) permits us to differentiate y_p (once!) as if u_1 and u_2 are constants. Differentiating (4) yields

$$y_p'' = u_1 y_1'' + u_2 y_2'' + u_1' y_1' + u_2' y_2'. \tag{7}$$

(Notice that there are no terms involving u_1'' and u_2'' here, as there would be if we hadn't required (5).) Substituting (3), (6), and (7) into (1) and collecting the coefficients of u_1 and u_2 yields

$$u_1(P_0y_1'' + P_1y_1' + P_2y_1) + u_2(P_0y_2'' + P_1y_2' + P_2y_2) + P_0(u_1'y_1' + u_2'y_2') = F.$$

As in the derivation of the method of reduction of order, the coefficients of u_1 and u_2 here are both zero because y_1 and y_2 satisfy the complementary equation. Hence we can rewrite the last equation as

$$P_0(u_1'y_1' + u_2'y_2') = F. \tag{8}$$

Therefore, y_p in (3) satisfies (1) if

$$\begin{aligned} u_1'y_1 + u_2'y_2 &= 0 \\ u_1'y_1' + u_2'y_2' &= \frac{F}{P_0}, \end{aligned} \tag{9}$$

where the first equation is the same as (5) and the second is from (8).

We will now show that you can always solve (9) for u_1' and u_2'. (The method that we use here will always work, but simpler methods usually work when you're dealing with specific equations.) To obtain u_1', multiply the first equation in (9) by y_2' and the second equation by y_2. This yields

$$\begin{aligned} u_1'y_1y_2' + u_2'y_2y_2' &= 0 \\ u_1'y_1'y_2 + u_2'y_2'y_2 &= \frac{Fy_2}{P_0}. \end{aligned}$$

Subtracting the second equation from the first yields

$$u_1'(y_1y_2' - y_1'y_2) = -\frac{Fy_2}{P_0}. \tag{10}$$

Since $\{y_1, y_2\}$ is a fundamental set of solutions of (2) on (a,b), Theorem 5.1.6 implies that the Wronskian $y_1y_2' - y_1'y_2$ has no zeros on (a,b). Therefore we can solve (10) for u_1', to obtain

$$u_1' = -\frac{Fy_2}{P_0(y_1y_2' - y_1'y_2)}. \tag{11}$$

We leave it to you to start from (9) and show by a similar argument that

$$u_2' = \frac{Fy_1}{P_0(y_1y_2' - y_1'y_2)}. \tag{12}$$

We can now obtain u_1 and u_2 by integrating u_1' and u_2'. The constants of integration can be taken to be zero, since any choice of u_1 and u_2 in (3) will suffice.

You should not memorize (11) and (12). On the other hand, you don't want to rederive the whole procedure for every specific problem. We recommend the following compromise:

1. Write

$$y_p = u_1y_1 + u_2y_2 \tag{13}$$

to remind yourself of what you're doing.

2. Write the system

$$u_1' y_1 + u_2' y_2 = 0$$

$$u_1' y_1' + u_2' y_2' = \frac{F}{P_0} \tag{14}$$

for the specific problem you're trying to solve.
3. Solve (14) for u_1' and u_2' by any convenient method.
4. Obtain u_1 and u_2 by integrating u_1' and u_2', taking the constants of integration to be zero.
5. Substitute u_1 and u_2 into (13) to obtain y_p.

EXAMPLE 1   Find a particular solution y_p of

$$x^2 y'' - 2xy' + 2y = x^{9/2}, \tag{15}$$

given that $y_1 = x$ and $y_2 = x^2$ are solutions of the complementary equation

$$x^2 y'' - 2xy' + 2y = 0.$$

Then find the general solution of (15).

Solution   We set

$$y_p = u_1 x + u_2 x^2,$$

where

$$u_1' x + u_2' x^2 = 0$$

$$u_1' + 2u_2' x = \frac{x^{9/2}}{x^2} = x^{5/2}.$$

From the first equation, $u_1' = -u_2' x$. Substituting this into the second equation yields $u_2' x = x^{5/2}$, so $u_2' = x^{3/2}$, and therefore $u_1' = -u_2' x = -x^{5/2}$. Integrating and taking the constants of integration to be zero yields

$$u_1 = -\frac{2}{7} x^{7/2} \quad \text{and} \quad u_2 = \frac{2}{5} x^{5/2}.$$

Therefore

$$y_p = u_1 x + u_2 x^2 = -\frac{2}{7} x^{7/2} x + \frac{2}{5} x^{5/2} x^2 = \frac{4}{35} x^{9/2},$$

and the general solution of (15) is

$$y = \frac{4}{35} x^{9/2} + c_1 x + c_2 x^2.$$

EXAMPLE 2   Find a particular solution y_p of

$$(x - 1)y'' - xy' + y = (x - 1)^2, \tag{16}$$

given that $y_1 = x$ and $y_2 = e^x$ are solutions of the complementary equation

$$(x - 1)y'' - xy' + y = 0.$$

Then find the general solution of (16).

Solution   We set

$$y_p = u_1 x + u_2 e^x$$

where

$$u_1' x + u_2' e^x = 0$$

$$u_1' \quad + u_2' e^x = \frac{(x-1)^2}{x-1} = x - 1.$$

Subtracting the first equation from the second yields $-u_1'(x-1) = x - 1$, so $u_1' = -1$. From this and the first equation, $u_2' = -xe^{-x}u_1' = xe^{-x}$. Integrating and taking the constants of integration to be zero yields

$$u_1 = -x \quad \text{and} \quad u_2 = -(x+1)e^{-x}.$$

Therefore

$$y_p = u_1 x + u_2 e^x = (-x)x + (-(x+1)e^{-x})e^x = -x^2 - x - 1,$$

so the general solution of (16) is

$$y = y_p + c_1 x + c_2 e^x = -x^2 - x - 1 + c_1 x + c_2 e^x$$
$$= -x^2 - 1 + (c_1 - 1)x + c_2 e^x. \tag{17}$$

However, since c_1 is an arbitrary constant, so is $c_1 - 1$; therefore, we improve the appearance of this result by renaming the constant and writing the general solution as

$$y = -x^2 - 1 + c_1 x + c_2 e^x. \tag{18}$$

∎

There is nothing *wrong* with leaving the general solution of (16) in the form (17); however, we think you'll agree that (18) is preferable. We can also view the transition from (17) to (18) differently. In this example the particular solution $y_p = -x^2 - x - 1$ contained the term $-x$, which satisfies the complementary equation. We can drop this term and redefine $y_p = -x^2 - 1$, since $-x^2 - x - 1$ is a solution of (16) and x is a solution of the complementary equation; hence, $-x^2 - 1 = (-x^2 - x - 1) + x$ is also a solution of (16). In general, it is always legitimate to drop linear combinations of $\{y_1, y_2\}$ from particular solutions obtained by variation of parameters. (See Exercise 36 for a general discussion of this question.) We will do this in the following examples and in the answers to exercises that ask for a particular solution. Therefore, don't be concerned if your answer to such an exercise differs from ours only by a solution of the complementary equation.

EXAMPLE 3   Find a particular solution of

$$y'' + 3y' + 2y = \frac{1}{1+e^x}. \tag{19}$$

Then find the general solution.

Solution   The characteristic polynomial of the complementary equation

$$y'' + 3y' + 2y = 0 \tag{20}$$

is $p(r) = r^2 + 3r + 2 = (r+1)(r+2)$, so $y_1 = e^{-x}$ and $y_2 = e^{-2x}$ form a fundamental set of solutions of (20). We look for a particular solution of (19) in the form

$$y_p = u_1 e^{-x} + u_2 e^{-2x}$$

where

$$u_1' e^{-x} + u_2' e^{-2x} = 0$$

$$-u_1' e^{-x} - 2u_2' e^{-2x} = \frac{1}{1 + e^x}.$$

Adding these two equations yields

$$-u_2' e^{-2x} = \frac{1}{1 + e^x}, \qquad \text{so} \qquad u_2' = -\frac{e^{2x}}{1 + e^x}.$$

From the first equation,

$$u_1' = -u_2' e^{-x} = \frac{e^x}{1 + e^x}.$$

Integrating by means of the substitution $v = e^x$ and taking the constants of integration to be zero yields

$$u_1 = \int \frac{e^x}{1 + e^x} \, dx = \int \frac{dv}{1 + v} = \ln(1 + v) = \ln(1 + e^x)$$

and

$$u_2 = -\int \frac{e^{2x}}{1 + e^x} \, dx = -\int \frac{v}{1 + v} \, dv = \int \left[\frac{1}{1 + v} - 1 \right] dv$$

$$= \ln(1 + v) - v = \ln(1 + e^x) - e^x.$$

Therefore

$$y_p = u_1 e^{-x} + u_2 e^{-2x}$$

$$= [\ln(1 + e^x)]e^{-x} + [\ln(1 + e^x) - e^x]e^{-2x},$$

so

$$y_p = (e^{-x} + e^{-2x}) \ln(1 + e^x) - e^{-x}.$$

Since the last term on the right satisfies the complementary equation, we drop it and redefine

$$y_p = (e^{-x} + e^{-2x}) \ln(1 + e^x).$$

The general solution of (19) is

$$y = y_p + c_1 e^{-x} + c_2 e^{-2x} = (e^{-x} + e^{-2x}) \ln(1 + e^x) + c_1 e^{-x} + c_2 e^{-2x}.$$

EXAMPLE 4   Solve the initial value problem

$$(x^2 - 1)y'' + 4xy' + 2y = \frac{2}{x + 1}, \qquad y(0) = -1, \qquad y'(0) = -5, \quad (21)$$

given that

$$y_1 = \frac{1}{x - 1} \qquad \text{and} \qquad y_2 = \frac{1}{x + 1}$$

are solutions of the complementary equation

$$(x^2 - 1)y'' + 4xy' + 2y = 0.$$

Solution   We first use variation of parameters to find a particular solution of

$$(x^2 - 1)y'' + 4xy' + 2y = \frac{2}{x + 1}$$

on $(-1, 1)$ in the form

$$y_p = \frac{u_1}{x - 1} + \frac{u_2}{x + 1},$$

where

$$\frac{u_1'}{x - 1} + \frac{u_2'}{x + 1} = 0$$

$$-\frac{u_1'}{(x - 1)^2} - \frac{u_2'}{(x + 1)^2} = \frac{2}{(x + 1)(x^2 - 1)}. \tag{22}$$

Multiplying the first equation by $1/(x - 1)$ and adding the result to the second equation yields

$$\left[\frac{1}{x^2 - 1} - \frac{1}{(x + 1)^2} \right] u_2' = \frac{2}{(x + 1)(x^2 - 1)}. \tag{23}$$

Since

$$\left[\frac{1}{x^2 - 1} - \frac{1}{(x + 1)^2} \right] = \frac{(x + 1) - (x - 1)}{(x + 1)(x^2 - 1)} = \frac{2}{(x + 1)(x^2 - 1)},$$

(23) implies that $u_2' = 1$. From (22),

$$u_1' = -\frac{x - 1}{x + 1} u_2' = -\frac{x - 1}{x + 1}.$$

Integrating and taking the constants of integration to be zero yields

$$u_1 = -\int \frac{x - 1}{x + 1} \, dx = -\int \frac{x + 1 - 2}{x + 1} \, dx$$

$$= \int \left[\frac{2}{x + 1} - 1 \right] dx = 2 \ln(x + 1) - x$$

and

$$u_2 = \int dx = x.$$

Therefore

$$y_p = \frac{u_1}{x - 1} + \frac{u_2}{x + 1} = [2 \ln(x + 1) - x] \frac{1}{x - 1} + x \frac{1}{x + 1}$$

$$= \frac{2 \ln(x + 1)}{x - 1} + x \left[\frac{1}{x + 1} - \frac{1}{x - 1} \right] = \frac{2 \ln(x + 1)}{x - 1} - \frac{2x}{(x + 1)(x - 1)}.$$

However, since

$$\frac{2x}{(x + 1)(x - 1)} = \left[\frac{1}{x + 1} + \frac{1}{x - 1} \right]$$

is a solution of the complementary equation, we redefine

$$y_p = \frac{2\ln(x+1)}{x-1}.$$

Therefore the general solution of (24) is

$$y = \frac{2\ln(x+1)}{x-1} + \frac{c_1}{x-1} + \frac{c_2}{x+1}. \tag{24}$$

Differentiating this yields

$$y' = \frac{2}{x^2-1} - \frac{2\ln(x+1)}{(x-1)^2} - \frac{c_1}{(x-1)^2} - \frac{c_2}{(x+1)^2}.$$

Setting $x = 0$ in the last two equations and imposing the initial conditions $y(0) = -1$ and $y'(0) = -5$ yields the system

$$-c_1 + c_2 = -1$$
$$-2 - c_1 - c_2 = -5.$$

The solution of this system is $c_1 = 2, c_2 = 1$. Substituting these into (24) yields

$$y = \frac{2\ln(x+1)}{x-1} + \frac{2}{x-1} + \frac{1}{x+1}$$
$$= \frac{2\ln(x+1)}{x-1} + \frac{3x+1}{x^2-1}$$

as the solution of (21). The graph of this solution is shown in Figure 1. ■

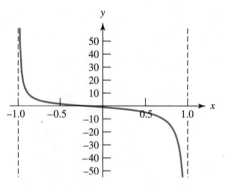

Figure 1　Graph of
$$y = \frac{2\ln(x+1)}{x-1} + \frac{3x+1}{x^2-1}$$

COMPARISON OF METHODS

We have now considered three methods for solving nonhomogeneous linear equations: undetermined coefficients, reduction of order, and variation of parameters. It is natural to ask which method is best for a given problem. The method of undetermined coefficients should be used for constant coefficient equations with forcing functions that are linear combinations of polynomials multiplied by functions of the form $e^{\alpha x}, e^{\lambda x}\cos \omega x$, or $e^{\lambda x}\sin \omega x$. Although the other two methods can be used to solve such problems, they will be more difficult except in the most trivial cases, because of the integrations involved.

If the equation is not a constant coefficient equation or the forcing function is not of the form just specified, then the method of undetermined coefficients does not apply and the choice is necessarily between the other two methods. The case could be made that reduction of order is better because it requires only one solution of the complementary equation while variation of parameters requires two. However, variation of parameters will probably be easier if you already know a fundamental set of solutions of the complementary equation.

5.7 EXERCISES

In Exercises 1–6 use variation of parameters to find a particular solution.

1. $y'' + 9y = \tan 3x$

2. $y'' + 4y = \sin 2x \sec^2 2x$

3. $y'' - 3y' + 2y = \dfrac{4}{1 + e^{-x}}$

4. $y'' - 2y' + 2y = 3e^x \sec x$

5. $y'' - 2y' + y = 14x^{3/2}e^x$

6. $y'' - y = \dfrac{4e^{-x}}{1 - e^{-2x}}$

In Exercises 7–29 use variation of parameters to find a particular solution, given the solutions y_1, y_2 of the complementary equation.

7. $x^2y'' + xy' - y = 2x^2 + 2; \quad y_1 = x, \quad y_2 = \dfrac{1}{x}$

8. $xy'' + (2 - 2x)y' + (x - 2)y = e^{2x}; \quad y_1 = e^x, \quad y_2 = \dfrac{e^x}{x}$

9. $4x^2y'' + (4x - 8x^2)y' + (4x^2 - 4x - 1)y = 4x^{1/2}e^x, \quad x > 0; \quad y_1 = x^{1/2}e^x, \quad y_2 = x^{-1/2}e^x$

10. $y'' + 4xy' + (4x^2 + 2)y = 4e^{-x(x+2)}; \quad y_1 = e^{-x^2}, \quad y_2 = xe^{-x^2}$

11. $x^2y'' - 4xy' + 6y = x^{5/2}, \quad x > 0; \quad y_1 = x^2, \quad y_2 = x^3$

12. $x^2y'' - 3xy' + 3y = 2x^4 \sin x; \quad y_1 = x, \quad y_2 = x^3$

13. $(2x + 1)y'' - 2y' - (2x + 3)y = (2x + 1)^2e^{-x}; \quad y_1 = e^{-x}, \quad y_2 = xe^x$

14. $4xy'' + 2y' + y = \sin \sqrt{x}; \quad y_1 = \cos \sqrt{x}, \quad y_2 = \sin \sqrt{x}$

15. $xy'' - (2x + 2)y' + (x + 2)y = 6x^3e^x; \quad y_1 = e^x, \quad y_2 = x^3e^x$

16. $x^2y'' - (2a - 1)xy' + a^2y = x^{a+1}; \quad y_1 = x^a, \quad y_2 = x^a \ln x$

17. $x^2y'' - 2xy' + (x^2 + 2)y = x^3 \cos x; \quad y_1 = x \cos x, \quad y_2 = x \sin x$

18. $xy'' - y' - 4x^3y = 8x^5; \quad y_1 = e^{x^2}, \quad y_2 = e^{-x^2}$

19. $(\sin x)y'' + (2 \sin x - \cos x)y' + (\sin x - \cos x)y = e^{-x}; \quad y_1 = e^{-x}, \quad y_2 = e^{-x} \cos x$

20. $4x^2y'' - 4xy' + (3 - 16x^2)y = 8x^{5/2}; \quad y_1 = \sqrt{x}e^{2x}, \quad y_2 = \sqrt{x}e^{-2x}$

21. $4x^2y'' - 4xy' + (4x^2 + 3)y = x^{7/2}; \quad y_1 = \sqrt{x} \sin x, \quad y_2 = \sqrt{x} \cos x$

22. $x^2y'' - 2xy' - (x^2 - 2)y = 3x^4; \quad y_1 = xe^x, \quad y_2 = xe^{-x}$

23. $x^2y'' - 2x(x + 1)y' + (x^2 + 2x + 2)y = x^3e^x; \quad y_1 = xe^x, \quad y_2 = x^2e^x$

24. $x^2y'' - xy' - 3y = x^{3/2}; \quad y_1 = 1/x, \quad y_2 = x^3$

25. $x^2y'' - x(x + 4)y' + 2(x + 3)y = x^4e^x; \quad y_1 = x^2, \quad y_2 = x^2e^x$

26. $x^2y'' - 2x(x + 2)y' + (x^2 + 4x + 6)y = 2xe^x; \quad y_1 = x^2e^x, \quad y_2 = x^3e^x$

27. $x^2y'' - 4xy' + (x^2 + 6)y = x^4; \quad y_1 = x^2 \cos x, \quad y_2 = x^2 \sin x$

28. $(x - 1)y'' - xy' + y = 2(x - 1)^2 e^x$; $y_1 = x$, $y_2 = e^x$

29. $4x^2 y'' - 4x(x + 1)y' + (2x + 3)y = x^{5/2} e^x$; $y_1 = \sqrt{x}$, $y_2 = \sqrt{x} e^x$

In Exercises 30–32 use variation of parameters to solve the initial value problem, given the solutions y_1, y_2 of the complementary equation.

30. $(3x - 1)y'' - (3x + 2)y' - (6x - 8)y = (3x - 1)^2 e^{2x}$, $y(0) = 1$, $y'(0) = 2$; $y_1 = e^{2x}$, $y_2 = xe^{-x}$

31. $(x - 1)^2 y'' - 2(x - 1)y' + 2y = (x - 1)^2$, $y(0) = 3$, $y'(0) = -6$; $y_1 = x - 1$, $y_2 = x^2 - 1$

32. $(x - 1)^2 y'' - (x^2 - 1)y' + (x + 1)y = (x - 1)^3 e^x$, $y(0) = 4$, $y'(0) = -6$; $y_1 = (x - 1)e^x$, $y_2 = x - 1$

In Exercises 33–35 use variation of parameters to solve the initial value problem and graph the solution, given the solutions y_1, y_2 of the complementary equation.

C   **33.** $(x^2 - 1)y'' + 4xy' + 2y = 2x$, $y(0) = 0$, $y'(0) = -2$; $y_1 = \dfrac{1}{x - 1}$, $y_2 = \dfrac{1}{x + 1}$

C   **34.** $x^2 y'' + 2xy' - 2y = -2x^2$, $y(1) = 1$, $y'(1) = -1$; $y_1 = x$, $y_2 = \dfrac{1}{x^2}$

C   **35.** $(x + 1)(2x + 3)y'' + 2(x + 2)y' - 2y = (2x + 3)^2$, $y(0) = 0$, $y'(0) = 0$; $y_1 = x + 2$, $y_2 = \dfrac{1}{x + 1}$

36. Suppose that

$$y_p = \bar{y} + a_1 y_1 + a_2 y_2$$

is a particular solution of

$$P_0(x)y'' + P_1(x)y' + P_2(x)y = F(x), \tag{A}$$

where y_1 and y_2 are solutions of the complementary equation

$$P_0(x)y'' + P_1(x)y' + P_2(x)y = 0.$$

Show that $\bar{y}$ is also a solution of (A).

37. Suppose that p, q, and f are continuous on (a, b) and let x_0 be in (a, b). Let y_1 and y_2 be the solutions of

$$y'' + p(x)y' + q(x)y = 0$$

such that

$$y_1(x_0) = 1, \qquad y_1'(x_0) = 0, \qquad y_2(x_0) = 0, \qquad y_2'(x_0) = 1.$$

Use variation of parameters to show that the solution of the initial value problem

$$y'' + p(x)y' + q(x)y = f(x), \qquad y(x_0) = k_0, \qquad y'(x_0) = k_1$$

is

$$y(x) = k_0 y_1(x) + k_1 y_2(x) + \int_{x_0}^{x} (y_1(t)y_2(x) - y_1(x)y_2(t)) f(t) \exp\left(\int_{x_0}^{t} p(s)\, ds \right) dt.$$

Hint: Use Abel's formula for the Wronskian of $\{y_1, y_2\}$, and integrate u_1' and u_2' from x_0 to x.

Show also that

$$y'(x) = k_0 y_1'(x) + k_1 y_2'(x) + \int_{x_0}^{x} (y_1(t)y_2'(x) - y_1'(x)y_2(t)) f(t) \exp\left(\int_{x_0}^{t} p(s)\, ds \right) dt.$$

38. Suppose that f is continuous on an open interval containing $x_0 = 0$. Use variation of parameters to find a formula for the solution of the initial value problem

$$y'' - y = f(x), \qquad y(0) = k_0, \qquad y'(0) = k_1.$$

39. Suppose that f is continuous on (a, ∞), where $a < 0$, so $x_0 = 0$ is in (a, ∞).

(a) Use variation of parameters to find a formula for the solution of the initial value problem

$$y'' + y = f(x), \qquad y(0) = k_0, \qquad y'(0) = k_1.$$

Hint: You will need the addition formulas for the sine and cosine:

$$\sin(A + B) = \sin A \cos B + \cos A \sin B$$
$$\cos(A + B) = \cos A \cos B - \sin A \sin B.$$

For the rest of this exercise assume that the improper integral $\int_0^\infty f(t)\, dt$ is absolutely convergent.

(b) Show that if y is a solution of

$$y'' + y = f(x) \tag{A}$$

on (a, ∞) then

$$\lim_{x \to \infty} (y(x) - A_0 \cos x - A_1 \sin x) = 0 \tag{B}$$

and

$$\lim_{x \to \infty} (y'(x) + A_0 \sin x - A_1 \cos x) = 0, \tag{C}$$

where

$$A_0 = k_0 - \int_0^\infty f(t) \sin t\, dt \quad \text{and} \quad A_1 = k_1 + \int_0^\infty f(t) \cos t\, dt.$$

Hint: Recall from calculus that if $\int_0^\infty f(t)\, dt$ converges absolutely, then $\lim_{x \to \infty} \int_x^\infty |f(t)|\, dt = 0.$

(c) Show that if A_0 and A_1 are arbitrary constants then there is a unique solution of $y'' + y = f(x)$ on (a, ∞) that satisfies (B) and (C).

6 Applications of Linear Second Order Equations

IN THIS CHAPTER we study applications of linear second order equations.

SECTIONS 6.1 AND 6.2 deal with spring–mass systems.

SECTION 6.3 discusses *RLC* circuits, the electrical analogs of spring–mass systems.

SECTION 6.4 is concerned with the motion of an object under a central force. This section is particularly relevant in the space age, since, for example, a satellite moving in orbit subject only to Earth's gravity is experiencing motion under a central force.

6.1 Spring Problems I

We consider the motion of an object of mass m, suspended from a spring of negligible mass. We say that the spring–mass system is in ***equilibrium*** when the object is at rest and the forces acting on it sum to zero. The position of the object in this case is the ***equilibrium position.*** We define y to be the displacement of the object from its equilibrium position (Figure 1), measured positive upward.

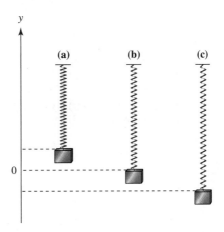

Figure 1 **(a)** $y > 0$; **(b)** $y = 0$; **(c)** $y < 0$

Our model accounts for the following kinds of forces acting on the object:

- The force $-mg$, due to gravity.
- A force F_s exerted by the spring resisting change in its length. The ***natural length*** of the spring is its length with no mass attached. We assume that the spring obeys ***Hooke's***[1] ***law:*** If the length of the spring is changed by an amount ΔL from its natural length, then the spring exerts a force $F_s = k\Delta L$, where k is a positive number called the ***spring constant.*** If the spring is stretched then $\Delta L > 0$ and $F_s > 0$, so the spring force is upward, while if the spring is compressed then $\Delta L < 0$ and $F_s < 0$, so the spring force is downward.
- A ***damping force*** $F_d = -cy'$ that resists the motion with a force proportional to the velocity of the object. It may be due to air resistance or friction in the spring. However, a convenient way to visualize a damping force is to assume that the object is rigidly attached to a piston with negligible mass immersed in a cylinder (called a ***dashpot***) filled with a viscous liquid (Figure 2). As the piston moves, the liquid exerts a damping force. We say that the motion is ***undamped*** if $c = 0$, or ***damped*** if $c > 0$.
- An external force F, other than the force due to gravity, that may vary with t, but is independent of displacement and velocity. We say that the motion is ***free*** if $F \equiv 0$, or ***forced*** if $F \not\equiv 0$.

From Newton's second law of motion,

$$my'' = -mg + F_d + F_s + F = -mg - cy' + F_s + F. \tag{1}$$

[1]Robert Hooke (1635–1703), an English scientist, was a contemporary and colleague of Isaac Newton.

Figure 2 A spring–mass system with damping

We must now relate F_s to y. In the absence of external forces the object stretches the spring by an amount Δl to assume its equilibrium position (Figure 3). Since the sum of the forces acting on the object is then zero, Hooke's Law implies that $mg = k\Delta l$. If the object is displaced y units from its equilibrium position then the total change in the length of the spring is $\Delta L = \Delta l - y$, so Hooke's law implies that

$$F_s = k\Delta L = k\Delta l - ky.$$

Substituting this into (1) yields

$$my'' = -mg - cy' + k\Delta L - ky + F.$$

Since $mg = k\Delta l$, this can be written as

$$my'' + cy' + ky = F. \tag{2}$$

We call this **the equation of motion.**

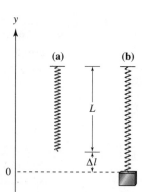

Figure 3 **(a)** Natural length of spring; **(b)** Spring stretched by mass

SIMPLE HARMONIC MOTION

Throughout the rest of this section we will consider spring–mass systems without damping; that is, $c = 0$. We will consider systems with damping in the next section.

We first consider the case where the motion is also free; that is, $F = 0$. We begin with an example.

EXAMPLE 1 An object stretches a spring 6 inches in equilibrium.

(a) Set up the equation of motion and find its general solution.

(b) Find the displacement of the object for $t > 0$ if it is initially displaced 18 inches above equilibrium and given a downward velocity of 3 ft/s.

Solution **(a)** Setting $c = 0$ and $F = 0$ in (2) yields the equation of motion

$$my'' + ky = 0,$$

which we rewrite as

$$y'' + \frac{k}{m}y = 0. \tag{3}$$

Although we would need the weight of the object to obtain k from the equation $mg = k\Delta l$, we can obtain k/m from Δl alone; thus, $k/m = g/\Delta l$. Consistent with the units used in the problem statement, we take $g = 32$ ft/s². Although Δl is stated in inches, we must convert it to feet to be consistent with this choice of g; that is, $\Delta l = 1/2$ ft. Therefore

$$\frac{k}{m} = \frac{32}{1/2} = 64$$

and (3) becomes

$$y'' + 64y = 0. \tag{4}$$

The characteristic equation of (4) is

$$r^2 + 64 = 0,$$

which has the zeros $r = \pm 8i$. Therefore, the general solution of (4) is

$$y = c_1 \cos 8t + c_2 \sin 8t. \tag{5}$$

Solution **(b)** The initial upward displacement of 18 inches is positive and must be expressed in feet. The initial downward velocity is negative; thus,

$$y(0) = \frac{3}{2} \quad \text{and} \quad y'(0) = -3.$$

Differentiating (5) yields

$$y' = -8c_1 \sin 8t + 8c_2 \cos 8t. \tag{6}$$

Setting $t = 0$ in (5) and (6) and imposing the initial conditions shows that $c_1 = 3/2$ and $c_2 = -3/8$. Therefore

$$y = \frac{3}{2} \cos 8t - \frac{3}{8} \sin 8t,$$

where y is in feet (Figure 4). ∎

We will now consider the equation

$$my'' + ky = 0$$

where m and k are arbitrary positive numbers. Dividing through by m and defining $\omega_0 = \sqrt{k/m}$ yields

$$y'' + \omega_0^2 y = 0.$$

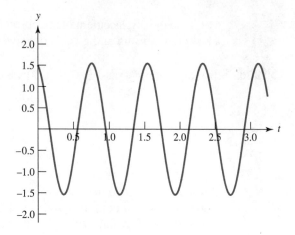

Figure 4 Graph of $y = \dfrac{3}{2} \cos 8t - \dfrac{3}{8} \sin 8t$

The general solution of this equation is

$$y = c_1 \cos \omega_0 t + c_2 \sin \omega_0 t. \tag{7}$$

We can rewrite this in a more useful form by defining

$$R = \sqrt{c_1^2 + c_2^2} \tag{8}$$

and

$$c_1 = R \cos \phi \qquad \text{and} \qquad c_2 = R \sin \phi. \tag{9}$$

Substituting from (9) into (7) and applying the identity

$$\cos \omega_0 t \cos \phi + \sin \omega_0 t \sin \phi = \cos(\omega_0 t - \phi)$$

yields

$$y = R \cos(\omega_0 t - \phi). \tag{10}$$

From (8) and (9) we see that the R and ϕ can be interpreted as polar coordinates of the point with rectangular coordinates (c_1, c_2) (Figure 5). Given c_1 and c_2, we can compute R from (8). From (8) and (9) we see that ϕ is related to c_1 and c_2 by

$$\cos \phi = \frac{c_1}{\sqrt{c_1^2 + c_2^2}} \qquad \text{and} \qquad \sin \phi = \frac{c_2}{\sqrt{c_1^2 + c_2^2}}.$$

There are infinitely many angles ϕ, differing by integer multiples of 2π, that satisfy these equations. We will always choose ϕ so that $-\pi \le \phi < \pi$.

The motion described by (7) or (10) is called **simple harmonic motion.** We see from either of these equations that the motion is periodic, with period

$$T = 2\pi/\omega_0.$$

This is the time required for the object to complete one full cycle of oscillation (for example, to move from its highest position to its lowest position and back to its highest position). Since the highest and lowest positions of the object are $y = R$ and $y = -R$, we say that R is the **amplitude** of the oscillation. The angle ϕ in (10) is called the **phase angle;** it is measured in radians. Equation (10) is called

Figure 5 $R = \sqrt{c_1^2 + c_2^2}$;
$c_1 = R \cos \phi$; $c_2 = R \sin \phi$

the *amplitude–phase form* of the displacement. If t is in seconds then ω_0 is in radians per second (rad/s); it is called the *frequency* of the motion. It is also called the *natural frequency* of the spring–mass system without damping.

EXAMPLE 2   We found the displacement of the object in Example 1 to be

$$y = \frac{3}{2} \cos 8t - \frac{3}{8} \sin 8t.$$

Find the frequency, period, amplitude R, and phase angle.

Solution   The frequency is $\omega_0 = 8$ rad/s, and the period is $T = 2\pi/\omega_0 = \pi/4$ s. Since $c_1 = 3/2$ and $c_2 = -3/8$, the amplitude is

$$R = \sqrt{c_1^2 + c_2^2} = \sqrt{\left(\frac{3}{2}\right)^2 + \left(\frac{3}{8}\right)^2} = \frac{3}{8}\sqrt{17},$$

and the phase angle is determined by

$$\cos \phi = \frac{\frac{3}{2}}{\frac{3}{8}\sqrt{17}} = \frac{4}{\sqrt{17}} \tag{11}$$

and

$$\sin \phi = \frac{-\frac{3}{8}}{\frac{3}{8}\sqrt{17}} = -\frac{1}{\sqrt{17}}. \tag{12}$$

Using a calculator, we see from (11) that

$$\phi \approx \pm .245 \text{ rad}.$$

Since $\sin \phi < 0$ (see (12)), the minus sign applies here; that is,

$$\phi \approx - .245 \text{ rad}.$$

EXAMPLE 3   The natural length of a spring is 1 m. An object is attached to it and the length of the spring increases to 102 cm when the object is in equilibrium. Then the object is initially displaced downward 1 cm and given an upward velocity of 14 cm/s. Find the displacement for $t > 0$. Also, find the natural frequency, period, amplitude, and phase angle of the resulting motion. Express the answers in cgs units.

Solution   In cgs units $g = 980$ cm/s^2. Since $\Delta l = 2$ cm, $\omega_0^2 = g/\Delta l = 490$. Therefore

$$y'' + 490y = 0, \qquad y(0) = -1, \qquad y'(0) = 14.$$

The general solution of the differential equation is

$$y = c_1 \cos 7\sqrt{10}\,t + c_2 \sin 7\sqrt{10}\,t,$$

so

$$y' = 7\sqrt{10}\left(-c_1 \sin 7\sqrt{10}\,t + c_2 \cos 7\sqrt{10}\,t\right).$$

Substituting the initial conditions into the last two equations yields $c_1 = -1$ and $c_2 = 2/\sqrt{10}$. Hence,

$$y = -\cos 7\sqrt{10}\,t + \frac{2}{\sqrt{10}} \sin 7\sqrt{10}\,t.$$

The frequency is $7\sqrt{10}$ rad/s, and the period is $T = 2\pi/(7\sqrt{10})$ s. The amplitude is

$$R = \sqrt{c_1^2 + c_2^2} = \sqrt{(-1)^2 + \left(\frac{2}{\sqrt{10}}\right)^2} = \sqrt{\frac{7}{5}} \text{ cm.}$$

The phase angle is determined by

$$\cos \phi = \frac{c_1}{R} = -\sqrt{\frac{5}{7}} \qquad \text{and} \qquad \sin \phi = \frac{c_2}{R} = \sqrt{\frac{2}{7}}.$$

Therefore ϕ is in the second quadrant, and

$$\phi = \cos^{-1}\left(-\sqrt{\frac{5}{7}}\right) \approx 2.58 \text{ rad.} \qquad \blacksquare$$

UNDAMPED FORCED OSCILLATION

In many mechanical problems a device is subjected to periodic external forces. For example, soldiers marching in cadence on a bridge cause periodic disturbances in the bridge, and the engines of a propeller-driven aircraft cause periodic disturbances in its wings. In the absence of sufficient damping forces, such disturbances—even if small in magnitude—can cause structural breakdown if they are at certain critical frequencies. To illustrate this we will consider the motion of an object in a spring–mass system without damping, subject to an external force

$$F(t) = F_0 \cos \omega t$$

where F_0 is a constant. In this case the equation of motion (2) is

$$my'' + ky = F_0 \cos \omega t,$$

which we rewrite as

$$y'' + \omega_0^2 y = \frac{F_0}{m} \cos \omega t \tag{13}$$

with $\omega_0 = \sqrt{k/m}$. We will see from the next two examples that the solutions of (13) with $\omega \neq \omega_0$ behave very differently from the solutions with $\omega = \omega_0$.

EXAMPLE 4 Solve the initial value problem

$$y'' + \omega_0^2 y = \frac{F_0}{m} \cos \omega t, \qquad y(0) = 0, \qquad y'(0) = 0, \tag{14}$$

given that $\omega \neq \omega_0$.

Solution We first obtain a particular solution of (13) by the method of undetermined coefficients. Since $\omega \neq \omega_0$, it follows that $\cos \omega t$ is not a solution of the complementary equation

$$y'' + \omega_0^2 y = 0.$$

Therefore (13) has a particular solution of the form

$$y_p = A \cos \omega t + B \sin \omega t.$$

Since

$$y_p'' = -\omega^2 (A \cos \omega t + B \sin \omega t),$$

we have

$$y_p'' + \omega_0^2 y_p = \frac{F_0}{m} \cos \omega t$$

if and only if

$$(\omega_0^2 - \omega^2)(A \cos \omega t + B \sin \omega t) = \frac{F_0}{m} \cos \omega t.$$

This holds if and only if

$$A = \frac{F_0}{m(\omega_0^2 - \omega^2)} \qquad \text{and} \qquad B = 0,$$

so

$$y_p = \frac{F_0}{m(\omega_0^2 - \omega^2)} \cos \omega t.$$

The general solution of (13) is

$$y = \frac{F_0}{m(\omega_0^2 - \omega^2)} \cos \omega t + c_1 \cos \omega_0 t + c_2 \sin \omega_0 t, \tag{15}$$

so

$$y' = \frac{-\omega F_0}{m(\omega_0^2 - \omega^2)} \sin \omega t + \omega_0(-c_1 \sin \omega_0 t + c_2 \cos \omega_0 t).$$

The initial conditions $y(0) = 0$ and $y'(0) = 0$ in (14) imply that

$$c_1 = -\frac{F_0}{m(\omega_0^2 - \omega^2)} \qquad \text{and} \qquad c_2 = 0.$$

Substituting these into (15) yields

$$y = \frac{F_0}{m(\omega_0^2 - \omega^2)} (\cos \omega t - \cos \omega_0 t). \tag{16}$$

It is revealing to write this in a different form. We start with the trigonometric identities

$$\cos(\alpha - \beta) = \cos \alpha \cos \beta + \sin \alpha \sin \beta$$
$$\cos(\alpha + \beta) = \cos \alpha \cos \beta - \sin \alpha \sin \beta.$$

Subtracting the second identity from the first yields

$$\cos(\alpha - \beta) - \cos(\alpha + \beta) = 2 \sin \alpha \sin \beta. \tag{17}$$

Now let

$$\alpha - \beta = \omega t \quad \text{and} \quad \alpha + \beta = \omega_0 t, \tag{18}$$

so that

$$\alpha = \frac{(\omega_0 + \omega)t}{2} \quad \text{and} \quad \beta = \frac{(\omega_0 - \omega)t}{2}. \tag{19}$$

Substituting (18) and (19) into (17) yields

$$\cos \omega t - \cos \omega_0 t = 2 \sin \frac{(\omega_0 - \omega)t}{2} \sin \frac{(\omega_0 + \omega)t}{2},$$

and substituting this into (16) yields

$$y = R(t) \sin \frac{(\omega_0 + \omega)t}{2}, \tag{20}$$

where

$$R(t) = \frac{2F_0}{m(\omega_0^2 - \omega^2)} \sin \frac{(\omega_0 - \omega)t}{2}. \tag{21}$$

From (20) we can regard y as a sinusoidal variation with frequency $(\omega_0 + \omega)/2$ and variable amplitude $|R(t)|$. In Figure 6 the dashed curve above the t-axis is $y = |R(t)|$, the dashed curve below the t-axis is $y = -|R(t)|$, and the displacement y appears as an oscillation bounded by them. The oscillation of y for t on an interval between successive zeros of $R(t)$ is called a **beat**.

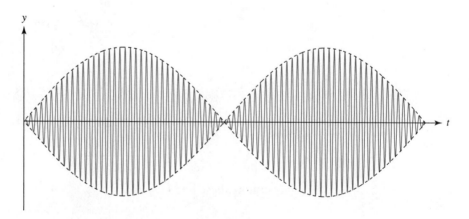

Figure 6 Undamped oscillation with beats

It can be seen from (20) and (21) that

$$|y(t)| \leq \frac{2|F_0|}{m|\omega_0^2 - \omega^2|};$$

moreover, if $\omega + \omega_0$ is sufficiently large compared with $\omega - \omega_0$, then $|y|$ assumes values close to (perhaps equal to) this upper bound during each beat. However, the oscillation remains bounded for all t. (This assumes that the spring can withstand deflections of this size and continue to obey Hooke's law.) The next example shows that this is not so if $\omega = \omega_0$.

EXAMPLE 5   Find the general solution of

$$y'' + \omega_0^2 y = \frac{F_0}{m} \cos \omega_0 t. \tag{22}$$

Solution   We first obtain a particular solution y_p of (22). Since $\cos \omega_0 t$ is a solution of the complementary equation, the form for y_p is

$$y_p = t(A \cos \omega_0 t + B \sin \omega_0 t). \tag{23}$$

Then

$$y_p' = A \cos \omega_0 t + B \sin \omega_0 t + \omega_0 t(-A \sin \omega_0 t + B \cos \omega_0 t)$$

and

$$y_p'' = 2\omega_0(-A \sin \omega_0 t + B \cos \omega_0 t) - \omega_0^2 t(A \cos \omega_0 t + B \sin \omega_0 t). \tag{24}$$

From (23) and (24) we see that y_p satisfies (22) if

$$-2A\omega_0 \sin \omega_0 t + 2B\omega_0 \cos \omega_0 t = \frac{F_0}{m} \cos \omega_0 t;$$

that is, if

$$A = 0 \quad \text{and} \quad B = \frac{F_0}{2m\omega_0}.$$

Therefore

$$y_p = \frac{F_0 t}{2m\omega_0} \sin \omega_0 t$$

is a particular solution of (22). The general solution of (22) is

$$y = \frac{F_0 t}{2m\omega_0} \sin \omega_0 t + c_1 \cos \omega_0 t + c_2 \sin \omega_0 t.$$

The graph of y_p is shown in Figure 7, where it can be seen that y_p oscillates between the dashed lines

$$y = \frac{F_0 t}{2m\omega_0} \quad \text{and} \quad y = -\frac{F_0 t}{2m\omega_0}$$

with increasing amplitude that approaches ∞ as $t \to \infty$. Of course, this means that the spring must eventually fail to obey Hooke's law or break.   ■

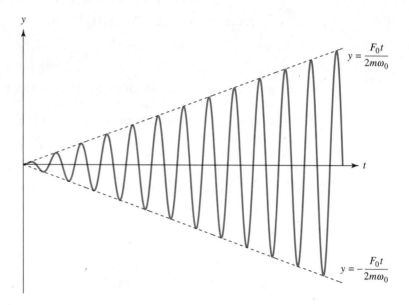

Figure 7   Unbounded displacement due to resonance

This phenomenon of unbounded displacements of a spring–mass system in response to a periodic forcing function at its natural frequency is called **reso-nance.** More complicated mechanical structures can also exhibit resonance-like phenomena. For example, rhythmic oscillations of a suspension bridge by wind forces or of an airplane wing by periodic vibrations of reciprocating engines can cause damage or even failure if the frequencies of the disturbances are close to critical frequencies determined by the parameters of the mechanical system in question.

6.1 EXERCISES

In the following exercises assume that there is no damping.

 1. An object stretches a spring 4 inches in equilibrium. Find and graph its displacement for $t > 0$ if it is initially displaced 36 inches above equilibrium and given a downward velocity of 2 ft/s.

2. An object stretches a string 1.2 inches in equilibrium. Find its displacement for $t > 0$ if it is initially displaced 3 inches below equilibrium and given a downward velocity of 2 ft/s.

3. A spring with natural length .5 m has length 50.5 cm with a mass of 2 g suspended from it. The mass is initially displaced 1.5 cm below equilibrium and released with zero velocity. Find its displacement for $t > 0$.

4. An object stretches a spring 6 inches in equilibrium. Find its displacement for $t > 0$ if it is initially displaced 3 inches above equilibrium and given a downward velocity of 6 inches/s. Find the frequency, period, amplitude, and phase angle of the motion.

 5. An object stretches a spring 5 cm in equilibrium. It is initially displaced 10 cm above equilibrium and given an upward velocity of .25 m/s. Find and graph its displacement for $t > 0$. Find the frequency, period, amplitude, and phase angle of the motion.

6. A 10-kg mass stretches a spring 70 cm in equilibrium. Suppose that a 2-kg mass is attached to the spring, initially displaced 25 cm below equilibrium, and given an upward velocity of 2 m/s. Find its displacement for $t > 0$. Find the frequency, period, amplitude, and phase angle of the motion.

7. A weight stretches a spring 1.5 inches in equilibrium. The weight is initially displaced 8 inches above equilibrium and given a downward velocity of 4 ft/s. Find its displacement for $t > 0$.

8. A weight stretches a spring 6 inches in equilibrium. The weight is initially displaced 6 inches above equilibrium and given a downward velocity of 3 ft/s. Find its displacement for $t > 0$.

9. A spring–mass system has natural frequency $7\sqrt{10}$ rad/s. The natural length of the spring is .7 m. What is the length of the spring when the mass is in equilibrium?

10. A 64-lb weight is attached to a spring with constant $k = 8$ lb/ft, and subjected to an external force $F(t) = 2 \sin t$. The weight is initially displaced 6 inches above equilibrium and given an upward velocity of 2 ft/s. Find its displacement for $t > 0$.

11. A unit mass hangs in equilibrium from a spring with constant $k = 1/16$. Starting at $t = 0$, a force $F(t) = 3 \sin t$ is applied to the mass. Find its displacement for $t > 0$.

C 12. A 4-lb weight stretches a spring 1 ft in equilibrium. An external force $F(t) = .25 \sin 8t$ lb is applied to the weight, which is initially displaced 4 inches above equilibrium and given a downward velocity of 1 ft/s. Find and graph its displacement for $t > 0$.

13. A 2-lb weight stretches a spring 6 inches in equilibrium. An external force $F(t) = \sin 8t$ lb is applied to the weight, which is released from rest 2 inches below equilibrium. Find its displacement for $t > 0$.

14. A 10-g mass suspended on a spring moves in simple harmonic motion with period 4 s. Find the period of the simple harmonic motion of a 20-g mass suspended from the same spring.

15. A 6-lb weight stretches a spring 6 inches in equilibrium. Suppose that an external force $F(t) = (3/16) \sin \omega t + (3/8) \cos \omega t$ lb is applied to the weight. For what value of ω will the displacement be unbounded? Find the displacement if ω has this value. Assume that the motion starts from equilibrium with zero initial velocity.

C 16. A 6-lb weight stretches a spring 4 inches in equilibrium. Suppose that an external force $F(t) = 4 \sin \omega t - 6 \cos \omega t$ lb is applied to the weight. For what value of ω will the displacement be unbounded? Find and graph the displacement if ω has this value. Assume that the motion starts from equilibrium with zero initial velocity.

17. A mass of 1 kg is attached to a spring with constant $k = 4$ N/m. An external force $F(t) = -\cos \omega t - 2 \sin \omega t$ N is applied to the mass. Find the displacement y for $t > 0$ if ω equals the natural frequency of the spring–mass system. Assume that the mass is initially displaced 3 m above equilibrium and given an upward velocity of 450 cm/s.

18. An object is in simple harmonic motion with frequency ω_0, with $y(0) = y_0$ and $y'(0) = v_0$. Find its displacement for $t > 0$. Also, find the amplitude of the oscillation and give formulas for the sine and cosine of the initial phase angle.

19. Two objects suspended from identical springs are set into motion. The period of one object is twice the period of the other. How are the weights of the two objects related?

20. Two objects suspended from identical springs are set into motion. The weight of one object is twice the weight of the other. How are the periods of the resulting motions related?

21. Two identical objects suspended from different springs are set into motion. The period of one motion is 3 times the period of the other. How are the two spring constants related?

6.2 Spring Problems II

FREE VIBRATIONS WITH DAMPING

In this section we consider the motion of an object in a spring–mass system with damping. We start with unforced motion, so the equation of motion is

$$my'' + cy' + ky = 0. \tag{1}$$

Now suppose that the object is displaced from equilibrium and given an initial velocity. Intuition suggests that if the damping force is sufficiently weak the resulting motion will be oscillatory, as in the undamped case considered in the previous section, while if it is sufficiently strong the object may just move slowly toward the equilibrium position without ever reaching it. We will now confirm these intuitive ideas mathematically. The characteristic equation of (1) is

$$mr^2 + cr + k = 0.$$

The roots of this equation are

$$r_1 = \frac{-c - \sqrt{c^2 - 4mk}}{2m} \quad \text{and} \quad r_2 = \frac{-c + \sqrt{c^2 - 4mk}}{2m}. \tag{2}$$

We saw in Section 5.2 that the form of the solution of (1) depends upon whether $c^2 - 4mk$ is positive, negative, or zero. We will now consider these three cases.

UNDERDAMPED MOTION

We say the motion is **underdamped** if $c < \sqrt{4mk}$. In this case r_1 and r_2 in (2) are complex conjugates, which we write as

$$r_1 = -\frac{c}{2m} - i\omega_1 \quad \text{and} \quad r_2 = -\frac{c}{2m} + i\omega_1,$$

where

$$\omega_1 = \frac{\sqrt{4mk - c^2}}{2m}.$$

The general solution of (1) in this case is

$$y = e^{-ct/2m}(c_1 \cos \omega_1 t + c_2 \sin \omega_1 t).$$

By the method used in Section 6.1 to derive the amplitude–phase form of the displacement of an object in simple harmonic motion, we can rewrite this equation as

$$y = Re^{-ct/2m} \cos(\omega_1 t - \phi), \tag{3}$$

where

$$R = \sqrt{c_1^2 + c_2^2}, \quad R \cos \phi = c_1, \quad \text{and} \quad R \sin \phi = c_2.$$

The factor $Re^{-ct/2m}$ in (3) is called the **time-varying amplitude** of the motion, the quantity ω_1 is called the **frequency,** and $T = 2\pi/\omega_1$ (which is the period of the cosine function in (3)) is called the **quasi-period.** A typical graph of (3) is shown in Figure 1. As illustrated in that figure, the graph of y oscillates between the dashed exponential curves $y = \pm Re^{-ct/2m}$.

OVERDAMPED MOTION

We say the motion is **overdamped** if $c > \sqrt{4mk}$. In this case the zeros r_1 and r_2 of the characteristic polynomial are real, with $r_1 < r_2 < 0$ (see (2)), and the general solution of (1) is

$$y = c_1 e^{r_1 t} + c_2 e^{r_2 t}.$$

Again $\lim_{t \to \infty} y(t) = 0$ as in the underdamped case, but the motion is not oscillatory, since y cannot equal zero for more than one value of t unless $c_1 = c_2 = 0$. (Exercise 23.)

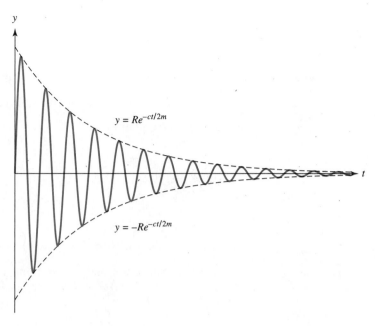

Figure 1 Underdamped motion

CRITICALLY DAMPED MOTION

We say the motion is ***critically damped*** if $c = \sqrt{4mk}$. In this case $r_1 = r_2 = -c/2m$ and the general solution of (1) is

$$y = e^{-ct/2m}(c_1 + c_2t).$$

Again $\lim_{t \to \infty} y(t) = 0$ and the motion is nonoscillatory, since y cannot equal zero for more than one value of t unless $c_1 = c_2 = 0$. (Exercise 22.)

EXAMPLE 1

Suppose that a 64-lb weight stretches a spring 6 inches in equilibrium and a dashpot provides a damping force of c lb for each ft/s of velocity.

(a) Write the equation of motion of the object and determine the value of c for which the motion is critically damped.
(b) Find the displacement y for $t > 0$ if the motion is critically damped and the initial conditions are $y(0) = 1$ and $y'(0) = 20$.
(c) Find the displacement y for $t > 0$ if the motion is critically damped and the initial conditions are $y(0) = 1$ and $y'(0) = -20$.

Solution

(a) Here $m = 2$ slugs and $k = 64/.5 = 128$ lb-s/ft. Therefore, the equation of motion (1) is

$$2y'' + cy' + 128y = 0. \qquad (4)$$

The characteristic equation is

$$2r^2 + cr + 128 = 0,$$

which has roots

$$r = \frac{-c \pm \sqrt{c^2 - 8 \cdot 128}}{4}.$$

Therefore, the damping is critical if

$$c = \sqrt{8 \cdot 128} = 32 \text{ lb-s/ft}.$$

Solution   **(b)** Setting $c = 32$ in (4) and canceling the common factor 2 yields

$$y'' + 16y + 64y = 0.$$

The characteristic equation is

$$r^2 + 16r + 64y = (r + 8)^2 = 0.$$

Hence, the general solution is

$$y = e^{-8t}(c_1 + c_2 t). \tag{5}$$

Differentiating this yields

$$y' = -8y + c_2 e^{-8t}. \tag{6}$$

Imposing the initial conditions $y(0) = 1$ and $y'(0) = 20$ in the last two equations shows that $1 = c_1$ and $20 = -8 + c_2$. Hence, the solution of the initial value problem is

$$y = e^{-8t}(1 + 28t).$$

Therefore, the object approaches equilibrium from above as $t \to \infty$. There is no oscillation.

Solution   **(c)** Imposing the initial conditions $y(0) = 1$ and $y'(0) = -20$ in (5) and (6) yields $1 = c_1$ and $-20 = -8 + c_2$. Hence, the solution of this initial value problem is

$$y = e^{-8t}(1 - 12t).$$

Therefore, the object moves downward through equilibrium just once, and then approaches equilibrium from below as $t \to \infty$. Again, there is no oscillation. The solutions of these two initial value problems are graphed in Figure 2.

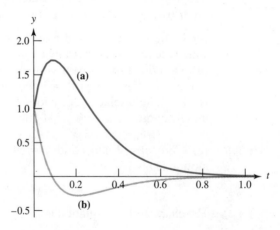

Figure 2   **(a)** $y = e^{-8t}(1 + 28t)$; **(b)** $y = e^{-8t}(1 - 12t)$

EXAMPLE 2   Find the displacement of the object in Example 1 if the damping constant is $c = 4$ lb-s/ft and the initial conditions are $y(0) = 1.5$ ft and $y'(0) = -3$ ft/s.

Solution With $c = 4$ the equation of motion (4) becomes

$$y'' + 2y' + 64y = 0 \tag{7}$$

after canceling the common factor 2. The characteristic equation

$$r^2 + 2r + 64 = 0$$

has complex conjugate roots

$$r = \frac{-2 \pm \sqrt{4 - 4 \cdot 64}}{2} = -1 \pm 3\sqrt{7}i.$$

Therefore, the motion is underdamped and the general solution of (7) is

$$y = e^{-t}\left(c_1 \cos 3\sqrt{7}t + c_2 \sin 3\sqrt{7}t\right).$$

Differentiating this yields

$$y' = -y + 3\sqrt{7}e^{-t}\left(-c_1 \sin 3\sqrt{7}t + c_2 \cos 3\sqrt{7}t\right).$$

Imposing the initial conditions $y(0) = 1.5$ and $y'(0) = -3$ in the last two equations yields $1.5 = c_1$ and $-3 = -1.5 + 3\sqrt{7}c_2$. Hence, the solution of the initial value problem is

$$y = e^{-t}\left(\frac{3}{2} \cos 3\sqrt{7}t - \frac{1}{2\sqrt{7}} \sin 3\sqrt{7}t\right). \tag{8}$$

The amplitude of the function in parentheses is

$$R = \sqrt{\left(\frac{3}{2}\right)^2 + \left(\frac{1}{2\sqrt{7}}\right)^2} = \sqrt{\frac{9}{4} + \frac{1}{4 \cdot 7}} = \sqrt{\frac{64}{4 \cdot 7}} = \frac{4}{\sqrt{7}}.$$

Therefore, we can rewrite (8) as

$$y = \frac{4}{\sqrt{7}}e^{-t} \cos\left(3\sqrt{7}t - \phi\right),$$

where

$$\cos \phi = \frac{3}{2R} = \frac{3\sqrt{7}}{8} \quad \text{and} \quad \sin \phi = -\frac{1}{2\sqrt{7}R} = -\frac{1}{8}.$$

Therefore $\phi \cong -.125$ radians.

EXAMPLE 3 Let the damping constant in Example 1 be $c = 40$ lb-s/ft. Find the displacement y for $t > 0$ if $y(0) = 1$ and $y'(0) = 1$.

Solution With $c = 40$ the equation of motion (4) reduces to

$$y'' + 20y' + 64y = 0 \tag{9}$$

after canceling the common factor 2. The characteristic equation

$$r^2 + 20r + 64 = (r + 16)(r + 4) = 0$$

has the roots $r_1 = -4$ and $r_2 = -16$. Therefore, the general solution of (9) is

$$y = c_1e^{-4t} + c_2e^{-16t}. \tag{10}$$

Differentiating this yields

$$y' = -4e^{-4t} - 16c_2e^{-16t}.$$

The last two equations and the initial conditions $y(0) = 1$ and $y'(0) = 1$ imply that

$$c_1 + c_2 = 1$$
$$-4c_1 - 16c_2 = 1.$$

The solution of this system is $c_1 = 17/12, c_2 = -5/12$. Substituting these into (10) yields

$$y = \frac{17}{12}e^{-4t} - \frac{5}{12}e^{-16t}$$

as the solution of the given initial value problem. Figure 3 shows the graph of the solution. ■

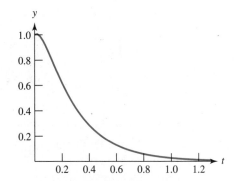

Figure 3   Graph of $y = \dfrac{17}{12}e^{-4t} - \dfrac{5}{12}e^{-16t}$

FORCED VIBRATIONS WITH DAMPING

Now we consider the motion of an object in a spring–mass system with damping, under the influence of a periodic forcing function $F(t) = F_0 \cos \omega t$, so that the equation of motion is

$$my'' + cy' + ky = F_0 \cos \omega t. \tag{11}$$

In Section 6.1 we considered this equation with $c = 0$ and found that the resulting displacement y assumed arbitrarily large values in the case of resonance (that is, when $\omega = \omega_0 = \sqrt{k/m}$). Here we will see that in the presence of damping the displacement remains bounded for all t, and the initial conditions have little effect on the motion as $t \to \infty$. In fact, we will see that for large t the displacement is closely approximated by a function of the form

$$y = R \cos(\omega t - \phi), \tag{12}$$

where the amplitude R depends upon m, c, k, F_0, and ω. We are interested in the following question:

QUESTION.   Assuming that m, c, k, and F_0 are held constant, what value of ω produces the largest amplitude R in (12), and what is this largest amplitude?

To answer this question we must solve (11) and determine R in terms of F_0, ω_0, ω, and c. We can obtain a particular solution of (11) by the method of undetermined coefficients. Since $\cos \omega t$ does not satisfy the complementary equation

$$my'' + cy' + ky = 0$$

we can obtain a particular solution of (11) in the form

$$y_p = A \cos \omega t + B \sin \omega t. \tag{13}$$

Differentiating this yields

$$y_p' = \omega(-A \sin \omega t + B \cos \omega t)$$

and

$$y_p'' = -\omega^2(A \cos \omega t + B \sin \omega t).$$

From the last three equations,

$$my_p'' + cy_p' + ky_p = (-m\omega^2 A + c\omega B + kA) \cos \omega t$$
$$+ (-m\omega^2 B - c\omega A + kB) \sin \omega t,$$

so y_p satisfies (11) if

$$(k - m\omega^2)A + c\omega B = F_0$$
$$-c\omega A + (k - m\omega^2)B = 0.$$

Solving for A and B and substituting the results into (13) yields

$$y_p = \frac{F_0}{(k - m\omega^2)^2 + c^2\omega^2}[(k - m\omega^2) \cos \omega t + c\omega \sin \omega t],$$

which can be written in amplitude–phase form as

$$y_p = \frac{F_0}{\sqrt{(k - m\omega^2)^2 + c^2\omega^2}} \cos(\omega t - \phi), \tag{14}$$

where

$$\cos \phi = \frac{k - m\omega^2}{\sqrt{(k - m\omega^2)^2 + c^2\omega^2}} \quad \text{and} \quad \sin \phi = \frac{c\omega}{\sqrt{(k - m\omega^2)^2 + c^2\omega^2}}. \tag{15}$$

To compare this with the undamped forced vibration that we considered in Section 6.1, it is useful to write

$$k - m\omega^2 = m\left(\frac{k}{m} - \omega^2\right) = m(\omega_0^2 - \omega^2), \tag{16}$$

where $\omega_0 = \sqrt{k/m}$ is the natural angular frequency of the undamped simple harmonic motion of an object with mass m on a spring with constant k. Substituting (16) into (14) yields

$$y_p = \frac{F_0}{\sqrt{m^2(\omega_0^2 - \omega^2)^2 + c^2\omega^2}} \cos(\omega t - \phi). \tag{17}$$

The solution of an initial value problem

$$my'' + cy' + ky = F_0 \cos \omega t, \quad y(0) = y_0, \quad y'(0) = v_0,$$

is of the form $y = y_c + y_p$, where y_c has one of the three forms

$$y_c = e^{-ct/2m}(c_1 \cos \omega_1 t + c_2 \sin \omega_1 t),$$
$$y_c = e^{-ct/2m}(c_1 + c_2 t),$$
$$y_c = c_1 e^{r_1 t} + c_2 e^{r_2 t} \quad (r_1, r_2 < 0).$$

In all three cases $\lim_{t\to\infty} y_c(t) = 0$ for any choice of c_1 and c_2. For this reason we say that y_c is the **transient component** of the solution y. The behavior of y for large t is determined by y_p, which we call the **steady state component** of y. Thus, for large t the motion is like simple harmonic motion at the frequency of the external force.

The amplitude of y_p in (17) is

$$R = \frac{F_0}{\sqrt{m^2(\omega_0^2 - \omega^2)^2 + c^2\omega^2}}, \tag{18}$$

which is finite for all ω; that is, the presence of damping precludes the phenomenon of resonance that we encountered in studying undamped vibrations under a periodic forcing function. We will now find the value ω_{max} of ω for which R is maximized. This is the value of ω for which the function

$$\rho(\omega) = m^2(\omega_0^2 - \omega^2)^2 + c^2\omega^2$$

in the denominator of (18) attains its minimum value. By rewriting this as

$$\rho(\omega) = m^2(\omega_0^4 + \omega^4) + (c^2 - 2m^2\omega_0^2)\omega^2 \tag{19}$$

it can be seen that ρ is a strictly increasing function of ω^2 if

$$c \geq \sqrt{2m^2\omega_0^2} = \sqrt{2mk}.$$

(Recall that $\omega_0^2 = k/m$.) Therefore, $\omega_{max} = 0$ if this inequality holds. From (15) it can be seen that $\phi = 0$ if $\omega = 0$. In this case (14) reduces to

$$y_p = \frac{F_0}{\sqrt{m^2\omega_0^4}} = \frac{F_0}{k},$$

which is consistent with Hooke's law: if the mass is subjected to a constant force F_0, then its displacement should approach a constant y_p such that $ky_p = F_0$. Now suppose that $c < \sqrt{2mk}$. Then, from (19),

$$\rho'(\omega) = 2\omega(2m^2\omega^2 + c^2 - 2m^2\omega_0^2),$$

and ω_{max} is the value of ω for which the expression in parentheses equals zero; that is,

$$\omega_{max} = \sqrt{\omega_0^2 - \frac{c^2}{2m^2}} = \sqrt{\frac{k}{m}\left(1 - \frac{c^2}{2km}\right)}.$$

(To see that $\rho(\omega_{max})$ is the minimum value of $\rho(\omega)$, notice that $\rho'(\omega) < 0$ if $\omega < \omega_{max}$ and $\rho'(\omega) > 0$ if $\omega > \omega_{max}$.) Substituting $\omega = \omega_{max}$ in (18) and simplifying shows that the maximum amplitude R_{max} is

$$R_{max} = \frac{2mF_0}{c\sqrt{4mk - c^2}} \quad \text{if} \quad c < \sqrt{2mk}.$$

We summarize our results as follows.

THEOREM 6.2.1

Suppose that we consider the amplitude R of the steady state component of the solution of

$$my'' + cy' + ky = F_0 \cos \omega t$$

as a function of ω.

(a) *If $c \geq \sqrt{2mk}$ then the maximum amplitude is $R_{max} = F_0/k$, and it is attained when $\omega = \omega_{max} = 0$.*

(b) *If $c < \sqrt{2mk}$ then the maximum amplitude is*

$$R_{max} = \frac{2mF_0}{c\sqrt{4mk - c^2}}, \tag{20}$$

and it is attained when

$$\omega = \omega_{max} = \sqrt{\frac{k}{m}\left(1 - \frac{c^2}{2km}\right)}. \tag{21}$$

Notice that R_{max} and ω_{max} are continuous functions of c, for $c \geq 0$, since (20) and (21) reduce to $R_{max} = F_0/k$ and $\omega_{max} = 0$ if $c = \sqrt{2km}$.

6.2 EXERCISES

1. A 64-lb object stretches a spring 4 ft in equilibrium. It is attached to a dashpot with damping constant $c = 8$ lb-s/ft. The object is initially displaced 18 inches above equilibrium and given a downward velocity of 4 ft/s. Find its displacement and time-varying amplitude for $t > 0$.

C 2. A 16-lb weight is attached to a spring with natural length 5 ft. With the weight attached the spring measures 8.2 ft. The weight is initially displaced 3 ft below equilibrium and given an upward velocity of 2 ft/s. Find and graph its displacement for $t > 0$ if the medium resists the motion with a force of 1 lb for each ft/s of velocity. Also, find its time-varying amplitude.

C 3. An 8-lb weight stretches a spring 1.5 inches. It is attached to a dashpot with damping constant $c = 8$ lb-s/ft. The weight is initially displaced 3 inches above equilibrium and given an upward velocity of 6 ft/s. Find and graph its displacement for $t > 0$.

4. A 96-lb weight stretches a spring 3.2 ft in equilibrium. It is attached to a dashpot with damping constant $c = 18$ lb-s/ft. The weight is initially displaced 15 inches below equilibrium and given a downward velocity of 12 ft/s. Find its displacement for $t > 0$.

5. A 16-lb weight stretches a spring 6 inches in equilibrium. It is attached to a damping mechanism with constant c. Find all values of c such that the free vibration of the weight has infinitely many oscillations.

6. An 8-lb weight stretches a spring .32 ft. The weight is initially displaced 6 inches above equilibrium and given an upward velocity of 4 ft/s. Find its displacement for $t > 0$ if the medium exerts a damping force of 1.5 lb for each ft/s of velocity.

7. A 32-lb weight stretches a spring 2 ft in equilibrium. It is attached to a dashpot with constant $c = 8$ lb-s/ft. The weight is initially displaced 8 inches below equilibrium and released from rest. Find its displacement for $t > 0$.

8. A mass of 20 g stretches a spring 5 cm. The spring is attached to a dashpot with damping constant 400 dyne-s/cm. Determine the displacement for $t > 0$ if the mass is initially displaced 9 cm above equilibrium and released from rest.

9. A 64-lb weight is suspended from a spring with constant $k = 25$ lb/ft. It is initially displaced 18 inches above equilibrium and released from rest. Find its displacement for $t > 0$ if the medium resists the motion with 6 lb of force for each ft/s of velocity.

10. A 32-lb weight stretches a spring 1 ft in equilibrium. The weight is initially displaced 6 inches above equilibrium and given a downward velocity of 3 ft/s. Find its displacement for $t > 0$ if the medium resists the motion with a force numerically equal to 3 times the speed in ft/s.

11. An 8-lb weight stretches a spring 2 inches. It is attached to a dashpot with damping constant $c = 4$ lb-s/ft. The weight is initially displaced 3 inches above equilibrium and given a downward velocity of 4 ft/s. Find its displacement for $t > 0$.

C **12.** A 2-lb weight stretches a spring .32 ft. The weight is initially displaced 4 inches below equilibrium and given an upward velocity of 5 ft/s. The medium provides damping with constant $c = (1/8)$ lb-s/ft. Find and graph the displacement for $t > 0$.

13. An 8-lb weight stretches a spring 8 inches in equilibrium. It is attached to a dashpot with damping constant $c = .5$ lb/ft and subjected to an external force $F(t) = 4 \cos 2t$ lb. Determine the steady state component of the displacement for $t > 0$.

14. A 32-lb weight stretches a spring 1 ft in equilibrium. It is attached to a dashpot with constant $c = 12$ lb-s/ft. The weight is initially displaced 8 inches above equilibrium and released from rest. Find its displacement for $t > 0$.

15. A mass of one kg stretches a spring 49 cm in equilibrium. A dashpot attached to the spring supplies a damping force of 4 N for each m/s of speed. The mass is initially displaced 10 cm above equilibrium and given a downward velocity of 1 m/s. Find its displacement for $t > 0$.

16. A mass of 100 g stretches a spring 98 cm in equilibrium. A dashpot attached to the spring supplies a damping force of 600 dynes for each cm/s of speed. The mass is initially displaced 10 cm above equilibrium and given a downward velocity of 1 m/s. Find its displacement for $t > 0$.

17. A 192-lb weight is suspended from a spring with constant $k = 6$ lb/ft and subjected to an external force $F(t) = 8 \cos 3t$ lb. Find the steady state component of the displacement for $t > 0$ if the medium resists the motion with a force numerically equal to 8 times the speed in ft/s.

18. A 2-g mass is attached to a spring with constant 20 dyne/cm. Find the steady state component of the displacement if the mass is subjected to an external force $F(t) = 3 \cos 4t - 5 \sin 4t$ dynes and a dashpot supplies 4 dynes of damping for each cm/s of velocity.

C **19.** A 96-lb weight is attached to a spring with constant 12 lb/ft. Find and graph the steady state component of the displacement if the mass is subjected to an external force $F(t) = 18 \cos t - 9 \sin t$ lb and a dashpot supplies 24 lb of damping for each ft/s of velocity.

20. A mass of one kg stretches a spring 49 cm in equilibrium. It is attached to a dashpot that supplies a damping force of 4 N for each m/s of speed. Find the steady state component of its displacement if it is subjected to an external force $F(t) = 8 \sin 2t - 6 \cos 2t$ N.

21. A mass m is suspended from a spring with constant k and subjected to an external force $F(t) = \alpha \cos \omega_0 t + \beta \sin \omega_0 t$, where ω_0 is the natural frequency of the spring–mass system without damping. Find the steady state component of the displacement if a dashpot with constant c supplies damping.

22. Show that if c_1 and c_2 are not both zero then the function

$$y = e^{r_1 t}(c_1 + c_2 t)$$

cannot equal zero for more than one value of t.

23. Show that if c_1 and c_2 are not both zero then the function

$$y = c_1 e^{r_1 t} + c_2 e^{r_2 t}$$

cannot equal zero for more than one value of t.

24. Find the solution of the initial value problem

$$my'' + cy' + ky = 0, \qquad y(0) = y_0, \qquad y'(0) = v_0,$$

given that the motion is underdamped, so that the general solution of the equation is of the form

$$y = e^{-ct/2m}(c_1 \cos \omega_1 t + c_2 \sin \omega_1 t).$$

25. Find the solution of the initial value problem

$$my'' + cy' + ky = 0, \qquad y(0) = y_0, \qquad y'(0) = v_0,$$

given that the motion is overdamped, so that the general solution of the equation is of the form

$$y = c_1 e^{r_1 t} + c_2 e^{r_2 t} \qquad (r_1, r_2 < 0).$$

26. Find the solution of the initial value problem

$$my'' + cy' + ky = 0, \qquad y(0) = y_0, \qquad y'(0) = v_0,$$

given that the motion is critically damped, so that the general solution of the equation is of the form

$$y = e^{r_1 t}(c_1 + c_2 t) \qquad (r_1 < 0).$$

6.3   The RLC Circuit

In this section we consider the ***RLC circuit,*** shown schematically in Figure 1. As we'll see, the *RLC* circuit is an electrical analog of a spring-mass system with damping.

Nothing happens while the switch is open (dashed line). When the switch is closed (solid line), we say that the ***circuit is closed.*** Differences in electrical potential in a closed circuit cause current to flow in the circuit. The battery or generator in Figure 1 creates a difference in electrical potential $E = E(t)$ between its two terminals, which we've marked arbitrarily as positive and negative. (We could just as well interchange the markings.) We'll say that $E(t) > 0$ if the potential at the positive terminal is greater than the potential at the negative terminal, $E(t) < 0$ if the potential at the positive terminal is less than the potential at the negative terminal, and $E(t) = 0$ if the potential is the same at the two terminals. We call E the ***impressed voltage.***

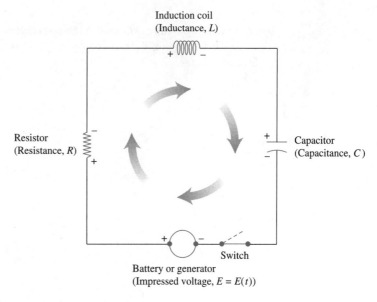

Figure 1   An *RLC* circuit

At any given time t the same current flows in all points of the circuit. We denote current by $I = I(t)$. We say that $I(t) > 0$ if the direction of flow is around the circuit from the positive terminal of the battery or generator back to the negative terminal, as indicated by the arrows in Figure 1; $I(t) < 0$ if the flow is in the opposite direction, and $I(t) = 0$ if no current flows at time t.

Differences in potential occur at the resistor, induction coil, and capacitor in Figure 1. Notice that the two sides of each of these components are also identified as positive and negative. The **voltage drop across** each component is defined to be the potential on the positive side of the component minus the potential on the negative side. This terminology is somewhat misleading, since "drop" suggests a decrease even though changes in potential are signed quantities and may therefore be increases. Nevertheless, we'll go along with tradition and call them voltage drops. The voltage drop across the resistor in Figure 1 is given by

$$V_R = IR, \tag{1}$$

where I is current and R is a positive constant, the **resistance** of the resistor. The voltage drop across the induction coil is given by

$$V_I = L\frac{dI}{dt} = LI', \tag{2}$$

where L is a positive constant, the **inductance** of the coil.

A capacitor stores electrical charge $Q = Q(t)$, which is related to the current in the circuit by the equation

$$Q(t) = Q_0 + \int_0^t I(\tau)\,d\tau, \tag{3}$$

where Q_0 is the charge on the capacitor at $t = 0$. The voltage drop across a capacitor is given by

$$V_C = \frac{Q}{C}, \tag{4}$$

where C is a positive constant, the **capacitance** of the capacitor.

Table 1 names the units for the quantities that we've discussed. The units are defined so that

$$1 \text{ volt} = 1 \text{ ampere} \cdot 1 \text{ ohm}$$
$$= 1 \text{ henry} \cdot 1 \text{ ampere/second}$$
$$= 1 \text{ coulomb/farad}$$

and

$$1 \text{ ampere} = 1 \text{ coulomb/second}.$$

TABLE I

Symbol	Name	Unit
E	impressed voltage	volt
I	current	ampere
Q	charge	coulomb
R	resistance	ohm
L	inductance	henry
C	capacitance	farad

According to **Kirchhoff's**[1] **law,** the sum of the voltage drops in a closed *RLC* circuit equals the impressed voltage. Therefore, from (1), (2), and (4),

$$LI' + RI + \frac{1}{C}Q = E(t). \tag{5}$$

This equation contains two unknowns, the current I in the circuit and the charge Q on the capacitor. However, (3) implies that $Q' = I$, so (5) can be converted into the second order equation

$$LQ'' + RQ' + \frac{1}{C}Q = E(t) \tag{6}$$

in Q. To find the current flowing in an *RLC* circuit, we solve (6) for Q and then differentiate the solution to obtain I.

In Sections 6.1 and 6.2 we encountered the equation

$$my'' + cy' + ky = F(t) \tag{7}$$

in connection with spring–mass systems. Except for notation this equation is the same as (6). The correspondence between electrical and mechanical quantities connected with (6) and (7) is shown in Table 2.

TABLE 2

Electrical		Mechanical	
charge	Q	displacement	y
current	I	velocity	y'
impressed voltage	$E(t)$	external force	$F(t)$
inductance	L	mass	m
resistance	R	damping	c
1/capacitance	$1/C$	spring constant	k

The equivalence between (6) and (7) is an example of how mathematics unifies fundamental similarities in diverse physical phenomena. Since we've already studied the properties of solutions of (7) in Sections 6.1 and 6.2, we can obtain results concerning solutions of (6) by simply changing notation, according to Table 1.

FREE OSCILLATIONS

We say that an *RLC* circuit is in *free oscillation* if $E(t) = 0$ for $t > 0$, so that (6) becomes

$$LQ'' + RQ' + \frac{1}{C}Q = 0. \tag{8}$$

The characteristic equation of (8) is

$$Lr^2 + Rr + \frac{1}{C} = 0,$$

---

[1]Gustav Kirchhoff (1824–1887) was a leading 19th-century physicist. He formulated the basic laws of electrical circuits from experiments performed while a student at the University of Königsberg.

with roots

$$r_1 = \frac{-R - \sqrt{R^2 - 4L/C}}{2L} \quad \text{and} \quad r_2 = \frac{-R + \sqrt{R^2 - 4L/C}}{2L}. \tag{9}$$

There are three cases to consider, all analogous to the cases considered in Section 6.2 for free vibrations of a damped spring–mass system.

CASE 1.   The oscillation is *underdamped* if $R < \sqrt{4L/C}$. In this case r_1 and r_2 in (9) are complex conjugates, which we write as

$$r_1 = -\frac{R}{2L} - i\omega_1 \quad \text{and} \quad r_2 = -\frac{R}{2L} + i\omega_1,$$

where

$$\omega_1 = \frac{\sqrt{4L/C - R^2}}{2L}.$$

The general solution of (8) is

$$Q = e^{-Rt/2L}(c_1 \cos \omega_1 t + c_2 \sin \omega_1 t),$$

which we can write as

$$Q = Ae^{-Rt/2L} \cos(\omega_1 t - \phi), \tag{10}$$

where

$$A = \sqrt{c_1^2 + c_2^2}, \quad A \cos \phi = c_1, \quad \text{and} \quad A \sin \phi = c_2.$$

In the idealized case where $R = 0$, the solution (10) reduces to

$$Q = A \cos\left(\frac{t}{\sqrt{LC}} - \phi\right),$$

which is analogous to the simple harmonic motion of an undamped spring–mass system in free vibration.

Actual *RLC* circuits are usually underdamped, so the case we've just considered is the most important. However, for completeness we'll consider the other two possibilities.

CASE 2.   The oscillation is *overdamped* if $R > \sqrt{4L/C}$. In this case the zeros r_1 and r_2 of the characteristic polynomial are real, with $r_1 < r_2 < 0$ (see (9)), and the general solution of (8) is

$$Q = c_1 e^{r_1 t} + c_2 e^{r_2 t}. \tag{11}$$

CASE 3.   The oscillation is *critically damped* if $R = \sqrt{4L/C}$. In this case $r_1 = r_2 = -R/2L$ and the general solution of (8) is

$$Q = e^{-Rt/2L}(c_1 + c_2 t). \tag{12}$$

If $R \neq 0$ then the exponentials in (10), (11), and (12) are negative, so the solution of any homogeneous initial value problem

$$LQ'' + RQ' + \frac{1}{C}Q = 0, \quad Q(0) = Q_0, \quad Q'(0) = I_0,$$

approaches zero exponentially as $t \to \infty$. Thus, all such solutions are *transient*, in the sense defined in Section 6.2, where we discussed forced vibrations of a spring–mass system with damping.

EXAMPLE 1   At $t = 0$ a current of 2 amperes flows in an RLC circuit with resistance $R = 40$ ohms, inductance $L = .2$ henrys, and capacitance $C = 10^{-5}$ farads. Find the current flowing in the circuit at $t > 0$ if the initial charge on the capacitor is 1 coulomb. Assume that $E(t) = 0$ for $t > 0$.

Solution   The equation for the charge Q is

$$\frac{1}{5} Q'' + 40 Q' + 10000 Q = 0,$$

or

$$Q'' + 200 Q' + 50000 Q = 0. \tag{13}$$

Therefore we must solve the initial value problem

$$Q'' + 200 Q' + 50000 Q = 0, \qquad Q(0) = 1, \qquad Q'(0) = 2. \tag{14}$$

The desired current is the derivative of the solution of this initial value problem.
    The characteristic equation of (13) is

$$r^2 + 200 r + 50000 = 0,$$

which has complex zeros $r = -100 \pm 200i$. Therefore, the general solution of (13) is

$$Q = e^{-100t}(c_1 \cos 200t + c_2 \sin 200t). \tag{15}$$

Differentiating this and collecting like terms yields

$$Q' = -e^{-100t}[(100c_1 - 200c_2) \cos 200t + (100c_2 + 200c_1) \sin 100t]. \tag{16}$$

To find the solution of the initial value problem (14) we set $t = 0$ in (15) and (16) to obtain

$$c_1 = Q(0) = 1 \qquad \text{and} \qquad -100c_1 + 200c_2 = Q'(0) = 2;$$

therefore $c_1 = 1$ and $c_2 = 51/100$, so

$$Q = e^{-100t}\left(\cos 200t + \frac{51}{100} \sin 200t\right)$$

is the solution of (14). Differentiating this yields

$$I = e^{-100t}(2 \cos 200t - 251 \sin 200t). \qquad \blacksquare$$

FORCED OSCILLATIONS WITH DAMPING

An initial value problem for (6) has the form

$$LQ'' + RQ' + \frac{1}{C}Q = E(t), \qquad Q(0) = Q_0, \qquad Q'(0) = I_0, \tag{17}$$

where Q_0 is the initial charge on the capacitor and I_0 is the initial current in the circuit. We've already seen that if $E \equiv 0$ then all solutions of (17) are transients. If $E \not\equiv 0$ then we know that the solution of (17) has the form $Q = Q_c + Q_p$, where Q_c satisfies the complementary equation, and approaches zero exponentially as $t \to \infty$ for any initial conditions, while Q_p depends only upon E and is independent of the initial conditions. As in the case of forced oscillations of a spring–mass system with damping, we call Q_p the **steady state charge** on the capacitor of the RLC circuit. Since $I = Q' = Q'_c + Q'_p$ and Q'_c also tends to zero exponentially as $t \to \infty$, we say that $I_c = Q'_c$ is the ***transient*** current and $I_p = Q'_p$ is the ***steady state*** current. In most applications we're interested only in the steady state charge and current.

EXAMPLE 2   Find the amplitude–phase form of the steady state current in the RLC circuit in Figure 1 if the impressed voltage, provided by an alternating current generator, is $E(t) = E_0 \cos \omega t$.

Solution   We'll first find the steady state charge on the capacitor as a particular solution of

$$LQ'' + RQ' + \frac{1}{C}Q = E_0 \cos \omega t.$$

To do this, we'll simply reinterpret a result obtained in Section 6.2, where we found that the steady state solution of

$$my'' + cy' + ky = F_0 \cos \omega t$$

is

$$y_p = \frac{F_0}{\sqrt{(k - m\omega^2)^2 + c^2\omega^2}} \cos(\omega t - \phi),$$

where

$$\cos \phi = \frac{k - m\omega^2}{\sqrt{(k - m\omega^2)^2 + c^2\omega^2}} \quad \text{and} \quad \sin \phi = \frac{c\omega}{\sqrt{(k - m\omega^2)^2 + c^2\omega^2}}.$$

(See Equations 14 and 15 of Section 6.2.) By making the appropriate changes in the symbols (according to Table 2) we obtain the steady state charge

$$Q_p = \frac{E_0}{\sqrt{(1/C - L\omega^2)^2 + R^2\omega^2}} \cos(\omega t - \phi),$$

where

$$\cos \phi = \frac{1/C - L\omega^2}{\sqrt{(1/C - L\omega^2)^2 + R^2\omega^2}} \quad \text{and} \quad \sin \phi = \frac{R\omega}{\sqrt{(1/C - L\omega^2)^2 + R^2\omega^2}}.$$

Therefore the steady state current in the circuit is

$$I_p = Q'_p = -\frac{\omega E_0}{\sqrt{(1/C - L\omega^2)^2 + R^2\omega^2}} \sin(\omega t - \phi). \qquad \blacksquare$$

6.3 EXERCISES

In Exercises 1–5 find the current in the given RLC circuit, assuming that $E(t) = 0$ for $t > 0$.

1. $R = 3$ ohms; $L = .1$ henrys; $C = .01$ farads; $Q_0 = 0$ coulombs; $I_0 = 2$ amperes

2. $R = 2$ ohms; $L = .05$ henrys; $C = .01$ farads; $Q_0 = 2$ coulombs; $I_0 = -2$ amperes

3. $R = 2$ ohms; $L = .1$ henrys; $C = .01$ farads; $Q_0 = 2$ coulombs; $I_0 = 0$ amperes

4. $R = 6$ ohms; $L = .1$ henrys; $C = .004$ farads; $Q_0 = 3$ coulombs; $I_0 = -10$ amperes

5. $R = 4$ ohms; $L = .05$ henrys; $C = .008$ farads; $Q_0 = -1$ coulombs; $I_0 = 2$ amperes

In Exercises 6–10 find the steady state current in the circuit described by the given equation.

6. $\dfrac{1}{10}Q'' + 3Q' + 100Q = 5 \cos 10t - 5 \sin 10t$

7. $\dfrac{1}{20}Q'' + 2Q' + 100Q = 10\cos 25t - 5\sin 25t$

8. $\dfrac{1}{10}Q'' + 2Q' + 100Q = 3\cos 50t - 6\sin 50t$

9. $\dfrac{1}{10}Q'' + 6Q' + 250Q = 10\cos 100t + 30\sin 100t$

10. $\dfrac{1}{20}Q'' + 4Q' + 125Q = 15\cos 30t - 30\sin 30t$

11. Show that if $E(t) = U\cos\omega t + V\sin\omega t$ where U and V are constants, then the steady state current in the *RLC* circuit shown in Figure 1 is

$$I_p = \frac{\omega^2 RE(t) + (1/C - L\omega^2)E'(t)}{\Delta},$$

where

$$\Delta = (1/C - L\omega^2)^2 + R^2\omega^2.$$

12. Find the amplitude of the steady state current I_p in the *RLC* circuit shown in Figure 1 if $E(t) = U\cos\omega t + V\sin\omega t$ where U and V are constants. Then find the value ω_0 of ω for which the amplitude of I_p is maximized, and find the maximum amplitude.

In Exercises 13–17 graph the amplitude of the steady state current against ω. Estimate the value of ω that maximizes the amplitude of the steady state current, and estimate this maximum amplitude.

Hint: *You can confirm your results by doing Exercise 12.*

L **13.** $\dfrac{1}{10}Q'' + 3Q' + 100Q = U\cos\omega t + V\sin\omega t$

L **14.** $\dfrac{1}{20}Q'' + 2Q' + 100Q = U\cos\omega t + V\sin\omega t$

L **15.** $\dfrac{1}{10}Q'' + 2Q' + 100Q = U\cos\omega t + V\sin\omega t$

L **16.** $\dfrac{1}{10}Q'' + 6Q' + 250Q = U\cos\omega t + V\sin\omega t$

L **17.** $\dfrac{1}{20}Q'' + 4Q' + 125Q = U\cos\omega t + V\sin\omega t$

6.4 Motion Under a Central Force

We will now study the motion of an object moving under the influence of a ***central force;*** that is, a force whose magnitude at any point P other than the origin depends only on the distance from P to the origin, and whose direction at P is parallel to the line connecting P and the origin, as indicated in Figure 1 for the case where the direction of the force at every point is toward the origin. Gravitational forces are central forces; for example, as mentioned in Section 4.3, if we assume that Earth is a perfect sphere with constant mass density, then Newton's law of gravitation asserts that the force exerted on an object by Earth's gravitational field is proportional to the mass of the object and inversely proportional to the square of its distance from the center of Earth, which we take to be the origin.

If the initial position and velocity vectors of an object moving under a central force are parallel, then the subsequent motion is along the line from the origin to the initial position. Here we will assume that the initial position and velocity vectors are not parallel; in this case the subsequent motion is in the plane determined by them. For convenience we take this to be the xy-plane. We will consider the problem of determining the curve traversed by the object. We call this curve the ***orbit.***

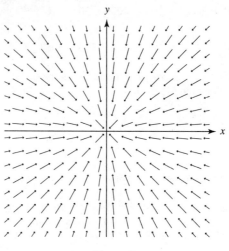

Figure 1

We can represent a central force in terms of polar coordinates

$$x = r \cos \theta, \qquad y = r \sin \theta$$

as

$$\mathbf{F}(r, \theta) = f(r)(\cos \theta \mathbf{i} + \sin \theta \mathbf{j}).$$

We assume that f is continuous for all $r > 0$. The magnitude of $\mathbf{F}$ at $(x, y) = (r \cos \theta, r \sin \theta)$ is $|f(r)|$, so it depends only on the distance r from the point to the origin; the direction of $\mathbf{F}$ is from the point to the origin if $f(r) < 0$, or from the origin to the point if $f(r) > 0$. We will show that the orbit of an object with mass m moving under this force is given by

$$r(\theta) = \frac{1}{u(\theta)},$$

where u is solution of the differential equation

$$\frac{d^2 u}{d\theta^2} + u = -\frac{1}{mh^2 u^2} f(1/u), \tag{1}$$

and h is a constant, defined below.

Newton's second law of motion ($\mathbf{F} = m\mathbf{a}$) says that the polar coordinates $r = r(t)$ and $\theta = \theta(t)$ of the particle satisfy the vector differential equation

$$m(r \cos \theta \mathbf{i} + r \sin \theta \mathbf{j})'' = f(r)(\cos \theta \mathbf{i} + \sin \theta \mathbf{j}). \tag{2}$$

To deal with this equation we introduce the unit vectors

$$\mathbf{e}_1 = \cos \theta \mathbf{i} + \sin \theta \mathbf{j} \qquad \text{and} \qquad \mathbf{e}_2 = -\sin \theta \mathbf{i} + \cos \theta \mathbf{j}.$$

Notice that $\mathbf{e}_1$ points in the direction of increasing r and $\mathbf{e}_2$ points in the direction of increasing θ (Figure 2); moreover,

$$\frac{d\mathbf{e}_1}{d\theta} = \mathbf{e}_2, \qquad \frac{d\mathbf{e}_2}{d\theta} = -\mathbf{e}_1, \tag{3}$$

and

$$\mathbf{e}_1 \cdot \mathbf{e}_2 = \cos\theta(-\sin\theta) + \sin\theta\cos\theta = 0,$$

so $\mathbf{e}_1$ and $\mathbf{e}_2$ are perpendicular. Recalling that the single prime (′) stands for differentiation with respect to t, we see from (3) and the chain rule that

$$\mathbf{e}_1' = \theta'\mathbf{e}_2 \quad \text{and} \quad \mathbf{e}_2' = -\theta'\mathbf{e}_1. \tag{4}$$

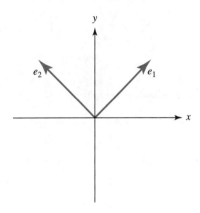

Figure 2

Now we can write (2) as

$$m(r\mathbf{e}_1)'' = f(r)\mathbf{e}_1. \tag{5}$$

But

$$(r\mathbf{e}_1)' = r'\mathbf{e}_1 + r\mathbf{e}_1' = r'\mathbf{e}_1 + r\theta'\mathbf{e}_2$$

(from (4)), and

$$\begin{aligned}
(r\mathbf{e}_1)'' &= (r'\mathbf{e}_1 + r\theta'\mathbf{e}_2)' \\
&= r''\mathbf{e}_1 + r'\mathbf{e}_1' + (r\theta'' + r'\theta')\mathbf{e}_2 + r\theta'\mathbf{e}_2' \\
&= r''\mathbf{e}_1 + r'\theta'\mathbf{e}_2 + (r\theta'' + r'\theta')\mathbf{e}_2 - r(\theta')^2\mathbf{e}_1 \quad \text{(from (4))} \\
&= (r'' - r(\theta')^2)\mathbf{e}_1 + (r\theta'' + 2r'\theta')\mathbf{e}_2.
\end{aligned}$$

Substituting this into (5) yields

$$m(r'' - r(\theta')^2)\mathbf{e}_1 + m(r\theta'' + 2r'\theta')\mathbf{e}_2 = f(r)\mathbf{e}_1.$$

By equating the coefficients of $\mathbf{e}_1$ and $\mathbf{e}_2$ on the two sides of this equation we see that

$$m(r'' - r(\theta')^2) = f(r) \tag{6}$$

and

$$r\theta'' + 2r'\theta' = 0.$$

Multiplying the last equation by r yields

$$r^2\theta'' + 2rr'\theta' = (r^2\theta')' = 0,$$

so

$$r^2\theta' = h, \tag{7}$$

where h is a constant that we can write in terms of the initial conditions as

$$h = r^2(0)\theta'(0).$$

Since the initial position and velocity vectors are

$$r(0)\mathbf{e}_1(0) \quad \text{and} \quad r'(0)\mathbf{e}_1(0) + r(0)\theta'(0)\mathbf{e}_2(0),$$

our assumption that these two vectors are not parallel implies that $\theta'(0) \neq 0$, so $h \neq 0$.

Now let $u = 1/r$. Then $u^2 = \theta'/h$ (from (7)) and

$$r' = -\frac{u'}{u^2} = -h\left(\frac{u'}{\theta'}\right),$$

which implies that

$$r' = -h\frac{du}{d\theta}, \tag{8}$$

since

$$\frac{u'}{\theta'} = \frac{du}{dt}\Big/\frac{d\theta}{dt} = \frac{du}{d\theta}.$$

Differentiating (8) with respect to t yields

$$r'' = -h\frac{d}{dt}\left(\frac{du}{d\theta}\right) = -h\frac{d^2u}{d\theta^2}\,\theta',$$

which implies that

$$r'' = -h^2u^2\,\frac{d^2u}{d\theta^2} \quad \text{since} \quad \theta' = hu^2.$$

Substituting from these equalities into (6) and recalling that $r = 1/u$ yields

$$-m\left(h^2u^2\,\frac{d^2u}{d\theta^2} + \frac{1}{u}h^2u^4\right) = f(1/u),$$

and dividing through by $-mh^2u^2$ yields (1).

Equation (7) has the following geometrical interpretation, which is known as **Kepler's[1] Second Law.**

THEOREM 6.4.1

The position vector of an object moving under a central force sweeps out equal areas in equal times; more precisely, if $\theta(t_1) \leq \theta(t_2)$ then the (signed) area of the sector

$$\{(x, y) = (r\cos\theta, r\sin\theta) : 0 \leq r \leq r(\theta),\ \theta(t_1) \leq \theta(t_2)\}$$

---

[1]The German astronomer Johannes Kepler (1571–1630) discovered three laws of planetary motion by studying observational data. After the invention of calculus it was shown that these laws could be deduced from Newton's laws of motion.

(*Figure* 3) *is given by*

$$A = \frac{h(t_2 - t_1)}{2},$$

where $h = r^2\theta'$, which we have shown to be constant.

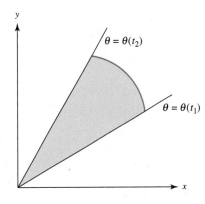

Figure 3

PROOF. Recall from calculus that the area of the shaded sector in Figure 3 is

$$A = \frac{1}{2} \int_{\theta(t_1)}^{\theta(t_2)} r^2(\theta) \, d\theta,$$

where $r = r(\theta)$ is the polar representation of the orbit. Making the change of variable $\theta = \theta(t)$ yields

$$A = \frac{1}{2} \int_{t_1}^{t_2} r^2(\theta(t))\theta'(t) \, dt. \tag{9}$$

But (7) and (9) imply that

$$A = \frac{1}{2} \int_{t_1}^{t_2} h \, dt = \frac{h(t_2 - t_1)}{2}. \qquad \square$$

MOTION UNDER AN INVERSE SQUARE LAW FORCE

In the special case where $f(r) = -mk/r^2 = -mku^2$, so **F** can be interpreted as a gravitational force, (1) becomes

$$\frac{d^2u}{d\theta^2} + u = \frac{k}{h^2}. \tag{10}$$

The general solution of the complementary equation

$$\frac{d^2u}{d\theta^2} + u = 0$$

can be written in amplitude–phase form as

$$u = A \cos(\theta - \phi),$$

where $A \geq 0$ and ϕ is a phase angle. Since $u_p = k/h^2$ is a particular solution of (10), the general solution of (10) is

$$u = A \cos(\theta - \phi) + \frac{k}{h^2};$$

hence, the orbit is given by

$$r = \left(A \cos(\theta - \phi) + \frac{k}{h^2} \right)^{-1},$$

which we rewrite as

$$r = \frac{\rho}{1 + e \cos(\theta - \phi)}, \tag{11}$$

where

$$\rho = \frac{h^2}{k} \quad \text{and} \quad e = A\rho.$$

A curve satisfying (11) is a conic section with a focus at the origin (Exercise 1). The nonnegative constant e is called the ***eccentricity*** of the orbit. If $e < 1$ the orbit is an ellipse (a circle if $e = 0$); if $e = 1$ the orbit is a parabola; if $e > 1$ the orbit is a branch of a hyperbola.

If the orbit is an ellipse the minimum and maximum values of r are given by

$$r_{\min} = \frac{\rho}{1 + e} \qquad \text{(the } \textbf{\textit{perihelion distance,}} \text{ attained when } \theta = \phi\text{)};$$

$$r_{\max} = \frac{\rho}{1 - e} \qquad \text{(the } \textbf{\textit{aphelion distance,}} \text{ attained when } \theta = \phi + \pi\text{)}.$$

A typical elliptic orbit is shown in Figure 4. The point P on the orbit where $r = r_{\min}$ is the ***perigee*** and the point A where $r = r_{\max}$ is the ***apogee.***

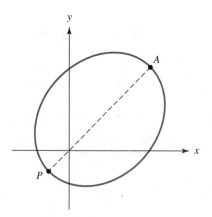

Figure 4

For example, Earth's orbit around the Sun is approximately an ellipse, with $e \approx .017$, $r_{\min} \approx 91 \times 10^6$ miles, and $r_{\max} \approx 95 \times 10^6$ miles. Halley's comet has a very elongated approximately elliptical orbit around the Sun, with $e \approx .967$,

$r_{\min} \approx 55 \times 10^6$ miles, and $r_{\max} \approx 33 \times 10^8$ miles. Some comets (the nonrecurring type) have parabolic or hyperbolic orbits.

6.4 EXERCISES

1. Find the equation of the curve

$$r = \frac{\rho}{1 + e \cos(\theta - \phi)} \tag{A}$$

in terms of $(X, Y) = (r \cos(\theta - \phi), r \sin(\theta - \phi))$, which are rectangular coordinates with respect to the axes shown in Figure 5. Use your results to verify that (A) is the equation of an ellipse if $0 < e < 1$, a parabola if $e = 1$, or a hyperbola if $e > 1$. If $e < 1$ leave your answer in the form

$$\frac{(X - X_0)^2}{a^2} + \frac{(Y - Y_0)^2}{b^2} = 1,$$

and show that the area of the ellipse is

$$A = \frac{\rho^2}{(1 - e^2)^{3/2}}.$$

Then use Theorem 6.4.1 to show that the time required for the object to traverse the entire orbit is

$$T = \frac{2\rho^2}{h(1 - e^2)^{3/2}}.$$

(This is **Kepler's Third Law;** T is called the **period** of the orbit.)

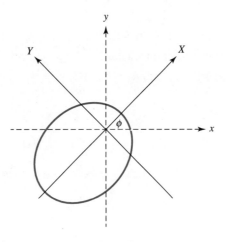

Figure 5

2. Suppose that an object with mass m moves in the xy-plane under the central force

$$\mathbf{F}(r, \theta) = -\frac{mk}{r^2}(\cos \theta \mathbf{i} + \sin \theta \mathbf{j}),$$

where k is a positive constant. As we have shown, the orbit of the object is given by

$$r = \frac{\rho}{1 + e \cos(\theta - \phi)}.$$

Determine ρ, e, and ϕ in terms of the initial conditions

$$r(0) = r_0, \qquad r'(0) = r'_0, \qquad \theta(0) = \theta_0, \qquad \theta'(0) = \theta'_0.$$

Assume that the initial position and velocity vectors are not collinear.

3. Suppose that we wish to put a satellite with mass m into an elliptical orbit around Earth. Assume that the only force acting on the object is Earth's gravity, given by

$$\mathbf{F}(r,\theta) = -mg\left(\frac{R^2}{r^2}\right)(\cos\theta\,\mathbf{i} + \sin\theta\,\mathbf{j}),$$

where R is Earth's radius, g is the acceleration due to gravity at Earth's surface, and r and θ are polar coordinates in the plane of the orbit, with the origin at Earth's center.

(a) Find the eccentricity required to make the aphelion and perihelion distances equal to $R\gamma_1$ and $R\gamma_2$, respectively, where $1 < \gamma_1 < \gamma_2$.

(b) Find the initial conditions

$$r(0) = r_0, \qquad r'(0) = r'_0, \qquad \theta(0) = \theta_0, \qquad \theta'(0) = \theta'_0$$

required to make the initial point the perigee, and the motion along the orbit in the direction of increasing θ.

Hint: Use the results of Exercise 2.

4. An object with mass m moves in a spiral orbit $r = c\theta^2$ under a central force $\mathbf{F}(r,\theta) = f(r)(\cos\theta\,\mathbf{i} + \sin\theta\,\mathbf{j})$. Find f.

5. An object with mass m moves in the orbit $r = r_0 e^{\gamma\theta}$ under a central force $\mathbf{F}(r,\theta) = f(r)(\cos\theta\,\mathbf{i} + \sin\theta\,\mathbf{j})$. Find f.

6. Suppose that an object with mass m moves under the central force

$$\mathbf{F}(r,\theta) = -\frac{mk}{r^3}(\cos\theta\,\mathbf{i} + \sin\theta\,\mathbf{j}),$$

with $r(0) = r_0, r'(0) = r'_0, \theta(0) = \theta_0$ and $\theta'(0) = \theta'_0$, where $h = r_0^2\theta'_0 \neq 0$.

(a) Set up a second order initial value problem for $u = 1/r$ as a function of θ.

(b) Determine $r = r(\theta)$ if (i) $h^2 < k$; (ii) $h^2 = k$; (iii) $h^2 > k$.

7 Series Solutions of Linear Second Order Equations

IN THIS CHAPTER we study a class of second order differential equations that occur in many applications, but can't be solved in closed form in terms of elementary functions. Here are some examples:

1. **Bessel's equation**

$$x^2 y'' + xy' + (x^2 - \nu^2)y = 0,$$

which occurs in problems displaying cylindrical symmetry, such as diffraction of light through a circular aperture, propagation of electromagnetic radiation through a coaxial cable, and vibrations of a circular drum head.

2. **Airy's[1] equation**

$$y'' - xy = 0,$$

which occurs in astronomy and quantum physics.

3. **Legendre's equation**

$$(1 - x^2)y'' - 2xy' + \alpha(\alpha + 1)y = 0,$$

which occurs in problems displaying spherical symmetry, particularly in electromagnetism.

These equations and others considered in this chapter can be written in the form

$$P_0(x)y'' + P_1(x)y' + P_2(x)y = 0, \tag{1}$$

where P_0, P_1, and P_2 are polynomials with no common factor. For most equations that occur in applications these polynomials are of degree two or less, so we'll impose this restriction, although the methods that we'll develop can be extended to the case where the coefficient functions are polynomials of arbitrary degree, or even power series that converge in some circle in the complex plane.

---

[1]Sir George Airy (1801–1892) was an English astronomer.

Since (1) does not in general have closed form solutions, we seek series representations for solutions. We'll see that if $P_0(0) \neq 0$ then solutions of (1) can be written as power series

$$y = \sum_{n=0}^{\infty} a_n x^n$$

that converge in an open interval centered at $x = 0$.

SECTION 7.1 reviews the properties of power series.

SECTIONS 7.2 AND 7.3 are devoted to finding power series solutions of (1) in the case where $P_0(0) \neq 0$. The situation is more complicated if $P_0(0) = 0$; however, if P_1 and P_2 satisfy certain assumptions that apply to most equations of interest, then we're able to use a modified series method to obtain solutions of (1).

SECTION 7.4 introduces the appropriate assumptions on P_1 and P_2 in the case where $P_0(0) = 0$, and deals with ***Euler's equation***

$$ax^2 y'' + bxy' + cy = 0,$$

in which a, b, and c are constants. This is the simplest equation that satisfies these assumptions.

SECTIONS 7.5, 7.6, AND 7.7 deal with three distinct cases satisfying the assumptions introduced in Section 7.4. In all three cases (1) has at least one solution of the form

$$y_1 = x^r \sum_{n=0}^{\infty} a_n x^n$$

where r need not be an integer. The problem is that there are three possibilities—each requiring a different approach—for the form of a second solution y_2 such that $\{y_1, y_2\}$ is a fundamental pair of solutions of (1).

7.1 Review of Power Series

Many applications give rise to differential equations with solutions that cannot be expressed in terms of elementary functions such as polynomials, rational functions, exponential and logarithmic functions, and trigonometric functions. The solutions of some of the most important of these equations can be expressed in terms of power series. We will study such equations in this chapter. In this section we review relevant properties of power series. We will omit proofs, which can be found in any standard calculus text.

DEFINITION 7.1.1

An infinite series of the form

$$\sum_{n=0}^{\infty} a_n(x - x_0)^n, \tag{1}$$

where x_0 and $a_0, a_1, \ldots, a_n, \ldots$ are constants, is called a ***power series in*** $\boldsymbol{x - x_0}$. We say that the power series (1) ***converges*** for a given x if the limit

$$\lim_{N \to \infty} \sum_{n=0}^{N} a_n(x - x_0)^n$$

exists; otherwise, we say that the power series ***diverges*** for the given x.

A power series in $x - x_0$ must converge if $x = x_0$, since the positive powers of $x - x_0$ are all zero in this case. This may be the only value of x for which the power series converges. However, the following theorem shows that if the power series converges for some $x \neq x_0$ then the set of all values of x for which it converges forms an interval.

THEOREM 7.1.2

For any power series

$$\sum_{n=0}^{\infty} a_n(x - x_0)^n$$

exactly one of the following statements is true:

  (i) *The power series converges only for $x = x_0$.*
  (ii) *The power series converges for all values of x.*
  (iii) *There is a positive number R such that the power series converges if $|x - x_0| < R$ and diverges if $|x - x_0| > R$.*

In case **(iii)** we say that R is the ***radius of convergence*** of the power series. For convenience, we include the other two cases in this definition by defining $R = 0$ in case **(i)** and $R = \infty$ in case **(ii)**. We define the ***open interval of convergence*** of $\sum_{n=0}^{\infty} a_n(x - x_0)^n$ to be

$$(x_0 - R, x_0 + R) \quad \text{if} \quad 0 < R < \infty, \qquad \text{or} \qquad (-\infty, \infty) \quad \text{if} \quad R = \infty.$$

If R is finite then no general statement can be made concerning convergence at the endpoints $x = x_0 \pm R$ of the open interval of convergence; the series may converge at one or both points, or diverge at both.

Recall from calculus that a series of constants $\sum_{n=0}^{\infty} \alpha_n$ is said to **converge absolutely** if the series of absolute values $\sum_{n=0}^{\infty} |\alpha_n|$ converges. It can be shown that a power series $\sum_{n=0}^{\infty} a_n(x - x_0)^n$ with a positive radius of convergence R converges absolutely in its open interval of convergence; that is, the series

$$\sum_{n=0}^{\infty} |a_n||x - x_0|^n$$

of absolute values converges if $|x - x_0| < R$. However, if $R < \infty$ the series may fail to converge absolutely at an endpoint $x_0 \pm R$, even if it converges there.

The following theorem provides a useful method for determining the radius of convergence of a power series. It is derived in calculus by applying the ratio test to the corresponding series of absolute values. For related theorems see Exercises 2 and 4.

THEOREM 7.1.3

Suppose that there is an integer N such that $a_n \neq 0$ if $n \geq N$, and

$$\lim_{n \to \infty} \left| \frac{a_{n+1}}{a_n} \right| = L,$$

where $0 \leq L \leq \infty$. Then the radius of convergence of $\sum_{n=0}^{\infty} a_n(x - x_0)^n$ is $R = 1/L$, which is interpreted to mean that $R = 0$ if $L = \infty$ or $R = \infty$ if $L = 0$.

EXAMPLE 1   Find the radius of convergence of the series:

(a) $\displaystyle\sum_{n=0}^{\infty} n! \, x^n$

(b) $\displaystyle\sum_{n=10}^{\infty} (-1)^n \frac{x^n}{n!}$

(c) $\displaystyle\sum_{n=0}^{\infty} 2^n n^2 (x - 1)^n.$

Solution   **(a)** Here $a_n = n!$, so

$$\lim_{n \to \infty} \left| \frac{a_{n+1}}{a_n} \right| = \lim_{n \to \infty} \frac{(n + 1)!}{n!} = \lim_{n \to \infty} (n + 1) = \infty.$$

Hence, $R = 0$.

Solution   **(b)** Here $a_n = (-1)^n/n!$ for $n \geq N = 10$, so

$$\lim_{n \to \infty} \left| \frac{a_{n+1}}{a_n} \right| = \lim_{n \to \infty} \frac{n!}{(n + 1)!} = \lim_{n \to \infty} \frac{1}{n + 1} = 0.$$

Hence, $R = \infty$.

Solution   **(c)** Here $a_n = 2^n n^2$, so

$$\lim_{n \to \infty} \left| \frac{a_{n+1}}{a_n} \right| = \lim_{n \to \infty} \frac{2^{n+1}(n + 1)^2}{2^n n^2} = 2 \lim_{n \to \infty} \left(1 + \frac{1}{n} \right)^2 = 2.$$

Hence $R = 1/2$.

TAYLOR SERIES

If a function f has derivatives of all orders at a point $x = x_0$ then the **Taylor series of f about x_0** is defined by

$$\sum_{n=0}^{\infty} \frac{f^{(n)}(x_0)}{n!}(x - x_0)^n.$$

In the special case where $x_0 = 0$ this series is also called the **Maclaurin series of f.**

Taylor series for most of the common elementary functions converge to the functions on their open intervals of convergence. For example, you are probably familiar with the following Maclaurin series:

$$e^x = \sum_{n=0}^{\infty} \frac{x^n}{n!}, \qquad -\infty < x < \infty, \tag{2}$$

$$\sin x = \sum_{n=0}^{\infty} (-1)^n \frac{x^{2n+1}}{(2n+1)!}, \qquad -\infty < x < \infty, \tag{3}$$

$$\cos x = \sum_{n=0}^{\infty} (-1)^n \frac{x^{2n}}{(2n)!}, \qquad -\infty < x < \infty, \tag{4}$$

$$\frac{1}{1-x} = \sum_{n=0}^{\infty} x^n, \qquad -1 < x < 1. \tag{5}$$

DIFFERENTIATION OF POWER SERIES

A power series with a positive radius of convergence defines a function

$$f(x) = \sum_{n=0}^{\infty} a_n(x - x_0)^n$$

on its open interval of convergence. We say that the series **represents** f on the open interval of convergence. A function f represented by a power series may be a familiar elementary function as in (2)–(5); however, it often happens that f is not a familiar function, so that the series actually *defines* f.

The following theorem shows that a function represented by a power series has derivatives of all orders on the open interval of convergence of the power series, and provides power series representations of the derivatives.

THEOREM 7.1.4

A power series

$$f(x) = \sum_{n=0}^{\infty} a_n(x - x_0)^n$$

with positive radius of convergence R has derivatives of all orders in its open interval of convergence, and successive derivatives can be obtained by repeatedly differentiating term by term; that is,

$$f'(x) = \sum_{n=1}^{\infty} na_n(x - x_0)^{n-1}, \tag{6}$$

$$f''(x) = \sum_{n=2}^{\infty} n(n-1)a_n(x-x_0)^{n-2}, \tag{7}$$

$$\vdots$$

$$f^{(k)}(x) = \sum_{n=k}^{\infty} n(n-1)\cdots(n-k+1)a_n(x-x_0)^{n-k}. \tag{8}$$

Moreover, all of these series have the same radius of convergence R.

EXAMPLE 2 Let $f(x) = \sin x$. From (3),

$$f(x) = \sum_{n=0}^{\infty} (-1)^n \frac{x^{2n+1}}{(2n+1)!}.$$

From (6),

$$f'(x) = \sum_{n=0}^{\infty} (-1)^n \frac{d}{dx}\left[\frac{x^{2n+1}}{(2n+1)!}\right] = \sum_{n=0}^{\infty} (-1)^n \frac{x^{2n}}{(2n)!},$$

which is the series (4) for $\cos x$. ∎

UNIQUENESS OF POWER SERIES

The following theorem shows that if f is *defined* by a power series in $x - x_0$ with a positive radius of convergence then the power series is the Taylor series of f about x_0.

THEOREM 7.1.5

If the power series

$$f(x) = \sum_{n=0}^{\infty} a_n(x-x_0)^n$$

has a positive radius of convergence, then

$$a_n = \frac{f^{(n)}(x_0)}{n!}; \tag{9}$$

that is, $\sum_{n=0}^{\infty} a_n(x-x_0)^n$ *is the Taylor series of f about* x_0.

This result can be obtained by setting $x = x_0$ in (8), which yields

$$f^{(k)}(x_0) = k(k-1)\cdots 1 \cdot a_k = k!\, a_k.$$

This implies that

$$a_k = \frac{f^{(k)}(x_0)}{k!}.$$

Except for notation this is the same as (9).

The following theorem lists two important properties of power series that follow from Theorem 7.1.5.

THEOREM 7.1.6

(a) *If*

$$\sum_{n=0}^{\infty} a_n(x - x_0)^n = \sum_{n=0}^{\infty} b_n(x - x_0)^n$$

for all x in an open interval containing x_0 then $a_n = b_n$ for $n = 0, 1, 2, \ldots$.
(b) *If*

$$\sum_{n=0}^{\infty} a_n(x - x_0)^n = 0$$

for all x in an open interval containing x_0 then $a_n = 0$ for $n = 0, 1, 2, \ldots$.

To obtain **(a)** we observe that the two series represent the same function f on the open interval; hence, Theorem 7.1.5 implies that

$$a_n = b_n = \frac{f^{(n)}(x_0)}{n!}, \qquad n = 0, 1, 2, \ldots.$$

Part **(b)** can be obtained from **(a)** by taking $b_n = 0$ for $n = 0, 1, 2, \ldots$.

TAYLOR POLYNOMIALS

If f has N derivatives at a point x_0 we say that

$$T_N(x) = \sum_{n=0}^{N} \frac{f^{(n)}(x_0)}{n!}(x - x_0)^n$$

is the **N-th Taylor polynomial of f about x_0.** From this definition and Theorem 7.1.5 it follows that if

$$f(x) = \sum_{n=0}^{\infty} a_n(x - x_0)^n$$

where the power series has a positive radius of convergence, then the Taylor polynomial T_N of f about x_0 is given by

$$T_N(x) = \sum_{n=0}^{N} a_n(x - x_0)^n.$$

In numerical applications we use the Taylor polynomials to approximate f on subintervals of the open interval of convergence of the power series. For example, from (2) the Taylor polynomial T_N of $f(x) = e^x$ is given by

$$T_N(x) = \sum_{n=0}^{N} \frac{x^n}{n!}.$$

The blue curve in Figure 1 is the graph of $y = e^x$ on the interval $[0, 5]$. The gray curves in Figure 1 are the graphs of the Taylor polynomials $T_1, \ldots, T_7$ of $y = e^x$ about $x_0 = 0$. From this figure we conclude that the accuracy of the approximation of $y = e^x$ by its Taylor polynomial T_N improves as N increases.

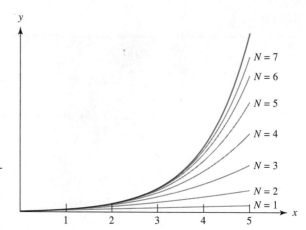

Figure 1 Approximation of $y = e^x$ by Taylor polynomials about $x = 0$

SHIFTING THE SUMMATION INDEX

In Definition 7.1.1 of a power series in $x - x_0$, the nth term is a constant multiple of $(x - x_0)^n$. This is not true in (6), (7), and (8), in which the general terms are constant multiples of $(x - x_0)^{n-1}$, $(x - x_0)^{n-2}$, and $(x - x_0)^{n-k}$, respectively. However, these series can all be rewritten so that their nth terms are constant multiples of $(x - x_0)^n$. For example, letting $n = k + 1$ in the series in (6) yields

$$f'(x) = \sum_{k=0}^{\infty} (k + 1)a_{k+1}(x - x_0)^k, \tag{10}$$

where we start the new summation index k from zero so that the first term in (10) (obtained by setting $k = 0$) is the same as the first term in (6) (obtained by setting $n = 1$). However, the sum of a series is independent of the symbol used to denote the summation index, just as the value of a definite integral is independent of the symbol used to denote the variable of integration. Therefore, we can replace k by n in (10) to obtain

$$f'(x) = \sum_{n=0}^{\infty} (n + 1)a_{n+1}(x - x_0)^n, \tag{11}$$

in which the general term is a constant multiple of $(x - x_0)^n$.

It is not really necessary to introduce the intermediate summation index k; we can obtain (11) directly from (6) by replacing n by $n + 1$ in the general term of (6) and subtracting 1 from the lower limit of (6). More generally, we have the following procedure for shifting indices.

Shifting the Summation Index in a Power Series

For any integer k the power series

$$\sum_{n=n_0}^{\infty} b_n(x - x_0)^{n-k}$$

can be rewritten as

$$\sum_{n=n_0-k}^{\infty} b_{n+k}(x - x_0)^n;$$

that is, replacing n by $n + k$ in the general term and subtracting k from the lower limit of summation leaves the series unchanged.

EXAMPLE 3 Rewrite the following power series from (7) and (8) so that the general term in each is a constant multiple of $(x - x_0)^n$:

(a) $\displaystyle\sum_{n=2}^{\infty} n(n - 1)a_n(x - x_0)^{n-2}$

(b) $\displaystyle\sum_{n=k}^{\infty} n(n - 1) \cdots (n - k + 1)a_n(x - x_0)^{n-k}$.

Solution **(a)** Replacing n by $n + 2$ in the general term and subtracting 2 from the lower limit of summation yields

$$\sum_{n=2}^{\infty} n(n - 1)a_n(x - x_0)^{n-2} = \sum_{n=0}^{\infty} (n + 2)(n + 1)a_{n+2}(x - x_0)^n.$$

Solution **(b)** Replacing n by $n + k$ in the general term and subtracting k from the lower limit of summation yields

$$\sum_{n=k}^{\infty} n(n - 1) \cdots (n - k + 1)a_n(x - x_0)^{n-k}$$

$$= \sum_{n=0}^{\infty} (n + k)(n + k - 1) \cdots (n + 1)a_{n+k}(x - x_0)^n.$$

EXAMPLE 4 Given that

$$f(x) = \sum_{n=0}^{\infty} a_n x^n,$$

write the function xf'' as a power series in which the general term is a constant multiple of x^n.

Solution From Theorem 7.1.4 with $x_0 = 0$,

$$f''(x) = \sum_{n=2}^{\infty} n(n - 1)a_n x^{n-2}.$$

Therefore

$$xf''(x) = \sum_{n=2}^{\infty} n(n - 1)a_n x^{n-1}.$$

Replacing n by $n + 1$ in the general term and subtracting 1 from the lower limit of summation yields

$$xf''(x) = \sum_{n=1}^{\infty} (n + 1)na_{n+1} x^n.$$

We can also write this as

$$xf''(x) = \sum_{n=0}^{\infty} (n + 1)na_{n+1} x^n,$$

since the first term in this last series is zero. (We will see later that it is sometimes useful to include zero terms at the beginning of a series.) ∎

LINEAR COMBINATIONS OF POWER SERIES

If a power series is multiplied by a constant then the constant can be placed inside the summation; that is,

$$c \sum_{n=0}^{\infty} a_n(x - x_0)^n = \sum_{n=0}^{\infty} ca_n(x - x_0)^n.$$

Two power series

$$f(x) = \sum_{n=0}^{\infty} a_n(x - x_0)^n \quad \text{and} \quad g(x) = \sum_{n=0}^{\infty} b_n(x - x_0)^n$$

with positive radii of convergence can be added term by term at points common to their open intervals of convergence; thus, if the first series converges for $|x - x_0| < R_1$ and the second converges for $|x - x_0| < R_2$, then

$$f(x) + g(x) = \sum_{n=0}^{\infty} (a_n + b_n)(x - x_0)^n$$

for $|x - x_0| < R$, where R is the smaller of R_1 and R_2. More generally, linear combinations of power series can be formed term by term; for example,

$$c_1 f(x) + c_2 f(x) = \sum_{n=0}^{\infty} (c_1 a_n + c_2 b_n)(x - x_0)^n.$$

EXAMPLE 5 Find the Maclaurin series for $\cosh x$ as a linear combination of the Maclaurin series for e^x and e^{-x}.

Solution By definition,

$$\cosh x = \frac{1}{2} e^x + \frac{1}{2} e^{-x}.$$

Since

$$e^x = \sum_{n=0}^{\infty} \frac{x^n}{n!} \quad \text{and} \quad e^{-x} = \sum_{n=0}^{\infty} (-1)^n \frac{x^n}{n!},$$

it follows that

$$\cosh x = \sum_{n=0}^{\infty} \frac{1}{2}[1 + (-1)^n] \frac{x^n}{n!}. \tag{12}$$

Since

$$\frac{1}{2}[1 + (-1)^n] = \begin{cases} 1 & \text{if } n = 2m, \text{ an even integer,} \\ 0 & \text{if } n = 2m + 1, \text{ an odd integer,} \end{cases}$$

we can rewrite (12) more simply as

$$\cosh x = \sum_{m=0}^{\infty} \frac{x^{2m}}{(2m)!}.$$

This result is valid on $(-\infty, \infty)$, since this is the open interval of convergence of the Maclaurin series for e^x and e^{-x}.

EXAMPLE 6 Suppose that

$$y = \sum_{n=0}^{\infty} a_n x^n$$

on an open interval I containing the origin.

(a) Express the function

$$(2 - x)y'' + 2y$$

as a power series in x on I.

(b) Use the result of **(a)** to find necessary and sufficient conditions on the coefficients $\{a_n\}$ for y to be a solution of the homogeneous equation

$$(2 - x)y'' + 2y = 0 \tag{13}$$

on I.

Solution **(a)** From (7) with $x_0 = 0$,

$$y'' = \sum_{n=2}^{\infty} n(n-1)a_n x^{n-2}.$$

Therefore

$$(2-x)y'' + 2y = 2y'' - xy' + 2y \qquad (14)$$

$$= \sum_{n=2}^{\infty} 2n(n-1)a_n x^{n-2} - \sum_{n=2}^{\infty} n(n-1)a_n x^{n-1} + \sum_{n=0}^{\infty} 2a_n x^n.$$

To combine the three series we shift indices in the first two to make their general terms constant multiples of x^n; thus,

$$\sum_{n=2}^{\infty} 2n(n-1))a_n x^{n-2} = \sum_{n=0}^{\infty} 2(n+2)(n+1)a_{n+2} x^n \qquad (15)$$

and

$$\sum_{n=2}^{\infty} n(n-1)a_n x^{n-1} = \sum_{n=1}^{\infty} (n+1)na_{n+1} x^n = \sum_{n=0}^{\infty} (n+1)na_{n+1} x^n, \qquad (16)$$

where we have added a zero term in the last series so that when we substitute from (15) and (16) into (14) all three series will start with $n = 0$; thus,

$$(2-x)y'' + 2y = \sum_{n=0}^{\infty} [2(n+2)(n+1)a_{n+2} - (n+1)na_{n+1} + 2a_n]x^n. \qquad (17)$$

Solution **(b)** From (17) we see that y satisfies (13) on I if

$$2(n+2)(n+1)a_{n+2} - (n+1)na_{n+1} + 2a_n = 0, \qquad n = 0,1,2,\ldots. \qquad (18)$$

Conversely, Theorem 7.1.6**(b)** implies that if $y = \sum_{n=0}^{\infty} a_n x^n$ satisfies (13) on I then (18) holds.

EXAMPLE 7 Suppose that

$$y = \sum_{n=0}^{\infty} a_n(x-1)^n$$

on an open interval I containing $x_0 = 1$. Express the function

$$(1+x)y'' + 2(x-1)^2 y' + 3y \qquad (19)$$

as a power series in $x - 1$ on I.

Solution Since we want a power series in $x - 1$, we rewrite the coefficient of y'' in (19) as $1 + x = 2 + (x - 1)$, so that (19) becomes

$$2y'' + (x-1)y'' + 2(x-1)^2 y' + 3y.$$

From (6) and (7) with $x_0 = 1$,

$$y' = \sum_{n=1}^{\infty} na_n(x-1)^{n-1} \qquad \text{and} \qquad y'' = \sum_{n=2}^{\infty} n(n-1)a_n(x-1)^{n-2}.$$

Therefore,

$$2y'' = \sum_{n=2}^{\infty} 2n(n-1)a_n(x-1)^{n-2},$$

$$(x-1)y'' = \sum_{n=2}^{\infty} n(n-1)a_n(x-1)^{n-1},$$

$$2(x-1)^2y' = \sum_{n=1}^{\infty} 2na_n(x-1)^{n+1},$$

$$3y = \sum_{n=0}^{\infty} 3a_n(x-1)^n.$$

Before adding these four series we shift indices in the first three so that their general terms become constant multiples of $(x-1)^n$. This yields

$$2y'' = \sum_{n=0}^{\infty} 2(n+2)(n+1)a_{n+2}(x-1)^n, \tag{20}$$

$$(x-1)y'' = \sum_{n=0}^{\infty} (n+1)na_{n+1}(x-1)^n, \tag{21}$$

$$2(x-1)^2y' = \sum_{n=1}^{\infty} 2(n-1)a_{n-1}(x-1)^n, \tag{22}$$

$$3y = \sum_{n=0}^{\infty} 3a_n(x-1)^n, \tag{23}$$

where we have added initial zero terms to the series in (21) and (22). Adding (20)–(23) yields

$$(1+x)y'' + 2(x-1)^2y' + 3y = 2y'' + (x-1)y'' + 2(x-1)^2y' + 3y$$
$$= \sum_{n=0}^{\infty} b_n(x-1)^n,$$

where

$$b_0 = 4a_2 + 3a_0, \tag{24}$$
$$b_n = 2(n+2)(n+1)a_{n+2} + (n+1)na_{n+1} + 2(n-1)a_{n-1} + 3a_n, \quad n \geq 1. \tag{25}$$

Notice that the formula (24) for b_0 can't be obtained by setting $n = 0$ in (25), since the summation in (22) begins with $n = 1$ while those in (20), (21), and (23) begin with $n = 0$. ∎

7.1 EXERCISES

---

1. For each power series use Theorem 7.1.3 to find the radius of convergence R. If $R > 0$ find the open interval of convergence.

(a) $\displaystyle\sum_{n=0}^{\infty} \frac{(-1)^n}{2^n n}(x-1)^n$

(b) $\displaystyle\sum_{n=0}^{\infty} 2^n n(x-2)^n$

(c) $\displaystyle\sum_{n=0}^{\infty} \frac{n!}{9^n} x^n$

(d) $\displaystyle\sum_{n=0}^{\infty} \frac{n(n+1)}{16^n}(x-2)^n$

(e) $\displaystyle\sum_{n=0}^{\infty} (-1)^n \frac{7^n}{n!} x^n$

(f) $\displaystyle\sum_{n=0}^{\infty} \frac{3^n}{4^{n+1}(n+1)^2}(x+7)^n$

2. Suppose that there is an integer M such that $b_m \neq 0$ for $m \geq M$, and

$$\lim_{m \to \infty} \left| \frac{b_{m+1}}{b_m} \right| = L,$$

where $0 \leq L \leq \infty$. Show that the radius of convergence of

$$\sum_{m=0}^{\infty} b_m (x - x_0)^{2m}$$

is $R = 1/\sqrt{L}$, which is interpreted to mean that $R = 0$ if $L = \infty$ or $R = \infty$ if $L = 0$.

Hint: *Apply Theorem 7.1.3 to the series $\sum_{m=0}^{\infty} b_m z^m$ and then let $z = (x - x_0)^2$.*

3. For each power series use the result of Exercise 2 to find the radius of convergence R. If $R > 0$ find the open interval of convergence.

(a) $\displaystyle\sum_{m=0}^{\infty} (-1)^m (3m + 1)(x - 1)^{2m+1}$

(b) $\displaystyle\sum_{m=0}^{\infty} (-1)^m \frac{m(2m + 1)}{2^m} (x + 2)^{2m}$

(c) $\displaystyle\sum_{m=0}^{\infty} \frac{m!}{(2m)!} (x - 1)^{2m}$

(d) $\displaystyle\sum_{m=0}^{\infty} (-1)^m \frac{m!}{9^m} (x + 8)^{2m}$

(e) $\displaystyle\sum_{m=0}^{\infty} (-1)^m \frac{(2m - 1)}{3^m} x^{2m+1}$

(f) $\displaystyle\sum_{m=0}^{\infty} (x - 1)^{2m}$

4. Suppose that there is an integer M such that $b_m \neq 0$ for $m \geq M$, and

$$\lim_{m \to \infty} \left| \frac{b_{m+1}}{b_m} \right| = L,$$

where $0 \leq L \leq \infty$. Let k be a positive integer. Show that the radius of convergence of

$$\sum_{m=0}^{\infty} b_m (x - x_0)^{km}$$

is $R = 1/\sqrt[k]{L}$, which is interpreted to mean that $R = 0$ if $L = \infty$ or $R = \infty$ if $L = 0$.

Hint: *Apply Theorem 7.1.3 to the series $\sum_{m=0}^{\infty} b_m z^m$ and then let $z = (x - x_0)^k$.*

5. For each power series use the result of Exercise 4 to find the radius of convergence R. If $R > 0$ find the open interval of convergence.

(a) $\displaystyle\sum_{m=0}^{\infty} \frac{(-1)^m}{(27)^m} (x - 3)^{3m+2}$

(b) $\displaystyle\sum_{m=0}^{\infty} \frac{x^{7m+6}}{m}$

(c) $\displaystyle\sum_{m=0}^{\infty} \frac{9^m (m + 1)}{(m + 2)} (x - 3)^{4m+2}$

(d) $\displaystyle\sum_{m=0}^{\infty} (-1)^m \frac{2^m}{m!} x^{4m+3}$

(e) $\displaystyle\sum_{m=0}^{\infty} \frac{m!}{(26)^m} (x + 1)^{4m+3}$

(f) $\displaystyle\sum_{m=0}^{\infty} \frac{(-1)^m}{8^m m(m + 1)} (x - 1)^{3m+1}$

L **6.** Graph $y = \sin x$ and the Taylor polynomial

$$T_{2M+1}(x) = \sum_{n=0}^{M} \frac{(-1)^n x^{2n+1}}{(2n + 1)!}$$

on the interval $(-2\pi, 2\pi)$ for $M = 1, 2, 3, \ldots$, until you find a value of M for which there is no perceptible difference between the two graphs.

L **7.** Graph $y = \cos x$ and the Taylor polynomial

$$T_{2M}(x) = \sum_{n=0}^{M} \frac{(-1)^n x^{2n}}{(2n)!}$$

on the interval $(-2\pi, 2\pi)$ for $M = 1, 2, 3, \ldots$, until you find a value of M for which there is no perceptible difference between the two graphs.

[L] 8. Graph $y = 1/(1 - x)$ and the Taylor polynomial

$$T_N(x) = \sum_{n=0}^{N} x^n$$

on the interval $[0, .95]$ for $N = 1, 2, 3, \ldots$, until you find a value of N for which there is no perceptible difference between the two graphs. Choose the scale on the y-axis so that $0 \le y \le 20$.

[L] 9. Graph $y = \cosh x$ and the Taylor polynomial

$$T_{2M}(x) = \sum_{n=0}^{M} \frac{x^{2n}}{(2n)!}$$

on the interval $(-5, 5)$ for $M = 1, 2, 3, \ldots$, until you find a value of M for which there is no perceptible difference between the two graphs. Choose the scale on the y-axis so that $0 \le y \le 75$.

[L] 10. Graph $y = \sinh x$ and the Taylor polynomial

$$T_{2M+1}(x) = \sum_{n=0}^{M} \frac{x^{2n+1}}{(2n + 1)!}$$

on the interval $(-5, 5)$ for $M = 0, 1, 2, \ldots$, until you find a value of M for which there is no perceptible difference between the two graphs. Choose the scale on the y-axis so that $-75 \le y \le 75$.

In Exercises 11–15 suppose that $y(x) = \sum_{n=0}^{\infty} a_n x^n$ on an open interval containing the origin, and find a power series in x for the given expression.

11. $(2 + x)y'' + xy' + 3y$

12. $(1 + 3x^2)y'' + 3x^2 y' - 2y$

13. $(1 + 2x^2)y'' + (2 - 3x)y' + 4y$

14. $(1 + x^2)y'' + (2 - x)y' + 3y$

15. $(1 + 3x^2)y'' - 2xy' + 4y$

16. Suppose that $y(x) = \sum_{n=0}^{\infty} a_n(x + 1)^n$ on an open interval containing $x_0 = -1$. Find a power series in $x + 1$ for

$$xy'' + (4 + 2x)y' + (2 + x)y.$$

17. Suppose that $y(x) = \sum_{n=0}^{\infty} a_n(x - 2)^n$ on an open interval containing $x_0 = 2$. Find a power series in $x - 2$ for

$$x^2 y'' + 2xy' - 3xy.$$

[L] 18. Do the following experiment for various choices of real numbers a_0 and a_1.

(a) Use differential equations software to solve the initial value problem

$$(2 - x)y'' + 2y = 0, \qquad y(0) = a_0, \qquad y'(0) = a_1$$

numerically on $(-1.95, 1.95)$. Choose the most accurate method your software provides. (See Section 10.1 for a brief discussion of one such method.)

(b) For $N = 2, 3, 4, \ldots$, compute $a_2, \ldots, a_N$ from Eqn. (18) and graph

$$T_N(x) = \sum_{n=0}^{N} a_n x^n$$

and the solution obtained in **(a)** on the same axes. Continue increasing N until it is obvious that there's no point in continuing. (This sounds vague, but you'll know when to stop.)

[L] 19. Follow the directions of Exercise 18 for the initial value problem

$$(1 + x)y'' + 2(x - 1)^2 y' + 3y = 0, \qquad y(1) = a_0, \qquad y'(1) = a_1$$

on the interval $(0, 2)$. Use Eqns. (24) and (25) to compute $\{a_n\}$.

20. Suppose that the series $\sum_{n=0}^{\infty} a_n x^n$ converges on an open interval $(-R, R)$, let r be an arbitrary real number, and define

$$y(x) = x^r \sum_{n=0}^{\infty} a_n x^n = \sum_{n=0}^{\infty} a_n x^{n+r}$$

on $(0, R)$. Use Theorem 7.1.4 and the rule for differentiating the product of two functions to show that

$$y'(x) = \sum_{n=0}^{\infty} (n+r) a_n x^{n+r-1},$$

$$y''(x) = \sum_{n=0}^{\infty} (n+r)(n+r-1) a_n x^{n+r-2},$$

$$\vdots$$

$$y^{(k)}(x) = \sum_{n=0}^{\infty} (n+r)(n+r-1) \cdots (n+r-k) a_n x^{n+r-k}$$

on $(0, R)$.

In Exercises 21–26 let y be as defined in Exercise 20, and write the given expression in the form $x^r \sum_{n=0}^{\infty} b_n x^n$.

21. $x^2(1-x)y'' + x(4+x)y' + (2-x)y$

22. $x^2(1+x)y'' + x(1+2x)y' - (4+6x)y$

23. $x^2(1+x)y'' - x(1-6x-x^2)y' + (1+6x+x^2)y$

24. $x^2(1+3x)y'' + x(2+12x+x^2)y' + 2x(3+x)y$

25. $x^2(1+2x^2)y'' + x(4+2x^2)y' + 2(1-x^2)y$

26. $x^2(2+x^2)y'' + 2x(5+x^2)y' + 2(3-x^2)y$

7.2   Series Solutions near an Ordinary Point I

Many physical applications give rise to second order homogeneous linear differential equations of the form

$$P_0(x)y'' + P_1(x)y' + P_2(x)y = 0, \tag{1}$$

where P_0, P_1, and P_2 are polynomials. Usually the solutions of these equations cannot be expressed in terms of familiar elementary functions. Therefore we will consider the problem of representing solutions of (1) by means of series.

We assume throughout that P_0, P_1, and P_2 have no common factors. Then we say that x_0 is an **ordinary point** of (1) if $P_0(x_0) \neq 0$, or a **singular point** if $P_0(x_0) = 0$. For Legendre's equation,

$$(1 - x^2)y'' - 2xy' + \alpha(\alpha + 1)y = 0, \tag{2}$$

$x_0 = 1$ and $x_0 = -1$ are singular points and all other points are ordinary points. For Bessel's equation,

$$x^2 y'' + xy' + (x^2 - \nu^2)y = 0,$$

$x_0 = 0$ is a singular point and all other points are ordinary points. If P_0 is a nonzero constant as in Airy's equation,

$$y'' - xy = 0, \tag{3}$$

then every point is an ordinary point.

Since polynomials are continuous everywhere, P_1/P_0 and P_2/P_0 are continuous at any point x_0 that is not a zero of P_0. Therefore, if x_0 is an ordinary point of (1) and a_0 and a_1 are arbitrary real numbers, then the initial value problem

$$P_0(x)y'' + P_1(x)y' + P_2(x)y = 0, \qquad y(x_0) = a_0, \qquad y'(x_0) = a_1 \qquad (4)$$

has a unique solution on the largest open interval that contains x_0 and does not contain any zeros of P_0. To see this, we rewrite the differential equation in (4) as

$$y'' + \frac{P_1(x)}{P_0(x)}y' + \frac{P_2(x)}{P_0(x)}y = 0$$

and apply Theorem 5.1.1 with $p = P_1/P_0$ and $q = P_2/P_0$. In this section and the next we consider the problem of representing solutions of (1) by power series that converge for values of x near an ordinary point x_0.

We state the following theorem without proof.

THEOREM 7.2.1

Suppose that P_0, P_1, and P_2 are polynomials with no common factor and P_0 is not identically zero. Let x_0 be a point such that $P_0(x_0) \neq 0$ and let ρ be the distance from x_0 to the nearest zero of P_0 in the complex plane. (If P_0 is constant then $\rho = \infty$.) Then every solution of

$$P_0(x)y'' + P_1(x)y' + P_2(x)y = 0 \qquad (5)$$

can be represented by a power series

$$y = \sum_{n=0}^{\infty} a_n(x - x_0)^n \qquad (6)$$

that converges at least on the open interval $(x_0 - \rho, x_0 + \rho)$. (If P_0 is nonconstant, so that ρ is necessarily finite, then the open interval of convergence of (6) may be larger than $(x_0 - \rho, x_0 + \rho)$; if P_0 is constant then $\rho = \infty$ and $(x_0 - \rho, x_0 + \rho) = (-\infty, \infty)$.)

We call (6) a ***power series solution in $x - x_0$*** of (5). We will now develop a method for finding power series solutions of (5). For this purpose we write (5) as $Ly = 0$, where

$$Ly = P_0 y'' + P_1 y' + P_2 y. \qquad (7)$$

Theorem 7.2.1 implies that every solution of $Ly = 0$ on $(x_0 - \rho, x_0 + \rho)$ can be written as

$$y = \sum_{n=0}^{\infty} a_n(x - x_0)^n.$$

Setting $x = x_0$ in this series and in the series

$$y' = \sum_{n=1}^{\infty} na_n(x - x_0)^{n-1}$$

shows that $y(x_0) = a_0$ and $y'(x_0) = a_1$. Since every initial value problem (4) has a unique solution, it follows that a_0 and a_1 can be chosen arbitrarily, and that $a_2, a_3, \ldots$ must be uniquely determined by them.

To find $a_2, a_3, \ldots$, we write P_0, P_1, and P_2 in powers of $x - x_0$, substitute

$$y = \sum_{n=0}^{\infty} a_n(x - x_0)^n,$$

$$y' = \sum_{n=1}^{\infty} na_n(x - x_0)^{n-1},$$

$$y'' = \sum_{n=2}^{\infty} n(n - 1)a_n(x - x_0)^{n-2}$$

into (7), and collect the coefficients of like powers of $x - x_0$. This yields

$$Ly = \sum_{n=0}^{\infty} b_n(x - x_0)^n, \tag{8}$$

where $\{b_0, b_1, \ldots, b_n, \ldots\}$ are expressed in terms of $\{a_0, a_1, \ldots, a_n, \ldots\}$ and the coefficients of P_0, P_1, and P_2, written in powers of $x - x_0$. Since (8) and Theorem 7.1.6 imply that $Ly = 0$ if and only if $b_n = 0$ for $n \geq 0$, it follows that all power series solutions in $x - x_0$ of $Ly = 0$ can be obtained by choosing a_0 and a_1 arbitrarily and computing $a_2, a_3, \ldots$ successively so that $b_n = 0$ for $n \geq 0$. For simplicity we call the power series obtained in this way the ***power series in $x - x_0$ for the general solution of $Ly = 0$,*** without explicitly identifying the open interval of convergence of the series.

EXAMPLE 1   Let x_0 be an arbitrary real number. Find the power series in $x - x_0$ for the general solution of

$$y'' + y = 0. \tag{9}$$

Solution   Here

$$Ly = y'' + y.$$

If

$$y = \sum_{n=0}^{\infty} a_n(x - x_0)^n$$

then

$$y'' = \sum_{n=2}^{\infty} n(n - 1)a_n(x - x_0)^{n-2},$$

so

$$Ly = \sum_{n=2}^{\infty} n(n - 1)a_n(x - x_0)^{n-2} + \sum_{n=0}^{\infty} a_n(x - x_0)^n.$$

To collect coefficients of like powers of $x - x_0$ we shift the summation index in the first sum. This yields

$$Ly = \sum_{n=0}^{\infty} (n + 2)(n + 1)a_{n+2}(x - x_0)^n + \sum_{n=0}^{\infty} a_n(x - x_0)^n = \sum_{n=0}^{\infty} b_n(x - x_0)^n$$

with

$$b_n = (n + 2)(n + 1)a_{n+2} + a_n.$$

Therefore $Ly = 0$ if and only if

$$a_{n+2} = \frac{-a_n}{(n + 2)(n + 1)}, \qquad n \geq 0, \tag{10}$$

where a_0 and a_1 are arbitrary. Since the indices on the left and right sides of (10) differ by two, we write (10) separately for n even ($n = 2m$) and n odd ($n = 2m + 1$). This yields

$$a_{2m+2} = \frac{-a_{2m}}{(2m + 2)(2m + 1)}, \qquad m \geq 0, \tag{11}$$

and

$$a_{2m+3} = \frac{-a_{2m+1}}{(2m + 3)(2m + 2)}, \qquad m \geq 0. \tag{12}$$

Computing the coefficients of the even powers of $x - x_0$ from (11) yields

$$a_2 = -\frac{a_0}{2 \cdot 1},$$

$$a_4 = -\frac{a_2}{4 \cdot 3} = -\frac{1}{4 \cdot 3}\left(-\frac{a_0}{2 \cdot 1}\right) = \frac{a_0}{4 \cdot 3 \cdot 2 \cdot 1},$$

$$a_6 = -\frac{a_4}{6 \cdot 5} = -\frac{1}{6 \cdot 5}\left(\frac{a_0}{4 \cdot 3 \cdot 2 \cdot 1}\right) = -\frac{a_0}{6 \cdot 5 \cdot 4 \cdot 3 \cdot 2 \cdot 1},$$

and, in general,

$$a_{2m} = (-1)^m \frac{a_0}{(2m)!}, \qquad m \geq 0. \tag{13}$$

Computing the coefficients of the odd powers of $x - x_0$ from (12) yields

$$a_3 = -\frac{a_1}{3 \cdot 2},$$

$$a_5 = -\frac{a_3}{5 \cdot 4} = -\frac{1}{5 \cdot 4}\left(-\frac{a_1}{3 \cdot 2}\right) = \frac{a_1}{5 \cdot 4 \cdot 3 \cdot 2},$$

$$a_7 = -\frac{a_5}{7 \cdot 6} = -\frac{1}{7 \cdot 6}\left(\frac{a_1}{5 \cdot 4 \cdot 3 \cdot 2}\right) = -\frac{a_1}{7 \cdot 6 \cdot 5 \cdot 4 \cdot 3 \cdot 2},$$

and, in general,

$$a_{2m+1} = \frac{(-1)^m a_1}{(2m + 1)!} \qquad m \geq 0. \tag{14}$$

Thus, the general solution of (9) can be written as

$$y = \sum_{m=0}^{\infty} a_{2m}(x - x_0)^{2m} + \sum_{m=0}^{\infty} a_{2m+1}(x - x_0)^{2m+1},$$

or, from (13) and (14), as

$$y = a_0 \sum_{m=0}^{\infty} (-1)^m \frac{(x - x_0)^{2m}}{(2m)!} + a_1 \sum_{m=0}^{\infty} (-1)^m \frac{(x - x_0)^{2m+1}}{(2m + 1)!}. \tag{15}$$

If we recall from calculus that

$$\sum_{m=0}^{\infty} (-1)^m \frac{(x - x_0)^{2m}}{(2m)!} = \cos(x - x_0)$$

and

$$\sum_{m=0}^{\infty} (-1)^m \frac{(x - x_0)^{2m+1}}{(2m + 1)!} = \sin(x - x_0),$$

then (15) becomes

$$y = a_0 \cos(x - x_0) + a_1 \sin(x - x_0),$$

which should look familiar.   ■

Equations like (10), (11), and (12), which define a given coefficient in the sequence $\{a_n\}$ in terms of one or more coefficients with lesser indices, are called **recurrence relations.** When we use a recurrence relation to compute terms of a sequence we are computing **recursively.**

In the remainder of this section we consider the problem of finding power series solutions in $x - x_0$ for equations of the form

$$(1 + \alpha(x - x_0)^2)y'' + \beta(x - x_0)y' + \gamma y = 0. \tag{16}$$

Many important equations that arise in applications are of this form with $x_0 = 0$, including Legendre's equation (2), Airy's equation (3), **Chebyshev's**[1] **equation,**

$$(1 - x^2)y'' - xy' + \alpha^2 y = 0,$$

and **Hermite's**[2] **equation,**

$$y'' - 2xy' + 2\alpha y = 0.$$

Since

$$P_0(x) = 1 + \alpha(x - x_0)^2$$

in (16), the point x_0 is an ordinary point of (16), and Theorem 7.2.1 implies that the solutions of (16) can be written as power series in $x - x_0$ that converge on the interval $(x_0 - 1/\sqrt{|\alpha|}, x_0 + 1/\sqrt{|\alpha|})$ if $\alpha \neq 0$, or on $(-\infty, \infty)$ if $\alpha = 0$. We will see that the coefficients in these power series can be obtained by methods similar to the one used in Example 1. To simplify finding the coefficients we introduce the following notation for products:

$$\prod_{j=r}^{s} b_j = b_r b_{r+1} \cdots b_s \qquad \text{if} \qquad s \geq r.$$

Thus,

$$\prod_{j=2}^{7} b_j = b_2 b_3 b_4 b_5 b_6 b_7,$$

$$\prod_{j=0}^{4} (2j + 1) = (1)(3)(5)(7)(9) = 945,$$

and

$$\prod_{j=2}^{2} j^2 = 2^2 = 4.$$

[1] Pafnuty L. Chebyshev (1821–1894) was a leading Russian mathematician of the 19th century. The polynomial solutions of the equation named after him (which occur when $\alpha = n$) are called **Chebyshev polynomials.** His interest in them stemmed from their application to the problem of approximating general continuous functions by polynomials.

[2] The French mathematician Charles Hermite (1822–1901) was the first to show that e is not a zero of any polynomial with integer coefficients. He strongly influenced the development of analysis, algebra, and number theory, and was the teacher of Henri Poincaré. **Hermitian matrices** are also named in his honor.

We define

$$\prod_{j=r}^{s} b_j = 1 \qquad \text{if} \qquad s < r,$$

no matter what the form of b_j.

EXAMPLE 2 Find the power series in x for the general solution of

$$(1 + 2x^2)y'' + 6xy' + 2y = 0. \tag{17}$$

Solution Here

$$Ly = (1 + 2x^2)y'' + 6xy' + 2y.$$

If

$$y = \sum_{n=0}^{\infty} a_n x^n$$

then

$$y' = \sum_{n=1}^{\infty} na_n x^{n-1} \qquad \text{and} \qquad y'' = \sum_{n=2}^{\infty} n(n-1)a_n x^{n-2},$$

so

$$Ly = (1 + 2x^2)\sum_{n=2}^{\infty} n(n-1)a_n x^{n-2} + 6x\sum_{n=1}^{\infty} na_n x^{n-1} + 2\sum_{n=0}^{\infty} a_n x^n$$

$$= \sum_{n=2}^{\infty} n(n-1)a_n x^{n-2} + \sum_{n=0}^{\infty} [2n(n-1) + 6n + 2]a_n x^n$$

$$= \sum_{n=2}^{\infty} n(n-1)a_n x^{n-2} + 2\sum_{n=0}^{\infty} (n+1)^2 a_n x^n.$$

To collect coefficients of x^n we shift the summation index in the first sum. This yields

$$Ly = \sum_{n=0}^{\infty} (n+2)(n+1)a_{n+2}x^n + 2\sum_{n=0}^{\infty} (n+1)^2 a_n x^n = \sum_{n=0}^{\infty} b_n x^n,$$

with

$$b_n = (n+2)(n+1)a_{n+2} + 2(n+1)^2 a_n, \qquad n \geq 0.$$

To obtain solutions of (17) we set $b_n = 0$ for $n \geq 0$. This is equivalent to the recurrence relation

$$a_{n+2} = -2\frac{n+1}{n+2}a_n, \qquad n \geq 0. \tag{18}$$

Since the indices on the left and right differ by two, we write (18) separately for $n = 2m$ and $n = 2m + 1$, as in Example 1. This yields

$$a_{2m+2} = -2\frac{2m+1}{2m+2}a_{2m} = -\frac{2m+1}{m+1}a_{2m}, \qquad m \geq 0, \tag{19}$$

and

$$a_{2m+3} = -2\frac{2m+2}{2m+3}a_{2m+1} = -4\frac{m+1}{2m+3}a_{2m+1}, \qquad m \geq 0. \tag{20}$$

Computing the coefficients of even powers of x from (19) yields

$$a_2 = -\frac{1}{1}a_0,$$

$$a_4 = -\frac{3}{2}a_2 = \left(-\frac{3}{2}\right)\left(-\frac{1}{1}\right)a_0 = \frac{1\cdot 3}{1\cdot 2}a_0,$$

$$a_6 = -\frac{5}{3}a_4 = -\frac{5}{3}\left(\frac{1\cdot 3}{1\cdot 2}\right)a_0 = -\frac{1\cdot 3\cdot 5}{1\cdot 2\cdot 3}a_0,$$

$$a_8 = -\frac{7}{4}a_6 = -\frac{7}{4}\left(-\frac{1\cdot 3\cdot 5}{1\cdot 2\cdot 3}\right)a_0 = \frac{1\cdot 3\cdot 5\cdot 7}{1\cdot 2\cdot 3\cdot 4}a_0.$$

In general,

$$a_{2m} = (-1)^m \frac{\prod_{j=1}^m (2j-1)}{m!}a_0, \qquad m \geq 0. \tag{21}$$

(Notice that (21) is correct for $m = 0$ because we have defined $\prod_{j=1}^0 b_j = 1$ for any b_j.)

Computing the coefficients of odd powers of x from (20) yields

$$a_3 = -4\frac{1}{3}a_1,$$

$$a_5 = -4\frac{2}{5}a_3 = -4\frac{2}{5}\left(-4\frac{1}{3}\right)a_1 = 4^2\frac{1\cdot 2}{3\cdot 5}a_1,$$

$$a_7 = -4\frac{3}{7}a_5 = -4\frac{3}{7}\left(4^2\frac{1\cdot 2}{3\cdot 5}\right)a_1 = -4^3\frac{1\cdot 2\cdot 3}{3\cdot 5\cdot 7}a_1,$$

$$a_9 = -4\frac{4}{9}a_7 = -4\frac{4}{9}\left(4^3\frac{1\cdot 2\cdot 3}{3\cdot 5\cdot 7}\right)a_1 = 4^4\frac{1\cdot 2\cdot 3\cdot 4}{3\cdot 5\cdot 7\cdot 9}a_1.$$

In general,

$$a_{2m+1} = \frac{(-1)^m 4^m m!}{\prod_{j=1}^m (2j+1)}a_1, \qquad m \geq 0. \tag{22}$$

From (21) and (22) we see that

$$y = a_0 \sum_{m=0}^\infty (-1)^m \frac{\prod_{j=1}^m (2j-1)}{m!}x^{2m} + a_1 \sum_{m=0}^\infty (-1)^m \frac{4^m m!}{\prod_{j=1}^m (2j+1)}x^{2m+1}$$

is the power series in x for the general solution of (17). Since $P_0(x) = 1 + 2x^2$ has no real zeros, Theorem 5.1.1 implies that every solution of (17) is defined on $(-\infty, \infty)$. However, since $P_0(\pm i/\sqrt{2}) = 0$, Theorem 7.2.1 implies only that the power series converges in $(-1/\sqrt{2}, 1/\sqrt{2})$ for any choice of a_0 and a_1. The blue curve in Figure 1 shows the solution on this interval (obtained by a highly accurate numerical method) corresponding to $a_0 = 1, a_1 = 0$. The gray curves are the graphs of the Taylor polynomials

$$T_{2M}(x) = \sum_{m=0}^M (-1)^m \frac{\prod_{j=1}^m (2j-1)}{m!}x^{2m}$$

for $M = 1, \ldots, 6$. ∎

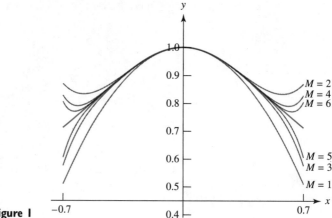

Figure I

The results in Examples 1 and 2 are consequences of the following general theorem.

THEOREM 7.2.2

The coefficients $\{a_n\}$ in any solution $y = \sum_{n=0}^{\infty} a_n(x - x_0)^n$ of

$$(1 + \alpha(x - x_0)^2)y'' + \beta(x - x_0)y' + \gamma y = 0 \qquad (23)$$

satisfy the recurrence relation

$$a_{n+2} = -\frac{p(n)}{(n + 2)(n + 1)} a_n, \qquad n \geq 0, \qquad (24)$$

where

$$p(n) = \alpha n(n - 1) + \beta n + \gamma. \qquad (25)$$

Moreover, the coefficients of the even and odd powers of $x - x_0$ can be computed separately as

$$a_{2m+2} = -\frac{p(2m)}{(2m + 2)(2m + 1)} a_{2m}, \qquad m \geq 0 \qquad (26)$$

and

$$a_{2m+3} = -\frac{p(2m + 1)}{(2m + 3)(2m + 2)} a_{2m+1}, \qquad m \geq 0, \qquad (27)$$

where a_0 and a_1 are arbitrary.

PROOF. Here

$$Ly = (1 + \alpha(x - x_0)^2)y'' + \beta(x - x_0)y' + \gamma y.$$

If

$$y = \sum_{n=0}^{\infty} a_n(x - x_0)^n$$

then

$$y' = \sum_{n=1}^{\infty} na_n(x - x_0)^{n-1} \quad \text{and} \quad y'' = \sum_{n=2}^{\infty} n(n-1)a_n(x - x_0)^{n-2}.$$

Hence,

$$Ly = \sum_{n=2}^{\infty} n(n-1)a_n(x - x_0)^{n-2} + \sum_{n=0}^{\infty} [\alpha n(n-1) + \beta n + \gamma]a_n(x - x_0)^n$$

$$= \sum_{n=2}^{\infty} n(n-1)a_n(x - x_0)^{n-2} + \sum_{n=0}^{\infty} p(n)a_n(x - x_0)^n,$$

from (25). To collect coefficients of powers of $x - x_0$ we shift the summation index in the first sum. This yields

$$Ly = \sum_{n=0}^{\infty} [(n+2)(n+1)a_{n+2} + p(n)a_n](x - x_0)^n.$$

Thus, $Ly = 0$ if and only if

$$(n+2)(n+1)a_{n+2} + p(n)a_n = 0, \quad n \geq 0,$$

which is equivalent to (24). Writing (24) separately for the cases where $n = 2m$ and $n = 2m + 1$ yields (26) and (27). $\qquad\square$

EXAMPLE 3 Find the power series in $x - 1$ for the general solution of

$$(2 + 4x - 2x^2)y'' - 12(x - 1)y' - 12y = 0. \tag{28}$$

Solution We must first write the coefficient $P_0(x) = 2 + 4x - x^2$ in powers of $x - 1$. To do this we write $x = (x - 1) + 1$ in $P_0(x)$ and then expand the terms, collecting powers of $x - 1$; thus,

$$2 + 4x - 2x^2 = 2 + 4[(x - 1) + 1] - 2[(x - 1) + 1]^2$$
$$= 4 - 2(x - 1)^2.$$

Therefore, we can rewrite (28) as

$$(4 - 2(x - 1)^2)y'' - 12(x - 1)y' - 12y = 0,$$

or, equivalently

$$\left(1 - \frac{1}{2}(x - 1)^2\right)y'' - 3(x - 1)y' - 3y = 0.$$

This is of the form (23) with $\alpha = -1/2$, $\beta = -3$, and $\gamma = -3$. Therefore, from (25),

$$p(n) = -\frac{n(n-1)}{2} - 3n - 3 = -\frac{(n+2)(n+3)}{2}.$$

Hence, Theorem 7.2.2 implies that

$$a_{2m+2} = -\frac{p(2m)}{(2m+2)(2m+1)}a_{2m}$$
$$= \frac{(2m+2)(2m+3)}{2(2m+2)(2m+1)}a_{2m} = \frac{2m+3}{2(2m+1)}a_{2m}, \quad m \geq 0,$$

and

$$a_{2m+3} = -\frac{p(2m+1)}{(2m+3)(2m+2)}a_{2m+1}$$
$$= \frac{(2m+3)(2m+4)}{2(2m+3)(2m+2)}a_{2m+1} = \frac{m+2}{2(m+1)}a_{2m+1}, \qquad m \geq 0.$$

We leave it to you to show that

$$a_{2m} = \frac{2m+1}{2^m}a_0 \qquad \text{and} \qquad a_{2m+1} = \frac{m+1}{2^m}a_1, \qquad m \geq 0,$$

which implies that the power series in $x - 1$ for the general solution of (28) is

$$y = a_0 \sum_{m=0}^{\infty} \frac{2m+1}{2^m}(x-1)^{2m} + a_1 \sum_{m=0}^{\infty} \frac{m+1}{2^m}(x-1)^{2m+1}. \qquad \blacksquare$$

In the examples considered so far we were able to obtain closed formulas for coefficients in the power series solutions. In some cases this is impossible, and we must settle for computing a finite number of terms in the series. The following example illustrates this with an initial value problem.

EXAMPLE 4   Compute $a_0, a_1, \ldots, a_7$ in the series solution $y = \sum_{n=0}^{\infty} a_n x^n$ of the initial value problem

$$(1 + 2x^2)y'' + 10xy' + 8y = 0, \qquad y(0) = 2, \qquad y'(0) = -3. \qquad (29)$$

Solution   Since $\alpha = 2, \beta = 10$, and $\gamma = 8$ in (29), we have

$$p(n) = 2n(n-1) + 10n + 8 = 2(n+2)^2.$$

Therefore

$$a_{n+2} = -2\frac{(n+2)^2}{(n+2)(n+1)}a_n = -2\frac{n+2}{n+1}a_n, \qquad n \geq 0.$$

Writing this equation separately for $n = 2m$ and $n = 2m + 1$ yields

$$a_{2m+2} = -2\frac{2m+2}{2m+1}a_{2m} = -4\frac{m+1}{2m+1}a_{2m}, \qquad m \geq 0 \qquad (30)$$

and

$$a_{2m+3} = -2\frac{2m+3}{2m+2}a_{2m+1} = -\frac{2m+3}{m+1}a_{2m+1}, \qquad m \geq 0. \qquad (31)$$

Starting with $a_0 = y(0) = 2$, we compute a_2, a_4, and a_6 from (30):

$$a_2 = -4\frac{1}{1}2 = -8,$$

$$a_4 = -4\frac{2}{3}(-8) = \frac{64}{3},$$

$$a_6 = -4\frac{3}{5}\left(\frac{64}{3}\right) = -\frac{256}{5}.$$

Starting with $a_1 = y'(0) = -3$, we compute a_3, a_5, and a_7 from (31):

$$a_3 = -\frac{3}{1}(-3) = 9,$$

$$a_5 = -\frac{5}{2}9 = -\frac{45}{2},$$

$$a_7 = -\frac{7}{3}\left(-\frac{45}{2}\right) = \frac{105}{2}.$$

Therefore the solution of (29) is

$$y = 2 - 3x - 8x^2 + 9x^3 + \frac{64}{3}x^4 - \frac{45}{2}x^5 - \frac{256}{5}x^6 + \frac{105}{2}x^7 + \cdots. \qquad \blacksquare$$

USING TECHNOLOGY

Computing coefficients recursively as in Example 4 is tedious. We recommend that you do this kind of computation by writing a short program to implement the appropriate recurrence relation on a calculator or computer. You may wish to do this in verifying examples and doing exercises (identified by the symbol $\boxed{\text{C}}$) in this chapter that call for numerical computation of the coefficients in series solutions. We obtained the answers to these exercises by using software that can produce answers in the form of rational numbers. However, it is perfectly acceptable—and more practical—to get your answers in decimal form. You can always check them by converting our fractions to decimals.

If you're interested in actually using series to compute numerical approximations to solutions of a differential equation, then whether or not there is a simple closed form for the coefficients is essentially irrelevant. For computational purposes it is usually more efficient to start with the given coefficients $a_0 = y(x_0)$ and $a_1 = y'(x_0)$, compute $a_2, \ldots, a_N$ recursively, and then compute approximate values of the solution from the Taylor polynomial

$$T_N(x) = \sum_{n=0}^{N} a_n(x - x_0)^n.$$

The trick is to decide how to choose N so the approximation $y(x) \approx T_N(x)$ is sufficiently accurate on the subinterval of the interval of convergence in which you're interested. In the computational exercises in this and the next two sections you will often be asked to obtain the solution of a given problem by numerical integration with software of your choice (see Section 10.1 for a brief discussion of one such method), and to compare the solution obtained in this way with the approximations obtained with T_N for various values of N. This is a typical textbook kind of exercise, designed to give you insight into how the accuracy of the approximation $y(x) \approx T_N(x)$ behaves as a function of N and the interval on which you are working. In real life you would choose one or the other of the two methods (numerical integration or series solution). If you choose the method of series solution, then a practical procedure for determining a suitable value of N is to continue increasing N until the maximum of $|T_N - T_{N-1}|$ on the interval of interest is within the margin of error that you're willing to accept.

In doing computational problems that call for numerical solution of differential equations you should choose the most accurate numerical integration procedure your software supports, and experiment with the step size until you're confident that the numerical results are sufficiently accurate for the problem at hand.

7.2 EXERCISES

In Exercises 1–8 find the power series in x for the general solution.

1. $(1 + x^2)y'' + 6xy' + 6y = 0$

2. $(1 + x^2)y'' + 2xy' - 2y = 0$

3. $(1 + x^2)y'' - 8xy' + 20y = 0$

4. $(1 - x^2)y'' - 8xy' - 12y = 0$

5. $(1 + 2x^2)y'' + 7xy' + 2y = 0$

6. $(1 + x^2)y'' + 2xy' + \dfrac{1}{4}y = 0$

7. $(1 - x^2)y'' - 5xy' - 4y = 0$

8. $(1 + x^2)y'' - 10xy' + 28y = 0$

L **9. (a)** Find the power series in x for the general solution of $y'' + xy' + 2y = 0$.

    (b) For several choices of a_0 and a_1 use differential equations software to solve the initial value problem

$$y'' + xy' + 2y = 0, \qquad y(0) = a_0, \qquad y'(0) = a_1 \tag{A}$$

    numerically on $(-5, 5)$.

    (c) For fixed r in $\{1, 2, 3, 4, 5\}$ graph

$$T_N(x) = \sum_{n=0}^{N} a_n x^n$$

    and the solution obtained in **(a)** on $(-r, r)$. Continue increasing N until there is no perceptible difference between the two graphs.

L **10.** Follow the directions of Exercise 9 for the differential equation

$$y'' + 2xy' + 3y = 0.$$

In Exercises 11–13 find $a_0, \dots, a_N$ for N at least 7 in the power series solution $y = \sum_{n=0}^{\infty} a_n x^n$ of the initial value problem.

C **11.** $(1 + x^2)y'' + xy' + y = 0$, $y(0) = 2$, $y'(0) = -1$

C **12.** $(1 + 2x^2)y'' - 9xy' - 6y = 0$, $y(0) = 1$, $y'(0) = -1$

C **13.** $(1 + 8x^2)y'' + 2y = 0$, $y(0) = 2$, $y'(0) = -1$

L **14.** Do the following experiment for various choices of real numbers a_0, a_1, and r, with $0 < r < 1/\sqrt{2}$.

    (a) Use differential equations software to solve the initial value problem

$$(1 - 2x^2)y'' - xy' + 3y = 0, \qquad y(0) = a_0, \qquad y'(0) = a_1 \tag{A}$$

    numerically on $(-r, r)$.

    (b) For $N = 2, 3, 4, \dots$, compute $a_2, \dots, a_N$ in the power series solution $y = \sum_{n=0}^{\infty} a_n x^n$ of (A), and graph

$$T_N(x) = \sum_{n=0}^{N} a_n x^n$$

    and the solution obtained in **(a)** on $(-r, r)$. Continue increasing N until there is no perceptible difference between the two graphs.

L **15.** Do **(a)** and **(b)** for several values of r in $(0, 1)$:

    (a) Use differential equations software to solve the initial value problem

$$(1 + x^2)y'' + 10xy' + 14y = 0, \qquad y(0) = 5, \qquad y'(0) = 1 \tag{A}$$

    numerically on $(-r, r)$.

    (b) For $N = 2, 3, 4, \dots$, compute $a_2, \dots, a_N$ in the power series solution $y = \sum_{n=0}^{\infty} a_n x^n$ of (A), and graph

$$T_N(x) = \sum_{n=0}^{N} a_n x^n$$

    and the solution obtained in **(a)** on $(-r, r)$. Continue increasing N until there is no perceptible difference between the two graphs. What happens to the required N as $r \to 1$?

(c) Try **(a)** and **(b)** with $r = 1.2$. Explain your results.

In Exercises 16–20 find the power series in $x - x_0$ for the general solution.

16. $y'' - y = 0$; $x_0 = 3$ **17.** $y'' - (x - 3)y' - y = 0$; $x_0 = 3$

18. $(1 - 4x + 2x^2)y'' + 10(x - 1)y' + 6y = 0$; $x_0 = 1$

19. $(11 - 8x + 2x^2)y'' - 16(x - 2)y' + 36y = 0$; $x_0 = 2$

20. $(5 + 6x + 3x^2)y'' + 9(x + 1)y' + 3y = 0$; $x_0 = -1$

In Exercises 21–26 find $a_0, \ldots, a_N$ for N at least 7 in the power series $y = \sum_{n=0}^{\infty} a_n(x - x_0)^n$ for the solution of the initial value problem. Take x_0 to be the point at which the initial conditions are imposed.

Ⓒ **21.** $(x^2 - 4)y'' - xy' - 3y = 0$, $y(0) = -1$, $y'(0) = 2$

Ⓒ **22.** $y'' + (x - 3)y' + 3y = 0$, $y(3) = -2$, $y'(3) = 3$

Ⓒ **23.** $(5 - 6x + 3x^2)y'' + 18(x - 1)y' + 12y = 0$, $y(1) = -1$, $y'(1) = 1$

Ⓒ **24.** $(4x^2 - 24x + 37)y'' + y = 0$, $y(3) = 4$, $y'(3) = -6$

Ⓒ **25.** $(x^2 - 8x + 14)y'' - 8(x - 4)y' + 20y = 0$, $y(4) = 3$, $y'(4) = -4$

Ⓒ **26.** $(2x^2 + 4x + 5)y'' - 20(x + 1)y' + 60y = 0$, $y(-1) = 3$, $y'(-1) = -3$

27. (a) Find a power series in x for the general solution of

$$(1 + x^2)y'' + 4xy' + 2y = 0. \tag{A}$$

(b) Use **(a)** and the formula

$$\frac{1}{1 - r} = 1 + r + r^2 + \cdots + r^n + \cdots \qquad (-1 < r < 1)$$

for the sum of a geometric series to find a closed form expression for the general solution of (A) on $(-1, 1)$.

(c) Show that the expression obtained in **(b)** is actually the general solution of (A) on $(-\infty, \infty)$.

28. Use Theorem 7.2.2 to show that the power series in x for the general solution of

$$(1 + \alpha x^2)y'' + \beta xy' + \gamma y = 0$$

is

$$y = a_0 \sum_{m=0}^{\infty} (-1)^m \left[\prod_{j=0}^{m-1} p(2j) \right] \frac{x^{2m}}{(2m)!} + a_1 \sum_{m=0}^{\infty} (-1)^m \left[\prod_{j=0}^{m-1} p(2j + 1) \right] \frac{x^{2m+1}}{(2m + 1)!}.$$

29. Use Exercise 28 to show that all solutions of

$$(1 + \alpha x^2)y'' + \beta xy' + \gamma y = 0$$

are polynomials if and only if

$$\alpha n(n - 1) + \beta n + \gamma = \alpha(n - 2r)(n - 2s - 1),$$

where r and s are nonnegative integers.

30. (a) Use Exercise 28 to show that the power series in x for the general solution of

$$(1 - x^2)y'' - 2bxy' + \alpha(\alpha + 2b - 1)y = 0$$

is $y = a_0 y_1 + a_1 y_2$, where

$$y_1 = \sum_{m=0}^{\infty} \left[\prod_{j=0}^{m-1} (2j - \alpha)(2j + \alpha + 2b - 1) \right] \frac{x^{2m}}{(2m)!}$$

and

$$y_2 = \sum_{m=0}^{\infty} \left[\prod_{j=0}^{m-1} (2j + 1 - \alpha)(2j + \alpha + 2b) \right] \frac{x^{2m+1}}{(2m + 1)!}.$$

(b) Suppose that $2b$ is not a negative integer and k is a nonnegative integer. Show that y_1 is a polynomial of degree $2k$ such that $y_1(-x) = y_1(x)$ if $\alpha = 2k$, while y_2 is a polynomial of degree $2k + 1$ such that $y_2(-x) = -y_2(-x)$ if $\alpha = 2k + 1$. Conclude that if n is a nonnegative integer then there is a polynomial P_n of degree n such that $P_n(-x) = (-1)^n P_n(x)$ and

$$(1 - x^2)P_n'' - 2bxP_n' + n(n + 2b - 1)P_n = 0. \tag{A}$$

(c) Show that (A) implies that

$$[(1 - x^2)^b P_n']' = -n(n + 2b - 1)(1 - x^2)^{b-1}P_n,$$

and use this to show that if m and n are nonnegative integers then

$$[(1 - x^2)^b P_n']'P_m - [(1 - x^2)^b P_m']'P_n \\ = [m(m + 2b - 1) - n(n + 2b - 1)](1 - x^2)^{b-1}P_m P_n. \tag{B}$$

(d) Now suppose that $b > 0$. Use (B) and integration by parts to show that if $m \neq n$ then

$$\int_{-1}^{1} (1 - x^2)^{b-1}P_m(x)P_n(x)\ dx = 0.$$

(We say that P_m and P_n are **orthogonal on $(-1, 1)$ with respect to the weighting function $(1 - x^2)^{b-1}$.**)

31. (a) Use Exercise 28 to show that the power series in x for the general solution of Hermite's equation

$$y'' - 2xy' + 2\alpha y = 0$$

is $y = a_0 y_1 + a_1 y_1$, where

$$y_1 = \sum_{m=0}^{\infty} \left[\prod_{j=0}^{m-1} (2j - \alpha) \right] \frac{2^m x^{2m}}{(2m)!}$$

and

$$y_2 = \sum_{m=0}^{\infty} \left[\prod_{j=0}^{m-1} (2j + 1 - \alpha) \right] \frac{2^m x^{2m+1}}{(2m + 1)!}.$$

(b) Suppose that k is a nonnegative integer. Show that y_1 is a polynomial of degree $2k$ such that $y_1(-x) = y_1(x)$ if $\alpha = 2k$, while y_2 is a polynomial of degree $2k + 1$ such that $y_2(-x) = -y_2(-x)$ if $\alpha = 2k + 1$. Conclude that if n is a nonnegative integer then there is a polynomial P_n of degree n such that $P_n(-x) = (-1)^n P_n(x)$ and

$$P_n'' - 2xP_n' + 2nP_n = 0. \tag{A}$$

(c) Show that (A) implies that

$$[e^{-x^2} P_n']' = -2ne^{-x^2}P_n,$$

and use this to show that if m and n are nonnegative integers then

$$[e^{-x^2} P_n']'P_m - [e^{-x^2} P_m']'P_n = 2(m - n)e^{-x^2}P_m P_n. \tag{B}$$

(d) Use (B) and integration by parts to show that if $m \neq n$ then

$$\int_{-\infty}^{\infty} e^{-x^2}P_m(x)P_n(x)\ dx = 0.$$

(We say that P_m and P_n are **orthogonal on $(-\infty, \infty)$ with respect to the weighting function e^{-x^2}.**)

32. Consider the equation

$$(1 + \alpha x^3)y'' + \beta x^2 y' + \gamma xy = 0, \tag{A}$$

and let $p(n) = \alpha n(n - 1) + \beta n + \gamma$. (The special case $y'' - xy = 0$ of (A) is Airy's equation.)

(a) Modify the argument used to prove Theorem 7.2.2 to show that

$$y = \sum_{n=0}^{\infty} a_n x^n$$

is a solution of (A) if and only if $a_2 = 0$ and

$$a_{n+3} = -\frac{p(n)}{(n+3)(n+2)} a_n, \qquad n \geq 0.$$

(b) Show from **(a)** that $a_n = 0$ unless $n = 3m$ or $n = 3m + 1$ for some nonnegative integer m, and that

$$a_{3m+3} = -\frac{p(3m)}{(3m+3)(3m+2)} a_{3m}, \qquad m \geq 0,$$

and

$$a_{3m+4} = -\frac{p(3m+1)}{(3m+4)(3m+3)} a_{3m+1}, \qquad m \geq 0,$$

where a_0 and a_1 may be specified arbitrarily.

(c) Conclude from **(b)** that the power series in x for the general solution of (A) is

$$y = a_0 \sum_{m=0}^{\infty} (-1)^m \left[\prod_{j=0}^{m-1} \frac{p(3j)}{3j+2} \right] \frac{x^{3m}}{3^m m!} + a_1 \sum_{m=0}^{\infty} (-1)^m \left[\prod_{j=0}^{m-1} \frac{p(3j+1)}{3j+4} \right] \frac{x^{3m+1}}{3^m m!}.$$

In Exercises 33–37 use the method of Exercise 32 to find the power series in x for the general solution.

33. $y'' - xy = 0$ **34.** $(1 - 2x^3)y'' - 10x^2 y' - 8xy = 0$

35. $(1 + x^3)y'' + 7x^2 y' + 9xy = 0$ **36.** $(1 - 2x^3)y'' + 6x^2 y' + 24xy = 0$

37. $(1 - x^3)y'' + 15x^2 y' - 63xy = 0$

38. Consider the equation

$$(1 + \alpha x^{k+2})y'' + \beta x^{k+1} y' + \gamma x^k y = 0, \tag{A}$$

where k is a positive integer, and let $p(n) = \alpha n(n-1) + \beta n + \gamma$.

(a) Modify the argument used to prove Theorem 7.2.2 to show that

$$y = \sum_{n=0}^{\infty} a_n x^n$$

is a solution of (A) if and only if $a_n = 0$ for $2 \leq n \leq k + 1$ and

$$a_{n+k+2} = -\frac{p(n)}{(n+k+2)(n+k+1)} a_n, \qquad n \geq 0.$$

(b) Show from **(a)** that $a_n = 0$ unless $n = (k+2)m$ or $n = (k+2)m + 1$ for some nonnegative integer m, and that

$$a_{(k+2)(m+1)} = -\frac{p((k+2)m)}{(k+2)(m+1)[(k+2)(m+1)-1]} a_{(k+2)m}, \qquad m \geq 0,$$

and

$$a_{(k+2)(m+1)+1} = -\frac{p((k+2)m+1)}{[(k+2)(m+1)+1](k+2)(m+1)} a_{(k+2)m+1}, \qquad m \geq 0,$$

where a_0 and a_1 may be specified arbitrarily.

(c) Conclude from **(b)** that the power series in x for the general solution of (A) is

$$y = a_0 \sum_{m=0}^{\infty} (-1)^m \left[\prod_{j=0}^{m-1} \frac{p((k+2)j)}{(k+2)(j+1)-1} \right] \frac{x^{(k+2)m}}{(k+2)^m m!}$$

$$+ a_1 \sum_{m=0}^{\infty} (-1)^m \left[\prod_{j=0}^{m-1} \frac{p((k+2)j+1)}{(k+2)(j+1)+1} \right] \frac{x^{(k+2)m+1}}{(k+2)^m m!}.$$

In Exercises 39–44 use the method of Exercise 38 to find the power series in x for the general solution.

39. $(1 + 2x^5)y'' + 14x^4 y' + 10x^3 y = 0$ **40.** $y'' + x^2 y = 0$

41. $y'' + x^6 y' + 7x^5 y = 0$ **42.** $(1 + x^8)y'' - 16x^7 y' + 72x^6 y = 0$

43. $(1 - x^6)y'' - 12x^5 y' - 30x^4 y = 0$ **44.** $y'' + x^5 y' + 6x^4 y = 0$

7.3   Series Solutions near an Ordinary Point II

In this section we continue to find series solutions

$$y = \sum_{n=0}^{\infty} a_n(x - x_0)^n$$

of initial value problems

$$P_0(x)y'' + P_1(x)y' + P_2(x)y = 0, \qquad y(x_0) = a_0, \qquad y'(x_0) = a_1, \qquad (1)$$

where P_0, P_1, and P_2 are polynomials and $P_0(x_0) \neq 0$, so x_0 is an ordinary point of (1). However, here we consider cases where the differential equation in (1) is not of the form

$$(1 + \alpha(x - x_0)^2)y'' + \beta(x - x_0)y' + \gamma y = 0,$$

so Theorem 7.2.2 does not apply, and the computation of the coefficients $\{a_n\}$ is more complicated. For the equations considered here it is difficult or impossible to obtain an explicit formula for a_n in terms of n. Nevertheless, we can calculate as many coefficients as we wish. The following three examples illustrate this.

EXAMPLE 1   Find the coefficients $a_0, \ldots, a_7$ in the series solution $y = \sum_{n=0}^{\infty} a_n x^n$ of the initial value problem

$$(1 + x + 2x^2)y'' + (1 + 7x)y' + 2y = 0, \qquad y(0) = -1, \qquad y'(0) = -2. \qquad (2)$$

Solution   Here

$$Ly = (1 + x + 2x^2)y'' + (1 + 7x)y' + 2y.$$

The zeros $(-1 \pm i\sqrt{7})/4$ of $P_0(x) = 1 + x + 2x^2$ have absolute value $1/\sqrt{2}$, so Theorem 7.2.1 implies that the series solution converges to the solution of (2) on $(-1/\sqrt{2}, 1/\sqrt{2})$. Since

$$y = \sum_{n=0}^{\infty} a_n x^n, \qquad y' = \sum_{n=1}^{\infty} n a_n x^{n-1} \qquad \text{and} \qquad y'' = \sum_{n=2}^{\infty} n(n-1)a_n x^{n-2},$$

we have

$$Ly = \sum_{n=2}^{\infty} n(n-1)a_n x^{n-2} + \sum_{n=2}^{\infty} n(n-1)a_n x^{n-1} + 2\sum_{n=2}^{\infty} n(n-1)a_n x^n$$

$$+ \sum_{n=1}^{\infty} n a_n x^{n-1} + 7\sum_{n=1}^{\infty} n a_n x^n + 2\sum_{n=0}^{\infty} a_n x^n.$$

Shifting indices so the general term in each series is a constant multiple of x^n yields

$$Ly = \sum_{n=0}^{\infty} (n + 2)(n + 1)a_{n+2}x^n + \sum_{n=0}^{\infty} (n + 1)na_{n+1}x^n + 2 \sum_{n=0}^{\infty} n(n - 1)a_nx^n$$

$$+ \sum_{n=0}^{\infty} (n + 1)a_{n+1}x^n + 7 \sum_{n=0}^{\infty} na_nx^n + 2 \sum_{n=0}^{\infty} a_nx^n = \sum_{n=0}^{\infty} b_nx^n,$$

where

$$b_n = (n + 2)(n + 1)a_{n+2} + (n + 1)^2a_{n+1} + (n + 2)(2n + 1)a_n.$$

Therefore, $y = \sum_{n=0}^{\infty} a_nx^n$ is a solution of $Ly = 0$ if and only if

$$a_{n+2} = -\frac{n + 1}{n + 2}a_{n+1} - \frac{2n + 1}{n + 1}a_n, \qquad n \geq 0. \tag{3}$$

From the initial conditions in (2), $a_0 = y(0) = -1$ and $a_1 = y'(0) = -2$. Setting $n = 0$ in (3), we obtain

$$a_2 = -\frac{1}{2}a_1 - a_0 = -\frac{1}{2}(-2) - (-1) = 2.$$

Setting $n = 1$ in (3), we obtain

$$a_3 = -\frac{2}{3}a_2 - \frac{3}{2}a_1 = -\frac{2}{3}(2) - \frac{3}{2}(-2) = \frac{5}{3}.$$

We leave it to you to compute a_4, a_5, a_6, a_7 from (3) and show that

$$y = -1 - 2x + 2x^2 + \frac{5}{3}x^3 - \frac{55}{12}x^4 + \frac{3}{4}x^5 + \frac{61}{8}x^6 - \frac{443}{56}x^7 + \cdots.$$

We also leave it to you (Exercise 13) to verify numerically that the Taylor polynomials $T_N(x) = \sum_{n=0}^{N} a_nx^n$ converge to the solution of (2) on $(-1/\sqrt{2}, 1/\sqrt{2})$.

EXAMPLE 2   Find the coefficients $a_0, \ldots, a_5$ in the series solution

$$y = \sum_{n=0}^{\infty} a_n(x + 1)^n$$

of the initial value problem

$$(3 + x)y'' + (1 + 2x)y' - (2 - x)y = 0, \qquad y(-1) = 2, \qquad y'(-1) = -3. \tag{4}$$

Solution   Since the desired series is in powers of $x + 1$ we rewrite the differential equation in (4) as $Ly = 0$, with

$$Ly = (2 + (x + 1))y'' - (1 - 2(x + 1))y' - (3 - (x + 1))y.$$

Since

$$y = \sum_{n=0}^{\infty} a_n(x + 1)^n, \qquad y' = \sum_{n=1}^{\infty} na_n(x + 1)^{n-1}$$

and

$$y'' = \sum_{n=2}^{\infty} n(n - 1)a_n(x + 1)^{n-2},$$

we have

$$Ly = 2 \sum_{n=2}^{\infty} n(n-1)a_n(x+1)^{n-2} + \sum_{n=2}^{\infty} n(n-1)a_n(x+1)^{n-1}$$

$$- \sum_{n=1}^{\infty} na_n(x+1)^{n-1} + 2 \sum_{n=1}^{\infty} na_n(x+1)^n$$

$$- 3 \sum_{n=0}^{\infty} a_n(x+1)^n + \sum_{n=0}^{\infty} a_n(x+1)^{n+1}.$$

Shifting indices so that the general term in each series is a constant multiple of $(x+1)^n$ yields

$$Ly = 2 \sum_{n=0}^{\infty} (n+2)(n+1)a_{n+2}(x+1)^n + \sum_{n=0}^{\infty} (n+1)na_{n+1}(x+1)^n$$

$$- \sum_{n=0}^{\infty} (n+1)a_{n+1}(x+1)^n + \sum_{n=0}^{\infty} (2n-3)a_n(x+1)^n + \sum_{n=1}^{\infty} a_{n-1}(x+1)^n$$

$$= \sum_{n=0}^{\infty} b_n(x+1)^n,$$

where

$$b_0 = 4a_2 - a_1 - 3a_0$$

and

$$b_n = 2(n+2)(n+1)a_{n+2} + (n^2-1)a_{n+1} + (2n-3)a_n + a_{n-1}, \qquad n \geq 1.$$

Therefore, $y = \sum_{n=0}^{\infty} a_n(x+1)^n$ is a solution of $Ly = 0$ if and only if

$$a_2 = \frac{1}{4}(a_1 + 3a_0) \tag{5}$$

and

$$a_{n+2} = -\frac{1}{2(n+2)(n+1)}[(n^2-1)a_{n+1} + (2n-3)a_n + a_{n-1}], \qquad n \geq 1. \tag{6}$$

From the initial conditions in (4), $a_0 = y(-1) = 2$ and $a_1 = y'(-1) = -3$. We leave it to you to compute $a_2, \ldots, a_5$ by means of (5) and (6) and show that the solution of (4) is

$$y = -2 - 3(x+1) + \frac{3}{4}(x+1)^2 - \frac{5}{12}(x+1)^3$$

$$+ \frac{7}{48}(x+1)^4 - \frac{1}{60}(x+1)^5 + \cdots.$$

We also leave it to you (Exercise 14) to verify numerically that the Taylor polynomials $T_N(x) = \sum_{n=0}^{N} a_n x^n$ converge to the solution of (4) on the interval of convergence of the power series solution.

EXAMPLE 3   Find the coefficients $a_0, \ldots, a_5$ in the series solution $y = \sum_{n=0}^{\infty} a_n x^n$ of the initial value problem

$$y'' + 3xy' + (4 + 2x^2)y = 0, \qquad y(0) = 2, \qquad y'(0) = -3. \tag{7}$$

Solution   Here

$$Ly = y'' + 3xy' + (4 + 2x^2)y.$$

Since

$$y = \sum_{n=0}^{\infty} a_n x^n, \qquad y' = \sum_{n=1}^{\infty} n a_n x^{n-1}, \qquad \text{and} \qquad y'' = \sum_{n=2}^{\infty} n(n-1) a_n x^{n-2},$$

we have

$$Ly = \sum_{n=2}^{\infty} n(n-1) a_n x^{n-2} + 3 \sum_{n=1}^{\infty} n a_n x^n + 4 \sum_{n=0}^{\infty} a_n x^n + 2 \sum_{n=0}^{\infty} a_n x^{n+2}.$$

Shifting indices so that the general term in each series is a constant multiple of x^n yields

$$Ly = \sum_{n=0}^{\infty} (n+2)(n+1) a_{n+2} x^n + \sum_{n=0}^{\infty} (3n+4) a_n x^n + 2 \sum_{n=2}^{\infty} a_{n-2} x^n = \sum_{n=0}^{\infty} b_n x^n$$

where

$$b_0 = 2a_2 + 4a_0, \qquad b_1 = 6a_3 + 7a_1,$$

and

$$b_n = (n+2)(n+1) a_{n+2} + (3n+4) a_n + 2a_{n-2}, \qquad n \geq 2.$$

Therefore, $y = \sum_{n=0}^{\infty} a_n x^n$ is a solution of $Ly = 0$ if and only if

$$a_2 = -2a_0, \qquad a_3 = -\frac{7}{6} a_1, \tag{8}$$

and

$$a_{n+2} = -\frac{1}{(n+2)(n+1)} [(3n+4) a_n + 2a_{n-2}], \qquad n \geq 2. \tag{9}$$

From the initial conditions in (7), $a_0 = y(0) = 2$ and $a_1 = y'(0) = -3$. We leave it to you to compute $a_2, \ldots, a_5$ by means of (8) and (9) and show that the solution of (7) is

$$y = 2 - 3x - 4x^2 + \frac{7}{2} x^3 + 3x^4 - \frac{79}{40} x^5 + \cdots.$$

We also leave it to you (Exercise 15) to verify numerically that the Taylor polynomials $T_N(x) = \sum_{n=0}^{N} a_n x^n$ converge to the solution of (9) on the interval of convergence of the power series solution.   ■

7.3 EXERCISES

In Exercises 1–12 find the coefficients $a_0, \ldots, a_N$ for N at least 7 in the series solution $y = \sum_{n=0}^{\infty} a_n x^n$ of the initial value problem.

C **1.** $(1 + 3x)y'' + xy' + 2y = 0, \quad y(0) = 2, \quad y'(0) = -3$

C **2.** $(1 + x + 2x^2)y'' + (2 + 8x)y' + 4y = 0, \quad y(0) = -1, \quad y'(0) = 2$

C **3.** $(1 - 2x^2)y'' + (2 - 6x)y' - 2y = 0, \quad y(0) = 1, \quad y'(0) = 0$

C **4.** $(1 + x + 3x^2)y'' + (2 + 15x)y' + 12y = 0, \quad y(0) = 0, \quad y'(0) = 1$

C 5. $(2 + x)y'' + (1 + x)y' + 3y = 0$, $y(0) = 4$, $y'(0) = 3$

C 6. $(3 + 3x + x^2)y'' + (6 + 4x)y' + 2y = 0$, $y(0) = 7$, $y'(0) = 3$

C 7. $(4 + x)y'' + (2 + x)y' + 2y = 0$, $y(0) = 2$, $y'(0) = 5$

C 8. $(2 - 3x + 2x^2)y'' - (4 - 6x)y' + 2y = 0$, $y(1) = 1$, $y'(1) = -1$

C 9. $(3x + 2x^2)y'' + 10(1 + x)y' + 8y = 0$, $y(-1) = 1$, $y'(-1) = -1$

C 10. $(1 - x + x^2)y'' - (1 - 4x)y' + 2y = 0$, $y(1) = 2$, $y'(1) = -1$

C 11. $(2 + x)y'' + (2 + x)y' + y = 0$, $y(-1) = -2$, $y'(-1) = 3$

C 12. $x^2y'' - (6 - 7x)y' + 8y = 0$, $y(1) = 1$, $y'(1) = -2$

L 13. Do the following experiment for various choices of real numbers a_0, a_1, and r, with $0 < r < 1/\sqrt{2}$.

   (a) Use differential equations software to solve the initial value problem

   $$(1 + x + 2x^2)y'' + (1 + 7x)y' + 2y = 0, \qquad y(0) = a_0, \qquad y'(0) = a_1 \tag{A}$$

   numerically on $(-r, r)$. (See Example 1.)

   (b) For $N = 2, 3, 4, \ldots$, compute $a_2, \ldots, a_N$ in the power series solution $y = \sum_{n=0}^{\infty} a_n x^n$ of (A), and graph

   $$T_N(x) = \sum_{n=0}^{N} a_n x^n$$

   and the solution obtained in **(a)** on $(-r, r)$. Continue increasing N until there is no perceptible difference between the two graphs.

L 14. Do the following experiment for various choices of real numbers a_0, a_1, and r, with $0 < r < 2$.

   (a) Use differential equations software to solve the initial value problem

   $$(3 + x)y'' + (1 + 2x)y' - (2 - x)y = 0, \qquad y(-1) = a_0, \qquad y'(-1) = a_1, \tag{A}$$

   numerically on $(-1 - r, -1 + r)$. (See Example 2. Why this interval?)

   (b) For $N = 2, 3, 4, \ldots$, compute $a_2, \ldots, a_N$ in the power series solution

   $$y = \sum_{n=0}^{\infty} a_n(x + 1)^n$$

   of (A), and graph

   $$T_N(x) = \sum_{n=0}^{N} a_n(x + 1)^n$$

   and the solution obtained in **(a)** on $(-1 - r, -1 + r)$. Continue increasing N until there is no perceptible difference between the two graphs.

L 15. Do the following experiment for several choices of a_0, a_1, and r, with $r > 0$.

   (a) Use differential equations software to solve the initial value problem

   $$y'' + 3xy' + (4 + 2x^2)y = 0, \qquad y(0) = a_0, \qquad y'(0) = a_1 \tag{A}$$

   numerically on $(-r, r)$. (See Example 3.)

   (b) Find the coefficients $a_0, a_1, \ldots, a_N$ in the power series solution $y = \sum_{n=0}^{\infty} a_n x^n$ of (A), and graph

   $$T_N(x) = \sum_{n=0}^{N} a_n x^n$$

   and the solution obtained in **(a)** on $(-r, r)$. Continue increasing N until there is no perceptible difference between the two graphs.

L 16. Do the following experiment for several choices of a_0, a_1, and r, with $0 < r < 1$.

(a) Use differential equations software to solve the initial value problem

$$(1 - x)y'' - (2 - x)y' + y = 0, \qquad y(0) = a_0, \qquad y'(0) = a_1 \qquad \text{(A)}$$

numerically on $(-r, r)$.

(b) Find the coefficients $a_0, a_1, \ldots, a_N$ in the power series solution $y = \sum_{n=0}^{N} a_n x^n$ of (A), and graph

$$T_N(x) = \sum_{n=0}^{N} a_n x^n$$

and the solution obtained in **(a)** on $(-r, r)$. Continue increasing N until there is no perceptible difference between the two graphs. What happens as you let $r \to 1$?

L **17.** Follow the directions of Exercise 16 for the initial value problem

$$(1 + x)y'' + 3y' + 32y = 0, \qquad y(0) = a_0, \qquad y'(0) = a_1.$$

L **18.** Follow the directions of Exercise 16 for the initial value problem

$$(1 + x^2)y'' + y' + 2y = 0, \qquad y(0) = a_0, \qquad y'(0) = a_1.$$

In Exercises 19–28 find the coefficients $a_0, \ldots, a_N$ for N at least 5 in the series solution

$$y = \sum_{n=0}^{\infty} a_n(x - x_0)^n$$

of the initial value problem. Take x_0 to be the point at which the initial conditions are imposed.

C **19.** $(2 + 4x)y'' - 4y' - (6 + 4x)y = 0, \quad y(0) = 2, \quad y'(0) = -7$

C **20.** $(1 + 2x)y'' - (1 - 2x)y' - (3 - 2x)y = 0, \quad y(1) = 1, \quad y'(1) = -2$

C **21.** $(5 + 2x)y'' - y' + (5 + x)y = 0, \quad y(-2) = 2, \quad y'(-2) = -1$

C **22.** $(4 + x)y'' - (4 + 2x)y' + (6 + x)y = 0, \quad y(-3) = 2, \quad y'(-3) = -2$

C **23.** $(2 + 3x)y'' - xy' + 2xy = 0, \quad y(0) = -1, \quad y'(0) = 2$

C **24.** $(3 + 2x)y'' + 3y' - xy = 0, \quad y(-1) = 2, \quad y'(-1) = -3$

C **25.** $(3 + 2x)y'' - 3y' - (2 + x)y = 0, \quad y(-2) = -2, \quad y'(-2) = 3$

C **26.** $(10 - 2x)y'' + (1 + x)y = 0, \quad y(2) = 2, \quad y'(2) = -4$

C **27.** $(7 + x)y'' + (8 + 2x)y' + (5 + x)y = 0, \quad y(-4) = 1, \quad y'(-4) = 2$

C **28.** $(6 + 4x)y'' + (1 + 2x)y = 0, \quad y(-1) = -1, \quad y'(-1) = 2$

29. Show that the coefficients in the power series in x for the general solution of

$$(1 + \alpha x + \beta x^2)y'' + (\gamma + \delta x)y' + \epsilon y = 0$$

satisfy the recurrrence relation

$$a_{n+2} = -\frac{\gamma + \alpha n}{n + 2} a_{n+1} - \frac{\beta n(n - 1) + \delta n + \epsilon}{(n + 2)(n + 1)} a_n.$$

30. (a) Let α and β be constants, with $\beta \neq 0$. Show that $y = \sum_{n=0}^{\infty} a_n x^n$ is a solution of

$$(1 + \alpha x + \beta x^2)y'' + (2\alpha + 4\beta x)y' + 2\beta y = 0 \qquad \text{(A)}$$

if and only if

$$a_{n+2} + \alpha a_{n+1} + \beta a_n = 0, \qquad n \geq 0. \qquad \text{(B)}$$

An equation of this form is called a **second order homogeneous linear difference equation**. The polynomial $p(r) = r^2 + \alpha r + \beta$ is called the **characteristic polynomial** of (B). Notice that if r_1 and r_2 are the zeros of p, then $1/r_1$ and $1/r_2$ are the zeros of

$$P_0(x) = 1 + \alpha x + \beta x^2.$$

(b) Suppose that $p(r) = (r - r_1)(r - r_2)$ where r_1 and r_2 are real and distinct, and let ρ be the smaller of the two numbers $\{1/|r_1|, 1/|r_2|\}$. Show that if c_1 and c_2 are constants then the sequence

$$a_n = c_1 r_1^n + c_2 r_2^n, \qquad n \geq 0,$$

satisfies (B). Conclude from this that any function of the form

$$y = \sum_{n=0}^{\infty} (c_1 r_1^n + c_2 r_2^n) x^n$$

is a solution of (A) on $(-\rho, \rho)$.

(c) Use **(b)** and the formula for the sum of a geometric series to show that the functions

$$y_1 = \frac{1}{1 - r_1 x} \qquad \text{and} \qquad y_2 = \frac{1}{1 - r_2 x}$$

form a fundamental set of solutions of (A) on $(-\rho, \rho)$.

(d) Show that $\{y_1, y_2\}$ is a fundamental set of solutions of (A) on any interval that does not contain either $1/r_1$ or $1/r_2$.

(e) Suppose that $p(r) = (r - r_1)^2$, and let $\rho = 1/|r_1|$. Show that if c_1 and c_2 are constants then the sequence

$$a_n = (c_1 + c_2 n) r_1^n, \qquad n \geq 0,$$

satisfies (B). Conclude from this that any function of the form

$$y = \sum_{n=0}^{\infty} (c_1 + c_2 n) r_1^n x^n$$

is a solution of (A) on $(-\rho, \rho)$.

(f) Use **(e)** and the formula for the sum of a geometric series to show that the functions

$$y_1 = \frac{1}{1 - r_1 x} \qquad \text{and} \qquad y_2 = \frac{x}{(1 - r_1 x)^2}$$

form a fundamental set of solutions of (A) on $(-\rho, \rho)$.

(g) Show that $\{y_1, y_2\}$ is a fundamental set of solutions of (A) on any interval that does not contain $1/r_1$.

31. Use the results of Exercise 30 to find the general solution of the given equation on any interval on which the polynomial multiplying y'' has no zeros.

(a) $(1 + 3x + 2x^2)y'' + (6 + 8x)y' + 4y = 0$

(b) $(1 - 5x + 6x^2)y'' - (10 - 24x)y' + 12y = 0$

(c) $(1 - 4x + 4x^2)y'' - (8 - 16x)y' + 8y = 0$

(d) $(4 + 4x + x^2)y'' + (8 + 4x)y' + 2y = 0$

(e) $(4 + 8x + 3x^2)y'' + (16 + 12x)y' + 6y = 0$

In Exercises 32–38 find the coefficients $a_0, \ldots, a_N$ for N at least 5 in the series solution $y = \sum_{n=0}^{\infty} a_n x^n$ of the initial value problem.

C **32.** $y'' + 2xy' + (3 + 2x^2)y = 0, \quad y(0) = 1, \quad y'(0) = -2$

C **33.** $y'' - 3xy' + (5 + 2x^2)y = 0, \quad y(0) = 1, \quad y'(0) = -2$

C **34.** $y'' + 5xy' - (3 - x^2)y = 0, \quad y(0) = 6, \quad y'(0) = -2$

C **35.** $y'' - 2xy' - (2 + 3x^2)y = 0, \quad y(0) = 2, \quad y'(0) = -5$

C **36.** $y'' - 3xy' + (2 + 4x^2)y = 0, \quad y(0) = 3, \quad y'(0) = 6$

C **37.** $2y'' + 5xy' + (4 + 2x^2)y = 0, \quad y(0) = 3, \quad y'(0) = -2$

C **38.** $3y'' + 2xy' + (4 - x^2)y = 0, \quad y(0) = -2, \quad y'(0) = 3$

C **39.** Find power series in x for the solutions y_1 and y_2 of

$$y'' + 4xy' + (2 + 4x^2)y = 0$$

such that $y_1(0) = 1$, $y_1'(0) = 0$, $y_2(0) = 0$, $y_2'(0) = 1$, and identify y_1 and y_2 in terms of familiar elementary functions.

In Exercises 40–49 find the coefficients $a_0, \dots, a_N$ for N at least 5 in the series solution

$$y = \sum_{n=0}^{\infty} a_n(x - x_0)^n$$

of the initial value problem. Take x_0 to be the point at which the initial conditions are imposed.

C **40.** $(1 + x)y'' + x^2y' + (1 + 2x)y = 0, \quad y(0) = -2, \quad y'(0) = 3$

C **41.** $y'' + (1 + 2x + x^2)y' + 2y = 0, \quad y(0) = 2, \quad y'(0) = 3$

C **42.** $(1 + x^2)y'' + (2 + x^2)y' + xy = 0, \quad y(0) = -3, \quad y'(0) = 5$

C **43.** $(1 + x)y'' + (1 - 3x + 2x^2)y' - (x - 4)y = 0, \quad y(1) = -2, \quad y'(1) = 3$

C **44.** $y'' + (13 + 12x + 3x^2)y' + (5 + 2x)y, \quad y(-2) = 2, \quad y'(-2) = -3$

C **45.** $(1 + 2x + 3x^2)y'' + (2 - x^2)y' + (1 + x)y = 0, \quad y(0) = 1, \quad y'(0) = -2$

C **46.** $(3 + 4x + x^2)y'' - (5 + 4x - x^2)y' - (2 + x)y = 0, \quad y(-2) = 2, \quad y'(-2) = -1$

C **47.** $(1 + 2x + x^2)y'' + (1 - x)y = 0, \quad y(0) = 2, \quad y'(0) = -1$

C **48.** $(x - 2x^2)y'' + (1 + 3x - x^2)y' + (2 + x)y = 0, \quad y(1) = 1, \quad y'(1) = 0$

C **49.** $(16 - 11x + 2x^2)y'' + (10 - 6x + x^2)y' - (2 - x)y, \quad y(3) = 1, \quad y'(3) = -2$

7.4   Regular Singular Points; Euler Equations

This section sets the stage for Sections 7.5, 7.6, and 7.7. If you're not interested in those sections, but wish to learn about Euler equations, then omit the introductory paragraphs and start reading at Definition 7.4.2.

In the next three sections we will continue to study equations of the form

$$P_0(x)y'' + P_1(x)y' + P_2(x)y = 0 \tag{1}$$

where P_0, P_1, and P_2 are polynomials, but the emphasis will be different from that of Sections 7.2 and 7.3, where we obtained solutions of (1) near an ordinary point x_0 in the form of power series in $x - x_0$. If x_0 is a singular point of (1) (that is, if $P(x_0) = 0$), then the solutions cannot in general be represented by power series in $x - x_0$. Nevertheless, it is often necessary in physical applications to study the behavior of solutions of (1) near a singular point. Although this can be difficult in the absence of some sort of assumption on the nature of the singular point, equations that satisfy the requirements of the following definition can be solved by series methods discussed in the next three sections. Fortunately, many equations arising in applications satisfy these requirements.

DEFINITION 7.4.1   Let P_0, P_1, and P_2 be polynomials with no common factor and suppose that $P_0(x_0) = 0$. Then x_0 is a ***regular singular point*** of the equation

$$P_0(x)y'' + P_1(x)y' + P_2(x)y = 0 \tag{2}$$

if (2) can be written as

$$(x - x_0)^2 A(x)y'' + (x - x_0)B(x)y' + C(x)y = 0 \qquad (3)$$

where A, B, and C are polynomials and $A(x_0) \neq 0$; otherwise, x_0 is an *irregular singular point* of (2).

EXAMPLE 1   Bessel's equation,

$$x^2 y'' + xy' + (x^2 - \nu^2)y = 0, \qquad (4)$$

has the singular point $x_0 = 0$. Since this equation is of the form (3) with $x_0 = 0$, $A(x) = 1$, $B(x) = 1$, and $C(x) = x^2 - \nu^2$, it follows that $x_0 = 0$ is a regular singular point of (4).

EXAMPLE 2   Legendre's equation,

$$(1 - x^2)y'' - 2xy' + \alpha(\alpha + 1)y = 0, \qquad (5)$$

has the singular points $x_0 = \pm 1$. Mutiplying through by $1 - x$ yields

$$(x - 1)^2(x + 1)y'' + 2x(x - 1)y' - \alpha(\alpha + 1)(x - 1)y = 0,$$

which is of the form (3) with $x_0 = 1$, $A(x) = x + 1$, $B(x) = 2x$, and $C(x) = -\alpha(\alpha + 1)(x - 1)$. Therefore $x_0 = 1$ is a regular singular point of (5). We leave it to you to show that $x_0 = -1$ is also a regular singular point of (5).

EXAMPLE 3   The equation

$$x^3 y'' + xy' + y = 0$$

has an irregular singular point at $x_0 = 0$. (Verify.)   ∎

For convenience we restrict our attention to the case where $x_0 = 0$ is a regular singular point of (2). This is not really a restriction, since if $x_0 \neq 0$ is a regular singular point of (2), then introducing the new independent variable $t = x - x_0$ and the new unknown $Y(t) = y(t + x_0)$ leads to a differential equation with polynomial coefficients that has a regular singular point at $t_0 = 0$. This is illustrated in Exercise 22 for Legendre's equation, and in Exercise 23 for the general case.

EULER EQUATIONS

The simplest kind of equation with a regular singular point at $x_0 = 0$ is the Euler equation, defined as follows.

DEFINITION 7.4.2

An *Euler equation* is an equation that can be written in the form

$$ax^2 y'' + bxy' + cy = 0, \qquad (6)$$

where a, b, and c are real constants and $a \neq 0$.

Theorem 5.1.1 implies that (6) has solutions defined on $(0, \infty)$ and $(-\infty, 0)$, since (6) can be rewritten as

$$ay'' + \frac{b}{x}y' + \frac{c}{x^2}y = 0.$$

For convenience we'll restrict our attention to the interval $(0, \infty)$. (Exercise 19 deals with solutions of (6) on $(-\infty, 0)$.) The key to finding solutions on $(0, \infty)$ is to notice that if $x > 0$ then x^r is defined as a real-valued function on $(0, \infty)$ for all values of r, and substituting $y = x^r$ into (6) produces

$$ax^2(x^r)'' + bx(x^r)' + cx^r = ax^2r(r-1)x^{r-2} + bxrx^{r-1} + cx^r$$
$$= [ar(r-1) + br + c]x^r. \tag{7}$$

The polynomial

$$p(r) = ar(r-1) + br + c$$

is called the **indicial polynomial** of (6), and $p(r) = 0$ is the **indicial equation.** From (7) we can see that $y = x^r$ is a solution of (6) on $(0, \infty)$ if and only if $p(r) = 0$. Therefore, if the indicial equation has distinct real roots r_1 and r_2, then $y_1 = x^{r_1}$ and $y_2 = x^{r_2}$ form a fundamental set of solutions of (6) on $(0, \infty)$, since $y_2/y_1 = x^{r_2-r_1}$ is nonconstant. In this case

$$y = c_1x^{r_1} + c_2x^{r_2}$$

is the general solution of (6) on $(0, \infty)$.

EXAMPLE 4   Find the general solution of

$$x^2y'' - xy' - 8y = 0 \tag{8}$$

on $(0, \infty)$.

Solution   The indicial polynomial of (8) is

$$p(r) = r(r-1) - r - 8 = (r-4)(r+2).$$

Therefore $y_1 = x^4$ and $y_2 = x^{-2}$ are solutions of (8) on $(0, \infty)$, and the general solution on $(0, \infty)$ is

$$y = c_1x^4 + \frac{c_2}{x^2}.$$

EXAMPLE 5   Find the general solution of

$$6x^2y'' + 5xy' - y = 0 \tag{9}$$

on $(0, \infty)$.

Solution   The indicial polynomial of (9) is

$$p(r) = 6r(r-1) + 5r - 1 = (2r-1)(3r+1).$$

Therefore the general solution of (9) on $(0, \infty)$ is

$$y = c_1x^{1/2} + c_2x^{-1/3}. \qquad \blacksquare$$

If the indicial equation has a repeated root r_1, then $y_1 = x^{r_1}$ is a solution of

$$ax^2y'' + bxy' + cy = 0, \tag{10}$$

on $(0, \infty)$, but (10) has no other solution of the form $y = x^r$. If the indicial equation has complex conjugate zeros then (10) has no real-valued solutions of the form $y = x^r$. Fortunately, we can use the results of Section 5.2 for constant coefficient equations to solve (10) in any case.

THEOREM 7.4.3

Suppose that the roots of the indicial equation

$$ar(r - 1) + br + c = 0 \tag{11}$$

are r_1 and r_2. Then the general solution of the Euler equation

$$ax^2y'' + bxy' + cy = 0 \tag{12}$$

on $(0, \infty)$ is

$y = c_1x^{r_1} + c_2x^{r_2}$ *if r_1 and r_2 are distinct real numbers;*

$y = x^{r_1}(c_1 + c_2 \ln x)$ *if $r_1 = r_2$;*

$y = x^\lambda[c_1 \cos(\omega \ln x) + c_2 \sin(\omega \ln x)]$ *if $r_1, r_2 = \lambda \pm i\omega$ with $\omega > 0$.*

PROOF. We first show that $y = y(x)$ satisfies (12) on $(0, \infty)$ if and only if $Y(t) = y(e^t)$ satisfies the constant coefficient equation

$$a\frac{d^2Y}{dt^2} + (b - a)\frac{dY}{dt} + cY = 0 \tag{13}$$

on $(-\infty, \infty)$. To do this it is convenient to write $x = e^t$, or, equivalently, $t = \ln x$; thus, $Y(t) = y(x)$, where $x = e^t$. From the chain rule,

$$\frac{dY}{dt} = \frac{dy}{dx}\frac{dx}{dt}$$

and, since

$$\frac{dx}{dt} = e^t = x,$$

it follows that

$$\frac{dY}{dt} = x\frac{dy}{dx}. \tag{14}$$

Differentiating this with respect to t and using the chain rule again yields

$$\begin{aligned}
\frac{d^2Y}{dt^2} &= \frac{d}{dt}\left(\frac{dY}{dt}\right) = \frac{d}{dt}\left(x\frac{dy}{dx}\right) \\
&= \frac{dx}{dt}\frac{dy}{dx} + x\frac{d^2y}{dx^2}\frac{dx}{dt} \\
&= x\frac{dy}{dx} + x^2\frac{d^2y}{dx^2} \quad \left(\text{since } \frac{dx}{dt} = x\right).
\end{aligned}$$

From this and (14),

$$x^2\frac{d^2y}{dx^2} = \frac{d^2Y}{dt^2} - \frac{dY}{dt}.$$

Substituting this and (14) into (12) yields (13). Since (11) is the characteristic equation of (13), Theorem 5.2.1 implies that the general solution of (13) on $(-\infty, \infty)$ is

$$Y(t) = c_1 e^{r_1 t} + c_2 e^{r_2 t} \quad \text{if } r_1 \text{ and } r_2 \text{ are distinct real numbers;}$$
$$Y(t) = e^{r_1 t}(c_1 + c_2 t) \quad \text{if } r_1 = r_2;$$
$$Y(t) = e^{\lambda t}(c_1 \cos \omega t + c_2 \sin \omega t) \quad \text{if } r_1, r_2 = \lambda \pm i\omega \quad \text{with } \omega \neq 0.$$

Since $Y(t) = y(e^t)$, substituting $t = \ln x$ in the last three equations shows that the general solution of (12) on $(0, \infty)$ has the form stated in the theorem. $\square$

EXAMPLE 6   Find the general solution of

$$x^2 y'' - 5xy' + 9y = 0 \tag{15}$$

on $(0, \infty)$.

Solution   The indicial polynomial of (15) is

$$p(r) = r(r - 1) - 5r + 9 = (r - 3)^2.$$

Therefore the general solution of (15) on $(0, \infty)$ is

$$y = x^3(c_1 + c_2 \ln x).$$

EXAMPLE 7   Find the general solution of

$$x^2 y'' + 3xy' + 2y = 0 \tag{16}$$

on $(0, \infty)$.

Solution   The indicial polynomial of (16) is

$$p(r) = r(r - 1) + 3r + 2 = (r + 1)^2 + 1.$$

The roots of the indicial equation are $r = -1 \pm i$, and the general solution of (16) on $(0, \infty)$ is

$$y = \frac{1}{x}[c_1 \cos(\ln x) + c_2 \sin(\ln x)]. \quad \blacksquare$$

7.4 EXERCISES

In Exercises 1–18 find the general solution of the given Euler equation on $(0, \infty)$.

1. $x^2 y'' + 7xy' + 8y = 0$ **2.** $x^2 y'' - 7xy' + 7y = 0$

3. $x^2 y'' - xy' + y = 0$ **4.** $x^2 y'' + 5xy' + 4y = 0$

5. $x^2 y'' + xy' + y = 0$ **6.** $x^2 y'' - 3xy' + 13y = 0$

7. $x^2 y'' + 3xy' - 3y = 0$ **8.** $12x^2 y'' - 5xy'' + 6y = 0$

9. $4x^2 y'' + 8xy' + y = 0$ **10.** $3x^2 y'' - xy' + y = 0$

11. $2x^2 y'' - 3xy' + 2y = 0$ **12.** $x^2 y'' + 3xy' + 5y = 0$

13. $9x^2 y'' + 15xy' + y = 0$ **14.** $x^2 y'' - xy' + 10y = 0$

15. $x^2 y'' - 6y = 0$ **16.** $2x^2 y'' + 3xy' - y = 0$

17. $x^2 y'' - 3xy' + 4y = 0$ **18.** $2x^2 y'' + 10xy' + 9y = 0$

19. **(a)** Adapt the proof of Theorem 7.4.3 to show that $y = y(x)$ satisfies the Euler equation

$$ax^2 y'' + bxy' + cy = 0 \tag{A}$$

on $(-\infty, 0)$ if and only if $Y(t) = y(-e^t)$ satisfies

$$a\frac{d^2Y}{dt^2} + (b-a)\frac{dy}{dt} + cY = 0$$

on $(-\infty, \infty)$.

(b) Use **(a)** to show that the general solution of (A) on $(-\infty, 0)$ is

$$y = c_1|x|^{r_1} + c_2|x|^{r_2} \quad \text{if } r_1 \text{ and } r_2 \text{ are distinct real numbers;}$$
$$y = |x|^{r_1}(c_1 + c_2 \ln |x|) \quad \text{if } r_1 = r_2;$$
$$y = |x|^{\lambda}[c_1 \cos(\omega \ln |x|) + c_2 \sin(\omega \ln |x|)] \quad \text{if } r_1, r_2 = \lambda \pm i\omega \quad \text{with } \omega > 0.$$

20. Use reduction of order to show that if

$$ar(r-1) + br + c = 0$$

has a repeated root r_1 then $y = x^{r_1}(c_1 + c_2 \ln x)$ is the general solution of

$$ax^2y'' + bxy' + cy = 0$$

on $(0, \infty)$.

21. A nontrivial solution of

$$P_0(x)y'' + P_1(x)y' + P_2(x)y = 0$$

is said to be **oscillatory** on an interval (a,b) if it has infinitely many zeros on (a,b). Otherwise y is said to be **nonoscillatory** on (a,b). Show that the equation

$$x^2y'' + ky = 0 \qquad (k = \text{constant})$$

has oscillatory solutions on $(0, \infty)$ if and only if $k > 1/4$.

22. We saw in Example 2 that $x_0 = 1$ and $x_0 = -1$ are regular singular points of Legendre's equation

$$(1 - x^2)y'' - 2xy' + \alpha(\alpha + 1)y = 0. \tag{A}$$

(a) Introduce the new variables $t = x - 1$ and $Y(t) = y(t + 1)$, and show that y is a solution of (A) if and only if Y is a solution of

$$t(2 + t)\frac{d^2Y}{dt^2} + 2(1 + t)\frac{dY}{dt} - \alpha(\alpha + 1)Y = 0,$$

which has a regular singular point at $t_0 = 0$.

(b) Introduce the new variables $t = x + 1$ and $Y(t) = y(t - 1)$, and show that y is a solution of (A) if and only if Y is a solution of

$$t(2 - t)\frac{d^2Y}{dt^2} + 2(1 - t)\frac{dY}{dt} + \alpha(\alpha + 1)Y = 0,$$

which has a regular singular point at $t_0 = 0$.

23. Let $P_0, P_1,$ and P_2 be polynomials with no common factor, and suppose that $x_0 \neq 0$ is a singular point of

$$P_0(x)y'' + P_1(x)y' + P_2(x)y = 0. \tag{A}$$

Let $t = x - x_0$ and $Y(t) = y(t + x_0)$.

(a) Show that y is a solution of (A) if and only if Y is a solution of

$$R_0(t)\frac{d^2Y}{dt^2} + R_1(t)\frac{dY}{dt} + R_2(t)Y = 0 \tag{B}$$

where

$$R_i(t) = P_i(t + x_0), \qquad i = 0, 1, 2.$$

(b) Show that $R_0, R_1,$ and R_2 are polynomials in t with no common factors, and $R_0(0) = 0$; thus, $t_0 = 0$ is a singular point of (B).

7.5 The Method of Frobenius I

In this section we begin to study series solutions of a homogeneous linear second order differential equation with a regular singular point at $x_0 = 0$, so it can be written as

$$x^2 A(x)y'' + xB(x)y' + C(x)y = 0, \qquad (1)$$

where A, B, and C are polynomials and $A(0) \neq 0$.

We'll see that (1) always has at least one solution of the form

$$y = x^r \sum_{n=0}^{\infty} a_n x^n$$

where $a_0 \neq 0$ and r is a suitably chosen number. The method we will use to find solutions of this form and other forms that we will encounter in the next two sections is called the ***method of Frobenius,***[1] and we will call them ***Frobenius solutions.***

It can be shown that the power series $\sum_{n=0}^{\infty} a_n x^n$ in a Frobenius solution of (1) converges on some open interval $(-\rho, \rho)$, where $0 < \rho \leq \infty$. However, since x^r may be complex for negative x or undefined if $x = 0$, we will consider solutions defined for positive values of x. Easy modifications of our results yield solutions defined for negative values of x (Exercise 54).

We will restrict our attention to the case where A, B, and C are polynomials of degree not greater than two, so (1) becomes

$$x^2(\alpha_0 + \alpha_1 x + \alpha_2 x^2)y'' + x(\beta_0 + \beta_1 x + \beta_2 x^2)y' + (\gamma_0 + \gamma_1 x + \gamma_2 x^2)y = 0 \quad (2)$$

where α_i, β_i, and γ_i are real constants and $\alpha_0 \neq 0$. Most equations that arise in applications can be written this way. Some examples are

$$\alpha x^2 y'' + \beta x y' + \gamma y = 0 \qquad \text{(Euler's equation),}$$
$$x^2 y'' + xy' + (x^2 - \nu^2)y = 0 \qquad \text{(Bessel's equation),}$$

and

$$xy'' + (1 - x)y' + \lambda y = 0 \qquad \text{(Laguerre's}[2] \text{ equation),}$$

where we would multiply the last equation through by x to put it in the form (2). However, the method of Frobenius can be extended to the case where A, B, and C are functions that can be represented by power series in x on some interval containing zero, and $A_0(0) \neq 0$ (Exercises 57 and 58).

The next two theorems will enable us to develop systematic methods for finding Frobenius solutions of (2).

---

[1]Ferdinand Georg Frobenius (1849–1917) was a professor at the University of Berlin. He developed his method of series solution of differential equations in his doctoral thesis, written under the direction of Karl Weierstrass, one of the founders of the theory of functions of a complex variable. However, Frobenius is best known for his outstanding contributions to group theory.

[2]Edmond Nicholas Laguerre (1834–1886) was a French geometer and analyst who made extensive contributions to the theory of infinite series.

THEOREM 7.5.1

Let

$$Ly = x^2(\alpha_0 + \alpha_1 x + \alpha_2 x^2)y'' + x(\beta_0 + \beta_1 x + \beta_2 x^2)y'$$
$$+ (\gamma_0 + \gamma_1 x + \gamma_2 x^2)y$$

and define

$$p_0(r) = \alpha_0 r(r-1) + \beta_0 r + \gamma_0,$$
$$p_1(r) = \alpha_1 r(r-1) + \beta_1 r + \gamma_1,$$
$$p_2(r) = \alpha_2 r(r-1) + \beta_2 r + \gamma_2.$$

Suppose that the series

$$y = \sum_{n=0}^{\infty} a_n x^{n+r} \tag{3}$$

converges on $(0, \rho)$. *Then*

$$Ly = \sum_{n=0}^{\infty} b_n x^{n+r} \tag{4}$$

on $(0, \rho)$, *where*

$$b_0 = p_0(r)a_0,$$
$$b_1 = p_0(r+1)a_1 + p_1(r)a_0, \tag{5}$$
$$b_n = p_0(n+r)a_n + p_1(n+r-1)a_{n-1} + p_2(n+r-2)a_{n-2}, \quad n \geq 2.$$

PROOF. We begin by showing that if y is given by (3) and α, β, and γ are constants, then

$$\alpha x^2 y'' + \beta xy' + \gamma y = \sum_{n=0}^{\infty} p(n+r)a_n x^{n+r}, \tag{6}$$

where

$$p(r) = \alpha r(r-1) + \beta r + \gamma.$$

Differentiating (3) twice yields

$$y' = \sum_{n=0}^{\infty} (n+r)a_n x^{n+r-1} \tag{7}$$

and

$$y'' = \sum_{n=0}^{\infty} (n+r)(n+r-1)a_n x^{n+r-2}. \tag{8}$$

Multiplying (7) by x and (8) by x^2 yields

$$xy' = \sum_{n=0}^{\infty} (n+r)a_n x^{n+r}$$

and

$$x^2 y'' = \sum_{n=0}^{\infty} (n+r)(n+r-1)a_n x^{n+r}.$$

Therefore,

$$\alpha x^2 y'' + \beta xy' + \gamma y = \sum_{n=0}^{\infty} [\alpha(n + r)(n + r - 1) + \beta(n + r) + \gamma]a_n x^{n+r}$$

$$= \sum_{n=0}^{\infty} p(n + r)a_n x^{n+r},$$

which proves (6).

Multiplying (6) by x yields

$$x(\alpha x^2 y'' + \beta xy' + \gamma y) = \sum_{n=0}^{\infty} p(n + r)a_n x^{n+r+1} = \sum_{n=1}^{\infty} p(n + r - 1)a_{n-1} x^{n+r}.$$

(9)

Multiplying (6) by x^2 yields

$$x^2(\alpha x^2 y'' + \beta xy' + \gamma y) = \sum_{n=0}^{\infty} p(n + r)a_n x^{n+r+2} = \sum_{n=2}^{\infty} p(n + r - 2)a_{n-2} x^{n+r}.$$

(10)

To use these results we rewrite

$$Ly = x^2(\alpha_0 + \alpha_1 x + \alpha_2 x^2)y'' + x(\beta_0 + \beta_1 x + \beta_2 x^2)y' + (\gamma_0 + \gamma_1 x + \gamma_2 x^2)y$$

as

$$Ly = (\alpha_0 x^2 y'' + \beta_0 xy' + \gamma_0 y) + x(\alpha_1 x^2 y'' + \beta_1 xy' + \gamma_1 y)$$
$$+ x^2(\alpha_2 x^2 y'' + \beta_2 xy' + \gamma_2 y).$$

(11)

From (6) with $p = p_0$,

$$\alpha_0 x^2 y'' + \beta_0 xy' + \gamma_0 y = \sum_{n=0}^{\infty} p_0(n + r)a_n x^{n+r}.$$

From (9) with $p = p_1$,

$$x(\alpha_1 x^2 y'' + \beta_1 xy' + \gamma_1 y) = \sum_{n=1}^{\infty} p_1(n + r - 1)a_{n-1} x^{n+r}.$$

From (10) with $p = p_2$,

$$x^2(\alpha_2 x^2 y'' + \beta_2 xy' + \gamma_2 y) = \sum_{n=2}^{\infty} p_2(n + r - 2)a_{n-2} x^{n+r}.$$

Therefore we can rewrite (11) as

$$Ly = \sum_{n=0}^{\infty} p_0(n + r)a_n x^{n+r} + \sum_{n=1}^{\infty} p_1(n + r - 1)a_{n-1} x^{n+r}$$

$$+ \sum_{n=2}^{\infty} p_2(n + r - 2)a_{n-2} x^{n+r},$$

or

$$Ly = p_0(r)a_0 x^r + [p_0(r + 1)a_1 + p_1(r)a_2]x^{r+1}$$

$$+ \sum_{n=1}^{\infty} [p_0(n + r)a_n + p_1(n + r - 1)a_{n-1} + p_2(n + r - 2)a_{n-2}]x^{n+r},$$

which implies (4) with $\{b_n\}$ defined as in (5). $\square$

THEOREM 7.5.2

Let

$$Ly = x^2(\alpha_0 + \alpha_1 x + \alpha_2 x^2)y'' + x(\beta_0 + \beta_1 x + \beta_2 x^2)y'$$
$$+ (\gamma_0 + \gamma_1 x + \gamma_2 x^2)y$$

where $\alpha_0 \neq 0$, and define

$$p_0(r) = \alpha_0 r(r - 1) + \beta_0 r + \gamma_0,$$
$$p_1(r) = \alpha_1 r(r - 1) + \beta_1 r + \gamma_1,$$
$$p_2(r) = \alpha_2 r(r - 1) + \beta_2 r + \gamma_2.$$

Suppose that r is a real number such that $p_0(n + r)$ is nonzero for all positive integers n. Define

$$a_0(r) = 1,$$

$$a_1(r) = -\frac{p_1(r)}{p_0(r + 1)}, \tag{12}$$

$$a_n(r) = -\frac{p_1(n + r - 1)a_{n-1}(r) + p_2(n + r - 2)a_{n-2}(r)}{p_0(n + r)}, \qquad n \geq 2.$$

Then the Frobenius series

$$y(x, r) = x^r \sum_{n=0}^{\infty} a_n(r)x^n \tag{13}$$

converges and satisfies

$$Ly(x, r) = p_0(r)x^r \tag{14}$$

on the interval $(0, \rho)$, where ρ is the distance from the origin to the nearest zero of $A(x) = \alpha_0 + \alpha_1 x + \alpha_2 x^2$ in the complex plane. (If A is constant then $\rho = \infty$.)

PARTIAL PROOF. If $\{a_n(r)\}$ is determined by the recurrence relation (12), then substituting $a_n = a_n(r)$ into (5) yields $b_0 = p_0(r)$ and $b_n = 0$ for $n \geq 1$, so (4) reduces to (14). We omit the proof that the series (13) converges on $(0, \rho)$. $\square$

If $\alpha_i = \beta_i = \gamma_i = 0$ for $i = 1, 2$, then $Ly = 0$ reduces to the Euler equation

$$\alpha_0 x^2 y'' + \beta_0 x y' + \gamma_0 y = 0.$$

Theorem 7.4.3 shows that the solutions of this equation are determined by the zeros of the indicial polynomial

$$p_0(r) = \alpha_0 r(r - 1) + \beta_0 r + \gamma_0.$$

Since (14) implies that this is also true for the solutions of $Ly = 0$, we will also say that p_0 is the **indicial polynomial** of (2), and that $p_0(r) = 0$ is the **indicial equation** of $Ly = 0$. We will consider only cases where the indicial equation has real roots r_1 and r_2, with $r_1 \geq r_2$.

THEOREM 7.5.3

Let L and $\{a_n(r)\}$ be as in Theorem 7.5.2, and suppose that the indicial equation $p_0(r) = 0$ of $Ly = 0$ has real roots r_1 and r_2, where $r_1 \geq r_2$. Then

$$y_1(x) = y(x, r_1) = x^{r_1} \sum_{n=0}^{\infty} a_n(r_1) x^n$$

is a Frobenius solution of $Ly = 0$. Moreover, if $r_1 - r_2$ is not an integer then

$$y_2(x) = y(x, r_2) = x^{r_2} \sum_{n=0}^{\infty} a_n(r_2) x^n$$

is also a Frobenius solution of $Ly = 0$, and $\{y_1, y_2\}$ is a fundamental set of solutions.

PROOF. Since r_1 and r_2 are roots of $p_0(r) = 0$, the indicial polynomial can be factored as

$$p_0(r) = \alpha_0(r - r_1)(r - r_2). \tag{15}$$

Therefore,

$$p_0(n + r_1) = n\alpha_0(n + r_1 - r_2),$$

which is nonzero if $n > 0$, since $r_1 - r_2 \geq 0$. Therefore, the assumptions of Theorem 7.5.2 hold with $r = r_1$, and (14) implies that $Ly_1 = p_0(r_1)x^{r_1} = 0$.

Now suppose that $r_1 - r_2$ is not an integer. From (15),

$$p_0(n + r_2) = n\alpha_0(n - r_1 + r_2) \neq 0 \qquad \text{if} \qquad n = 1, 2, \ldots .$$

Hence, the assumptions of Theorem 7.5.2 hold with $r = r_2$, and (14) implies that $Ly_2 = p_0(r_2)x^{r_2} = 0$. We leave the proof that $\{y_1, y_2\}$ is a fundamental set of solutions as an exercise (Exercise 52). □

It is not always possible to obtain explicit formulas for the coefficients in Frobenius solutions. However, we can always set up the recurrence relations and use them to compute as many coefficients as we want. The following example illustrates this.

EXAMPLE 1   Find a fundamental set of Frobenius solutions of

$$2x^2(1 + x + x^2)y'' + x(9 + 11x + 11x^2)y' + (6 + 10x + 7x^2)y = 0. \tag{16}$$

Compute just the first six coefficients $a_0, \ldots, a_5$ in each solution.

Solution   For the given equation the polynomials defined in Theorem 7.5.2 are

$$\begin{aligned} p_0(r) &= 2r(r - 1) + 9r + 6 &&= (2r + 3)(r + 2), \\ p_1(r) &= 2r(r - 1) + 11r + 10 &&= (2r + 5)(r + 2), \\ p_2(r) &= 2r(r - 1) + 11r + 7 &&= (2r + 7)(r + 1). \end{aligned}$$

The zeros of the indicial polynomial p_0 are $r_1 = -3/2$ and $r_2 = -2$, so $r_1 - r_2 = 1/2$. Therefore Theorem 7.5.3 implies that

$$y_1 = x^{-3/2} \sum_{n=0}^{\infty} a_n(-3/2) x^n \qquad \text{and} \qquad y_2 = x^{-2} \sum_{n=0}^{\infty} a_n(-2) x^n \tag{17}$$

form a fundamental set of Frobenius solutions of (16). To find the coefficients in these series we use the recurrence relation of Theorem 7.5.2; thus,

$$a_0(r) = 1,$$

$$a_1(r) = -\frac{p_1(r)}{p_0(r+1)} = -\frac{(2r+5)(r+2)}{(2r+5)(r+3)} = -\frac{r+2}{r+3},$$

$$a_n(r) = -\frac{p_1(n+r-1)a_{n-1} + p_2(n+r-2)a_{n-2}}{p_0(n+r)}$$

$$= -\frac{(n+r+1)(2n+2r+3)a_{n-1}(r) + (n+r-1)(2n+2r+3)a_{n-2}(r)}{(n+r+2)(2n+2r+3)}$$

$$= -\frac{(n+r+1)a_{n-1}(r) + (n+r-1)a_{n-2}(r)}{n+r+2}, \qquad n \geq 2.$$

Setting $r = -3/2$ in these equations yields

$$a_0(-3/2) = 1,$$
$$a_1(-3/2) = -1/3, \tag{18}$$
$$a_n(-3/2) = -\frac{(2n-1)a_{n-1}(-3/2) + (2n-5)a_{n-2}(-3/2)}{2n+1}, \qquad n \geq 2,$$

and setting $r = -2$ yields

$$a_0(-2) = 1,$$
$$a_1(-2) = 0, \tag{19}$$
$$a_n(-2) = -\frac{(n-1)a_{n-1}(-2) + (n-3)a_{n-2}(-2)}{n}, \qquad n \geq 2.$$

Calculating with (18) and (19) and substituting the results into (17) yields the fundamental set of Frobenius solutions

$$y_1 = x^{-3/2}\left(1 - \frac{1}{3}x + \frac{2}{5}x^2 - \frac{5}{21}x^3 + \frac{7}{135}x^4 + \frac{76}{1155}x^5 + \cdots\right),$$

$$y_2 = x^{-2}\left(1 + \frac{1}{2}x^2 - \frac{1}{3}x^3 + \frac{1}{8}x^4 + \frac{1}{30}x^5 + \cdots\right). \qquad \blacksquare$$

SPECIAL CASES WITH TWO TERM RECURRENCE RELATIONS

For $n \geq 2$ the recurrence relation (12) of Theorem 7.5.2 involves the three coefficients $a_n(r)$, $a_{n-1}(r)$, and $a_{n-2}(r)$. We will now consider some special cases in which (12) reduces to a two term recurrence relation; that is, a relation involving only $a_n(r)$ and $a_{n-1}(r)$ or only $a_n(r)$ and $a_{n-2}(r)$. This simplification often makes it possible to obtain explicit formulas for the coefficents of Frobenius solutions.

We first consider equations of the form

$$x^2(\alpha_0 + \alpha_1 x)y'' + x(\beta_0 + \beta_1 x)y' + (\gamma_0 + \gamma_1 x)y = 0$$

with $\alpha_0 \neq 0$. For this equation we have $\alpha_2 = \beta_2 = \gamma_2 = 0$, so $p_2 \equiv 0$ and the recurrence relations in Theorem 7.5.2 simplify to

$$a_0(r) = 1,$$
$$a_n(r) = -\frac{p_1(n+r-1)}{p_0(n+r)}a_{n-1}(r), \qquad n \geq 1. \tag{20}$$

EXAMPLE 2   Find a fundamental set of Frobenius solutions of

$$x^2(3 + x)y'' + 5x(1 + x)y' - (1 - 4x)y = 0. \tag{21}$$

Give explicit formulas for the coefficients in the solutions.

Solution   For the given equation the polynomials defined in Theorem 7.5.2 are

$$p_0(r) = 3r(r - 1) + 5r - 1 = (3r - 1)(r + 1),$$
$$p_1(r) = r(r - 1) + 5r + 4 = (r + 2)^2,$$
$$p_2(r) = \qquad 0$$

The zeros of the indicial polynomial p_0 are $r_1 = 1/3$ and $r_2 = -1$, so $r_1 - r_2 = 4/3$. Therefore Theorem 7.5.3 implies that

$$y_1 = x^{1/3} \sum_{n=0}^{\infty} a_n(1/3)x^n \quad \text{and} \quad y_2 = x^{-1} \sum_{n=0}^{\infty} a_n(-1)x^n$$

form a fundamental set of Frobenius solutions of (21). To find the coefficients in these series we use the recurrence relations (20); thus,

$$a_0(r) = 1,$$
$$a_n(r) = -\frac{p_1(n + r - 1)}{p_0(n + r)} a_{n-1}(r)$$
$$= -\frac{(n + r + 1)^2}{(3n + 3r - 1)(n + r + 1)} a_{n-1}(r) \tag{22}$$
$$= -\frac{n + r + 1}{3n + 3r - 1} a_{n-1}(r), \qquad n \geq 1.$$

Setting $r = 1/3$ in (22) yields

$$a_0(1/3) = 1,$$
$$a_n(1/3) = -\frac{3n + 4}{9n} a_{n-1}(1/3), \qquad n \geq 1.$$

By using the product notation introduced in Section 7.2 and proceeding as we did in the examples in that section, we obtain

$$a_n(1/3) = \frac{(-1)^n \prod_{j=1}^{n} (3j + 4)}{9^n n!}, \qquad n \geq 0.$$

Therefore

$$y_1 = x^{1/3} \sum_{n=0}^{\infty} \frac{(-1)^n \prod_{j=1}^{n} (3j + 4)}{9^n n!} x^n$$

is a Frobenius solution of (21).

Setting $r = -1$ in (22) yields

$$a_0(-1) = 1,$$
$$a_n(-1) = -\frac{n}{3n - 4} a_{n-1}(-1), \qquad n \geq 1,$$

so

$$a_n(-1) = \frac{(-1)^n n!}{\prod_{j=1}^{n} (3j - 4)}.$$

Therefore

$$y_2 = x^{-1} \sum_{n=0}^{\infty} \frac{(-1)^n n!}{\prod_{j=1}^{n} (3j - 4)} x^n$$

is a Frobenius solution of (21), and $\{y_1, y_2\}$ is a fundamental set of solutions. ∎

We now consider equations of the form

$$x^2(\alpha_0 + \alpha_2 x^2)y'' + x(\beta_0 + \beta_2 x^2)y' + (\gamma_0 + \gamma_2 x^2)y = 0 \qquad (23)$$

with $\alpha_0 \neq 0$. For this equation we have $\alpha_1 = \beta_1 = \gamma_1 = 0$, so $p_1 \equiv 0$ and the recurrence relations in Theorem 7.5.2 simplify to

$$a_0(r) = 1,$$
$$a_1(r) = 0,$$
$$a_n(r) = -\frac{p_2(n + r - 2)}{p_0(n + r)} a_{n-2}(r), \qquad n \geq 2.$$

Since $a_1(r) = 0$ the last equation implies that $a_n(r) = 0$ if n is odd, so the Frobenius solutions are of the form

$$y(x,r) = x^r \sum_{m=0}^{\infty} a_{2m}(r)x^{2m},$$

where

$$a_0(r) = 1,$$
$$a_{2m}(r) = -\frac{p_2(2m + r - 2)}{p_0(2m + r)} a_{2m-2}(r), \qquad m \geq 1. \qquad (24)$$

EXAMPLE 3 Find a fundamental set of Frobenius solutions of

$$x^2(2 - x^2)y'' - x(3 + 4x^2)y' + (2 - 2x^2)y = 0. \qquad (25)$$

Give explicit formulas for the coefficients in the solutions.

Solution For the given equation the polynomials defined in Theorem 7.5.2 are

$$\begin{aligned}
p_0(r) &= 2r(r - 1) - 3r + 2 = (r - 2)(2r - 1), \\
p_1(r) &= 0 \\
p_2(r) &= -[r(r - 1) + 4r + 2] = -(r + 1)(r + 2).
\end{aligned}$$

The zeros of the indicial polynomial p_0 are $r_1 = 2$ and $r_2 = 1/2$, so $r_1 - r_2 = 3/2$. Therefore Theorem 7.5.3 implies that

$$y_1 = x^2 \sum_{m=0}^{\infty} a_{2m}(2)x^{2m} \qquad \text{and} \qquad y_2 = x^{1/2} \sum_{m=0}^{\infty} a_{2m}(1/2)x^{2m}$$

form a fundamental set of Frobenius solutions of (25). To find the coefficients in these series we use the recurrence relation (24); thus,

$$\begin{aligned}
a_0(r) &= 1, \\
a_{2m}(r) &= -\frac{p_2(2m + r - 2)}{p_0(2m + r)} a_{2m-2}(r) \qquad (26) \\
&= \frac{(2m + r)(2m + r - 1)}{(2m + r - 2)(4m + 2r - 1)} a_{2m-2}(r), \qquad m \geq 1.
\end{aligned}$$

Setting $r = 2$ in (26) yields

$$a_0(2) = 1,$$

$$a_{2m}(2) = \frac{(m+1)(2m+1)}{m(4m+3)} a_{2m-2}(2), \quad m \geq 1,$$

so

$$a_{2m}(2) = (m+1) \prod_{j=1}^{m} \frac{2j+1}{4j+3}.$$

Therefore

$$y_1 = x^2 \sum_{m=0}^{\infty} (m+1) \left(\prod_{j=1}^{m} \frac{2j+1}{4j+3} \right) x^{2m}$$

is a Frobenius solution of (25).

Setting $r = 1/2$ in (26) yields

$$a_0(1/2) = 1,$$

$$a_{2m}(1/2) = \frac{(4m-1)(4m+1)}{8m(4m-3)} a_{2m-2}(1/2), \quad m \geq 1,$$

so

$$a_{2m}(1/2) = \frac{1}{8^m m!} \prod_{j=1}^{m} \frac{(4j-1)(4j+1)}{4j-3}.$$

Therefore

$$y_2 = x^{1/2} \sum_{m=0}^{\infty} \frac{1}{8^m m!} \left(\prod_{j=1}^{m} \frac{(4j-1)(4j+1)}{4j-3} \right) x^{2m}$$

is a Frobenius solution of (25), and $\{y_1, y_2\}$ is a fundamental set of solutions.   ∎

REMARK. Thus far we have considered only the case where the indicial equation has real roots that do not differ by an integer, which allows us to apply Theorem 7.5.3. However, for equations of the form (23) the sequence $\{a_{2m}(r)\}$ in (24) is defined for $r = r_2$ if $r_1 - r_2$ is not an *even* integer. It can be shown (Exercise 56) that in this case

$$y_1 = x^{r_1} \sum_{m=0}^{\infty} a_{2m}(r_1) x^{2m} \quad \text{and} \quad y_2 = x^{r_2} \sum_{m=0}^{\infty} a_{2m}(r_2) x^{2m}$$

form a fundamental set Frobenius solutions of (23).

USING TECHNOLOGY

As we said at the end of Section 7.2, if you're interested in actually using series to compute numerical approximations to solutions of a differential equation, then whether or not there is a simple closed form for the coefficients is essentially irrelevant; recursive computation is usually more efficient. Since it is also laborious, we encourage you to write short programs to implement recurrence relations on a calculator or computer, even in exercises where this is not specifically required.

In practical use of the method of Frobenius when $x_0 = 0$ is a regular singular point we are interested in how well the functions

$$y_N(x, r_i) = x^{r_i} \sum_{n=0}^{N} a_n(r_i)x^n, \quad i = 1, 2,$$

approximate solutions to a given equation when r_i is a zero of the indicial polynomial. In dealing with the corresponding problem for the case where $x_0 = 0$ is an ordinary point, we used numerical integration to solve the differential equation subject to initial conditions $y(0) = a_0$, $y'(0) = a_1$, and compared the result with values of the Taylor polynomial

$$T_N(x) = \sum_{n=0}^{N} a_n x^n.$$

We can't do that here, since in general we can't prescribe arbitrary initial values for solutions of a differential equation at a singular point. Therefore, motivated by Theorem 7.5.2 (specifically, (14)), we suggest the following procedure.

Verification Procedure

Let L and $y_N(x; r_i)$ be defined by

$$Ly = x^2(\alpha_0 + \alpha_1 x + \alpha_2 x^2)y'' + x(\beta_0 + \beta_1 x + \beta_2 x^2)y'$$
$$+ (\gamma_0 + \gamma_1 x + \gamma_2 x^2)y$$

and

$$y_N(x, r_i) = x^{r_i} \sum_{n=0}^{N} a_n(r_i)x^n,$$

where the coefficients $\{a_n(r_i)\}_{n=0}^{N}$ are computed as in (12), Theorem 7.5.2. Compute the error

$$E_N(x; r_i) = x^{-r_i} Ly_N(x; r_i)/\alpha_0 \tag{27}$$

for various values of N and various values of x in the interval $(0, \rho)$, with ρ as defined in Theorem 7.5.2.

The multiplier x^{-r_i}/α_0 on the right of (27) eliminates the effects of small or large values of x^{r_i} near $x = 0$, and of multiplication by an arbitrary constant. In some exercises you will be asked to estimate the maximum value of $E_N(x; r_i)$ on an interval $(0, \delta]$ by computing $E_N(x_m, r_i)$ at the M points $x_m = m\delta/M$, $m = 1, 2, \ldots, M$, and finding the maximum of the absolute values:

$$\sigma_N(\delta) = \max\{|E_N(x_m, r_i)|, m = 1, 2, \ldots, M\}. \tag{28}$$

(For simplicity this notation ignores the dependence of the right side of the equation on i and M.)

To implement this procedure you'll have to write a computer program to calculate $\{a_n(r_i)\}$ from the applicable recurrence relation, and to evaluate $E_N(x; r_i)$.

7.5 EXERCISES

This set contains exercises specifically identified by **L** *that ask you to implement the verification procedure. These particular exercises were chosen arbitrarily; you can just as well formulate such laboratory problems for any of the equations in Exercises 1–10, 14–25, and 28–51.*

In Exercises 1–10 find a fundamental set of Frobenius solutions. Compute $a_0, \ldots, a_N$ for N at least 3 in each solution.

C **1.** $2x^2(1 + x + x^2)y'' + x(3 + 3x + 5x^2)y' - y = 0$

C **2.** $3x^2y'' + 2x(1 + x - 2x^2)y' + (2x - 8x^2)y = 0$

C **3.** $x^2(3 + 3x + x^2)y'' + x(5 + 8x + 7x^2)y' - (1 - 2x - 9x^2)y = 0$

C **4.** $4x^2y'' + x(7 + 2x + 4x^2)y' - (1 - 4x - 7x^2)y = 0$

C **5.** $12x^2(1 + x)y'' + x(11 + 35x + 3x^2)y' - (1 - 10x - 5x^2)y = 0$

C **6.** $x^2(5 + x + 10x^2)y'' + x(4 + 3x + 48x^2)y' + (x + 36x^2)y = 0$

C **7.** $8x^2y'' - 2x(3 - 4x - x^2)y' + (3 + 6x + x^2)y = 0$

C **8.** $18x^2(1 + x)y'' + 3x(5 + 11x + x^2)y' - (1 - 2x - 5x^2)y = 0$

C **9.** $x(3 + x + x^2)y'' + (4 + x - x^2)y' + xy = 0$

C **10.** $10x^2(1 + x + 2x^2)y'' + x(13 + 13x + 66x^2)y' - (1 + 4x + 10x^2)y = 0$

L **11.** The Frobenius solutions of

$$2x^2(1 + x + x^2)y'' + x(9 + 11x + 11x^2)y' + (6 + 10x + 7x^2)y = 0$$

obtained in Example 1 are defined on $(0, \rho)$, where ρ is defined in Theorem 7.5.2. Find ρ. Then do the following experiments for each Frobenius solution, with $M = 20$ and $\delta = .5\rho, .7\rho$, and $.9\rho$ in the verification procedure described at the end of this section.

(a) Compute $\sigma_N(\delta)$ (see Eqn. (28)) for $N = 5, 10, 15, \ldots, 50$.

(b) Find N such that $\sigma_N(\delta) < 10^{-5}$.

(c) Find N such that $\sigma_N(\delta) < 10^{-10}$.

L **12.** By Theorem 7.5.2 the Frobenius solutions of the equation in Exercise 4 are defined on $(0, \infty)$. Do experiments **(a)**, **(b)**, and **(c)** of Exercise 11 for each Frobenius solution, with $M = 20$ and $\delta = 1, 2$, and 3 in the verification procedure described at the end of this section.

L **13.** The Frobenius solutions of the equation in Exercise 6 are defined on $(0, \rho)$, where ρ is defined in Theorem 7.5.2. Find ρ and do experiments **(a)**, **(b)**, and **(c)** of Exercise 11 for each Frobenius solution, with $M = 20$ and $\delta = .3\rho, .4\rho$, and $.5\rho$, in the verification procedure described at the end of this section.

In Exercises 14–25 find a fundamental set of Frobenius solutions. Give explicit formulas for the coefficients in each solution.

14. $2x^2y'' + x(3 + 2x)y' - (1 - x)y = 0$

15. $x^2(3 + x)y'' + x(5 + 4x)y' - (1 - 2x)y = 0$

16. $2x^2y'' + x(5 + x)y' - (2 - 3x)y = 0$

17. $3x^2y'' + x(1 + x)y' - y = 0$

18. $2x^2y'' - xy' + (1 - 2x)y = 0$

19. $9x^2y'' + 9xy' - (1 + 3x)y = 0$

20. $3x^2y'' + x(1 + x)y' - (1 + 3x)y = 0$

21. $2x^2(3 + x)y'' + x(1 + 5x)y' + (1 + x)y = 0$

22. $x^2(4 + x)y'' - x(1 - 3x)y' + y = 0$

23. $2x^2y'' + 5xy' + (1 + x)y = 0$

24. $x^2(3 + 4x)y'' + x(5 + 18x)y' - (1 - 12x)y = 0$

25. $6x^2y'' + x(10 - x)y' - (2 + x)y = 0$

L **26.** By Theorem 7.5.2 the Frobenius solutions of the equation in Exercise 17 are defined on $(0, \infty)$. Do experiments **(a)**, **(b)**, and **(c)** of Exercise 11 for each Frobenius solution, with $M = 20$ and $\delta = 3, 6, 9$, and 12 in the verification procedure described at the end of this section.

L **27.** The Frobenius solutions of the equation in Exercise 22 are defined on $(0,\rho)$, where ρ is defined in Theorem 7.5.2. Find ρ and do experiments **(a)**, **(b)**, and **(c)** of Exercise 11 for each Frobenius solution, with $M = 20$ and $\delta = .25\rho, .5\rho$, and $.75\rho$ in the verification procedure described at the end of this section.

In Exercises 28–32 find a fundamental set of Frobenius solutions. Compute coefficients $a_0, \dots, a_N$ for N at least 5 in each solution.

C **28.** $x^2(8 + x)y'' + x(2 + 3x)y' + (1 + x)y = 0$ **C** **29.** $x^2(3 + 4x)y'' + x(11 + 4x)y' - (3 + 4x)y = 0$

C **30.** $2x^2(2 + 3x)y'' + x(4 + 11x)y' - (1 - x)y = 0$ **C** **31.** $x^2(2 + x)y'' + 5x(1 - x)y' - (2 - 8x)y$

C **32.** $x^2(6 + x)y'' + x(11 + 4x)y' + (1 + 2x)y = 0$

In Exercises 33–46 find a fundamental set of Frobenius solutions. Give explicit formulas for the coefficients in each solution.

33. $8x^2y'' + x(2 + x^2)y' + y = 0$

34. $8x^2(1 - x^2)y'' + 2x(1 - 13x^2)y' + (1 - 9x^2)y = 0$

35. $x^2(1 + x^2)y'' - 2x(2 - x^2)y' + 4y = 0$ **36.** $x(3 + x^2)y'' + (2 - x^2)y' - 8xy = 0$

37. $4x^2(1 - x^2)y'' + x(7 - 19x^2)y' - (1 + 14x^2)y = 0$

38. $3x^2(2 - x^2)y'' + x(1 - 11x^2)y' + (1 - 5x^2)y = 0$

39. $2x^2(2 + x^2)y'' - x(12 - 7x^2)y' + (7 + 3x^2)y = 0$

40. $2x^2(2 + x^2)y'' + x(4 + 7x^2)y' - (1 - 3x^2)y = 0$

41. $2x^2(1 + 2x^2)y'' + 5x(1 + 6x^2)y' - (2 - 40x^2)y = 0$

42. $3x^2(1 + x^2)y'' + 5x(1 + x^2)y' - (1 + 5x^2)y = 0$

43. $x(1 + x^2)y'' + (4 + 7x^2)y' + 8xy = 0$ **44.** $x^2(2 + x^2)y'' + x(3 + x^2)y' - y = 0$

45. $2x^2(1 + x^2)y'' + x(3 + 8x^2)y' - (3 - 4x^2)y = 0$

46. $9x^2y'' + 3x(3 + x^2)y' - (1 - 5x^2)y = 0$

In Exercises 47–51 find a fundamental set of Frobenius solutions. Compute the coefficients $a_0, \dots, a_{2M}$ for M at least 2 in each solution.

C **47.** $6x^2y'' + x(1 + 6x^2)y' + (1 + 9x^2)y = 0$ **C** **48.** $x^2(8 + x^2)y'' + 7x(2 + x^2)y' - (2 - 9x^2)y = 0$

C **49.** $9x^2(1 + x^2)y'' + 3x(3 + 13x^2)y' - (1 - 25x^2)y = 0$

C **50.** $4x^2(1 + x^2)y'' + 4x(1 + 6x^2)y' - (1 - 25x^2)y = 0$

C **51.** $8x^2(1 + 2x^2)y'' + 2x(5 + 34x^2)y' - (1 - 30x^2)y = 0$

52. Suppose that $r_1 > r_2$, $a_0 = b_0 = 1$, and the Frobenius series

$$y_1 = x^{r_1} \sum_{n=0}^{\infty} a_n x^n \quad \text{and} \quad y_2 = x^{r_2} \sum_{n=0}^{\infty} b_n x^n$$

both converge on an interval $(0,\rho)$.

(a) Show that y_1 and y_2 are linearly independent on $(0,\rho)$.

Hint: Show that if c_1 and c_2 are constants such that $c_1y_1 + c_2y_2 \equiv 0$ on $(0,\rho)$, then

$$c_1 x^{r_1 - r_2} \sum_{n=0}^{\infty} a_n x^n + c_2 \sum_{n=0}^{\infty} b_n x^n = 0, \quad 0 < x < \rho.$$

Then let $x \to 0+$ to conclude that $c_2 = 0$.

(b) Use the result of **(a)** to complete the proof of Theorem 7.5.3.

53. The equation

$$x^2y'' + xy' + (x^2 - \nu^2)y = 0 \tag{A}$$

is known as ***Bessel's equation of order ν.*** (Here ν is a parameter and this use of "order" should not be confused with its usual use, as in "the order of the equation.") The solutions of (A) are known as ***Bessel functions of order ν.***

(a) Assuming that ν is not an integer, find a fundamental set of Frobenius solutions of (A).

(b) If $\nu = 1/2$ then the solutions of (A) reduce to familiar elementary functions. Identify these functions.

54. (a) Verify that

$$\frac{d}{dx}\left(|x|^r x^n\right) = (n + r)|x|^r x^{n-1} \quad \text{and} \quad \frac{d^2}{dx^2}\left(|x|^r x^n\right) = (n + r)(n + r - 1)|x|^r x^{n-2}$$

if $x \neq 0$.

(b) Let

$$Ly = x^2(\alpha_0 + \alpha_1 x + \alpha_2 x^2)y'' + x(\beta_0 + \beta_1 x + \beta_2 x^2)y' + (\gamma_0 + \gamma_1 x + \gamma_2 x^2)y = 0.$$

Show that if $x^r \sum_{n=0}^{\infty} a_n x^n$ is a solution of $Ly = 0$ on $(0, \rho)$ then $|x|^r \sum_{n=0}^{\infty} a_n x^n$ is a solution on $(-\rho, 0)$ and $(0, \rho)$.

55. (a) Deduce from Eqn. (20) that

$$a_n(r) = (-1)^n \prod_{j=1}^{n} \frac{p_1(j + r - 1)}{p_0(j + r)}.$$

(b) Conclude that if $p_0(r) = \alpha_0(r - r_1)(r - r_2)$ where $r_1 - r_2$ is not an integer then

$$y_1 = x^{r_1} \sum_{n=0}^{\infty} a_n(r_1)x^n \quad \text{and} \quad y_2 = x^{r_2} \sum_{n=0}^{\infty} a_n(r_2)x^n$$

form a fundamental set of Frobenius solutions of

$$x^2(\alpha_0 + \alpha_1 x)y'' + x(\beta_0 + \beta_1 x)y' + (\gamma_0 + \gamma_1 x)y = 0.$$

(c) Show that if $p_0(r) = \alpha_0 r(r - 1) + \beta_0 r + \gamma_0$ satisfies the hypotheses of **(b)** then

$$y_1 = x^{r_1} \sum_{n=0}^{\infty} \frac{(-1)^n}{n! \, \prod_{j=1}^{n} (j + r_1 - r_2)} \left(\frac{\gamma_1}{\alpha_0}\right)^n x^n$$

and

$$y_2 = x^{r_2} \sum_{n=0}^{\infty} \frac{(-1)^n}{n! \, \prod_{j=1}^{n} (j + r_2 - r_1)} \left(\frac{\gamma_1}{\alpha_0}\right)^n x^n$$

form a fundamental set of Frobenius solutions of

$$\alpha_0 x^2 y'' + \beta_0 xy' + (\gamma_0 + \gamma_1 x)y = 0.$$

56. Let

$$Ly = x^2(\alpha_0 + \alpha_2 x^2)y'' + x(\beta_0 + \beta_2 x^2)y' + (\gamma_0 + \gamma_2 x^2)y = 0$$

and define

$$p_0(r) = \alpha_0 r(r - 1) + \beta_0 r + \gamma_0 \quad \text{and} \quad p_2(r) = \alpha_2 r(r - 1) + \beta_2 r + \gamma_2.$$

(a) Use Theorem 7.5.2 to show that if

$$a_0(r) = 1,$$

$$p_0(2m + r)a_{2m}(r) + p_2(2m + r - 2)a_{2m-2}(r) = 0, \quad m \geq 1, \tag{A}$$

then the Frobenius series $y(x, r) = x^r \sum_{m=0}^{\infty} a_{2m}(r)x^{2m}$ satisfies $Ly(x, r) = p_0(r)x^r$.

(b) Deduce from (A) that if $p_0(2m + r)$ is nonzero for every positive integer m then

$$a_{2m}(r) = (-1)^m \prod_{j=1}^{m} \frac{p_2(2j + r - 2)}{p_0(2j + r)}.$$

(c) Conclude that if $p_0(r) = \alpha_0(r - r_1)(r - r_2)$ where $r_1 - r_2$ is not an even integer then

$$y_1 = x^{r_1} \sum_{m=0}^{\infty} a_{2m}(r_1)x^{2m} \quad \text{and} \quad y_2 = x^{r_2} \sum_{m=0}^{\infty} a_{2m}(r_2)x^{2m}$$

form a fundamental set of Frobenius solutions of $Ly = 0$.

(d) Show that if p_0 satisfies the hypotheses of **(c)** then

$$y_1 = x^{r_1} \sum_{m=0}^{\infty} \frac{(-1)^m}{2^m m! \, \prod_{j=1}^{m} (2j + r_1 - r_2)} \left(\frac{\gamma_2}{\alpha_0}\right)^m x^{2m}$$

and

$$y_2 = x^{r_2} \sum_{m=0}^{\infty} \frac{(-1)^m}{2^m m! \, \prod_{j=1}^{m} (2j + r_2 - r_1)} \left(\frac{\gamma_2}{\alpha_0}\right)^m x^{2m}$$

form a fundamental set of Frobenius solutions of

$$\alpha_0 x^2 y'' + \beta_0 xy' + (\gamma_0 + \gamma_2 x^2)y = 0.$$

57. Let

$$Ly = x^2 q_0(x)y'' + x q_1(x)y' + q_2(x)y,$$

where

$$q_0(x) = \sum_{j=0}^{\infty} \alpha_j x^j, \quad q_1(x) = \sum_{j=0}^{\infty} \beta_j x^j, \quad q_2(x) = \sum_{j=0}^{\infty} \gamma_j x^j,$$

and define

$$p_j(r) = \alpha_j r(r-1) + \beta_j r + \gamma_j, \quad j = 0, 1, \dots.$$

Let $y = x^r \sum_{n=0}^{\infty} a_n x^n$. Show that

$$Ly = x^r \sum_{n=0}^{\infty} b_n x^n,$$

where

$$b_n = \sum_{j=0}^{n} p_j(n + r - j)a_{n-j}.$$

58. (a) Let L be as in Exercise 57. Show that if

$$y(x,r) = x^r \sum_{n=0}^{\infty} a_n(r)x^n$$

where

$$a_0(r) = 1,$$

$$a_n(r) = -\frac{1}{p_0(n+r)} \sum_{j=1}^{n} p_j(n + r - j)a_{n-j}(r), \quad n \geq 1,$$

then

$$Ly(x,r) = p_0(r)x^r.$$

(b) Conclude that if

$$p_0(r) = \alpha_0(r - r_1)(r - r_2)$$

where $r_1 - r_2$ is not an integer, then $y_1 = y(x,r_1)$ and $y_2 = y(x,r_2)$ are solutions of $Ly = 0$.

59. Let

$$Ly = x^2(\alpha_0 + \alpha_q x^q)y'' + x(\beta_0 + \beta_q x^q)y' + (\gamma_0 + \gamma_q x^q)y$$

where q is a positive integer, and define

$$p_0(r) = \alpha_0 r(r-1) + \beta_0 r + \gamma_0 \quad \text{and} \quad p_q(r) = \alpha_q r(r-1) + \beta_q r + \gamma_q.$$

(a) Show that if

$$y(x,r) = x^r \sum_{m=0}^{\infty} a_{qm}(r)x^{qm}$$

where

$$a_0(r) = 1,$$

(A)

$$a_{qm}(r) = -\frac{p_q(q(m-1)+r)}{p_0(qm+r)}a_{q(m-1)}(r), \quad m \geq 1,$$

then

$$Ly(x,r) = p_0(r)x^r.$$

(b) Deduce from (A) that

$$a_{qm}(r) = (-1)^m \prod_{j=1}^{m} \frac{p_q(q(j-1)+r)}{p_0(qj+r)}.$$

(c) Conclude that if $p_0(r) = \alpha_0(r-r_1)(r-r_2)$ where $r_1 - r_2$ is not an integer multiple of q then

$$y_1 = x^{r_1} \sum_{m=0}^{\infty} a_{qm}(r_1)x^{qm} \quad \text{and} \quad y_2 = x^{r_2} \sum_{m=0}^{\infty} a_{qm}(r_2)x^{qm}$$

form a fundamental set of Frobenius solutions of $Ly = 0$.

(d) Show that if p_0 satisfies the hypotheses of **(c)** then

$$y_1 = x^{r_1} \sum_{m=0}^{\infty} \frac{(-1)^m}{q^m m! \, \prod_{j=1}^{m}(qj + r_1 - r_2)} \left(\frac{\gamma_q}{\alpha_0}\right)^m x^{qm}$$

and

$$y_2 = x^{r_2} \sum_{m=0}^{\infty} \frac{(-1)^m}{q^m m! \, \prod_{j=1}^{m}(qj + r_2 - r_1)} \left(\frac{\gamma_q}{\alpha_0}\right)^m x^{qm}$$

form a fundamental set of Frobenius solutions of

$$\alpha_0 x^2 y'' + \beta_0 xy' + (\gamma_0 + \gamma_q x^q)y = 0.$$

60. (a) Suppose that α_0, α_1, and α_2 are real numbers with $\alpha_0 \neq 0$, and the sequence $\{a_n\}_{n=0}^{\infty}$ satisfies

$$\alpha_0 a_1 + \alpha_1 a_0 = 0$$

and

$$\alpha_0 a_n + \alpha_1 a_{n-1} + \alpha_2 a_{n-2} = 0, \quad n \geq 2.$$

Show that

$$(\alpha_0 + \alpha_1 x + \alpha_2 x^2) \sum_{n=0}^{\infty} a_n x^n = \alpha_0 a_0,$$

and infer that

$$\sum_{n=0}^{\infty} a_n x^n = \frac{\alpha_0 a_0}{\alpha_0 + \alpha_1 x + \alpha_2 x^2}.$$

(b) With α_0, α_1, and α_2 as in **(a)**, consider the equation

$$x^2(\alpha_0 + \alpha_1 x + \alpha_2 x^2)y'' + x(\beta_0 + \beta_1 x + \beta_2 x^2)y' + (\gamma_0 + \gamma_1 x + \gamma_2 x^2)y = 0, \tag{A}$$

and define

$$p_j(r) = \alpha_j r(r-1) + \beta_j r + \gamma_j, \quad j = 0,1,2.$$

Suppose that

$$\frac{p_1(r-1)}{p_0(r)} = \frac{\alpha_1}{\alpha_0} \quad \text{and} \quad \frac{p_2(r-2)}{p_0(r)} = \frac{\alpha_2}{\alpha_0} \quad \text{for all } r,$$

and

$$p_0(r) = \alpha_0(r - r_1)(r - r_2),$$

where $r_1 > r_2$. Show that

$$y_1 = \frac{x^{r_1}}{\alpha_0 + \alpha_1 x + \alpha_2 x^2} \quad \text{and} \quad y_2 = \frac{x^{r_2}}{\alpha_0 + \alpha_1 x + \alpha_2 x^2}$$

form a fundamental set of Frobenius solutions of (A) on any interval $(0, \rho)$ on which $\alpha_0 + \alpha_1 x + \alpha_2 x^2$ has no zeros.

In Exercises 61–68 use the method suggested by Exercise 60 to find a fundamental set of Frobenius solutions.

61. $2x^2(1 + x)y'' - x(1 - 3x)y' + y = 0$

62. $6x^2(1 + 2x^2)y'' + x(1 + 50x^2)y' + (1 + 30x^2)y = 0$

63. $28x^2(1 - 3x)y'' - 7x(5 + 9x)y' + 7(2 + 9x)y = 0$

64. $9x^2(5 + x)y'' + 9x(5 + 3x)y' - (5 - 8x)y = 0$

65. $8x^2(2 - x^2)y'' + 2x(10 - 21x^2)y' - (2 + 35x^2)y = 0$

66. $4x^2(1 + 3x + x^2)y'' - 4x(1 - 3x - 3x^2)y' + 3(1 - x + x^2)y = 0$

67. $3x^2(1 + x)^2 y'' - x(1 - 10x - 11x^2)y' + (1 + 5x^2)y = 0$

68. $4x^2(3 + 2x + x^2)y'' - x(3 - 14x - 15x^2)y' + (3 + 7x^2)y = 0$

7.6   The Method of Frobenius II

In this section we discuss a method for finding two linearly independent Frobenius solutions of a homogeneous linear second order equation near a regular singular point in the case where the indicial equation has a repeated real root. As in the preceding section, we consider equations that can be written as

$$x^2(\alpha_0 + \alpha_1 x + \alpha_2 x^2)y'' + x(\beta_0 + \beta_1 x + \beta_2 x^2)y' + (\gamma_0 + \gamma_1 x + \gamma_2 x^2)y = 0 \tag{1}$$

where $\alpha_0 \ne 0$. We assume that the indicial equation $p_0(r) = 0$ has a repeated real root r_1. In this case Theorem 7.5.3 implies that (1) has one solution of the form

$$y_1 = x^{r_1} \sum_{n=0}^{\infty} a_n x^n,$$

but does not provide a second solution y_2 such that $\{y_1, y_2\}$ is a fundamental set of solutions. The following extension of Theorem 7.5.2 provides a way to find a second solution.

THEOREM 7.6.1

Let

$$Ly = x^2(\alpha_0 + \alpha_1 x + \alpha_2 x^2)y'' + x(\beta_0 + \beta_1 x + \beta_2 x^2)y'$$
$$+ (\gamma_0 + \gamma_1 x + \gamma_2 x^2)y \tag{2}$$

where $\alpha_0 \neq 0$, and define

$$p_0(r) = \alpha_0 r(r-1) + \beta_0 r + \gamma_0,$$
$$p_1(r) = \alpha_1 r(r-1) + \beta_1 r + \gamma_1,$$
$$p_2(r) = \alpha_2 r(r-1) + \beta_2 r + \gamma_2.$$

Suppose that r is a real number such that $p_0(n + r)$ is nonzero for all positive integers n, and define

$$a_0(r) = 1,$$

$$a_1(r) = -\frac{p_1(r)}{p_0(r+1)},$$

$$a_n(r) = -\frac{p_1(n+r-1)a_{n-1}(r) + p_2(n+r-2)a_{n-2}(r)}{p_0(n+r)}, \qquad n \geq 2.$$

Then the Frobenius series

$$y(x,r) = x^r \sum_{n=0}^{\infty} a_n(r)x^n \tag{3}$$

satisfies

$$Ly(x,r) = p_0(r)x^r. \tag{4}$$

Moreover,

$$\frac{\partial y}{\partial r}(x,r) = y(x,r) \ln x + x^r \sum_{n=1}^{\infty} a_n'(r)x^n, \tag{5}$$

and

$$L\left(\frac{\partial y}{\partial r}(x,r)\right) = p_0'(r)x^r + x^r p_0(r) \ln x. \tag{6}$$

PROOF. Theorem 7.5.2 implies (4). Differentiating formally with respect to r in (3) yields

$$\frac{\partial y}{\partial r}(x,r) = \frac{\partial}{\partial r}(x^r) \sum_{n=0}^{\infty} a_n(r)x^n + x^r \sum_{n=1}^{\infty} a_n'(r)x^n$$

$$= x^r \ln x \sum_{n=0}^{\infty} a_n(r)x^n + x^r \sum_{n=1}^{\infty} a_n'(r)x^n$$

$$= y(x,r) \ln x + x^r \sum_{n=1}^{\infty} a_n'(r)x^n,$$

which proves (5).

To prove that $\partial y(x,r)/\partial r$ satisfies (6), we view y in (2) as a function $y = y(x,r)$ of two variables, where the prime indicates partial differentiation with respect to x; thus,

$$y' = y'(x,r) = \frac{\partial y}{\partial x}(x,r) \quad \text{and} \quad y'' = y''(x,r) = \frac{\partial^2 y}{\partial x^2}(x,r).$$

With this notation we can use (2) to rewrite (4) as

$$x^2 q_0(x) \frac{\partial^2 y}{\partial x^2}(x,r) + x q_1(x) \frac{\partial y}{\partial x}(x,r) + q_2(x) y(x,r) = p_0(r)x^r, \tag{7}$$

where

$$q_0(x) = \alpha_0 + \alpha_1 x + \alpha_2 x^2,$$
$$q_1(x) = \beta_0 + \beta_1 x + \beta_2 x^2,$$
$$q_2(x) = \gamma_0 + \gamma_1 x + \gamma_2 x^2.$$

Differentiating both sides of (7) with respect to r yields

$$x^2 q_0(x) \frac{\partial^3 y}{\partial r\, \partial x^2}(x,r) + x q_1(x) \frac{\partial^2 y}{\partial r\, \partial x}(x,r) + q_2(x) \frac{\partial y}{\partial r}(x,r)$$

$$= p_0'(r)x^r + p_0(r)x^r \ln x.$$

By changing the order of differentiation in the first two terms on the left we can rewrite this as

$$x^2 q_0(x) \frac{\partial^3 y}{\partial x^2\, \partial r}(x,r) + x q_1(x) \frac{\partial^2 y}{\partial x\, \partial r}(x,r) + q_2(x) \frac{\partial y}{\partial r}(x,r)$$

$$= p_0'(r)x^r + p_0(r)x^r \ln x,$$

or

$$x^2 q_0(x) \frac{\partial^2}{\partial x^2}\left(\frac{\partial y}{\partial r}(x,r)\right) + x q_1(x) \frac{\partial}{\partial r}\left(\frac{\partial y}{\partial x}(x,r)\right) + q_2(x) \frac{\partial y}{\partial r}(x,r)$$

$$= p_0'(r)x^r + p_0(r)x^r \ln x,$$

which is equivalent to (6). $\square$

THEOREM 7.6.2

Let L be as in Theorem 7.6.1 and suppose that the indicial equation $p_0(r) = 0$ has a repeated real root r_1. Then

$$y_1(x) = y(x,r_1) = x^{r_1} \sum_{n=0}^{\infty} a_n(r_1)x^n$$

and

$$y_2(x) = \frac{\partial y}{\partial r}(x,r_1) = y_1(x) \ln x + x^{r_1} \sum_{n=1}^{\infty} a_n'(r_1)x^n \tag{8}$$

form a fundamental set of solutions of $Ly = 0$.

PARTIAL PROOF. Since r_1 is a repeated root of $p_0(r) = 0$, the indicial polynomial can be factored as

$$p_0(r) = \alpha_0(r - r_1)^2,$$

so

$$p_0(n + r_1) = \alpha_0 n^2,$$

which is nonzero if $n > 0$. Therefore the assumptions of Theorem 7.6.1 hold with $r = r_1$, and (4) implies that $Ly_1 = p_0(r_1)x^{r_1} = 0$. Since

$$p_0'(r) = 2\alpha(r - r_1)$$

it follows that $p_0'(r_1) = 0$, so (6) implies that

$$Ly_2 = p_0'(r_1)x^{r_1} + x^{r_1}p_0(r_1) \ln x = 0.$$

This proves that y_1 and y_2 are both solutions of $Ly = 0$. We leave the proof that $\{y_1, y_2\}$ is a fundamental set as an exercise (Exercise 53).   ☐

EXAMPLE I   Find a fundamental set of solutions of

$$x^2(1 - 2x + x^2)y'' - x(3 + x)y' + (4 + x)y = 0. \tag{9}$$

Compute just the terms involving x^{n+r_1}, where $0 \le n \le 4$ and r_1 is the root of the indicial equation.

Solution   For the given equation the polynomials defined in Theorem 7.6.1 are

$$\begin{aligned}
p_0(r) &= r(r - 1) - 3r + 4 = (r - 2)^2, \\
p_1(r) &= -2r(r - 1) - r + 1 = -(r - 1)(2r + 1), \\
p_2(r) &= r(r - 1).
\end{aligned}$$

Since $r_1 = 2$ is a repeated root of the indicial polynomial p_0, Theorem 7.6.2 implies that

$$y_1 = x^2 \sum_{n=0}^{\infty} a_n(2)x^n \quad \text{and} \quad y_2 = y_1 \ln x + x^2 \sum_{n=1}^{\infty} a_n'(2)x^n \tag{10}$$

form a fundamental set of Frobenius solutions of (9). To find the coefficients in these series we use the recurrence formulas from Theorem 7.6.1:

$$a_0(r) = 1,$$

$$a_1(r) = -\frac{p_1(r)}{p_0(r + 1)} = -\frac{(r - 1)(2r + 1)}{(r - 1)^2} = \frac{2r + 1}{r - 1},$$

$$\begin{aligned}
a_n(r) &= -\frac{p_1(n + r - 1)a_{n-1}(r) + p_2(n + r - 2)a_{n-2}(r)}{p_0(n + r)} \tag{11} \\
&= \frac{(n + r - 2)[(2n + 2r - 1)a_{n-1}(r) - (n + r - 3)a_{n-2}(r)]}{(n + r - 2)^2} \\
&= \frac{(2n + 2r - 1)}{(n + r - 2)}a_{n-1}(r) - \frac{(n + r - 3)}{(n + r - 2)}a_{n-2}(r), \quad n \ge 2.
\end{aligned}$$

Differentiating yields

$$a_1'(r) = -\frac{3}{(r-1)^2},$$

$$a_n'(r) = \frac{2n+2r-1}{n+r-2}a_{n-1}'(r) - \frac{n+r-3}{n+r-2}a_{n-2}'(r) \tag{12}$$

$$- \frac{3}{(n+r-2)^2}a_{n-1}(r) - \frac{1}{(n+r-2)^2}a_{n-2}(r), \quad n \geq 2.$$

Setting $r = 2$ in (11) and (12) yields

$$a_0(2) = 1,$$
$$a_1(2) = 5, \tag{13}$$
$$a_n(2) = \frac{(2n+3)}{n}a_{n-1}(2) - \frac{(n-1)}{n}a_{n-2}(2), \quad n \geq 2,$$

and

$$a_1'(2) = -3, \tag{14}$$
$$a_n'(2) = \frac{2n+3}{n}a_{n-1}'(2) - \frac{n-1}{n}a_{n-2}'(2) - \frac{3}{n^2}a_{n-1}(2) - \frac{1}{n^2}a_{n-2}(2), \quad n \geq 2.$$

Computing recursively with (13) and (14) yields

$$a_0(2) = 1, \quad a_1(2) = 5, \quad a_2(2) = 17, \quad a_3(2) = \frac{143}{3}, \quad a_4(2) = \frac{355}{3},$$

and

$$a_1'(2) = -3, \quad a_2'(2) = -\frac{29}{2}, \quad a_3'(2) = -\frac{859}{18}, \quad a_4'(2) = -\frac{4693}{36}.$$

Substituting these coefficients into (10) yields

$$y_1 = x^2\left(1 + 5x + 17x^2 + \frac{143}{3}x^3 + \frac{355}{3}x^4 + \cdots\right)$$

and

$$y_2 = y_1 \ln x - x^3\left(3 + \frac{29}{2}x + \frac{859}{18}x^2 + \frac{4693}{36}x^3 + \cdots\right). \qquad \blacksquare$$

Since the recurrence formula (11) involves three terms it is not possible to obtain a simple explicit formula for the coefficients in the Frobenius solutions of (9). However, as we saw in the preceding sections, the recurrrence formula for $\{a_n(r)\}$ involves only two terms if either $\alpha_1 = \beta_1 = \gamma_1 = 0$ or $\alpha_2 = \beta_2 = \gamma_2 = 0$ in (1). In this case it is often possible to find explicit formulas for the coefficients. The next two examples illustrate this.

EXAMPLE 2   Find a fundamental set of Frobenius solutions of

$$2x^2(2+x)y'' + 5x^2y' + (1+x)y = 0. \tag{15}$$

Give explicit formulas for the coefficients in the solutions.

Solution   For the given equation the polynomials defined in Theorem 7.6.1 are

$$\begin{aligned} p_0(r) &= 4r(r-1) + 1 = (2r-1)^2, \\ p_1(r) &= 2r(r-1) + 5r + 1 = (r+1)(2r+1), \\ p_2(r) &= 0. \end{aligned}$$

Since $r_1 = 1/2$ is a repeated zero of the indicial polynomial p_0, Theorem 7.6.2 implies that

$$y_1 = x^{1/2} \sum_{n=0}^{\infty} a_n(1/2) x^n \tag{16}$$

and

$$y_2 = y_1 \ln x + x^{1/2} \sum_{n=1}^{\infty} a_n'(1/2) x^n \tag{17}$$

form a fundamental set of Frobenius solutions of (15). Since $p_2 \equiv 0$ the recurrence formulas in Theorem 7.6.1 reduce to

$$a_0(r) = 1,$$

$$a_n(r) = -\frac{p_1(n + r - 1)}{p_0(n + r)} a_{n-1}(r),$$

$$= -\frac{(n + r)(2n + 2r - 1)}{(2n + 2r - 1)^2} a_{n-1}(r),$$

$$= -\frac{n + r}{2n + 2r - 1} a_{n-1}(r), \qquad n \geq 0.$$

We leave it to you to show that

$$a_n(r) = (-1)^n \prod_{j=1}^{n} \frac{j + r}{2j + 2r - 1}, \qquad n \geq 1. \tag{18}$$

Setting $r = 1/2$ yields

$$a_n(1/2) = (-1)^n \prod_{j=1}^{n} \frac{j + 1/2}{2j} = (-1)^n \prod_{j=1}^{n} \frac{2j + 1}{4j},$$

$$= \frac{(-1)^n \prod_{j=1}^{n} (2j + 1)}{4^n n!}, \qquad n \geq 0. \tag{19}$$

Substituting this into (16) yields

$$y_1 = x^{1/2} \sum_{n=0}^{\infty} \frac{(-1)^n \prod_{j=1}^{n} (2j + 1)}{4^n n!} x^n.$$

To obtain y_2 in (17) we must compute $a_n'(1/2)$ for $n = 1, 2, \ldots$. We will do this by logarithmic differentiation. From (18),

$$|a_n(r)| = \prod_{j=1}^{n} \frac{|j + r|}{|2j + 2r - 1|}, \qquad n \geq 1.$$

Therefore,

$$\ln |a_n(r)| = \sum_{j=1}^{n} (\ln |j + r| - \ln |2j + 2r - 1|).$$

Differentiating with respect to r yields

$$\frac{a_n'(r)}{a_n(r)} = \sum_{j=1}^{n} \left(\frac{1}{j + r} - \frac{2}{2j + 2r - 1} \right).$$

Therefore

$$a'_n(r) = a_n(r) \sum_{j=1}^{n} \left(\frac{1}{j+r} - \frac{2}{2j+2r-1} \right).$$

Setting $r = 1/2$ here and recalling (19) yields

$$a'_n(1/2) = \frac{(-1)^n \prod_{j=1}^{n} (2j+1)}{4^n n!} \left(\sum_{j=1}^{n} \frac{1}{j+1/2} - \sum_{j=1}^{n} \frac{1}{j} \right). \tag{20}$$

Since

$$\frac{1}{j+1/2} - \frac{1}{j} = \frac{j-j-1/2}{j(j+1/2)} = -\frac{1}{j(2j+1)},$$

so (20) can be rewritten as

$$a'_n(1/2) = -\frac{(-1)^n \prod_{j=1}^{n} (2j+1)}{4^n n!} \sum_{j=1}^{n} \frac{1}{j(2j+1)}.$$

Therefore, from (17),

$$y_2 = y_1 \ln x - x^{1/2} \sum_{n=1}^{\infty} \frac{(-1)^n \prod_{j=1}^{n} (2j+1)}{4^n n!} \left(\sum_{j=1}^{n} \frac{1}{j(2j+1)} \right) x^n.$$

EXAMPLE 3 Find a fundamental set of Frobenius solutions of

$$x^2(2 - x^2)y'' - 2x(1 + 2x^2)y' + (2 - 2x^2)y = 0. \tag{21}$$

Give explicit formulas for the coefficients in the solutions.

Solution For (21) the polynomials defined in Theorem 7.6.1 are

$$\begin{aligned} p_0(r) &= 2r(r-1) - 2r + 2 = 2(r-1)^2, \\ p_1(r) &= 0, \\ p_2(r) &= -r(r-1) - 4r - 2 = -(r+1)(r+2). \end{aligned}$$

As we saw in Section 7.5, since $p_1 \equiv 0$ the recurrence formulas of Theorem 7.6.1 imply that $a_n(r) = 0$ if n is odd, and

$$a_0(r) = 1,$$

$$\begin{aligned} a_{2m}(r) &= -\frac{p_2(2m+r-2)}{p_0(2m+r)} a_{2m-2}(r) \\ &= \frac{(2m+r-1)(2m+r)}{2(2m+r-1)^2} a_{2m-2}(r) \\ &= \frac{2m+r}{2(2m+r-1)} a_{2m-2}(r), \quad m \geq 1. \end{aligned}$$

Since $r_1 = 1$ is a repeated root of the indicial polynomial p_0, Theorem 7.6.2 implies that

$$y_1 = x \sum_{m=0}^{\infty} a_{2m}(1)x^{2m} \tag{22}$$

and

$$y_2 = y_1 \ln x + x \sum_{m=1}^{\infty} a_{2m}'(1)x^{2m} \tag{23}$$

form a fundamental set of Frobenius solutions of (21). We leave it to you to show that

$$a_{2m}(r) = \frac{1}{2^m} \prod_{j=1}^{m} \frac{2j + r}{2j + r - 1}. \tag{24}$$

Setting $r = 1$ yields

$$a_{2m}(1) = \frac{1}{2^m} \prod_{j=1}^{m} \frac{2j + 1}{2j} = \frac{\prod_{j=1}^{m} (2j + 1)}{4^m m!}, \tag{25}$$

and substituting this into (22) yields

$$y_1 = x \sum_{m=0}^{\infty} \frac{\prod_{j=1}^{m} (2j + 1)}{4^m m!} x^{2m}.$$

To obtain y_2 in (23) we must compute $a_{2m}'(1)$ for $m = 1, 2, \ldots$. Again we use logarithmic differentiation. From (24),

$$|a_{2m}(r)| = \frac{1}{2^m} \prod_{j=1}^{m} \frac{|2j + r|}{|2j + r - 1|}.$$

Taking logarithms yields

$$\ln |a_{2m}(r)| = -m \ln 2 + \sum_{j=1}^{m} (\ln |2j + r| - \ln |2j + r - 1|).$$

Differentiating with respect to r yields

$$\frac{a_{2m}'(r)}{a_{2m}(r)} = \sum_{j=1}^{m} \left(\frac{1}{2j + r} - \frac{1}{2j + r - 1} \right).$$

Therefore

$$a_{2m}'(r) = a_{2m}(r) \sum_{j=1}^{m} \left(\frac{1}{2j + r} - \frac{1}{2j + r - 1} \right).$$

Setting $r = 1$ and recalling (25) yields

$$a_{2m}'(1) = \frac{\prod_{j=1}^{m} (2j + 1)}{4^m m!} \sum_{j=1}^{m} \left(\frac{1}{2j + 1} - \frac{1}{2j} \right). \tag{26}$$

Since

$$\frac{1}{2j + 1} - \frac{1}{2j} = - \frac{1}{2j(2j + 1)},$$

(26) can be rewritten as

$$a_{2m}'(1) = - \frac{\prod_{j=1}^{m} (2j + 1)}{2 \cdot 4^m m!} \sum_{j=1}^{m} \frac{1}{j(2j + 1)}.$$

Substituting this into (23) yields

$$y_2 = y_1 \ln x - \frac{x}{2} \sum_{m=1}^{\infty} \frac{\prod_{j=1}^{m} (2j + 1)}{4^m m!} \left(\sum_{j=1}^{m} \frac{1}{j(2j + 1)} \right) x^{2m}.$$

If the solution $y_1 = y(x, r_1)$ of $Ly = 0$ reduces to a finite sum then there is a difficulty in using logarithmic differentiation to obtain the coefficients $\{a'_n(r_1)\}$ in the second solution. The following example illustrates this difficulty and shows how to overcome it.

EXAMPLE 4   Find a fundamental set of Frobenius solutions of

$$x^2y'' - x(5 - x)y' + (9 - 4x)y = 0. \tag{27}$$

Give explicit formulas for the coefficients in the solutions.

Solution   For (27) the polynomials defined in Theorem 7.6.1 are

$$p_0(r) = r(r - 1) - 5r + 9 = (r - 3)^2,$$
$$p_1(r) = \quad\quad r - 4,$$
$$p_2(r) = \quad\quad\quad 0.$$

Since $r_1 = 3$ is a repeated zero of the indicial polynomial p_0, Theorem 7.6.2 implies that

$$y_1 = x^3 \sum_{n=0}^{\infty} a_n(3)x^n \tag{28}$$

and

$$y_2 = y_1 \ln x + x^3 \sum_{n=1}^{\infty} a'_n(3)x^n \tag{29}$$

are linearly independent Frobenius solutions of (27). To find the coefficients in (28) we use the recurrence formulas

$$a_0(r) = 1,$$
$$a_n(r) = -\frac{p_1(n + r - 1)}{p_0(n + r)} a_{n-1}(r)$$
$$= -\frac{n + r - 5}{(n + r - 3)^2} a_{n-1}(r), \qquad n \geq 1.$$

We leave it to you to show that

$$a_n(r) = (-1)^n \prod_{j=1}^{n} \frac{j + r - 5}{(j + r - 3)^2}. \tag{30}$$

Setting $r = 3$ here yields

$$a_n(3) = (-1)^n \prod_{j=1}^{n} \frac{j - 2}{j^2},$$

so $a_1(3) = 1$ and $a_n(3) = 0$ if $n \geq 2$. Substituting these coefficients into (28) yields

$$y_1 = x^3(1 + x).$$

To obtain y_2 in (29) we must compute $a'_n(3)$ for $n = 1, 2, \ldots$ Let's first try logarithmic differentiation. From (30),

$$|a_n(r)| = \prod_{j=1}^{n} \frac{|j + r - 5|}{|j + r - 3|^2}, \qquad n \geq 1,$$

so

$$\ln |a_n(r)| = \sum_{j=1}^{n} (\ln |j + r - 5| - 2 \ln |j + r - 3|).$$

Differentiating with respect to r yields

$$\frac{a_n'(r)}{a_n(r)} = \sum_{j=1}^{n} \left(\frac{1}{j + r - 5} - \frac{2}{j + r - 3} \right).$$

Therefore

$$a_n'(r) = a_n(r) \sum_{j=1}^{n} \left(\frac{1}{j + r - 5} - \frac{2}{j + r - 3} \right). \tag{31}$$

However, we cannot simply set $r = 3$ here if $n \geq 2$, since the bracketed expression in the sum corresponding to $j = 2$ contains the term $1/(r - 3)$. In fact, since $a_n(3) = 0$ for $n \geq 2$, the formula (31) for $a_n'(r)$ is actually an indeterminate form at $r = 3$.

We overcome this difficulty as follows. From (30) with $n = 1$,

$$a_1(r) = -\frac{r - 4}{(r - 2)^2}.$$

Therefore,

$$a_1'(r) = \frac{r - 6}{(r - 2)^3},$$

so

$$a_1'(3) = -3. \tag{32}$$

From (30) with $n \geq 2$,

$$a_n(r) = (-1)^n (r - 4)(r - 3) \frac{\prod_{j=3}^{n} (j + r - 5)}{\prod_{j=1}^{n} (j + r - 3)^2} = (r - 3)c_n(r),$$

where

$$c_n(r) = (-1)^n (r - 4) \frac{\prod_{j=3}^{n} (j + r - 5)}{\prod_{j=1}^{n} (j + r - 3)^2}, \qquad n \geq 2.$$

Therefore,

$$a_n'(r) = c_n(r) + (r - 3)c_n'(r), \qquad n \geq 2,$$

which implies that $a_n'(3) = c_n(3)$ if $n \geq 3$. We leave it to you to verify that

$$a_n'(3) = c_n(3) = \frac{(-1)^{n+1}}{n(n - 1)n!}, \qquad n \geq 2.$$

Substituting this and (32) into (29) yields

$$y_2 = x^3(1 + x) \ln x - 3x^4 - x^3 \sum_{n=2}^{\infty} \frac{(-1)^n}{n(n - 1)n!} x^n.$$

∎

7.6 EXERCISES

In Exercises 1–11 find a fundamental set of Frobenius solutions. Compute the terms involving x^{n+r_1}, where $0 \le n \le N$ (N at least 3) and r_1 is the root of the indicial equation. Optionally, write a computer program to implement the applicable recurrence formulas and take $N > 3$.

C   **1.** $x^2 y'' - x(1 - x)y' + (1 - x^2)y = 0$

C   **2.** $x^2(1 + x + 2x^2)y' + x(3 + 6x + 7x^2)y' + (1 + 6x - 3x^2)y = 0$

C   **3.** $x^2(1 + 2x + x^2)y'' + x(1 + 3x + 4x^2)y' - x(1 - 2x)y = 0$

C   **4.** $4x^2(1 + x + x^2)y'' + 12x^2(1 + x)y' + (1 + 3x + 3x^2)y = 0$

C   **5.** $x^2(1 + x + x^2)y'' - x(1 - 4x - 2x^2)y' + y = 0$

C   **6.** $9x^2 y'' + 3x(5 + 3x - 2x^2)y' + (1 + 12x - 14x^2)y = 0$

C   **7.** $x^2 y'' + x(1 + x + x^2)y' + x(2 - x)y = 0$

C   **8.** $x^2(1 + 2x)y'' + x(5 + 14x + 3x^2)y' + (4 + 18x + 12x^2)y = 0$

C   **9.** $4x^2 y'' + 2x(4 + x + x^2)y' + (1 + 5x + 3x^2)y = 0$

C   **10.** $16x^2 y'' + 4x(6 + x + 2x^2)y' + (1 + 5x + 18x^2)y = 0$

C   **11.** $9x^2(1 + x)y'' + 3x(5 + 11x - x^2)y' + (1 + 16x - 7x^2)y = 0$

In Exercises 12–22 find a fundamental set of Frobenius solutions. Give explicit formulas for the coefficients.

12. $4x^2 y'' + (1 + 4x)y = 0$

13. $36x^2(1 - 2x)y'' + 24x(1 - 9x)y' + (1 - 70x)y = 0$

14. $x^2(1 + x)y'' - x(3 - x)y' + 4y = 0$ **15.** $x^2(1 - 2x)y'' - x(5 - 4x)y' + (9 - 4x)y = 0$

16. $25x^2 y'' + x(15 + x)y' + (1 + x)y = 0$ **17.** $2x^2(2 + x)y'' + x^2 y' + (1 - x)y = 0$

18. $x^2(9 + 4x)y'' + 3xy' + (1 + x)y = 0$ **19.** $x^2 y'' - x(3 - 2x)y' + (4 + 3x)y = 0$

20. $x^2(1 - 4x)y'' + 3x(1 - 6x)y' + (1 - 12x)y = 0$ **21.** $x^2(1 + 2x)y'' + x(3 + 5x)y' + (1 - 2x)y = 0$

22. $2x^2(1 + x)y'' - x(6 - x)y' + (8 - x)y = 0$

In Exercises 23–27 find a fundamental set of Frobenius solutions. Compute the terms involving x^{n+r_1}, where $0 \le n \le N$ (N at least 3) and r_1 is the root of the indicial equation. Optionally, write a computer program to implement the applicable recurrence formulas and take $N > 3$.

C   **23.** $x^2(1 + 2x)y'' + x(5 + 9x)y' + (4 + 3x)y = 0$ **C**   **24.** $x^2(1 - 2x)y'' - x(5 + 4x)y' + (9 + 4x)y = 0$

C   **25.** $x^2(1 + 4x)y'' - x(1 - 4x)y' + (1 + x)y = 0$ **C**   **26.** $x^2(1 + x)y'' + x(1 + 2x)y' + xy = 0$

C   **27.** $x^2(1 - x)y'' + x(7 + x)y' + (9 - x)y = 0$

In Exercises 28–38 find a fundamental set of Frobenius solutions. Give explicit formulas for the coefficients.

28. $x^2 y'' - x(1 - x^2)y' + (1 + x^2)y = 0$ **29.** $x^2(1 + x^2)y'' - 3x(1 - x^2)y' + 4y = 0$

30. $4x^2 y'' + 2x^3 y' + (1 + 3x^2)y = 0$ **31.** $x^2(1 + x^2)y'' - x(1 - 2x^2)y' + y = 0$

32. $2x^2(2 + x^2)y'' + 7x^3 y' + (1 + 3x^2)y = 0$ **33.** $x^2(1 + x^2)y'' - x(1 - 4x^2)y' + (1 + 2x^2)y = 0$

34. $4x^2(4 + x^2)y'' + 3x(8 + 3x^2)y' + (1 - 9x^2)y = 0$

35. $3x^2(3 + x^2)y'' + x(3 + 11x^2)y' + (1 + 5x^2)y = 0$

36. $4x^2(1 + 4x^2)y'' + 32x^3 y' + y = 0$ **37.** $9x^2 y'' - 3x(7 - 2x^2)y' + (25 + 2x^2)y = 0$

38. $x^2(1 + 2x^2)y'' + x(3 + 7x^2)y' + (1 - 3x^2)y = 0$

In Exercises 39–43 find a fundamental set of Frobenius solutions. Compute the terms involving x^{2m+r_1}, where $0 \le m \le M$ (M at least 3) and r_1 is the root of the indicial equation. Optionally, write a computer program to implement the applicable recurrence formulas and take $M > 3$.

C   **39.** $x^2(1 + x^2)y'' + x(3 + 8x^2)y' + (1 + 12x^2)y = 0$

C **40.** $x^2y'' - x(1 - x^2)y' + (1 + x^2)y = 0$

C **41.** $x^2(1 - 2x^2)y'' + x(5 - 9x^2)y' + (4 - 3x^2)y = 0$

C **42.** $x^2(2 + x^2)y'' + x(14 - x^2)y' + 2(9 + x^2)y = 0$

C **43.** $x^2(1 + x^2)y'' + x(3 + 7x^2)y' + (1 + 8x^2)y = 0$

In Exercises 44–52 find a fundamental set of Frobenius solutions. Give explicit formulas for the coefficients.

44. $x^2(1 - 2x)y'' + 3xy' + (1 + 4x)y = 0$

45. $x(1 + x)y'' + (1 - x)y' + y = 0$

46. $x^2(1 - x)y'' + x(3 - 2x)y' + (1 + 2x)y = 0$

47. $4x^2(1 + x)y'' - 4x^2y' + (1 - 5x)y = 0$

48. $x^2(1 - x)y'' - x(3 - 5x)y' + (4 - 5x)y = 0$

49. $x^2(1 + x^2)y'' - x(1 + 9x^2)y' + (1 + 25x^2)y = 0$

50. $9x^2y'' + 3x(1 - x^2)y' + (1 + 7x^2)y = 0$

51. $x(1 + x^2)y'' + (1 - x^2)y' - 8xy = 0$

52. $4x^2y'' + 2x(4 - x^2)y' + (1 + 7x^2)y = 0$

53. Under the assumptions of Theorem 7.6.2, suppose that the power series

$$\sum_{n=0}^{\infty} a_n(r_1)x^n \quad \text{and} \quad \sum_{n=1}^{\infty} a_n'(r_1)x^n$$

converge on $(-\rho, \rho)$.

(a) Show that

$$y_1 = x^{r_1} \sum_{n=0}^{\infty} a_n(r_1)x^n \quad \text{and} \quad y_2 = y_1 \ln x + x^{r_1} \sum_{n=1}^{\infty} a_n'(r_1)x^n$$

are linearly independent on $(0, \rho)$.

Hint: Show that if c_1 and c_2 are constants such that $c_1 y_1 + c_2 y_2 \equiv 0$ on $(0, \rho)$ then

$$(c_1 + c_2 \ln x) \sum_{n=0}^{\infty} a_n(r_1)x^n + c_2 \sum_{n=1}^{\infty} a_n'(r_1)x^n = 0, \quad 0 < x < \rho.$$

Then let $x \to 0+$ to conclude that $c_2 = 0$.

(b) Use the result of **(a)** to complete the proof of Theorem 7.6.2.

54. Let

$$Ly = x^2(\alpha_0 + \alpha_1 x)y'' + x(\beta_0 + \beta_1 x)y' + (\gamma_0 + \gamma_1 x)y$$

and define

$$p_0(r) = \alpha_0 r(r - 1) + \beta_0 r + \gamma_0 \quad \text{and} \quad p_1(r) = \alpha_1 r(r - 1) + \beta_1 r + \gamma_1.$$

Theorem 7.6.1 and Exercise 55**(a)** of Section 7.5 imply that if

$$y(x, r) = x^r \sum_{n=0}^{\infty} a_n(r)x^n,$$

where

$$a_n(r) = (-1)^n \prod_{j=1}^{n} \frac{p_1(j + r - 1)}{p_0(j + r)},$$

then

$$Ly(x, r) = p_0(r)x^r.$$

Now suppose that $p_0(r) = \alpha_0(r - r_1)^2$ and $p_1(k + r_1) \neq 0$ if k is a nonnegative integer.

(a) Show that $Ly = 0$ has the solution

$$y_1 = x^{r_1} \sum_{n=0}^{\infty} a_n(r_1)x^n,$$

where

$$a_n(r_1) = \frac{(-1)^n}{\alpha_0^n(n!)^2} \prod_{j=1}^{n} p_1(j + r_1 - 1).$$

(b) Show that $Ly = 0$ has the second solution

$$y_2 = y_1 \ln x + x^{r_1} \sum_{n=1}^{\infty} a_n(r_1)J_n x^n,$$

where

$$J_n = \sum_{j=1}^{n} \frac{p_1'(j + r_1 - 1)}{p_1(j + r_1 - 1)} - 2 \sum_{j=1}^{n} \frac{1}{j}.$$

(c) Conclude from **(a)** and **(b)** that if $\gamma_1 \neq 0$ then

$$y_1 = x^{r_1} \sum_{n=0}^{\infty} \frac{(-1)^n}{(n!)^2} \left(\frac{\gamma_1}{\alpha_0}\right)^n x^n$$

and

$$y_2 = y_1 \ln x - 2x^{r_1} \sum_{n=1}^{\infty} \frac{(-1)^n}{(n!)^2} \left(\frac{\gamma_1}{\alpha_0}\right)^n \left(\sum_{j=1}^{n} \frac{1}{j}\right) x_n$$

are solutions of

$$\alpha_0 x^2 y'' + \beta_0 xy' + (\gamma_0 + \gamma_1 x)y = 0.$$

(The conclusion is also valid if $\gamma_1 = 0$. Why?)

55. Let

$$Ly = x^2(\alpha_0 + \alpha_q x^q)y'' + x(\beta_0 + \beta_q x^q)y' + (\gamma_0 + \gamma_q x^q)y$$

where q is a positive integer, and define

$$p_0(r) = \alpha_0 r(r - 1) + \beta_0 r + \gamma_0 \quad \text{and} \quad p_q(r) = \alpha_q r(r - 1) + \beta_q r + \gamma_q.$$

Suppose that $p_0(r) = \alpha_0(r - r_1)^2$ and $p_q(qk + r_1) \neq 0$ if k is a nonnegative integer.

(a) Recall from Exercise 59 of Section 7.5 that $Ly = 0$ has the solution

$$y_1 = x^{r_1} \sum_{m=0}^{\infty} a_{qm}(r_1)x^{qm},$$

where

$$a_{qm}(r_1) = \frac{(-1)^m}{(q^2\alpha_0)^m(m!)^2} \prod_{j=1}^{m} p_q(q(j - 1) + r_1).$$

(b) Show that $Ly = 0$ has the second solution

$$y_2 = y_1 \ln x + x^{r_1} \sum_{m=1}^{\infty} a'_{qm}(r_1)J_m x^{qm},$$

where

$$J_m = \sum_{j=1}^{m} \frac{p_q'(q(j-1) + r_1)}{p_q(q(j-1) + r_1)} - \frac{2}{q} \sum_{j=1}^{m} \frac{1}{j}.$$

(c) Conclude from **(a)** and **(b)** that if $\gamma_q \neq 0$ then

$$y_1 = x^{r_1} \sum_{m=0}^{\infty} \frac{(-1)^m}{(m!)^2} \left(\frac{\gamma_q}{q^2 \alpha_0} \right)^m x^{qm}$$

and

$$y_2 = y_1 \ln x - \frac{2}{q} x^{r_1} \sum_{m=1}^{\infty} \frac{(-1)^m}{(m!)^2} \left(\frac{\gamma_q}{q^2 \alpha_0} \right)^m \left(\sum_{j=1}^{m} \frac{1}{j} \right) x^{qm}$$

are solutions of

$$\alpha_0 x^2 y'' + \beta_0 xy' + (\gamma_0 + \gamma_q x^q)y = 0.$$

56. The equation

$$xy'' + y' + xy = 0$$

is known as **Bessel's equation of order zero.** (See Exercise 53 of Section 7.5.) Find two linearly independent Frobenius solutions of this equation.

57. Suppose that the assumptions of Exercise 60 of Section 7.5 hold, except that

$$p_0(r) = \alpha_0 (r - r_1)^2.$$

Show that

$$y_1 = \frac{x^{r_1}}{\alpha_0 + \alpha_1 x + \alpha_2 x^2} \quad \text{and} \quad y_2 = \frac{x^{r_1} \ln x}{\alpha_0 + \alpha_1 x + \alpha_2 x^2}$$

are linearly independent Frobenius solutions of

$$x^2(\alpha_0 + \alpha_1 x + \alpha_2 x^2)y'' + x(\beta_0 + \beta_1 x + \beta_2 x^2)y' + (\gamma_0 + \gamma_1 x + \gamma_2 x^2)y = 0$$

on any interval $(0, \rho)$ on which $\alpha_0 + \alpha_1 x + \alpha_2 x^2$ has no zeros.

In Exercises 58–65 use the method suggested by Exercise 57 to find the general solution on some interval $(0, \rho)$.

58. $4x^2(1 + x)y'' + 8x^2 y' + (1 + x)y = 0$ **59.** $9x^2(3 + x)y'' + 3x(3 + 7x)y' + (3 + 4x)y = 0$

60. $x^2(2 - x^2)y'' - x(2 + 3x^2)y' + (2 - x^2)y = 0$

61. $16x^2(1 + x^2)y'' + 8x(1 + 9x^2)y' + (1 + 49x^2)y = 0$

62. $x^2(4 + 3x)y'' - x(4 - 3x)y' + 4y = 0$

63. $4x^2(1 + 3x + x^2)y'' + 8x^2(3 + 2x)y' + (1 + 3x + 9x^2)y = 0$

64. $x^2(1 - x)^2 y'' - x(1 + 2x - 3x^2)y' + (1 + x^2)y = 0$

65. $9x^2(1 + x + x^2)y'' + 3x(1 + 7x + 13x^2)y' + (1 + 4x + 25x^2)y = 0$

66. (a) Let L and $y(x, r)$ be as in Exercises 57 and 58 of Section 7.5. Extend Theorem 7.6.2 by showing that

$$L \left(\frac{\partial y}{\partial r}(x, r) \right) = p_0'(r)x^r + x^r p_0(r) \ln x.$$

(b) Show that if

$$p_0(r) = \alpha_0 (r - r_1)^2$$

then

$$y_1 = y(x, r_1) \quad \text{and} \quad y_2 = \frac{\partial y}{\partial r}(x, r_1)$$

are solutions of $Ly = 0$.

7.7   The Method of Frobenius III

In Sections 7.5 and 7.6 we discussed methods for finding Frobenius solutions of a homogeneous linear second order equation near a regular singular point in the case where the indicial equation has a repeated root or distinct real roots that do not differ by an integer. In this section we consider the case where the indicial equation has distinct real roots that differ by an integer. We will limit our discussion to equations that can be written as

$$x^2(\alpha_0 + \alpha_1 x)y'' + x(\beta_0 + \beta_1 x)y' + (\gamma_0 + \gamma_1 x)y = 0 \tag{1}$$

or

$$x^2(\alpha_0 + \alpha_2 x^2)y'' + x(\beta_0 + \beta_2 x^2)y' + (\gamma_0 + \gamma_2 x^2)y = 0,$$

where the roots of the indicial equation differ by a positive integer.

We begin with a theorem that provides a fundamental set of solutions of equations of the form (1).

THEOREM 7.7.1

Let

$$Ly = x^2(\alpha_0 + \alpha_1 x)y'' + x(\beta_0 + \beta_1 x)y' + (\gamma_0 + \gamma_1 x)y$$

where $\alpha_0 \neq 0$, and define

$$p_0(r) = \alpha_0 r(r - 1) + \beta_0 r + \gamma_0,$$
$$p_1(r) = \alpha_1 r(r - 1) + \beta_1 r + \gamma_1.$$

Suppose that r is a real number such that $p_0(n + r)$ is nonzero for all positive integers n, and define

$$a_0(r) = 1,$$
$$a_n(r) = -\frac{p_1(n + r - 1)}{p_0(n + r)} a_{n-1}(r), \qquad n \geq 1. \tag{2}$$

Let r_1 and r_2 be the roots of the indicial equation $p_0(r) = 0$, and suppose that $r_1 = r_2 + k$, where k is a positive integer. Then

$$y_1 = x^{r_1} \sum_{n=0}^{\infty} a_n(r_1)x^n$$

is a Frobenius solution of $Ly = 0$. Moreover, if we define

$$a_0(r_2) = 1,$$
$$a_n(r_2) = -\frac{p_1(n + r_2 - 1)}{p_0(n + r_2)} a_{n-1}(r_2), \qquad 1 \leq n \leq k - 1, \tag{3}$$

and

$$C = -\frac{p_1(r_1 - 1)}{k\alpha_0} a_{k-1}(r_2), \tag{4}$$

then

$$y_2 = x^{r_2} \sum_{n=0}^{k-1} a_n(r_2)x^n + C\left(y_1 \ln x + x^{r_1} \sum_{n=1}^{\infty} a'_n(r_1)x^n \right) \tag{5}$$

is also a solution of $Ly = 0$, and $\{y_1, y_2\}$ is a fundamental set of solutions.

PROOF. Theorem 7.5.3 implies that $Ly_1 = 0$. We will now show that $Ly_2 = 0$. Since L is a linear operator this is equivalent to showing that

$$L\left(x^{r_2} \sum_{n=0}^{k-1} a_n(r_2)x^n \right) + CL\left(y_1 \ln x + x^{r_1} \sum_{n=1}^{\infty} a'_n(r_1)x^n \right) = 0. \tag{6}$$

To verify this we'll show that

$$L\left(x^{r_2} \sum_{n=0}^{k-1} a_n(r_2)x^n \right) = p_1(r_1 - 1)a_{k-1}(r_2)x^{r_1} \tag{7}$$

and

$$L\left(y_1 \ln x + x^{r_1} \sum_{n=1}^{\infty} a'_n(r_1)x^n \right) = k\alpha_0 x^{r_1}. \tag{8}$$

This will imply that $Ly_2 = 0$, since substituting (7) and (8) into (6) and using (4) yields

$$Ly_2 = [p_1(r_1 - 1)a_{k-1}(r_2) + Ck\alpha_0]x^{r_1}$$
$$= [p_1(r_1 - 1)a_{k-1}(r_2) - p_1(r_1 - 1)a_{k-1}(r_2)]x^{r_1} = 0.$$

We will prove (8) first. From Theorem 7.6.1,

$$L\left(y(x,r) \ln x + x^r \sum_{n=1}^{\infty} a'_n(r)x^n \right) = p'_0(r)x^r + x^r p_0(r) \ln x.$$

Setting $r = r_1$ and recalling that $p_0(r_1) = 0$ and $y_1 = y(x,r_1)$ yields

$$L\left(y_1 \ln x + x^{r_1} \sum_{n=1}^{\infty} a'_n(r_1)x^n \right) = p'_0(r_1)x^{r_1}. \tag{9}$$

Since r_1 and r_2 are the roots of the indicial equation, the indicial polynomial can be written as

$$p_0(r) = \alpha_0(r - r_1)(r - r_2) = \alpha_0[r^2 - (r_1 + r_2)r + r_1 r_2].$$

Differentiating this yields

$$p'_0(r) = \alpha_0(2r - r_1 - r_2).$$

Therefore $p'_0(r_1) = \alpha_0(r_1 - r_2) = k\alpha_0$, so (9) implies (8).

Before proving (7) we first note $a_n(r_2)$ is well defined by (3) for $1 \le n \le k - 1$, since $p_0(n + r_2) \ne 0$ for these values of n. However, we cannot define $a_n(r_2)$ for $n \ge k$ by means of (3), since $p_0(k + r_2) = p_0(r_1) = 0$. For convenience we define $a_n(r_2) = 0$ for $n \ge k$. Then, from Theorem 7.5.1,

$$L\left(x^{r_2} \sum_{n=0}^{k-1} a_n(r_2)x^n \right) = L\left(x^{r_2} \sum_{n=0}^{\infty} a_n(r_2)x^n \right) = x^{r_2} \sum_{n=0}^{\infty} b_n x^n, \tag{10}$$

where $b_0 = p_0(r_2) = 0$ and

$$b_n = p_0(n + r_2)a_n(r_2) + p_1(n + r_2 - 1)a_{n-1}(r_2), \qquad n \ge 1.$$

If $1 \le n \le k - 1$ then (3) implies that $b_n = 0$. If $n \ge k + 1$ then $b_n = 0$ because $a_{n-1}(r_2) = a_n(r_2) = 0$. Therefore (10) reduces to

$$L\left(x^{r_2} \sum_{n=0}^{k-1} a_n(r_2)x^n \right) = [p_0(k + r_2)a_k(r_2) + p_1(k + r_2 - 1)a_{k-1}(r_2)]x^{k+r_2}.$$

Since $a_k(r_2) = 0$ and $k + r_2 = r_1$, this implies (7).

We leave the proof that $\{y_1, y_2\}$ is a fundamental set as an exercise (Exercise 41). $\qquad \square$

EXAMPLE 1   Find a fundamental set of Frobenius solutions of

$$2x^2(2 + x)y'' - x(4 - 7x)y' - (5 - 3x)y = 0.$$

Give explicit formulas for the coefficients in the solutions.

Solution   For the given equation the polynomials defined in Theorem 7.7.1 are

$$p_0(r) = 4r(r - 1) - 4r - 5 = (2r + 1)(2r - 5),$$
$$p_1(r) = 2r(r - 1) + 7r + 3 = (r + 1)(2r + 3).$$

The roots of the indicial equation are $r_1 = 5/2$ and $r_2 = -1/2$, so $k = r_1 - r_2 = 3$. Therefore Theorem 7.7.1 implies that

$$y_1 = x^{5/2} \sum_{n=0}^{\infty} a_n(5/2)x^n \tag{11}$$

and

$$y_2 = x^{-1/2} \sum_{n=0}^{2} a_n(-1/2)x^n + C\left(y_1 \ln x + x^{5/2} \sum_{n=1}^{\infty} a_n'(5/2)x^n \right) \tag{12}$$

(with C as in (4)) form a fundamental set of solutions of $Ly = 0$. The recurrence formula (2) is

$$a_0(r) = 1,$$
$$a_n(r) = -\frac{p_1(n + r - 1)}{p_0(n + r)}a_{n-1}(r)$$
$$= -\frac{(n + r)(2n + 2r + 1)}{(2n + 2r + 1)(2n + 2r - 5)}a_{n-1}(r), \tag{13}$$
$$= -\frac{n + r}{2n + 2r - 5}a_{n-1}(r), \qquad n \ge 1,$$

which implies that

$$a_n(r) = (-1)^n \prod_{j=1}^{n} \frac{j + r}{2j + 2r - 5}, \qquad n \ge 0. \tag{14}$$

Therefore,

$$a_n(5/2) = \frac{(-1)^n \prod_{j=1}^{n} (2j + 5)}{4^n n!}. \tag{15}$$

Substituting this into (11) yields

$$y_1 = x^{5/2} \sum_{n=0}^{\infty} \frac{(-1)^n \prod_{j=1}^{n} (2j + 5)}{4^n n!} x^n.$$

To compute the coefficients $a_0(-1/2)$, $a_1(-1/2)$, and $a_2(-1/2)$ in y_2, we set $r = -1/2$ in (13) and apply the resulting recurrence formula for $n = 1, 2$; thus,

$$a_0(-1/2) = 1,$$

$$a_n(-1/2) = -\frac{2n - 1}{4(n - 3)} a_{n-1}(-1/2), \qquad n = 1, 2.$$

The last formula yields

$$a_1(-1/2) = 1/8, \qquad a_2(-1/2) = 3/32.$$

Substituting $r_1 = 5/2$, $r_2 = -1/2$, $k = 3$, and $\alpha_0 = 4$ into (4) yields $C = -15/128$. Therefore, from (12),

$$y_2 = x^{-1/2}\left(1 + \frac{1}{8}x + \frac{3}{32}x^2\right) - \frac{15}{128}\left(y_1 \ln x + x^{5/2} \sum_{n=1}^{\infty} a_n'(5/2)x^n\right). \quad (16)$$

To obtain $a_n'(r)$ we use logarithmic differentiation. From (14),

$$|a_n(r)| = \prod_{j=1}^{n} \frac{|j + r|}{|2j + 2r - 5|}, \qquad n \geq 1.$$

Therefore

$$\ln|a_n(r)| = \sum_{j=1}^{n} (\ln|j + r| - \ln|2j + 2r - 5|).$$

Differentiating with respect to r yields

$$\frac{a_n'(r)}{a_n(r)} = \sum_{j=1}^{n} \left(\frac{1}{j + r} - \frac{2}{2j + 2r - 5}\right).$$

Therefore

$$a_n'(r) = a_n(r) \sum_{j=1}^{n} \left(\frac{1}{j + r} - \frac{2}{2j + 2r - 5}\right).$$

Setting $r = 5/2$ here and recalling (15) yields

$$a_n'(5/2) = \frac{(-1)^n \prod_{j=1}^{n} (2j + 5)}{4^n n!} \sum_{j=1}^{n} \left(\frac{1}{j + 5/2} - \frac{1}{j}\right). \quad (17)$$

Since

$$\frac{1}{j + 5/2} - \frac{1}{j} = -\frac{5}{j(2j + 5)},$$

we can rewrite (17) as

$$a_n'(5/2) = -5\frac{(-1)^n \prod_{j=1}^{n} (2j + 5)}{4^n n!}\left(\sum_{j=1}^{n} \frac{1}{j(2j + 5)}\right).$$

Substituting this into (16) yields

$$y_2 = x^{-1/2}\left(1 + \frac{1}{8}x + \frac{3}{32}x^2\right) - \frac{15}{128}y_1 \ln x$$

$$+ \frac{75}{128} x^{5/2} \sum_{n=1}^{\infty} \frac{(-1)^n \prod_{j=1}^{n} (2j + 5)}{4^n n!}\left(\sum_{j=1}^{n} \frac{1}{j(2j + 5)}\right)x^n.$$ ■

If $C = 0$ in (4) then there is no need to compute

$$y_1 \ln x + x^{r_1} \sum_{n=1}^{\infty} a'_n(r_1)x^n$$

in the formula (5) for y_2. Therefore, it is best to compute C before computing $\{a'_n(r_1)\}_{n=1}^{\infty}$. This is illustrated in the next example. (See also Exercises 44 and 45.)

EXAMPLE 2 Find a fundamental set of Frobenius solutions of

$$x^2(1 - 2x)y'' + x(8 - 9x)y' + (6 - 3x)y = 0.$$

Give explicit formulas for the coefficients in the solutions.

Solution For the given equation the polynomials defined in Theorem 7.7.1 are

$$p_0(r) = r(r - 1) + 8r + 6 = (r + 1)(r + 6),$$
$$p_1(r) = -2r(r - 1) - 9r - 3 = -(r + 3)(2r + 1).$$

The roots of the indicial equation are $r_1 = -1$ and $r_2 = -6$, so $k = r_1 - r_2 = 5$. Therefore Theorem 7.7.1 implies that

$$y_1 = x^{-1} \sum_{n=0}^{\infty} a_n(-1)x^n \tag{18}$$

and

$$y_2 = x^{-6} \sum_{n=0}^{4} a_n(-6)x^n + C\left(y_1 \ln x + x^{-1} \sum_{n=1}^{\infty} a'_n(-1)x^n\right) \tag{19}$$

(with C as in (4)) form a fundamental set of solutions of $Ly = 0$. The recurrence formula (2) is

$$a_0(r) = 1,$$
$$a_n(r) = -\frac{p_1(n + r - 1)}{p_0(n + r)}a_{n-1}(r) \tag{20}$$
$$= \frac{(n + r + 2)(2n + 2r - 1)}{(n + r + 1)(n + r + 6)}a_{n-1}(r), \quad n \geq 1,$$

which implies that

$$a_n(r) = \prod_{j=1}^{n} \frac{(j + r + 2)(2j + 2r - 1)}{(j + r + 1)(j + r + 6)} \tag{21}$$
$$= \left(\prod_{j=1}^{n} \frac{j + r + 2}{j + r + 1}\right)\left(\prod_{j=1}^{n} \frac{2j + 2r - 1}{j + r + 6}\right).$$

Since

$$\prod_{j=1}^{n} \frac{j + r + 2}{j + r + 1} = \frac{(r + 3)(r + 4) \cdots (n + r + 2)}{(r + 2)(r + 3) \cdots (n + r + 1)} = \frac{n + r + 2}{r + 2}$$

because of cancellations, (21) simplifies to

$$a_n(r) = \frac{n + r + 2}{r + 2} \prod_{j=1}^{n} \frac{2j + 2r - 1}{j + r + 6}.$$

Therefore

$$a_n(-1) = (n + 1) \prod_{j=1}^{n} \frac{2j - 3}{j + 5}.$$

Substituting this into (18) yields

$$y_1 = x^{-1} \sum_{n=0}^{\infty} (n + 1) \left(\prod_{j=1}^{n} \frac{2j - 3}{j + 5} \right) x^n.$$

To compute the coefficients $a_0(-6), \ldots, a_4(-6)$ in y_2, we set $r = -6$ in (20) and apply the resulting recurrence formula for $n = 1, 2, 3, 4$; thus,

$$a_0(-6) = 1,$$

$$a_n(-6) = \frac{(n - 4)(2n - 13)}{n(n - 5)} a_{n-1}(-6), \qquad n = 1, 2, 3, 4.$$

The last formula yields

$$a_1(-6) = -\frac{33}{4}, \qquad a_2(-6) = \frac{99}{4}, \qquad a_3(-6) = -\frac{231}{8}, \qquad a_4(-6) = 0.$$

Since $a_4(-6) = 0$, it follows from (4) that the constant C in (19) is zero. Therefore (19) reduces to

$$y_2 = x^{-6} \left(1 - \frac{33}{4} x + \frac{99}{4} x^2 - \frac{231}{8} x^3 \right). \qquad \blacksquare$$

We now consider equations of the form

$$x^2(\alpha_0 + \alpha_2 x^2) y'' + x(\beta_0 + \beta_2 x^2) y' + (\gamma_0 + \gamma_2 x^2) y = 0,$$

where the roots of the indicial equation are real and differ by an even integer. The case where the roots are real and differ by an odd integer can be handled by the method discussed in Exercise 56 of Section 7.5.

The proof of the following theorem is similar to the proof of Theorem 7.7.1 (Exercise 43).

THEOREM 7.7.2

Let

$$Ly = x^2(\alpha_0 + \alpha_2 x^2) y'' + x(\beta_0 + \beta_2 x^2) y' + (\gamma_0 + \gamma_2 x^2) y$$

where $\alpha_0 \neq 0$, and define

$$p_0(r) = \alpha_0 r(r - 1) + \beta_0 r + \gamma_0,$$
$$p_2(r) = \alpha_2 r(r - 1) + \beta_2 r + \gamma_2.$$

Suppose that r is a real number such that $p_0(2m + r)$ is nonzero for all positive integers m, and define

$$a_0(r) = 1,$$

$$a_{2m}(r) = -\frac{p_2(2m + r - 2)}{p_0(2m + r)} a_{2m-2}(r), \qquad m \geq 1. \qquad (22)$$

Let r_1 and r_2 be the roots of the indicial equation $p_0(r) = 0$, and suppose that $r_1 = r_2 + 2k$, where k is a positive integer. Then

$$y_1 = x^{r_1} \sum_{m=0}^{\infty} a_{2m}(r_1)x^{2m}$$

is a Frobenius solution of $Ly = 0$. Moreover, if we define

$$a_0(r_2) = 1,$$

$$a_{2m}(r_2) = -\frac{p_2(2m + r_2 - 2)}{p_0(2m + r_2)}a_{2m-2}(r_2), \qquad 1 \le m \le k - 1,$$

and

$$C = -\frac{p_2(r_1 - 2)}{2k\alpha_0}a_{2k-2}(r_2), \tag{23}$$

then

$$y_2 = x^{r_2} \sum_{m=0}^{k-1} a_{2m}(r_2)x^{2m} + C\left(y_1 \ln x + x^{r_1} \sum_{m=1}^{\infty} a'_{2m}(r_1)x^{2m}\right) \tag{24}$$

is also a solution of $Ly = 0$, and $\{y_1, y_2\}$ is a fundamental set of solutions.

EXAMPLE 3   Find a fundamental set of Frobenius solutions of

$$x^2(1 + x^2)y'' + x(3 + 10x^2)y' - (15 - 14x^2)y = 0.$$

Give explicit formulas for the coefficients in the solutions.

Solution   For the given equation the polynomials defined in Theorem 7.7.2 are

$$p_0(r) = r(r - 1) + 3r - 15 = (r - 3)(r + 5),$$
$$p_2(r) = r(r - 1) + 10r + 14 = (r + 2)(r + 7).$$

The roots of the indicial equation are $r_1 = 3$ and $r_2 = -5$, so $k = (r_1 - r_2)/2 = 4$. Therefore Theorem 7.7.2 implies that

$$y_1 = x^3 \sum_{m=0}^{\infty} a_{2m}(3)x^{2m} \tag{25}$$

and

$$y_2 = x^{-5} \sum_{m=0}^{3} a_{2m}(-5)x^{2m} + C\left(y_1 \ln x + x^3 \sum_{m=1}^{\infty} a'_{2m}(3)x^{2m}\right)$$

(with C as in (23)) form a fundamental set of solutions of $Ly = 0$. The recurrence formula (22) is

$$a_0(r) = 1,$$

$$a_{2m}(r) = -\frac{p_2(2m + r - 2)}{p_0(2m + r)}a_{2m-2}(r)$$

$$= -\frac{(2m + r)(2m + r + 5)}{(2m + r - 3)(2m + r + 5)}a_{2m-2}(r) \tag{26}$$

$$= -\frac{2m + r}{2m + r - 3}a_{2m-2}(r), \qquad m \ge 1,$$

which implies that

$$a_{2m}(r) = (-1)^m \prod_{j=1}^{m} \frac{2j + r}{2j + r - 3}, \qquad m \ge 0. \tag{27}$$

Therefore

$$a_{2m}(3) = \frac{(-1)^m \prod_{j=1}^{m} (2j + 3)}{2^m m!}. \tag{28}$$

Substituting this into (25) yields

$$y_1 = x^3 \sum_{m=0}^{\infty} \frac{(-1)^m \prod_{j=1}^{m} (2j + 3)}{2^m m!} x^{2m}.$$

To compute the coefficients $a_2(-5)$, $a_4(-5)$, and $a_6(-5)$ in y_2, we set $r = -5$ in (26) and apply the resulting recurrence formula for $m = 1, 2, 3$; thus,

$$a_{2m}(-5) = -\frac{2m - 5}{2(m - 4)} a_{2m-2}(-5), \qquad m = 1, 2, 3.$$

This yields

$$a_2(-5) = -\frac{1}{2}, \qquad a_4(-5) = \frac{1}{8}, \qquad a_6(-5) = \frac{1}{16}.$$

Substituting $r_1 = 3$, $r_2 = -5$, $k = 4$, and $\alpha_0 = 1$ into (23) yields $C = -3/16$. Therefore, from (24),

$$y_2 = x^{-5}\left(1 - \frac{1}{2}x^2 + \frac{1}{8}x^4 + \frac{1}{16}x^6\right) - \frac{3}{16}\left(y_1 \ln x + x^3 \sum_{m=1}^{\infty} a'_{2m}(3)x^{2m}\right). \tag{29}$$

To obtain $a'_{2m}(r)$ we use logarithmic differentiation. From (27),

$$|a_{2m}(r)| = \prod_{j=1}^{m} \frac{|2j + r|}{|2j + r - 3|}, \qquad m \geq 1.$$

Therefore,

$$\ln |a_{2m}(r)| = \sum_{j=1}^{m} (\ln |2j + r| - \ln |2j + r - 3|).$$

Differentiating with respect to r yields

$$\frac{a'_{2m}(r)}{a_{2m}(r)} = \sum_{j=1}^{m} \left(\frac{1}{2j + r} - \frac{1}{2j + r - 3}\right).$$

Therefore

$$a'_{2m}(r) = a_{2m}(r) \sum_{j=1}^{m} \left(\frac{1}{2j + r} - \frac{1}{2j + r - 3}\right).$$

Setting $r = 3$ here and recalling (28) yields

$$a'_{2m}(3) = \frac{(-1)^m \prod_{j=1}^{m} (2j + 3)}{2^m m!} \sum_{j=1}^{m} \left(\frac{1}{2j + 3} - \frac{1}{2j}\right). \tag{30}$$

Since

$$\frac{1}{2j + 3} - \frac{1}{2j} = -\frac{3}{2j(2j + 3)},$$

we can rewrite (30) as

$$a'_{2m}(3) = -\frac{3}{2} \frac{(-1)^m \prod_{j=1}^{m} (2j + 3)}{2^m m!} \left(\sum_{j=1}^{m} \frac{1}{j(2j + 3)}\right).$$

Substituting this into (29) yields

$$y_2 = x^{-5}\left(1 - \frac{1}{2}x^2 + \frac{1}{8}x^4 + \frac{1}{16}x^6\right) - \frac{3}{16}y_1 \ln x$$
$$+ \frac{9}{32}x^3 \sum_{m=1}^{\infty} \frac{(-1)^m \prod_{j=1}^m (2j + 3)}{2^m m!} \left(\sum_{j=1}^m \frac{1}{j(2j + 3)}\right)x^{2m}.$$

EXAMPLE 4 Find a fundamental set of Frobenius solutions of

$$x^2(1 - 2x^2)y'' + x(7 - 13x^2)y' - 14x^2y = 0.$$

Give explicit formulas for the coefficients in the solutions.

Solution For the given equation the polynomials defined in Theorem 7.7.2 are

$$p_0(r) = \quad r(r - 1) + 7r \quad = \quad r(r + 6),$$
$$p_2(r) = -2r(r - 1) - 13r - 14 = -(r + 2)(2r + 7).$$

The roots of the indicial equation are $r_1 = 0$ and $r_2 = -6$, so $k = (r_1 - r_2)/2 = 3$. Therefore Theorem 7.7.2 implies that

$$y_1 = \sum_{m=0}^{\infty} a_{2m}(0)x^{2m} \tag{31}$$

and

$$y_2 = x^{-6} \sum_{m=0}^{2} a_{2m}(-6)x^{2m} + C\left(y_1 \ln x + \sum_{m=1}^{\infty} a'_{2m}(0)x^{2m}\right) \tag{32}$$

(with C as in (23)) form a fundamental set of solutions of $Ly = 0$. The recurrence formulas (22) are

$$a_0(r) = 1,$$
$$a_{2m}(r) = -\frac{p_2(2m + r - 2)}{p_0(2m + r)}a_{2m-2}(r)$$
$$= \frac{(2m + r)(4m + 2r + 3)}{(2m + r)(2m + r + 6)}a_{2m-2}(r) \tag{33}$$
$$= \frac{4m + 2r + 3}{2m + r + 6}a_{2m-2}(r), \quad m \geq 1,$$

which implies that

$$a_{2m}(r) = \prod_{j=1}^m \frac{4j + 2r + 3}{2j + r + 6}.$$

Setting $r = 0$ yields

$$a_{2m}(0) = 6\frac{\prod_{j=1}^m (4j + 3)}{2^m(m + 3)!}.$$

Substituting this into (31) yields

$$y_1 = 6\sum_{m=0}^{\infty} \frac{\prod_{j=1}^m (4j + 3)}{2^m(m + 3)!}x^{2m}.$$

To compute the coefficients $a_0(-6)$, $a_2(-6)$, and $a_4(-6)$ in y_2, we set $r = -6$ in (33) and apply the resulting recurrence formula for $m = 1, 2$; thus,

$$a_0(-6) = 1,$$

$$a_{2m}(-6) = \frac{4m - 9}{2m} a_{2m-2}(-6), \qquad m = 1, 2.$$

The last formula yields

$$a_2(-6) = -\frac{5}{2}, \qquad a_4(-6) = \frac{5}{8}.$$

Since $p_2(-2) = 0$ the constant C in (23) is zero. Therefore (32) reduces to

$$y_2 = x^{-6}\left(1 - \frac{5}{2}x^2 + \frac{5}{8}x^4\right). \qquad \blacksquare$$

7.7 EXERCISES

In Exercises 1–40 find a fundamental set of Frobenius solutions. Give explicit formulas for the coefficients.

1. $x^2y'' - 3xy' + (3 + 4x)y = 0$

2. $xy'' + y = 0$

3. $4x^2(1 + x)y'' + 4x(1 + 2x)y' - (1 + 3x)y = 0$

4. $xy'' + xy' + y = 0$

5. $2x^2(2 + 3x)y'' + x(4 + 21x)y' - (1 - 9x)y = 0$

6. $x^2y'' + x(2 + x)y' - (2 - 3x)y = 0$

7. $4x^2y'' + 4xy' - (9 - x)y = 0$

8. $x^2y'' + 10xy' + (14 + x)y = 0$

9. $4x^2(1 + x)y'' + 4x(3 + 8x)y' - (5 - 49x)y = 0$

10. $x^2(1 + x)y'' - x(3 + 10x)y' + 30xy = 0$

11. $x^2y'' + x(1 + x)y' - 3(3 + x)y = 0$

12. $x^2y'' + x(1 - 2x)y' - (4 + x)y = 0$

13. $x(1 + x)y'' - 4y' - 2y = 0$

14. $x^2(1 + 2x)y'' + x(9 + 13x)y' + (7 + 5x)y = 0$

15. $4x^2y'' - 2x(4 - x)y' - (7 + 5x)y = 0$

16. $3x^2(3 + x)y'' - x(15 + x)y' - 20y = 0$

17. $x^2(1 + x)y'' + x(1 - 10x)y' - (9 - 10x)y = 0$

18. $x^2(1 + x)y'' + 3x^2y' - (6 - x)y = 0$

19. $x^2(1 + 2x)y'' - 2x(3 + 14x)y' + (6 + 100x)y = 0$

20. $x^2(1 + x)y'' - x(6 + 11x)y' + (6 + 32x)y = 0$

21. $4x^2(1 + x)y'' + 4x(1 + 4x)y' - (49 + 27x)y = 0$

22. $x^2(1 + 2x)y'' - x(9 + 8x)y' - 12xy = 0$

23. $x^2(1 + x^2)y'' - x(7 - 2x^2)y' + 12y = 0$

24. $x^2y'' - x(7 - x^2)y' + 12y = 0$

25. $xy'' - 5y' + xy = 0$

26. $x^2y'' + x(1 + 2x^2)y' - (1 - 10x^2)y = 0$

27. $x^2y'' - xy' - (3 - x^2)y = 0$

28. $4x^2y'' + 2x(8 + x^2)y' + (5 + 3x^2)y = 0$

29. $x^2y'' + x(1 + x^2)y' - (1 - 3x^2)y = 0$

30. $x^2y'' + x(1 - 2x^2)y' - 4(1 + 2x^2)y = 0$

31. $4x^2y'' + 8xy' - (35 - x^2)y = 0$

32. $9x^2y'' - 3x(11 + 2x^2)y' + (13 + 10x^2)y = 0$

33. $x^2y'' + x(1 - 2x^2)y' - 4(1 - x^2)y = 0$

34. $x^2y'' + x(1 - 3x^2)y' - 4(1 - 3x^2)y = 0$

35. $x^2(1 + x^2)y'' + x(5 + 11x^2)y' + 24x^2y = 0$

36. $4x^2(1 + x^2)y'' + 8xy' - (35 - x^2)y = 0$

37. $x^2(1 + x^2)y'' - x(5 - x^2)y' - (7 + 25x^2)y = 0$

38. $x^2(1 + x^2)y'' + x(5 + 2x^2)y' - 21y = 0$

39. $x^2(1 + 2x^2)y'' - x(3 + x^2)y' - 2x^2y = 0$

40. $4x^2(1 + x^2)y'' + 4x(2 + x^2)y' - (15 + x^2)y = 0$

41. (a) Under the assumptions of Theorem 7.7.1 show that

$$y_1 = x^{r_1} \sum_{n=0}^{\infty} a_n(r_1)x^n$$

and

$$y_2 = x^{r_2} \sum_{n=0}^{k-1} a_n(r_2)x^n + C\left(y_1 \ln x + x^{r_1} \sum_{n=1}^{\infty} a_n'(r_1)x^n\right)$$

are linearly independent.

Hint: *If c_1 and c_2 are constants such that $c_1y_1 + c_2y_2 \equiv 0$ on an interval $(0,\rho)$ then*

$$x^{-r_2}(c_1y_1(x) + c_2y_2(x)) = 0, \qquad 0 < x < \rho.$$

Let $x \to 0+$ to conclude that $c_2 = 0$.

(b) Use the result of **(a)** to complete the proof of Theorem 7.7.1.

42. Find a fundamental set of Frobenius solutions of Bessel's equation

$$x^2y'' + xy' + (x^2 - \nu^2)y = 0$$

in the case where ν is a positive integer.

43. Prove Theorem 7.7.2.

44. Under the assumptions of Theorem 7.7.1 show that $C = 0$ if and only if $p_1(r_2 + l) = 0$ for some integer l in $\{0, \ldots, k - 1\}$.

45. Under the assumptions of Theorem 7.7.2 show that $C = 0$ if and only if $p_2(r_2 + 2l) = 0$ for some integer l in $\{0, \ldots, k - 1\}$.

46. Let

$$Ly = \alpha_0 x^2 y'' + \beta_0 xy' + (\gamma_0 + \gamma_1 x)y$$

and define

$$p_0(r) = \alpha_0 r(r - 1) + \beta_0 r + \gamma_0.$$

Show that if

$$p_0(r) = \alpha_0(r - r_1)(r - r_2)$$

where $r_1 - r_2 = k$, a positive integer, then $Ly = 0$ has the solutions

$$y_1 = x^{r_1} \sum_{n=0}^{\infty} \frac{(-1)^n}{n! \, \prod_{j=1}^{n} (j + k)} \left(\frac{\gamma_1}{\alpha_0}\right)^n x^n$$

and

$$y_2 = x^{r_2} \sum_{n=0}^{k-1} \frac{(-1)^n}{n! \, \prod_{j=1}^{n} (j - k)} \left(\frac{\gamma_1}{\alpha_0}\right)^n x^n$$

$$- \frac{1}{k!(k - 1)!} \left(\frac{\gamma_1}{\alpha_0}\right)^k \left(y_1 \ln x - x^{r_1} \sum_{n=1}^{\infty} \frac{(-1)^n}{n! \, \prod_{j=1}^{n} (j + k)} \left(\frac{\gamma_1}{\alpha_0}\right)^n \left(\sum_{j=1}^{n} \frac{2j + k}{j(j + k)}\right) x^n\right).$$

47. Let

$$Ly = \alpha_0 x^2 y'' + \beta_0 xy' + (\gamma_0 + \gamma_2 x^2)y$$

and define

$$p_0(r) = \alpha_0 r(r - 1) + \beta_0 r + \gamma_0.$$

Show that if

$$p_0(r) = \alpha_0(r - r_1)(r - r_2)$$

where $r_1 - r_2 = 2k$, an even positive integer, then $Ly = 0$ has the solutions

$$y_1 = x^{r_1} \sum_{m=0}^{\infty} \frac{(-1)^m}{4^m m! \, \prod_{j=1}^{m} (j + k)} \left(\frac{\gamma_2}{\alpha_0}\right)^m x^{2m}$$

and

$$y_2 = x^{r_2} \sum_{m=0}^{k-1} \frac{(-1)^m}{4^m m! \ \prod_{j=1}^{m} (j - k)} \left(\frac{\gamma_2}{\alpha_0}\right)^m x^{2m}$$

$$- \frac{2}{4^k k! (k - 1)!} \left(\frac{\gamma_2}{\alpha_0}\right)^k \left(y_1 \ln x - \frac{x^{r_1}}{2} \sum_{m=1}^{\infty} \frac{(-1)^m}{4^m m! \ \prod_{j=1}^{m} (j + k)} \left(\frac{\gamma_2}{\alpha_0}\right)^m \left(\sum_{j=1}^{m} \frac{2j + k}{j(j + k)}\right) x^{2m}\right).$$

48. Let L be as in Exercises 57 and 58 of Section 7.5, and suppose that the indicial polynomial of $Ly = 0$ is

$$p_0(r) = \alpha_0(r - r_1)(r - r_2),$$

with $r_1 - r_2 = k$, where k is a positive integer. Define $a_0(r) = 1$ for all r. If r is a real number such that $p_0(n + r)$ is nonzero for all positive integers n, define

$$a_n(r) = -\frac{1}{p_0(n + r)} \sum_{j=1}^{n} p_j(n + r - j)a_{n-j}(r), \qquad n \geq 1,$$

and let

$$y_1 = x^{r_1} \sum_{n=1}^{\infty} a_n(r_1)x^n.$$

Define

$$a_n(r_2) = -\frac{1}{p_0(n + r_2)} \sum_{j=1}^{n} p_j(n + r_2 - j)a_{n-j}(r_2) \qquad \text{if } n \geq 1 \text{ and } n \neq k,$$

and let $a_k(r_2)$ be arbitrary.

(a) Conclude from Exercise 66 of Section 7.6 that

$$L\left(y_1 \ln x + x^{r_1} \sum_{n=1}^{\infty} a_n'(r_1)x^n\right) = k\alpha_0 x^{r_1}.$$

(b) Conclude from Exercise 57 of Section 7.5 that

$$L\left(x^{r_2} \sum_{n=0}^{\infty} a_n(r_2)x^n\right) = Ax^{r_1},$$

where

$$A = \sum_{j=1}^{k} p_j(r_1 - j)a_{k-j}(r_2).$$

(c) Show that y_1 and

$$y_2 = x^{r_2} \sum_{n=0}^{\infty} a_n(r_2)x^n - \frac{A}{k\alpha_0}\left(y_1 \ln x + x^{r_1} \sum_{n=1}^{\infty} a_n'(r_1)x^n\right)$$

form a fundamental set of Frobenius solutions of $Ly = 0$.

(d) Show that choosing the arbitrary quantity $a_k(r_2)$ to be nonzero merely adds a multiple of y_1 to y_2. Conclude that we may as well take $a_k(r_2) = 0$.

8 Laplace Transforms

IN THIS CHAPTER we study the method of Laplace[1] transforms, which illustrates one of the basic problem solving techniques in mathematics: transform a difficult problem into an easier one, solve the latter, and then use its solution to obtain a solution of the original problem. The method discussed here transforms an initial value problem for a constant coefficient equation into an algebraic equation whose solution can then be used to solve the initial value problem. In some cases this method is merely an alternative procedure for solving problems that can be solved equally well by methods that we have considered previously; however, in other cases the method of Laplace transforms is more efficient than the methods previously discussed. This is especially true in physical problems dealing with discontinuous forcing functions.

SECTION 8.1 defines the Laplace transform and develops its properties.

SECTION 8.2 deals with the problem of finding a function that has a given Laplace transform.

SECTION 8.3 applies the Laplace transform to solve initial value problems for constant coefficient second order differential equations on $(0, \infty)$.

SECTION 8.4 introduces the unit step function.

SECTION 8.5 uses the unit step function to solve constant coefficient equations with piecewise continuous forcing functions.

SECTION 8.6 deals with the convolution theorem, an important theoretical property of the Laplace transform.

SECTION 8.7 introduces the idea of impulsive force, and treats constant coefficient equations with impulsive forcing functions.

---

[1]Pierre Simon de Laplace (1749–1827) was a professor in Paris. He made important contributions to celestial mechanics, astronomy, special functions, probability, and mathematical physics. Although the integral known as the Laplace transform bears his name, it was not used to solve differential equations until long after his death.

8.1   Introduction to the Laplace Transform

DEFINITION OF THE LAPLACE TRANSFORM

To define the Laplace transform we first recall the definition of an improper integral from calculus. If g is integrable over the interval $[a, T]$ for every $T > a$, then the **improper integral of g over** $[a, \infty)$ is defined as

$$\int_a^\infty g(t)\, dt = \lim_{T \to \infty} \int_a^T g(t)\, dt. \tag{1}$$

We say that the improper integral **converges** if the limit in (1) exists; otherwise we say that the improper integral **diverges** or **does not exist.** The Laplace transform of a function f is defined as follows.

DEFINITION 8.1.1

> Let f be defined for $t \geq 0$ and let s be a real number. Then the **Laplace transform** of f is the function F defined by
>
> $$F(s) = \int_0^\infty e^{-st} f(t)\, dt, \tag{2}$$
>
> for those values of s for which the improper integral converges.

It is important to keep in mind that the variable of integration in (2) is t, while s is a parameter independent of t. We use t as the independent variable for f because in applications the Laplace transform is usually applied to functions of time.

The Laplace transform can be viewed as an operator $\mathcal{L}$ that transforms the function $f = f(t)$ into the function $F = F(s)$. Thus, (2) can be expressed as

$$F = \mathcal{L}(f).$$

The functions f and F form a **transform pair,** which we'll sometimes denote by

$$f(t) \leftrightarrow F(s).$$

It can be shown that if $F(s)$ is defined for $s = s_0$ then it is defined for all $s > s_0$ (Exercise 14**(b)**).

COMPUTATION OF SOME SIMPLE LAPLACE TRANSFORMS

EXAMPLE 1   Find the Laplace transform of $f(t) = 1$.

Solution   From (2) with $f(t) = 1$,

$$F(s) = \int_0^\infty e^{-st}\, dt = \lim_{T \to \infty} \int_0^T e^{-st}\, dt.$$

If $s \neq 0$ then

$$\int_0^T e^{-st}\, dt = -\frac{1}{s} e^{-st} \Big|_0^T = \frac{1 - e^{-sT}}{s}. \tag{3}$$

Therefore

$$\lim_{T\to\infty}\int_0^T e^{-st}\,dt = \begin{cases} \dfrac{1}{s}, & s > 0, \\ \infty, & s < 0. \end{cases} \tag{4}$$

If $s = 0$ the integrand reduces to the constant 1, and

$$\lim_{T\to\infty}\int_0^T 1\,dt = \lim_{T\to\infty}\int_0^T 1\,dt = \lim_{T\to\infty} T = \infty.$$

Therefore $F(0)$ is undefined, and

$$F(s) = \int_0^\infty e^{-st}\,dt = \frac{1}{s}, \qquad s > 0.$$

This result can be written in operator notation as

$$\mathcal{L}(1) = \frac{1}{s}, \qquad s > 0,$$

or as the transform pair

$$1 \leftrightarrow \frac{1}{s}, \qquad s > 0. \qquad\blacksquare$$

REMARK. It is convenient to combine the steps of integrating from 0 to T and letting $T \to \infty$. Therefore, instead of writing (3) and (4) as separate steps we write

$$\int_0^\infty e^{-st}\,dt = -\frac{1}{s}e^{-st}\bigg|_0^\infty = \begin{cases} \dfrac{1}{s}, & s > 0, \\ \infty, & s < 0. \end{cases}$$

We'll follow this practice throughout this chapter.

EXAMPLE 2   Find the Laplace transform of $f(t) = t$.

Solution   From (2) with $f(t) = t$,

$$F(s) = \int_0^\infty e^{-st} t\,dt. \tag{5}$$

If $s \neq 0$ integrating by parts yields

$$\int_0^\infty e^{-st} t\,dt = -\frac{te^{-st}}{s}\bigg|_0^\infty + \frac{1}{s}\int_0^\infty e^{-st}\,dt = -\left[\frac{t}{s} + \frac{1}{s^2}\right]e^{-st}\bigg|_0^\infty$$

$$= \begin{cases} \dfrac{1}{s^2}, & s > 0, \\ \infty, & s < 0. \end{cases}$$

If $s = 0$ the integral in (5) becomes

$$\int_0^\infty t \, dt = \frac{t^2}{2} \Big|_0^\infty = \infty.$$

Therefore $F(0)$ is undefined and

$$F(s) = \frac{1}{s^2}, \qquad s > 0.$$

This result can also be written as

$$\mathcal{L}(t) = \frac{1}{s^2}, \qquad s > 0,$$

or as the transform pair

$$t \leftrightarrow \frac{1}{s^2}, \qquad s > 0.$$

EXAMPLE 3 Find the Laplace transform of $f(t) = e^{at}$, where a is a constant.

Solution From (2) with $f(t) = e^{at}$,

$$F(s) = \int_0^\infty e^{-st} e^{at} \, dt.$$

Combining the exponentials yields

$$F(s) = \int_0^\infty e^{-(s-a)t} \, dt.$$

However, we know from Example 1 that

$$\int_0^\infty e^{-st} \, dt = \frac{1}{s}, \qquad s > 0.$$

Replacing s by $s - a$ here shows that

$$F(s) = \frac{1}{s - a}, \qquad s > a.$$

This can also be written as

$$\mathcal{L}(e^{at}) = \frac{1}{s - a}, \qquad s > a, \qquad \text{or} \qquad e^{at} \leftrightarrow \frac{1}{s - a}, \qquad s > a.$$

EXAMPLE 4 Find the Laplace transforms of $f(t) = \sin \omega t$ and $g(t) = \cos \omega t$, where ω is a constant.

Solution Define

$$F(s) = \int_0^\infty e^{-st} \sin \omega t \, dt \tag{6}$$

and

$$G(s) = \int_0^\infty e^{-st} \cos \omega t \, dt. \tag{7}$$

If $s > 0$ then integrating (6) by parts yields

$$F(s) = -\frac{e^{-st}}{s}\sin \omega t \Big|_0^\infty + \frac{\omega}{s}\int_0^\infty e^{-st}\cos \omega t \, dt,$$

so

$$F(s) = \frac{\omega}{s}G(s). \tag{8}$$

If $s > 0$ then integrating (7) by parts yields

$$G(s) = -\frac{e^{-st}\cos \omega t}{s}\Big|_0^\infty - \frac{\omega}{s}\int_0^\infty e^{-st}\sin \omega t \, dt,$$

so

$$G(s) = \frac{1}{s} - \frac{\omega}{s}F(s).$$

Now substitute from (8) into this to obtain

$$G(s) = \frac{1}{s} - \frac{\omega^2}{s^2}G(s).$$

Solving this for $G(s)$ yields

$$G(s) = \frac{s}{s^2 + \omega^2}, \qquad s > 0.$$

This and (8) imply that

$$F(s) = \frac{\omega}{s^2 + \omega^2}, \qquad s > 0. \qquad\blacksquare$$

TABLES OF LAPLACE TRANSFORMS

Extensive tables of Laplace transforms have been compiled and are commonly used in applications. The brief table of Laplace transforms on the inside front cover of this text will be adequate for our purposes.

EXAMPLE 5   Use the table of Laplace transforms to find $\mathcal{L}(t^3 e^{4t})$.

Solution   The table includes the transform pair

$$t^n e^{at} \leftrightarrow \frac{n!}{(s - a)^{n+1}}.$$

Setting $n = 3$ and $a = 4$ here yields

$$\mathcal{L}(t^3 e^{4t}) = \frac{3!}{(s - 4)^4} = \frac{6}{(s - 4)^4}. \qquad\blacksquare$$

We'll sometimes write Laplace transforms of specific functions without explicitly stating how they are obtained. In such cases you should refer to the table of Laplace transforms.

LINEARITY OF THE LAPLACE TRANSFORM

The following theorem presents an important property of the Laplace transform.

THEOREM 8.1.2

(Linearity Property)

Suppose that $\mathcal{L}(f_i)$ is defined for $s > s_i$ $(1 \leq i \leq n)$. Let s_0 be the largest of the numbers $s_1, \ldots, s_n$ and let $c_1, c_2, \ldots, c_n$ be constants. Then

$$\mathcal{L}(c_1 f_1 + c_2 f_2 + \cdots + c_n f_n)$$
$$= c_1 \mathcal{L}(f_1) + c_2 \mathcal{L}(f_2) + \cdots + c_n \mathcal{L}(f_n) \qquad \text{for } s > s_0.$$

PROOF. We give the proof for the case where $n = 2$. If $s > s_0$ then

$$\mathcal{L}(c_1 f_1 + c_2 f_2) = \int_0^\infty e^{-st}(c_1 f_1(t) + c_2 f_2(t))\, dt$$
$$= c_1 \int_0^\infty e^{-st} f_1(t)\, dt + c_2 \int_0^\infty e^{-st} f_2(t)\, dt$$
$$= c_1 \mathcal{L}(f_1) + c_2 \mathcal{L}(f_2). \qquad \square$$

EXAMPLE 6 Use Theorem 8.1.2 and the known Laplace transform

$$\mathcal{L}(e^{at}) = \frac{1}{s - a}$$

to find $\mathcal{L}(\cosh bt)$ $(b \neq 0)$.

Solution By definition,

$$\cosh bt = \frac{e^{bt} + e^{-bt}}{2}.$$

Therefore

$$\mathcal{L}(\cosh bt) = \mathcal{L}\left(\frac{1}{2}e^{bt} + \frac{1}{2}e^{-bt}\right)$$
$$= \frac{1}{2}\mathcal{L}(e^{bt}) + \frac{1}{2}\mathcal{L}(e^{-bt}) \qquad \text{(linearity property)} \qquad (9)$$
$$= \frac{1}{2}\frac{1}{s - b} + \frac{1}{2}\frac{1}{s + b},$$

where the first transform on the right is defined for $s > b$ and the second for $s > -b$; hence, both are defined for $s > |b|$. Simplifying the last expression in (9) yields

$$\mathcal{L}(\cosh bt) = \frac{s}{s^2 - b^2}, \qquad s > |b|. \qquad \blacksquare$$

THE FIRST SHIFTING THEOREM

The next theorem enables us to start with known transform pairs and derive others. (For other results of this kind, see Exercises 6, 12, and 13.)

THEOREM 8.1.3

(First Shifting Theorem)   *If*

$$F(s) = \int_0^\infty e^{-st}f(t)\, dt \tag{10}$$

is the Laplace transform of $f(t)$ for $s > s_0$, then $F(s - a)$ is the Laplace transform of $e^{at}f(t)$ for $s > s_0 + a$.

PROOF. Replacing s by $s - a$ in (10) yields

$$F(s - a) = \int_0^\infty e^{-(s-a)t}f(t)\, dt \tag{11}$$

if $s - a > s_0$; that is, if $s > s_0 + a$. However, (11) can be rewritten as

$$F(s - a) = \int_0^\infty e^{-st}(e^{at}f(t))\, dt,$$

which implies the conclusion.   □

EXAMPLE 7   Use Theorem 8.1.3 and the known Laplace transforms of $1, t, \cos \omega t$, and $\sin \omega t$ to find

$$\mathcal{L}(e^{at}), \qquad \mathcal{L}(te^{at}), \qquad \mathcal{L}(e^{\lambda t} \sin \omega t), \qquad \text{and} \qquad \mathcal{L}(e^{\lambda t} \cos \omega t).$$

Solution   In the following table the known transform pairs are listed on the left and the required transform pairs listed on the right are obtained by applying Theorem 8.1.3.   ■

$f(t) \leftrightarrow F(s)$	$e^{at}f(t) \leftrightarrow F(s - a)$
$1 \leftrightarrow \dfrac{1}{s}, \quad s > 0$	$e^{at} \leftrightarrow \dfrac{1}{(s - a)}, \quad s > a$
$t \leftrightarrow \dfrac{1}{s^2}, \quad s > 0$	$te^{at} \leftrightarrow \dfrac{1}{(s - a)^2}, \quad s > a$
$\sin \omega t \leftrightarrow \dfrac{\omega}{s^2 + \omega^2}, \quad s > 0$	$e^{\lambda t} \sin \omega t \leftrightarrow \dfrac{\omega}{(s - \lambda)^2 + \omega^2}, \quad s > \lambda$
$\cos \omega t \leftrightarrow \dfrac{s}{s^2 + \omega^2}, \quad s > 0$	$e^{\lambda t} \sin \omega t \leftrightarrow \dfrac{s - \lambda}{(s - \lambda)^2 + \omega^2}, \quad s > \lambda$

EXISTENCE OF LAPLACE TRANSFORMS

Not every function has a Laplace transform. For example, it can be shown (Exercise 3) that

$$\int_0^\infty e^{-st}e^{t^2}\, dt = \infty$$

for every real number s. Hence, the function $f(t) = e^{t^2}$ does not have a Laplace transform.

Our next objective is to establish conditions that ensure the existence of the Laplace transform of a function. We first review some relevant definitions from calculus.

Recall that a limit

$$\lim_{t \to t_0} f(t)$$

exists if and only if the one-sided limits

$$\lim_{t \to t_0-} f(t) \quad \text{and} \quad \lim_{t \to t_0+} f(t)$$

both exist and are equal, in which case

$$\lim_{t \to t_0} f(t) = \lim_{t \to t_0-} f(t) = \lim_{t \to t_0+} f(t).$$

Recall also that f is continuous at a point t_0 in an open interval (a, b) if and only if

$$\lim_{t \to t_0} f(t) = f(t_0),$$

which is equivalent to

$$\lim_{t \to t_0+} f(t) = \lim_{t \to t_0-} f(t) = f(t_0). \tag{12}$$

For simplicity we define

$$f(t_0+) = \lim_{t \to t_0+} f(t) \quad \text{and} \quad f(t_0-) = \lim_{t \to t_0-} f(t),$$

so (12) can be expressed as

$$f(t_0+) = f(t_0-) = f(t_0).$$

If $f(t_0+)$ and $f(t_0-)$ have finite but distinct values, then we say that f has a ***jump discontinuity*** at t_0, and

$$f(t_0+) - f(t_0-)$$

is called the ***jump in f at t_0*** (Figure 1).

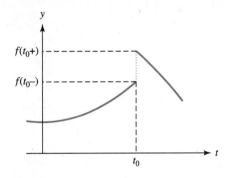

Figure 1 A jump discontinuity

If $f(t_0+)$ and $f(t_0-)$ are finite and equal, but either f is not defined at t_0 or it is defined but

$$f(t_0) \neq f(t_0+) = f(t_0-),$$

then we say that f has a ***removable discontinuity*** at t_0 (Figure 2). This terminology is appropriate since a function f with a removable discontinuity at t_0 can be made continuous at t_0 by defining (or redefining)

$$f(t_0) = f(t_0+) = f(t_0-).$$

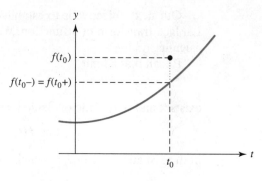

Figure 2

REMARK. We know from calculus that a definite integral is not affected by changing the values of its integrand at isolated points. Therefore, redefining a function f to make it continuous at removable discontinuities does not change $\mathcal{L}(f)$.

We make the following definition.

DEFINITION 8.1.4

> **(i)** A function f is said to be *piecewise continuous* on a finite closed interval $[0, T]$ if $f(0+)$ and $f(T-)$ are finite and f is continuous on the open interval $(0, T)$ except possibly at finitely many points, where f may have jump discontinuities or removable discontinuities.
> **(ii)** A function f is said to be *piecewise continuous* on the infinite interval $[0, \infty)$ if it is piecewise continuous on $[0, T]$ for every $T > 0$.

Figure 3 shows the graph of a typical piecewise continuous function.

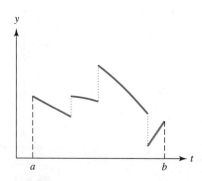

Figure 3 A piecewise continuous function on $[a, b]$

It is shown in calculus that if a function is piecewise continuous on a finite closed interval, then it is integrable on that interval. But if f is piecewise continuous on $[0, \infty)$, then so is $e^{-st} f(t)$, and therefore

$$\int_0^T e^{-st} f(t)\, dt$$

exists for every $T > 0$. However, piecewise continuity alone does not guarantee that the improper integral

$$\int_0^\infty e^{-st}f(t)\, dt = \lim_{T\to\infty} \int_0^T e^{-st}f(t)\, dt \qquad (13)$$

converges for s in some interval (s_0, ∞). For example, we noted earlier that (13) diverges for all s if $f(t) = e^{t^2}$. Stated informally, this occurs because e^{t^2} increases too rapidly as $t \to \infty$. The following definition provides a constraint on the growth of a function that guarantees convergence of its Laplace transform for s in some interval (s_0, ∞).

DEFINITION 8.1.5

A function f is said to be ***of exponential order s_0*** if there are constants M and t_0 such that

$$|f(t)| \le Me^{s_0 t}, \qquad t \ge t_0. \qquad (14)$$

In situations where the specific value of s_0 is irrelevant we say simply that f is ***of exponential order.***

The following theorem gives useful sufficient conditions for a function f to have a Laplace transform. The proof is sketched in Exercise 10.

THEOREM 8.1.6

If f is piecewise continuous on $[0, \infty)$ and of exponential order s_0 for some real number s_0, then $\mathcal{L}(f)$ is defined for $s > s_0$.

REMARK. We emphasize that the conditions of Theorem 8.1.6 are sufficient, but *not necessary*, for f to have a Laplace transform. For example, Exercise 14**(c)** shows that f may have a Laplace transform even though f is not of exponential order.

EXAMPLE 8 If f is bounded on some interval $[t_0, \infty)$, say

$$|f(t)| \le M, \qquad t \ge t_0,$$

then (14) holds with $s_0 = 0$, so f is of exponential order zero. Thus, for example, $\sin \omega t$ and $\cos \omega t$ are of exponential order zero, and Theorem 8.1.6 implies that $\mathcal{L}(\sin \omega t)$ and $\mathcal{L}(\cos \omega t)$ exist for $s > 0$. This is consistent with the conclusion of Example 4.

EXAMPLE 9 It can be shown that if $\lim_{t\to\infty} e^{-s_0 t}f(t)$ exists and is finite, then f is of exponential order s_0 (Exercise 9). If α is any real number and $s_0 > 0$, then $f(t) = t^\alpha$ is of exponential order s_0, since

$$\lim_{t\to\infty} e^{-s_0 t} t^\alpha = 0,$$

by L'Hôpital's rule. If $\alpha \ge 0$ then f is also continuous on $[0, \infty)$. Therefore, Exercise 9 and Theorem 8.1.6 imply that $\mathcal{L}(t^\alpha)$ exists for $s \ge s_0$. However, since s_0 is an arbitrary positive number, this really implies that $\mathcal{L}(t^\alpha)$ exists for all $s > 0$. This is consistent with the results of Example 2 and Exercises 6 and 8.

EXAMPLE 10   Find the Laplace transform of the piecewise continuous function

$$f(t) = \begin{cases} 1, & 0 \le t < 1, \\ -3e^{-t}, & t \ge 1. \end{cases}$$

Solution   Since f is defined by different formulas on $[0,1)$ and $[1,\infty)$, we write

$$F(s) = \int_0^\infty e^{-st}f(t)\, dt = \int_0^1 e^{-st}(1)\, dt + \int_1^\infty e^{-st}(-3e^{-t})\, dt.$$

Since

$$\int_0^1 e^{-st}\, dt = \begin{cases} \dfrac{1-e^{-s}}{s}, & s \ne 0, \\ 1, & s = 0, \end{cases}$$

and

$$\int_1^\infty e^{-st}(-3e^{-t})\, dt = -3\int_1^\infty e^{-(s+1)t}\, dt = -\frac{3e^{-(s+1)}}{s+1}, \qquad s > -1,$$

we have

$$F(s) = \begin{cases} \dfrac{1-e^{-s}}{s} - 3\dfrac{e^{-(s+1)}}{s+1}, & s > -1, \quad s \ne 0, \\ 1 - \dfrac{3}{e}, & s = 0. \end{cases}$$

This is consistent with Theorem 8.1.6, since

$$|f(t)| \le 3e^{-t}, \qquad t \ge 1,$$

and therefore f is of exponential order $s_0 = -1$.   ∎

REMARK.   In Section 8.4 we will develop a more efficient method for finding Laplace transforms of piecewise continuous functions.

EXAMPLE 11   We stated earlier that

$$\int_0^\infty e^{-st}e^{t^2}\, dt = \infty$$

for all s, so Theorem 8.1.6 implies that $f(t) = e^{t^2}$ cannot be of exponential order. To verify this, observe that for any choice of s_0 and $M > 0$

$$\lim_{t\to\infty} \frac{e^{t^2}}{Me^{s_0 t}} = \lim_{t\to\infty} \frac{1}{M}e^{t^2-s_0 t} = \infty,$$

so

$$e^{t^2} > Me^{s_0 t}$$

for sufficiently large values of t.   ∎

8.1 EXERCISES

1. Find the Laplace transforms of the following functions by evaluating the integral $F(s) = \int_0^\infty e^{-st}f(t)\,dt$.

    (a) t **(b)** te^{-t} **(c)** $\sinh bt$

    (d) $e^{2t} - 3e^t$ **(e)** t^2

2. Use the table of Laplace transforms to find the Laplace transforms of the following functions.

    (a) $\cosh t \sin t$ **(b)** $\sin^2 t$ **(c)** $\cos^2 2t$

    (d) $\cosh^2 t$ **(e)** $t \sinh 2t$ **(f)** $\sin t \cos t$

    (g) $\sin\left(t + \dfrac{\pi}{4}\right)$ **(h)** $\cos 2t - \cos 3t$ **(i)** $\sin 2t + \cos 4t$

3. Show that

$$\int_0^\infty e^{-st}e^{t^2}\,dt = \infty$$

    for every real number s.

4. Graph the following piecewise continuous functions and evaluate $f(t+)$, $f(t-)$, and $f(t)$ at each point of discontinuity.

    (a) $f(t) = \begin{cases} -t, & 0 \le t < 2, \\ t - 4, & 2 \le t < 3, \\ 1, & t \ge 3. \end{cases}$ **(b)** $f(t) = \begin{cases} t^2 + 2, & 0 \le t < 1, \\ 4, & t = 1, \\ t, & t > 1. \end{cases}$

    (c) $f(t) = \begin{cases} \sin t, & 0 \le t < \pi/2, \\ 2\sin t, & \pi/2 \le t < \pi, \\ \cos t, & t \ge \pi. \end{cases}$ **(d)** $f(t) = \begin{cases} t, & 0 \le t < 1, \\ 2, & t = 1, \\ 2 - t, & 1 \le t < 2, \\ 3, & t = 2, \\ 6, & t > 2. \end{cases}$

5. Find the Laplace transform.

    (a) $f(t) = \begin{cases} e^{-t}, & 0 \le t < 1, \\ e^{-2t}, & t \ge 1. \end{cases}$ **(b)** $f(t) = \begin{cases} 1, & 0 \le t < 4, \\ t, & t \ge 4. \end{cases}$

    (c) $f(t) = \begin{cases} t, & 0 \le t < 1, \\ 1, & t \ge 1. \end{cases}$ **(d)** $f(t) = \begin{cases} te^t, & 0 \le t < 1, \\ e^t, & t \ge 1. \end{cases}$

6. Prove that if $f(t) \leftrightarrow F(s)$ then $t^k f(t) \leftrightarrow (-1)^k F^{(k)}(s)$.

    Hint: *Assume that it is permissible to differentiate the integral* $\displaystyle\int_0^\infty e^{-st}f(t)\,dt$ *with respect to s under the integral sign.*

7. Use the known Laplace transforms

$$\mathcal{L}(e^{\lambda t}\sin \omega t) = \frac{\omega}{(s - \lambda)^2 + \omega^2} \quad \text{and} \quad \mathcal{L}(e^{\lambda t}\cos \omega t) = \frac{s - \lambda}{(s - \lambda)^2 + \omega^2}$$

    and the result of Exercise 6 to find $\mathcal{L}(te^{\lambda t}\cos \omega t)$ and $\mathcal{L}(te^{\lambda t}\sin \omega t)$.

8. Use the known Laplace transform $\mathcal{L}(1) = 1/s$ and the result of Exercise 6 to show that

$$\mathcal{L}(t^n) = \frac{n!}{s^{n+1}} \quad (n = \text{integer})$$

9. **(a)** Show that if $\lim_{t\to\infty} e^{-s_0 t}f(t)$ exists and is finite, then f is of exponential order s_0.

    (b) Show that if f is of exponential order s_0, then $\lim_{t\to\infty} e^{-st}f(t) = 0$ for all $s > s_0$.

(c) Show that if f is of exponential order s_0 and $g(t) = f(t + \tau)$ where $\tau > 0$, then g is also of exponential order s_0.

10. Recall the following theorem from calculus.

THEOREM A *Let g be integrable on $[0,T]$ for every $T > 0$. Suppose that there is a function w defined on some interval $[\tau, \infty)$ (with $\tau \geq 0$) such that $|g(t)| \leq w(t)$ for $t \geq \tau$ and $\int_\tau^\infty w(t)\, dt$ converges. Then $\int_0^\infty g(t)\, dt$ converges.*

Use Theorem A to show that if f is piecewise continuous on $[0, \infty)$ and of exponential order s_0, then f has a Laplace transform $F(s)$ defined for $s > s_0$.

11. Prove: If f is piecewise continuous and of exponential order, then $\lim_{t \to \infty} F(s) = 0$.

12. Prove: If f is continuous on $[0, \infty)$ and of exponential order $s_0 > 0$, then

$$\mathcal{L}\left(\int_0^t f(\tau)\, d\tau \right) = \frac{1}{s} \mathcal{L}(f), \qquad s > s_0.$$

Hint: Use integration by parts to evaluate the transform on the left.

13. Suppose that f is piecewise continuous and of exponential order, and that $\lim_{t \to 0+} f(t)/t$ exists. Show that

$$\mathcal{L}\left(\frac{f(t)}{t} \right) = \int_s^\infty F(r)\, dr.$$

Hint: Use the results of Exercises 6 and 12.

14. Suppose that f is piecewise continuous on $[0, \infty)$.

(a) Prove: If the integral $g(t) = \int_0^t e^{-s_0\tau} f(\tau)\, d\tau$ satisfies the inequality $|g(t)| \leq M$ $(t \geq 0)$, then f has a Laplace transform $F(s)$ defined for $s > s_0$.

Hint: Use integration by parts to show that

$$\int_0^T e^{-st} f(t)\, dt = e^{-(s-s_0)T} g(T) + (s - s_0) \int_0^T e^{-(s-s_0)t} g(t)\, dt.$$

(b) Show that if $\mathcal{L}(f)$ exists for $s = s_0$, then it exists for $s > s_0$.

(c) Show that the function

$$f(t) = te^{t^2} \cos(e^{t^2})$$

has a Laplace transform defined for $s > 0$, even though f is not of exponential order.

15. Use the table of Laplace transforms and the result of Exercise 13 to find the Laplace transforms of the following functions.

(a) $\dfrac{\sin \omega t}{t}$ $(\omega > 0)$

(b) $\dfrac{\cos \omega t - 1}{t}$ $(\omega > 0)$

(c) $\dfrac{e^{at} - e^{bt}}{t}$

(d) $\dfrac{\cosh t - 1}{t}$

(e) $\dfrac{\sinh^2 t}{t}$

16. The *gamma function* is defined by the integral

$$\Gamma(\alpha) = \int_0^\infty x^{\alpha-1} e^{-x}\, dx,$$

which can be shown to converge if $\alpha > 0$.

(a) Use integration by parts to show that

$$\Gamma(\alpha + 1) = \alpha \Gamma(\alpha), \qquad \alpha > 0.$$

(b) Show that $\Gamma(n + 1) = n!$ if $n = 1, 2, 3, \ldots$.

(c) From **(b)** and the table of Laplace transforms,

$$\mathcal{L}(t^{\alpha}) = \frac{\Gamma(\alpha + 1)}{s^{\alpha+1}}, \qquad s > 0,$$

if α is a nonnegative integer. Show that this formula is valid for any $\alpha > -1$.

Hint: *Change the variable of integration in the integral for $\Gamma(\alpha + 1)$.*

17. Suppose that f is continuous on $[0, T]$ and $f(t + T) = f(t)$ for all $t \geq 0$. (We say in this case that f is **periodic with period T.**)

(a) Conclude from Theorem 8.1.6 that f has a Laplace transform $F(s)$ defined for $s > 0$.

Hint: *Since f is continuous on $[0, T]$ and periodic with period T, it is bounded on $[0, \infty)$.*

(b) Show that

$$F(s) = \frac{1}{1 - e^{-sT}} \int_0^T e^{-st} f(t) \, dt, \qquad s > 0.$$

Hint: *Write*

$$F(s) = \sum_{n=0}^{\infty} \int_{nT}^{(n+1)T} e^{-st} f(t) \, dt.$$

Then show that

$$\int_{nT}^{(n+1)T} e^{-st} f(t) \, dt = e^{-nsT} \int_0^T e^{-st} f(t) \, dt,$$

and recall the formula for the sum of a geometric series.

18. Use the formula given in Exercise 17**(b)** to find the Laplace transforms of the given periodic functions:

(a) $f(t) = \begin{cases} t, & 0 \leq t < 1, \\ 2 - t, & 1 \leq t < 2, \end{cases} \quad f(t + 2) = f(t), \quad t \geq 0$

(b) $f(t) = \begin{cases} 1, & 0 \leq t < \frac{1}{2}, \\ -1, & \frac{1}{2} \leq t < 1, \end{cases} \quad f(t + 1) = f(t), \quad t \geq 0$

(c) $f(t) = |\sin t|$

(d) $f(t) = \begin{cases} \sin t, & 0 \leq t < \pi, \\ 0, & \pi \leq t < 2\pi, \end{cases} \quad f(t + 2\pi) = f(t)$

8.2   The Inverse Laplace Transform

DEFINITION OF THE INVERSE LAPLACE TRANSFORM

In Section 8.1 we defined the Laplace transform of f by

$$F(s) = \mathcal{L}(f) = \int_0^{\infty} e^{-st} f(t) \, dt.$$

We will also say that f is an **inverse Laplace transform** of F, and write

$$f = \mathcal{L}^{-1}(F).$$

To solve differential equations by means of the Laplace transform we must be able to obtain f from its transform F. There is a formula for doing this, but we can't use it because it requires the theory of functions of a complex variable. Fortunately, we can use the table of Laplace transforms to find inverse transforms that we'll need.

EXAMPLE 1   Use the table of Laplace transforms to find

$$\textbf{(a)} \; \mathcal{L}^{-1}\left(\frac{1}{s^2 - 1}\right) \quad \text{and} \quad \textbf{(b)} \; \mathcal{L}^{-1}\left(\frac{s}{s^2 + 9}\right).$$

Solution   **(a)** Setting $b = 1$ in the transform pair

$$\sinh bt \leftrightarrow \frac{b}{s^2 - b^2}$$

shows that

$$\mathcal{L}^{-1}\left(\frac{1}{s^2 - 1}\right) = \sinh t.$$

Solution   **(b)** Setting $\omega = 3$ in the transform pair

$$\cos \omega t \leftrightarrow \frac{s}{s^2 + \omega^2}$$

shows that

$$\mathcal{L}^{-1}\left(\frac{s}{s^2 + 9}\right) = \cos 3t.$$ ∎

The following theorem enables us to find inverse transforms of linear combinations of transforms in the table. We omit the proof.

THEOREM 8.2.1

(Linearity Property)

> If $F_1, F_2, \ldots, F_n$ are Laplace transforms and $c_1, c_2, \ldots, c_n$ are constants, then
>
> $$\mathcal{L}^{-1}(c_1 F_1 + c_2 F_2 + \cdots + c_n F_n) = c_1 \mathcal{L}^{-1}(F_1) + c_2 \mathcal{L}^{-1}(F_2) + \cdots + c_n \mathcal{L}^{-1} F_n.$$

EXAMPLE 2   Find

$$\mathcal{L}^{-1}\left(\frac{8}{s + 5} + \frac{7}{s^2 + 3}\right).$$

Solution   From the table of Laplace transforms,

$$e^{at} \leftrightarrow \frac{1}{s - a} \quad \text{and} \quad \sin \omega t \leftrightarrow \frac{\omega}{s^2 + \omega^2}.$$

Taking $a = -5$ and $\omega = \sqrt{3}$ using Theorem 8.2.1 yields

$$\mathcal{L}^{-1}\left(\frac{8}{s + 5} + \frac{7}{s^2 + 3}\right) = 8\mathcal{L}^{-1}\left(\frac{1}{s + 5}\right) + 7\mathcal{L}^{-1}\left(\frac{1}{s^2 + 3}\right)$$

$$= 8\mathcal{L}^{-1}\left(\frac{1}{s + 5}\right) + \frac{7}{\sqrt{3}}\mathcal{L}^{-1}\left(\frac{\sqrt{3}}{s^2 + 3}\right)$$

$$= 8e^{-5t} + \frac{7}{\sqrt{3}} \sin \sqrt{3}t.$$

EXAMPLE 3 Find

$$\mathcal{L}^{-1}\left(\frac{3s + 8}{s^2 + 2s + 5}\right).$$

Solution Completing the square in the denominator yields

$$\frac{3s + 8}{s^2 + 2s + 5} = \frac{3s + 8}{(s + 1)^2 + 4}.$$

Because of the form of the denominator, we consider the transform pairs

$$e^{-t}\cos 2t \leftrightarrow \frac{s + 1}{(s + 1)^2 + 4} \quad \text{and} \quad e^{-t}\sin 2t \leftrightarrow \frac{2}{(s + 1)^2 + 4},$$

and write

$$\mathcal{L}^{-1}\left(\frac{3s + 8}{(s + 1)^2 + 4}\right) = \mathcal{L}^{-1}\left(\frac{3s + 3}{(s + 1)^2 + 4}\right) + \mathcal{L}^{-1}\left(\frac{5}{(s + 1)^2 + 4}\right)$$

$$= 3\mathcal{L}^{-1}\left(\frac{s + 1}{(s + 1)^2 + 4}\right) + \frac{5}{2}\mathcal{L}^{-1}\left(\frac{2}{(s + 1)^2 + 4}\right)$$

$$= e^{-t}\left(3\cos 2t + \frac{5}{2}\sin 2t\right). \qquad \blacksquare$$

REMARK. We will often write inverse Laplace transforms of specific functions without explicitly stating how they are obtained. In such cases you should refer to the table of Laplace transforms.

INVERSE LAPLACE TRANSFORMS OF RATIONAL FUNCTIONS

In solving differential equations by means of Laplace transforms it is often necessary to find the inverse transform of a rational function

$$F(s) = \frac{P(s)}{Q(s)}$$

where P and Q are polynomials in s with no common factors. Since it can be shown that $\lim_{s\to\infty} F(s) = 0$ if F is a Laplace transform, we need only consider the case where degree$(P) <$ degree(Q). To obtain $\mathcal{L}^{-1}(F)$ we find the partial fraction expansion of F, obtain inverse transforms of the individual terms in the expansion from the table of Laplace transforms, and use the linearity property of the inverse transform. The next two examples illustrate this.

EXAMPLE 4 Find the inverse Laplace transform of

$$F(s) = \frac{3s + 2}{s^2 - 3s + 2}. \tag{1}$$

Solution (Method 1) Factoring the denominator in (1) yields

$$F(s) = \frac{3s + 2}{(s - 1)(s - 2)}. \tag{2}$$

The form for the partial fraction expansion is

$$\frac{3s + 2}{(s - 1)(s - 2)} = \frac{A}{s - 1} + \frac{B}{s - 2}. \tag{3}$$

Multiplying this by $(s - 1)(s - 2)$ yields

$$3s + 2 = (s - 2)A + (s - 1)B.$$

Setting $s = 2$ yields $B = 8$ and setting $s = 1$ yields $A = -5$. Therefore

$$F(s) = -\frac{5}{s - 1} + \frac{8}{s - 2}$$

and

$$\mathcal{L}^{-1}(F) = -5\mathcal{L}^{-1}\left(\frac{1}{s - 1}\right) + 8\mathcal{L}^{-1}\left(\frac{1}{s - 2}\right) = -5e^t + 8e^{2t}.$$

Solution   (Method 2) We don't really have to multiply (3) by $(s - 1)(s - 2)$ to compute A and B. We can obtain A by simply ignoring the factor $s - 1$ in the denominator of (2) and setting $s = 1$ elsewhere; thus

$$A = \frac{3s + 2}{s - 2}\bigg|_{s=1} = \frac{3 \cdot 1 + 2}{1 - 2} = -5. \tag{4}$$

Similarly, we can obtain B by ignoring the factor $s - 2$ in the denominator of (2) and setting $s = 2$ elsewhere; thus

$$B = \frac{3s + 2}{s - 1}\bigg|_{s=2} = \frac{3 \cdot 2 + 2}{2 - 1} = 8. \tag{5}$$

To justify this, we observe that multiplying (3) by $s - 1$ yields

$$\frac{3s + 2}{s - 2} = A + (s - 1)\frac{B}{s - 2},$$

and setting $s = 1$ leads to (4). Similarly, multiplying (3) by $s - 2$ yields

$$\frac{3s + 2}{s - 1} = (s - 2)\frac{A}{s - 2} + B$$

and setting $s = 2$ leads to (5). (It isn't necesary to write the last two equations. We wrote them only to justify the shortcut procedure indicated in (4) and (5).)   ∎

The shortcut employed in the second solution of Example 4 is called ***Heaviside's*** [1] ***method.*** The following theorem states Heaviside's method formally. For a proof and an extension of this theorem, see Exercise 10.

THEOREM 8.2.2

Suppose that

$$F(s) = \frac{P(s)}{(s - s_1)(s - s_2) \cdots (s - s_n)}, \tag{6}$$

[1] Oliver Heaviside (1850–1925) was a British electrical engineer who probably deserves most of the credit for formulating the method of Laplace transforms as we know it. However, he was a controversial figure whose techniques were viewed with skepticism, so he was forced to publish his results privately and did not receive appropriate recognition during his lifetime.

where $s_1, s_2, \ldots, s_n$ are distinct and P is a polynomial of degree less than n. Then

$$F(s) = \frac{A_1}{s - s_1} + \frac{A_2}{s - s_2} + \cdots + \frac{A_n}{s - s_n},$$

where A_i can be computed from (6) by ignoring the factor $s - s_i$ and setting $s = s_i$ elsewhere.

EXAMPLE 5   Find the inverse Laplace transform of

$$F(s) = \frac{6 + (s + 1)(s^2 - 5s + 11)}{s(s - 1)(s - 2)(s + 1)}. \tag{7}$$

Solution   The partial fraction expansion of (7) is of the form

$$F(s) = \frac{A}{s} + \frac{B}{s - 1} + \frac{C}{s - 2} + \frac{D}{s + 1}. \tag{8}$$

To find A we ignore the factor s in the denominator of (7) and set $s = 0$ elsewhere. This yields

$$A = \frac{6 + (1)(11)}{(-1)(-2)(1)} = \frac{17}{2}.$$

Similarly, the other coefficients are given by

$$B = \frac{6 + (2)(7)}{(1)(-1)(2)} = -10,$$

$$C = \frac{6 + 3(5)}{2(1)(3)} = \frac{7}{2},$$

and

$$D = \frac{6}{(-1)(-2)(-3)} = -1.$$

Therefore

$$F(s) = \frac{17}{2}\frac{1}{s} - \frac{10}{s - 1} + \frac{7}{2}\frac{1}{s - 2} - \frac{1}{s + 1}$$

and

$$\mathcal{L}^{-1}(F) = \frac{17}{2}\mathcal{L}^{-1}\left(\frac{1}{s}\right) - 10\mathcal{L}^{-1}\left(\frac{1}{s - 1}\right) + \frac{7}{2}\mathcal{L}^{-1}\left(\frac{1}{s - 2}\right) - \mathcal{L}^{-1}\left(\frac{1}{s + 1}\right)$$

$$= \frac{17}{2} - 10e^t + \frac{7}{2}e^{2t} - e^{-t}. \qquad \blacksquare$$

REMARK.   Notice that we didn't expand the numerator in (7) before computing the coefficients in (8). There would be no point to this, since it wouldn't simplify the computations.

EXAMPLE 6   Find the inverse Laplace transform of

$$F(s) = \frac{8 - (s + 2)(4s + 10)}{(s + 1)(s + 2)^2}. \tag{9}$$

Solution   The form for the partial fraction expansion is

$$F(s) = \frac{A}{s + 1} + \frac{B}{s + 2} + \frac{C}{(s + 2)^2}. \tag{10}$$

Because of the repeated factor $(s + 2)^2$ in (9), Heaviside's method cannot be used. Instead, we find a common denominator in (10). This yields

$$F(s) = \frac{A(s + 2)^2 + B(s + 1)(s + 2) + C(s + 1)}{(s + 1)(s + 2)^2}. \tag{11}$$

If (9) and (11) are to be equivalent then

$$A(s + 2)^2 + B(s + 1)(s + 2) + C(s + 1) = 8 - (s + 2)(4s + 10). \tag{12}$$

The two sides of this equation are polynomials of degree two. From a theorem of algebra, they will be equal for all s if they are equal for any three distinct values of s. We may determine $A, B,$ and C by choosing convenient values of s.

The left side of (12) suggests that we take $s = -2$ to obtain $C = -8$, and $s = -1$ to obtain $A = 2$. We can now choose any third value of s to determine B. Taking $s = 0$ yields $4A + 2B + C = -12$. Since $A = 2$ and $C = -8$ this implies that $B = -6$. Therefore

$$F(s) = \frac{2}{s + 1} - \frac{6}{s + 2} - \frac{8}{(s + 2)^2}$$

and

$$\mathcal{L}^{-1}(F) = 2\mathcal{L}^{-1}\left(\frac{1}{s + 1}\right) - 6\mathcal{L}^{-1}\left(\frac{1}{s + 2}\right) - 8\mathcal{L}^{-1}\left(\frac{1}{(s + 2)^2}\right)$$

$$= 2e^{-t} - 6e^{-2t} - 8te^{-2t}.$$

EXAMPLE 7   Find the inverse Laplace transform of

$$F(s) = \frac{s^2 - 5s + 7}{(s + 2)^3}.$$

Solution   The form for the partial fraction expansion is

$$F(s) = \frac{A}{s + 2} + \frac{B}{(s + 2)^2} + \frac{C}{(s + 2)^3}.$$

The easiest way to obtain $A, B,$ and C is to expand the numerator in powers of $s + 2$. This yields

$$s^2 - 5s + 7 = [(s + 2) - 2]^2 - 5[(s + 2) - 2] + 7$$
$$= (s + 2)^2 - 9(s + 2) + 21.$$

Therefore

$$F(s) = \frac{(s + 2)^2 - 9(s + 2) + 21}{(s + 2)^3}$$

$$= \frac{1}{s+2} - \frac{9}{(s+2)^2} + \frac{21}{(s+2)^3}$$

and

$$\mathcal{L}^{-1}(F) = \mathcal{L}^{-1}\left(\frac{1}{s+2}\right) - 9\mathcal{L}^{-1}\left(\frac{1}{(s+2)^2}\right) + \frac{21}{2}\mathcal{L}^{-1}\left(\frac{2}{(s+2)^3}\right)$$

$$= e^{-2t}\left(1 - 9t + \frac{21}{2}t^2\right).$$

EXAMPLE 8   Find the inverse Laplace transform of

$$F(s) = \frac{1 - s(5 + 3s)}{s[(s+1)^2 + 1]}. \tag{13}$$

Solution   One form for the partial fraction expansion of F is

$$F(s) = \frac{A}{s} + \frac{Bs + C}{(s+1)^2 + 1}. \tag{14}$$

However, we see from the table of Laplace transforms that the inverse transform of the second fraction on the right of (14) will be a linear combination of the inverse transforms

$$e^{-t}\cos t \quad \text{and} \quad e^{-t}\sin t$$

of

$$\frac{s+1}{(s+1)^2 + 1} \quad \text{and} \quad \frac{1}{(s+1)^2 + 1},$$

respectively. Therefore, instead of (14) we write

$$F(s) = \frac{A}{s} + \frac{B(s+1) + C}{(s+1)^2 + 1}. \tag{15}$$

Finding a common denominator yields

$$F(s) = \frac{A[(s+1)^2 + 1] + B(s+1)s + Cs}{s[(s+1)^2 + 1]}. \tag{16}$$

If (13) and (16) are to be equivalent we must have

$$A[(s+1)^2 + 1] + B(s+1)s + Cs = 1 - s(5 + 3s).$$

This will hold for all s if it holds for three distinct values of s. Choosing $s = 0, -1,$ and 1 yields the system

$$2A = 1$$
$$A - C = 3$$
$$5A + 2B + C = -7.$$

Solving this system yields

$$A = \frac{1}{2}, \quad B = -\frac{7}{2}, \quad C = -\frac{5}{2}.$$

Hence, from (15),

$$F(s) = \frac{1}{2s} - \frac{7}{2}\frac{s+1}{(s+1)^2+1} - \frac{5}{2}\frac{1}{(s+1)^2+1}.$$

Therefore

$$\mathcal{L}^{-1}(F) = \frac{1}{2}\mathcal{L}^{-1}\left(\frac{1}{s}\right) - \frac{7}{2}\mathcal{L}^{-1}\left(\frac{s+1}{(s+1)^2+1}\right) - \frac{5}{2}\mathcal{L}^{-1}\left(\frac{1}{(s+1)^2+1}\right)$$

$$= \frac{1}{2} - \frac{7}{2}e^{-t}\cos t - \frac{5}{2}e^{-t}\sin t.$$

EXAMPLE 9   Find the inverse Laplace transform of

$$F(s) = \frac{8+3s}{(s^2+1)(s^2+4)}. \tag{17}$$

Solution   The form for the partial fraction expansion is

$$F(s) = \frac{A+Bs}{s^2+1} + \frac{C+Ds}{s^2+4}.$$

The coefficients $A, B, C,$ and D can be obtained by finding a common denominator and equating the resulting numerator to the numerator in (17). However, since there is no first power of s in the denominator of (17), there is an easier way: the expansion of

$$F_1(s) = \frac{1}{(s^2+1)(s^2+4)}$$

can be obtained quickly by using Heaviside's method to expand

$$\frac{1}{(x+1)(x+4)} = \frac{1}{3}\left(\frac{1}{x+1} - \frac{1}{x+4}\right)$$

and then setting $x = s^2$ to obtain

$$\frac{1}{(s^2+1)(s^2+4)} = \frac{1}{3}\left(\frac{1}{s^2+1} - \frac{1}{s^2+4}\right).$$

Multiplying this by $8 + 3s$ yields

$$F(s) = \frac{8+3s}{(s^2+1)(s^2+4)} = \frac{1}{3}\left(\frac{8+3s}{s^2+1} - \frac{8+3s}{s^2+4}\right).$$

Therefore,

$$\mathcal{L}^{-1}(F) = \frac{8}{3}\sin t + \cos t - \frac{4}{3}\sin 2t - \cos 2t. \qquad \blacksquare$$

USING TECHNOLOGY

Some software packages that do symbolic algebra can find partial fraction expansions very easily. We recommend that you use such a package if one is available to you, but only after you have done enough partial fraction expansions on your own to master the technique.

8.2 EXERCISES

1. Use the table of Laplace transforms to find the inverse Laplace transform.

(a) $\dfrac{3}{(s-7)^4}$

(b) $\dfrac{2s-4}{s^2-4s+13}$

(c) $\dfrac{1}{s^2+4s+20}$

(d) $\dfrac{2}{s^2+9}$

(e) $\dfrac{s^2-1}{(s^2+1)^2}$

(f) $\dfrac{1}{(s-2)^2-4}$

(g) $\dfrac{12s-24}{(s^2-4s+85)^2}$

(h) $\dfrac{2}{(s-3)^2-9}$

(i) $\dfrac{s^2-4s+3}{(s^2-4s+5)^2}$

2. Use Theorem 8.2.1 and the table of Laplace transforms to find the inverse Laplace transform.

(a) $\dfrac{2s+3}{(s-7)^4}$

(b) $\dfrac{s^2-1}{(s-2)^6}$

(c) $\dfrac{s+5}{s^2+6s+18}$

(d) $\dfrac{2s+1}{s^2+9}$

(e) $\dfrac{s}{s^2+2s+1}$

(f) $\dfrac{s+1}{s^2-9}$

(g) $\dfrac{s^3+2s^2-s-3}{(s+1)^4}$

(h) $\dfrac{2s+3}{(s-1)^2+4}$

(i) $\dfrac{1}{s}-\dfrac{s}{s^2+1}$

(j) $\dfrac{3s+4}{s^2-1}$

(k) $\dfrac{3}{s-1}+\dfrac{4s+1}{s^2+9}$

(l) $\dfrac{3}{(s+2)^2}-\dfrac{2s+6}{s^2+4}$

3. Use Heaviside's method to find the inverse Laplace transform.

(a) $\dfrac{3-(s+1)(s-2)}{(s+1)(s+2)(s-2)}$

(b) $\dfrac{7+(s+4)(18-3s)}{(s-3)(s-1)(s+4)}$

(c) $\dfrac{2+(s-2)(3-2s)}{(s-2)(s+2)(s-3)}$

(d) $\dfrac{3-(s-1)(s+1)}{(s+4)(s-2)(s-1)}$

(e) $\dfrac{3+(s-2)(10-2s-s^2)}{(s-2)(s+2)(s-1)(s+3)}$

(f) $\dfrac{3+(s-3)(2s^2+s-21)}{(s-3)(s-1)(s+4)(s-2)}$

4. Find the inverse Laplace transform.

(a) $\dfrac{2+3s}{(s^2+1)(s+2)(s+1)}$

(b) $\dfrac{3s^2+2s+1}{(s^2+1)(s^2+2s+2)}$

(c) $\dfrac{3s+2}{(s-2)(s^2+2s+5)}$

(d) $\dfrac{3s^2+2s+1}{(s-1)^2(s+2)(s+3)}$

(e) $\dfrac{2s^2+s+3}{(s-1)^2(s+2)^2}$

(f) $\dfrac{3s+2}{(s^2+1)(s-1)^2}$

5. Use the method of Example 9 to find the inverse Laplace transform.

(a) $\dfrac{3s+2}{(s^2+4)(s^2+9)}$

(b) $\dfrac{-4s+1}{(s^2+1)(s^2+16)}$

(c) $\dfrac{5s+3}{(s^2+1)(s^2+4)}$

(d) $\dfrac{-s+1}{(4s^2+1)(s^2+1)}$

(e) $\dfrac{17s-34}{(s^2+16)(16s^2+1)}$

(f) $\dfrac{2s-1}{(4s^2+1)(9s^2+1)}$

6. Find the inverse Laplace transform.

(a) $\dfrac{17s-15}{(s^2-2s+5)(s^2+2s+10)}$

(b) $\dfrac{8s+56}{(s^2-6s+13)(s^2+2s+5)}$

(c) $\dfrac{s+9}{(s^2+4s+5)(s^2-4s+13)}$

(d) $\dfrac{3s-2}{(s^2-4s+5)(s^2-6s+13)}$

(e) $\dfrac{3s - 1}{(s^2 - 2s + 2)(s^2 + 2s + 5)}$

(f) $\dfrac{20s + 40}{(4s^2 - 4s + 5)(4s^2 + 4s + 5)}$

7. Find the inverse Laplace transform.

(a) $\dfrac{1}{s(s^2 + 1)}$

(b) $\dfrac{1}{(s - 1)(s^2 - 2s + 17)}$

(c) $\dfrac{3s + 2}{(s - 2)(s^2 + 2s + 10)}$

(d) $\dfrac{34 - 17s}{(2s - 1)(s^2 - 2s + 5)}$

(e) $\dfrac{s + 2}{(s - 3)(s^2 + 2s + 5)}$

(f) $\dfrac{2s - 2}{(s - 2)(s^2 + 2s + 10)}$

8. Find the inverse Laplace transform.

(a) $\dfrac{2s + 1}{(s^2 + 1)(s - 1)(s - 3)}$

(b) $\dfrac{s + 2}{(s^2 + 2s + 2)(s^2 - 1)}$

(c) $\dfrac{2s - 1}{(s^2 - 2s + 2)(s + 1)(s - 2)}$

(d) $\dfrac{s - 6}{(s^2 - 1)(s^2 + 4)}$

(e) $\dfrac{2s - 3}{s(s - 2)(s^2 - 2s + 5)}$

(f) $\dfrac{5s - 15}{(s^2 - 4s + 13)(s - 2)(s - 1)}$

9. Given that $f(t) \leftrightarrow F(s)$, find the inverse Laplace transform of $F(as - b)$, where $a > 0$.

10. (a) If $s_1, s_2, \ldots, s_n$ are distinct and P is a polynomial of degree less than n, then

$$\frac{P(s)}{(s - s_1)(s - s_2)\cdots(s - s_n)} = \frac{A_1}{s - s_1} + \frac{A_2}{s - s_2} + \cdots + \frac{A_n}{s - s_n}.$$

Multiply through by $s - s_i$ to show that A_i can be obtained by ignoring the factor $s - s_i$ on the left and setting $s = s_i$ elsewhere.

(b) Suppose that P and Q_1 are polynomials such that degree$(P) \leq$ degree(Q_1) and $Q_1(s_1) \neq 0$. Show that the coefficient of $1/(s - s_1)$ in the partial fraction expansion of

$$F(s) = \frac{P(s)}{(s - s_1)Q_1(s)}$$

is $P(s_1)/Q_1(s_1)$.

(c) Explain how the results of **(a)** and **(b)** are related.

8.3   Solution of Initial Value Problems

LAPLACE TRANSFORMS OF DERIVATIVES

In the rest of this chapter we will use the Laplace transform to solve initial value problems for constant coefficient second order equations. To do this we must know how the Laplace transform of a derivative f' is related to the Laplace transform of f. The following theorem answers this question.

THEOREM 8.3.1

Suppose that f is continuous on $[0, \infty)$ and of exponential order s_0, and f' is piecewise continuous on $[0, \infty)$. Then f and f' have Laplace transforms for $s > s_0$, and

$$\mathcal{L}(f') = s\mathcal{L}(f) - f(0). \tag{1}$$

PROOF. We know from Theorem 8.1.6 that $\mathcal{L}(f)$ is defined for $s > s_0$. We first consider the case where f' is continuous on $[0, \infty)$. Integration by parts yields

$$\int_0^T e^{-st} f'(t)\, dt = e^{-st} f(t) \Big|_0^T + s \int_0^T e^{-st} f(t)\, dt$$

$$= e^{-sT} f(T) - f(0) + s \int_0^T e^{-st} f(t)\, dt \tag{2}$$

for any $T > 0$. Since f is of exponential order s_0, it follows that $\lim_{T \to \infty} e^{-sT} f(T) = 0$ and the last integral in (2) converges as $T \to \infty$ if $s > s_0$. Therefore

$$\int_0^\infty e^{-st} f'(t)\, dt = -f(0) + s \int_0^\infty e^{-st} f(t)\, dt$$

$$= -f(0) + s\mathcal{L}(f),$$

which proves (1). Now suppose that $T > 0$ and f' is only piecewise continuous on $[0, T]$, with discontinuities at $t_1 < t_2 < \cdots < t_{n-1}$. For convenience, let $t_0 = 0$ and $t_n = T$. Integrating by parts yields

$$\int_{t_{i-1}}^{t_i} e^{-st} f'(t)\, dt = e^{-st} f(t) \Big|_{t_{i-1}}^{t_i} + s \int_{t_{i-1}}^{t_i} e^{-st} f(t)\, dt$$

$$= e^{-st_i} f(t_i) - e^{-st_{i-1}} f(t_{i-1}) + s \int_{t_{i-1}}^{t_i} e^{-st} f(t)\, dt.$$

Summing both sides of this equation from $i = 1$ to n and noting that

$$(e^{-st_1} f(t_1) - e^{-st_0} f(t_0)) + (e^{-st_2} f(t_2) - e^{-st_1} f(t_1))$$

$$+ \cdots + (e^{-st_N} f(t_N) - e^{-st_{N-1}} f(t_{N-1}))$$

$$= e^{-st_N} f(t_N) - e^{-st_0} f(t_0) = e^{-sT} f(T) - f(0)$$

yields (2), so (1) follows as before. $\qquad\square$

EXAMPLE 1 In Example 4 of Section 8.1 we saw that

$$\mathcal{L}(\cos \omega t) = \frac{s}{s^2 + \omega^2}.$$

Applying (1) with $f(t) = \cos \omega t$ shows that

$$\mathcal{L}(-\omega \sin \omega t) = s\, \frac{s}{s^2 + \omega^2} - 1 = -\frac{\omega^2}{s^2 + \omega^2},$$

and therefore

$$\mathcal{L}(\sin \omega t) = \frac{\omega}{s^2 + \omega^2},$$

which agrees with the corresponding result obtained in Example 4 of Section 8.1. ∎

In Section 2.1 we showed that the solution of the initial value problem

$$y' = ay, \qquad y(0) = y_0 \tag{3}$$

is $y = y_0 e^{at}$. We will now obtain this result by using the Laplace transform.

Let $Y(s) = \mathcal{L}(y)$ be the Laplace transform of the unknown solution of (3). Taking Laplace transforms of both sides of (3) yields

$$\mathcal{L}(y') = \mathcal{L}(ay),$$

which, by Theorem 8.3.1, can be rewritten as

$$s\mathcal{L}(y) - y(0) = a\mathcal{L}(y),$$

or

$$sY(s) - y_0 = aY(s).$$

Solving for $Y(s)$ yields

$$Y(s) = \frac{y_0}{s - a},$$

so

$$y = \mathcal{L}^{-1}(Y(s)) = \mathcal{L}^{-1}\left(\frac{y_0}{s - a}\right) = y_0\mathcal{L}^{-1}\left(\frac{1}{s - a}\right) = y_0 e^{at},$$

which agrees with the known result.

We need the following theorem to solve second order differential equations using the Laplace transform.

THEOREM 8.3.2

Suppose that f and f' are continuous on $[0, \infty)$ and of exponential order s_0, and that f'' is piecewise continuous on $[0, \infty)$. Then f, f', and f'' have Laplace transforms for $s > s_0$,

$$\mathcal{L}(f') = s\mathcal{L}(f) - f(0), \tag{4}$$

and

$$\mathcal{L}(f'') = s^2\mathcal{L}(f) - f'(0) - sf(0). \tag{5}$$

PROOF. Theorem 8.3.1 implies that $\mathcal{L}(f')$ exists and satisfies (4) for $s > s_0$. To prove that $\mathcal{L}(f'')$ exists and satisfies (5) for $s > s_0$, we first apply Theorem 8.3.1 to $g = f'$. Since g satisfies the hypotheses of Theorem 8.3.1, we conclude that $\mathcal{L}(g')$ is defined and satisfies

$$\mathcal{L}(g') = s\mathcal{L}(g) - g(0)$$

for $s > s_0$. However, since $g' = f''$, this can be rewritten as

$$\mathcal{L}(f'') = s\mathcal{L}(f') - f'(0),$$

and substituting (4) into this yields (5). $\square$

SOLVING SECOND ORDER EQUATIONS WITH THE LAPLACE TRANSFORM

We will now use the Laplace transform to solve initial value problems for second order equations.

EXAMPLE 2   Use the Laplace transform to solve the initial value problem

$$y'' - 6y' + 5y = 3e^{2t}, \qquad y(0) = 2, \qquad y'(0) = 3. \tag{6}$$

Solution   Taking Laplace transforms of both sides of the differential equation in (6) yields

$$\mathcal{L}(y'' - 6y' + 5y) = \mathcal{L}(3e^{2t}) = \frac{3}{s-2},$$

which we rewrite as

$$\mathcal{L}(y'') - 6\mathcal{L}(y') + 5\mathcal{L}(y) = \frac{3}{s-2}. \tag{7}$$

Now denote $\mathcal{L}(y) = Y(s)$. Theorem 8.3.2 and the initial conditions in (6) imply that

$$\mathcal{L}(y') = sY(s) - y(0) = sY(s) - 2$$

and

$$\mathcal{L}(y'') = s^2 Y(s) - y'(0) - sy(0) = s^2 Y(s) - 3 - 2s.$$

Substituting from the last two equations into (7) yields

$$(s^2 Y(s) - 3 - 2s) - 6(sY(s) - 2) + 5Y(s) = \frac{3}{s-2}.$$

Therefore,

$$(s^2 - 6s + 5)Y(s) = \frac{3}{s-2} + (3 + 2s) + 6(-2), \tag{8}$$

so

$$(s - 5)(s - 1)Y(s) = \frac{3 + (s-2)(2s-9)}{s-2},$$

and

$$Y(s) = \frac{3 + (s-2)(2s-9)}{(s-2)(s-5)(s-1)}.$$

Heaviside's method yields the partial fraction expansion

$$Y(s) = -\frac{1}{s-2} + \frac{1}{2}\frac{1}{s-5} + \frac{5}{2}\frac{1}{s-1}$$

and taking the inverse transform of this yields

$$y = -e^{2t} + \frac{1}{2}e^{5t} + \frac{5}{2}e^t$$

as the solution of (6).   ∎

It isn't necessary to write all the steps that we used to obtain (8). To see how to avoid this, let's apply the method of Example 2 to the general initial value problem

$$ay'' + by' + cy = f(t), \qquad y(0) = k_0, \qquad y'(0) = k_1. \tag{9}$$

Taking Laplace transforms of both sides of the differential equation in (9) yields

$$a\mathcal{L}(y'') + b\mathcal{L}(y') + c\mathcal{L}(y) = F(s). \tag{10}$$

Now let $Y(s) = \mathcal{L}(y)$. Theorem 8.3.2 and the initial conditions in (9) imply that

$$\mathcal{L}(y') = sY(s) - k_0 \quad \text{and} \quad \mathcal{L}(y'') = s^2 Y(s) - k_1 - k_0 s.$$

Substituting these into (10) yields

$$a(s^2 Y(s) - k_1 - k_0 s) + b(sY(s) - k_0) + cY(s) = F(s). \tag{11}$$

Notice that the coefficient of $Y(s)$ on the left is the characteristic polynomial

$$p(s) = as^2 + bs + c$$

of the complementary equation for (9). Using this and moving the terms involving k_0 and k_1 to the right side of (11) yields

$$p(s)Y(s) = F(s) + a(k_1 + k_0 s) + bk_0. \tag{12}$$

This equation corresponds to (8) of Example 2. Having established the form of this equation in the general case, it is preferable to go directly from the initial value problem to this equation. You may find it easier to remember (12) rewritten as

$$p(s)Y(s) = F(s) + a(y'(0) + sy(0)) + by(0). \tag{13}$$

EXAMPLE 3   Use the Laplace transform to solve the initial value problem

$$2y'' + 3y' + y = 8e^{-2t}, \quad y(0) = -4, \quad y'(0) = 2. \tag{14}$$

Solution   The characteristic polynomial is

$$p(s) = 2s^2 + 3s + 1 = (2s + 1)(s + 1)$$

and

$$F(s) = \mathcal{L}(8e^{-2t}) = \frac{8}{s + 2},$$

so (13) becomes

$$(2s + 1)(s + 1)Y(s) = \frac{8}{s + 2} + 2(2 - 4s) + 3(-4).$$

Solving for $Y(s)$ yields

$$Y(s) = \frac{4(1 - (s + 2)(s + 1))}{(s + 1/2)(s + 1)(s + 2)}.$$

Heaviside's method yields the partial fraction expansion

$$Y(s) = \frac{4}{3} \frac{1}{s + 1/2} - \frac{8}{s + 1} + \frac{8}{3} \frac{1}{s + 2},$$

so the solution of (14) is

$$y = \mathcal{L}^{-1}(Y(s)) = \frac{4}{3}e^{-t/2} - 8e^{-t} + \frac{8}{3}e^{-2t}.$$

The graph of the solution is shown in Figure 1. ■

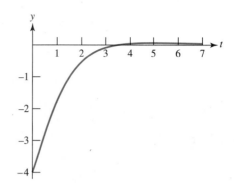

Figure 1 Graph of
$$y = \frac{4}{3}e^{-t/2} - 8e^{-t} + \frac{8}{3}e^{-2t}$$

EXAMPLE 4 Solve the initial value problem
$$y'' + 2y' + 2y = 1, \qquad y(0) = -3, \qquad y'(0) = 1. \tag{15}$$

Solution The characteristic polynomial is
$$p(s) = s^2 + 2s + 2 = (s + 1)^2 + 1$$

and
$$F(s) = \mathcal{L}(1) = \frac{1}{s},$$

so (13) becomes
$$[(s + 1)^2 + 1]Y(s) = \frac{1}{s} + 1 \cdot (1 - 3s) + 2(-3).$$

Solving for $Y(s)$ yields
$$Y(s) = \frac{1 - s(5 + 3s)}{s[(s + 1)^2 + 1]}.$$

In Example 8 of Section 8.2 we found the inverse transform of this function to be
$$y = \frac{1}{2} - \frac{7}{2}e^{-t}\cos t - \frac{5}{2}e^{-t}\sin t$$

which is therefore the solution of (15). The graph of the solution is shown in Figure 2. ∎

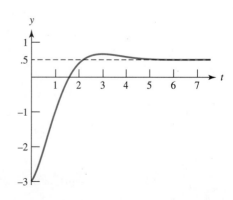

Figure 2 Graph of
$$y = \frac{1}{2} - \frac{7}{2}e^{-t}\cos t - \frac{5}{2}e^{-t}\sin t$$

REMARK.   In our examples we applied Theorems 8.3.1 and 8.3.2 without verifying that the unknown function y satisfies their hypotheses. This is characteristic of the formal manipulative way in which the Laplace transform is used to solve differential equations. Any doubts about the validity of the method for solving a given equation can be resolved by verifying that the resulting function y is in fact the solution of the given problem.

8.3 EXERCISES

In Exercises 1–31 use the Laplace transform to solve the initial value problem.

1. $y'' + 3y' + 2y = e^t$, $y(0) = 1$, $y'(0) = -6$

2. $y'' - y' - 6y = 2$, $y(0) = 1$, $y'(0) = 0$

3. $y'' + y' - 2y = 2e^{3t}$, $y(0) = -1$, $y'(0) = 4$

4. $y'' - 4y = 2e^{3t}$, $y(0) = 1$, $y'(0) = -1$

5. $y'' + y' - 2y = e^{3t}$, $y(0) = 1$, $y'(0) = -1$

6. $y'' + 3y' + 2y = 6e^t$, $y(0) = 1$, $y'(0) = -1$

7. $y'' + y = \sin 2t$, $y(0) = 0$, $y'(0) = 1$

8. $y'' - 3y' + 2y = 2e^{3t}$, $y(0) = 1$, $y'(0) = -1$

9. $y'' - 3y' + 2y = e^{4t}$, $y(0) = 1$, $y'(0) = -2$

10. $y'' - 3y' + 2y = e^{3t}$, $y(0) = -1$, $y'(0) = -4$

11. $y'' + 3y' + 2y = 2e^t$, $y(0) = 0$, $y'(0) = -1$

12. $y'' + y' - 2y = -4$, $y(0) = 2$, $y'(0) = 3$

13. $y'' + 4y = 4$, $y(0) = 0$, $y'(0) = 1$

14. $y'' - y' - 6y = 2$, $y(0) = 1$, $y'(0) = 0$

15. $y'' + 3y' + 2y = e^t$, $y(0) = 0$, $y'(0) = 1$

16. $y'' - y = 1$, $y(0) = 1$, $y'(0) = 0$

17. $y'' + 4y = 3 \sin t$, $y(0) = 1$, $y'(0) = -1$

18. $y'' + y' = 2e^{3t}$, $y(0) = -1$, $y'(0) = 4$

19. $y'' + y = 1$, $y(0) = 2$, $y'(0) = 0$

20. $y'' + y = t$, $y(0) = 0$, $y'(0) = 2$

21. $y'' + y = t - 3 \sin 2t$, $y(0) = 1$, $y'(0) = -3$

22. $y'' + 5y' + 6y = 2e^{-t}$, $y(0) = 1$, $y'(0) = 3$

23. $y'' + 2y' + y = 6 \sin t - 4 \cos t$, $y(0) = -1$, $y'(0) = 1$

24. $y'' - 2y' - 3y = 10 \cos t$, $y(0) = 2$, $y'(0) = 7$

25. $y'' + y = 4 \sin t + 6 \cos t$, $y(0) = -6$, $y'(0) = 2$

26. $y'' + 4y = 8 \sin 2t + 9 \cos t$, $y(0) = 1$, $y'(0) = 0$

27. $y'' - 5y' + 6y = 10e^t \cos t$, $y(0) = 2$, $y'(0) = 1$

28. $y'' + 2y' + 2y = 2t$, $y(0) = 2$, $y'(0) = -7$

29. $y'' - 2y' + 2y = 5 \sin t + 10 \cos t$, $y(0) = 1$, $y'(0) = 2$

30. $y'' + 4y' + 13y = 10e^{-t} - 36e^t$, $y(0) = 0$, $y'(0) = -16$

31. $y'' + 4y' + 5y = e^{-t}(\cos t + 3 \sin t)$, $y(0) = 0$, $y'(0) = 4$

32. $2y'' - 3y' - 2y = 4e^t$, $y(0) = 1$, $y'(0) = -2$

33. $6y'' - y' - y = 3e^{2t}$, $y(0) = 0$, $y'(0) = 0$

34. $2y'' + 2y' + y = 2t$, $y(0) = 1$, $y'(0) = -1$

35. $4y'' - 4y' + 5y = 4 \sin t - 4 \cos t$, $y(0) = 0$, $y'(0) = 1$

36. $4y'' + 4y' + y = 3 \sin t + \cos t$, $y(0) = 2$, $y'(0) = -1$

37. $9y'' + 6y' + y = 3e^{3t}$, $y(0) = 0$, $y'(0) = -3$

38. Suppose that a, b, and c are constants and $a \neq 0$. Let

$$y_1 = \mathcal{L}^{-1}\left(\frac{as + b}{as^2 + bs + c} \right) \quad \text{and} \quad y_2 = \mathcal{L}^{-1}\left(\frac{a}{as^2 + bs + c} \right).$$

Show that

$$y_1(0) = 1, \qquad y_1'(0) = 0 \quad \text{and} \quad y_2(0) = 0, \qquad y_2'(0) = 1.$$

Hint: Use the Laplace transform to solve the initial value problems

$$ay'' + by' + cy = 0, \qquad y(0) = 1, \qquad y'(0) = 0$$
$$ay'' + by' + cy = 0, \qquad y(0) = 0, \qquad y'(0) = 1.$$

8.4   The Unit Step Function

In the next section we will consider initial value problems

$$ay'' + by' + cy = f(t), \qquad y(0) = k_0, \qquad y'(0) = k_1,$$

where $a, b,$ and c are constants and f is a piecewise continuous function. In this section we will develop procedures for using the table of Laplace transforms to find Laplace transforms of piecewise continuous functions, and to find the piecewise continuous inverses of Laplace transforms. We begin with an example.

EXAMPLE 1   Use the table of Laplace transforms to find the Laplace transform of

$$f(t) = \begin{cases} 2t + 1, & 0 \le t < 2, \\ 3t, & t \ge 2 \end{cases} \tag{1}$$

(Figure 1).

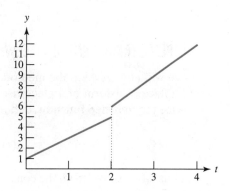

Figure 1   Graph of the piecewise continuous function (1)

Solution   Since the formula for f changes at $t = 2$ we write

$$\mathcal{L}(f) = \int_0^\infty e^{-st} f(t) \, dt$$
$$= \int_0^2 e^{-st}(2t + 1) \, dt + \int_2^\infty e^{-st}(3t) \, dt. \tag{2}$$

To relate the first term to a Laplace transform we add and subtract

$$\int_2^\infty e^{-st}(2t + 1) \, dt$$

in (2) to obtain

$$\mathcal{L}(f) = \int_0^\infty e^{-st}(2t + 1)\, dt + \int_2^\infty e^{-st}(3t - 2t - 1)\, dt$$

$$= \int_0^\infty e^{-st}(2t + 1)\, dt + \int_2^\infty e^{-st}(t - 1)\, dt \tag{3}$$

$$= \mathcal{L}(2t + 1) + \int_2^\infty e^{-st}(t - 1)\, dt.$$

To relate the last integral to a Laplace transform we make the change of variable $x = t - 2$ and rewrite the integral as

$$\int_2^\infty e^{-st}(t - 1)\, dt = \int_0^\infty e^{-s(x+2)}(x + 1)\, dx = e^{-2s}\int_0^\infty e^{-sx}(x + 1)\, dx.$$

Since the symbol used for the variable of integration has no effect on the value of a definite integral, we can now replace x by the more standard t and write

$$\int_2^\infty e^{-st}(t - 1)\, dt = e^{-2s}\int_0^\infty e^{-st}(t + 1)\, dt = e^{-2s}\mathcal{L}(t + 1).$$

This and (3) imply that

$$\mathcal{L}(f) = \mathcal{L}(2t + 1) + e^{-2s}\mathcal{L}(t + 1),$$

and now we can use the table of Laplace transforms to find that

$$\mathcal{L}(f) = \frac{2}{s^2} + \frac{1}{s} + e^{-2s}\left(\frac{1}{s^2} + \frac{1}{s}\right). \qquad\blacksquare$$

LAPLACE TRANSFORMS OF PIECEWISE CONTINUOUS FUNCTIONS

We will now develop the method of Example 1 into a systematic way to find the Laplace transform of a piecewise continuous function. It is convenient to introduce the **unit step function,** defined as

$$u(t) = \begin{cases} 0, & t < 0 \\ 1, & t \geq 0. \end{cases} \tag{4}$$

Thus, $u(t)$ "steps" from the constant value 0 to the constant value 1 at $t = 0$. If we replace t by $t - \tau$ in (4), then we have

$$u(t - \tau) = \begin{cases} 0, & t < \tau, \\ 1, & t \geq \tau; \end{cases}$$

that is, the step now occurs at $t = \tau$ (Figure 2).

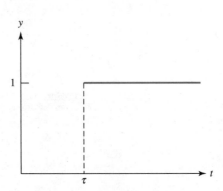

Figure 2   Graph of $y = u(t - \tau)$

The step function enables us to represent piecewise continuous functions conveniently. For example, consider the function

$$f(t) = \begin{cases} f_0(t), & 0 \le t < t_1, \\ f_1(t), & t \ge t_1, \end{cases} \tag{5}$$

where we assume that f_0 and f_1 are defined on $[0, \infty)$, even though they equal f only on the indicated intervals. This assumption enables us to rewrite (5) as

$$f(t) = f_0(t) + u(t - t_1)(f_1(t) - f_0(t)). \tag{6}$$

To verify this, notice that if $t < t_1$, then $u(t - t_1) = 0$ and (6) becomes

$$f(t) = f_0(t) + 0(f_1(t) - f_0(t)) = f_0(t).$$

If $t \ge t_1$, then $u(t - t_1) = 1$ and (6) becomes

$$f(t) = f_0(t) + (1)(f_1(t) - f_0(t)) = f_1(t).$$

We will now show how (6) can be used to find $\mathcal{L}(f)$. For this purpose we need the following theorem.

THOEREM 8.4.1

> Let g be defined on $[0, \infty)$. Suppose that $\tau \ge 0$ and $\mathcal{L}(g(t + \tau))$ exists for $s > s_0$. Then $\mathcal{L}(u(t - \tau)g(t))$ exists for $s > s_0$ and
>
> $$\mathcal{L}(u(t - \tau)g(t)) = e^{-s\tau}\mathcal{L}(g(t + \tau)).$$

PROOF. By definition,

$$\mathcal{L}(u(t - \tau)g(t)) = \int_0^\infty e^{-st}u(t - \tau)g(t)\, dt.$$

From this and the definition of $u(t - \tau)$,

$$\mathcal{L}(u(t - \tau)g(t)) = \int_0^\tau e^{-st}(0)\, dt + \int_\tau^\infty e^{-st}g(t)\, dt.$$

The first integral on the right equals zero. Introducing the new variable of integration $x = t - \tau$ in the second integral yields

$$\mathcal{L}(u(t - \tau)g(t)) = \int_0^\infty e^{-s(x+\tau)}g(x + \tau)\, dx = e^{-s\tau}\int_0^\infty e^{-sx}g(x + \tau)\, dx.$$

Changing the name of the variable of integration in the last integral from x to t yields

$$\mathcal{L}(u(t - \tau)g(t)) = e^{-s\tau}\int_0^\infty e^{-st}g(t + \tau)\, dt = e^{-s\tau}\mathcal{L}(g(t + \tau)). \qquad \square$$

EXAMPLE 2   Find

$$\mathcal{L}(u(t - 1)(t^2 + 1)).$$

Solution   Here $\tau = 1$ and $g(t) = t^2 + 1$, so

$$g(t + 1) = (t + 1)^2 + 1 = t^2 + 2t + 2.$$

Since

$$\mathcal{L}(g(t + 1)) = \frac{2}{s^3} + \frac{2}{s^2} + \frac{2}{s},$$

Theorem 8.4.1 implies that

$$\mathcal{L}(u(t - 1)(t^2 + 1)) = e^{-s}\left(\frac{2}{s^3} + \frac{2}{s^2} + \frac{2}{s}\right).$$

EXAMPLE 3   Use Theorem 8.4.1 to find the Laplace transform of the function

$$f(t) = \begin{cases} 2t + 1, & 0 \le t < 2, \\ 3t, & t \ge 2, \end{cases}$$

from Example 1.

Solution   We first write f in the form (6) as

$$f(t) = 2t + 1 + u(t - 2)(t - 1).$$

Therefore,

$$\begin{aligned} \mathcal{L}(f) &= \mathcal{L}(2t + 1) + \mathcal{L}(u(t - 2)(t - 1)) \\ &= \mathcal{L}(2t + 1) + e^{-2s}\mathcal{L}(t + 1) \qquad \text{(from Theorem 8.4.1)} \\ &= \frac{2}{s^2} + \frac{1}{s} + e^{-2s}\left(\frac{1}{s^2} + \frac{1}{s}\right), \end{aligned}$$

which is the result obtained in Example 1.   ◼

Formula (6) can be extended to more general piecewise continuous functions. For example, we can write

$$f(t) = \begin{cases} f_0(t), & 0 \le t < t_1, \\ f_1(t), & t_1 \le t < t_2, \\ f_2(t), & t \ge t_2, \end{cases}$$

as

$$f(t) = f_0(t) + u(t - t_1)(f_1(t) - f_0(t)) + u(t - t_2)(f_2(t) - f_1(t))$$

if $f_0, f_1,$ and f_2 are all defined on $[0, \infty)$.

EXAMPLE 4   Find the Laplace transform of

$$f(t) = \begin{cases} 1, & 0 \le t < 2, \\ -2t + 1, & 2 \le t < 3, \\ 3t, & 3 \le t < 5, \\ t - 1, & t \ge 5 \end{cases} \tag{7}$$

(Figure 3).

Solution   In terms of step functions,

$$\begin{aligned} f(t) = {} &1 + u(t - 2)(-2t + 1 - 1) + u(t - 3)(3t + 2t - 1) \\ &+ u(t - 5)(t - 1 - 3t), \end{aligned}$$

or

$$f(t) = 1 - 2u(t - 2)t + u(t - 3)(5t - 1) - u(t - 5)(2t + 1).$$

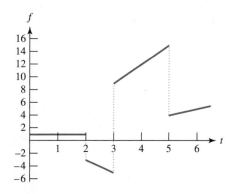

Figure 3   Graph of the piecewise continuous function (7)

Now Theorem 8.4.1 implies that

$$\mathcal{L}(f) = \mathcal{L}(1) - 2e^{-2s}\mathcal{L}(t + 2) + e^{-3s}\mathcal{L}(5(t + 3) - 1) - e^{-5s}\mathcal{L}(2(t + 5) + 1)$$
$$= \mathcal{L}(1) - 2e^{-2s}\mathcal{L}(t + 2) + e^{-3s}\mathcal{L}(5t + 14) - e^{-5s}\mathcal{L}(2t + 11)$$
$$= \frac{1}{s} - 2e^{-2s}\left(\frac{1}{s^2} + \frac{2}{s}\right) + e^{-3s}\left(\frac{5}{s^2} + \frac{14}{s}\right) - e^{-5s}\left(\frac{2}{s^2} + \frac{11}{s}\right).$$

The trigonometric identities

$$\sin(A + B) = \sin A \cos B + \cos A \sin B \tag{8}$$
$$\cos(A + B) = \cos A \cos B - \sin A \sin B \tag{9}$$

are useful in problems that involve shifting the arguments of trigonometric functions. We will use these identities in the following example.

EXAMPLE 5   Find the Laplace transform of

$$f(t) = \begin{cases} \sin t, & 0 \le t < \dfrac{\pi}{2}, \\[2mm] \cos t - 3 \sin t, & \dfrac{\pi}{2} \le t < \pi, \\[2mm] 3 \cos t, & t \ge \pi \end{cases} \tag{10}$$

(Figure 4).

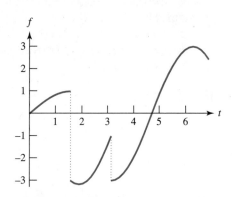

Figure 4   Graph of the piecewise continuous function (10)

Solution   In terms of step functions,

$$f(t) = \sin t + u(t - \pi/2)(\cos t - 4 \sin t) + u(t - \pi)(2 \cos t + 3 \sin t).$$

Now Theorem 8.4.1 implies that

$$\mathcal{L}(f) = \mathcal{L}(\sin t) + e^{-\pi s/2}\mathcal{L}\left(\cos\left(t + \frac{\pi}{2}\right) - 4 \sin\left(t + \frac{\pi}{2}\right)\right)$$

$$+ e^{-\pi s}\mathcal{L}(2 \cos(t + \pi) + 3 \sin(t + \pi)). \tag{11}$$

Since

$$\cos\left(t + \frac{\pi}{2}\right) - 4 \sin\left(t + \frac{\pi}{2}\right) = -\sin t - 4 \cos t$$

and

$$2 \cos(t + \pi) + 3 \sin(t + \pi) = -2 \cos t - 3 \sin t,$$

we see from (11) that

$$\mathcal{L}(f) = \mathcal{L}(\sin t) - e^{-\pi s/2}\mathcal{L}(\sin t + 4 \cos t) - e^{-\pi s}\mathcal{L}(2 \cos t + 3 \sin t)$$

$$= \frac{1}{s^2 + 1} - e^{-\pi s/2}\left(\frac{1 + 4s}{s^2 + 1}\right) - e^{-\pi s}\left(\frac{3 + 2s}{s^2 + 1}\right). \qquad ■$$

THE SECOND SHIFTING THEOREM

Replacing $g(t)$ by $g(t - \tau)$ in Theorem 8.4.1 yields the following theorem.

THEOREM 8.4.2

(Second Shifting Theorem)   *Suppose that $\tau \geq 0$ and $\mathcal{L}(g)$ exists for $s > s_0$. Then $\mathcal{L}(u(t - \tau)g(t - \tau))$ exists for $s > s_0$ and*

$$\mathcal{L}(u(t - \tau)g(t - \tau)) = e^{-s\tau}\mathcal{L}(g(t)),$$

or, equivalently,

if $g(t) \leftrightarrow G(s)$ *then* $u(t - \tau)g(t - \tau) \leftrightarrow e^{-s\tau}G(s).$ (12)

REMARK.   Recall that the First Shifting Theorem (Theorem 8.1.3) states that multiplying a function of time by e^{at} corresponds to shifting the argument of its transform by a units. Theorem 8.4.2 states that multiplying a Laplace transform by the exponential $e^{-\tau s}$ corresponds to shifting the argument of the inverse transform by τ units.

EXAMPLE 6   Use (12) to find

$$\mathcal{L}^{-1}\left(\frac{e^{-2s}}{s^2}\right).$$

Solution   To apply (12) we let $\tau = 2$ and $G(s) = 1/s^2$. Then $g(t) = t$ and (12) implies that

$$\mathcal{L}^{-1}\left(\frac{e^{-2s}}{s^2}\right) = u(t - 2)(t - 2).$$

EXAMPLE 7   Find the inverse Laplace transform h of

$$H(s) = \frac{1}{s^2} - e^{-s}\left(\frac{1}{s^2} + \frac{2}{s}\right) + e^{-4s}\left(\frac{4}{s^3} + \frac{1}{s}\right),$$

and find distinct formulas for h on appropriate intervals.

Solution   Let

$$G_0(s) = \frac{1}{s^2}, \qquad G_1(s) = \frac{1}{s^2} + \frac{2}{s}, \qquad G_2(s) = \frac{4}{s^3} + \frac{1}{s}.$$

Then

$$g_0(t) = t, \qquad g_1(t) = t + 2, \qquad g_2(t) = 2t^2 + 1.$$

Hence, (12) and the linearity of $\mathcal{L}^{-1}$ imply that

$$\begin{aligned}
h(t) &= \mathcal{L}^{-1}(G_0(s)) - \mathcal{L}^{-1}(e^{-s}G_1(s)) + \mathcal{L}^{-1}(e^{-4s}G_2(s)) \\
&= t - u(t-1)[(t-1) + 2] + u(t-4)[2(t-4)^2 + 1] \\
&= t - u(t-1)(t+1) + u(t-4)(2t^2 - 16t + 33),
\end{aligned}$$

which can also be written as

$$h(t) = \begin{cases} t, & 0 \le t < 1, \\ -1, & 1 \le t < 4, \\ 2t^2 - 16t + 32, & t \ge 4. \end{cases}$$

EXAMPLE 8   Find the inverse transform of

$$H(s) = \frac{2s}{s^2 + 4} - e^{-\pi s/2}\frac{3s + 1}{s^2 + 9} + e^{-\pi s}\frac{s + 1}{s^2 + 6s + 10}.$$

Solution   Let

$$G_0(s) = \frac{2s}{s^2 + 4}, \qquad G_1(s) = -\frac{(3s + 1)}{s^2 + 9},$$

and

$$G_2(s) = \frac{s + 1}{s^2 + 6s + 10} = \frac{(s + 3) - 2}{(s + 3)^2 + 1}.$$

Then

$$g_0(t) = 2\cos 2t, \qquad g_1(t) = -3\cos 3t - \frac{1}{3}\sin 3t$$

and

$$g_2(t) = e^{-3t}(\cos t - 2\sin t).$$

Therefore, (12) and the linearity of $\mathcal{L}^{-1}$ imply that

$$\begin{aligned}
h(t) = 2\cos 2t - u(t - \pi/2)\left[3\cos 3(t - \pi/2) + \frac{1}{3}\sin 3\left(t - \frac{\pi}{2}\right)\right] \\
+ u(t - \pi)e^{-3(t-\pi)}[\cos(t - \pi) - 2\sin(t - \pi)].
\end{aligned}$$

By using the trigonometric identities (8) and (9) we can rewrite this as

$$h(t) = 2 \cos 2t + u(t - \pi/2)\left(3 \sin 3t - \frac{1}{3} \cos 3t \right)$$
$$-u(t - \pi)e^{-3(t - \pi)}(\cos t - 2 \sin t) \tag{13}$$

(Figure 5).

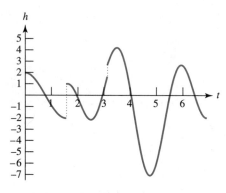

Figure 5 Graph of the piecewise continuous function (13)

8.4 EXERCISES

In Exercises 1–6 find the Laplace transform by the method of Example 1. Then express the given function f in terms of unit step functions as in Eqn. (6), and use Theorem 8.4.1 to find $\mathcal{L}(f)$. Where indicated by **C**, *graph f.*

1. $f(t) = \begin{cases} 1, & 0 \le t < 4, \\ t, & t \ge 4. \end{cases}$

2. $f(t) = \begin{cases} t, & 0 \le t < 1, \\ 1, & t \ge 1. \end{cases}$

C 3. $f(t) = \begin{cases} 2t - 1, & 0 \le t < 2, \\ t, & t \ge 2. \end{cases}$

C 4. $f(t) = \begin{cases} 1, & 0 \le t < 1, \\ t + 2, & t \ge 1. \end{cases}$

5. $f(t) = \begin{cases} t - 1, & 0 \le t < 2, \\ 4, & t \ge 2. \end{cases}$

6. $f(t) = \begin{cases} t^2, & 0 \le t < 1, \\ 0, & t \ge 1. \end{cases}$

In Exercises 7–18 express the given function f in terms of unit step functions and use Theorem 8.4.1 to find $\mathcal{L}(f)$. Where indicated by **C**, *graph f.*

7. $f(t) = \begin{cases} 0, & 0 \le t < 2, \\ t^2 + 3t, & t \ge 2. \end{cases}$

8. $f(t) = \begin{cases} t^2 + 2, & 0 \le t < 1, \\ t, & t \ge 1. \end{cases}$

9. $f(t) = \begin{cases} te^t, & 0 \le t < 1, \\ e^t, & t \ge 1. \end{cases}$

10. $f(t) = \begin{cases} e^{-t}, & 0 \le t < 1, \\ e^{-2t}, & t \ge 1. \end{cases}$

11. $f(t) = \begin{cases} -t, & 0 \le t < 2, \\ t - 4, & 2 \le t < 3, \\ 1, & t \ge 3. \end{cases}$

12. $f(t) = \begin{cases} 0, & 0 \le t < 1, \\ t, & 1 \le t < 2, \\ 0, & t \ge 2. \end{cases}$

13. $f(t) = \begin{cases} t, & 0 \le t < 1, \\ t^2, & 1 \le t < 2, \\ 0, & t \ge 2. \end{cases}$

14. $f(t) = \begin{cases} t, & 0 \le t < 1, \\ 2 - t, & 1 \le t < 2, \\ 6, & t > 2. \end{cases}$

C **15.** $f(t) = \begin{cases} \sin t, & 0 \le t < \dfrac{\pi}{2}, \\ 2\sin t, & \dfrac{\pi}{2} \le t < \pi, \\ \cos t, & t \ge \pi. \end{cases}$

C **16.** $f(t) = \begin{cases} 2, & 0 \le t < 1, \\ -2t + 2, & 1 \le t < 3, \\ 3t, & t \ge 3. \end{cases}$

C **17.** $f(t) = \begin{cases} 3, & 0 \le t < 2, \\ 3t + 2, & 2 \le t < 4, \\ 4t, & t \ge 4. \end{cases}$

C **18.** $f(t) = \begin{cases} (t+1)^2, & 0 \le t < 1, \\ (t+2)^2, & t \ge 1. \end{cases}$

In Exercises 19–28 use Theorem 8.4.2 to express the inverse transforms in terms of step functions, and then find distinct formulas for the inverse transforms on the appropriate intervals, as in Example 7. Where indicated by **C**, *graph the inverse transform.*

19. $H(s) = \dfrac{e^{-2s}}{s - 2}$

20. $H(s) = \dfrac{e^{-s}}{s(s + 1)}$

C **21.** $H(s) = \dfrac{e^{-s}}{s^3} + \dfrac{e^{-2s}}{s^2}$

C **22.** $H(s) = \left(\dfrac{2}{s} + \dfrac{1}{s^2}\right) + e^{-s}\left(\dfrac{3}{s} - \dfrac{1}{s^2}\right) + e^{-3s}\left(\dfrac{1}{s} + \dfrac{1}{s^2}\right)$

23. $H(s) = \left(\dfrac{5}{s} - \dfrac{1}{s^2}\right) + e^{-3s}\left(\dfrac{6}{s} + \dfrac{7}{s^2}\right) + \dfrac{3e^{-6s}}{s^3}$

24. $H(s) = \dfrac{e^{-\pi s}(1 - 2s)}{s^2 + 4s + 5}$

C **25.** $H(s) = \left(\dfrac{1}{s} - \dfrac{s}{s^2 + 1}\right) + e^{-\pi s/2}\left(\dfrac{3s - 1}{s^2 + 1}\right)$

26. $H(s) = e^{-2s}\left[\dfrac{3(s - 3)}{(s + 1)(s - 2)} - \dfrac{s + 1}{(s - 1)(s - 2)}\right]$

27. $H(s) = \dfrac{1}{s} + \dfrac{1}{s^2} + e^{-s}\left(\dfrac{3}{s} + \dfrac{2}{s^2}\right) + e^{-3s}\left(\dfrac{4}{s} + \dfrac{3}{s^2}\right)$

28. $H(s) = \dfrac{1}{s} - \dfrac{2}{s^3} + e^{-2s}\left(\dfrac{3}{s} - \dfrac{1}{s^3}\right) + \dfrac{e^{-4s}}{s^2}$

29. Find $\mathcal{L}(u(t - \tau))$.

30. Let $\{t_m\}_{m=0}^{\infty}$ be a sequence of points such that $t_0 = 0$, $t_{m+1} > t_m$, and $\lim_{m \to \infty} t_m = \infty$. For each nonnegative integer m let f_m be continuous on $[t_m, \infty)$, and let f be defined on $[0, \infty)$ by

$$f(t) = f_m(t), \qquad t_m \le t < t_{m+1} \qquad (m = 0, 1, \ldots).$$

Show that f is piecewise continuous on $[0, \infty)$ and that it has the step function representation

$$f(t) = f_0(t) + \sum_{m=1}^{\infty} u(t - t_m)(f_m(t) - f_{m-1}(t)), \qquad 0 \le t < \infty.$$

How do we know that the series on the right converges for all t in $[0, \infty)$?

31. In addition to the assumptions of Exercise 30, assume that

$$|f_m(t)| \le Me^{s_0 t}, \qquad t \ge t_m, \qquad m = 0, 1, \ldots, \tag{A}$$

and that the series

$$\sum_{m=0}^{\infty} e^{-\rho t_m} \tag{B}$$

converges for some $\rho > 0$. By means of the steps listed below, show that $\mathcal{L}(f)$ is defined for $s > s_0$ and

$$\mathcal{L}(f) = \mathcal{L}(f_0) + \sum_{m=1}^{\infty} e^{-s t_m} \mathcal{L}(g_m) \tag{C}$$

for $s > s_0 + \rho$, where

$$g_m(t) = f_m(t + t_m) - f_{m-1}(t + t_m).$$

(a) Use (A) and Theorem 8.1.6 to show that

$$\mathcal{L}(f) = \sum_{m=0}^{\infty} \int_{t_m}^{t_{m+1}} e^{-st} f_m(t) \, dt \qquad \text{(D)}$$

is defined for $s > s_0$.

(b) Show that (D) can be rewritten as

$$\mathcal{L}(f) = \sum_{m=0}^{\infty} \left(\int_{t_m}^{\infty} e^{-st} f_m(t) \, dt - \int_{t_{m+1}}^{\infty} e^{-st} f_m(t) \, dt \right). \qquad \text{(E)}$$

(c) Use (A), the assumed convergence of (B), and the comparison test to show that the series

$$\sum_{m=0}^{\infty} \int_{t_m}^{\infty} e^{-st} f_m(t) \, dt \qquad \text{and} \qquad \sum_{m=0}^{\infty} \int_{t_{m+1}}^{\infty} e^{-st} f_m(t) \, dt$$

both converge (absolutely) if $s > s_0 + \rho$.

(d) Show that (E) can be rewritten as

$$\mathcal{L}(f) = \mathcal{L}(f_0) + \sum_{m=1}^{\infty} \int_{t_m}^{\infty} e^{-st}(f_m(t) - f_{m-1}(t)) \, dt$$

if $s > s_0 + \rho$.

(e) Complete the proof of (C).

32. Suppose that $\{t_m\}_{m=0}^{\infty}$ and $\{f_m\}_{m=0}^{\infty}$ satisfy the assumptions of Exercises 30 and 31, and there is a positive constant K such that $t_m \geq Km$ for m sufficiently large. Show that the series (B) of Exercise 31 converges for any $\rho > 0$, and conclude from this that (C) of Exercise 31 holds for $s > s_0$.

In Exercises 33–36 find the step function representation of f and use the result of Exercise 32 to find $\mathcal{L}(f)$.

Hint: You will need formulas related to the formula for the sum of a geometric series.

33. $f(t) = m + 1, \quad m \leq t < m + 1 \quad (m = 0, 1, 2, \dots)$

34. $f(t) = (-1)^m, \quad m \leq t < m + 1 \quad (m = 0, 1, 2, \dots)$

35. $f(t) = (m + 1)^2, \quad m \leq t < m + 1 \quad (m = 0, 1, 2, \dots)$

36. $f(t) = (-1)^m m, \quad m \leq t < m + 1 \quad (m = 0, 1, 2, \dots)$

8.5 Constant Coefficient Equations with Piecewise Continuous Forcing Functions

We will now consider initial value problems of the form

$$ay'' + by' + cy = f(t), \qquad y(0) = k_0, \qquad y'(0) = k_1, \qquad \text{(1)}$$

where a, b, and c are constants ($a \neq 0$) and f is piecewise continuous on $[0, \infty)$. Problems of this kind occur in situations where the input to a physical system undergoes instantaneous changes, as when a switch is turned on or off, or the forces acting on the system change abruptly.

It can be shown (Exercises 23 and 24) that the differential equation in (1) does not have any solutions on an open interval containing a jump discontinuity of f. Therefore, we must define what we mean by a solution of (1) on $[0, \infty)$ in the case where f has jump discontinuities. The following theorem motivates our definition. We omit the proof.

THEOREM 8.5.1

Suppose that a, b, and c are constants ($a \neq 0$), and f is piecewise continuous on $[0, \infty)$, with jump discontinuities at $t_1, \ldots, t_n$, where

$$0 < t_1 < \cdots < t_n.$$

Let k_0 and k_1 be arbitrary real numbers. Then there is a unique function y defined on $[0, \infty)$ with the following properties:

(a) *$y(0) = k_0$ and $y'(0) = k_1$.*
(b) *y and y' are continuous on $[0, \infty)$.*
(c) *y'' is defined on every open subinterval of $[0, \infty)$ that does not contain any of the points $t_1, \ldots, t_n$, and*

$$ay'' + by' + cy = f(t)$$

on every such subinterval.
(d) *y'' has limits from the right and left at $t_1, \ldots, t_n$.*

We define the function y of Theorem 8.5.1 to be the solution of the initial value problem (1).

We begin by considering initial value problems of the form

$$ay'' + by' + cy = \begin{cases} f_0(t), & 0 \le t < t_1, \\ f_1(t), & t \ge t_1, \end{cases} \qquad y(0) = k_0, \quad y'(0) = k_1, \qquad (2)$$

in which the forcing function has a single jump discontinuity at t_1.

We can solve (2) by means of the following steps:

Step 1. Find the solution y_0 of the initial value problem

$$ay'' + by' + cy = f_0(t), \qquad y(0) = k_0, \qquad y'(0) = k_1.$$

Step 2. Compute $c_0 = y_0(t_1)$ and $c_1 = y_0'(t_1)$.
Step 3. Find the solution y_1 of the initial value problem

$$ay'' + by' + cy = f_1(t), \qquad y(t_1) = c_0, \qquad y'(t_1) = c_1.$$

Step 4. Obtain the solution y of (2) as

$$y = \begin{cases} y_0(t), & 0 \le t < t_1, \\ y_1(t), & t \ge t_1. \end{cases}$$

It is shown in Exercise 23 that y' exists and is continuous at t_1. The following example illustrates this procedure.

EXAMPLE 1   Solve the initial value problem

$$y'' + y = f(t), \qquad y(0) = 2, \qquad y'(0) = -1, \qquad (3)$$

where

$$f(t) = \begin{cases} 1, & 0 \le t < \dfrac{\pi}{2}, \\[2mm] -1, & t \ge \dfrac{\pi}{2}. \end{cases}$$

Solution   The initial value problem in Step 1 is

$$y'' + y = 1, \qquad y(0) = 2, \qquad y'(0) = -1.$$

We leave it to you to verify that its solution is

$$y_0 = 1 + \cos t - \sin t.$$

Doing Step 2 yields $y_0(\pi/2) = 0$ and $y_0'(\pi/2) = -1$, so the second initial value problem is

$$y'' + y = -1, \qquad y\left(\frac{\pi}{2}\right) = 0, \qquad y'\left(\frac{\pi}{2}\right) = -1.$$

We leave it to you to verify that the solution of this problem is

$$y_1 = -1 + \cos t + \sin t.$$

Hence, the solution of (3) is

$$y = \begin{cases} 1 + \cos t - \sin t, & 0 \le t < \dfrac{\pi}{2}, \\[2mm] -1 + \cos t + \sin t, & t \ge \dfrac{\pi}{2}. \end{cases} \tag{4}$$

The graph of y is shown in Figure 1.   ∎

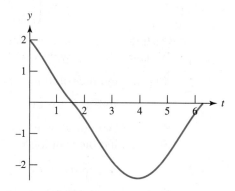

Figure 1   Graph of (4)

If f_0 and f_1 are defined on $[0, \infty)$ we can rewrite (2) as

$$ay'' + by' + cy = f_0(t) + u(t - t_1)(f_1(t) - f_0(t)), \qquad y(0) = k_0, \qquad y'(0) = k_1,$$

and apply the method of Laplace transforms. We will now solve the problem considered in Example 1 by this method.

EXAMPLE 2   Use the Laplace transform to solve the initial value problem

$$y'' + y = f(t), \qquad y(0) = 2, \qquad y'(0) = -1, \tag{5}$$

where

$$f(t) = \begin{cases} 1, & 0 \le t < \dfrac{\pi}{2}, \\[2mm] -1, & t \ge \dfrac{\pi}{2}. \end{cases}$$

Solution Here

$$f(t) = 1 - 2u\left(t - \frac{\pi}{2}\right),$$

so Theorem 8.4.1 (with $g(t) = 1$) implies that

$$\mathcal{L}(f) = \frac{1 - 2e^{-\pi s/2}}{s}.$$

Therefore, transforming (5) yields

$$(s^2 + 1)Y(s) = \frac{1 - 2e^{-\pi s/2}}{s} - 1 + 2s,$$

so

$$Y(s) = (1 - 2e^{-\pi s/2})G(s) + \frac{2s - 1}{s^2 + 1}, \tag{6}$$

with

$$G(s) = \frac{1}{s(s^2 + 1)}.$$

The form for the partial fraction expansion of G is

$$\frac{1}{s(s^2 + 1)} = \frac{A}{s} + \frac{Bs + C}{s^2 + 1}. \tag{7}$$

Multiplying through by $s(s^2 + 1)$ yields

$$A(s^2 + 1) + (Bs + C)s = 1,$$

or

$$(A + B)s^2 + Cs + A = 1.$$

Equating coefficients of like powers of s on the two sides of this equation shows that $A = 1$, $B = -A = -1$, and $C = 0$. Hence, from (7),

$$G(s) = \frac{1}{s} - \frac{s}{s^2 + 1}.$$

Therefore

$$g(t) = 1 - \cos t.$$

From this, (6), and Theorem 8.4.2,

$$y = 1 - \cos t - 2u\left(t - \frac{\pi}{2}\right)\left(1 - \cos\left(t - \frac{\pi}{2}\right)\right) + 2\cos t - \sin t.$$

Simplifying this (recalling that $\cos(t - \pi/2) = \sin t$) yields

$$y = 1 + \cos t - \sin t - 2u\left(t - \frac{\pi}{2}\right)(1 - \sin t),$$

or

$$y = \begin{cases} 1 + \cos t - \sin t, & 0 \le t < \dfrac{\pi}{2}, \\ -1 + \cos t + \sin t, & t \ge \dfrac{\pi}{2}, \end{cases}$$

which is the result obtained in Example 1.   ∎

REMARK.   It isn't obvious that using the Laplace transform to solve (2) as we did in Example 2 yields a function y with the properties stated in Theorem 8.5.1; that is, such that y and y' are continuous on $[0, \infty)$ and y'' has limits from the right and left at t_1. However, this is in fact the case if f_0 and f_1 are continuous and of exponential order on $[0, \infty)$. A proof of this assertion is sketched in Exercises 11–13 of Section 8.6.

EXAMPLE 3   Solve the initial value problem

$$y'' - y = f(t), \qquad y(0) = -1, \qquad y'(0) = 2, \tag{8}$$

where

$$f(t) = \begin{cases} t, & 0 \le t < 1, \\ 1, & t \ge 1. \end{cases}$$

Solution   Here

$$f(t) = t - u(t - 1)(t - 1),$$

so

$$\begin{aligned} \mathcal{L}(f) &= \mathcal{L}(t) - \mathcal{L}(u(t - 1)(t - 1)) \\ &= \mathcal{L}(t) - e^{-s}\mathcal{L}(t) \quad \text{(from Theorem 8.4.1)} \\ &= \frac{1}{s^2} - \frac{e^{-s}}{s^2}. \end{aligned}$$

Since transforming (8) yields

$$(s^2 - 1)Y(s) = \mathcal{L}(f) + 2 - s,$$

we see that

$$Y(s) = (1 - e^{-s})H(s) + \frac{2 - s}{s^2 - 1}, \tag{9}$$

where

$$H(s) = \frac{1}{s^2(s^2 - 1)} = \frac{1}{s^2 - 1} - \frac{1}{s^2}$$

and therefore

$$h(t) = \sinh t - t. \tag{10}$$

Since

$$\mathcal{L}^{-1}\left(\frac{2 - s}{s^2 - 1}\right) = 2 \sinh t - \cosh t,$$

we conclude from (9), (10), and Theorem 8.4.2 that

$$y = \sinh t - t - u(t - 1)(\sinh(t - 1) - t + 1) + 2 \sinh t - \cosh t,$$

or

$$y = 3 \sinh t - \cosh t - t - u(t - 1)(\sinh(t - 1) - t + 1) \qquad (11)$$

We leave it to you to verify that y and y' are continuous and y'' has limits from the right and left at $t_1 = 1$.

EXAMPLE 4 Solve the initial value problem

$$y'' + y = f(t), \qquad y(0) = 0, \qquad y'(0) = 0, \qquad (12)$$

where

$$f(t) = \begin{cases} 0, & 0 \le t < \dfrac{\pi}{4}, \\ \cos 2t, & \dfrac{\pi}{4} \le t < \pi, \\ 0, & t \ge \pi. \end{cases}$$

Solution Here

$$f(t) = u(t - \pi/4) \cos 2t - u(t - \pi) \cos 2t,$$

so

$$\begin{aligned} \mathcal{L}(f) &= \mathcal{L}(u(t - \pi/4) \cos 2t) - \mathcal{L}(u(t - \pi) \cos 2t) \\ &= e^{-\pi s/4} \mathcal{L}(\cos 2(t + \pi/4)) - e^{-\pi s} \mathcal{L}(\cos 2(t + \pi)) \\ &= -e^{-\pi s/4} \mathcal{L}(\sin 2t) - e^{-\pi s} \mathcal{L}(\cos 2t) \\ &= -\frac{2e^{-\pi s/4}}{s^2 + 4} - \frac{se^{-\pi s}}{s^2 + 4}. \end{aligned}$$

Since transforming (12) yields

$$(s^2 + 1)Y(s) = \mathcal{L}(f),$$

we see that

$$Y(s) = e^{-\pi s/4} H_1(s) + e^{-\pi s} H_2(s), \qquad (13)$$

where

$$H_1(s) = -\frac{2}{(s^2 + 1)(s^2 + 4)} \qquad \text{and} \qquad H_2(s) = -\frac{s}{(s^2 + 1)(s^2 + 4)}. \qquad (14)$$

To simplify the required partial fraction expansions we first write

$$\frac{1}{(x + 1)(x + 4)} = \frac{1}{3}\left[\frac{1}{x + 1} - \frac{1}{x + 4}\right].$$

Setting $x = s^2$ and substituting the result in (14) yields

$$H_1(s) = -\frac{2}{3}\left[\frac{1}{s^2 + 1} - \frac{1}{s^2 + 4}\right] \qquad \text{and} \qquad H_2(s) = -\frac{1}{3}\left[\frac{s}{s^2 + 1} - \frac{s}{s^2 + 4}\right].$$

The inverse transforms are

$$h_1(t) = -\frac{2}{3} \sin t + \frac{1}{3} \sin 2t \qquad \text{and} \qquad h_2(t) = -\frac{1}{3} \cos t + \frac{1}{3} \cos 2t.$$

From (13) and Theorem 8.4.2,

$$y = u\left(t - \frac{\pi}{4}\right)h_1\left(t - \frac{\pi}{4}\right) + u(t - \pi)h_2(t - \pi). \tag{15}$$

Since

$$h_1\left(t - \frac{\pi}{4}\right) = -\frac{2}{3}\sin\left(t - \frac{\pi}{4}\right) + \frac{1}{3}\sin 2\left(t - \frac{\pi}{4}\right)$$

$$= -\frac{\sqrt{2}}{3}(\sin t - \cos t) - \frac{1}{3}\cos 2t$$

and

$$h_2(t - \pi) = -\frac{1}{3}\cos(t - \pi) + \frac{1}{3}\cos 2(t - \pi)$$

$$= \frac{1}{3}\cos t + \frac{1}{3}\cos 2t,$$

(15) can be rewritten as

$$y = -\frac{1}{3}u\left(t - \frac{\pi}{4}\right)(\sqrt{2}\,(\sin t - \cos t) + \cos 2t)$$

$$+ \frac{1}{3}u(t - \pi)(\cos t + \cos 2t)$$

or

$$y = \begin{cases} 0, & 0 \le t < \dfrac{\pi}{4}, \\[2mm] -\dfrac{\sqrt{2}}{3}(\sin t - \cos t) - \dfrac{1}{3}\cos 2t, & \dfrac{\pi}{4} \le t < \pi, \\[2mm] -\dfrac{\sqrt{2}}{3}\sin t + \dfrac{1 + \sqrt{2}}{3}\cos t, & t \ge \pi. \end{cases} \tag{16}$$

We leave it to you to verify that y and y' are continuous and y'' has limits from the right and left at $t_1 = \pi/4$ and $t_2 = \pi$. The graph of y is shown in Figure 2. ∎

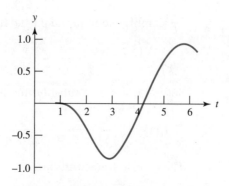

Figure 2   Graph of (16)

8.5 EXERCISES

In Exercises 1–20 use the Laplace transform to solve the initial value problem. Where indicated by **C** *, graph the solution.*

1. $y'' + y = \begin{cases} 3, & 0 \le t < \pi, \\ 0, & t \ge \pi, \end{cases}$ $y(0) = 0, \quad y'(0) = 0$

2. $y'' + y = \begin{cases} 3, & 0 \le t < 4, \\ 2t - 5, & t > 4, \end{cases}$ $y(0) = 1, \quad y'(0) = 0$

3. $y'' - 2y' = \begin{cases} 4, & 0 \le t < 1, \\ 6, & t \ge 1, \end{cases}$ $y(0) = -6, \quad y'(0) = 1$

4. $y'' - y = \begin{cases} e^{2t}, & 0 \le t < 2, \\ 1 & t \ge 2, \end{cases}$ $y(0) = 3, \quad y'(0) = -1$

5. $y'' - 3y' + 2y = \begin{cases} 0, & 0 \le t < 1, \\ 1, & 1 \le t < 2, \\ -1, & t \ge 2, \end{cases}$ $y(0) = -3, \quad y'(0) = 1$

C **6.** $y'' + 4y = \begin{cases} |\sin t|, & 0 \le t < 2\pi, \\ 0, & t \ge 2\pi, \end{cases}$ $y(0) = -3, \quad y'(0) = 1$

7. $y'' - 5y' + 4y = \begin{cases} 1, & 0 \le t < 1, \\ -1, & 1 \le t < 2, \\ 0, & t \ge 2, \end{cases}$ $y(0) = 3, \quad y'(0) = -5$

8. $y'' + 9y = \begin{cases} \cos t, & 0 \le t < \dfrac{3\pi}{2}, \\[2mm] \sin t, & t \ge \dfrac{3\pi}{2}, \end{cases}$ $y(0) = 0, \quad y'(0) = 0$

C **9.** $y'' + 4y = \begin{cases} t, & 0 \le t < \dfrac{\pi}{2}, \\[2mm] \pi, & t \ge \dfrac{\pi}{2}, \end{cases}$ $y(0) = 0, \quad y'(0) = 0$

10. $y'' + y = \begin{cases} t, & 0 \le t < \pi, \\ -t, & t \ge \pi, \end{cases}$ $y(0) = 0, \quad y'(0) = 0$

11. $y'' - 3y' + 2y = \begin{cases} 0, & 0 \le t < 2, \\ 2t - 4, & t \ge 2, \end{cases}$ $y(0) = 0, \quad y'(0) = 0$

12. $y'' + y = \begin{cases} t, & 0 \le t < 2\pi, \\ -2t, & t \ge 2\pi, \end{cases}$ $y(0) = 1, \quad y'(0) = 2$

C **13.** $y'' + 3y' + 2y = \begin{cases} 1, & 0 \le t < 2, \\ -1, & t \ge 2, \end{cases}$ $y(0) = 0, \quad y'(0) = 0$

14. $y'' - 4y' + 3y = \begin{cases} -1, & 0 \le t < 1, \\ 1, & t \ge 1, \end{cases}$ $y(0) = 0, \quad y'(0) = 0$

15. $y'' + 2y' + y = \begin{cases} e^t, & 0 \le t < 1, \\ e^t - 1, & t \ge 1, \end{cases}$ $y(0) = 3, \quad y'(0) = -1$

16. $y'' + 2y' + y = \begin{cases} 4e^t, & 0 \le t < 1, \\ 0, & t \ge 1, \end{cases}$ $y(0) = 0, \quad y'(0) = 0$

17. $y'' + 3y' + 2y = \begin{cases} e^{-t}, & 0 \le t < 1, \\ 0, & t \ge 1, \end{cases}$ $\qquad y(0) = 1, \quad y'(0) = -1$

18. $y'' - 4y' + 4y = \begin{cases} e^{2t}, & 0 \le t < 2, \\ -e^{2t}, & t \ge 2, \end{cases}$ $\qquad y(0) = 0, \quad y'(0) = -1$

C **19.** $y'' = \begin{cases} t^2, & 0 \le t < 1, \\ -t, & 1 \le t < 2, \\ t + 1, & t \ge 2, \end{cases}$ $\quad y(0) = 1, \quad y'(0) = 0$

20. $y'' + 2y' + 2y = \begin{cases} 1, & 0 \le t < 2\pi, \\ t, & 2\pi \le t < 3\pi, \\ -1, & t \ge 3\pi, \end{cases}$ $\quad y(0) = 2, \quad y'(0) = -1$

21. Solve the initial value problem

$$y'' = f(t), \qquad y(0) = 0, \qquad y'(0) = 0,$$

where

$$f(t) = m + 1, \qquad m \le t < m + 1 \qquad (m = 0, 1, 2, \dots).$$

22. Solve the given initial value problem and find a formula that does not involve step functions and represents y on each interval of continuity of f.

(a) $y'' + y = f(t), \quad y(0) = 0, \quad y'(0) = 0;$
$f(t) = m + 1, \quad m\pi \le t < (m + 1)\pi \quad (m = 0, 1, 2, \dots)$

(b) $y'' + y = f(t), \quad y(0) = 0, \quad y'(0) = 0;$
$f(t) = (m + 1)t, \quad 2m\pi \le t < 2(m + 1)\pi \quad (m = 0, 1, 2, \dots)$

Hint: *You will need the formula*

$$1 + 2 + \cdots + m = \frac{m(m + 1)}{2}.$$

(c) $y'' + y = f(t), \quad y(0) = 0, \quad y'(0) = 0;$
$f(t) = (-1)^m, \quad m\pi \le t < (m + 1)\pi \quad (m = 0, 1, 2, \dots)$

(d) $y'' - y = f(t), \quad y(0) = 0, \quad y'(0) = 0;$
$f(t) = m + 1, \quad m \le t < m + 1 \quad (m = 0, 1, 2, \dots)$

Hint: *You will need the formula*

$$1 + r + \cdots + r^m = \frac{1 - r^{m+1}}{1 - r} \qquad (r \ne 1).$$

(e) $y'' + 2y' + 2y = f(t), \quad y(0) = 0, \quad y'(0) = 0;$
$f(t) = (m + 1)(\sin t + 2 \cos t), \quad 2m\pi \le t < 2(m + 1)\pi \quad (m = 0, 1, 2, \dots)$ (See the hint in **(d)**.)

(f) $y'' - 3y' + 2y = f(t), \quad y(0) = 0, \quad y'(0) = 0;$
$f(t) = m + 1, \quad m \le t < m + 1 \quad (m = 0, 1, 2, \dots)$ (See the hints in **(b)** and **(d)**.)

23. (a) Let g be continuous on the interval (α, β) and differentiable on the subintervals (α, t_0) and (t_0, β), and suppose that $A = \lim_{t \to t_0-} g'(t)$ and $B = \lim_{t \to t_0+} g'(t)$ both exist. Use the mean value theorem to show that

$$\lim_{t \to t_0-} \frac{g(t) - g(t_0)}{t - t_0} = A \qquad \text{and} \qquad \lim_{t \to t_0+} \frac{g(t) - g(t_0)}{t - t_0} = B.$$

(b) Conclude from (a) that $g'(t_0)$ exists and g' is continuous at t_0 if $A = B$.

(c) Conclude from (a) that if g is differentiable on (α, β) then g' cannot have a jump discontinuity on (α, β).

24. (a) Let $a, b,$ and c be constants, with $a \ne 0$. Let f be piecewise continuous on an interval (α, β), with a single jump discontinuity at a point t_0 in (α, β). Suppose that y and y' are continuous on (α, β) and y'' is continuous on (α, t_0) and (t_0, β). Suppose also that

$$ay'' + by' + cy = f(t) \tag{A}$$

on (α, t_0) and (t_0, β). Show that

$$y''(t_0+) - y''(t_0-) = \frac{f(t_0+) - f(t_0-)}{a} \neq 0.$$

(b) Use (a) and Exercise 23(c) to show that (A) does not have solutions on any interval (α, β) containing a jump discontinuity of f.

25. Suppose that P_0, P_1, and P_2 are continuous and P_0 has no zeros on an open interval (a, b), and that F has a jump discontinuity at a point t_0 in (a, b). Show that the differential equation

$$P_0(t)y'' + P_1(t)y' + P_2(t)y = F(t)$$

has no solutions on (a, b).

Hint: Generalize the result of Exercise 24 and use Exercise 23(c).

26. Let $0 = t_0 < t_1 < \cdots < t_n$. Suppose that f_m is continuous on $[t_m, \infty)$ for $m = 1, \ldots, n$. Let

$$f(t) = \begin{cases} f_m(t), & t_m \leq t < t_{m+1}, \quad m = 1, \ldots, n-1. \\ f_n(t), & t \geq t_n. \end{cases}$$

Show that the solution of

$$ay'' + by' + cy = f(t), \qquad y(0) = k_0, \qquad y'(0) = k_1,$$

as defined following Theorem 8.5.1, is given by

$$y = \begin{cases} z_0(t), & 0 \leq t < t_1, \\ z_0(t) + z_1(t), & t_1 \leq t < t_2, \\ \quad \vdots \\ z_0(t) + \cdots + z_{n-1}(t), & t_{n-1} \leq t < t_n, \\ z_0(t) + \cdots + z_n(t), & t \geq t_n, \end{cases}$$

where z_0 is the solution of

$$az'' + bz' + cz = f_0(t), \qquad z(0) = k_0, \qquad z'(0) = k_1$$

and z_m is the solution of

$$az'' + bz' + cz = f_m(t) - f_{m-1}(t), \qquad z(t_m) = 0, \qquad z'(t_m) = 0$$

for $m = 1, \ldots, n$.

8.6   Convolution

In this section we consider the problem of finding the inverse Laplace transform of a product $H(s) = F(s)G(s)$, where F and G are the Laplace transforms of known functions f and g. To motivate our interest in this problem, consider the initial value problem

$$ay'' + by' + cy = f(t), \qquad y(0) = 0, \qquad y'(0) = 0.$$

Taking Laplace transforms yields

$$(as^2 + bs + c)Y(s) = F(s),$$

so

$$Y(s) = F(s)G(s), \tag{1}$$

where

$$G(s) = \frac{1}{as^2 + bs + c}.$$

Until now we haven't been interested in the factorization indicated in (1), since we have dealt only with differential equations with specific forcing functions. Hence, we could simply do the indicated multiplication in (1) and use the table of Laplace transforms to find $y = \mathcal{L}^{-1}(Y)$. However, this is not possible if we want a *formula* for y in terms of f, which may be unspecified.

To motivate the formula for $\mathcal{L}^{-1}(FG)$, consider the initial value problem

$$y' - ay = f(t), \qquad y(0) = 0, \tag{2}$$

which we first solve without using the Laplace transform. The solution of the differential equation in (2) is of the form $y = ue^{at}$ where

$$u' = e^{-at}f(t).$$

Integrating this from 0 to t and imposing the initial condition $u(0) = y(0) = 0$ yields

$$u = \int_0^t e^{-a\tau}f(\tau)\,d\tau,$$

and therefore

$$y(t) = e^{at}\int_0^t e^{-a\tau}f(\tau)\,d\tau = \int_0^t e^{a(t-\tau)}f(\tau)\,d\tau. \tag{3}$$

Now we will solve (2) by means of the Laplace transform and compare the result to (3). Taking Laplace transforms in (2) yields

$$(s - a)Y(s) = F(s),$$

so

$$Y(s) = F(s)\frac{1}{s - a},$$

which implies that

$$y(t) = \mathcal{L}^{-1}\left(F(s)\frac{1}{s - a}\right). \tag{4}$$

If we now let $g(t) = e^{at}$, so that

$$G(s) = \frac{1}{s - a},$$

then (3) and (4) can be written as

$$y(t) = \int_0^t f(\tau)g(t - \tau)\,d\tau$$

and

$$y = \mathcal{L}^{-1}(FG),$$

respectively. Therefore

$$\mathcal{L}^{-1}(FG) = \int_0^t f(\tau)g(t - \tau)\, d\tau \tag{5}$$

in this case.

This motivates the following definition.

DEFINITION 8.6.1

The **convolution** *f* ∗ *g* of two functions *f* and *g* is defined by

$$(f * g)(t) = \int_0^t f(\tau)g(t - \tau)\, d\tau.$$

It can be shown (Exercise 6) that *f* ∗ *g* = *g* ∗ *f*; that is,

$$\int_0^t f(t - \tau)g(\tau)\, d\tau = \int_0^t f(\tau)g(t - \tau)\, d\tau.$$

Equation (5) shows that $\mathcal{L}^{-1}(FG) = f * g$ in the special case where $g(t) = e^{at}$. This following theorem states that this is true in general.

THEOREM 8.6.2

(The Convolution
Theorem)

If $\mathcal{L}(f) = F$ and $\mathcal{L}(g) = G$ then

$$\mathcal{L}(f * g) = FG.$$

A complete proof of the convolution theorem is beyond the scope of this book. However, we will assume that *f* ∗ *g* has a Laplace transform and verify the conclusion of the theorem in a purely computational way. By the definition of the Laplace transform,

$$\mathcal{L}(f * g) = \int_0^\infty e^{-st}(f * g)(t)\, dt = \int_0^\infty e^{-st} \int_0^t f(\tau)g(t - \tau)\, d\tau\, dt.$$

This iterated integral equals a double integral over the region shown in Figure 1. Reversing the order of integration yields

$$\mathcal{L}(f * g) = \int_0^\infty f(\tau) \int_\tau^\infty e^{-st}g(t - \tau)\, dt\, d\tau. \tag{6}$$

However, the substitution $x = t - \tau$ shows that

$$\int_\tau^\infty e^{-st}g(t - \tau)\, dt = \int_0^\infty e^{-s(x+\tau)}g(x)\, dx$$

$$= e^{-s\tau} \int_0^\infty e^{-sx}g(x)\, dx = e^{-s\tau}G(s).$$

Substituting this into (6) and noting that $G(s)$ is independent of τ yields

$$\mathcal{L}(f * g) = \int_0^\infty e^{-s\tau}f(\tau)G(s)\, d\tau$$

$$= G(s) \int_0^\infty e^{-st}f(\tau)\, d\tau = F(s)G(s).$$

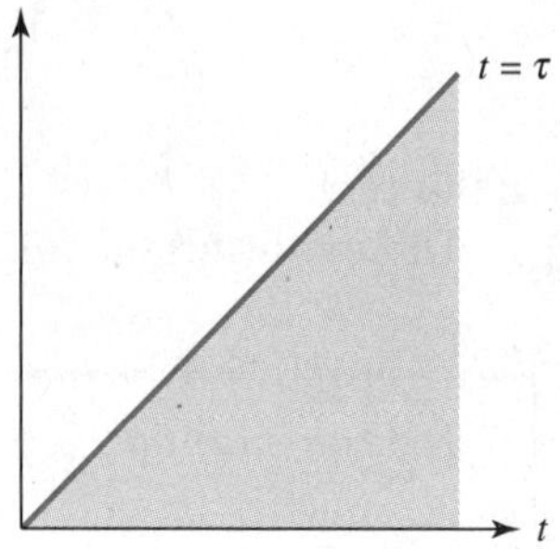

Figure 1

EXAMPLE 1   Let

$$f(t) = e^{at} \quad \text{and} \quad g(t) = e^{bt} \quad (a \neq b).$$

Verify that $\mathcal{L}(f * g) = \mathcal{L}(f)\mathcal{L}(g)$, as implied by the convolution theorem.

Solution   We first compute

$$(f * g)(t) = \int_0^t e^{a\tau} e^{b(t-\tau)} \, d\tau = e^{bt} \int_0^t e^{(a-b)\tau} \, d\tau$$

$$= e^{bt} \frac{e^{(a-b)\tau}}{a-b} \Big|_0^t = \frac{e^{bt}[e^{(a-b)t} - 1]}{a-b}$$

$$= \frac{e^{at} - e^{bt}}{a-b}.$$

Since

$$e^{at} \leftrightarrow \frac{1}{s-a} \quad \text{and} \quad e^{bt} \leftrightarrow \frac{1}{s-b},$$

it follows that

$$\mathcal{L}(f * g) = \frac{1}{a-b}\left[\frac{1}{s-a} - \frac{1}{s-b}\right]$$

$$= \frac{1}{(s-a)(s-b)}$$

$$= \mathcal{L}(e^{at})\mathcal{L}(e^{bt}) = \mathcal{L}(f)\mathcal{L}(g). \qquad \blacksquare$$

A FORMULA FOR THE SOLUTION OF AN INITIAL VALUE PROBLEM

The convolution theorem can be used to provide a formula for the solution of an initial value problem for a linear constant coefficient second order equation in which the forcing function is unspecified. The next three examples illustrate this.

EXAMPLE 2   Find a formula for the solution of the initial value problem

$$y'' - 2y' + y = f(t), \qquad y(0) = k_0, \quad y'(0) = k_1. \tag{7}$$

Solution   Taking Laplace transforms in (7) yields

$$(s^2 - 2s + 1)Y(s) = F(s) + (k_1 + k_0 s) - 2k_0.$$

Therefore

$$Y(s) = \frac{1}{(s-1)^2} F(s) + \frac{k_1 + k_0 s - 2k_0}{(s-1)^2}$$

$$= \frac{1}{(s-1)^2} F(s) + \frac{k_0}{s-1} + \frac{k_1 - k_0}{(s-1)^2}.$$

From the table of Laplace transforms,

$$\mathcal{L}^{-1}\left(\frac{k_0}{s-1} + \frac{k_1 - k_0}{(s-1)^2}\right) = e^t(k_0 + (k_1 - k_0)t).$$

Since

$$\frac{1}{(s-1)^2} \leftrightarrow te^t \qquad \text{and} \qquad F(s) \leftrightarrow f(t),$$

the convolution theorem implies that

$$\mathcal{L}^{-1}\left(\frac{1}{(s-1)^2} F(s)\right) = \int_0^t \tau e^\tau f(t - \tau) \, d\tau.$$

Therefore the solution of (7) is

$$y(t) = e^t(k_0 + (k_1 - k_0)t) + \int_0^t \tau e^\tau f(t - \tau) \, d\tau.$$

EXAMPLE 3 Find a formula for the solution of the initial value problem

$$y'' + 4y = f(t), \qquad y(0) = k_0, \qquad y'(0) = k_1. \tag{8}$$

Solution Taking Laplace transforms in (8) yields

$$(s^2 + 4)Y(s) = F(s) + k_1 + k_0 s.$$

Therefore

$$Y(s) = \frac{1}{(s^2 + 4)} F(s) + \frac{k_1 + k_0 s}{s^2 + 4}.$$

From the table of Laplace transforms,

$$\mathcal{L}^{-1}\left(\frac{k_1 + k_0 s}{s^2 + 4}\right) = k_0 \cos 2t + \frac{k_1}{2} \sin 2t.$$

Since

$$\frac{1}{(s^2 + 4)} \leftrightarrow \frac{1}{2} \sin 2t \qquad \text{and} \qquad F(s) \leftrightarrow f(t),$$

the convolution theorem implies that

$$\mathcal{L}^{-1}\left(\frac{1}{(s^2 + 4)} F(s)\right) = \frac{1}{2} \int_0^t f(t - \tau) \sin 2\tau \, d\tau.$$

Therefore the solution of (8) is

$$y(t) = k_0 \cos 2t + \frac{k_1}{2} \sin 2t + \frac{1}{2} \int_0^t f(t - \tau) \sin 2\tau \, d\tau.$$

EXAMPLE 4 Find a formula for the solution of the initial value problem

$$y'' + 2y' + 2y = f(t), \qquad y(0) = k_0, \qquad y'(0) = k_1. \tag{9}$$

Solution Taking Laplace transforms in (9) yields

$$(s^2 + 2s + 2)Y(s) = F(s) + k_1 + k_0 s + 2k_0.$$

Therefore

$$Y(s) = \frac{1}{(s+1)^2+1}F(s) + \frac{k_1 + k_0 s + 2k_0}{(s+1)^2+1}$$

$$= \frac{1}{(s+1)^2+1}F(s) + \frac{(k_1+k_0) + k_0(s+1)}{(s+1)^2+1}.$$

From the table of Laplace transforms,

$$\mathcal{L}^{-1}\left(\frac{(k_1+k_0)+k_0(s+1)}{(s+1)^2+1}\right) = e^{-t}((k_1+k_0)\sin t + k_0\cos t).$$

Since

$$\frac{1}{(s+1)^2+1} \leftrightarrow e^{-t}\sin t \qquad \text{and} \qquad F(s) \leftrightarrow f(t),$$

the convolution theorem implies that

$$\mathcal{L}^{-1}\left(\frac{1}{(s+1)^2+1}F(s)\right) = \int_0^t f(t-\tau)e^{-\tau}\sin\tau \, d\tau.$$

Therefore the solution of (9) is

$$y(t) = e^{-t}((k_1+k_0)\sin t + k_0\cos t) + \int_0^t f(t-\tau)e^{-\tau}\sin\tau \, d\tau. \tag{10}$$

■

EVALUATING CONVOLUTION INTEGRALS

We will say that an integral of the form $\int_0^t u(\tau)v(t-\tau)\,d\tau$ is a **convolution integral.** The convolution theorem provides a convenient way to evaluate convolution integrals.

EXAMPLE 5   Evaluate the convolution integral

$$h(t) = \int_0^t (t-\tau)^5 \tau^7 \, d\tau.$$

Solution   We could evaluate this integral by expanding $(t-\tau)^5$ in powers of τ and then integrating. However, the convolution theorem provides an easier way. The integral is the convolution of $f(t) = t^5$ and $g(t) = t^7$. Since

$$t^5 \leftrightarrow \frac{5!}{s^6} \qquad \text{and} \qquad t^7 \leftrightarrow \frac{7!}{s^8},$$

the convolution theorem implies that

$$h(t) \leftrightarrow \frac{5!\,7!}{s^{14}} = \frac{5!\,7!}{13!}\frac{13!}{s^{14}},$$

where we have written the second equality because

$$\frac{13!}{s^{14}} \leftrightarrow t^{13}.$$

Hence,

$$h(t) = \frac{5!\,7!}{13!}t^{13}.$$

EXAMPLE 6 Use the convolution theorem and a partial fraction expansion to evaluate the convolution integral

$$h(t) = \int_0^t \sin a(t - \tau) \cos b\tau \, d\tau \qquad (|a| \neq |b|).$$

Solution Since

$$\sin at \leftrightarrow \frac{a}{s^2 + a^2} \qquad \text{and} \qquad \cos bt \leftrightarrow \frac{s}{s^2 + b^2},$$

the convolution theorem implies that

$$H(s) = \frac{a}{s^2 + a^2} \frac{s}{s^2 + b^2}.$$

Expanding this in a partial fraction expansion yields

$$H(s) = \frac{a}{b^2 - a^2}\left[\frac{s}{s^2 + a^2} - \frac{s}{s^2 + b^2}\right],$$

and therefore

$$h(t) = \frac{a}{b^2 - a^2}(\cos at - \cos bt). \qquad \blacksquare$$

VOLTERRA INTEGRAL EQUATIONS

An equation of the form

$$y(t) = f(t) + \int_0^t k(t - \tau)y(\tau) \, d\tau \tag{11}$$

is called a ***Volterra***[1] ***integral equation.*** Here f and k are given functions and y is unknown. Since the integral on the right is a convolution integral, the convolution theorem provides a convenient formula for solving (11). Taking Laplace transforms in (11) yields

$$Y(s) = F(s) + K(s)Y(s),$$

and solving this for $Y(s)$ yields

$$Y(s) = \frac{F(s)}{1 - K(s)}.$$

We then obtain the solution of (11) as $y = \mathcal{L}^{-1}(Y)$.

EXAMPLE 7 Solve the integral equation

$$y(t) = 1 + 2 \int_0^t e^{-2(t - \tau)}y(\tau) \, d\tau. \tag{12}$$

Solution Taking Laplace transforms in (12) yields

$$Y(s) = \frac{1}{s} + \frac{2}{s + 2}Y(s),$$

---

[1]The Italian mathematician Vito Volterra (1860–1940) developed the theory of the integral equations bearing his name in connection with his research on mathematical models for population growth.

and solving this for $Y(s)$ yields

$$Y(s) = \frac{1}{s} + \frac{2}{s^2}.$$

Hence,

$$y(t) = 1 + 2t. \qquad \blacksquare$$

TRANSFER FUNCTIONS

The following theorem presents a formula for the solution of the general initial value problem

$$ay'' + by' + cy = f(t), \qquad y(0) = k_0, \qquad y'(0) = k_1,$$

where we assume for simplicity that f is continuous on $[0,\infty)$ and that $\mathcal{L}(f)$ exists. In Exercises 11–14 it is shown that the formula is valid under much weaker conditions on f.

THEOREM 8.6.3

Suppose that f is continuous on $[0,\infty)$ and has a Laplace transform. Then the solution of the initial value problem

$$ay'' + by' + cy = f(t), \qquad y(0) = k_0, \qquad y'(0) = k_1 \qquad (13)$$

is

$$y(t) = k_0 y_1(t) + k_1 y_2(t) + \int_0^t w(\tau) f(t - \tau) \, d\tau, \qquad (14)$$

where y_1 and y_2 satisfy

$$ay_1'' + by_1' + cy_1 = 0, \qquad y_1(0) = 1, \qquad y_1'(0) = 0 \qquad (15)$$

and

$$ay_2'' + by_2' + cy_2 = 0, \qquad y_2(0) = 0, \qquad y_2'(0) = 1, \qquad (16)$$

and

$$w(t) = \frac{1}{a} y_2(t). \qquad (17)$$

PROOF. Taking Laplace transforms in (13) yields

$$p(s)Y(s) = F(s) + a(k_1 + k_0 s) + bk_0,$$

where

$$p(s) = as^2 + bs + c.$$

Hence

$$Y(s) = W(s)F(s) + V(s), \qquad (18)$$

with

$$W(s) = \frac{1}{p(s)} \qquad (19)$$

and

$$V(s) = \frac{a(k_1 + k_0 s) + b k_0}{p(s)}. \tag{20}$$

Taking Laplace transforms in (15) and (16) shows that

$$p(s)Y_1(s) = as + b \quad \text{and} \quad p(s)Y_2(s) = a.$$

Therefore

$$Y_1(s) = \frac{as + b}{p(s)}$$

and

$$Y_2(s) = \frac{a}{p(s)}. \tag{21}$$

Hence, (20) can be rewritten as

$$V(s) = k_0 Y_1(s) + k_1 Y_2(s).$$

Substituting this into (18) yields

$$Y(s) = k_0 Y_1(s) + k_1 Y_2(s) + \frac{1}{a} Y_2(s) F(s).$$

Taking inverse transforms and invoking the convolution theorem yields (14). Finally, (19) and (21) imply (17). ☐

It is useful to note from (14) that y is of the form

$$y = v + h,$$

where

$$v(t) = k_0 y_1(t) + k_1 y_2(t)$$

depends on the initial conditions and is independent of the forcing function, while

$$h(t) = \int_0^t w(\tau) f(t - \tau) \, d\tau$$

depends on the forcing function and is independent of the initial conditions. If the zeros of the characteristic polynomial

$$p(s) = as^2 + bs + c$$

of the complementary equation have negative real parts, then y_1 and y_2 both approach zero as $t \to \infty$, so $\lim_{t \to \infty} v(t) = 0$ for any choice of initial conditions. Moreover, the value of $h(t)$ is essentially independent of the values of $f(t - \tau)$ for large τ, since $\lim_{\tau \to \infty} w(\tau) = 0$. In this case we say that v and h are **transient** and **steady state components,** respectively, of the solution y of (13). These definitions apply to the initial value problem of Example 4, where the zeros of

$$p(s) = s^2 + 2s + 2 = (s + 1)^2 + 1$$

are $-1 \pm i$. From (10) we see that the solution of the general initial value problem of Example 4 is $y = v + h$, where

$$v(t) = e^{-t}((k_1 + k_0) \sin t + k_0 \cos t)$$

is the transient component of the solution and

$$h(t) = \int_0^t f(t - \tau)e^{-\tau} \sin \tau \, d\tau$$

is the steady state component. The definitions do not apply to the initial value problems considered in Examples 2 and 3, since the zeros of the characteristic polynomials in these two examples do not have negative real parts.

In physical applications where the input f and the output y of a device are related by (13), the zeros of the characteristic polynomial usually do have negative real parts. Then $W = \mathcal{L}(w)$ is called the **transfer function** of the device. Since

$$H(s) = W(s)F(s),$$

we see that

$$W(s) = \frac{H(s)}{F(s)}$$

is the ratio of the transform of the steady state output to the transform of the input.

Because of the form of

$$h(t) = \int_0^t w(\tau)f(t - \tau) \, d\tau,$$

w is sometimes called the **weighting function** of the device, since it assigns weights to past values of the input f. It is also called the **impulse response** of the device, for reasons discussed in the next section.

Formula (14) is given in more detail in Exercises 8–10 for the three possible cases where the zeros of $p(s)$ are real and distinct, real and repeated, or complex conjugates.

8.6 EXERCISES

1. Express the inverse transform as an integral.

(a) $\dfrac{1}{s^2(s^2 + 4)}$

(b) $\dfrac{s}{(s + 2)(s^2 + 9)}$

(c) $\dfrac{s}{(s^2 + 4)(s^2 + 9)}$

(d) $\dfrac{s}{(s^2 + 1)^2}$

(e) $\dfrac{1}{s(s - a)}$

(f) $\dfrac{1}{(s + 1)(s^2 + 2s + 2)}$

(g) $\dfrac{1}{(s + 1)^2(s^2 + 4s + 5)}$

(h) $\dfrac{1}{(s - 1)^3(s + 2)^2}$

(i) $\dfrac{s - 1}{s^2(s^2 - 2s + 2)}$

(j) $\dfrac{s(s + 3)}{(s^2 + 4)(s^2 + 6s + 10)}$

(k) $\dfrac{1}{(s - 3)^5 s^6}$

(l) $\dfrac{1}{(s - 1)^3(s^2 + 4)}$

(m) $\dfrac{1}{s^2(s - 2)^3}$

(n) $\dfrac{1}{s^7(s - 2)^6}$

2. Find the Laplace transform.

(a) $\int_0^t \sin a\tau \cos b(t - \tau) \, d\tau$

(b) $\int_0^t e^\tau \sin a(t - \tau) \, d\tau$

(c) $\int_0^t \sinh a\tau \cosh a(t - \tau) \, d\tau$

(d) $\int_0^t \tau(t - \tau) \sin \omega\tau \cos \omega(t - \tau) \, d\tau$

(e) $e^t \int_0^t \sin \omega\tau \cos \omega(t - \tau) \, d\tau$

(f) $e^t \int_0^t \tau^2(t - \tau)e^\tau \, d\tau$

(g) $e^{-t} \int_0^t e^{-\tau}\tau \cos \omega(t - \tau) \, d\tau$

(h) $e^t \int_0^t e^{2\tau} \sinh(t - \tau) \, d\tau$

(i) $\int_0^t \tau e^{2\tau} \sin 2(t - \tau) \, d\tau$

(j) $\int_0^t (t - \tau)^3 e^\tau \, d\tau$

(k) $\int_0^t \tau^6 e^{-(t-\tau)} \sin 3(t - \tau) \, d\tau$

(l) $\int_0^t \tau^2(t - \tau)^3 \, d\tau$

(m) $\int_0^t (t - \tau)^7 e^{-\tau} \sin 2\tau \, d\tau$

(n) $\int_0^t (t - \tau)^4 \sin 2\tau \, d\tau$

3. Find a formula for the solution of the initial value problem.

(a) $y'' + 3y' + y = f(t)$, $y(0) = 0$, $y'(0) = 0$

(b) $y'' + 4y = f(t)$, $y(0) = 0$, $y'(0) = 0$

(c) $y'' + 2y' + y = f(t)$, $y(0) = 0$, $y'(0) = 0$

(d) $y'' + k^2 y = f(t)$, $y(0) = 1$, $y'(0) = -1$

(e) $y'' + 6y' + 9y = f(t)$, $y(0) = 0$, $y'(0) = -2$

(f) $y'' - 4y = f(t)$, $y(0) = 0$, $y'(0) = 3$

(g) $y'' - 5y' + 6y = f(t)$, $y(0) = 1$, $y'(0) = 3$

(h) $y'' + \omega^2 y = f(t)$, $y(0) = k_0$, $y'(0) = k_1$

4. Solve the integral equation.

(a) $y(t) = t - \int_0^t (t - \tau)y(\tau) \, d\tau$

(b) $y(t) = \sin t - 2 \int_0^t \cos(t - \tau)y(\tau) \, d\tau$

(c) $y(t) = 1 + 2 \int_0^t y(\tau) \cos(t - \tau) \, d\tau$

(d) $y(t) = t + \int_0^t y(\tau)e^{-(t-\tau)} \, d\tau$

(e) $y'(t) = t + \int_0^t y(\tau) \cos(t - \tau) \, d\tau$, $y(0) = 4$

(f) $y(t) = \cos t - \sin t + \int_0^t y(\tau) \sin(t - \tau) \, d\tau$

5. Use the convolution theorem to evaluate the integral.

(a) $\int_0^t (t - \tau)^7 \tau^8 \, d\tau$

(b) $\int_0^t (t - \tau)^{13} \tau^7 \, d\tau$

(c) $\int_0^t (t - \tau)^6 \tau^7 \, d\tau$

(d) $\int_0^t e^{-\tau} \sin(t - \tau) \, d\tau$

(e) $\int_0^t \sin \tau \cos 2(t - \tau) \, d\tau$

6. Show that

$$\int_0^t f(t - \tau)g(\tau) \, d\tau = \int_0^t f(\tau)g(t - \tau) \, d\tau$$

by introducing the new variable of integration $x = t - \tau$ in the first integral.

7. Use the convolution theorem to show that if $f(t) \leftrightarrow F(s)$ then

$$\int_0^t f(\tau) \, d\tau \leftrightarrow \frac{F(s)}{s}.$$

8. Show that if $p(s) = as^2 + bs + c$ has distinct real zeros r_1 and r_2, then the solution of

$$ay'' + by' + cy = f(t), \qquad y(0) = k_0, \qquad y'(0) = k_1$$

is

$$y(t) = k_0 \frac{r_2 e^{r_1 t} - r_1 e^{r_2 t}}{r_2 - r_1} + k_1 \frac{e^{r_2 t} - e^{r_1 t}}{r_2 - r_1} + \frac{1}{a(r_2 - r_1)} \int_0^t (e^{r_2 \tau} - e^{r_1 \tau}) f(t - \tau) \, d\tau.$$

9. Show that if $p(s) = as^2 + bs + c$ has a repeated real zero r_1, then the solution of

$$ay'' + by' + cy = f(t), \qquad y(0) = k_0, \qquad y'(0) = k_1$$

is

$$y(t) = k_0(1 - r_1 t) e^{r_1 t} + k_1 t e^{r_1 t} + \frac{1}{a} \int_0^t \tau e^{r_1 \tau} f(t - \tau) \, d\tau.$$

10. Show that if $p(s) = as^2 + bs + c$ has complex conjugate zeros $\lambda \pm i\omega$, then the solution of

$$ay'' + by' + cy = f(t), \qquad y(0) = k_0, \qquad y'(0) = k_1$$

is

$$y(t) = e^{\lambda t} \left[k_0 \left(\cos \omega t - \frac{\lambda}{\omega} \sin \omega t \right) + \frac{k_1}{\omega} \sin \omega t \right] + \frac{1}{a\omega} \int_0^t e^{\lambda t} f(t - \tau) \sin \omega \tau \, d\tau.$$

11. Let

$$w = \mathcal{L}^{-1} \left(\frac{1}{as^2 + bs + c} \right),$$

where $a, b,$ and c are constants and $a \neq 0$.

(a) Show that w is the solution of

$$aw'' + bw' + cw = 0, \qquad w(0) = 0, \qquad w'(0) = \frac{1}{a}.$$

(b) Let f be continuous on $[0, \infty)$ and define

$$h(t) = \int_0^t w(t - \tau) f(\tau) \, d\tau.$$

Use Leibniz's rule for differentiating an integral with respect to a parameter to show that h is the solution of

$$ah'' + bh' + ch = f, \qquad h(0) = 0, \qquad h'(0) = 0.$$

(c) Show that the function y in Eqn. (14) is the solution of Eqn. (13) provided that f is continuous on $[0, \infty)$; thus, it is not necessary to assume that f has a Laplace transform.

12. Consider the initial value problem

$$ay'' + by' + cy = f(t), \qquad y(0) = 0, \qquad y'(0) = 0, \tag{A}$$

where $a, b,$ and c are constants, $a \neq 0$, and

$$f(t) = \begin{cases} f_0(t), & 0 \leq t < t_1, \\ f_1(t), & t \geq t_1. \end{cases}$$

Assume that f_0 is continuous and of exponential order on $[0, \infty)$ and f_1 is continuous and of exponential order on $[t_1, \infty)$. Let

$$p(s) = as^2 + bs + c.$$

(a) Show that the Laplace transform of the solution of (A) is

$$Y(s) = \frac{F_0(s) + e^{-st_1}G(s)}{p(s)}$$

where $g(t) = f_1(t + t_1) - f_0(t + t_1)$.

(b) Let w be as in Exercise 11. Use Theorem 8.4.2 and the convolution theorem to show that the solution of (A) is

$$y(t) = \int_0^t w(t - \tau)f_0(\tau) \, d\tau + u(t - t_1) \int_0^{t-t_1} w(t - t_1 - \tau)g(\tau) \, d\tau$$

for $t > 0$.

(c) Henceforth assume only that f_0 is continuous on $[0, \infty)$ and f_1 is continuous on $[t_1, \infty)$. Use **(a)** and **(b)** of Exercise 11 to show that

$$y'(t) = \int_0^t w'(t - \tau)f_0(\tau) \, d\tau + u(t - t_1) \int_0^{t-t_1} w'(t - t_1 - \tau)g(\tau) \, d\tau$$

for $t > 0$ and

$$y''(t) = \frac{f(t)}{a} + \int_0^t w''(t - \tau)f_0(\tau) \, d\tau + u(t - t_1) \int_0^{t-t_1} w''(t - t_1 - \tau)g(\tau) \, d\tau$$

for $0 < t < t_1$ and $t > t_1$, and that y satisfies the differential equation in (A) on $(0, t_1)$ and (t_1, ∞).

(d) Show that y and y' are continuous on $[0, \infty)$.

13. Suppose that

$$f(t) = \begin{cases} f_0(t), & 0 \le t < t_1, \\ f_1(t), & t_1 \le t < t_2, \\ \quad\vdots \\ f_{k-1}(t), & t_{k-1} \le t < t_k, \\ f_k(t), & t \ge t_k, \end{cases}$$

where f_m is continuous on $[t_m, \infty)$ for $m = 0, \dots, k$ (let $t_0 = 0$), and define

$$g_m(t) = f_m(t + t_m) - f_{m-1}(t + t_m), \qquad m = 1, \dots, k.$$

Extend the results of Exercise 12 to show that the solution of

$$ay'' + by' + cy = f(t), \qquad y(0) = 0, \qquad y'(0) = 0$$

is

$$y(t) = \int_0^t w(t - \tau)f_0(\tau) \, d\tau + \sum_{m=1}^k u(t - t_m) \int_0^{t-t_m} w(t - t_m - \tau)g_m(\tau) \, d\tau.$$

14. Let $\{t_m\}_{m=0}^\infty$ be a sequence of points such that $t_0 = 0$, $t_{m+1} > t_m$, and $\lim_{m \to \infty} t_m = \infty$. For each nonnegative integer m let f_m be continuous on $[t_m, \infty)$, and let f be defined on $[0, \infty)$ by

$$f(t) = f_m(t), \qquad t_m \le t < t_{m+1} \qquad (m = 0, 1, \dots).$$

Let

$$g_m(t) = f_m(t + t_m) - f_{m-1}(t + t_m), \qquad m = 1, \dots, k.$$

Extend the results of Exercise 13 to show that the solution of

$$ay'' + by' + cy = f(t), \qquad y(0) = 0, \qquad y'(0) = 0$$

is

$$y(t) = \int_0^t w(t - \tau)f_0(\tau)\, d\tau + \sum_{m=1}^{\infty} u(t - t_m) \int_0^{t-t_m} w(t - t_m - \tau)g_m(\tau)\, d\tau.$$

Hint: *See Exercise 30 of Section 8.4.*

8.7   Constant Coefficient Equations with Impulses

So far in this chapter we've considered initial value problems for the constant coefficient equation

$$ay'' + by' + cy = f(t)$$

where f is continuous or piecewise continuous on $[0, \infty)$. In this section we consider initial value problems in which f represents a force that is very large for a short time and zero otherwise. We say that such forces are ***impulsive.*** Impulsive forces occur, for example, when two objects collide. Since it is not feasible to represent such forces as continuous or piecewise continuous functions, we must construct a different mathematical model to deal with them.

If f is an integrable function such that $f(t) = 0$ for t outside of the interval $[t_0, t_0 + h]$, then $\int_{t_0}^{t_0+h} f(t)\, dt$ is called the ***total impulse*** of f. We are interested in the idealized situation in which h is so small that the total impulse can be assumed to be applied instantaneously at $t = t_0$. We say in this case that f is an ***impulse function.*** In particular, we denote by $\delta(t - t_0)$ the impulse function with total impulse equal to 1, applied at $t = t_0$. (The impulse function $\delta(t)$ obtained by setting $t_0 = 0$ is called the ***Dirac***[1] ***delta function.***) It must be understood, however, that $\delta(t - t_0)$ is not a function in the standard sense, since our "definition" implies that $\delta(t - t_0) = 0$ if $t \neq t_0$ while

$$\int_{t_0}^{t_0} \delta(t - t_0)\, dt = 1.$$

From calculus we know that no function can have these properties; nevertheless, there is a branch of mathematics known as the ***theory of distributions*** in which the definition can be made rigorous. Since the theory of distributions is beyond the scope of this book, we will take an intuitive approach to impulse functions.

Our first task is to define what we mean by the solution of the initial value problem

$$ay'' + by' + cy = \delta(t - t_0), \qquad y(0) = 0, \qquad y'(0) = 0,$$

where t_0 is a fixed nonnegative number. The following theorem will motivate our definition.

---

[1]The English mathematical physicist Paul A. M. Dirac (1902–1984) created the delta function in connection with his work on quantum mechanics, for which he was a joint winner of the 1933 Nobel Prize in physics.

THEOREM 8.7.1

Suppose that $t_0 \geq 0$. For each positive number h let y_h be the solution of the initial value problem

$$ay_h'' + by_h' + cy_h = f_h(t), \qquad y_h(0) = 0, \qquad y_h'(0) = 0, \qquad (1)$$

where

$$f_h(t) = \begin{cases} 0, & 0 \leq t < t_0, \\ 1/h, & t_0 \leq t < t_0 + h, \\ 0, & t \geq t_0 + h, \end{cases} \qquad (2)$$

so that f_h has unit total impulse equal to the area of the shaded rectangle in Figure 1. Then

$$\lim_{h \to 0+} y_h(t) = u(t - t_0)w(t - t_0), \qquad (3)$$

where

$$w = \mathcal{L}^{-1}\left(\frac{1}{as^2 + bs + c}\right).$$

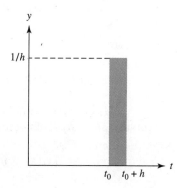

Figure 1   Graph of $y = f_h(t)$

PROOF. Taking Laplace transforms in (1) yields

$$(as^2 + bs + c)Y_h(s) = F_h(s),$$

so

$$Y_h(s) = \frac{F_h(s)}{as^2 + bs + c}.$$

The convolution theorem implies that

$$y_h(t) = \int_0^t w(t - \tau)f_h(\tau)\, d\tau.$$

Therefore, (2) implies that

$$y_h(t) = \begin{cases} 0, & 0 \leq t < t_0, \\ \dfrac{1}{h}\displaystyle\int_{t_0}^t w(t - \tau)\, d\tau, & t_0 \leq t \leq t_0 + h, \\ \dfrac{1}{h}\displaystyle\int_{t_0}^{t_0+h} w(t - \tau)\, d\tau, & t > t_0 + h. \end{cases} \qquad (4)$$

Since $y_h(t) = 0$ for all h if $0 \leq t \leq t_0$, it follows that

$$\lim_{h \to 0+} y_h(t) = 0 \qquad \text{if} \qquad 0 \leq t \leq t_0. \tag{5}$$

We will now show that

$$\lim_{h \to 0+} y_h(t) = w(t - t_0) \qquad \text{if} \qquad t > t_0. \tag{6}$$

Suppose that t is fixed and $t > t_0$. From (4),

$$y_h(t) = \frac{1}{h} \int_{t_0}^{t_0+h} w(t - \tau)\, d\tau \qquad \text{if} \qquad h < t - t_0. \tag{7}$$

Since

$$\frac{1}{h} \int_{t_0}^{t_0+h} d\tau = 1 \tag{8}$$

we can write

$$w(t - t_0) = \frac{1}{h} w(t - t_0) \int_{t_0}^{t_0+h} d\tau = \frac{1}{h} \int_{t_0}^{t_0+h} w(t - t_0)\, d\tau.$$

From this and (7),

$$y_h(t) - w(t - t_0) = \frac{1}{h} \int_{t_0}^{t_0+h} \left(w(t - \tau) - w(t - t_0)\right) d\tau.$$

Therefore

$$|y_h(t) - w(t - t_0)| \leq \frac{1}{h} \int_{t_0}^{t_0+h} |w(t - \tau) - w(t - t_0)|\, d\tau. \tag{9}$$

Now let M_h be the maximum value of $|w(t - \tau) - w(t - t_0)|$ as τ varies over the interval $[t_0, t_0 + h]$. (Remember that t and t_0 are fixed.) Then (8) and (9) imply that

$$|y_h(t) - w(t - t_0)| \leq \frac{1}{h} M_h \int_{t_0}^{t_0+h} d\tau = M_h. \tag{10}$$

But $\lim_{h \to 0+} M_h = 0$, since w is continuous. Therefore (10) implies (6). This and (5) imply (3). $\square$

Theorem 8.7.1 motivates the following definition.

DEFINITION 8.7.2

If $t_0 > 0$ then the solution of the initial value problem

$$ay'' + by' + cy = \delta(t - t_0), \qquad y(0) = 0, \qquad y'(0) = 0 \tag{11}$$

is defined to be

$$y = u(t - t_0)w(t - t_0),$$

where

$$w = \mathcal{L}^{-1}\left(\frac{1}{as^2 + bs + c}\right).$$

In physical applications where the input f and the output y of a device are related by the differential equation

$$ay'' + by' + cy = f(t),$$

w is called the **impulse response** of the device. Notice that w is the solution of the initial value problem

$$aw'' + bw' + cw = 0, \qquad w(0) = 0, \qquad w'(0) = 1/a, \qquad (12)$$

as can be seen by using the Laplace transform to solve this problem. (Verify.) On the other hand, we can solve (12) by the methods of Section 5.2 and show that w is defined on $(-\infty, \infty)$ by

$$w = \frac{e^{r_2 t} - e^{r_1 t}}{a(r_2 - r_1)}, \qquad w = \frac{1}{a} t e^{r_1 t}, \qquad \text{or} \qquad w = \frac{1}{a\omega} e^{\lambda t} \sin \omega t, \qquad (13)$$

depending upon whether the polynomial $p(r) = ar^2 + br + c$ has distinct real zeros r_1 and r_2, a repeated zero r_1, or complex conjugate zeros $\lambda \pm i\omega$. (In most physical applications the zeros of the characteristic polynomial have negative real parts, so $\lim_{t \to \infty} w(t) = 0$.) This means that $y = u(t - t_0)w(t - t_0)$ is defined on $(-\infty, \infty)$ and has the following properties:

$$y(t) = 0, \qquad t < t_0,$$
$$ay'' + by' + cy = 0 \qquad \text{on } (-\infty, t_0) \text{ and } (t_0, \infty),$$
$$y'_-(t_0) = 0, \qquad y'_+(t_0) = 1/a \qquad (14)$$

(remember that $y'_-(t_0)$ and $y'_+(t_0)$ are derivatives from the right and left, respectively), and $y'(t_0)$ does not exist. Thus, even though we have defined $y = u(t - t_0)w(t - t_0)$ to be the solution of (11), the fact is that this function *doesn't satisfy* the differential equation in (11) *at* t_0, since it isn't differentiable there; in fact (14) indicates that an impulse causes a jump discontinuity in velocity. (To see that this is reasonable, think of what happens if you strike a ball with a bat.) This means that the initial value problem (11) doesn't make sense if $t_0 = 0$, since $y'(0)$ doesn't exist in this case. However, $y = u(t)w(t)$ can be defined to be the solution of the modified initial value problem

$$ay'' + by' + cy = \delta(t), \qquad y(0) = 0, \qquad y'_-(0) = 0,$$

where the condition on the derivative at $t = 0$ has been replaced by a condition on the derivative from the left.

Figure 2 illustrates Theorem 8.7.1 for the case where the impulse response w is the first expression in (13) and r_1 and r_2 are distinct and both negative. The blue curve in the figure is the graph of w. The gray curves are solutions of (1) for various values of h. As h decreases the graph of y_h moves to the left toward the graph of w.

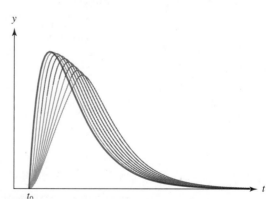

Figure 2   An illustration of Theorem 8.7.1

EXAMPLE 1   Find the solution of the initial value problem

$$y'' - 2y' + y = \delta(t - t_0), \qquad y(0) = 0, \qquad y'(0) = 0, \tag{15}$$

where $t_0 > 0$. Then interpret the solution for the case where $t_0 = 0$.

Solution   Here

$$w = \mathcal{L}^{-1}\left(\frac{1}{s^2 - 2s + 1}\right) = \mathcal{L}^{-1}\left(\frac{1}{(s-1)^2}\right) = te^{-t},$$

so Definition 8.7.2 yields

$$y = u(t - t_0)(t - t_0)e^{-(t-t_0)}$$

as the solution of (15) if $t_0 > 0$. If $t_0 = 0$ then (15) doesn't have a solution; however, $y = u(t)te^{-t}$ (which we usually write simply as $y = te^{-t}$) is the solution of the modified initial value problem

$$y'' - 2y' + y = \delta(t), \qquad y(0) = 0, \qquad y'_-(0) = 0.$$

The graph of $y = u(t - t_0)(t - t_0)e^{-(t-t_0)}$ is shown in Figure 3.   ∎

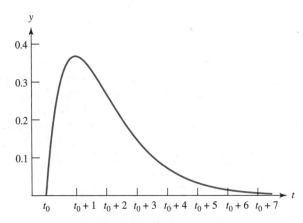

Figure 3   Graph of $y = u(t - t_0)(t - t_0)e^{-(t-t_0)}$

Definition 8.7.2 and the principle of superposition motivate the following definition.

DEFINITION 8.7.3

Suppose that α is a nonzero constant and f is piecewise continuous on $[0, \infty)$. If $t_0 > 0$ then the solution of the initial value problem

$$ay'' + by' + cy = f(t) + \alpha\delta(t - t_0), \qquad y(0) = k_0, \qquad y'(0) = k_1$$

is defined to be

$$y(t) = \hat{y}(t) + \alpha u(t - t_0)w(t - t_0),$$

where $\hat{y}$ is the solution of

$$ay'' + by' + cy = f(t), \qquad y(0) = k_0, \qquad y'(0) = k_1.$$

This definition also applies if $t_0 = 0$, provided that the initial condition $y'(0) = k_1$ is replaced by $y'_-(0) = k_1$.

EXAMPLE 2   Solve the initial value problem

$$y'' + 6y' + 5y = 3e^{-2t} + 2\delta(t-1), \qquad y(0) = -3, \qquad y'(0) = 2. \quad (16)$$

Solution   We leave it to you to show that the solution of

$$y'' + 6y' + 5y = 3e^{-2t}, \qquad y(0) = -3, \qquad y'(0) = 2$$

is

$$\hat{y} = -e^{-2t} + \frac{1}{2}e^{-5t} - \frac{5}{2}e^{-t}.$$

Since

$$w(t) = \mathcal{L}^{-1}\left(\frac{1}{s^2 + 6s + 5}\right) = \mathcal{L}^{-1}\left(\frac{1}{(s+1)(s+5)}\right)$$

$$= \frac{1}{4}\mathcal{L}^{-1}\left(\frac{1}{s+1} - \frac{1}{s+5}\right) = \frac{e^{-t} - e^{-5t}}{4},$$

the solution of (16) is

$$y = -e^{-2t} + \frac{1}{2}e^{-5t} - \frac{5}{2}e^{-t} + u(t-1)\frac{e^{-(t-1)} - e^{-5(t-1)}}{2} \quad (17)$$

(Figure 4).   ■

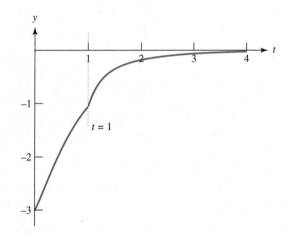

Figure 4   Graph of (17)

Definition 8.7.3 can be extended in the obvious way to cover the case where the forcing function contains more than one impulse.

EXAMPLE 3   Solve the initial value problem

$$y'' + y = 1 + 2\delta(t - \pi) - 3\delta(t - 2\pi), \qquad y(0) = -1, \qquad y'(0) = 2. \quad (18)$$

Solution   We leave it to you to show that

$$\hat{y} = 1 - 2\cos t + 2\sin t$$

is the solution of

$$y'' + y = 1, \qquad y(0) = -1, \qquad y'(0) = 2.$$

Since

$$w = \mathcal{L}^{-1}\left(\frac{1}{s^2 + 1}\right) = \sin t,$$

the solution of (18) is

$$y = 1 - 2 \cos t + 2 \sin t + 2u(t - \pi) \sin(t - \pi) - 3u(t - 2\pi) \sin(t - 2\pi)$$
$$= 1 - 2 \cos t + 2 \sin t - 2u(t - \pi) \sin t - 3u(t - 2\pi) \sin t,$$

or

$$y = \begin{cases} 1 - 2 \cos t + 2 \sin t, & 0 \le t < \pi, \\ 1 - 2 \cos t, & \pi \le t < 2\pi, \\ 1 - 2 \cos t - 3 \sin t, & t \ge 2\pi \end{cases} \tag{19}$$

(Figure 5). ■

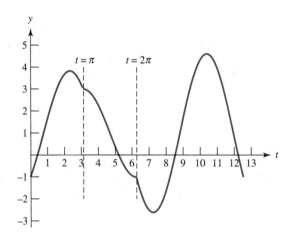

Figure 5 Graph of (19)

8.7 EXERCISES

In Exercises 1–20 solve the initial value problem. Where indicated by **C**, *graph the solution.*

1. $y'' + 3y' + 2y = 6e^{2t} + 2\delta(t - 1)$, $y(0) = 2$, $y'(0) = -6$

C **2.** $y'' + y' - 2y = -10e^{-t} + 5\delta(t - 1)$, $y(0) = 7$, $y'(0) = -9$

3. $y'' - 4y = 2e^{-t} + 5\delta(t - 1)$, $y(0) = -1$, $y'(0) = 2$

C **4.** $y'' + y = \sin 3t + 2\delta(t - \pi/2)$, $y(0) = 1$, $y'(0) = -1$

5. $y'' + 4y = 4 + \delta(t - 3\pi)$, $y(0) = 0$, $y'(0) = 1$

6. $y'' - y = 8 + 2\delta(t - 2)$, $y(0) = -1$, $y'(0) = 1$

7. $y'' + y' = e^t + 3\delta(t - 6)$, $y(0) = -1$, $y'(0) = 4$

8. $y'' + 4y = 8e^{2t} + \delta(t - \pi/2)$, $y(0) = 8$, $y'(0) = 0$

C **9.** $y'' + 3y' + 2y = 1 + \delta(t - 1)$, $y(0) = 1$, $y'(0) = -1$

10. $y'' + 2y' + y = e^t + 2\delta(t - 2)$, $y(0) = -1$, $y'(0) = 2$

C **11.** $y'' + 4y = \sin t + \delta(t - \pi/2)$, $y(0) = 0$, $y'(0) = 2$

12. $y'' + 2y' + 2y = \delta(t - \pi) - 3\delta(t - 2\pi)$, $y(0) = -1$, $y'(0) = 2$

13. $y'' + 4y' + 13y = \delta(t - \pi/6) + 2\delta(t - \pi/3)$, $y(0) = 1$, $y'(0) = 2$

14. $2y'' - 3y' - 2y = 1 + \delta(t - 2)$, $y(0) = -1$, $y'(0) = 2$

15. $4y'' - 4y' + 5y = 4 \sin t - 4 \cos t + \delta(t - \pi/2) - \delta(t - \pi)$, $y(0) = 1$, $y'(0) = 1$

16. $y'' + y = \cos 2t + 2\delta(t - \pi/2) - 3\delta(t - \pi)$, $y(0) = 0$, $y'(0) = -1$

C **17.** $y'' - y = 4e^{-t} - 5\delta(t - 1) + 3\delta(t - 2)$, $y(0) = 0$, $y'(0) = 0$

18. $y'' + 2y' + y = e^t - \delta(t - 1) + 2\delta(t - 2)$, $y(0) = 0$, $y'(0) = -1$

19. $y'' + y = f(t) + \delta(t - 2\pi)$, $y(0) = 0$, $y'(0) = 1$; $f(t) = \begin{cases} \sin 2t, & 0 \le t < \pi, \\ 0, & t \ge \pi. \end{cases}$

20. $y'' + 4y = f(t) + \delta(t - \pi) - 3\delta(t - 3\pi/2)$, $y(0) = 1$, $y'(0) = -1$; $f(t) = \begin{cases} 1, & 0 \le t < \pi/2, \\ 2, & t \ge \pi/2 \end{cases}$

21. $y'' + y = \delta(t)$, $y(0) = 1$, $y'_-(0) = -2$ **22.** $y'' - 4y = 3\delta(t)$, $y(0) = -1$, $y'_-(0) = 7$

23. $y'' + 3y' + 2y = -5\delta(t)$, $y(0) = 0$, $y'_-(0) = 0$ **24.** $y'' + 4y' + 4y = -\delta(t)$, $y(0) = 1$, $y'_-(0) = 5$

25. $4y'' + 4y' + y = 3\delta(t)$, $y(0) = 1$, $y'_-(0) = -6$

In Exercises 26–28 solve the initial value problem

$$ay''_h + by'_h + cy_h = \begin{cases} 0, & 0 \le t < t_0, \\ 1/h, & t_0 \le t < t_0 + h, \\ 0, & t \ge t_0 + h, \end{cases} \qquad y_h(0) = 0, \quad y'_h(0) = 0,$$

where $t_0 > 0$ and $h > 0$. Then find

$$w = \mathcal{L}^{-1}\left(\frac{1}{as^2 + bs + c}\right)$$

and verify Theorem 8.7.1 by graphing w and y_h on the same axes, for small positive values of h.

L **26.** $y'' + 2y' + 2y = f_h(t)$, $y(0) = 0$, $y'(0) = 0$

L **27.** $y'' + 2y' + y = f_h(t)$, $y(0) = 0$, $y'(0) = 0$

L **28.** $y'' + 3y' + 2y = f_h(t)$, $y(0) = 0$, $y'(0) = 0$

29. Recall from Section 6.2 that the displacement of an object of mass m in a spring–mass system in free damped oscillation is

$$my'' + cy' + ky = 0, \quad y(0) = y_0, \quad y'(0) = v_0,$$

and that y can be written as

$$y = Re^{-ct/2m} \cos(\omega_1 t - \phi)$$

if the motion is underdamped. Suppose that $y(\tau) = 0$. Find the impulse that would have to be applied to the object at $t = \tau$ to put it in equilibrium.

30. Solve the initial value problem. Find a formula that does not involve step functions and represents y on each subinterval of $[0, \infty)$ on which the forcing function is zero.

(a) $y'' - y = \sum\limits_{k=1}^{\infty} \delta(t - k)$, $y(0) = 0$, $y'(0) = 1$

(b) $y'' + y = \sum\limits_{k=1}^{\infty} \delta(t - 2k\pi)$, $y(0) = 0$, $y'(0) = 1$

(c) $y'' - 3y' + 2y = \sum\limits_{k=1}^{\infty} \delta(t - k)$, $y(0) = 0$, $y'(0) = 1$

(d) $y'' + y = \sum\limits_{k=1}^{\infty} \delta(t - k\pi)$, $y(0) = 0$, $y'(0) = 0$

9

Linear Higher Order Equations

IN THIS CHAPTER we extend the results obtained in Chapter 5 for linear second order equations to linear higher order equations.

SECTION 9.1 presents a theoretical introduction to linear higher order equations.

SECTION 9.2 discusses higher order constant coefficient homogeneous equations.

SECTION 9.3 presents the method of undetermined coefficients for higher order equations.

SECTION 9.4 extends the method of variation of parameters to higher order equations.

9.1 Introduction to Linear Higher Order Equations

An *n*th order differential equation is said to be *linear* if it can be written in the form

$$y^{(n)} + p_1(x)y^{(n-1)} + \cdots + p_n(x)y = f(x). \tag{1}$$

We considered equations of this form with $n = 1$ in Section 2.1 and with $n = 2$ in Chapter 5. In this chapter *n* is an arbitrary positive integer.

In this section we sketch the general theory of linear *n*th order equations. Since this theory has already been discussed for $n = 2$ in Sections 5.1 and 5.3, we will omit proofs.

For convenience we consider linear differential equations written as

$$P_0(x)y^{(n)} + P_1(x)y^{(n-1)} + \cdots + P_n(x)y = F(x), \tag{2}$$

which can be rewritten as (1) on any interval on which P_0 has no zeros, with $p_1 = P_1/P_0, \ldots, p_n = P_n/P_0$ and $f = F/P_0$. For simplicity, throughout this chapter we will abbreviate the left side of (2) by Ly; that is,

$$Ly = P_0y^{(n)} + P_1y^{(n-1)} + \cdots + P_ny.$$

We say that the equation $Ly = F$ is *normal* on (a,b) if $P_0, P_1, \ldots, P_n$ and F are continuous on (a,b) and P_0 has no zeros on (a,b). If this is so then $Ly = F$ can be written as (1) with $p_1, \ldots, p_n$ and f continuous on (a,b).

The following theorem is analogous to Theorem 5.3.1.

THEOREM 9.1.1

Suppose that $Ly = F$ is normal on (a,b), let x_0 be a point in (a,b), and let $k_0, k_1, \ldots, k_{n-1}$ be arbitrary real numbers. Then the initial value problem

$$Ly = F, \qquad y(x_0) = k_0, \qquad y'(x_0) = k_1, \ldots, y^{(n-1)}(x_0) = k_{n-1}$$

has a unique solution on (a,b).

HOMOGENEOUS EQUATIONS

Equation (2) is said to be *homogeneous* if $F \equiv 0$ and *nonhomogeneous* otherwise. Since $y \equiv 0$ is obviously a solution of $Ly = 0$ we call it the *trivial* solution. Any other solution is *nontrivial*.

If $y_1, y_2, \ldots, y_n$ are defined on (a,b) and $c_1, c_2 \ldots, c_n$ are constants, then

$$y = c_1y_1 + c_2y_2 + \cdots + c_ny_n \tag{3}$$

is a *linear combination* of $\{y_1, y_2, \ldots, y_n\}$. It is straightforward to show that if $y_1, y_2, \ldots, y_n$ are solutions of $Ly = 0$ on (a,b) then so is any linear combination of $\{y_1, y_2, \ldots, y_n\}$. (See the proof of Theorem 5.1.2.) We say that $\{y_1, y_2, \ldots, y_n\}$ is a *fundamental set of solutions* of $Ly = 0$ on (a,b) if every solution of $Ly = 0$ on (a,b) can be written as a linear combination of $\{y_1, y_2, \ldots, y_n\}$, as in (3). In this case we say that (3) is the *general solution* of $Ly = 0$ on (a,b).

It can be shown (Exercises 14 and 15) that if the equation $Ly = 0$ is normal on (a,b) then it has infinitely many fundamental sets of solutions on (a,b). The following definition will help to identify fundamental sets of solutions of $Ly = 0$.

We say that $\{y_1, y_2, \ldots, y_n\}$ is ***linearly independent*** on (a, b) if the only constants $c_1, c_2, \ldots, c_n$ such that

$$c_1 y_1(x) + c_2 y_2(x) + \cdots + c_n y_n(x) = 0, \qquad a < x < b, \tag{4}$$

are $c_1 = c_2 = \cdots = c_n = 0$. If (4) holds for some set of constants $c_1, c_2, \ldots, c_n$ that are not all zero, then $\{y_1, y_2, \ldots, y_n\}$ is ***linearly dependent*** on (a, b).

The following theorem is analogous to Theorem 5.1.3.

THEOREM 9.1.2

Suppose that $Ly = 0$ is normal on (a, b). Then a set $\{y_1, y_2, \ldots, y_n\}$ of n solutions of $Ly = 0$ on (a, b) is a fundamental set if and only if it is linearly independent on (a, b).

EXAMPLE 1   The equation

$$x^3 y''' - x^2 y'' - 2xy' + 6y = 0 \tag{5}$$

is normal and has the solutions $y_1 = x^2$, $y_2 = x^3$, and $y_3 = 1/x$ on $(-\infty, 0)$ and $(0, \infty)$. Show that $\{y_1, y_2, y_3\}$ is linearly independent on $(-\infty, 0)$ and $(0, \infty)$. Then find the general solution of (5) on $(-\infty, 0)$ and $(0, \infty)$.

Solution   Suppose that

$$c_1 x^2 + c_2 x^3 + \frac{c_3}{x} = 0 \tag{6}$$

on $(0, \infty)$. We must show that $c_1 = c_2 = c_3 = 0$. Differentiating (6) twice yields the system

$$c_1 x^2 + c_2 x^3 + \frac{c_3}{x} = 0$$

$$2c_1 x + 3c_2 x^2 - \frac{c_3}{x^2} = 0 \tag{7}$$

$$2c_1 + 6c_2 x + \frac{2c_3}{x^3} = 0.$$

If (7) holds for all x in $(0, \infty)$ then it certainly holds at $x = 1$; therefore,

$$\begin{aligned} c_1 + c_2 + c_3 &= 0 \\ 2c_1 + 3c_2 - c_3 &= 0 \\ 2c_1 + 6c_2 + 2c_3 &= 0. \end{aligned} \tag{8}$$

By solving this system directly you can verify that it has only the trivial solution $c_1 = c_2 = c_3 = 0$; however, for our purposes it is more useful to recall from linear algebra that a homogeneous linear system of n equations in n unknowns has only the trivial solution if its determinant is nonzero. Since the determinant of (8) is

$$\begin{vmatrix} 1 & 1 & 1 \\ 2 & 3 & -1 \\ 2 & 6 & 2 \end{vmatrix} = \begin{vmatrix} 1 & 0 & 0 \\ 2 & 1 & -3 \\ 2 & 4 & 0 \end{vmatrix} = 12,$$

it follows that (8) has only the trivial solution, so $\{y_1, y_2, y_3\}$ is linearly independent on $(0, \infty)$. Now Theorem 9.1.2 implies that

$$y = c_1 x^2 + c_2 x^3 + \frac{c_3}{x}$$

is the general solution of (5) on $(0,\infty)$. To see that this is also true on $(-\infty,0)$, assume that (6) holds on $(-\infty,0)$. Setting $x = -1$ in (7) yields

$$
\begin{aligned}
c_1 - c_2 - c_3 &= 0 \\
-2c_1 + 3c_2 - c_3 &= 0 \\
2c_1 - 6c_2 - 2c_3 &= 0.
\end{aligned}
$$

Since the determinant of this system is

$$
\begin{vmatrix} 1 & -1 & -1 \\ -2 & 3 & -1 \\ 2 & -6 & -2 \end{vmatrix} = \begin{vmatrix} 1 & 0 & 0 \\ -2 & 1 & -3 \\ 2 & -4 & 0 \end{vmatrix} = -12,
$$

it follows that $c_1 = c_2 = c_3 = 0$; that is, $\{y_1, y_2, y_3\}$ is linearly independent on $(-\infty,0)$.

EXAMPLE 2 The equation

$$
y^{(4)} + y''' - 7y'' - y' + 6y = 0 \tag{9}
$$

is normal and has the solutions $y_1 = e^x$, $y_2 = e^{-x}$, $y_3 = e^{2x}$, and $y_4 = e^{-3x}$ on $(-\infty,\infty)$. (Verify.) Show that $\{y_1, y_2, y_3, y_4\}$ is linearly independent on $(-\infty,\infty)$. Then find the general solution of (9).

Solution Suppose that c_1, c_2, c_3, and c_4 are constants such that

$$
c_1 e^x + c_2 e^{-x} + c_3 e^{2x} + c_4 e^{-3x} = 0 \tag{10}
$$

for all x. We must show that $c_1 = c_2 = c_3 = c_4 = 0$. Differentiating (10) three times yields the system

$$
\begin{aligned}
c_1 e^x + c_2 e^{-x} + c_3 e^{2x} + c_4 e^{-3x} &= 0 \\
c_1 e^x - c_2 e^{-x} + 2c_3 e^{2x} - 3c_4 e^{-3x} &= 0 \\
c_1 e^x + c_2 e^{-x} + 4c_3 e^{2x} + 9c_4 e^{-3x} &= 0 \\
c_1 e^x - c_2 e^{-x} + 8c_3 e^{2x} - 27c_4 e^{-3x} &= 0.
\end{aligned} \tag{11}
$$

If (11) holds for all x then it certainly holds for $x = 0$. Therefore,

$$
\begin{aligned}
c_1 + c_2 + c_3 + c_4 &= 0 \\
c_1 - c_2 + 2c_3 - 3c_4 &= 0 \\
c_1 + c_2 + 4c_3 + 9c_4 &= 0 \\
c_1 - c_2 + 8c_3 - 27c_4 &= 0.
\end{aligned}
$$

The determinant of this system is

$$
\begin{vmatrix} 1 & 1 & 1 & 1 \\ 1 & -1 & 2 & -3 \\ 1 & 1 & 4 & 9 \\ 1 & -1 & 8 & -27 \end{vmatrix} = \begin{vmatrix} 1 & 1 & 1 & 1 \\ 0 & -2 & 1 & -4 \\ 0 & 0 & 3 & 8 \\ 0 & -2 & 7 & -28 \end{vmatrix} = \begin{vmatrix} -2 & 1 & -4 \\ 0 & 3 & 8 \\ -2 & 7 & -28 \end{vmatrix}
$$

$$
= \begin{vmatrix} -2 & 1 & -4 \\ 0 & 3 & 8 \\ 0 & 6 & -24 \end{vmatrix} = -2 \begin{vmatrix} 3 & 8 \\ 6 & -24 \end{vmatrix} = 240, \tag{12}
$$

so the system has only the trivial solution $c_1 = c_2 = c_3 = c_4 = 0$. Now Theorem 9.1.2 implies that

$$y = c_1 e^x + c_2 e^{-x} + c_3 e^{2x} + c_4 e^{-3x}$$

is the general solution of (9). ∎

THE WRONSKIAN

We can use the method used in Examples 1 and 2 to test a set of n solutions $\{y_1, y_2, \ldots, y_n\}$ of any nth order equation $Ly = 0$ for linear independence on an interval (a, b) on which the equation is normal. Thus, if $c_1, c_2, \ldots, c_n$ are constants such that

$$c_1 y_1 + c_2 y_2 + \cdots + c_n y_n = 0, \qquad a < x < b,$$

then differentiating $n - 1$ times leads to the $n \times n$ system of equations

$$
\begin{aligned}
c_1 y_1(x) + c_2 y_2(x) + \cdots + c_n y_n(x) &= 0 \\
c_1 y_1'(x) + c_2 y_2'(x) + \cdots + c_n y_n'(x) &= 0 \\
&\ \ \vdots \\
c_1 y_1^{(n-1)}(x) + c_2 y_2^{(n-1)}(x) + \cdots + c_n y_n^{(n-1)}(x) &= 0
\end{aligned}
\tag{13}
$$

for $c_1, c_2, \ldots, c_n$. For a fixed x the determinant of this system is

$$
W(x) = \begin{vmatrix}
y_1(x) & y_2(x) & \cdots & y_n(x) \\
y_1'(x) & y_2'(x) & \cdots & y_n'(x) \\
\vdots & \vdots & \ddots & \vdots \\
y_1^{(n-1)}(x) & y_2^{(n-1)}(x) & \cdots & y_n^{(n-1)}(x)
\end{vmatrix}.
$$

We call this determinant the **Wronskian** of $\{y_1, y_2, \ldots, y_n\}$. If $W(x) \neq 0$ for some x in (a, b) then (13) has only the trivial solution $c_1 = c_2 = \cdots = c_n = 0$, and Theorem 9.1.2 implies that

$$y = c_1 y_1 + c_2 y_2 + \cdots + c_n y_n$$

is the general solution of $Ly = 0$ on (a, b).

The following theorem generalizes Theorem 5.1.4. The proof is sketched in Exercises 17–20.

THEOREM 9.1.3

Suppose that the homogeneous linear nth order equation

$$P_0(x) y'' + P_1(x) y' + \cdots + P_n(x) y = 0 \tag{14}$$

is normal on (a, b), let $y_1, y_2, \ldots, y_n$ be solutions of (14) on (a, b), and let x_0 be in (a, b). Then the Wronskian of $\{y_1, y_2, \ldots, y_n\}$ is given by

$$W(x) = W(x_0) \exp\left\{ -\int_{x_0}^{x} \frac{P_1(t)}{P_0(t)}\, dt \right\}, \qquad a < x < b. \tag{15}$$

Therefore, either W has no zeros in (a, b) or $W \equiv 0$ on (a, b).

Formula (15) is **Abel's formula**.

The following theorem is analogous to Theorem 5.1.6.

THEOREM 9.1.4

Suppose that $Ly = 0$ is normal on (a,b) and let $y_1, y_2, \ldots, y_n$ be n solutions of $Ly = 0$ on (a,b). Then the following statements are equivalent; that is, they are either all true or all false:

(a) *The general solution of $Ly = 0$ on (a,b) is $y = c_1 y_1 + c_2 y_2 + \cdots + c_n y_n$.*
(b) *$\{y_1, y_2, \ldots, y_n\}$ is a fundamental set of solutions of $Ly = 0$ on (a,b).*
(c) *$\{y_1, y_2, \ldots, y_n\}$ is linearly independent on (a,b).*
(d) *The Wronskian of $\{y_1, y_2, \ldots, y_n\}$ is nonzero at some point in (a,b).*
(e) *The Wronskian of $\{y_1, y_2, \ldots, y_n\}$ is nonzero at all points in (a,b).*

EXAMPLE 3   We saw in Example 1 that the solutions $y_1 = x^2$, $y_2 = x^3$, and $y_3 = 1/x$ of

$$x^3 y''' - x^2 y'' - 2xy' + 6y = 0$$

are linearly independent on $(-\infty, 0)$ and $(0, \infty)$. Calculate the Wronskian of $\{y_1, y_2, y_3\}$.

Solution   If $x \neq 0$ then

$$W(x) = \begin{vmatrix} x^2 & x^3 & \dfrac{1}{x} \\ 2x & 3x^2 & -\dfrac{1}{x^2} \\ 2 & 6x & \dfrac{2}{x^3} \end{vmatrix} = 2x^3 \begin{vmatrix} 1 & x & \dfrac{1}{x^3} \\ 2 & 3x & -\dfrac{1}{x^3} \\ 1 & 3x & \dfrac{1}{x^3} \end{vmatrix},$$

where we factored x^2, x, and 2 out of the first, second, and third rows of $W(x)$, respectively. Adding the second row of the last determinant to the first and third rows yields

$$W(x) = 2x^3 \begin{vmatrix} 3 & 4x & 0 \\ 2 & 3x & -\dfrac{1}{x^3} \\ 3 & 6x & 0 \end{vmatrix} = 2x^3 \left(\dfrac{1}{x^3} \right) \begin{vmatrix} 3 & 4x \\ 3 & 6x \end{vmatrix} = 12x.$$

Therefore $W(x) \neq 0$ on $(-\infty, 0)$ and $(0, \infty)$.

EXAMPLE 4   We saw in Example 2 that the solutions $y_1 = e^x$, $y_2 = e^{-x}$, $y_3 = e^{2x}$, and $y_4 = e^{-3x}$ of

$$y^{(4)} + y''' - 7y'' - y' + 6y = 0$$

are linearly independent on every open interval. Calculate the Wronskian of $\{y_1, y_2, y_3, y_4\}$.

Solution   For all x,

$$W(x) = \begin{vmatrix} e^x & e^{-x} & e^{2x} & e^{-3x} \\ e^x & -e^{-x} & 2e^{2x} & -3e^{-3x} \\ e^x & e^{-x} & 4e^{2x} & 9e^{-3x} \\ e^x & -e^{-x} & 8e^{2x} & -27e^{-3x} \end{vmatrix}.$$

Factoring the exponential common factor from each row yields

$$W(x) = e^{-x} \begin{vmatrix} 1 & 1 & 1 & 1 \\ 1 & -1 & 2 & -3 \\ 1 & 1 & 4 & 9 \\ 1 & -1 & 8 & -27 \end{vmatrix} = 240e^{-x},$$

from (12).

REMARK. Under the assumptions of Theorem 9.1.4, it isn't necessary to obtain a formula for $W(x)$. Just evaluate $W(x)$ at a convenient point in (a,b), as we did in Examples 1 and 2.

GENERAL SOLUTION OF A NONHOMOGENEOUS EQUATION

The following theorem is analogous to Theorem 5.3.2. It shows how to find the general solution of $Ly = F$ if we know a particular solution of $Ly = F$ and a fundamental set of solutions of the **complementary equation** $Ly = 0$.

THEOREM 9.1.5

Suppose that $Ly = F$ is normal on (a,b). Let y_p be a particular solution of $Ly = F$ on (a,b), and let $\{y_1, y_2, \ldots, y_n\}$ be a fundamental set of solutions of the complementary equation $Ly = 0$ on (a,b). Then y is a solution of $Ly = F$ on (a,b) if and only if

$$y = y_p + c_1 y_1 + c_2 y_2 + \cdots + c_n y_n,$$

where $c_1, c_2, \ldots, c_n$ are constants.

The following theorem is analogous to 5.3.3.

THEOREM 9.1.6

(The Principle of Superposition)

Suppose that for each $i = 1, 2, \ldots, r$ the function y_{p_i} is a particular solution of $Ly = F_i$ on (a,b). Then

$$y_p = y_{p_1} + y_{p_2} + \cdots + y_{p_r}$$

is a particular solution of

$$Ly = F_1(x) + F_2(x) + \cdots + F_r(x)$$

on (a,b).

We will apply Theorems 9.1.5 and 9.1.6 throughout the rest of this chapter.

9.1 EXERCISES

1. Verify that the given function is the solution of the initial value problem.

   (a) $x^3y''' - 3x^2y'' + 6xy' - 6y = -\dfrac{24}{x}$, $\quad y(-1) = 0$, $\quad y'(-1) = 0$, $\quad y''(-1) = 0$;

   $$y = -6x - 8x^2 - 3x^3 + \frac{1}{x}$$

   (b) $y''' - \dfrac{1}{x}y'' - y' + \dfrac{1}{x}y = \dfrac{x^2 - 4}{x^4}$, $\quad y(1) = \dfrac{3}{2}$, $\quad y'(1) = \dfrac{1}{2}$, $\quad y''(1) = 1$; $\quad y = x + \dfrac{1}{2x}$

   (c) $xy''' - y'' - xy' + y = x^2$, $\quad y(1) = 2$, $\quad y'(1) = 5$, $\quad y''(1) = -1$;

   $$y = -x^2 - 2 + 2e^{(x-1)} - e^{-(x-1)} + 4x$$

   (d) $4x^3y''' + 4x^2y'' - 5xy' + 2y = 30x^2$, $\quad y(1) = 5$, $\quad y'(1) = \dfrac{17}{2}$, $\quad y''(1) = \dfrac{63}{4}$;

   $$y = 2x^2 \ln x - x^{1/2} + 2x^{-1/2} + 4x^2$$

   (e) $x^4y^{(4)} - 4x^3y''' + 12x^2y'' - 24xy' + 24y = 6x^4$, $\quad y(1) = -2$, $\quad y'(1) = -9$, $\quad y''(1) = -27$,

   $\quad y'''(1) = -52$; $\quad y = x^4 \ln x + x - 2x^2 + 3x^3 - 4x^4$

   (f) $xy^{(4)} - y''' - 4xy'' + 4y' = 96x^2$, $\quad y(1) = -5$, $\quad y'(1) = -24$, $\quad y''(1) = -36$,

   $\quad y'''(1) = -48$; $\quad y = 9 - 12x + 6x^2 - 8x^3$

2. Solve the initial value problem

   $$x^3y''' - x^2y'' - 2xy' + 6y = 0, \quad y(-1) = 4, \quad y'(-1) = -14, \quad y''(-1) = 20.$$

   Hint: *See Example 1.*

3. Solve the initial value problem

   $$y^{(4)} + y''' - 7y'' - y' + 6y = 0, \quad y(0) = 5, \quad y'(0) = -6, \quad y''(0) = 10, \quad y'''(0) = -36.$$

   Hint: *See Example 2.*

4. Find solutions $y_1, y_2, \ldots, y_n$ of the equation $y^{(n)} = 0$ that satisfy the initial conditions

   $$y_i^{(j)}(x_0) = \begin{cases} 0, & j \neq i - 1, \\ 1, & j = i - 1, \end{cases} \quad 1 \leq i \leq n.$$

5. (a) Verify that the function

   $$y = c_1x^3 + c_2x^2 + \frac{c_3}{x}$$

   satisfies

   $$x^3y''' - x^2y'' - 2xy' + 6y = 0 \tag{A}$$

   if c_1, c_2, and c_3 are constants.

   (b) Use (a) to find solutions y_1, y_2, and y_3 of (A) such that

   $$\begin{aligned}
   y_1(1) &= 1, & y_1'(1) &= 0, & y_1''(1) &= 0; \\
   y_2(1) &= 0, & y_2'(1) &= 1, & y_2''(1) &= 0; \\
   y_3(1) &= 0, & y_3'(1) &= 0, & y_3''(1) &= 1.
   \end{aligned}$$

   (c) Use (b) to find the solution of (A) such that

   $$y(1) = k_0; \quad y'(1) = k_1, \quad y''(1) = k_2.$$

6. Verify that the given functions are solutions of the given equation, and show that they form a fundamental set of solutions of the equation on any interval on which the equation is normal.

   (a) $y''' + y'' - y' - y = 0$; $\quad \{e^x, e^{-x}, xe^{-x}\}$

(b) $y''' - 3y'' + 7y' - 5y = 0;$ $\{e^x, e^x \cos 2x, e^x \sin 2x\}$

(c) $xy''' - y'' - xy' + y = 0;$ $\{e^x, e^{-x}, x\}$

(d) $x^2y''' + 2xy'' - (x^2 + 2)y = 0;$ $\{e^x/x, e^{-x}/x, 1\}$

(e) $(x^2 - 2x + 2)y''' - x^2y'' + 2xy' - 2y = 0;$ $\{x, x^2, e^x\}$

(f) $(2x - 1)y^{(4)} - 4xy''' + (5 - 2x)y'' + 4xy' - 4y = 0;$ $\{x, e^x, e^{-x}, e^{2x}\}$

(g) $xy^{(4)} - y''' - 4xy' + 4y' = 0;$ $\{1, x^2, e^{2x}, e^{-2x}\}$

7. Find the Wronskian W of a set of three solutions of

$$y''' + 2xy'' + e^x y' - y = 0,$$

given that $W(0) = 2$.

8. Find the Wronskian W of a set of four solutions of

$$y^{(4)} + (\tan x)y''' + x^2 y'' + 2xy = 0,$$

given that $W(\pi/4) = K$.

9. **(a)** Evaluate the Wronskian W of $\{e^x, xe^x, x^2e^x\}$. Evaluate $W(0)$.

(b) Verify that $y_1, y_2,$ and y_3 satisfy

$$y''' - 3y'' + 3y' - y = 0. \tag{A}$$

(c) Use $W(0)$ from **(a)** and Abel's formula to calculate $W(x)$.

(d) What is the general solution of (A)?

10. Compute the Wronskian of the given set of functions.

(a) $\{1, e^x, e^{-x}\}$

(b) $\{e^x, e^x \sin x, e^x \cos x\}$

(c) $\{2, x + 1, x^2 + 2\}$

(d) $\{x, x \ln x, 1/x\}$

(e) $\left\{1, x, \dfrac{x^2}{2!}, \dfrac{x^3}{3!}, \ldots, \dfrac{x^n}{n!}\right\}$

(f) $\{e^x, e^{-x}, x\}$

(g) $\{e^x/x, e^{-x}/x, 1\}$

(h) $\{x, x^2, e^x\}$

(i) $\{x, x^3, 1/x, 1/x^2\}$

(j) $\{e^x, e^{-x}, x, e^{2x}\}$

(k) $\{e^{2x}, e^{-2x}, 1, x^2\}$

11. Suppose that the linear nth order equation $Ly = 0$ is normal on (a, b) and x_0 is in (a, b). Use Theorem 9.1.1 to show that $y \equiv 0$ is the only solution of the initial value problem

$$Ly = 0, \quad y(x_0) = 0, \quad y'(x_0) = 0, \ldots, y^{(n-1)}(x_0) = 0$$

on (a, b).

12. Prove: If $y_1, y_2, \ldots, y_n$ are solutions of $Ly = 0$ and the functions

$$z_i = \sum_{j=1}^{n} a_{ij} y_j, \quad 1 \le i \le n,$$

form a fundamental set of solutions of $Ly = 0$, then so do $y_1, y_2, \ldots, y_n$.

13. Prove: If

$$y = c_1 y_1 + c_2 y_2 + \cdots + c_k y_k + y_p$$

is a solution of a linear equation $Ly = F$ for every choice of the constants $c_1, c_2, \ldots, c_k$, then $Ly_i = 0$ for $1 \le i \le k$.

14. Suppose that the linear nth order equation $Ly = 0$ is normal on (a, b) and let x_0 be in (a, b). For $1 \le i \le n$, let y_i be the solution of the initial value problem

$$Ly_i = 0, \quad y_i^{(j)}(x_0) = \begin{cases} 0, & j \ne i - 1, \\ 1, & j = i - 1, \end{cases}$$

where x_0 is an arbitrary point in (a, b). Show that any solution of $Ly = 0$ on (a, b) can be written as

$$y = c_1 y_1 + c_2 y_2 + \cdots + c_n y_n,$$

with $c_j = y^{(j-1)}(x_0)$.

15. Suppose that $\{y_1, y_2, \ldots, y_n\}$ is a fundamental set of solutions of

$$P_0(x) y^{(n)} + P_1(x) y^{(n-1)} + \cdots + P_n(x) y = 0$$

on (a, b) and let

$$z_1 = a_{11} y_1 + a_{12} y_2 + \cdots + a_{1n} y_n$$
$$z_2 = a_{21} y_1 + a_{22} y_2 + \cdots + a_{2n} y_n$$
$$\vdots$$
$$z_n = a_{n1} y_1 + a_{n2} y_2 + \cdots + a_{nn} y_n,$$

where the $\{a_{ij}\}$ are constants. Show that $\{z_1, z_2, \ldots, z_n\}$ is a fundamental set of solutions of (A) if and only if the determinant

$$\begin{vmatrix} a_{11} & a_{12} & \cdots & a_{1n} \\ a_{21} & a_{22} & \cdots & a_{2n} \\ \vdots & \vdots & \ddots & \vdots \\ a_{n1} & a_{n2} & \cdots & a_{nn} \end{vmatrix}$$

is nonzero.

Hint: *The determinant of a product of $n \times n$ matrices equals the product of the determinants.*

16. Show that $\{y_1, y_2, \ldots, y_n\}$ is linearly dependent on (a, b) if and only if at least one of the functions $y_1, y_2, \ldots, y_n$ can be written as a linear combination of the others on (a, b).

Take the following as a hint in Exercises 17–19.

Hint: *By the definiton of determinant,*

$$\begin{vmatrix} a_{11} & a_{12} & \cdots & a_{1n} \\ a_{21} & a_{22} & \cdots & a_{2n} \\ \vdots & \vdots & \ddots & \vdots \\ a_{n1} & a_{n2} & \cdots & a_{nn} \end{vmatrix} = \sum \pm a_{1i_1}, a_{2i_2}, \ldots, a_{ni_n},$$

where the sum is over all permutations $(i_1, i_2, \ldots, i_n)$ of $(1, 2, \ldots, n)$ and the choice of $+$ or $-$ in each term depends only on the permutation associated with that term.

17. Prove: If

$$A(u_1, u_2, \ldots, u_n) = \begin{vmatrix} a_{11} & a_{12} & \cdots & a_{1n} \\ a_{21} & a_{22} & \cdots & a_{2n} \\ \vdots & \vdots & \ddots & \vdots \\ a_{n-1,1} & a_{n-1,2} & \cdots & a_{n-1,n} \\ u_1 & u_2 & \cdots & u_n \end{vmatrix},$$

then

$$A(u_1 + v_1, u_2 + v_2, \ldots, u_n + v_n) = A(u_1, u_2, \ldots, u_n) + A(v_1, v_2, \ldots, v_n).$$

18. Let

$$F = \begin{vmatrix} f_{11} & f_{12} & \cdots & f_{1n} \\ f_{21} & f_{22} & \cdots & f_{2n} \\ \vdots & \vdots & \ddots & \vdots \\ f_{n1} & f_{n2} & \cdots & f_{nn} \end{vmatrix},$$

where f_{ij} $(1 \le i, j \le n)$ is differentiable. Show that

$$F' = F_1 + F_2 + \cdots + F_n,$$

where F_i is the determinant obtained by differentiating just the ith row of F.

19. Use Exercise 18 to show that if W is the Wronskian of the n-times differentiable functions $y_1, y_2, \ldots, y_n$ then

$$W' = \begin{vmatrix} y_1 & y_2 & \cdots & y_n \\ y_1' & y_2' & \cdots & y_n' \\ \vdots & \vdots & \ddots & \vdots \\ y_1^{(n-2)} & y_2^{(n-2)} & \cdots & y_n^{(n-2)} \\ y_1^{(n)} & y_2^{(n)} & \cdots & y_n^{(n)} \end{vmatrix}.$$

20. Use Exercises 17 and 19 to show that if W is the Wronskian of solutions $\{y_1, y_2, \ldots, y_n\}$ of the normal equation

$$P_0(x)y^{(n)} + P_1(x)y^{(n-1)} + \cdots + P_n(x)y = 0, \tag{A}$$

then $W' = -P_1 W / P_0$. Derive Abel's formula (Eqn. (15)) from this.

Hint: *Use* (A) *to write* $y^{(n)}$ *in terms of* $y, y', \ldots, y^{(n-1)}$.

21. Prove Theorem 9.1.5.

22. Prove Theorem 9.1.6.

23. Show that if the Wronskian of the n-times continuously differentiable functions $\{y_1, y_2, \ldots, y_n\}$ has no zeros in (a, b), then the differential equation obtained by expanding the determinant

$$\begin{vmatrix} y & y_1 & y_2 & \cdots & y_n \\ y' & y_1' & y_2' & \cdots & y_n' \\ \vdots & \vdots & \vdots & \ddots & \vdots \\ y^{(n)} & y_1^{(n)} & y_2^{(n)} & \cdots & y_n^{(n)} \end{vmatrix} = 0$$

in cofactors of its first column is normal and has $\{y_1, y_2, \ldots, y_n\}$ as a fundamental set of solutions on (a, b).

24. Use the method suggested by Exercise 23 to find a linear homogeneous equation for which the given set of functions is a fundamental set of solutions on intervals on which the Wronskian of the set has no zeros.

(a) $\{x, x^2 - 1, x^2 + 1\}$

(b) $\{e^x, e^{-x}, x\}$

(c) $\{e^x, xe^{-x}, 1\}$

(d) $\{x, x^2, e^x\}$

(e) $\{x, x^2, 1/x\}$

(f) $\{x + 1, e^x, e^{3x}\}$

(g) $\{x, x^3, 1/x, 1/x^2\}$

(h) $\{x, x \ln x, 1/x, x^2\}$

(i) $\{e^x, e^{-x}, x, e^{2x}\}$

(j) $\{e^{2x}, e^{-2x}, 1, x^2\}$

9.2 Higher Order Constant Coefficient Homogeneous Equations

If $a_0, a_1, \ldots, a_n$ are constants and $a_0 \neq 0$, then

$$a_0 y^{(n)} + a_1 y^{(n-1)} + \cdots + a_n y = F(x)$$

is said to be a **constant coefficient equation.** In this section we consider the homogeneous constant coefficient equation

$$a_0 y^{(n)} + a_1 y^{(n-1)} + \cdots + a_n y = 0. \qquad (1)$$

Since (1) is normal on $(-\infty, \infty)$, the theorems in Section 9.1 all apply with $(a, b) = (-\infty, \infty)$.

As in Section 5.2, we call

$$p(r) = a_0 r^n + a_1 r^{n-1} + \cdots + a_n \qquad (2)$$

the **characteristic polynomial** of (1). We saw in Section 5.2 that when $n = 2$ the solutions of (1) are determined by the zeros of the characteristic polynomial. This is also true when $n > 2$, but the situation is more complicated in this case. Consequently, we take a different approach here than in Section 5.2.

If k is a positive integer, let D^k stand for the kth derivative operator; that is,

$$D^k y = y^{(k)}.$$

If

$$q(r) = b_0 r^m + b_1 r^{m-1} + \cdots + b_m$$

is an arbitrary polynomial then define the operator

$$q(D) = b_0 D^m + b_1 D^{m-1} + \cdots + b_m$$

such that

$$q(D)y = (b_0 D^m + b_1 D^{m-1} + \cdots + b_m)y = b_0 y^m + b_1 y^{(m-1)} + \cdots + b_m y$$

whenever y is a function with m derivatives. We call $q(D)$ a **polynomial operator.**

With p as in (2) we have

$$p(D) = a_0 D^n + a_1 D^{n-1} + \cdots + a_n,$$

so (1) can be written as $p(D)y = 0$. If r is a constant then

$$p(D)e^{rx} = (a_0 D^n e^{rx} + a_1 D^{n-1} e^{rx} + \cdots + a_n e^{rx})$$
$$= (a_0 r^n + a_1 r^{n-1} + \cdots + a_n)e^{rx};$$

that is,

$$p(D)(e^{rx}) = p(r)e^{rx}.$$

This shows that $y = e^{rx}$ is a solution of (1) if $p(r) = 0$. In the simplest case, where p has n distinct real zeros $r_1, r_2, \ldots, r_n$, this argument yields n solutions

$$y_1 = e^{r_1 x}, \qquad y_2 = e^{r_2 x}, \ldots, y_n = e^{r_n x}.$$

It can be shown (Exercise 39) that the Wronskian of $\{e^{r_1 x}, e^{r_2 x}, \ldots, e^{r_n x}\}$ is nonzero if $r_1, r_2, \ldots, r_n$ are distinct; hence, $\{e^{r_1 x}, e^{r_2 x}, \ldots, e^{r_n x}\}$ is a fundamental set of solutions of $p(D)y = 0$ in this case.

EXAMPLE 1 **(a)** Find the general solution of

$$y''' - 6y'' + 11y' - 6y = 0. \qquad (3)$$

(b) Solve the initial value problem

$$y''' - 6y'' + 11y' - 6y = 0, \qquad y(0) = 4, \qquad y'(0) = 5, \qquad y''(0) = 9. \quad (4)$$

Solution   **(a)** The characteristic polynomial of (3) is

$$p(r) = r^3 - 6r^2 + 11r - 6 = (r-1)(r-2)(r-3).$$

Therefore $\{e^x, e^{2x}, e^{3x}\}$ is a set of solutions of (3). It is a fundamental set, since its Wronskian is

$$W(x) = \begin{vmatrix} e^x & e^{2x} & e^{3x} \\ e^x & 2e^{2x} & 3e^{3x} \\ e^x & 4e^{2x} & 9e^{3x} \end{vmatrix} = e^{6x} \begin{vmatrix} 1 & 1 & 1 \\ 1 & 2 & 3 \\ 1 & 4 & 9 \end{vmatrix} = 2e^{6x} \neq 0.$$

Therefore the general solution of (3) is

$$y = c_1 e^x + c_2 e^{2x} + c_3 e^{3x}. \quad (5)$$

Solution   **(b)** We must determine c_1, c_2, and c_3 in (5) so that y satisfies the initial conditions in (4). Differentiating (5) twice yields

$$\begin{aligned} y' &= c_1 e^x + 2c_2 e^{2x} + 3c_3 e^{3x} \\ y'' &= c_1 e^x + 4c_2 e^{2x} + 9c_3 e^{3x}. \end{aligned} \quad (6)$$

Setting $x = 0$ in (5) and (6) and imposing the initial conditions yields

$$\begin{aligned} c_1 + c_2 + c_3 &= 4 \\ c_1 + 2c_2 + 3c_3 &= 5 \\ c_1 + 4c_2 + 9c_3 &= 9. \end{aligned}$$

The solution of this system is $c_1 = 4, c_2 = -1, c_3 = 1$. Therefore the solution of (4) is

$$y = 4e^x - e^{2x} + e^{3x}$$

(Figure 1). ■

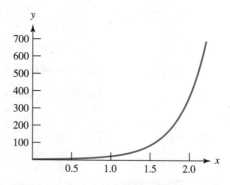

Figure 1   Graph of $y = 4e^x - e^{2x} + e^{3x}$

Now we consider the case where the characteristic polynomial (2) does not have n distinct real zeros. For this purpose it is useful to define what we mean by a factorization of a polynomial operator. We begin with an example.

EXAMPLE 2   Consider the polynomial

$$p(r) = r^3 - r^2 + r - 1$$

and the associated polynomial operator

$$p(D) = D^3 - D^2 + D - 1.$$

Since $p(r)$ can be factored as

$$p(r) = (r - 1)(r^2 + 1) = (r^2 + 1)(r - 1)$$

it is reasonable to expect that $p(D)$ can be factored as

$$p(D) = (D - 1)(D^2 + 1) = (D^2 + 1)(D - 1). \tag{7}$$

However, before we can make this assertion we must *define* what we mean by saying that two operators are equal, and what we mean by the products of operators in (7). We say that two operators are equal if they apply to the same functions and always produce the same result. The definitions of the products in (7) are as follows: if y is any three-times differentiable function then

(a) $(D - 1)(D^2 + 1)y$ is the function obtained by first applying $D^2 + 1$ to y and then applying $D - 1$ to the resulting function;
(b) $(D^2 + 1)(D - 1)y$ is the function obtained by first applying $D - 1$ to y and then applying $D^2 + 1$ to the resulting function.

From **(a)**,

$$\begin{aligned}
(D - 1)(D^2 + 1)y &= (D - 1)[(D^2 + 1)y] \\
&= (D - 1)(y'' + y) = D(y'' + y) - (y'' + y) \\
&= (y''' + y') - (y'' + y) \\
&= y''' - y'' + y' - y = (D^3 - D^2 + D - 1)y.
\end{aligned} \tag{8}$$

From this we conclude that

$$(D - 1)(D^2 + 1) = (D^3 - D^2 + D - 1).$$

From **(b)**,

$$\begin{aligned}
(D^2 + 1)(D - 1)y &= (D^2 + 1)[(D - 1)y] \\
&= (D^2 - 1)(y' - y) = D^2(y' - y) + (y' - y) \\
&= (y''' - y'') + (y' - y) \\
&= y''' - y'' + y' - y = (D^3 - D^2 + D - 1)y,
\end{aligned} \tag{9}$$

so

$$(D^2 + 1)(D - 1) = (D^3 - D^2 + D - 1),$$

which completes the justification of (7).

EXAMPLE 3 Use the result of Example 2 to find the general solution of

$$y''' - y'' + y' - y = 0. \tag{10}$$

Solution From (8), we can rewrite (10) as

$$(D - 1)(D^2 + 1)y = 0,$$

which implies that any solution of $(D^2 + 1)y = 0$ is a solution of (10). Therefore $y_1 = \cos x$ and $y_2 = \sin x$ are solutions of (10).

From (9), we can rewrite (10) as

$$(D^2 + 1)(D - 1)y = 0,$$

which implies that any solution of $(D - 1)y = 0$ is a solution of (10). Therefore $y_3 = e^x$ is solution of (10).

The Wronskian of $\{\cos x, \sin x, e^x\}$ is

$$W(x) = \begin{vmatrix} \cos x & \sin x & e^x \\ -\sin x & \cos x & e^x \\ -\cos x & -\sin x & e^x \end{vmatrix}.$$

Since

$$W(0) = \begin{vmatrix} 1 & 0 & 1 \\ 0 & 1 & 1 \\ -1 & 0 & 1 \end{vmatrix} = 2,$$

it follows that $\{\cos x, \sin x, e^x\}$ is linearly independent, so the general solution of (10) is

$$y = c_1 \cos x + c_2 \sin x + c_3 e^x.$$

EXAMPLE 4   Find the general solution of

$$y^{(4)} - 16y = 0. \tag{11}$$

Solution   The characteristic polynomial of (11) is

$$p(r) = r^4 - 16 = (r^2 - 4)(r^2 + 4) = (r - 2)(r + 2)(r^2 + 4).$$

By arguments similar to those used in Examples 2 and 3 it can be shown that (11) can be written as

$$(D^2 + 4)(D + 2)(D - 2)y = 0$$

or

$$(D^2 + 4)(D - 2)(D + 2)y = 0$$

or

$$(D - 2)(D + 2)(D^2 + 4)y = 0;$$

therefore, y is a solution of (11) if it is a solution of any of the three equations

$$(D - 2)y = 0, \qquad (D + 2)y = 0, \qquad (D^2 + 4)y = 0.$$

Hence, $\{e^{2x}, e^{-2x}, \cos 2x, \sin 2x\}$ is a set of solutions of (11). The Wronskian of this set is

$$W(x) = \begin{vmatrix} e^{2x} & e^{-2x} & \cos 2x & \sin 2x \\ 2e^{2x} & -2e^{-2x} & -2\sin 2x & 2\cos 2x \\ 4e^{2x} & 4e^{-2x} & -4\cos 2x & -4\sin 2x \\ 8e^{2x} & -8e^{-2x} & 8\sin 2x & -8\cos 2x \end{vmatrix}.$$

Since

$$W(0) = \begin{vmatrix} 1 & 1 & 1 & 0 \\ 2 & -2 & 0 & 2 \\ 4 & 4 & -4 & 0 \\ 8 & -8 & 0 & -8 \end{vmatrix} = -512,$$

it follows that $\{e^{2x}, e^{-2x}, \cos 2x, \sin 2x\}$ is linearly independent, so the general solution of (11) is

$$y_1 = c_1 e^{2x} + c_2 e^{-2x} + c_3 \cos 2x + c_4 \sin 2x. \qquad \blacksquare$$

It is known from algebra that every polynomial

$$p(r) = a_0 r^n + a_1 r^{n-1} + \cdots + a_n$$

with real coefficients can be factored as

$$p(r) = a_0 p_1(r) p_2(r) \cdots p_k(r),$$

where no pair of the polynomials $p_1, p_2, \ldots, p_k$ has a common factor and each is either of the form

$$p_j(r) = (r - r_j)^{m_j} \qquad (12)$$

where r_j is real and m_j is a positive integer, or

$$p_j(r) = [(r - \lambda_j)^2 + \omega_j^2]^{m_j} \qquad (13)$$

where λ_j and ω_j are real, $\omega_j \neq 0$, and m_j is a positive integer. If (12) holds then r_j is a real zero of p, while if (13) holds then $\lambda + i\omega$ and $\lambda - i\omega$ are complex conjugate zeros of p. In either case m_j is the **multiplicity** of the zero(s).

By arguments similar to those used in our examples, it can be shown that

$$p(D) = a_0 p_1(D) p_2(D) \cdots p_k(D) \qquad (14)$$

and that the order of the factors on the right can be chosen arbitrarily. It follows that if $p_j(D)y = 0$ for some j, then $p(D)y = 0$. To see this, we simply rewrite (14) so that $p_j(D)$ is applied first. Therefore the problem of finding solutions of $p(D)y = 0$ with p as in (14) reduces to finding solutions of each of the equations

$$p_j(D)y = 0, \qquad 1 \leq j \leq k,$$

where p_j is a power of a first degree term or of an irreducible quadratic. To find a fundamental set of solutions $\{y_1, y_2, \ldots, y_n\}$ of $p(D)y = 0$, we find a fundamental set of solutions of each of these equations and take $\{y_1, y_2, \ldots, y_n\}$ to be the set of all functions in these separate fundamental sets. In Exercise 40 we sketch the proof that $\{y_1, y_2, \ldots, y_n\}$ is linearly independent, and therefore a fundamental set of solutions of $p(D)y = 0$.

To apply this procedure to general homogeneous constant coefficient equations we must be able to find fundamental sets of solutions of equations of the form

$$(D - a)^m y = 0$$

and

$$[(D - \lambda)^2 + \omega^2]^m y = 0$$

where m is an arbitrary positive integer. The next two theorems show how to do this.

THEOREM 9.2.1

If m is a positive integer then

$$\{e^{ax}, xe^{ax}, \ldots, x^{m-1}e^{ax}\} \qquad (15)$$

is a fundamental set of solutions of

$$(D - a)^m y = 0. \qquad (16)$$

PROOF. We will show that if

$$f(x) = c_1 + c_2 x + \cdots + c_m x^{m-1}$$

is an arbitrary polynomial of degree $\leq m - 1$ then $y = e^{ax}f$ is a solution of (16). First note that if g is any differentiable function then

$$(D - a)e^{ax}g = De^{ax}g - ae^{ax}g = ae^{ax}g + e^{ax}g' - ae^{ax}g,$$

so

$$(D - a)e^{ax}g = e^{ax}g'. \tag{17}$$

Therefore

$$(D - a)e^{ax}f \; = e^{ax}f' \qquad\qquad \text{(from (17) with } g = f)$$
$$(D - a)^2 e^{ax}f = (D - a)e^{ax}f' = e^{ax}f'' \qquad \text{(from (17) with } g = f')$$
$$(D - a)^3 e^{ax}f = (D - a)e^{ax}f'' = e^{ax}f''' \qquad \text{(from (17) with } g = f'')$$
$$\vdots$$
$$(D - a)^m e^{ax}f = (D - a)e^{ax}f^{(m-1)} = e^{ax}f^{(m)} \quad \text{(from (17) with } g = f^{(m-1)}).$$

Since $f^{(m)} = 0$, the last equation implies that $y = e^{ax}f$ is a solution of (16) if f is any polynomial of degree $\leq m - 1$. In particular, each function in (15) is a solution of (16). To see that (15) is linearly independent (and therefore a fundamental set of solutions of (16)), note that if

$$c_1 e^{ax} + c_2 x e^{ax} + \cdots + c_{m-1} x^{m-1} e^{ax} = 0$$

for all x in some interval (a, b), then

$$c_1 + c_2 x + \cdots + c_{m-1} x^{m-1} = 0$$

for all x in (a, b). However, we know from algebra that if this polynomial has more than $m - 1$ zeros then $c_1 = c_2 = \cdots = c_n = 0$. $\qquad \square$

EXAMPLE 5   Find the general solution of

$$y''' + 3y'' + 3y' + y = 0. \tag{18}$$

Solution   The characteristic polynomial of (18) is

$$p(r) = r^3 + 3r^2 + 3r + 1 = (r + 1)^3.$$

Therefore (18) can be written as

$$(D + 1)^3 y = 0,$$

so Theorem 9.2.1 implies that the general solution of (18) is

$$y = e^{-x}(c_1 + c_2 x + c_3 x^2). \qquad \blacksquare$$

The proof of the following theorem is sketched in Exercise 41.

THEOREM 9.2.2

If $\omega \neq 0$ and m is a positive integer, then

$$\{e^{\lambda x} \cos \omega x, x e^{\lambda x} \cos \omega x, \ldots, x^{m-1} e^{\lambda x} \cos \omega x,$$
$$e^{\lambda x} \sin \omega x, x e^{\lambda x} \sin \omega x, \ldots, x^{m-1} e^{\lambda x} \sin \omega x\}$$

is a fundamental set of solutions of

$$[(D - \lambda)^2 + \omega^2]^m y = 0.$$

EXAMPLE 6 Find the general solution of

$$(D^2 + 4D + 13)^3 y = 0. \tag{19}$$

Solution The characteristic polynomial of (19) is

$$p(r) = (r^2 + 4r + 13)^3 = ((r + 2)^2 + 9)^3.$$

Therefore (19) can be be written as

$$[(D + 2)^2 + 9]^3 y = 0,$$

so Theorem 9.2.2 implies that the general solution of (19) is

$$y = (a_1 + a_2 x + a_3 x^2)e^{-2x} \cos 3x + (b_1 + b_2 x + b_3 x^2)e^{-2x} \sin 3x.$$

EXAMPLE 7 Find the general solution of

$$y^{(4)} + 4y''' + 6y'' + 4y' = 0. \tag{20}$$

Solution The characteristic polynomial of (20) is

$$\begin{aligned}
p(r) &= r^4 + 4r^3 + 6r^2 + 4r \\
&= r(r^3 + 4r^2 + 6r + 4) \\
&= r(r + 2)(r^2 + 2r + 2) \\
&= r(r + 2)[(r + 1)^2 + 1].
\end{aligned}$$

Therefore (20) can be written as

$$[(D + 1)^2 + 1](D + 2)\, Dy = 0.$$

Fundamental sets of solutions of

$$[(D + 1)^2 + 1]y = 0, \qquad (D + 2)y = 0, \qquad \text{and} \qquad Dy = 0$$

are given by

$$\{e^{-x} \cos x, e^{-x} \sin x\}, \qquad \{e^{-2x}\}, \qquad \text{and} \qquad \{1\},$$

respectively. Therefore the general solution of (20) is

$$y = e^{-x}(c_1 \cos x + c_2 \sin x) + c_3 e^{-2x} + c_4.$$

EXAMPLE 8 Find a fundamental set of solutions of

$$[(D + 1)^2 + 1]^2(D - 1)^3(D + 1)D^2 y = 0. \tag{21}$$

Solution A fundamental set of solutions of (21) can be obtained by combining fundamental sets of solutions of

$$[(D + 1)^2 + 1]^2 y = 0, \qquad (D - 1)^3 y = 0, \qquad (D + 1)y = 0, \qquad \text{and} \qquad D^2 y = 0.$$

Fundamental sets of solutions of these equations are given by

$$\{e^{-x} \cos x, xe^{-x} \cos x, e^{-x} \sin x, xe^{-x} \sin x\}, \quad \{e^x, xe^x, x^2 e^x\}, \quad \{e^{-x}\}, \quad \text{and} \quad \{1, x\},$$

respectively. These ten functions form a fundamental set of solutions of (21). ■

9.2 EXERCISES

In Exercises 1–14 find the general solution.

1. $y''' - 3y'' + 3y' - y = 0$

2. $y^{(4)} + 8y'' - 9y = 0$

3. $y''' - y'' + 16y' - 16y = 0$

4. $2y''' + 3y'' - 2y' - 3y = 0$

5. $y''' + 5y'' + 9y' + 5y = 0$

6. $4y''' - 8y'' + 5y' - y = 0$

7. $27y''' + 27y'' + 9y' + y = 0$

8. $y^{(4)} + y'' = 0$

9. $y^{(4)} - 16y = 0$

10. $y^{(4)} + 12y'' + 36y = 0$

11. $16y^{(4)} - 72y'' + 81y = 0$

12. $6y^{(4)} + 5y''' + 7y'' + 5y' + y = 0$

13. $4y^{(4)} + 12y''' + 3y'' - 13y' - 6y = 0$

14. $y^{(4)} - 4y''' + 7y'' - 6y' + 2y = 0$

In Exercises 15–27 solve the initial value problem. Where indicated by $\boxed{C}$, graph the solution.

15. $y''' - 2y'' + 4y' - 8y = 0,\quad y(0) = 2,\quad y'(0) = -2,\quad y''(0) = 0$

16. $y''' + 3y'' - y' - 3y = 0,\quad y(0) = 0,\quad y'(0) = 14,\quad y''(0) = -40$

$\boxed{C}$ **17.** $y''' - y'' - y' + y = 0,\quad y(0) = -2,\quad y'(0) = 9,\quad y''(0) = 4$

$\boxed{C}$ **18.** $y''' - 2y' - 4y = 0,\quad y(0) = 6,\quad y'(0) = 3,\quad y''(0) = 22$

$\boxed{C}$ **19.** $3y''' - y'' - 7y' + 5y = 0,\quad y(0) = \dfrac{14}{5},\quad y'(0) = 0,\quad y''(0) = 10$

20. $y''' - 6y'' + 12y' - 8y = 0,\quad y(0) = 1,\quad y'(0) = -1,\quad y''(0) = -4$

21. $2y''' - 11y'' + 12y' + 9y = 0,\quad y(0) = 6,\quad y'(0) = 3,\quad y''(0) = 13$

22. $8y''' - 4y'' - 2y' + y = 0,\quad y(0) = 4,\quad y'(0) = -3,\quad y''(0) = -1$

23. $y^{(4)} - 16y = 0,\quad y(0) = 2,\quad y'(0) = 2,\quad y''(0) = -2,\quad y'''(0) = 0$

24. $y^{(4)} - 6y''' + 7y'' + 6y' - 8y = 0,\quad y(0) = -2,\quad y'(0) = -8,\quad y''(0) = -14,\quad y'''(0) = -62$

25. $4y^{(4)} - 13y'' + 9y = 0,\quad y(0) = 1,\quad y'(0) = 3,\quad y''(0) = 1,\quad y'''(0) = 3$

26. $y^{(4)} + 2y''' - 2y'' - 8y' - 8y = 0,\quad y(0) = 5,\quad y'(0) = -2,\quad y''(0) = 6,\quad y'''(0) = 8$

$\boxed{C}$ **27.** $4y^{(4)} + 8y''' + 19y'' + 32y' + 12y = 0,\quad y(0) = 3,\quad y'(0) = -3,\quad y''(0) = -\dfrac{7}{2},\quad y'''(0) = \dfrac{31}{4}$

28. Find a fundamental set of solutions of the given equation, and verify that it is a fundamental set by evaluating its Wronskian at $x = 0$.

    (a) $(D - 1)^2(D - 2)y = 0$

    (b) $(D^2 + 4)(D - 3)y = 0$

    (c) $(D^2 + 2D + 2)(D - 1)y = 0$

    (d) $D^3(D - 1)y = 0$

    (e) $(D^2 - 1)(D^2 + 1)y = 0$

    (f) $(D^2 - 2D + 2)(D^2 + 1)y = 0$

In Exercises 29–38 find a fundamental set of solutions of the given equation.

29. $(D^2 + 6D + 13)(D - 2)^2 D^3 y = 0$

30. $(D - 1)^2(2D - 1)^3(D^2 + 1)y = 0$

31. $(D^2 + 9)^3 D^2 y = 0$

32. $(D - 2)^3(D + 1)^2 Dy = 0$

33. $(D^2 + 1)(D^2 + 9)^2(D - 2)y = 0$

34. $(D^4 - 16)^2 y = 0$

35. $(4D^2 + 4D + 9)^3 y = 0$

36. $D^3(D - 2)^2(D^2 + 4)^2 y = 0$

37. $(4D^2 + 1)^2(9D^2 + 4)^3 y = 0$

38. $[(D - 1)^4 - 16]y = 0$

39. It can be shown that

$$\begin{vmatrix} 1 & 1 & \cdots & 1 \\ a_1 & a_2 & \cdots & a_n \\ a_1^2 & a_2^2 & \cdots & a_n^2 \\ \vdots & \vdots & \ddots & \vdots \\ a_1^{n-1} & a_2^{n-1} & \cdots & a_n^{n-1} \end{vmatrix} = \prod_{1 \le i < j \le n} (a_j - a_i), \tag{A}$$

where the left side is called the ***Vandermonde determinant*** and the right side is the product of all factors of the form $(a_j - a_i)$ with i and j between 1 and n and $i < j$.

(a) Verify (A) for $n = 2$ and $n = 3$.

(b) Find the Wronskian of $\{e^{a_1 x}, e^{a_2 x}, \ldots, e^{a_n x}\}$.

40. A theorem from algebra says that if P_1 and P_2 are polynomials with no common factors, then there are polynomials Q_1 and Q_2 such that

$$Q_1 P_1 + Q_2 P_2 = 1.$$

This implies that

$$Q_1(D)P_1(D)y + Q_2(D)P_2(D)y = y$$

for every function y with enough derivatives for the left side to be defined.

(a) Use this to show that if P_1 and P_2 have no common factors and

$$P_1(D)y = P_2(D)y = 0$$

then $y = 0$.

(b) Suppose that P_1 and P_2 are polynomials with no common factors. Let $u_1, \ldots, u_r$ be linearly independent solutions of $P_1(D)y = 0$ and let $v_1, \ldots, v_s$ be linearly independent solutions of $P_2(D)y = 0$. Use **(a)** to show that $\{u_1, \ldots, u_r, v_1, \ldots, v_s\}$ is a linearly independent set.

(c) Suppose that the characteristic polynomial of the constant coefficient equation

$$a_0 y^{(n)} + a_1 y^{(n-1)} + \cdots + a_n y = 0 \tag{A}$$

has the factorization

$$p(r) = a_0 p_1(r) p_2(r) \cdots p_k(r),$$

where each p_j is of the form

$$p_j(r) = (r - r_j)^{n_j} \quad \text{or} \quad p_j(r) = [(r - \lambda_j)^2 + \omega_j^2]^{m_j} \quad (\omega_j > 0)$$

and no two of the polynomials $p_1, p_2, \ldots, p_k$ have a common factor. Show that we can find a fundamental set of solutions $\{y_1, y_2, \ldots, y_n\}$ of (A) by finding a fundamental set of solutions of each of the equations

$$p_j(D)y = 0, \quad 1 \le j \le k,$$

and taking $\{y_1, y_2, \ldots, y_n\}$ to be the set of all functions in these separate fundamental sets.

41. (a) Show that if

$$z = p(x) \cos \omega x + q(x) \sin \omega x \tag{A}$$

where p and q are polynomials of degree $\le k$, then

$$(D^2 + \omega^2)z = p_1(x) \cos \omega x + q_1(x) \sin \omega x,$$

where p_1 and q_1 are polynomials of degree $\le k - 1$.

(b) Apply **(a)** m times to show that if z is of the form (A) where p and q are polynomials of degree $\le m - 1$ then

$$(D^2 + \omega^2)^m z = 0. \tag{B}$$

(c) Use Eqn. (17) to show that if $y = e^{\lambda x} z$ then

$$[(D - \lambda)^2 + \omega^2]^m y = e^{\lambda x}(D^2 + \omega^2)^m z.$$

(d) Conclude from **(b)** and **(c)** that if p and q are arbitrary polynomials of degree $\le m - 1$ then

$$y = e^{\lambda x}(p(x) \cos \omega x + q(x) \sin \omega x)$$

is a solution of

$$[(D - \lambda)^2 + \omega^2]^m y = 0. \tag{C}$$

Conclude from **(d)** that the functions

$$e^{\lambda x} \cos \omega x, xe^{\lambda x} \cos \omega x, \ldots, x^{m-1}e^{\lambda x} \cos \omega x,$$

$$e^{\lambda x} \sin \omega x, xe^{\lambda x} \sin \omega x, \ldots, x^{m-1}e^{\lambda x} \sin \omega x \tag{D}$$

are all solutions of (C).

(f) Complete the proof of Theorem 9.2.2 by showing that the functions in (D) are linearly independent.

42. (a) Use the trigonometric identities

$$\cos(A + B) = \cos A \cos B - \sin A \sin B$$

$$\sin(A + B) = \cos A \sin B + \sin A \cos B$$

to show that

$$(\cos A + i \sin A)(\cos B + i \sin B) = \cos(A + B) + i \sin(A + B).$$

(b) Apply **(a)** repeatedly to show that if n is a positive integer then

$$\prod_{k=1}^{n} (\cos A_k + i \sin A_k) = \cos(A_1 + A_2 + \cdots + A_n) + i \sin(A_1 + A_2 + \cdots + A_n).$$

(c) Infer from **(b)** that if n is a positive integer then

$$(\cos \theta + i \sin \theta)^n = \cos n\theta + i \sin n\theta. \tag{A}$$

(d) Show that (A) also holds if $n = 0$ or a negative integer.

Hint: Verify by direct calculation that

$$(\cos \theta + i \sin \theta)^{-1} = (\cos \theta - i \sin \theta).$$

Then replace θ by $-\theta$ in (A).

(e) Now suppose that n is a positive integer. Infer from (A) that if

$$z_k = \cos\left(\frac{2k\pi}{n}\right) + i \sin\left(\frac{2k\pi}{n}\right), \qquad k = 0, 1, \ldots, n - 1$$

and

$$\zeta_k = \cos\left(\frac{(2k + 1)\pi}{n}\right) + i \sin\left(\frac{(2k + 1)\pi}{n}\right), \qquad k = 0, 1, \ldots, n - 1,$$

then

$$z_k^n = 1 \quad \text{and} \quad \zeta_k^n = -1, \quad k = 0, 1, \ldots, n - 1.$$

(Why don't we also consider other integer values for k ?)

(f) Let ρ be a positive number. Use **(e)** to show that

$$z^n - \rho = (z - \rho^{1/n}z_0)(z - \rho^{1/n}z_1) \cdots (z - \rho^{1/n}z_{n-1})$$

and

$$z^n + \rho = (z - \rho^{1/n}\zeta_0)(z - \rho^{1/n}\zeta_1) \cdots (z - \rho^{1/n}\zeta_{n-1}).$$

43. Use **(e)** of Exercise 42 to find a fundamental set of solutions of the given equation.

(a) $y''' - y = 0$ **(b)** $y''' + y = 0$

(c) $y^{(4)} + 64y = 0$ **(d)** $y^{(6)} - y = 0$

(e) $y^{(6)} + 64y = 0$ **(f)** $[(D - 1)^6 - 1]y = 0$

(g) $y^{(5)} + y^{(4)} + y''' + y'' + y' + y = 0$

44. An equation of the form

$$a_0 x^n y^{(n)} + a_1 x^{n-1} y^{(n-1)} + \cdots + a_{n-1} xy' + a_n y = 0, \qquad x > 0, \tag{A}$$

where $a_0, a_1, \ldots, a_n$ are constants, is called an **Euler** or **equidimensional** equation.

Show that if

$$x = e^t \qquad \text{and} \qquad Y(t) = y(x(t)), \tag{B}$$

then

$$x \frac{dy}{dx} = \frac{dY}{dt}$$

$$x^2 \frac{d^2 y}{dx^2} = \frac{d^2 Y}{dt^2} - \frac{dY}{dt}$$

$$x^3 \frac{d^3 y}{dx^3} = \frac{d^3 Y}{dt^3} - 3\frac{d^2 Y}{dt^2} + 2\frac{dY}{dt}.$$

In general it can be shown that if r is any integer ≥ 2 then

$$x^r \frac{d^r y}{dx^r} = \frac{d^r Y}{dt^r} + A_{1r} \frac{d^{r-1} Y}{dt^{r-1}} + \cdots + A_{r-1,r} \frac{dY}{dt}$$

where $A_{1r}, \ldots, A_{r-1,r}$ are integers. Use these results to show that the substitution (B) transforms (A) into a constant coefficient equation for Y as a function of t.

45. Use Exercise 44 to show that a function $y = y(x)$ satisfies the equation

$$a_0 x^3 y''' + a_1 x^2 y'' + a_2 xy' + a_3 y = 0 \tag{A}$$

on $(0, \infty)$ if and only if the function $Y(t) = y(e^t)$ satisfies

$$a_0 \frac{d^3 Y}{dt^3} + (a_1 - 3a_0) \frac{d^2 Y}{dt^2} + (a_2 - a_1 + 2a_0) \frac{dY}{dt} + a_3 Y = 0.$$

Assuming that a_0, a_1, a_2, a_3 are real and $a_0 \neq 0$, find the possible forms for the general solution of (A).

9.3   Undetermined Coefficients for Higher Order Equations

In this section we consider the constant coefficient equation

$$a_0 y^{(n)} + a_1 y^{(n-1)} + \cdots + a_n y = F(x) \tag{1}$$

where $n \geq 3$ and F is a linear combination of functions of the form

$$e^{\alpha x}(p_0 + p_1 x + \cdots + p_k x^k)$$

or

$$e^{\lambda x}[(p_0 + p_1 x + \cdots + p_k x^k) \cos \omega x + (q_0 + q_1 x + \cdots + q_k x^k) \sin \omega x].$$

From Theorem 9.1.5 the general solution of (1) is $y = y_p + y_c$, where y_p is a particular solution of (1) and y_c is the general solution of the complementary equation

$$a_0 y^{(n)} + a_1 y^{(n-1)} + \cdots + a_n y = 0.$$

In Section 9.2 we learned how to find y_c. Here we will learn how to find y_p when the forcing function has the form stated above. The procedure that we will use is a generalization of the method that we used in Sections 5.4 and 5.5, and is called the ***method of undetermined coefficients,*** as it was there. Since the underlying ideas are the same as those considered in those two sections, we'll confine ourselves to an informal presentation based on examples.

FORCING FUNCTIONS OF THE FORM $e^{\alpha x}(p_0 + p_1 x + \cdots + p_k x^k)$

We first consider equations of the form

$$a_0 y^{(n)} + a_1 y^{(n-1)} + \cdots + a_n y = e^{\alpha x}(p_0 + p_1 x + \cdots + p_k x^k).$$

EXAMPLE 1   Find a particular solution of

$$y''' + 3y'' + 2y' - y = e^x(21 + 24x + 28x^2 + 5x^3). \tag{2}$$

Solution   Substituting

$$y = ue^x,$$
$$y' = e^x(u' + u),$$
$$y'' = e^x(u'' + 2u' + u),$$
$$y''' = e^x(u''' + 3u'' + 3u' + u)$$

into (2) and canceling e^x yields

$$(u''' + 3u'' + 3u' + u) + 3(u'' + 2u' + u) + 2(u' + u) - u$$
$$= 21 + 24x + 28x^2 + 5x^3,$$

or

$$u''' + 6u'' + 11u' + 5u = 21 + 24x + 28x^2 + 5x^3. \tag{3}$$

Since the unknown u appears on the left, we can see that (3) has a particular solution of the form

$$u_p = A + Bx + Cx^2 + Dx^3.$$

Then

$$u_p' = B + 2Cx + 3Dx^2$$
$$u_p'' = 2C + 6Dx$$
$$u_p''' = 6D.$$

Substituting from the last four equations into the left side of (3) yields

$$u_p''' + 6u_p'' + 11u_p' + 5u_p = 6D + 6(2C + 6Dx) + 11(B + 2Cx + 3Dx^2)$$
$$+ 5(A + Bx + Cx^2 + Dx^3)$$
$$= (5A + 11B + 12C + 6D) + (5B + 22C + 36D)x$$
$$+ (5C + 33D)x^2 + 5Dx^3.$$

Comparing coefficients of like powers of x on the right sides of this equation and (3) shows that u_p satisfies (3) if

$$5D = 5$$
$$5C + 33D = 28$$
$$5B + 22C + 36D = 24$$
$$5A + 11B + 12C + 6D = 21.$$

Solving these equations successively yields $D = 1$, $C = -1$, $B = 2$, $A = 1$. Therefore

$$u_p = 1 + 2x - x^2 + x^3$$

is a particular solution of (3), so

$$y_p = e^x u_p = e^x(1 + 2x - x^2 + x^3)$$

is a particular solution of (2). (See Figure 1.)

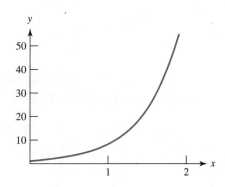

Figure 1   Graph of
$$y_p = e^x(1 + 2x - x^2 + x^3)$$

EXAMPLE 2   Find a particular solution of

$$y^{(4)} - y''' - 6y'' + 4y' + 8y = e^{2x}(4 + 19x + 6x^2). \qquad (4)$$

Solution   Substituting

$$\begin{aligned}
y &= ue^{2x}, \\
y' &= e^{2x}(u' + 2u), \\
y'' &= e^{2x}(u'' + 4u' + 4u), \\
y''' &= e^{2x}(u''' + 6u'' + 12u' + 8u), \\
y^{(4)} &= e^{2x}(u^{(4)} + 8u''' + 24u'' + 32u' + 16u)
\end{aligned}$$

into (4) and canceling e^{2x} yields

$$(u^{(4)} + 8u''' + 24u'' + 32u' + 16u) - (u''' + 6u'' + 12u' + 8u)$$
$$-6(u'' + 4u' + 4u) + 4(u' + 2u) + 8u = 4 + 19x + 6x^2,$$

or

$$u^{(4)} + 7u''' + 12u'' = 4 + 19x + 6x^2. \qquad (5)$$

Since neither u nor u' appear on the left, we can see that (5) has a particular solution of the form

$$u_p = Ax^2 + Bx^3 + Cx^4. \qquad (6)$$

Then

$$\begin{aligned}
u'_p &= 2Ax + 3Bx^2 + 4Cx^3 \\
u''_p &= 2A + 6Bx + 12Cx^2
\end{aligned}$$

$$u_p''' = 6B + 24Cx$$
$$u_p^{(4)} = 24C.$$

Substituting u_p'', u_p''', and $u_p^{(4)}$ into the left side of (5) yields

$$u_p^{(4)} + 7u_p''' + 12u_p'' = 24C + 7(6B + 24Cx) + 12(2A + 6Bx + 12Cx^2)$$
$$= (24A + 42B + 24C) + (72B + 168C)x + 144Cx^2.$$

Comparing coefficients of like powers of x on the right sides of this equation and (5) shows that u_p satisfies (5) if

$$144C = 6$$
$$72B + 168C = 19$$
$$24A + 42B + 24C = 4.$$

Solving these equations successively yields $C = 1/24, B = 1/6, A = -1/6$. Substituting these into (6) shows that

$$u_p = \frac{x^2}{24}(-4 + 4x + x^2)$$

is a particular solution of (5), so

$$y_p = e^{2x}u_p = \frac{x^2 e^{2x}}{24}(-4 + 4x + x^2)$$

is a particular solution of (4). (See Figure 2.) ■

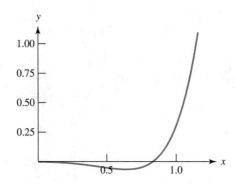

Figure 2   Graph of
$$y_p = \frac{x^2 e^{2x}}{24}(-4 + 4x + x^2)$$

FORCING FUNCTIONS OF THE FORM $e^{\lambda x}(P(x)\cos \omega x + Q(x)\sin \omega x)$

We now consider equations of the form

$$a_0 y^{(n)} + a_1 y^{(n-1)} + \cdots + a_n y = e^{\lambda x}(P(x)\cos \omega x + Q(x)\sin \omega x)$$

where P and Q are polynomials.

EXAMPLE 3   Find a particular solution of

$$y''' + y'' - 4y' - 4y = e^x[(5 - 5x)\cos x + (2 + 5x)\sin x]. \tag{7}$$

Solution   Substituting

$$y = ue^x,$$

$$y' = e^x(u' + u),$$
$$y'' = e^x(u'' + 2u' + u),$$
$$y''' = e^x(u''' + 3u'' + 3u' + u)$$

into (7) and canceling e^x yields

$$(u''' + 3u'' + 3u' + u) + (u'' + 2u' + u) - 4(u' + u) - 4u$$
$$= (5 - 5x) \cos x + (2 + 5x) \sin x,$$

or

$$u''' + 4u'' + u' - 6u = (5 - 5x) \cos x + (2 + 5x) \sin x. \tag{8}$$

Since $\cos x$ and $\sin x$ are not solutions of the complementary equation

$$u''' + 4u'' + u' - 6u = 0,$$

a theorem analogous to Theorem 5.5.1 implies that (8) has a particular solution of the form

$$u_p = (A_0 + A_1 x) \cos x + (B_0 + B_1 x) \sin x. \tag{9}$$

Then

$$u_p' = (A_1 + B_0 + B_1 x) \cos x + (B_1 - A_0 - A_1 x) \sin x,$$
$$u_p'' = (2B_1 - A_0 - A_1 x) \cos x - (2A_1 + B_0 + B_1 x) \sin x,$$
$$u_p''' = -(3A_1 + B_0 + B_1 x) \cos x - (3B_1 - A_0 - A_1 x) \sin x,$$

so

$$u_p''' + 4u_p'' + u_p' - 6u_p = -[10A_0 + 2A_1 - 8B_1 + 10A_1 x] \cos x$$
$$- [10B_0 + 2B_1 + 8A_1 + 10B_1 x] \sin x.$$

Comparing the coefficients of $x \cos x, x \sin x, \cos x$, and $\sin x$ here with the corresponding coefficients in (8) shows that u_p is a solution of (8) if

$$-10A_1 = -5$$
$$-10B_1 = 5$$
$$-10A_0 - 2A_1 + 8B_1 = 5$$
$$-10B_0 - 2B_1 - 8A_1 = 2.$$

Solving the first two equations yields $A_1 = 1/2, B_1 = -1/2$. Substituting these into the last two equations yields

$$-10A_0 = 5 + 2A_1 - 8B_1 = 10$$
$$-10B_0 = 2 + 2B_1 + 8A_1 = 5,$$

so $A_0 = -1, B_0 = -1/2$. Substituting $A_0 = -1, A_1 = 1/2, B_0 = -1/2$, and $B_1 = -1/2$ into (9) shows that

$$u_p = -\frac{1}{2}[(2 - x) \cos x + (1 + x) \sin x]$$

is a particular solution of (8), so

$$y_p = e^x u_p = -\frac{e^x}{2}[(2 - x) \cos x + (1 + x) \sin x]$$

is a particular solution of (7). (See Figure 3.)

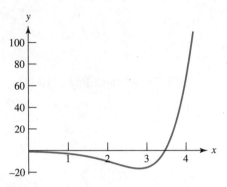

Figure 3 Graph of
$$y_p = -\frac{e^x}{2}[(2 - x)\cos x + (1 + x)\sin x]$$

EXAMPLE 4 Find a particular solution of
$$y''' + 4y'' + 6y' + 4y = e^{-x}[(1 - 6x)\cos x - (3 + 2x)\sin x]. \tag{10}$$

Solution Substituting
$$y = ue^{-x},$$
$$y' = e^{-x}(u' - u),$$
$$y'' = e^{-x}(u'' - 2u' + u),$$
$$y''' = e^{-x}(u''' - 3u'' + 3u' - u)$$

into (10) and canceling e^{-x} yields
$$(u''' - 3u'' + 3u' - u) + 4(u'' - 2u' + u) + 6(u' - u) + 4u$$
$$= (1 - 6x)\cos x - (3 + 2x)\sin x,$$

or
$$u''' + u'' + u' + u = (1 - 6x)\cos x - (3 + 2x)\sin x. \tag{11}$$

Since $\cos x$ and $\sin x$ are solutions of the complementary equation
$$u''' + u'' + u' + u = 0,$$

a theorem analogous to Theorem 5.5.1 implies that (11) has a particular solution of the form
$$u_p = (A_0x + A_1x^2)\cos x + (B_0x + B_1x^2)\sin x. \tag{12}$$

Then
$$u_p' = [A_0 + (2A_1 + B_0)x + B_1x^2]\cos x + [B_0 + (2B_1 - A_0)x - A_1x^2]\sin x,$$
$$u_p'' = [2A_1 + 2B_0 - (A_0 - 4B_1)x - A_1x^2]\cos x$$
$$\quad + [2B_1 - 2A_0 - (B_0 + 4A_1)x - B_1x^2]\sin x,$$
$$u_p''' = -[3A_0 - 6B_1 + (6A_1 + B_0)x + B_1x^2]\cos x$$
$$\quad - [3B_0 + 6A_1 + (6B_1 - A_0)x - A_1x^2]\sin x,$$

so
$$u_p''' + u_p'' + u_p' + u_p = -[2A_0 - 2B_0 - 2A_1 - 6B_1 + (4A_1 - 4B_1)x]\cos x$$
$$\quad - [2B_0 + 2A_0 - 2B_1 + 6A_1 + (4B_1 + 4A_1)x]\sin x.$$

Comparing the coefficients of $x \cos x, x \sin x, \cos x$, and $\sin x$ here with the corresponding coefficients in (11) shows that u_p is a solution of (11) if

$$-4A_1 + 4B_1 = -6$$
$$-4A_1 - 4B_1 = -2$$
$$-2A_0 + 2B_0 + 2A_1 + 6B_1 = 1$$
$$-2A_0 - 2B_0 - 6A_1 + 2B_1 = -3.$$

Solving the first two equations yields $A_1 = 1, B_1 = -1/2$. Substituting these into the last two equations yields

$$-2A_0 + 2B_0 = 1 - 2A_1 - 6B_1 = 2$$
$$-2A_0 - 2B_0 = -3 + 6A_1 - 2B_1 = 4,$$

so $A_0 = -3/2, B_0 = -1/2$. Substituting $A_0 = -3/2, A_1 = 1, B_0 = -1/2, B_1 = -1/2$ into (12) shows that

$$u_p = -\frac{x}{2}[(3 - 2x) \cos x + (1 + x) \sin x]$$

is a particular solution of (11), so

$$y_p = e^{-x}u_p = -\frac{xe^{-x}}{2}[(3 - 2x) \cos x + (1 + x) \sin x]$$

is a particular solution of (10). (See Figure 4.) ∎

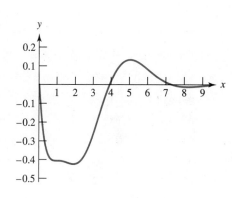

Figure 4 Graph of
$$y_p = -\frac{xe^{-x}}{2}[(3 - 2x) \cos x + (1 + x) \sin x]$$

9.3 EXERCISES

In Exercises 1–59 find a particular solution.

1. $y''' - 6y'' + 11y' - 6y = -e^{-x}(4 + 76x - 24x^2)$

2. $y''' - 2y'' - 5y' + 6y = e^{-3x}(32 - 23x + 6x^2)$

3. $4y''' + 8y'' - y' - 2y = -e^x(4 + 45x + 9x^2)$

4. $y''' + 3y'' - y' - 3y = e^{-2x}(2 - 17x + 3x^2)$

5. $y''' + 3y'' - y' - 3y = e^x(-1 + 2x + 24x^2 + 16x^3)$

6. $y''' + y'' - 2y = e^x(14 + 34x + 15x^2)$

7. $4y''' + 8y'' - y' - 2y = -e^{-2x}(1 - 15x)$

8. $y''' - y'' - y' + y = e^x(7 + 6x)$

9. $2y''' - 7y'' + 4y' + 4y = e^{2x}(17 + 30x)$

10. $y''' - 5y'' + 3y' + 9y = 2e^{3x}(11 - 24x^2)$

11. $y''' - 7y'' + 8y' + 16y = 2e^{4x}(13 + 15x)$

12. $8y''' - 12y'' + 6y' - y = e^{x/2}(1 + 4x)$

13. $y^{(4)} + 3y''' - 3y'' - 7y' + 6y = -e^{-x}(12 + 8x - 8x^2)$

14. $y^{(4)} + 3y''' + y'' - 3y' - 2y = -3e^{2x}(11 + 12x)$

15. $y^{(4)} + 8y''' + 24y'' + 32y' = -16e^{-2x}(1 + x + x^2 - x^3)$

16. $4y^{(4)} - 11y'' - 9y' - 2y = -e^x(1 - 6x)$ **17.** $y^{(4)} - 2y''' + 3y' - y = e^x(3 + 4x + x^2)$

18. $y^{(4)} - 4y''' + 6y'' - 4y' + 2y = e^x(24 + x + x^4)$ **19.** $2y^{(4)} + 5y''' - 5y' - 2y = 18e^x(5 + 2x)$

20. $y^{(4)} + y''' - 2y'' - 6y' - 4y = -e^{2x}(4 + 28x + 15x^2)$

21. $2y^{(4)} + y''' - 2y' - y = 3e^{-x/2}(1 - 6x)$ **22.** $y^{(4)} - 5y'' + 4y = e^x(3 + x - 3x^2)$

23. $y^{(4)} - 2y''' - 3y'' + 4y' + 4y = e^{2x}(13 + 33x + 18x^2)$

24. $y^{(4)} - 3y''' + 4y' = e^{2x}(15 + 26x + 12x^2)$ **25.** $y^{(4)} - 2y''' + 2y' - y = e^x(1 + x)$

26. $2y^{(4)} - 5y''' + 3y'' + y' - y = e^x(11 + 12x)$ **27.** $y^{(4)} + 3y''' + 3y'' + y' = e^{-x}(5 - 24x + 10x^2)$

28. $y^{(4)} - 7y''' + 18y'' - 20y' + 8y = e^{2x}(3 - 8x - 5x^2)$

29. $y''' - y'' - 4y' + 4y = e^{-x}[(16 + 10x)\cos x + (30 - 10x)\sin x]$

30. $y''' + y'' - 4y' - 4y = e^{-x}[(1 - 22x)\cos 2x - (1 + 6x)\sin 2x]$

31. $y''' - y'' + 2y' - 2y = e^{2x}[(27 + 5x - x^2)\cos x + (2 + 13x + 9x^2)\sin x]$

32. $y''' - 2y'' + y' - 2y = -e^x[(9 - 5x + 4x^2)\cos 2x - (6 - 5x - 3x^2)\sin 2x]$

33. $y''' + 3y'' + 4y' + 12y = 8\cos 2x - 16\sin 2x$

34. $y''' - y'' + 2y = e^x[(20 + 4x)\cos x - (12 + 12x)\sin x]$

35. $y''' - 7y'' + 20y' - 24y = -e^{2x}[(13 - 8x)\cos 2x - (8 - 4x)\sin 2x]$

36. $y''' - 6y'' + 18y' = -e^{3x}[(2 - 3x)\cos 3x - (3 + 3x)\sin 3x]$

37. $y^{(4)} + 2y''' - 2y'' - 8y' - 8y = e^x(8\cos x + 16\sin x)$

38. $y^{(4)} - 3y''' + 2y'' + 2y' - 4y = e^x(2\cos 2x - \sin 2x)$

39. $y^{(4)} - 8y''' + 24y'' - 32y' + 15y = e^{2x}(15x\cos 2x + 32\sin 2x)$

40. $y^{(4)} + 6y''' + 13y'' + 12y' + 4y = e^{-x}[(4 - x)\cos x - (5 + x)\sin x]$

41. $y^{(4)} + 3y''' + 2y'' - 2y' - 4y = -e^{-x}(\cos x - \sin x)$

42. $y^{(4)} - 5y''' + 13y'' - 19y' + 10y = e^x(\cos 2x + \sin 2x)$

43. $y^{(4)} + 8y''' + 32y'' + 64y' + 39y = e^{-2x}[(4 - 15x)\cos 3x - (4 + 15x)\sin 3x]$

44. $y^{(4)} - 5y''' + 13y'' - 19y' + 10y = e^x[(7 + 8x)\cos 2x + (8 - 4x)\sin 2x]$

45. $y^{(4)} + 4y''' + 8y'' + 8y' + 4y = -2e^{-x}(\cos x - 2\sin x)$

46. $y^{(4)} - 8y''' + 32y'' - 64y' + 64y = e^{2x}(\cos 2x - \sin 2x)$

47. $y^{(4)} - 8y''' + 26y'' - 40y' + 25y = e^{2x}[3\cos x - (1 + 3x)\sin x]$

48. $y''' - 4y'' + 5y' - 2y = e^{2x} - 4e^x - 2\cos x + 4\sin x$

49. $y''' - y'' + y' - y = 5e^{2x} + 2e^x - 4\cos x + 4\sin x$

50. $y''' - y' = -2(1 + x) + 4e^x - 6e^{-x} + 96e^{3x}$ **51.** $y''' - 4y'' + 9y' - 10y = 10e^{2x} + 20e^x\sin 2x - 10$

52. $y''' + 3y'' + 3y' + y = 12e^{-x} + 9\cos 2x - 13\sin 2x$

53. $y''' + y'' - y' - y = 4e^{-x}(1 - 6x) - 2x\cos x + 2(1 + x)\sin x$

54. $y^{(4)} - 5y'' + 4y = -12e^x + 6e^{-x} + 10\cos x$

55. $y^{(4)} - 4y''' + 11y'' - 14y' + 10y = -e^x(\sin x + 2\cos 2x)$

56. $y^{(4)} + 2y''' - 3y'' - 4y' + 4y = 2e^x(1 + x) + e^{-2x}$ **57.** $y^{(4)} + 4y = \sinh x\cos x - \cosh x\sin x$

58. $y^{(4)} + 5y''' + 9y'' + 7y' + 2y = e^{-x}(30 + 24x) - e^{-2x}$

59. $y^{(4)} - 4y''' + 7y'' - 6y' + 2y = e^x(12x - 2\cos x + 2\sin x)$

In Exercises 60–68 find the general solution.

60. $y''' - y'' - y' + y = e^{2x}(10 + 3x)$

61. $y''' + y'' - 2y = -e^{3x}(9 + 67x + 17x^2)$

62. $y''' - 6y'' + 11y' - 6y = e^{2x}(5 - 4x - 3x^2)$

63. $y''' + 2y'' + y' = -2e^{-x}(7 - 18x + 6x^2)$

64. $y''' - 3y'' + 3y' - y = e^x(1 + x)$

65. $y^{(4)} - 2y'' + y = -e^{-x}(4 - 9x + 3x^2)$

66. $y''' + 2y'' - y' - 2y = e^{-2x}[(23 - 2x)\cos x + (8 - 9x)\sin x]$

67. $y^{(4)} - 3y''' + 4y'' - 2y' = e^x[(28 + 6x)\cos 2x + (11 - 12x)\sin 2x]$

68. $y^{(4)} - 4y''' + 14y'' - 20y' + 25y = e^x[(2 + 6x)\cos 2x + 3\sin 2x]$

In Exercises 69–74 solve the initial value problem and graph the solution.

C **69.** $y''' - 2y'' - 5y' + 6y = 2e^x(1 - 6x), \quad y(0) = 2, \quad y'(0) = 7, \quad y''(0) = 9$

C **70.** $y''' - y'' - y' + y = -e^{-x}(4 - 8x), \quad y(0) = 2, \quad y'(0) = 0, \quad y''(0) = 0$

C **71.** $4y''' - 3y' - y = e^{-x/2}(2 - 3x), \quad y(0) = -1, \quad y'(0) = 15, \quad y''(0) = -17$

C **72.** $y^{(4)} + 2y''' + 2y'' + 2y' + y = e^{-x}(20 - 12x), \quad y(0) = 3, \quad y'(0) = -4, \quad y''(0) = 7, \quad y'''(0) = -22$

C **73.** $y''' + 2y'' + y' + 2y = 30\cos x - 10\sin x, \quad y(0) = 3, \quad y'(0) = -4, \quad y''(0) = 16$

C **74.** $y^{(4)} - 3y''' + 4y'' - 2y' = -2e^x(\cos x - \sin x), \quad y(0) = 2, \quad y'(0) = 0, \quad y''(0) = -1, \quad y'''(0) = -5$

75. Prove: A function y is a solution of the constant coefficient nonhomogeneous equation

$$a_0 y^{(n)} + a_1 y^{(n-1)} + \cdots + a_n y = e^{\alpha x} G(x) \tag{A}$$

if and only if $y = u e^{\alpha x}$, where u satisfies the differential equation

$$a_0 u^{(n)} + \frac{p^{(n-1)}(\alpha)}{(n-1)!} u^{(n-1)} + \frac{p^{(n-2)}(\alpha)}{(n-2)!} u^{(n-2)} + \cdots + p(\alpha)u = G(x), \tag{B}$$

in which

$$p(r) = a_0 r^n + a_1 r^{n-1} + \cdots + a_n$$

is the characteristic polynomial of the complementary equation

$$a_0 y^{(n)} + a_1 y^{(n-1)} + \cdots + a_n y = 0.$$

9.4 Variation of Parameters for Higher Order Equations

DERIVATION OF THE METHOD

We assume throughout this section that the nonhomogeneous linear equation

$$P_0(x)y^{(n)} + P_1(x)y^{(n-1)} + \cdots + P_n(x)y = F(x) \tag{1}$$

is normal on an interval (a,b). We will abbreviate this equation as $Ly = F$; that is,

$$Ly = P_0(x)y^{(n)} + P_1(x)y^{(n-1)} + \cdots + P_n(x)y.$$

When we speak of solutions of this equation and its complementary equation $Ly = 0$ we mean solutions on (a,b). We will show how to use the method of variation of parameters to find a particular solution of $Ly = F$ provided that we know a fundamental set of solutions $\{y_1, y_2, \ldots, y_n\}$ of $Ly = 0$.

We seek a particular solution of $Ly = F$ in the form

$$y_p = u_1 y_1 + u_2 y_2 + \cdots + u_n y_n \tag{2}$$

where $\{y_1, y_2, \ldots, y_n\}$ is a known fundamental set of solutions of the complementary equation

$$P_0(x)y^{(n)} + P_1(x)y^{(n-1)} + \cdots + P_n(x)y = 0$$

and $u_1, u_2, \ldots, u_n$ are functions to be determined. We begin by imposing the following $n - 1$ conditions on $u_1, u_2, \ldots, u_n$:

$$
\begin{aligned}
u_1' y_1 + u_2' y_2 + \cdots + u_n' y_n &= 0 \\
u_1' y_1' + u_2' y_2' + \cdots + u_n' y_n' &= 0 \\
&\vdots \\
u_1' y_1^{(n-2)} + u_2' y_2^{(n-2)} + \cdots + u_n' y_n^{(n-2)} &= 0.
\end{aligned}
\tag{3}
$$

These conditions lead to simple formulas for the first $n - 1$ derivatives of y_p; namely,

$$y_p^{(r)} = u_1 y_1^{(r)} + u_2 y_2^{(r)} + \cdots + u_n y_n^{(r)}, \qquad 0 \le r \le n - 1. \tag{4}$$

These formulas are easy to remember, since they look as though we obtained them by differentiating (2) $n - 1$ times while treating $u_1, u_2, \ldots, u_n$ as constants. To see that (3) implies (4), we first differentiate (2) to obtain

$$y_p' = u_1 y_1' + u_2 y_2' + \cdots + u_n y_n' + u_1' y_1 + u_2' y_2 + \cdots + u_n' y_n,$$

which reduces to

$$y_p' = u_1 y_1' + u_2 y_2' + \cdots + u_n y_n'$$

because of the first equation in (3). Differentiating this yields

$$y_p'' = u_1 y_1'' + u_2 y_2'' + \cdots + u_n y_n'' + u_1' y_1' + u_2' y_2' + \cdots + u_n' y_n',$$

which reduces to

$$y_p'' = u_1 y_1'' + u_2 y_2'' + \cdots + u_n y_n''$$

because of the second equation in (3). Continuing in this way we obtain (4).

The last equation in (4) is

$$y_p^{(n-1)} = u_1 y_1^{(n-1)} + u_2 y_2^{(n-1)} + \cdots + u_n y_n^{(n-1)}.$$

Differentiating this yields

$$y_p^{(n)} = u_1 y_1^{(n)} + u_2 y_2^{(n)} + \cdots + u_n y_n^{(n)} + u_1' y_1^{(n-1)} + u_2' y_2^{(n-1)} + \cdots + u_n' y_n^{(n-1)}.$$

Substituting this and (4) into (1) yields

$$u_1 L y_1 + u_2 L y_2 + \cdots + u_n L y_n$$
$$+ P_0(x)(u_1' y_1^{(n-1)} + u_2' y_2^{(n-1)} + \cdots + u_n' y_n^{(n-1)}) = F(x).$$

Since $Ly_i = 0$ $(1 \le i \le n)$, this reduces to

$$u_1' y_1^{(n-1)} + u_2' y_2^{(n-1)} + \cdots + u_n' y_n^{(n-1)} = \frac{F(x)}{P_0(x)}.$$

Combining this equation with (3) shows that

$$y_p = u_1 y_1 + u_2 y_2 + \cdots + u_n y_n$$

is a solution of (1) if

$$u_1'y_1 + u_2'y_2 + \cdots + u_n'y_n = 0$$
$$u_1'y_1' + u_2'y_2' + \cdots + u_n'y_n' = 0$$
$$\vdots$$
$$u_1'y_1^{(n-2)} + u_2'y_2^{(n-2)} + \cdots + u_n'y_n^{(n-2)} = 0$$
$$u_1'y_1^{(n-1)} + u_2'y_2^{(n-1)} + \cdots + u_n'y_n^{(n-1)} = F/P_0,$$

which can be written in matrix form as

$$
\begin{bmatrix}
y_1 & y_2 & \cdots & y_n \\
y_1' & y_2' & \cdots & y_n' \\
\vdots & \vdots & \ddots & \vdots \\
y_1^{(n-2)} & y_2^{(n-2)} & \cdots & y_n^{(n-2)} \\
y_1^{(n-1)} & y_2^{(n-1)} & \cdots & y_n^{(n-1)}
\end{bmatrix}
\begin{bmatrix}
u_1' \\
u_2' \\
\vdots \\
u_{n-1}' \\
u_n'
\end{bmatrix}
=
\begin{bmatrix}
0 \\
0 \\
\vdots \\
0 \\
F/P_0
\end{bmatrix}.
\tag{5}
$$

The determinant of this system is the Wronskian W of the fundamental set of solutions $\{y_1, y_2, \ldots, y_n\}$, which has no zeros on (a, b), by Theorem 9.1.4. Solving (5) by Cramer's rule yields

$$u_j' = (-1)^{n-j} \frac{FW_j}{P_0 W}, \qquad 1 \leq j \leq n, \tag{6}$$

where W_j is the Wronskian of the set of functions obtained by deleting y_j from $\{y_1, y_2, \ldots, y_n\}$ and keeping the remaining functions in the same order. Equivalently, W_j is the determinant obtained by deleting the last row and jth column of W.

Having obtained $u_1', u_2', \ldots, u_n'$, we can integrate to obtain $u_1, u_2, \ldots, u_n$. As in Section 5.7, we take the constants of integration to be zero, and we drop any linear combination of $\{y_1, y_2, \ldots, y_n\}$ that may appear in y_p.

REMARK.   For efficiency it is best to compute $W_1, W_2, \ldots, W_n$ first, and then compute W by expanding in cofactors of the last row; thus,

$$W = \sum_{j=1}^{n} (-1)^{n-j} y_j^{(n-1)} W_j.$$

THIRD ORDER EQUATIONS

If $n = 3$ then

$$
W =
\begin{vmatrix}
y_1 & y_2 & y_3 \\
y_1' & y_2' & y_3' \\
y_1'' & y_2'' & y_3''
\end{vmatrix}.
$$

Therefore,

$$
W_1 =
\begin{vmatrix}
y_2 & y_3 \\
y_2' & y_3'
\end{vmatrix},
\qquad
W_2 =
\begin{vmatrix}
y_1 & y_3 \\
y_1' & y_3'
\end{vmatrix},
\qquad
W_3 =
\begin{vmatrix}
y_1 & y_2 \\
y_1' & y_2'
\end{vmatrix},
$$

and (6) becomes

$$u_1' = \frac{FW_1}{P_0 W}, \qquad u_2' = -\frac{FW_2}{P_0 W}, \qquad u_3' = \frac{FW_3}{P_0 W}. \tag{7}$$

EXAMPLE 1   Find a particular solution of

$$xy''' - y'' - xy' + y = 8x^2e^x, \tag{8}$$

given that $y_1 = x$, $y_2 = e^x$, and $y_3 = e^{-x}$ form a fundamental set of solutions of the complementary equation. Then find the general solution of (8).

Solution   We seek a particular solution of (8) of the form

$$y_p = u_1x + u_2e^x + u_3e^{-x}.$$

The Wronskian of $\{y_1, y_2, y_3\}$ is

$$W(x) = \begin{vmatrix} x & e^x & e^{-x} \\ 1 & e^x & -e^{-x} \\ 0 & e^x & e^{-x} \end{vmatrix},$$

so

$$W_1 = \begin{vmatrix} e^x & e^{-x} \\ e^x & -e^{-x} \end{vmatrix} = -2,$$

$$W_2 = \begin{vmatrix} x & e^{-x} \\ 1 & -e^{-x} \end{vmatrix} = -e^{-x}(x + 1),$$

$$W_3 = \begin{vmatrix} x & e^x \\ 1 & e^x \end{vmatrix} = e^x(x - 1).$$

Expanding W by cofactors of the last row yields

$$W = 0W_1 - e^xW_2 + e^{-x}W_3 = 0(-2) - e^x(-e^{-x}(x + 1)) + e^{-x}e^x(x - 1) = 2x.$$

Since $F(x) = 8x^2e^x$ and $P_0(x) = x$, we have

$$\frac{F}{P_0W} = \frac{8x^2e^x}{x \cdot 2x} = 4e^x.$$

Therefore, from (7)

$$\begin{aligned} u_1' &= 4e^xW_1 = 4e^x(-2) = -8e^x, \\ u_2' &= -4e^xW_2 = -4e^x(-e^{-x}(x + 1)) = 4(x + 1), \\ u_3' &= 4e^xW_3 = 4e^x(e^x(x - 1)) = 4e^{2x}(x - 1). \end{aligned}$$

Integrating these and taking the constants of integration to be zero yields

$$u_1 = -8e^x, \qquad u_2 = 2(x + 1)^2, \qquad u_3 = e^{2x}(2x - 3).$$

Hence,

$$\begin{aligned} y_p &= u_1y_1 + u_2y_2 + u_3y_3 \\ &= (-8e^x)x + e^x(2(x + 1)^2) + e^{-x}(e^{2x}(2x - 3)) \\ &= e^x(2x^2 - 2x - 1). \end{aligned}$$

Since $-e^x$ is a solution of the complementary equation, we redefine

$$y_p = 2xe^x(x - 1).$$

Therefore the general solution of (8) is

$$y = 2xe^x(x - 1) + c_1x + c_2e^x + c_3e^{-x}.$$ ∎

FOURTH ORDER EQUATIONS

If $n = 4$ then

$$W = \begin{vmatrix} y_1 & y_2 & y_3 & y_4 \\ y_1' & y_2' & y_3' & y_4' \\ y_1'' & y_2'' & y_3'' & y_4'' \\ y_1''' & y_2''' & y_3''' & y_4''' \end{vmatrix}.$$

Therefore,

$$W_1 = \begin{vmatrix} y_2 & y_3 & y_4 \\ y_2' & y_3' & y_4' \\ y_2'' & y_3'' & y_4'' \end{vmatrix}, \qquad W_2 = \begin{vmatrix} y_1 & y_3 & y_4 \\ y_1' & y_3' & y_4' \\ y_1'' & y_3'' & y_4'' \end{vmatrix},$$

$$W_3 = \begin{vmatrix} y_1 & y_2 & y_4 \\ y_1' & y_2' & y_4' \\ y_1'' & y_2'' & y_4'' \end{vmatrix}, \qquad W_4 = \begin{vmatrix} y_1 & y_2 & y_3 \\ y_1' & y_2' & y_3' \\ y_1'' & y_2'' & y_3'' \end{vmatrix},$$

and (6) becomes

$$u_1' = -\frac{FW_1}{P_0W}, \qquad u_2' = \frac{FW_2}{P_0W}, \qquad u_3' = -\frac{FW_3}{P_0W}, \qquad u_4' = \frac{FW_4}{P_0W}. \qquad (9)$$

EXAMPLE 2   Find a particular solution of

$$x^4y^{(4)} + 6x^3y''' + 2x^2y'' - 4xy' + 4y = 12x^2, \qquad (10)$$

given that $y_1 = x$, $y_2 = x^2$, $y_3 = 1/x$, and $y_4 = 1/x^2$ form a fundamental set of solutions of the complementary equation. Then find the general solution of (10) on $(-\infty, 0)$ and $(0, \infty)$.

Solution   We seek a particular solution of (10) of the form

$$y_p = u_1x + u_2x^2 + \frac{u_3}{x} + \frac{u_4}{x^2}.$$

The Wronskian of $\{y_1, y_2, y_3, y_4\}$ is

$$W(x) = \begin{vmatrix} x & x^2 & \dfrac{1}{x} & \dfrac{1}{x^2} \\ 1 & 2x & -\dfrac{1}{x^2} & -\dfrac{2}{x^3} \\ 0 & 2 & \dfrac{2}{x^3} & \dfrac{6}{x^4} \\ 0 & 0 & -\dfrac{6}{x^4} & -\dfrac{24}{x^5} \end{vmatrix},$$

so

$$W_1 = \begin{vmatrix} x^2 & \dfrac{1}{x} & \dfrac{1}{x^2} \\ 2x & -\dfrac{1}{x^2} & -\dfrac{2}{x^3} \\ 2 & \dfrac{2}{x^3} & \dfrac{6}{x^4} \end{vmatrix} = -\dfrac{12}{x^4},$$

$$W_2 = \begin{vmatrix} x & \dfrac{1}{x} & \dfrac{1}{x^2} \\ 1 & -\dfrac{1}{x^2} & -\dfrac{2}{x^3} \\ 0 & \dfrac{2}{x^3} & \dfrac{6}{x^4} \end{vmatrix} = -\dfrac{6}{x^5},$$

$$W_3 = \begin{vmatrix} x & x^2 & \dfrac{1}{x^2} \\ 1 & 2x & -\dfrac{2}{x^3} \\ 0 & 2 & \dfrac{6}{x^4} \end{vmatrix} = \dfrac{12}{x^2},$$

$$W_4 = \begin{vmatrix} x & x^2 & \dfrac{1}{x} \\ 1 & 2x & -\dfrac{1}{x^2} \\ 0 & 2 & \dfrac{2}{x^3} \end{vmatrix} = \dfrac{6}{x}.$$

Expanding W by cofactors of the last row yields

$$W = -0W_1 + 0W_2 - \left(-\dfrac{6}{x^4}\right)W_3 + \left(-\dfrac{24}{x^5}\right)W_4$$

$$= \dfrac{6}{x^4}\dfrac{12}{x^2} - \dfrac{24}{x^5}\dfrac{6}{x} = -\dfrac{72}{x^6}.$$

Since $F(x) = 12x^2$ and $P_0(x) = x^4$, we have

$$\dfrac{F}{P_0 W} = \dfrac{12x^2}{x^4}\left(-\dfrac{x^6}{72}\right) = -\dfrac{x^4}{6}.$$

Therefore, from (9),

$$u_1' = -\left(-\dfrac{x^4}{6}\right)W_1 = \dfrac{x^4}{6}\left(-\dfrac{12}{x^4}\right) = -2,$$

$$u_2' = -\dfrac{x^4}{6}W_2 = -\dfrac{x^4}{6}\left(-\dfrac{6}{x^5}\right) = \dfrac{1}{x},$$

$$u_3' = -\left(-\dfrac{x^4}{6}\right)W_3 = \dfrac{x^4}{6}\dfrac{12}{x^2} = 2x^2,$$

$$u_4' = -\dfrac{x^4}{6}W_4 = -\dfrac{x^4}{6}\dfrac{6}{x} = -x^3.$$

Integrating these and taking the constants of integration to be zero yields

$$u_1 = -2x, \qquad u_2 = \ln|x|, \qquad u_3 = \frac{2x^3}{3}, \qquad u_4 = -\frac{x^4}{4}.$$

Hence,

$$y_p = u_1 y_1 + u_2 y_2 + u_3 y_3 + u_4 y_4$$

$$= (-2x)x + (\ln|x|)x^2 + \frac{2x^3}{3}\frac{1}{x} + \left(-\frac{x^4}{4}\right)\frac{1}{x^2}$$

$$= x^2 \ln|x| - \frac{19x^2}{12}.$$

Since $-19x^2/12$ is a solution of the complementary equation, we redefine

$$y_p = x^2 \ln|x|.$$

Therefore

$$y = x^2 \ln|x| + c_1 x + c_2 x^2 + \frac{c_3}{x} + \frac{c_4}{x^2}$$

is the general solution of (10) on $(-\infty, 0)$ and $(0, \infty)$.

9.4 EXERCISES

In Exercises 1–21 find a particular solution, given the fundamental set of solutions of the complementary equation.

1. $x^3 y''' - x^2(x+3)y'' + 2x(x+3)y' - 2(x+3)y = -4x^4;$ $\{x, x^2, xe^x\}$

2. $y''' + 6xy'' + (6+12x^2)y' + (12x+8x^3)y = x^{1/2}e^{-x^2};$ $\{e^{-x^2}, xe^{-x^2}, x^2 e^{-x^2}\}$

3. $x^3 y''' - 3x^2 y'' + 6xy' - 6y = 2x;$ $\{x, x^2, x^3\}$

4. $x^2 y''' + 2xy'' - (x^2 + 2)y' = 2x^2;$ $\{1, e^x/x, e^{-x}/x\}$

5. $x^3 y''' - 3x^2(x+1)y'' + 3x(x^2 + 2x + 2)y' - (x^3 + 3x^2 + 6x + 6)y = x^4 e^{-3x};$ $\{xe^x, x^2 e^x, x^3 e^x\}$

6. $x(x^2-2)y''' + (x^2-6)y'' + x(2-x^2)y' + (6-x^2)y = 2(x^2-2)^2;$ $\{e^x, e^{-x}, 1/x\}$

7. $xy''' - (x-3)y'' - (x+2)y' + (x-1)y = -4e^{-x};$ $\{e^x, e^x/x, e^{-x}/x\}$

8. $4x^3 y''' + 4x^2 y'' - 5xy' + 2y = 30x^2;$ $\{\sqrt{x}, 1/\sqrt{x}, x^2\}$

9. $x(x^2-1)y''' + (5x^2+1)y'' + 2xy' - 2y = 12x^2;$ $\{x, 1/(x-1), 1/(x+1)\}$

10. $x(1-x)y''' + (x^2-3x+3)y'' + xy' - y = 2(x-1)^2;$ $\{x, 1/x, e^x/x\}$

11. $x^3 y''' + x^2 y'' - 2xy' + 2y = x^2;$ $\{x, x^2, 1/x\}$

12. $xy''' - y'' - xy' + y = x^2;$ $\{x, e^x, e^{-x}\}$

13. $xy^{(4)} + 4y''' = 6\ln|x|;$ $\{1, x, x^2, 1/x\}$

14. $16x^4 y^{(4)} + 96x^3 y''' + 72x^2 y'' - 24xy' + 9y = 96x^{5/2};$ $\{\sqrt{x}, 1/\sqrt{x}, x^{3/2}, x^{-3/2}\}$

15. $x(x^2-6)y^{(4)} + 2(x^2-12)y''' + x(6-x^2)y'' + 2(12-x^2)y' = 2(x^2-6)^2;$ $\{1, 1/x, e^x, e^{-x}\}$

16. $x^4 y^{(4)} - 4x^3 y''' + 12x^2 y'' - 24xy' + 24y = x^4;$ $\{x, x^2, x^3, x^4\}$

17. $x^4 y^{(4)} - 4x^3 y''' + 2x^2(6-x^2)y'' + 4x(x^2-6)y' + (x^4 - 4x^2 + 24)y = 4x^5 e^x;$ $\{xe^x, x^2 e^x, xe^{-x}, x^2 e^{-x}\}$

18. $x^4 y^{(4)} + 6x^3 y''' + 2x^2 y'' - 4xy' + 4y = 12x^2;$ $\{x, x^2, 1/x, 1/x^2\}$

19. $xy^{(4)} + 4y''' - 2xy'' - 4y' + xy = 4e^x;$ $\{e^x, e^{-x}, e^x/x, e^{-x}/x\}$

20. $xy^{(4)} + (4-6x)y''' + (13x-18)y'' + (26-12x)y' + (4x-12)y = 3e^x;$ $\{e^x, e^{2x}, e^x/x, e^{2x}/x\}$

21. $x^4 y^{(4)} - 4x^3 y''' + x^2(12-x^2)y'' + 2x(x^2-12)y' + 2(12-x^2)y = 2x^5;$ $\{x, x^2, xe^x, xe^{-x}\}$

In Exercises 22–33 solve the initial value problem, given the fundamental set of solutions of the complementary equation. Where indicated by C *, graph the solution.*

C **22.** $x^3y''' - 2x^2y'' + 3xy' - 3y = 4x$, $y(1) = 4$, $y'(1) = 4$, $y''(1) = 2$; $\{x, x^3, x \ln x\}$

23. $x^3y''' - 5x^2y'' + 14xy' - 18y = x^3$, $y(1) = 0$, $y'(1) = 1$, $y''(1) = 7$; $\{x^2, x^3, x^3 \ln x\}$

24. $(5 - 6x)y''' + (12x - 4)y'' + (6x - 23)y' + (22 - 12x)y = -(6x - 5)^2e^x$,

$y(0) = -4$, $y'(0) = -\dfrac{3}{2}$, $y''(0) = -19$; $\{e^x, e^{2x}, xe^{-x}\}$

25. $x^3y''' - 6x^2y'' + 16xy' - 16y = 9x^4$, $y(1) = 2$, $y'(1) = 1$, $y''(1) = 5$; $\{x, x^4, x^4 \ln x\}$

C **26.** $(x^2 - 2x + 2)y''' - x^2y'' + 2xy' - 2y = (x^2 - 2x + 2)^2$, $y(0) = 0$, $y'(0) = 5$, $y''(0) = 0$; $\{x, x^2, e^x\}$

27. $x^3y''' + x^2y'' - 2xy' + 2y = x(x + 1)$, $y(-1) = -6$, $y'(-1) = \dfrac{43}{6}$, $y''(-1) = -\dfrac{5}{2}$; $\{x, x^2, 1/x\}$

28. $(3x - 1)y''' - (12x - 1)y'' + 9(x + 1)y' - 9y = 2e^x(3x - 1)^2$, $y(0) = \dfrac{3}{4}$, $y'(0) = \dfrac{5}{4}$, $y''(0) = \dfrac{1}{4}$;

$\{x + 1, e^x, e^{3x}\}$

C **29.** $(x^2 - 2)y''' - 2xy'' + (2 - x^2)y' + 2xy = 2(x^2 - 2)^2$, $y(0) = 1$, $y'(0) = -5$, $y''(0) = 5$; $\{x^2, e^x, e^{-x}\}$

C **30.** $x^4y^{(4)} + 3x^3y''' - x^2y'' + 2xy' - 2y = 9x^2$, $y(1) = -7$, $y'(1) = -11$, $y''(1) = -5$, $y'''(1) = 6$;

$\{x, x^2, 1/x, x \ln x\}$

31. $(2x - 1)y^{(4)} - 4xy''' + (5 - 2x)y'' + 4xy' - 4y = 6(2x - 1)^2$, $y(0) = \dfrac{55}{4}$, $y'(0) = 0$, $y''(0) = 13$, **F**

$y'''(0) = 1$; $\{x, e^x, e^{-x}, e^{2x}\}$

32. $4x^4y^{(4)} + 24x^3y''' + 23x^2y'' - xy' + y = 6x$, $y(1) = 2$, $y'(1) = 0$, $y''(1) = 4$, $y'''(1) = -\dfrac{37}{4}$;

$\{x, \sqrt{x}, 1/x, 1/\sqrt{x}\}$

33. $x^4y^{(4)} + 5x^3y''' - 3x^2y'' - 6xy' + 6y = 40x^3$, $y(-1) = -1$, $y'(-1) = -7$, $y''(-1) = -1$,

$y'''(-1) = -31$; $\{x, x^3, 1/x, 1/x^2\}$

34. Suppose that the equation

$$P_0(x)y^{(n)} + P_1(x)y^{(n-1)} + \cdots + P_n(x)y = F(x) \tag{A}$$

is normal on an interval (a, b). Let $\{y_1, y_2, \ldots, y_n\}$ be a fundamental set of solutions of the complementary equation on (a, b), let W be the Wronskian of $\{y_1, y_2, \ldots, y_n\}$ and let W_j be the determinant obtained by deleting the last row and the jth column of W. Suppose that x_0 is in (a, b), let

$$u_j(x) = (-1)^{(n-j)} \int_{x_0}^x \frac{F(t)W_j(t)}{P_0(t)W(t)} \, dt, \qquad 1 \le j \le n,$$

and define

$$y_p = u_1y_1 + u_2y_2 + \cdots + u_ny_n.$$

(a) Show that y_p is a solution of (A) and that

$$y_p^{(r)} = u_1y_1^{(r)} + u_2y_2^{(r)} + \cdots + u_ny_n^{(r)}, \qquad 1 \le r \le n - 1,$$

and

$$y_p^{(n)} = u_1y_1^{(n)} + u_2y_2^{(n)} + \cdots + u_ny_n^{(n)} + \frac{F}{P_0}.$$

Hint: See the derivation of the method of variation of parameters at the beginning of the section.

(b) Show that y_p is the solution of the initial value problem

$$P_0(x)y^{(n)} + P_1(x)y^{(n-1)} + \cdots + P_n(x)y = F(x), \quad y(x_0) = 0, \quad y'(x_0) = 0, \ldots, \quad y^{(n-1)}(x_0) = 0.$$

(c) Show that y_p can be written as

$$y_p(x) = \int_{x_0}^x G(x,t)F(t)\, dt,$$

where

$$G(x,t) = \frac{1}{P_0(t)W(t)} \begin{vmatrix} y_1(t) & y_2(t) & \cdots & y_n(t) \\ y_1'(t) & y_2'(t) & \cdots & y_n'(t) \\ \vdots & \vdots & \ddots & \vdots \\ y_1^{(n-2)}(t) & y_2^{(n-2)}(t) & \cdots & y_n^{(n-2)}(t) \\ y_1(x) & y_2(x) & \cdots & y_n(x) \end{vmatrix},$$

which is called **the Green's function** for (A).

(d) Show that

$$\frac{\partial^j G(x,t)}{\partial x^j} = \frac{1}{P_0(t)W(t)} \begin{vmatrix} y_1(t) & y_2(t) & \cdots & y_n(t) \\ y_1'(t) & y_2'(t) & \cdots & y_n'(t) \\ \vdots & \vdots & \ddots & \vdots \\ y_1^{(n-2)}(t) & y_2^{(n-2)}(t) & \cdots & y_n^{(n-2)}(t) \\ y_1^{(j)}(x) & y_2^{(j)}(x) & \cdots & y_n^{(j)}(x) \end{vmatrix}, \quad 1 \le j \le n.$$

(e) Show that if $a < t < b$ then

$$\left. \frac{\partial^j G(x,t)}{\partial x^j} \right|_{x=t} = \begin{cases} 0, & 0 \le j \le n-2, \\ \dfrac{1}{P_0(t)}, & j = n-1. \end{cases}$$

(f) Show that

$$y_p^{(j)}(x) = \begin{cases} \displaystyle\int_{x_0}^x \frac{\partial^j G(x,t)}{\partial x^j} F(t)\, dt, & 0 \le j \le n-1, \\[3mm] \dfrac{F(x)}{P_0(x)} + \displaystyle\int_{x_0}^x \frac{\partial^{(n)} G(x,t)}{\partial x^n} F(t)\, dt, & j = n. \end{cases}$$

In Exercises 35–42 use the method suggested by Exercise 34 to find a particular solution in the form $y_p = \int_{x_0}^x G(x,t)F(t)\, dt$, given the indicated fundamental set of solutions. Assume that x and x_0 are in an interval on which the equation is normal.

35. $y''' + 2y'' - y' - 2y = F(x); \quad \{e^x, e^{-x}, e^{-2x}\}$

36. $x^3y''' + x^2y'' - 2xy' + 2y = F(x); \quad \{x, x^2, 1/x\}$

37. $x^3y''' - x^2(x+3)y'' + 2x(x+3)y' - 2(x+3)y = F(x); \quad \{x, x^2, xe^x\}$

38. $x(1-x)y''' + (x^2 - 3x + 3)y'' + xy' - y = F(x); \quad \{x, 1/x, e^x/x\}$

39. $y^{(4)} - 5y'' + 4y = F(x); \quad \{e^x, e^{-x}, e^{2x}, e^{-2x}\}$

40. $xy^{(4)} + 4y''' = F(x); \quad \{1, x, x^2, 1/x\}$

41. $x^4y^{(4)} + 6x^3y''' + 2x^2y'' - 4xy' + 4y = F(x); \quad \{x, x^2, 1/x, 1/x^2\}$

42. $xy^{(4)} - y''' - 4xy'' + 4y' = F(x); \quad \{1, x^2, e^{2x}, e^{-2x}\}$

10 Linear Systems of Differential Equations

IN THIS CHAPTER we consider systems of differential equations involving more than one unknown function. Such systems arise in many physical applications.

SECTION 10.1 presents examples of physical situations that lead to systems of differential equations.

SECTION 10.2 discusses linear systems of differential equations.

SECTION 10.3 deals with the basic theory of homogeneous linear systems.

SECTIONS 10.4, 10.5, AND 10.6 present the theory of constant coefficient homogeneous systems.

SECTION 10.7 presents the method of variation of parameters for nonhomogeneous linear systems.

10.1   Introduction to Systems of Differential Equations

Many situations are modeled by systems of n differential equations in n unknown functions, where $n \geq 2$. The following three examples illustrate physical problems that lead to systems of differential equations. In these examples and throughout this chapter we will denote the independent variable by t.

EXAMPLE 1   Tanks T_1 and T_2 contain 100 gallons and 300 gallons of salt solutions, respectively. Salt solutions are simultaneously added to both tanks from external sources, pumped from each tank to the other, and drained from both tanks (Figure 1). A solution with 1 pound of salt per gallon is pumped into T_1 from an external source at 5 gal/min, and a solution with 2 pounds of salt per gallon is pumped into T_2 from an external source at 4 gal/min. The solution from T_1 is pumped into T_2 at 2 gal/min, and the solution from T_2 is pumped into T_1 at 3 gal/min. T_1 is drained at 6 gal/min and T_2 is drained at 3 gal/min. Let $Q_1(t)$ and $Q_2(t)$ be the number of pounds of salt in T_1 and T_2, respectively, at time $t > 0$. Derive a system of differential equations for Q_1 and Q_2. Assume that both mixtures are well stirred.

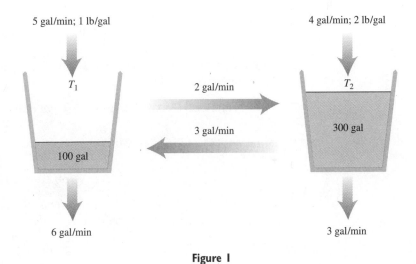

Figure 1

Solution   As in Section 4.2, let *rate in* and *rate out* denote the rates (lb/min) at which salt enters and leaves a tank; thus,

$$Q_1' = (\text{rate in})_1 - (\text{rate out})_1,$$
$$Q_2' = (\text{rate in})_2 - (\text{rate out})_2.$$

Notice that the volumes of the solutions in T_1 and T_2 remain constant at 100 gallons and 300 gallons, respectively.

T_1 receives salt from the external source at the rate of

$$(1 \text{ lb/gal}) \times (5 \text{ gal/min}) = 5 \text{ lb/min},$$

and from T_2 at the rate of

$$\text{(lb/gal in } T_2) \times (3 \text{ gal/min}) = \frac{1}{300}Q_2 \times 3 = \frac{1}{100}Q_2 \text{ lb/min}.$$

Therefore

$$\text{(rate in)}_1 = 5 + \frac{1}{100}Q_2. \tag{1}$$

Solution leaves T_1 at the rate of 8 gal/min, since 6 gal/min are drained and 2 gal/min are pumped to T_2; hence

$$\text{(rate out)}_1 = \text{(lb/gal in } T_1) \times (8 \text{ gal/min}) = \frac{1}{100}Q_1 \times 8 = \frac{2}{25}Q_1. \tag{2}$$

Equations (1) and (2) imply that

$$Q_1' = 5 + \frac{1}{100}Q_2 - \frac{2}{25}Q_1. \tag{3}$$

T_2 receives salt from the external source at the rate of

$$(2 \text{ lb/gal}) \times (4 \text{ gal/min}) = 8 \text{ lb/min},$$

and from T_1 at the rate of

$$\text{(lb/gal in } T_1) \times (2 \text{ gal/min}) = \frac{1}{100}Q_1 \times 2 = \frac{1}{50}Q_1 \text{ lb/min}.$$

Therefore

$$\text{(rate in)}_2 = 8 + \frac{1}{50}Q_1. \tag{4}$$

Solution leaves T_2 at the rate of 6 gal/min, since 3 gal/min are drained and 3 gal/min are pumped to T_1; hence

$$\text{(rate out)}_2 = \text{(lb/gal in } T_2) \times (6 \text{ gal/min}) = \frac{1}{300}Q_2 \times 6 = \frac{1}{50}Q_2. \tag{5}$$

Equations (4) and (5) imply that

$$Q_2' = 8 + \frac{1}{50}Q_1 - \frac{1}{50}Q_2. \tag{6}$$

We say that (3) and (6) form a *system of two first order equations in two unknowns*, and write them together as

$$Q_1' = 5 - \frac{2}{25}Q_1 + \frac{1}{100}Q_2$$

$$Q_2' = 8 + \frac{1}{50}Q_1 - \frac{1}{50}Q_2.$$

EXAMPLE 2 A mass m_1 is suspended from a rigid support on a spring S_1 and a second mass m_2 is suspended from the first on a spring S_2 (Figure 2). The springs obey Hooke's law, with spring constants k_1 and k_2. Internal friction causes the springs to exert damping forces proportional to the rates of change of their lengths, with damping constants c_1 and c_2. Let $y_1 = y_1(t)$ and $y_2 = y_2(t)$ be the displacements of the two masses from their equilibrium positions at time t, measured positive upward. Derive a system of differential equations for y_1 and y_2, assuming that the masses of the springs are negligible and that vertical external forces F_1 and F_2 also act on the objects.

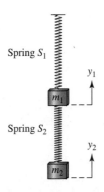

Figure 2

Solution   In equilibrium S_1 supports both m_1 and m_2 and S_2 supports only m_2. Therefore, if Δl_1 and Δl_2 are the elongations of the springs in equilibrium, then

$$(m_1 + m_2)g = k_1 \Delta l_1 \qquad \text{and} \qquad m_2 g = k_2 \Delta l_2. \tag{7}$$

Let H_1 be the Hooke's law force acting on m_1, and let D_1 be the damping force on m_1. Similarly, let H_2 and D_2 be the Hooke's law and damping forces acting on m_2. According to Newton's second law of motion,

$$
\begin{aligned}
m_1 y_1'' &= -m_1 g + H_1 + D_1 + F_1 \\
m_2 y_2'' &= -m_2 g + H_2 + D_2 + F_2.
\end{aligned} \tag{8}
$$

When the displacements are y_1 and y_2 the change in length of S_1 is $-y_1 + \Delta l_1$ and the change in length of S_2 is $-y_2 + y_1 + \Delta l_2$. Both springs exert Hooke's law forces on m_1, while only S_2 exerts a Hooke's law force on m_2. These forces are in directions that tend to restore the springs to their natural lengths. Therefore

$$
\begin{aligned}
H_1 &= k_1(-y_1 + \Delta l_1) - k_2(-y_2 + y_1 + \Delta l_2) \qquad \text{and} \\
H_2 &= k_2(-y_2 + y_1 + \Delta l_2).
\end{aligned} \tag{9}
$$

When the velocities are y_1' and y_2', S_1 and S_2 are changing length at the rates $-y_1'$ and $-y_2' + y_1'$, respectively. Both springs exert damping forces on m_1, while only S_2 exerts a damping force on m_2. Since the force due to damping exerted by a spring is proportional to the rate of change of length of the spring and in a direction that opposes the change, it follows that

$$D_1 = -c_1 y_1' + c_2(y_2' - y_1') \qquad \text{and} \qquad D_2 = -c_2(y_2' - y_1'). \tag{10}$$

From (8), (9), and (10),

$$
\begin{aligned}
m_1 y_1'' &= -m_1 g + k_1(-y_1 + \Delta l_1) - k_2(-y_2 + y_1 + \Delta l_2) \\
&\quad - c_1 y_1' + c_2(y_2' - y_1') + F_1 \\
&= -(m_1 g - k_1 \Delta l_1 + k_2 \Delta l_2) - k_1 y_1 + k_2(y_2 - y_1) \\
&\quad - c_1 y_1' + c_2(y_2' - y_1') + F_1
\end{aligned} \tag{11}
$$

and

$$
\begin{aligned}
m_2 y_2'' &= -m_2 g + k_2(-y_2 + y_1 + \Delta l_2) - c_2(y_2' - y_1') + F_2 \\
&= -(m_2 g - k_2 \Delta l_2) - k_2(y_2 - y_1) - c_2(y_2' - y_1') + F_2.
\end{aligned} \tag{12}
$$

Because of (7),

$$m_1 g - k_1 \Delta l_1 + k_2 \Delta l_2 = m_2 g - k_2 \Delta l_2 = 0.$$

Therefore, we can rewrite (11) and (12) as

$$m_1 y_1'' = -(c_1 + c_2)y_1' + c_2 y_2' - (k_1 + k_2)y_1 + k_2 y_2 + F_1$$
$$m_2 y_2'' = c_2 y_1' - c_2 y_2' + k_2 y_1 - k_2 y_2 + F_2.$$

EXAMPLE 3   Let $\mathbf{X} = \mathbf{X}(t) = x(t)\mathbf{i} + y(t)\mathbf{j} + z(t)\mathbf{k}$ be the position vector at time t of an object with mass m, relative to a rectangular coordinate system with origin at Earth's center (Figure 3). According to Newton's law of gravitation, Earth's gravitational force $\mathbf{F} = \mathbf{F}(x, y, z)$ on the object is inversely proportional to the square of the distance of the object from Earth's center, and directed toward the center; thus

$$\mathbf{F} = \frac{K}{\|\mathbf{X}\|^2}\left(-\frac{\mathbf{X}}{\|\mathbf{X}\|}\right) = -K\frac{x\mathbf{i} + y\mathbf{j} + z\mathbf{k}}{(x^2 + y^2 + z^2)^{3/2}}, \tag{13}$$

where K is a constant. To determine K we observe that the magnitude of $\mathbf{F}$ is

$$\|\mathbf{F}\| = K\frac{\|\mathbf{X}\|}{\|\mathbf{X}\|^3} = \frac{K}{\|\mathbf{X}\|^2} = \frac{K}{(x^2 + y^2 + z^2)}.$$

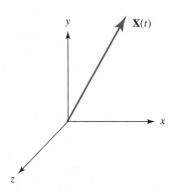

Figure 3

Let R be Earth's radius. Since $\|\mathbf{F}\| = mg$ when the object is at Earth's surface, we see that

$$mg = \frac{K}{R^2}, \quad \text{so} \quad K = mgR^2.$$

Therefore we can rewrite (13) as

$$\mathbf{F} = -mgR^2\frac{x\mathbf{i} + y\mathbf{j} + z\mathbf{k}}{(x^2 + y^2 + z^2)^{3/2}}.$$

Now suppose that $\mathbf{F}$ is the only force acting on the object. According to Newton's second law of motion, $\mathbf{F} = m\mathbf{X}''$; that is,

$$m(x''\mathbf{i} + y''\mathbf{j} + z''\mathbf{k}) = -mgR^2\frac{x\mathbf{i} + y\mathbf{j} + z\mathbf{k}}{(x^2 + y^2 + z^2)^{3/2}}.$$

Canceling the common factor m and equating components on the two sides of this equation yields the system

$$x'' = -\frac{gR^2 x}{(x^2 + y^2 + z^2)^{3/2}}$$

$$y'' = -\frac{gR^2 y}{(x^2 + y^2 + z^2)^{3/2}} \qquad (14)$$

$$z'' = -\frac{gR^2 z}{(x^2 + y^2 + z^2)^{3/2}}.$$

■

REWRITING HIGHER ORDER SYSTEMS AS FIRST ORDER SYSTEMS

A system of the form

$$\begin{aligned}
y_1' &= g_1(t, y_1, y_2, \ldots, y_n) \\
y_2' &= g_2(t, y_1, y_2, \ldots, y_n) \\
&\;\;\vdots \\
y_n' &= g_n(t, y_1, y_2, \ldots, y_n)
\end{aligned} \qquad (15)$$

is called a **first order system,** since the only derivatives occurring in it are first derivatives. Notice that the derivative of each of the unknowns may depend upon the independent variable and all the unknowns, but not on the derivatives of other unknowns. When we wish to emphasize the number of unknown functions in (15) we will say that (15) is an **$n \times n$ system.**

Systems involving higher order derivatives can often be reformulated as first order systems by introducing additional unknowns. The next two examples illustrate this.

EXAMPLE 4   Rewrite the system

$$\begin{aligned}
m_1 y_1'' &= -(c_1 + c_2)y_1' + c_2 y_2' - (k_1 + k_2)y_1 + k_2 y_2 + F_1 \\
m_2 y_2'' &= c_2 y_1' - c_2 y_2' + k_2 y_1 - k_2 y_2 + F_2
\end{aligned} \qquad (16)$$

derived in Example 2 as a system of first order equations.

Solution   If we define $v_1 = y_1'$ and $v_2 = y_2'$ then $v_1' = y_1''$ and $v_2' = y_2''$, so (16) becomes

$$\begin{aligned}
m_1 v_1' &= -(c_1 + c_2)v_1 + c_2 v_2 - (k_1 + k_2)y_1 + k_2 y_2 + F_1 \\
m_2 v_2' &= c_2 v_1 - c_2 v_2 + k_2 y_1 - k_2 y_2 + F_2.
\end{aligned}$$

Therefore $\{y_1, y_2, v_1, v_2\}$ satisfies the 4×4 first order system

$$y_1' = v_1$$

$$y_2' = v_2$$

$$v_1' = \frac{1}{m_1}[-(c_1 + c_2)v_1 + c_2 v_2 - (k_1 + k_2)y_1 + k_2 y_2 + F_1] \qquad (17)$$

$$v_2' = \frac{1}{m_2}[c_2 v_1 - c_2 v_2 + k_2 y_1 - k_2 y_2 + F_2].$$

■

REMARK.   The difference in form between (15) and (17), due to the way in which the unknowns are *denoted* in the two systems, is not important; (17) is a first order system in that each equation in (17) expresses the first derivative of one of the unknown functions in a way that does not involve derivatives of any of the other unknowns.

EXAMPLE 5 Rewrite the system

$$x'' = f(t, x, x', y, y', y'')$$
$$y''' = g(t, x, x', y, y'y'')$$

as a first order system.

Solution We regard $x, x', y, y',$ and y'' as unknown functions, and rename them

$$x = x_1, \qquad x' = x_2, \qquad y = y_1, \qquad y' = y_2, \qquad y'' = y_3.$$

These unknowns satisfy the system

$$x_1' = x_2$$
$$x_2' = f(t, x_1, x_2, y_1, y_2, y_3)$$
$$y_1' = y_2$$
$$y_2' = y_3$$
$$y_3' = g(t, x_1, x_2, y_1, y_2, y_3).$$ ∎

REWRITING SCALAR DIFFERENTIAL EQUATIONS AS SYSTEMS

In this chapter we will refer to differential equations involving only one unknown function as *scalar* differential equations. Scalar differential equations can be rewritten as systems of first order equations by the method illustrated in the next two examples.

EXAMPLE 6 Rewrite the equation

$$y^{(4)} + 4y''' + 6y'' + 4y' + y = 0 \tag{18}$$

as a 4×4 first order system.

Solution We regard $y, y', y'',$ and y''' as unknowns and rename them

$$y = y_1, \qquad y' = y_2, \qquad y'' = y_3, \qquad \text{and} \qquad y''' = y_4.$$

Then $y^{(4)} = y_4'$, so (18) can be written as

$$y_4' + 4y_4 + 6y_3 + 4y_2 + y_1 = 0.$$

Therefore, $\{y_1, y_2, y_3, y_4\}$ satisfies the system

$$y_1' = y_2$$
$$y_2' = y_3$$
$$y_3' = y_4$$
$$y_4' = -4y_4 - 6y_3 - 4y_2 - y_1.$$

EXAMPLE 7 Rewrite

$$x''' = f(t, x, x', x'')$$

as a system of first order equations.

Solution We regard $x, x',$ and x'' as unknowns and rename them

$$x = y_1, \qquad x' = y_2, \qquad \text{and} \qquad x'' = y_3.$$

Then

$$y_1' = x' = y_2, \qquad y_2' = x'' = y_3, \qquad \text{and} \qquad y_3' = x'''.$$

Therefore, $\{y_1, y_2, y_3\}$ satisfies the first order system

$$y_1' = y_2$$
$$y_2' = y_3$$
$$y_3' = f(t, y_1, y_2, y_3).$$

■

Since systems of differential equations involving higher derivatives can be rewritten as first order systems by the method used in Examples 5–7, we will consider only first order systems.

NUMERICAL SOLUTION OF SYSTEMS

The numerical methods that we studied in Chapter 3 can be extended to systems, and most differential equation software packages include programs to solve systems of equations. We won't go into detail on numerical methods for systems; however, for illustrative purposes we'll describe the Runge–Kutta method for the numerical solution of the initial value problem

$$y_1' = g_1(t, y_1, y_2), \qquad y_1(t_0) = y_{10},$$
$$y_2' = g_2(t, y_1, y_2), \qquad y_2(t_0) = y_{20}$$

at equally spaced points $t_0, t_1, \ldots, t_n = b$ in an interval $[t_0, b]$. Thus,

$$t_i = t_0 + ih, \qquad i = 0, 1, \ldots, n,$$

where

$$h = \frac{b - t_0}{n}.$$

We will denote the approximate values of y_1 and y_2 at these points by $y_{10}, y_{11}, \ldots, y_{1n}$ and $y_{20}, y_{21}, \ldots, y_{2n}$. The Runge–Kutta method computes these approximate values as follows: given y_{1i} and y_{2i}, compute

$$I_{1i} = g_1(t_i, y_{1i}, y_{2i}),$$
$$J_{1i} = g_2(t_i, y_{1i}, y_{2i}),$$
$$I_{2i} = g_1\left(t_i + \frac{h}{2}, y_{1i} + \frac{h}{2}I_{1i}, y_{2i} + \frac{h}{2}J_{1i}\right),$$
$$J_{2i} = g_2\left(t_i + \frac{h}{2}, y_{1i} + \frac{h}{2}I_{1i}, y_{2i} + \frac{h}{2}J_{1i}\right),$$
$$I_{3i} = g_1\left(t_i + \frac{h}{2}, y_{1i} + \frac{h}{2}I_{2i}, y_{2i} + \frac{h}{2}J_{2i}\right),$$
$$J_{3i} = g_2\left(t_i + \frac{h}{2}, y_{1i} + \frac{h}{2}I_{2i}, y_{2i} + \frac{h}{2}J_{2i}\right),$$
$$I_{4i} = g_1(t_i + h, y_{1i} + hI_{3i}, y_{2i} + hJ_{3i}),$$
$$J_{4i} = g_2(t_i + h, y_{1i} + hI_{3i}, y_{2i} + hJ_{3i}),$$

and

$$y_{1,i+1} = y_{1i} + \frac{h}{6}(I_{1i} + 2I_{2i} + 2I_{3i} + I_{4i}),$$

$$y_{2,i+1} = y_{2i} + \frac{h}{6}(J_{1i} + 2J_{2i} + 2J_{3i} + J_{4i})$$

for $i = 0, \ldots, n - 1$. Under appropriate conditions on g_1 and g_2 it can be shown that the global truncation error for the Runge–Kutta method is $O(h^4)$, as in the scalar case considered in Section 3.3. ■

10.1 EXERCISES

1. Tanks T_1 and T_2 contain 50 gallons and 100 gallons of salt solutions, respectively. A solution with 2 pounds of salt per gallon is pumped into T_1 from an external source at 1 gal/min, and a solution with 3 pounds of salt per gallon is pumped into T_2 from an external source at 2 gal/min. The solution from T_1 is pumped into T_2 at 3 gal/min, and the solution from T_2 is pumped into T_1 at 4 gal/min. T_1 is drained at 2 gal/min and T_2 is drained at 1 gal/min. Let $Q_1(t)$ and $Q_2(t)$ be the number of pounds of salt in T_1 and T_2, respectively, at time $t > 0$. Derive a system of differential equations for Q_1 and Q_2. Assume that both mixtures are well stirred.

2. Two 500-gallon tanks T_1 and T_2 initially contain 100 gallons each of salt solution. A solution with 2 pounds of salt per gallon is pumped into T_1 from an external source at 6 gal/min, and a solution with 1 pound of salt per gallon is pumped into T_2 from an external source at 5 gal/min. The solution from T_1 is pumped into T_2 at 2 gal/min, and the solution from T_2 is pumped into T_1 at 1 gal/min. Both tanks are drained at 3 gal/min. Let $Q_1(t)$ and $Q_2(t)$ be the number of pounds of salt in T_1 and T_2, respectively, at time $t > 0$. Derive a system of differential equations for Q_1 and Q_2 that is valid until a tank is about to overflow. Assume that both mixtures are well stirred.

3. A mass m_1 is suspended from a rigid support on a spring S_1 with spring constant k_1 and damping constant c_1. A second mass m_2 is suspended from the first on a spring S_2 with spring constant k_2 and damping constant c_2, and a third mass m_3 is suspended from the second on a spring S_3 with spring constant k_3 and damping constant c_3. Let $y_1 = y_1(t), y_2 = y_2(t)$, and $y_3 = y_3(t)$ be the displacements of the three masses from their equilibrium positions at time t, measured positive upward. Derive a system of differential equations for y_1, y_2, and y_3, assuming that the masses of the springs are negligible and that vertical external forces F_1, F_2, and F_3 also act on the masses.

4. Let $\mathbf{X} = x\mathbf{i} + y\mathbf{j} + z\mathbf{k}$ be the position vector of an object with mass m, expressed in terms of a rectangular coordinate system with origin at Earth's center (Figure 3). Derive a system of differential equations for x, y, and z, assuming that the object moves under Earth's gravitational force (given by Newton's law of gravitation, as in Example 3) and a resistive force proportional to the speed of the object. Let α be the constant of proportionality.

5. Rewrite the given system as a first order system.

   (a) $x''' = f(t, x, y, y')$
   $\quad\ y'' = g(t, y, y')$

   (b) $u' = f(t, u, v, v', w')$
   $\quad\ v'' = g(t, u, v, v', w)$
   $\quad\ w'' = h(t, u, v, v', w, w')$

   (c) $y''' = f(t, y, y', y'')$

   (d) $y^{(4)} = f(t, y)$

   (e) $x'' = f(t, x, y)$
   $\quad\ y'' = g(t, x, y)$

6. Rewrite the system in Eqn. (14) as a first order system.

7. Formulate a version of Euler's method (Section 3.1) for the numerical solution of the initial value problem

   $$y_1' = g_1(t, y_1, y_2), \qquad y_1(t_0) = y_{10},$$
   $$y_2' = g_2(t, y_1, y_2), \qquad y_2(t_0) = y_{20}$$

   on an interval $[t_0, b]$.

8. Formulate a version of the improved Euler method (Section 3.2) for the numerical solution of the initial value problem

$$y_1' = g_1(t, y_1, y_2), \qquad y_1(t_0) = y_{10},$$
$$y_2' = g_2(t, y_1, y_2), \qquad y_2(t_0) = y_{20}$$

on an interval $[t_0, b]$.

10.2 Linear Systems of Differential Equations

A first order system of differential equations that can be written in the form

$$y_1' = a_{11}(t)y_1 + a_{12}(t)y_2 + \cdots + a_{1n}(t)y_n + f_1(t)$$
$$y_2' = a_{21}(t)y_1 + a_{22}(t)y_2 + \cdots + a_{2n}(t)y_n + f_2(t)$$
$$\vdots$$
$$y_n' = a_{n1}(t)y_1 + a_{n2}(t)y_2 + \cdots + a_{nn}(t)y_n + f_n(t)$$

(1)

is called a **linear system**.

The linear system (1) can be written in matrix form as

$$\begin{bmatrix} y_1' \\ y_2' \\ \vdots \\ y_n' \end{bmatrix} = \begin{bmatrix} a_{11}(t) & a_{12}(t) & \cdots & a_{1n}(t) \\ a_{21}(t) & a_{22}(t) & \cdots & a_{2n}(t) \\ \vdots & \vdots & \ddots & \vdots \\ a_{n1}(t) & a_{n2}(t) & \cdots & a_{nn}(t) \end{bmatrix} \begin{bmatrix} y_1 \\ y_2 \\ \vdots \\ y_n \end{bmatrix} + \begin{bmatrix} f_1(t) \\ f_2(t) \\ \vdots \\ f_n(t) \end{bmatrix},$$

or more briefly as

$$\mathbf{y}' = A(t)\mathbf{y} + \mathbf{f}(t),$$

(2)

where

$$\mathbf{y} = \begin{bmatrix} y_1 \\ y_2 \\ \vdots \\ y_n \end{bmatrix}, \quad A(t) = \begin{bmatrix} a_{11}(t) & a_{12}(t) & \cdots & a_{1n}(t) \\ a_{21}(t) & a_{22}(t) & \cdots & a_{2n}(t) \\ \vdots & \vdots & \ddots & \vdots \\ a_{n1}(t) & a_{n2}(t) & \cdots & a_{nn}(t) \end{bmatrix}, \quad \text{and} \quad \mathbf{f}(t) = \begin{bmatrix} f_1(t) \\ f_2(t) \\ \vdots \\ f_n(t) \end{bmatrix}.$$

We call A the **coefficient matrix** of (2) and $\mathbf{f}$ the **forcing function**. We will say that A and $\mathbf{f}$ are **continuous** if their entries are continuous. If $\mathbf{f} = \mathbf{0}$ then (2) is **homogeneous;** otherwise (2) is **nonhomogeneous**.

An initial value problem for (2) consists of finding a solution of (2) that equals a given constant vector

$$\mathbf{k} = \begin{bmatrix} k_1 \\ k_2 \\ \vdots \\ k_n \end{bmatrix}$$

at some initial point t_0. We write this initial value problem as

$$\mathbf{y}' = A(t)\mathbf{y} + \mathbf{f}(t), \qquad \mathbf{y}(t_0) = \mathbf{k}.$$

The following theorem gives sufficient conditions for the existence of solutions of initial value problems for (2). We omit the proof.

THEOREM 10.2.1

Suppose that the coefficient matrix A and the forcing function **f** *are contin-uous on* (a,b)*, let* t_0 *be in* (a,b)*, and let* **k** *be an arbitrary constant n-vector. Then the initial value problem*

$$\mathbf{y}' = A(t)\mathbf{y} + \mathbf{f}(t), \qquad \mathbf{y}(t_0) = \mathbf{k}$$

has a unique solution on (a,b)*.*

EXAMPLE 1   **(a)** Write the system

$$\begin{aligned} y_1' &= y_1 + 2y_2 + 2e^{4t} \\ y_2' &= 2y_1 + y_2 + e^{4t} \end{aligned} \tag{3}$$

in matrix form and conclude from Theorem 10.2.1 that every initial value prob-lem for (3) has a unique solution on $(-\infty, \infty)$.

(b) Verify that

$$\mathbf{y} = \frac{1}{5}\begin{bmatrix} 8 \\ 7 \end{bmatrix} e^{4t} + c_1 \begin{bmatrix} 1 \\ 1 \end{bmatrix} e^{3t} + c_2 \begin{bmatrix} 1 \\ -1 \end{bmatrix} e^{-t} \tag{4}$$

is a solution of (3) for all values of the constants c_1 and c_2.

(c) Find the solution of the initial value problem

$$\mathbf{y}' = \begin{bmatrix} 1 & 2 \\ 2 & 1 \end{bmatrix} \mathbf{y} + \begin{bmatrix} 2 \\ 1 \end{bmatrix} e^{4t}, \qquad \mathbf{y}(0) = \frac{1}{5}\begin{bmatrix} 3 \\ 22 \end{bmatrix}. \tag{5}$$

Solution   **(a)** The system (3) can be written in matrix form as

$$\mathbf{y}' = \begin{bmatrix} 1 & 2 \\ 2 & 1 \end{bmatrix} \mathbf{y} + \begin{bmatrix} 2 \\ 1 \end{bmatrix} e^{4t}.$$

An initial value problem for (3) can be written as

$$\mathbf{y}' = \begin{bmatrix} 1 & 2 \\ 2 & 1 \end{bmatrix} \mathbf{y} + \begin{bmatrix} 2 \\ 1 \end{bmatrix} e^{4t}, \qquad y(t_0) = \begin{bmatrix} k_1 \\ k_2 \end{bmatrix}.$$

Since the coefficient matrix and the forcing function are both continuous on $(-\infty, \infty)$, Theorem 10.2.1 implies that this problem has a unique solution on $(-\infty, \infty)$.

Solution   **(b)** If **y** is given by (4) then

$$A\mathbf{y} + \mathbf{f} = \frac{1}{5}\begin{bmatrix} 1 & 2 \\ 2 & 1 \end{bmatrix}\begin{bmatrix} 8 \\ 7 \end{bmatrix} e^{4t} + c_1 \begin{bmatrix} 1 & 2 \\ 2 & 1 \end{bmatrix}\begin{bmatrix} 1 \\ 1 \end{bmatrix} e^{3t}$$

$$+ c_2 \begin{bmatrix} 1 & 2 \\ 2 & 1 \end{bmatrix}\begin{bmatrix} 1 \\ -1 \end{bmatrix} e^{-t} + \begin{bmatrix} 2 \\ 1 \end{bmatrix} e^{4t}$$

$$= \frac{1}{5}\begin{bmatrix} 22 \\ 23 \end{bmatrix} e^{4t} + c_1 \begin{bmatrix} 3 \\ 3 \end{bmatrix} e^{3t} + c_2 \begin{bmatrix} -1 \\ 1 \end{bmatrix} e^{-t} + \begin{bmatrix} 2 \\ 1 \end{bmatrix} e^{4t}$$

$$= \frac{1}{5}\begin{bmatrix} 32 \\ 28 \end{bmatrix} e^{4t} + 3c_1 \begin{bmatrix} 1 \\ 1 \end{bmatrix} e^{3t} - c_2 \begin{bmatrix} 1 \\ -1 \end{bmatrix} e^{-t} = \mathbf{y}'.$$

Solution **(c)** We must choose c_1 and c_2 in (4) so that

$$\frac{1}{5}\begin{bmatrix} 8 \\ 7 \end{bmatrix} + c_1 \begin{bmatrix} 1 \\ 1 \end{bmatrix} + c_2 \begin{bmatrix} 1 \\ -1 \end{bmatrix} = \frac{1}{5}\begin{bmatrix} 3 \\ 22 \end{bmatrix},$$

which is equivalent to

$$\begin{bmatrix} 1 & 1 \\ 1 & -1 \end{bmatrix}\begin{bmatrix} c_1 \\ c_2 \end{bmatrix} = \begin{bmatrix} -1 \\ 3 \end{bmatrix}.$$

Solving this system yields $c_1 = 1, c_2 = -2$, so

$$\mathbf{y} = \frac{1}{5}\begin{bmatrix} 8 \\ 7 \end{bmatrix} e^{4t} + \begin{bmatrix} 1 \\ 1 \end{bmatrix} e^{3t} - 2\begin{bmatrix} 1 \\ -1 \end{bmatrix} e^{-t}$$

is the solution of (5). ∎

REMARK. The theory of $n \times n$ linear systems of differential equations is analogous to the theory of the scalar nth order equation

$$P_0(t)y^{(n)} + P_1(t)y^{(n-1)} + \cdots + P_n(t)y = F(t), \tag{6}$$

as developed in Section 9.1. For example, by rewriting (6) as an equivalent linear system it can be shown that Theorem 10.2.1 implies Theorem 9.1.1 (Exercise 12).

10.2 EXERCISES

1. Rewrite the system in matrix form and verify that the given vector function satisfies the system for any choice of the constants c_1 and c_2.

(a) $y_1' = 2y_1 + 4y_2$
   $y_2' = 4y_1 + 2y_2$; $\mathbf{y} = c_1 \begin{bmatrix} 1 \\ 1 \end{bmatrix} e^{6t} + c_2 \begin{bmatrix} 1 \\ -1 \end{bmatrix} e^{-2t}$

(b) $y_1' = -2y_1 - 2y_2$
   $y_2' = -5y_1 + y_2$; $\mathbf{y} = c_1 \begin{bmatrix} 1 \\ 1 \end{bmatrix} e^{-4t} + c_2 \begin{bmatrix} -2 \\ 5 \end{bmatrix} e^{3t}$

(c) $y_1' = -4y_1 - 10y_2$
   $y_2' = 3y_1 + 7y_2$; $\mathbf{y} = c_1 \begin{bmatrix} -5 \\ 3 \end{bmatrix} e^{2t} + c_2 \begin{bmatrix} 2 \\ -1 \end{bmatrix} e^{t}$

(d) $y_1' = 2y_1 + y_2$
   $y_2' = y_1 + 2y_2$; $\mathbf{y} = c_1 \begin{bmatrix} 1 \\ 1 \end{bmatrix} e^{3t} + c_2 \begin{bmatrix} 1 \\ -1 \end{bmatrix} e^{t}$

2. Rewrite the system in matrix form and verify that the given vector function satisfies the system for any choice of the constants c_1, c_2, and c_3.

(a) $y_1' = -y_1 + 2y_2 + 3y_3$
   $y_2' = \qquad y_2 + 6y_3$; $\mathbf{y} = c_1 \begin{bmatrix} 1 \\ 1 \\ 0 \end{bmatrix} e^{t} + c_2 \begin{bmatrix} 1 \\ 0 \\ 0 \end{bmatrix} e^{-t} + c_3 \begin{bmatrix} 1 \\ -2 \\ 1 \end{bmatrix} e^{-2t}$
   $y_3' = \qquad - 2y_3$

(b) $\begin{aligned} y_1' &= 2y_2 + 2y_3 \\ y_2' &= 2y_1 + 2y_3 \;; \quad \mathbf{y} = c_1 \begin{bmatrix} -1 \\ 0 \\ 1 \end{bmatrix} e^{-2t} + c_2 \begin{bmatrix} 0 \\ -1 \\ 1 \end{bmatrix} e^{-2t} + c_3 \begin{bmatrix} 1 \\ 1 \\ 1 \end{bmatrix} e^{4t} \\ y_3' &= 2y_1 + 2y_2 \end{aligned}$

(c) $\begin{aligned} y_1' &= -y_1 + 2y_2 + 2y_3 \\ y_2' &= 2y_1 - y_2 + 2y_3 \;; \quad \mathbf{y} = c_1 \begin{bmatrix} -1 \\ 0 \\ 1 \end{bmatrix} e^{-3t} + c_2 \begin{bmatrix} 0 \\ -1 \\ 1 \end{bmatrix} e^{-3t} + c_3 \begin{bmatrix} 1 \\ 1 \\ 1 \end{bmatrix} e^{3t} \\ y_3' &= 2y_1 + 2y_2 - y_3 \end{aligned}$

(d) $\begin{aligned} y_1' &= 3y_1 - y_2 - y_3 \\ y_2' &= -2y_1 + 3y_2 + 2y_3 \;; \quad \mathbf{y} = c_1 \begin{bmatrix} 1 \\ 0 \\ 1 \end{bmatrix} e^{2t} + c_2 \begin{bmatrix} 1 \\ -1 \\ 1 \end{bmatrix} e^{3t} + c_3 \begin{bmatrix} 1 \\ -3 \\ 7 \end{bmatrix} e^{-t} \\ y_3' &= 4y_1 - y_2 - 2y_3 \end{aligned}$

3. Rewrite the initial value problem in matrix form and verify that the given vector function is a solution.

(a) $\begin{aligned} y_1' &= y_1 + y_2 \\ y_2' &= -2y_1 + 4y_2 \end{aligned}, \quad \begin{aligned} y_1(0) &= 1 \\ y_2(0) &= 0 \end{aligned}; \quad \mathbf{y} = 2 \begin{bmatrix} 1 \\ 1 \end{bmatrix} e^{2t} - \begin{bmatrix} 1 \\ 2 \end{bmatrix} e^{3t}$

(b) $\begin{aligned} y_1' &= 5y_1 + 3y_2 \\ y_2' &= -y_1 + y_2 \end{aligned}, \quad \begin{aligned} y_1(0) &= 12 \\ y_2(0) &= -6 \end{aligned}; \quad \mathbf{y} = 3 \begin{bmatrix} 1 \\ -1 \end{bmatrix} e^{2t} + 3 \begin{bmatrix} 3 \\ -1 \end{bmatrix} e^{4t}$

4. Rewrite the initial value problem in matrix form and verify that the given vector function is a solution.

(a) $\begin{aligned} y_1' &= 6y_1 + 4y_2 + 4y_3 \\ y_2' &= -7y_1 - 2y_2 - y_3 \;, \\ y_3' &= 7y_1 + 4y_2 + 3y_3 \end{aligned} \quad \begin{aligned} y_1(0) &= 3 \\ y_2(0) &= -6 \;; \\ y_3(0) &= 4 \end{aligned} \quad \mathbf{y} = \begin{bmatrix} 1 \\ -1 \\ 1 \end{bmatrix} e^{6t} + 2 \begin{bmatrix} 1 \\ -2 \\ 1 \end{bmatrix} e^{2t} + \begin{bmatrix} 0 \\ -1 \\ 1 \end{bmatrix} e^{-t}$

(b) $\begin{aligned} y_1' &= 8y_1 + 7y_2 + 7y_3 \\ y_2' &= -5y_1 - 6y_2 - 9y_3 \;, \\ y_3' &= 5y_1 + 7y_2 + 10y_3 \end{aligned} \quad \begin{aligned} y_1(0) &= 2 \\ y_2(0) &= -4 \;; \\ y_3(0) &= 3 \end{aligned} \quad \mathbf{y} = \begin{bmatrix} 1 \\ -1 \\ 1 \end{bmatrix} e^{8t} + \begin{bmatrix} 0 \\ -1 \\ 1 \end{bmatrix} e^{3t} + \begin{bmatrix} 1 \\ -2 \\ 1 \end{bmatrix} e^{t}$

5. Rewrite the system in matrix form and verify that the given vector function satisfies the system for any choice of the constants c_1 and c_2.

(a) $\begin{aligned} y_1' &= -3y_1 + 2y_2 + 3 - 2t \\ y_2' &= -5y_1 + 3y_2 + 6 - 3t \end{aligned} \;; \quad \mathbf{y} = c_1 \begin{bmatrix} 2\cos t \\ 3\cos t - \sin t \end{bmatrix} + c_2 \begin{bmatrix} 2\sin t \\ 3\sin t + \cos t \end{bmatrix} + \begin{bmatrix} 1 \\ t \end{bmatrix}$

(b) $\begin{aligned} y_1' &= 3y_1 + y_2 - 5e^t \\ y_2' &= -y_1 + y_2 + e^t \end{aligned} \;; \quad \mathbf{y} = c_1 \begin{bmatrix} -1 \\ 1 \end{bmatrix} e^{2t} + c_2 \begin{bmatrix} 1+t \\ -t \end{bmatrix} e^{2t} + \begin{bmatrix} 1 \\ 3 \end{bmatrix} e^t$

(c) $\begin{aligned} y_1' &= -y_1 - 4y_2 + 4e^t + 8te^t \\ y_2' &= -y_1 - y_2 + e^{3t} + (4t+2)e^t \end{aligned} \;; \quad \mathbf{y} = c_1 \begin{bmatrix} 2 \\ 1 \end{bmatrix} e^{-3t} + c_2 \begin{bmatrix} -2 \\ 1 \end{bmatrix} e^t + \begin{bmatrix} e^{3t} \\ 2te^t \end{bmatrix}$

(d) $\begin{aligned} y_1' &= -6y_1 - 3y_2 + 14e^{2t} + 12e^t \\ y_2' &= y_1 - 2y_2 + 7e^{2t} - 12e^t \end{aligned} \;; \quad \mathbf{y} = c_1 \begin{bmatrix} -3 \\ 1 \end{bmatrix} e^{-5t} + c_2 \begin{bmatrix} -1 \\ 1 \end{bmatrix} e^{-3t} + \begin{bmatrix} e^{2t} + 3e^t \\ 2e^{2t} - 3e^t \end{bmatrix}$

6. Convert the linear scalar equation

$$P_0(t)y^{(n)} + P_1(t)y^{(n-1)} + \cdots + P_n(t)y(t) = F(t) \tag{A}$$

into an equivalent $n \times n$ system

$$\mathbf{y}' = A(t)\mathbf{y} + \mathbf{f}(t),$$

and show that A and $\mathbf{f}$ are continuous on an interval (a,b) if (A) is normal on (a,b).

7. A matrix function

$$Q(t) = \begin{bmatrix} q_{11}(t) & q_{12}(t) & \cdots & q_{1s}(t) \\ q_{21}(t) & q_{22}(t) & \cdots & q_{2s}(t) \\ \vdots & \vdots & \ddots & \vdots \\ q_{r1}(t) & q_{r2}(t) & \cdots & q_{rs}(t) \end{bmatrix}$$

is said to be **differentiable** if its entries $\{q_{ij}\}$ are differentiable. Then the **derivative Q'** is defined by

$$Q'(t) = \begin{bmatrix} q'_{11}(t) & q'_{12}(t) & \cdots & q'_{1s}(t) \\ q'_{21}(t) & q'_{22}(t) & \cdots & q'_{2s}(t) \\ \vdots & \vdots & \ddots & \vdots \\ q'_{r1}(t) & q'_{r2}(t) & \cdots & q'_{rs}(t) \end{bmatrix}.$$

(a) Prove: If P and Q are differentiable matrices such that $P + Q$ is defined and if c_1 and c_2 are constants, then

$$(c_1 P + c_2 Q)' = c_1 P' + c_2 Q'.$$

(b) Prove: If P and Q are differentiable matrices such that PQ is defined, then

$$(PQ)' = P'Q + PQ'.$$

8. Verify that $Y' = AY$.

(a) $Y = \begin{bmatrix} e^{6t} & e^{-2t} \\ e^{6t} & -e^{-2t} \end{bmatrix}$, $A = \begin{bmatrix} 2 & 4 \\ 4 & 2 \end{bmatrix}$ **(b)** $Y = \begin{bmatrix} e^{-4t} & -2e^{3t} \\ e^{-4t} & 5e^{3t} \end{bmatrix}$, $A = \begin{bmatrix} -2 & -2 \\ -5 & 1 \end{bmatrix}$

(c) $Y = \begin{bmatrix} -5e^{2t} & 2e^{t} \\ 3e^{2t} & -e^{t} \end{bmatrix}$, $A = \begin{bmatrix} -4 & -10 \\ 3 & 7 \end{bmatrix}$ **(d)** $Y = \begin{bmatrix} e^{3t} & e^{t} \\ e^{3t} & -e^{t} \end{bmatrix}$, $A = \begin{bmatrix} 2 & 1 \\ 1 & 2 \end{bmatrix}$

(e) $Y = \begin{bmatrix} e^{t} & e^{-t} & e^{-2t} \\ e^{t} & 0 & -2e^{-2t} \\ 0 & 0 & e^{-2t} \end{bmatrix}$, $A = \begin{bmatrix} -1 & 2 & 3 \\ 0 & 1 & 6 \\ 0 & 0 & -2 \end{bmatrix}$

(f) $Y = \begin{bmatrix} -e^{-2t} & -e^{-2t} & e^{4t} \\ 0 & e^{-2t} & e^{4t} \\ e^{-2t} & 0 & e^{4t} \end{bmatrix}$, $A = \begin{bmatrix} 0 & 2 & 2 \\ 2 & 0 & 2 \\ 2 & 2 & 0 \end{bmatrix}$

(g) $Y = \begin{bmatrix} e^{3t} & e^{-3t} & 0 \\ e^{3t} & 0 & -e^{-3t} \\ e^{3t} & e^{-3t} & e^{-3t} \end{bmatrix}$, $A = \begin{bmatrix} -9 & 6 & 6 \\ -6 & 3 & 6 \\ -6 & 6 & 3 \end{bmatrix}$

(h) $Y = \begin{bmatrix} e^{2t} & e^{3t} & e^{-t} \\ 0 & -e^{3t} & -3e^{-t} \\ e^{2t} & e^{3t} & 7e^{-t} \end{bmatrix}$, $A = \begin{bmatrix} 3 & -1 & -1 \\ -2 & 3 & 2 \\ 4 & -1 & -2 \end{bmatrix}$

9. Suppose that

$$\mathbf{y}_1 = \begin{bmatrix} y_{11} \\ y_{21} \end{bmatrix} \quad \text{and} \quad \mathbf{y}_2 = \begin{bmatrix} y_{12} \\ y_{22} \end{bmatrix}$$

are solutions of the homogeneous system

$$\mathbf{y}' = A(t)\mathbf{y} \tag{A}$$

and define

$$Y = \begin{bmatrix} y_{11} & y_{12} \\ y_{21} & y_{22} \end{bmatrix}.$$

(a) Show that $Y' = AY$.

(b) Show that if $\mathbf{c}$ is a constant vector, then $\mathbf{y} = Y\mathbf{c}$ is a solution of (A).

(c) State generalizations of **(a)** and **(b)** for $n \times n$ systems.

10. Suppose that Y is a differentiable square matrix.

(a) Find a formula for the derivative of Y^2.

(b) Find a formula for the derivative of Y^n, where n is any positive integer.

(c) State how the results obtained in **(a)** and **(b)** are analogous to results from calculus concerning scalar functions.

11. It can be shown that if Y is a differentiable and invertible square matrix function, then Y^{-1} is differentiable.

(a) Show that $(Y^{-1})' = -Y^{-1}Y'Y^{-1}$.

Hint: Differentiate the identity $Y^{-1}Y = I$.

(b) Find the derivative of $Y^{-n} = (Y^{-1})^n$, where n is a positive integer.

(c) State how the results obtained in **(a)** and **(b)** are analogous to results from calculus concerning scalar functions.

12. Show that Theorem 10.2.1 implies 9.1.1.

Hint: Write the scalar equation

$$P_0(x)y^{(n)} + P_1(x)y^{(n-1)} + \cdots + P_n(x)y = F(x)$$

as an $n \times n$ system of linear equations.

13. Suppose that $\mathbf{y}$ is a solution of the $n \times n$ system $\mathbf{y}' = A(t)\mathbf{y}$ on (a,b), and that the $n \times n$ matrix P is invertible and differentiable on (a,b). Find a matrix B such that the function $\mathbf{x} = P\mathbf{y}$ is a solution of $\mathbf{x}' = B\mathbf{x}$ on (a,b).

10.3   Basic Theory of Homogeneous Linear Systems

In this section we consider homogeneous linear systems $\mathbf{y}' = A(t)\mathbf{y}$, in which $A = A(t)$ is a continuous $n \times n$ matrix function on an interval (a,b). The theory of homogeneous linear systems has much in common with the theory of linear homogeneous scalar equations, which we considered in Sections 2.1, 5.1, and 9.1.

Whenever we refer to solutions of $\mathbf{y}' = A(t)\mathbf{y}$ we will mean solutions on (a,b). Since $\mathbf{y} \equiv \mathbf{0}$ is obviously a solution of $\mathbf{y}' = A(t)\mathbf{y}$, we call it the ***trivial*** solution. Any other solution is ***nontrivial.***

If $\mathbf{y}_1, \mathbf{y}_2, \ldots, \mathbf{y}_n$ are vector functions defined on an interval (a,b) and $c_1, c_2, \ldots, c_n$ are constants, then

$$\mathbf{y} = c_1\mathbf{y}_1 + c_2\mathbf{y}_2 + \cdots + c_n\mathbf{y}_n \tag{1}$$

is a ***linear combination*** of $\mathbf{y}_1, \mathbf{y}_2, \ldots, \mathbf{y}_n$. It is straightforward to show that if $\mathbf{y}_1, \mathbf{y}_2, \ldots, \mathbf{y}_n$ are solutions of $\mathbf{y}' = A(t)\mathbf{y}$ on (a,b) then so is any linear combination of $\{\mathbf{y}_1, \mathbf{y}_2, \ldots, \mathbf{y}_n\}$ (Exercise 1). We say that $\{\mathbf{y}_1, \mathbf{y}_2, \ldots, \mathbf{y}_n\}$ is a ***fundamental set of solutions of*** $\mathbf{y}' = A(t)\mathbf{y}$ ***on*** (a,b) if every solution of $\mathbf{y}' = A(t)\mathbf{y}$ on (a,b) can be written as a linear combination of $\{\mathbf{y}_1, \mathbf{y}_2, \ldots, \mathbf{y}_n\}$, as in (1). In this case we say that (1) is the ***general solution of*** $\mathbf{y}' = A(t)\mathbf{y}$ ***on*** (a,b).

It can be shown that if A is continuous on (a,b) then $\mathbf{y}' = A(t)\mathbf{y}$ has infinitely many fundamental sets of solutions on (a,b) (Exercises 15 and 16). The following definition will help to characterize fundamental sets of solutions of $\mathbf{y}' = A(t)\mathbf{y}$.

We say that a set $\{\mathbf{y}_1, \mathbf{y}_2, \ldots, \mathbf{y}_n\}$ of n-vector functions is ***linearly independent on*** (a,b) if the only constants $c_1, c_2, \ldots, c_n$ such that

$$c_1\mathbf{y}_1(t) + c_2\mathbf{y}_2(t) + \cdots + c_n\mathbf{y}_n(t) = 0, \qquad a < t < b, \tag{2}$$

are $c_1 = c_2 = \cdots = c_n = 0$. If (2) holds for some set of constants $c_1, c_2, \ldots, c_n$ that are not all zero, then $\{\mathbf{y}_1, \mathbf{y}_2, \ldots, \mathbf{y}_n\}$ is ***linearly dependent on*** (a,b)

The following theorem is analogous to Theorems 5.1.3. and 9.1.2.

THEOREM 10.3.1

Suppose that the $n \times n$ *matrix* $A = A(t)$ *is continuous on* (a,b). *Then a set* $\{\mathbf{y}_1, \mathbf{y}_2, \ldots, \mathbf{y}_n\}$ *of* n *solutions of* $\mathbf{y}' = A(t)\mathbf{y}$ *on* (a,b) *is a fundamental set if and only if it is linearly independent on* (a,b).

EXAMPLE 1 Show that the vector functions

$$\mathbf{y}_1 = \begin{bmatrix} e^t \\ 0 \\ e^{-t} \end{bmatrix}, \qquad \mathbf{y}_2 = \begin{bmatrix} 0 \\ e^{3t} \\ 1 \end{bmatrix}, \qquad \text{and} \qquad \mathbf{y}_3 = \begin{bmatrix} e^{2t} \\ e^{3t} \\ 0 \end{bmatrix}$$

are linearly independent on every interval (a,b).

Solution Suppose that

$$c_1 \begin{bmatrix} e^t \\ 0 \\ e^{-t} \end{bmatrix} + c_2 \begin{bmatrix} 0 \\ e^{3t} \\ 1 \end{bmatrix} + c_3 \begin{bmatrix} e^{2t} \\ e^{3t} \\ 0 \end{bmatrix} = \begin{bmatrix} 0 \\ 0 \\ 0 \end{bmatrix}, \qquad a < t < b.$$

We must show that $c_1 = c_2 = c_3 = 0$. Rewriting this equation in matrix form yields

$$\begin{bmatrix} e^t & 0 & e^{2t} \\ 0 & e^{3t} & e^{3t} \\ e^{-t} & 1 & 0 \end{bmatrix} \begin{bmatrix} c_1 \\ c_2 \\ c_3 \end{bmatrix} = \begin{bmatrix} 0 \\ 0 \\ 0 \end{bmatrix}, \qquad a < t < b.$$

Expanding the determinant of this system in cofactors of the entries of the first row yields

$$\begin{vmatrix} e^t & 0 & e^{2t} \\ 0 & e^{3t} & e^{3t} \\ e^{-t} & 1 & 0 \end{vmatrix} = e^t \begin{vmatrix} e^{3t} & e^{3t} \\ 1 & 0 \end{vmatrix} - 0 \begin{vmatrix} 0 & e^{3t} \\ e^{-t} & 0 \end{vmatrix} + e^{2t} \begin{vmatrix} 0 & e^{3t} \\ e^{-t} & 1 \end{vmatrix}$$

$$= e^t(-e^{3t}) + e^{2t}(-e^{2t}) = -2e^{4t}.$$

Since this determinant is never zero it follows that $c_1 = c_2 = c_3 = 0$. ∎

We can use the method used in Example 1 to test n solutions $\{\mathbf{y}_1, \mathbf{y}_2, \dots, \mathbf{y}_n\}$ of any $n \times n$ system $\mathbf{y}' = A(t)\mathbf{y}$ for linear independence on an interval (a, b) on which A is continuous. To explain this (and for other purposes later), it is useful to write a linear combination of $\{\mathbf{y}_1, \mathbf{y}_2, \dots, \mathbf{y}_n\}$ in a different way. We first write the vector functions in terms of their components as

$$\mathbf{y}_1 = \begin{bmatrix} y_{11} \\ y_{21} \\ \vdots \\ y_{n1} \end{bmatrix}, \qquad \mathbf{y}_2 = \begin{bmatrix} y_{12} \\ y_{22} \\ \vdots \\ y_{n2} \end{bmatrix}, \dots, \qquad \mathbf{y}_n = \begin{bmatrix} y_{1n} \\ y_{2n} \\ \vdots \\ y_{nn} \end{bmatrix}.$$

If

$$\mathbf{y} = c_1\mathbf{y}_1 + c_2\mathbf{y}_2 + \cdots + c_n\mathbf{y}_n$$

then

$$\mathbf{y} = c_1 \begin{bmatrix} y_{11} \\ y_{21} \\ \vdots \\ y_{n1} \end{bmatrix} + c_2 \begin{bmatrix} y_{12} \\ y_{22} \\ \vdots \\ y_{n2} \end{bmatrix} + \cdots + c_n \begin{bmatrix} y_{1n} \\ y_{2n} \\ \vdots \\ y_{nn} \end{bmatrix}$$

$$= \begin{bmatrix} y_{11} & y_{12} & \cdots & y_{1n} \\ y_{21} & y_{22} & \cdots & y_{2n} \\ \vdots & \vdots & \ddots & \vdots \\ y_{n1} & y_{n2} & \cdots & y_{nn} \end{bmatrix} \begin{bmatrix} c_1 \\ c_2 \\ \vdots \\ c_n \end{bmatrix}.$$

This shows that

$$c_1\mathbf{y}_1 + c_2\mathbf{y}_2 + \cdots + c_n\mathbf{y}_n = Y\mathbf{c}, \tag{3}$$

where

$$\mathbf{c} = \begin{bmatrix} c_1 \\ c_2 \\ \vdots \\ c_n \end{bmatrix}$$

and

$$Y = [\mathbf{y}_1\, \mathbf{y}_2 \cdots \mathbf{y}_n] = \begin{bmatrix} y_{11} & y_{12} & \cdots & y_{1n} \\ y_{21} & y_{22} & \cdots & y_{2n} \\ \vdots & \vdots & \ddots & \vdots \\ y_{n1} & y_{n2} & \cdots & y_{nn} \end{bmatrix}; \tag{4}$$

that is, the columns of Y are the vector functions $\mathbf{y}_1, \mathbf{y}_2, \dots, \mathbf{y}_n$.

For reference below, note that

$$\begin{aligned} Y' &= [\mathbf{y}_1'\, \mathbf{y}_2' \cdots \mathbf{y}_n'] \\ &= [A\mathbf{y}_1\ A\mathbf{y}_2 \cdots A\mathbf{y}_n] \\ &= A[\mathbf{y}_1\ \mathbf{y}_2 \cdots \mathbf{y}_n] = AY; \end{aligned}$$

that is, Y satisfies the matrix differential equation

$$Y' = AY.$$

The determinant of Y,

$$W = \begin{vmatrix} y_{11} & y_{12} & \cdots & y_{1n} \\ y_{21} & y_{22} & \cdots & y_{2n} \\ \vdots & \vdots & \ddots & \vdots \\ y_{n1} & y_{n2} & \cdots & y_{nn} \end{vmatrix} \tag{5}$$

is called the **Wronskian** of $\{\mathbf{y}_1, \mathbf{y}_2, \ldots, \mathbf{y}_n\}$. It can be shown (Exercises 2 and 3) that this definition is analogous to definitions of the Wronskian of scalar functions given in Sections 5.1 and 9.1.

The following theorem is analogous to Theorems 5.1.4 and 9.1.3. The proof is sketched in Exercise 4 for $n = 2$ and in Exercise 5 for general n.

THEOREM 10.3.2

(Abel's Formula)

Suppose that the $n \times n$ matrix $A = A(t)$ is continuous on (a, b), let $\{\mathbf{y}_1, \mathbf{y}_2, \ldots, \mathbf{y}_n\}$ be solutions of $\mathbf{y}' = A(t)\mathbf{y}$ on (a, b), and let t_0 be in (a, b). Then the Wronskian of $\{\mathbf{y}_1, \mathbf{y}_2, \ldots, \mathbf{y}_n\}$ is given by

$$W(t) = W(t_0) \exp\left(\int_{t_0}^{t} [a_{11}(s) + a_{22}(s) + \cdots + a_{nn}(s)]\, ds \right), \tag{6}$$
$$a < t < b.$$

Therefore, either W has no zeros in (a, b) or $W \equiv 0$ on (a, b).

REMARK. The sum of the diagonal entries of a square matrix A is called the *trace* of A, denoted by $\operatorname{tr}(A)$. Thus, for an $n \times n$ matrix A,

$$\operatorname{tr}(A) = a_{11} + a_{22} + \cdots + a_{nn},$$

and (6) can be written as

$$W(t) = W(t_0) \exp\left(\int_{t_0}^{t} \operatorname{tr}(A(s))\, ds \right), \qquad a < t < b.$$

The following theorem is analogous to Theorems 5.1.6 and 9.1.4.

THEOREM 10.3.3

Suppose that the $n \times n$ matrix $A = A(t)$ is continuous on (a, b) and let $\{\mathbf{y}_1, \mathbf{y}_2, \ldots, \mathbf{y}_n\}$ be n solutions of $\mathbf{y}' = A(t)\mathbf{y}$ on (a, b). Then the following statements are equivalent; that is, they are either all true or all false:

(a) *The general solution of $\mathbf{y}' = A(t)\mathbf{y}$ on (a, b) is*

$$\mathbf{y} = c_1 \mathbf{y}_1 + c_2 \mathbf{y}_2 + \cdots + c_n \mathbf{y}_n$$

where $c_1, c_2, \ldots, c_n$ are arbitrary constants.
(b) *$\{\mathbf{y}_1, \mathbf{y}_2, \ldots, \mathbf{y}_n\}$ is a fundamental set of solutions of $\mathbf{y}' = A(t)\mathbf{y}$ on (a, b).*
(c) *$\{\mathbf{y}_1, \mathbf{y}_2, \ldots, \mathbf{y}_n\}$ is linearly independent on (a, b).*
(d) *The Wronskian of $\{\mathbf{y}_1, \mathbf{y}_2, \ldots, \mathbf{y}_n\}$ is nonzero at some point in (a, b).*
(e) *The Wronskian of $\{\mathbf{y}_1, \mathbf{y}_2, \ldots, \mathbf{y}_n\}$ is nonzero at all points in (a, b).*

We say that Y in (4) is a ***fundamental matrix*** for $\mathbf{y}' = A(t)\mathbf{y}$ if any (and therefore all) of the statements **(a)–(e)** of Theorem 10.3.3 are true for the columns of Y. In this case (3) implies that the general solution of $\mathbf{y}' = A(t)\mathbf{y}$ can be written as $\mathbf{y} = Y\mathbf{c}$, where $\mathbf{c}$ is an arbitrary constant n-vector.

EXAMPLE 2 The vector functions

$$\mathbf{y}_1 = \begin{bmatrix} -e^{2t} \\ 2e^{2t} \end{bmatrix} \quad \text{and} \quad \mathbf{y}_2 = \begin{bmatrix} -e^{-t} \\ e^{-t} \end{bmatrix}$$

are solutions of the constant coefficient system

$$\mathbf{y}' = \begin{bmatrix} -4 & -3 \\ 6 & 5 \end{bmatrix} \mathbf{y} \tag{7}$$

on $(-\infty, \infty)$. (Verify.)

(a) Compute the Wronskian of $\{\mathbf{y}_1, \mathbf{y}_2\}$ directly from the definition (5).
(b) Verify Abel's formula (6) for the Wronskian of $\{\mathbf{y}_1, \mathbf{y}_2\}$.
(c) Find the general solution of (7).
(d) Solve the initial value problem

$$\mathbf{y}' = \begin{bmatrix} -4 & -3 \\ 6 & 5 \end{bmatrix} \mathbf{y}, \quad \mathbf{y}(0) = \begin{bmatrix} 4 \\ -5 \end{bmatrix}. \tag{8}$$

Solution **(a)** From (5)

$$W(t) = \begin{vmatrix} -e^{2t} & -e^{-t} \\ 2e^{2t} & e^{-t} \end{vmatrix} = e^{2t}e^{-t} \begin{vmatrix} -1 & -1 \\ 2 & 1 \end{vmatrix} = e^{t}. \tag{9}$$

Solution **(b)** Here

$$A = \begin{bmatrix} -4 & -3 \\ 6 & 5 \end{bmatrix},$$

so $\text{tr}(A) = -4 + 5 = 1$. If t_0 is an arbitrary real number then (6) implies that

$$W(t) = W(t_0) \exp\left(\int_{t_0}^{t} 1\, ds \right) = \begin{vmatrix} -e^{2t_0} & -e^{-t_0} \\ 2e^{2t_0} & e^{-t_0} \end{vmatrix} e^{(t - t_0)} = e^{t_0} e^{t - t_0} = e^{t},$$

which is consistent with (9).

Solution **(c)** Since $W(t) \neq 0$, Theorem 10.3.3 implies that $\{\mathbf{y}_1, \mathbf{y}_2\}$ is a fundamental set of solutions of (7) and

$$Y = \begin{bmatrix} -e^{2t} & -e^{-t} \\ 2e^{2t} & e^{-t} \end{bmatrix}$$

is a fundamental matrix for (7). Therefore the general solution of (7) is

$$\mathbf{y} = c_1 \mathbf{y}_1 + c_2 \mathbf{y}_2 = c_1 \begin{bmatrix} -e^{2t} \\ 2e^{2t} \end{bmatrix} + c_2 \begin{bmatrix} -e^{-t} \\ e^{-t} \end{bmatrix} = \begin{bmatrix} -e^{2t} & -e^{-t} \\ 2e^{2t} & e^{-t} \end{bmatrix} \begin{bmatrix} c_1 \\ c_2 \end{bmatrix}. \tag{10}$$

Solution **(d)** Setting $t = 0$ in (10) and imposing the initial condition in (8) yields

$$c_1 \begin{bmatrix} -1 \\ 2 \end{bmatrix} + c_2 \begin{bmatrix} -1 \\ 1 \end{bmatrix} = \begin{bmatrix} 4 \\ -5 \end{bmatrix}.$$

Thus,

$$-c_1 - c_2 = 4$$
$$2c_1 + c_2 = -5.$$

The solution of this system is $c_1 = -1$, $c_2 = -3$. Substituting these values into (10) yields

$$\mathbf{y} = -\begin{bmatrix} -e^{2t} \\ 2e^{2t} \end{bmatrix} - 3\begin{bmatrix} -e^{-t} \\ e^{-t} \end{bmatrix} = \begin{bmatrix} e^{2t} + 3e^{-t} \\ -2e^{2t} - 3e^{-t} \end{bmatrix}$$

as the solution of (8). ∎

10.3 EXERCISES

1. Prove: If $\mathbf{y}_1, \mathbf{y}_2, \ldots, \mathbf{y}_n$ are solutions of $\mathbf{y}' = A(t)\mathbf{y}$ on (a,b) then any linear combination of $\mathbf{y}_1, \mathbf{y}_2, \ldots, \mathbf{y}_n$ is also a solution of $\mathbf{y}' = A(t)\mathbf{y}$ on (a,b).

2. In Section 5.1 the Wronskian of two solutions y_1 and y_2 of the scalar second order equation

$$P_0(x)y'' + P_1(x)y' + P_2(x)y = 0 \tag{A}$$

was defined to be

$$W = \begin{vmatrix} y_1 & y_2 \\ y_1' & y_2' \end{vmatrix}.$$

   (a) Rewrite (A) as a system of first order equations and show that W is the Wronskian (as defined in this section) of two solutions of this system.

   (b) Apply Eqn. (6) to the system derived in **(a)**, and show that

   $$W(x) = W(x_0) \exp\left\{ -\int_{x_0}^x \frac{P_1(s)}{P_0(s)}\, ds \right\},$$

   which is the form of Abel's formula given in Theorem 5.1.4.

3. In Section 9.1 the Wronskian of n solutions $y_1, y_2, \ldots, y_n$ of the nth order equation

$$P_0(x)y^{(n)} + P_1(x)y^{(n-1)} + \cdots + P_n(x)y = 0 \tag{A}$$

was defined to be

$$W = \begin{vmatrix} y_1 & y_2 & \cdots & y_n \\ y_1' & y_2' & \cdots & y_n' \\ \vdots & \vdots & \ddots & \vdots \\ y_1^{(n-1)} & y_2^{(n-1)} & \cdots & y_n^{(n-1)} \end{vmatrix}.$$

   (a) Rewrite (A) as a system of first order equations and show that W is the Wronskian (as defined in this section) of n solutions of this system.

   (b) Apply Eqn. (6) to the system derived in **(a)**, and show that

   $$W(x) = W(x_0) \exp\left\{ -\int_{x_0}^x \frac{P_1(s)}{P_0(s)}\, ds \right\},$$

   which is the form of Abel's formula given in Theorem 9.1.3.

4. Suppose that

$$\mathbf{y}_1 = \begin{bmatrix} y_{11} \\ y_{21} \end{bmatrix} \quad \text{and} \quad \mathbf{y}_2 = \begin{bmatrix} y_{12} \\ y_{22} \end{bmatrix}$$

are solutions of the 2×2 system $\mathbf{y}' = A\mathbf{y}$ on (a, b), and let

$$Y = \begin{bmatrix} y_{11} & y_{12} \\ y_{21} & y_{22} \end{bmatrix} \quad \text{and} \quad W = \begin{vmatrix} y_{11} & y_{12} \\ y_{21} & y_{22} \end{vmatrix};$$

thus, W is the Wronskian of $\{\mathbf{y}_1, \mathbf{y}_2\}$.

(a) Deduce from the definition of determinant that

$$W' = \begin{vmatrix} y'_{11} & y'_{12} \\ y_{21} & y_{22} \end{vmatrix} + \begin{vmatrix} y_{11} & y_{12} \\ y'_{21} & y'_{22} \end{vmatrix}.$$

(b) Use the equation $Y' = A(t)Y$ and the definition of matrix multiplication to show that

$$[y'_{11} \quad y'_{12}] = a_{11}[y_{11} \quad y_{12}] + a_{12}[y_{21} \quad y_{22}]$$

and

$$[y'_{21} \quad y'_{22}] = a_{21}[y_{11} \quad y_{12}] + a_{22}[y_{21} \quad y_{22}].$$

(c) Use properties of determinants to deduce from **(a)** and **(b)** that

$$\begin{vmatrix} y'_{11} & y'_{12} \\ y_{21} & y_{22} \end{vmatrix} = a_{11}W \quad \text{and} \quad \begin{vmatrix} y_{11} & y_{12} \\ y'_{21} & y'_{22} \end{vmatrix} = a_{22}W.$$

(d) Conclude from **(a)** and **(c)** that if $a < t_0 < b$ then

$$W' = (a_{11} + a_{22})W$$

and use this to show that

$$W(t) = W(t_0) \exp\left(\int_{t_0}^{t} [a_{11}(s) + a_{22}(s)] \, ds \right), \quad a < t < b.$$

5. Suppose that the $n \times n$ matrix $A = A(t)$ is continuous on (a, b). Let

$$Y = \begin{bmatrix} y_{11} & y_{12} & \cdots & y_{1n} \\ y_{21} & y_{22} & \cdots & y_{2n} \\ \vdots & \vdots & \ddots & \vdots \\ y_{n1} & y_{n2} & \cdots & y_{nn} \end{bmatrix}$$

where the columns of Y are solutions of $\mathbf{y}' = A(t)\mathbf{y}$. Let

$$\mathbf{r}_i = [y_{i1} \quad y_{i2} \quad \cdots \quad y_{in}]$$

be the ith row of Y, and let W be the determinant of Y.

(a) Deduce from the definition of determinant that

$$W' = W_1 + W_2 + \cdots + W_n,$$

where, for $1 \le m \le n$, the ith row of W_m is $\mathbf{r}_i$ if $i \ne m$, and $\mathbf{r}'_m$ if $i = m$.

(b) Use the equation $Y' = AY$ and the definition of matrix multiplication to show that

$$\mathbf{r}'_m = a_{m1}\mathbf{r}_1 + a_{m2}\mathbf{r}_2 + \cdots + a_{mn}\mathbf{r}_n.$$

(c) Use properties of determinants to deduce from (b) that

$$\det(W_m) = a_{mm}W.$$

(d) Conclude from (a) and (c) that

$$W' = (a_{11} + a_{22} + \cdots + a_{nn})W,$$

and use this to show that if $a < t_0 < b$ then

$$W(t) = W(t_0) \exp \int_{t_0}^{t} [a_{11}(s) + a_{22}(s) + \cdots + a_{nn}(s)]\, ds, \qquad a < t < b.$$

6. Suppose that the $n \times n$ matrix A is continuous on (a,b) and t_0 is a point in (a,b). Let Y be a fundamental matrix for $\mathbf{y}' = A(t)\mathbf{y}$ on (a,b).

(a) Show that $Y(t_0)$ is invertible.

(b) Show that if $\mathbf{k}$ is an arbitrary n-vector then the solution of the initial value problem

$$\mathbf{y}' = A(t)\mathbf{y}, \qquad \mathbf{y}(t_0) = \mathbf{k}$$

is

$$\mathbf{y} = Y(t)Y^{-1}(t_0)\mathbf{k}.$$

7. Let

$$A = \begin{bmatrix} 2 & 4 \\ 4 & 2 \end{bmatrix}, \qquad \mathbf{y}_1 = \begin{bmatrix} e^{6t} \\ e^{6t} \end{bmatrix}, \qquad \mathbf{y}_2 = \begin{bmatrix} e^{-2t} \\ -e^{-2t} \end{bmatrix}, \qquad \mathbf{k} = \begin{bmatrix} -3 \\ 9 \end{bmatrix}.$$

(a) Verify that $\{\mathbf{y}_1, \mathbf{y}_2\}$ is a fundamental set of solutions for $\mathbf{y}' = A\mathbf{y}$.

(b) Solve the initial value problem

$$\mathbf{y}' = A\mathbf{y}, \qquad \mathbf{y}(0) = \mathbf{k}. \tag{A}$$

(c) Use the result of Exercise 6(b) to find a formula for the solution of (A) for an arbitrary initial vector $\mathbf{k}$.

8. Repeat Exercise 7 with

$$A = \begin{bmatrix} -2 & -2 \\ -5 & 1 \end{bmatrix}, \qquad \mathbf{y}_1 = \begin{bmatrix} e^{-4t} \\ e^{-4t} \end{bmatrix}, \qquad \mathbf{y}_2 = \begin{bmatrix} -2e^{3t} \\ 5e^{3t} \end{bmatrix}, \qquad \mathbf{k} = \begin{bmatrix} 10 \\ -4 \end{bmatrix}.$$

9. Repeat Exercise 7 with

$$A = \begin{bmatrix} -4 & -10 \\ 3 & 7 \end{bmatrix}, \qquad \mathbf{y}_1 = \begin{bmatrix} -5e^{2t} \\ 3e^{2t} \end{bmatrix}, \qquad \mathbf{y}_2 = \begin{bmatrix} 2e^{t} \\ -e^{t} \end{bmatrix}, \qquad \mathbf{k} = \begin{bmatrix} -19 \\ 11 \end{bmatrix}.$$

10. Repeat Exercise 7 with

$$A = \begin{bmatrix} 2 & 1 \\ 1 & 2 \end{bmatrix}, \qquad \mathbf{y}_1 = \begin{bmatrix} e^{3t} \\ e^{3t} \end{bmatrix}, \qquad \mathbf{y}_2 = \begin{bmatrix} e^{t} \\ -e^{t} \end{bmatrix}, \qquad \mathbf{k} = \begin{bmatrix} 2 \\ 8 \end{bmatrix}.$$

11. Let

$$A = \begin{bmatrix} 3 & -1 & -1 \\ -2 & 3 & 2 \\ 4 & -1 & -2 \end{bmatrix}, \qquad \mathbf{y}_1 = \begin{bmatrix} e^{2t} \\ 0 \\ e^{2t} \end{bmatrix}, \qquad \mathbf{y}_2 = \begin{bmatrix} e^{3t} \\ -e^{3t} \\ e^{3t} \end{bmatrix}, \qquad \mathbf{y}_3 = \begin{bmatrix} e^{-t} \\ -3e^{-t} \\ 7e^{-t} \end{bmatrix}, \qquad \mathbf{k} = \begin{bmatrix} 2 \\ -7 \\ 20 \end{bmatrix}.$$

(a) Verify that $\{\mathbf{y}_1, \mathbf{y}_2, \mathbf{y}_3\}$ is a fundamental set of solutions for $\mathbf{y}' = A\mathbf{y}$.

(b) Solve the initial value problem

$$\mathbf{y}' = A\mathbf{y}, \qquad \mathbf{y}(0) = \mathbf{k}. \tag{A}$$

(c) Use the result of Exercise 6(b) to find a formula for the solution of (A) for an arbitrary initial vector $\mathbf{k}$.

12. Repeat Exercise 11 with

$$A = \begin{bmatrix} 0 & 2 & 2 \\ 2 & 0 & 2 \\ 2 & 2 & 0 \end{bmatrix}, \quad \mathbf{y}_1 = \begin{bmatrix} -e^{-2t} \\ 0 \\ e^{-2t} \end{bmatrix}, \quad \mathbf{y}_2 = \begin{bmatrix} -e^{-2t} \\ e^{-2t} \\ 0 \end{bmatrix}, \quad \mathbf{y}_3 = \begin{bmatrix} e^{4t} \\ e^{4t} \\ e^{4t} \end{bmatrix}, \quad \mathbf{k} = \begin{bmatrix} 0 \\ -9 \\ 12 \end{bmatrix}.$$

13. Repeat Exercise 11 with

$$A = \begin{bmatrix} -1 & 2 & 3 \\ 0 & 1 & 6 \\ 0 & 0 & -2 \end{bmatrix}, \quad \mathbf{y}_1 = \begin{bmatrix} e^{t} \\ e^{t} \\ 0 \end{bmatrix}, \quad \mathbf{y}_2 = \begin{bmatrix} e^{-t} \\ 0 \\ 0 \end{bmatrix}, \quad \mathbf{y}_3 = \begin{bmatrix} e^{-2t} \\ -2e^{-2t} \\ e^{-2t} \end{bmatrix}, \quad \mathbf{k} = \begin{bmatrix} 5 \\ 5 \\ -1 \end{bmatrix}.$$

14. Suppose that Y and Z are fundamental matrices for the $n \times n$ system $\mathbf{y}' = A(t)\mathbf{y}$. Then some of the four matrices $YZ^{-1}, Y^{-1}Z, Z^{-1}Y, ZY^{-1}$ are necessarily constant. Identify them and prove that they are constant.

15. Suppose that the columns of an $n \times n$ matrix Y are solutions of the $n \times n$ system $\mathbf{y}' = A\mathbf{y}$ and C is an $n \times n$ constant matrix.

(a) Show that the matrix $Z = YC$ satisfies the differential equation $Z' = AZ$.

(b) Show that Z is a fundamental matrix for $\mathbf{y}' = A(t)\mathbf{y}$ if and only if C is invertible and Y is a fundamental matrix for $\mathbf{y}' = A(t)\mathbf{y}$.

16. Suppose that the $n \times n$ matrix $A = A(t)$ is continuous on (a, b) and t_0 is in (a, b). For $i = 1, 2, \ldots, n$ let $\mathbf{y}_i$ be the solution of the initial value problem $\mathbf{y}_i' = A(t)\mathbf{y}_i$, $\mathbf{y}_i(t_0) = \mathbf{e}_i$, where

$$\mathbf{e}_1 = \begin{bmatrix} 1 \\ 0 \\ \vdots \\ 0 \end{bmatrix}, \quad \mathbf{e}_2 = \begin{bmatrix} 0 \\ 1 \\ \vdots \\ 0 \end{bmatrix}, \ldots, \quad \mathbf{e}_n = \begin{bmatrix} 0 \\ 0 \\ \vdots \\ 1 \end{bmatrix};$$

that is, the jth component of $\mathbf{e}_i$ is 1 if $j = i$, 0 if $j \neq i$.

(a) Show that $\{\mathbf{y}_1, \mathbf{y}_2, \ldots, \mathbf{y}_n\}$ is a fundamental set of solutions of $\mathbf{y}' = A(t)\mathbf{y}$ on (a, b).

(b) Conclude from (a) and Exercise 15 that $\mathbf{y}' = A(t)\mathbf{y}$ has infinitely many fundamental sets of solutions on (a, b).

17. Show that Y is a fundamental matrix for the system $\mathbf{y}' = A(t)\mathbf{y}$ if and only if Y^{-1} is a fundamental matrix for $\mathbf{y}' = -A^T(t)\mathbf{y}$, where A^T denotes the transpose of A.

Hint: See Exercise 11 of Section 10.2.

18. Let Z be the fundamental matrix for the constant coefficient system $\mathbf{y}' = A\mathbf{y}$ such that $Z(0) = I$.

(a) Show that $Z(t)Z(s) = Z(t + s)$ for all s and t.

Hint: For fixed s let $\Gamma_1(t) = Z(t)Z(s)$ and $\Gamma_2(t) = Z(t + s)$. Show that Γ_1 and Γ_2 are both solutions of the matrix initial value problem $\Gamma' = A\Gamma$, $\Gamma(0) = Z(s)$. Then conclude from Theorem 10.2.1 that $\Gamma_1 = \Gamma_2$.

(b) Show that $(Z(t))^{-1} = Z(-t)$.

(c) The matrix Z defined above is sometimes denoted by e^{tA}. Discuss the motivation for this notation.

10.4 Constant Coefficient Homogeneous Systems I

We will now begin our study of the homogeneous system

$$\mathbf{y}' = A\mathbf{y} \tag{1}$$

where A is an $n \times n$ constant matrix. Since A is continuous on $(-\infty, \infty)$, Theorem 10.2.1 implies that all solutions of (1) are defined on $(-\infty, \infty)$. Therefore, when we speak of solutions of $\mathbf{y}' = A\mathbf{y}$, we will mean solutions on $(-\infty, \infty)$.

In this section we assume that all the eigenvalues of A are real and that A has a set of n linearly independent eigenvectors. In the next two sections we consider the cases where some of the eigenvalues of A are complex, or where A does not have n linearly independent eigenvectors.

In Example 2 of Section 10.3 we showed that the vector functions

$$\mathbf{y}_1 = \begin{bmatrix} -e^{2t} \\ 2e^{2t} \end{bmatrix} \quad \text{and} \quad \mathbf{y}_2 = \begin{bmatrix} -e^{-t} \\ e^{-t} \end{bmatrix}$$

form a fundamental set of solutions of the system

$$\mathbf{y}' = \begin{bmatrix} -4 & -3 \\ 6 & 5 \end{bmatrix} \mathbf{y}, \tag{2}$$

but we did not show how we obtained $\mathbf{y}_1$ and $\mathbf{y}_2$ in the first place. To see how these solutions can be obtained we write (2) as

$$\begin{aligned} y_1' &= -4y_1 - 3y_2 \\ y_2' &= 6y_1 + 5y_2 \end{aligned} \tag{3}$$

and look for solutions of the form

$$y_1 = x_1 e^{\lambda t} \quad \text{and} \quad y_2 = x_2 e^{\lambda t}, \tag{4}$$

where x_1, x_2, and λ are constants to be determined. Differentiating (4) yields

$$y_1' = \lambda x_1 e^{\lambda t} \quad \text{and} \quad y_2' = \lambda x_2 e^{\lambda t}.$$

Substituting this and (4) into (3) and canceling the common factor $e^{\lambda t}$ yields

$$\begin{aligned} -4x_1 - 3x_2 &= \lambda x_1 \\ 6x_1 + 5x_2 &= \lambda x_2. \end{aligned}$$

For a given λ this is a homogeneous algebraic system, since it can be rewritten as

$$\begin{aligned} (-4 - \lambda)x_1 - 3x_2 &= 0 \\ 6x_1 + (5 - \lambda)x_2 &= 0. \end{aligned} \tag{5}$$

The trivial solution $x_1 = x_2 = 0$ of this system is not useful, since it corresponds to the trivial solution $y_1 \equiv y_2 \equiv 0$ of (3), which cannot be part of a fundamental set of solutions of (2). Therefore, we consider only those values of λ for which (5) has nontrivial solutions. These are the values of λ for which the determinant of (5) is zero; that is,

$$\begin{aligned} \begin{vmatrix} -4 - \lambda & -3 \\ 6 & 5 - \lambda \end{vmatrix} &= (-4 - \lambda)(5 - \lambda) + 18 \\ &= \lambda^2 - \lambda - 2 \\ &= (\lambda - 2)(\lambda + 1) = 0, \end{aligned}$$

which has the solutions $\lambda_1 = 2$ and $\lambda_2 = -1$.

Taking $\lambda = 2$ in (5) yields

$$-6x_1 - 3x_2 = 0$$
$$6x_1 + 3x_2 = 0,$$

which implies that $x_1 = -x_2/2$, where x_2 can be chosen arbitrarily. Choosing $x_2 = 2$ yields the solution $y_1 = -e^{2t}$, $y_2 = 2e^{2t}$ of (3). We can write this solution in vector form as

$$y_1 = \begin{bmatrix} -1 \\ 2 \end{bmatrix} e^{2t}. \tag{6}$$

Taking $\lambda = -1$ in (5) yields the system

$$-3x_1 - 3x_2 = 0$$
$$6x_1 + 6x_2 = 0,$$

so $x_1 = -x_2$. Taking $x_2 = 1$ here yields the solution $y_1 = -e^{-t}$, $y_2 = e^{-t}$ of (3). We can write this solution in vector form as

$$\mathbf{y}_2 = \begin{bmatrix} -1 \\ 1 \end{bmatrix} e^{-t}. \tag{7}$$

In (6) and (7) the constant coefficients in the arguments of the exponential functions are the eigenvalues of the coefficient matrix in (2), and the vector coefficients of the exponential functions are associated eigenvectors. This illustrates the following theorem.

THEOREM 10.4.1

Suppose that the $n \times n$ constant matrix A has n real (not necessarily distinct) eigenvalues $\lambda_1, \lambda_2, \ldots, \lambda_n$ with associated linearly independent eigenvectors $\mathbf{x}_1, \mathbf{x}_2, \ldots, \mathbf{x}_n$. Then the functions

$$\mathbf{y}_1 = \mathbf{x}_1 e^{\lambda_1 t}, \qquad \mathbf{y}_2 = \mathbf{x}_2 e^{\lambda_2 t}, \ldots, \qquad \mathbf{y}_n = \mathbf{x}_n e^{\lambda_n t}$$

form a fundamental set of solutions of $\mathbf{y}' = A\mathbf{y}$; that is, the general solution of this system is

$$\mathbf{y} = c_1 \mathbf{x}_1 e^{\lambda_1 t} + c_2 \mathbf{x}_2 e^{\lambda_2 t} + \cdots + c_n \mathbf{x}_n e^{\lambda_n t}.$$

PROOF. Differentiating $\mathbf{y}_i = \mathbf{x}_i e^{\lambda_i t}$ and recalling that $A\mathbf{x}_i = \lambda_i \mathbf{x}_i$ yields

$$\mathbf{y}_i' = \lambda_i \mathbf{x}_i e^{\lambda_i t} = A\mathbf{x}_i e^{\lambda_i t} = A\mathbf{y}_i.$$

This shows that $\mathbf{y}_i$ is a solution of $\mathbf{y}' = A\mathbf{y}$.

The Wronskian of $\{\mathbf{y}_1, \mathbf{y}_2, \ldots, \mathbf{y}_n\}$ is

$$\begin{vmatrix} x_{11}e^{\lambda_1 t} & x_{12}e^{\lambda_2 t} & \cdots & x_{1n}e^{\lambda_n t} \\ x_{21}e^{\lambda_1 t} & x_{22}e^{\lambda_2 t} & \cdots & x_{2n}e^{\lambda_n t} \\ \vdots & \vdots & \ddots & \vdots \\ x_{n1}e^{\lambda_1 t} & x_{n2}e^{\lambda_2 t} & \cdots & x_{nn}e^{\lambda_n t} \end{vmatrix} = e^{\lambda_1 t}e^{\lambda_2 t} \cdots e^{\lambda_n t} \begin{vmatrix} x_{11} & x_{12} & \cdots & x_{1n} \\ x_{21} & x_{22} & \cdots & x_{2n} \\ \vdots & \vdots & \ddots & \vdots \\ x_{n1} & x_{n2} & \cdots & x_{nn} \end{vmatrix}.$$

Since the columns of the determinant on the right are $\mathbf{x}_1, \mathbf{x}_2, \ldots, \mathbf{x}_n$, which are assumed to be linearly independent, the determinant is nonzero. Therefore Theorem 10.3.3 implies that $\{\mathbf{y}_1, \mathbf{y}_2, \ldots, \mathbf{y}_n\}$ is a fundamental set of solutions of $\mathbf{y}' = A\mathbf{y}$. $\qquad \square$

EXAMPLE 1 (a) Find the general solution of

$$\mathbf{y}' = \begin{bmatrix} 2 & 4 \\ 4 & 2 \end{bmatrix} \mathbf{y}. \tag{8}$$

(b) Solve the initial value problem

$$\mathbf{y}' = \begin{bmatrix} 2 & 4 \\ 4 & 2 \end{bmatrix} \mathbf{y}, \qquad \mathbf{y}(0) = \begin{bmatrix} 5 \\ -1 \end{bmatrix}. \tag{9}$$

Solution (a) The characteristic polynomial of the coefficient matrix A in (8) is

$$\begin{vmatrix} 2 - \lambda & 4 \\ 4 & 2 - \lambda \end{vmatrix} = (\lambda - 2)^2 - 16$$

$$= (\lambda - 2 - 4)(\lambda - 2 + 4)$$

$$= (\lambda - 6)(\lambda + 2).$$

Hence, $\lambda_1 = 6$ and $\lambda_2 = -2$ are eigenvalues of A. To obtain the eigenvectors we must solve the system

$$\begin{bmatrix} 2 - \lambda & 4 \\ 4 & 2 - \lambda \end{bmatrix} \begin{bmatrix} x_1 \\ x_2 \end{bmatrix} = \begin{bmatrix} 0 \\ 0 \end{bmatrix} \tag{10}$$

with $\lambda = 6$ and $\lambda = -2$. Setting $\lambda = 6$ in (10) yields

$$\begin{bmatrix} -4 & 4 \\ 4 & -4 \end{bmatrix} \begin{bmatrix} x_1 \\ x_2 \end{bmatrix} = \begin{bmatrix} 0 \\ 0 \end{bmatrix},$$

which implies that $x_1 = x_2$. Taking $x_2 = 1$ yields the eigenvector

$$\mathbf{x}_1 = \begin{bmatrix} 1 \\ 1 \end{bmatrix},$$

so

$$\mathbf{y}_1 = \begin{bmatrix} 1 \\ 1 \end{bmatrix} e^{6t}$$

is a solution of (8). Setting $\lambda = -2$ in (10) yields

$$\begin{bmatrix} 4 & 4 \\ 4 & 4 \end{bmatrix} \begin{bmatrix} x_1 \\ x_2 \end{bmatrix} = \begin{bmatrix} 0 \\ 0 \end{bmatrix},$$

which implies that $x_1 = -x_2$. Taking $x_2 = 1$ yields the eigenvector

$$\mathbf{x}_2 = \begin{bmatrix} -1 \\ 1 \end{bmatrix},$$

so

$$\mathbf{y}_2 = \begin{bmatrix} -1 \\ 1 \end{bmatrix} e^{-2t}$$

is a solution of (8). From Theorem 10.4.1 the general solution of (8) is

$$\mathbf{y} = c_1 \mathbf{y}_1 + c_2 \mathbf{y}_2 = c_1 \begin{bmatrix} 1 \\ 1 \end{bmatrix} e^{6t} + c_2 \begin{bmatrix} -1 \\ 1 \end{bmatrix} e^{-2t}. \tag{11}$$

Solution **(b)** To satisfy the initial condition in (9) we must choose c_1 and c_2 in (11) so that

$$c_1 \begin{bmatrix} 1 \\ 1 \end{bmatrix} + c_2 \begin{bmatrix} -1 \\ 1 \end{bmatrix} = \begin{bmatrix} 5 \\ -1 \end{bmatrix}.$$

This is equivalent to the system

$$\begin{aligned} c_1 - c_2 &= 5 \\ c_1 + c_2 &= -1, \end{aligned}$$

so $c_1 = 2$, $c_2 = -3$. Therefore the solution of (9) is

$$\mathbf{y} = 2 \begin{bmatrix} 1 \\ 1 \end{bmatrix} e^{6t} - 3 \begin{bmatrix} -1 \\ 1 \end{bmatrix} e^{-2t},$$

or, in terms of components,

$$y_1 = 2e^{6t} + 3e^{-2t}, \qquad y_2 = 2e^{6t} - 3e^{-2t}.$$

EXAMPLE 2 **(a)** Find the general solution of

$$\mathbf{y}' = \begin{bmatrix} 3 & -1 & -1 \\ -2 & 3 & 2 \\ 4 & -1 & -2 \end{bmatrix} \mathbf{y}. \tag{12}$$

(b) Solve the initial value problem

$$\mathbf{y}' = \begin{bmatrix} 3 & -1 & -1 \\ -2 & 3 & 2 \\ 4 & -1 & -2 \end{bmatrix} \mathbf{y}, \qquad \mathbf{y}(0) = \begin{bmatrix} 2 \\ -1 \\ 8 \end{bmatrix}. \tag{13}$$

Solution **(a)** The characteristic polynomial of the coefficient matrix A in (12) is

$$\begin{vmatrix} 3 - \lambda & -1 & -1 \\ -2 & 3 - \lambda & 2 \\ 4 & -1 & -2 - \lambda \end{vmatrix} = -(\lambda - 2)(\lambda - 3)(\lambda + 1).$$

Hence, the eigenvalues of A are $\lambda_1 = 2$, $\lambda_2 = 3$ and $\lambda_3 = -1$. To find the eigenvectors we must solve the system

$$\begin{bmatrix} 3 - \lambda & -1 & -1 \\ -2 & 3 - \lambda & 2 \\ 4 & -1 & -2 - \lambda \end{bmatrix} \begin{bmatrix} x_1 \\ x_2 \\ x_3 \end{bmatrix} = \begin{bmatrix} 0 \\ 0 \\ 0 \end{bmatrix} \tag{14}$$

with $\lambda = 2, 3, -1$. With $\lambda = 2$, the augmented matrix of (14) is

$$\begin{bmatrix} 1 & -1 & -1 & \vdots & 0 \\ -2 & 1 & 2 & \vdots & 0 \\ 4 & -1 & -4 & \vdots & 0 \end{bmatrix},$$

which is row equivalent to

$$\left[\begin{array}{ccc:c} 1 & 0 & -1 & 0 \\ 0 & 1 & 0 & 0 \\ 0 & 0 & 0 & 0 \end{array}\right].$$

Hence, we must have $x_1 = x_3$ and $x_2 = 0$. Taking $x_3 = 1$ yields

$$\mathbf{y}_1 = \left[\begin{array}{c} 1 \\ 0 \\ 1 \end{array}\right] e^{2t}$$

as a solution of (12). With $\lambda = 3$, the augmented matrix of (14) is

$$\left[\begin{array}{ccc:c} 0 & -1 & -1 & 0 \\ -2 & 0 & 2 & 0 \\ 4 & -1 & -5 & 0 \end{array}\right],$$

which is row equivalent to

$$\left[\begin{array}{ccc:c} 1 & 0 & -1 & 0 \\ 0 & 1 & 1 & 0 \\ 0 & 0 & 0 & 0 \end{array}\right].$$

Hence, we must have $x_1 = x_3$ and $x_2 = -x_3$. Taking $x_3 = 1$ yields

$$\mathbf{y}_2 = \left[\begin{array}{c} 1 \\ -1 \\ 1 \end{array}\right] e^{3t}$$

as a solution of (12). With $\lambda = -1$, the augmented matrix of (14) is

$$\left[\begin{array}{ccc:c} 4 & -1 & -1 & 0 \\ -2 & 4 & 2 & 0 \\ 4 & -1 & -1 & 0 \end{array}\right],$$

which is row equivalent to

$$\left[\begin{array}{ccc:c} 1 & 0 & -\frac{1}{7} & 0 \\ 0 & 1 & \frac{3}{7} & 0 \\ 0 & 0 & 0 & 0 \end{array}\right].$$

Hence $x_1 = x_3/7$ and $x_2 = -3x_3/7$. Taking $x_3 = 7$ yields

$$\mathbf{y}_3 = \left[\begin{array}{c} 1 \\ -3 \\ 7 \end{array}\right] e^{-t}$$

as a solution of (12). By Theorem 10.4.1 the general solution of (12) is

$$\mathbf{y} = c_1 \left[\begin{array}{c} 1 \\ 0 \\ 1 \end{array}\right] e^{2t} + c_2 \left[\begin{array}{c} 1 \\ -1 \\ 1 \end{array}\right] e^{3t} + c_3 \left[\begin{array}{c} 1 \\ -3 \\ 7 \end{array}\right] e^{-t},$$

which can also be written as

$$\mathbf{y} = \left[\begin{array}{ccc} e^{2t} & e^{3t} & e^{-t} \\ 0 & -e^{3t} & -3e^{-t} \\ e^{2t} & e^{3t} & 7e^{-t} \end{array}\right] \left[\begin{array}{c} c_1 \\ c_2 \\ c_3 \end{array}\right]. \tag{15}$$

Solution   **(b)** To satisfy the initial condition in (13) we must choose c_1, c_2, c_3 in (15) so that

$$\begin{bmatrix} 1 & 1 & 1 \\ 0 & -1 & -3 \\ 1 & 1 & 7 \end{bmatrix} \begin{bmatrix} c_1 \\ c_2 \\ c_3 \end{bmatrix} = \begin{bmatrix} 2 \\ -1 \\ 8 \end{bmatrix}.$$

Solving this system yields $c_1 = 3$, $c_2 = -2$, $c_3 = 1$. Hence, the solution of (13) is

$$\mathbf{y} = \begin{bmatrix} e^{2t} & e^{3t} & e^{-t} \\ 0 & -e^{3t} & -3e^{-t} \\ e^{2t} & e^{3t} & 7e^{-t} \end{bmatrix} \begin{bmatrix} 3 \\ -2 \\ 1 \end{bmatrix}$$

$$= 3 \begin{bmatrix} 1 \\ 0 \\ 1 \end{bmatrix} e^{2t} - 2 \begin{bmatrix} 1 \\ -1 \\ 1 \end{bmatrix} e^{3t} + \begin{bmatrix} 1 \\ -3 \\ 7 \end{bmatrix} e^{-t}.$$

EXAMPLE 3   Find the general solution of

$$\mathbf{y}' = \begin{bmatrix} -3 & 2 & 2 \\ 2 & -3 & 2 \\ 2 & 2 & -3 \end{bmatrix} \mathbf{y}. \tag{16}$$

Solution   The characteristic polynomial of the coefficient matrix A in (16) is

$$\begin{vmatrix} -3 - \lambda & 2 & 2 \\ 2 & -3 - \lambda & 2 \\ 2 & 2 & -3 - \lambda \end{vmatrix} = -(\lambda - 1)(\lambda + 5)^2.$$

Hence, $\lambda_1 = 1$ is an eigenvalue of multiplicity 1, while $\lambda_2 = -5$ is an eigenvalue of multiplicity 2. Eigenvectors associated with $\lambda_1 = 1$ are solutions of the system with augmented matrix

$$\begin{bmatrix} -4 & 2 & 2 & \vdots & 0 \\ 2 & -4 & 2 & \vdots & 0 \\ 2 & 2 & -4 & \vdots & 0 \end{bmatrix},$$

which is row equivalent to

$$\begin{bmatrix} 1 & 0 & -1 & \vdots & 0 \\ 0 & 1 & -1 & \vdots & 0 \\ 0 & 0 & 0 & \vdots & 0 \end{bmatrix}.$$

Hence $x_1 = x_2 = x_3$, and we choose $x_3 = 1$ to obtain the solution

$$\mathbf{y}_1 = \begin{bmatrix} 1 \\ 1 \\ 1 \end{bmatrix} e^t \tag{17}$$

of (16). Eigenvectors associated with $\lambda_2 = -5$ are solutions of the system with augmented matrix

$$\begin{bmatrix} 2 & 2 & 2 & \vdots & 0 \\ 2 & 2 & 2 & \vdots & 0 \\ 2 & 2 & 2 & \vdots & 0 \end{bmatrix}.$$

Hence, the components of these eigenvectors need only satisfy the single condition

$$x_1 + x_2 + x_3 = 0.$$

Since there is only one equation here we can choose x_2 and x_3 arbitrarily. We obtain one eigenvector by choosing $x_2 = 0$ and $x_3 = 1$, and another by choosing $x_2 = 1$ and $x_3 = 0$. In both cases $x_1 = -1$. Therefore

$$\begin{bmatrix} -1 \\ 0 \\ 1 \end{bmatrix} \quad \text{and} \quad \begin{bmatrix} -1 \\ 1 \\ 0 \end{bmatrix}$$

are linearly independent eigenvectors associated with $\lambda_2 = -5$, and the corresponding solutions of (16) are

$$\mathbf{y}_2 = \begin{bmatrix} -1 \\ 0 \\ 1 \end{bmatrix} e^{-5t} \quad \text{and} \quad \mathbf{y}_3 = \begin{bmatrix} -1 \\ 1 \\ 0 \end{bmatrix} e^{-5t}.$$

Because of this and (17), Theorem 10.4.1 implies that the general solution of (16) is

$$\mathbf{y} = c_1 \begin{bmatrix} 1 \\ 1 \\ 1 \end{bmatrix} e^t + c_2 \begin{bmatrix} -1 \\ 0 \\ 1 \end{bmatrix} e^{-5t} + c_3 \begin{bmatrix} -1 \\ 1 \\ 0 \end{bmatrix} e^{-5t}. \qquad \blacksquare$$

GEOMETRIC PROPERTIES OF SOLUTIONS WHEN $n = 2$

We will now consider the geometric properties of solutions of a 2×2 constant coefficient system

$$\begin{bmatrix} y_1' \\ y_2' \end{bmatrix} = \begin{bmatrix} a_{11} & a_{12} \\ a_{21} & a_{22} \end{bmatrix} \begin{bmatrix} y_1 \\ y_2 \end{bmatrix}. \qquad (18)$$

It is convenient to think of a $y_1 y_2$-plane, in which a point is identified by rectangular coordinates (y_1, y_2). If $\mathbf{y} = \begin{bmatrix} y_1 \\ y_2 \end{bmatrix}$ is a nonconstant solution of (18) then the point $(y_1(t), y_2(t))$ moves along a curve C in the $y_1 y_2$-plane as t varies from $-\infty$ to ∞. We call C the **trajectory** of $\mathbf{y}$. (We also say that C is a trajectory of the system (18).) It is important to note that C is the trajectory of infinitely many solutions of (18), since if τ is any real number then $\mathbf{y}(t - \tau)$ is a solution of (18) (Exercise 28**(b)**), and $(y_1(t - \tau), y_2(t - \tau))$ also moves along C as t varies from $-\infty$ to ∞. Moreover, Exercise 28**(c)** implies that distinct trajectories of (18) cannot intersect, and that two solutions $\mathbf{y}_1$ and $\mathbf{y}_2$ of (18) have the same trajectory if and only if $\mathbf{y}_2(t) = \mathbf{y}_1(t - \tau)$ for some τ.

From Exercise 28**(a)**, a trajectory of a nontrivial solution of (18) cannot contain $(0,0)$, which we define to be the trajectory of the trivial solution $\mathbf{y} \equiv \mathbf{0}$. More generally, if $\mathbf{y} = \begin{bmatrix} k_1 \\ k_2 \end{bmatrix} \neq \mathbf{0}$ is a constant solution of (18) (which could occur if zero is an eigenvalue of the matrix of (18)), we define the trajectory of $\mathbf{y}$ to be the single point (k_1, k_2).

To be specific, this is the question: What do the trajectories look like, and how are they traversed? In this section we'll answer this question assuming that the matrix

$$A = \begin{bmatrix} a_{11} & a_{12} \\ a_{21} & a_{22} \end{bmatrix}$$

of (18) has real eigenvalues λ_1 and λ_2 with associated linearly independent eigenvectors $\mathbf{x}_1$ and $\mathbf{x}_2$, so that the general solution of (18) is

$$\mathbf{y} = c_1 \mathbf{x}_1 e^{\lambda_1 t} + c_2 \mathbf{x}_2 e^{\lambda_2 t}. \tag{19}$$

We will consider other situations in the following two sections.

We leave it to you (Exercise 35) to classify the trajectories of (18) if zero is an eigenvalue of A. We'll confine our attention here to the case where both eigenvalues are nonzero. In this case the simplest situation is where $\lambda_1 = \lambda_2 \neq 0$, so (19) becomes

$$\mathbf{y} = (c_1 \mathbf{x}_1 + c_2 \mathbf{x}_2)\, e^{\lambda_1 t}.$$

Since $\mathbf{x}_1$ and $\mathbf{x}_2$ are linearly independent, an arbitrary vector $\mathbf{x}$ can be written as $\mathbf{x} = c_1 \mathbf{x}_1 + c_2 \mathbf{x}_2$. Therefore the general solution of (18) can be written as $\mathbf{y} = \mathbf{x} e^{\lambda_1 t}$ where $\mathbf{x}$ is an arbitrary 2-vector, and the trajectories of nontrivial solutions of (18) are half-lines through (but not including) the origin. The direction of motion is away from the origin if $\lambda_1 > 0$ (Figure 1), toward it if $\lambda_1 < 0$ (Figure 2). (In these and the following figures an arrow through a point indicates the direction of motion along the trajectory through the point.)

Now suppose that $\lambda_2 > \lambda_1$, and let L_1 and L_2 denote lines through the origin parallel to $\mathbf{x}_1$ and $\mathbf{x}_2$, respectively. By a ***half-line*** of L_1 (or L_2) we mean either of the rays obtained by removing the origin from L_1 (or L_2).

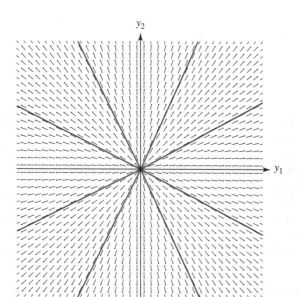

Figure 1   Trajectories of a 2 × 2 system with a repeated positive eigenvalue

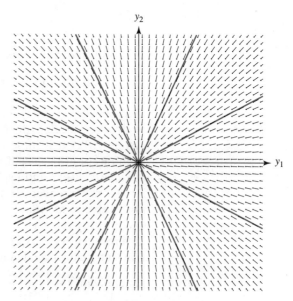

Figure 2   Trajectories of a 2 × 2 system with a repeated negative eigenvalue

Letting $c_2 = 0$ in (19) yields $\mathbf{y} = c_1\mathbf{x}_1e^{\lambda_1 t}$. If $c_1 \neq 0$ the trajectory defined by this solution is a half-line of L_1. The direction of motion is away from the origin if $\lambda_1 > 0$, toward the origin if $\lambda_1 < 0$. Similarly, the trajectory of $\mathbf{y} = c_2\mathbf{x}_2e^{\lambda_2 t}$ with $c_2 \neq 0$ is a half-line of L_2.

Henceforth we assume that c_1 and c_2 in (19) are both nonzero. In this case the trajectory of (19) cannot intersect L_1 or L_2, since every point on these lines is on the trajectory of a solution for which either $c_1 = 0$ or $c_2 = 0$. (Remember: distinct trajectories can't intersect!) Therefore, the trajectory of (19) must lie entirely in one of the four open sectors bounded by L_1 and L_2, but not containing any point on L_1 or L_2. Since the initial point $(y_1(0), y_2(0))$ defined by

$$\mathbf{y}(0) = c_1\mathbf{x}_1 + c_2\mathbf{x}_2$$

is on the trajectory, we can determine which sector contains the trajectory from the signs of c_1 and c_2, as shown in Figure 3.

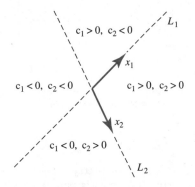

Figure 3 Four open sectors bounded by L_1 and L_2

The direction of $\mathbf{y}(t)$ in (19) is the same as that of

$$e^{-\lambda_2 t}\mathbf{y}(t) = c_1\mathbf{x}_1e^{-(\lambda_2 - \lambda_1)t} + c_2\mathbf{x}_2 \tag{20}$$

and of

$$e^{-\lambda_1 t}\mathbf{y}(t) = c_1\mathbf{x}_1 + c_2\mathbf{x}_2e^{(\lambda_2 - \lambda_1)t}. \tag{21}$$

Since the right side of (20) approaches $c_2\mathbf{x}_2$ as $t \to \infty$, the trajectory is asymptotically parallel to L_2 as $t \to \infty$. Since the right side of (21) approaches $c_1\mathbf{x}_1$ as $t \to -\infty$, the trajectory is asymptotically parallel to L_1 as $t \to -\infty$.

The shape and direction of traversal of the trajectory of (19) depend upon whether λ_1 and λ_2 are both positive, both negative, or of opposite signs. We'll now analyze these three cases.

In the following let $\|\mathbf{u}\|$ denote the length of the vector $\mathbf{u}$.

CASE 1: $\lambda_2 > \lambda_1 > 0$

Some typical trajectories are shown in Figure 4. In this case $\lim_{t \to -\infty} \|\mathbf{y}(t)\| = 0$, so the trajectory is not only asymptotically parallel to L_1 as $t \to -\infty$, but is actually asymptotically tangent to L_1 at the origin. On the other hand, $\lim_{t \to \infty} \|\mathbf{y}(t)\| = \infty$ and

$$\lim_{t \to \infty} \|\mathbf{y}(t) - c_2\mathbf{x}_2e^{\lambda_2 t}\| = \lim_{t \to \infty} \|c_1\mathbf{x}_1e^{\lambda_1 t}\| = \infty,$$

so although the trajectory is asymptotically parallel to L_2 as $t \to \infty$, it is not asymptotically tangent to L_2. The direction of motion along each trajectory is away from the origin.

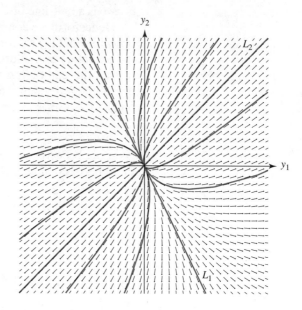

Figure 4 Two positive eigen-values; motion away from origin

CASE 2: $0 > \lambda_2 > \lambda_1$

Some typical trajectories are shown in Figure 5. In this case $\lim_{t \to \infty} \| \mathbf{y}(t) \| = 0$, so the trajectory is asymptotically tangent to L_2 at the origin as $t \to \infty$. On the other hand, $\lim_{t \to -\infty} \| \mathbf{y}(t) \| = \infty$ and

$$\lim_{t \to -\infty} \| \mathbf{y}(t) - c_1 \mathbf{x}_1 e^{\lambda_1 t} \| = \lim_{t \to -\infty} \| c_2 \mathbf{x}_2 e^{\lambda_2 t} \| = \infty,$$

so although the trajectory is asymptotically parallel to L_1 as $t \to -\infty$, it is not asymptotically tangent to it. The direction of motion along each trajectory is toward the origin.

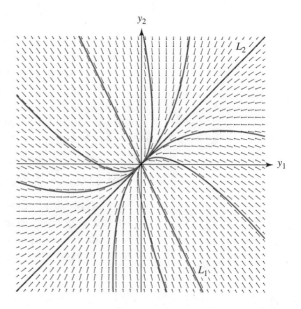

Figure 5 Two negative eigen-values; motion toward the origin

CASE 3: $\lambda_2 > 0 > \lambda_1$

Some typical trajectories are shown in Figure 6. In this case

$$\lim_{t\to\infty} \|\mathbf{y}(t)\| = \infty \qquad \text{and} \qquad \lim_{t\to\infty} \|\mathbf{y}(t) - c_2\mathbf{x}_2 e^{\lambda_2 t}\| = \lim_{t\to\infty} \|c_1\mathbf{x}_1 e^{\lambda_1 t}\| = 0,$$

so the trajectory is asymptotically tangent to L_2 as $t \to \infty$. Similarly,

$$\lim_{t\to-\infty} \|\mathbf{y}(t)\| = \infty \qquad \text{and} \qquad \lim_{t\to-\infty} \|\mathbf{y}(t) - c_1\mathbf{x}_1 e^{\lambda_1 t}\| = \lim_{t\to-\infty} \|c_2\mathbf{x}_2 e^{\lambda_2 t}\| = 0,$$

so the trajectory is asymptotically tangent to L_1 as $t \to -\infty$. The direction of motion is toward the origin on L_1 and away from the origin on L_2. The direction of motion along any other trajectory is away from L_1, toward L_2.

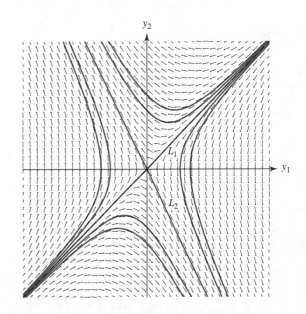

Figure 6 Eigenvalues of different signs

10.4 EXERCISES

In Exercises 1–15 find the general solution.

1. $\mathbf{y}' = \begin{bmatrix} 1 & 2 \\ 2 & 1 \end{bmatrix} \mathbf{y}$

2. $\mathbf{y}' = \dfrac{1}{4} \begin{bmatrix} -5 & 3 \\ 3 & -5 \end{bmatrix} \mathbf{y}$

3. $\mathbf{y}' = \dfrac{1}{5} \begin{bmatrix} -4 & 3 \\ -2 & -11 \end{bmatrix} \mathbf{y}$

4. $\mathbf{y}' = \begin{bmatrix} -1 & -4 \\ -1 & -1 \end{bmatrix} \mathbf{y}$

5. $\mathbf{y}' = \begin{bmatrix} 2 & -4 \\ -1 & -1 \end{bmatrix} \mathbf{y}$

6. $\mathbf{y}' = \begin{bmatrix} 4 & -3 \\ 2 & -1 \end{bmatrix} \mathbf{y}$

7. $\mathbf{y}' = \begin{bmatrix} -6 & -3 \\ 1 & -2 \end{bmatrix} \mathbf{y}$

8. $\mathbf{y}' = \begin{bmatrix} 1 & -1 & -2 \\ 1 & -2 & -3 \\ -4 & 1 & -1 \end{bmatrix} \mathbf{y}$

9. $\mathbf{y}' = \begin{bmatrix} -6 & -4 & -8 \\ -4 & 0 & -4 \\ -8 & -4 & -6 \end{bmatrix} \mathbf{y}$

10. $\mathbf{y}' = \begin{bmatrix} 3 & 5 & 8 \\ 1 & -1 & -2 \\ -1 & -1 & -1 \end{bmatrix} \mathbf{y}$

11. $\mathbf{y}' = \begin{bmatrix} 1 & -1 & 2 \\ 12 & -4 & 10 \\ -6 & 1 & -7 \end{bmatrix} \mathbf{y}$

12. $\mathbf{y}' = \begin{bmatrix} 4 & -1 & -4 \\ 4 & -3 & -2 \\ 1 & -1 & -1 \end{bmatrix} \mathbf{y}$

13. $\mathbf{y}' = \begin{bmatrix} -2 & 2 & 6 \\ 2 & 6 & 2 \\ -2 & -2 & 2 \end{bmatrix} \mathbf{y}$

14. $\mathbf{y}' = \begin{bmatrix} 3 & 2 & -2 \\ -2 & 7 & -2 \\ -10 & 10 & -5 \end{bmatrix} \mathbf{y}$

15. $\mathbf{y}' = \begin{bmatrix} 3 & 1 & -1 \\ 3 & 5 & 1 \\ -6 & 2 & 4 \end{bmatrix} \mathbf{y}$

In Exercises 16–27 solve the initial value problem.

16. $\mathbf{y}' = \begin{bmatrix} -7 & 4 \\ -6 & 7 \end{bmatrix} \mathbf{y}, \quad \mathbf{y}(0) = \begin{bmatrix} 2 \\ -4 \end{bmatrix}$

17. $\mathbf{y}' = \dfrac{1}{6} \begin{bmatrix} 7 & 2 \\ -2 & 2 \end{bmatrix} \mathbf{y}, \quad \mathbf{y}(0) = \begin{bmatrix} 0 \\ -3 \end{bmatrix}$

18. $\mathbf{y}' = \begin{bmatrix} 21 & -12 \\ 24 & -15 \end{bmatrix} \mathbf{y}, \quad \mathbf{y}(0) = \begin{bmatrix} 5 \\ 3 \end{bmatrix}$

19. $\mathbf{y}' = \begin{bmatrix} -7 & 4 \\ -6 & 7 \end{bmatrix} \mathbf{y}, \quad \mathbf{y}(0) = \begin{bmatrix} -1 \\ 7 \end{bmatrix}$

20. $\mathbf{y}' = \dfrac{1}{6} \begin{bmatrix} 1 & 2 & 0 \\ 4 & -1 & 0 \\ 0 & 0 & 3 \end{bmatrix} \mathbf{y}, \quad \mathbf{y}(0) = \begin{bmatrix} 4 \\ 7 \\ 1 \end{bmatrix}$

21. $\mathbf{y}' = \dfrac{1}{3} \begin{bmatrix} 2 & -2 & 3 \\ -4 & 4 & 3 \\ 2 & 1 & 0 \end{bmatrix} \mathbf{y}, \quad \mathbf{y}(0) = \begin{bmatrix} 1 \\ 1 \\ 5 \end{bmatrix}$

22. $\mathbf{y}' = \begin{bmatrix} 6 & -3 & -8 \\ 2 & 1 & -2 \\ 3 & -3 & -5 \end{bmatrix} \mathbf{y}, \quad \mathbf{y}(0) = \begin{bmatrix} 0 \\ -1 \\ -1 \end{bmatrix}$

23. $\mathbf{y}' = \dfrac{1}{3} \begin{bmatrix} 2 & 4 & -7 \\ 1 & 5 & -5 \\ -4 & 4 & -1 \end{bmatrix} \mathbf{y}, \quad \mathbf{y}(0) = \begin{bmatrix} 4 \\ 1 \\ 3 \end{bmatrix}$

24. $\mathbf{y}' = \begin{bmatrix} 3 & 0 & 1 \\ 11 & -2 & 7 \\ 1 & 0 & 3 \end{bmatrix} \mathbf{y}, \quad \mathbf{y}(0) = \begin{bmatrix} 2 \\ 7 \\ 6 \end{bmatrix}$

25. $\mathbf{y}' = \begin{bmatrix} -2 & -5 & -1 \\ -4 & -1 & 1 \\ 4 & 5 & 3 \end{bmatrix} \mathbf{y}, \quad \mathbf{y}(0) = \begin{bmatrix} 8 \\ -10 \\ -4 \end{bmatrix}$

26. $\mathbf{y}' = \begin{bmatrix} 3 & -1 & 0 \\ 4 & -2 & 0 \\ 4 & -4 & 2 \end{bmatrix} \mathbf{y}, \quad \mathbf{y}(0) = \begin{bmatrix} 7 \\ 10 \\ 2 \end{bmatrix}$

27. $\mathbf{y}' = \begin{bmatrix} -2 & 2 & 6 \\ 2 & 6 & 2 \\ -2 & -2 & 2 \end{bmatrix} \mathbf{y}, \quad \mathbf{y}(0) = \begin{bmatrix} 6 \\ -10 \\ 7 \end{bmatrix}$

28. Let A be an $n \times n$ constant matrix. Then Theorem 10.2.1 implies that the solutions of

$$\mathbf{y}' = A\mathbf{y} \tag{A}$$

are all defined on $(-\infty, \infty)$.

(a) Use Theorem 10.2.1 to show that the only solution of (A) that can ever equal the zero vector is $\mathbf{y} \equiv \mathbf{0}$.

(b) Suppose that $\mathbf{y}_1$ is a solution of (A) and $\mathbf{y}_2$ is defined by $\mathbf{y}_2(t) = \mathbf{y}_1(t - \tau)$, where τ is an arbitrary real number. Show that $\mathbf{y}_2$ is also a solution of (A).

(c) Suppose that $\mathbf{y}_1$ and $\mathbf{y}_2$ are solutions of (A) and there are real numbers t_1 and t_2 such that $\mathbf{y}_1(t_1) = \mathbf{y}_2(t_2)$. Show that $\mathbf{y}_2(t) = \mathbf{y}_1(t - \tau)$ for all t, where $\tau = t_2 - t_1$.

Hint: Show that $\mathbf{y}_1(t - \tau)$ and $\mathbf{y}_2(t)$ are solutions of the same initial value problem for (A) and apply the uniqueness assertion of Theorem 10.2.1.

In Exercises 29–34 describe and graph trajectories of the given system.

C **29.** $\mathbf{y}' = \begin{bmatrix} 1 & 1 \\ 1 & -1 \end{bmatrix} \mathbf{y}$

C **30.** $\mathbf{y}' = \begin{bmatrix} -4 & 3 \\ -2 & -11 \end{bmatrix} \mathbf{y}$

C **31.** $\mathbf{y}' = \begin{bmatrix} 9 & -3 \\ -1 & 11 \end{bmatrix} \mathbf{y}$

C **32.** $\mathbf{y}' = \begin{bmatrix} -1 & -10 \\ -5 & 4 \end{bmatrix} \mathbf{y}$

C **33.** $\mathbf{y}' = \begin{bmatrix} 5 & -4 \\ 1 & 10 \end{bmatrix} \mathbf{y}$

C **34.** $\mathbf{y}' = \begin{bmatrix} -7 & 1 \\ 3 & -5 \end{bmatrix} \mathbf{y}$

C **35.** Suppose that the eigenvalues of the 2×2 matrix A are $\lambda = 0$ and $\mu \neq 0$, with corresponding eigenvectors $\mathbf{x}_1$ and $\mathbf{x}_2$. Let L_1 be the line through the origin parallel to $\mathbf{x}_1$.

    (a) Show that every point on L_1 is the trajectory of a constant solution of $\mathbf{y}' = A\mathbf{y}$.

    (b) Show that the trajectories of nonconstant solutions of $\mathbf{y}' = A\mathbf{y}$ are half-lines parallel to $\mathbf{x}_2$ and on either side of L_1, and that the direction of motion along these trajectories is away from L_1 if $\mu > 0$, or toward L_1 if $\mu < 0$.

The matrices of the systems in Exercises 36–41 are singular. Describe and graph the trajectories of nonconstant solutions of the given systems.

C **36.** $\mathbf{y}' = \begin{bmatrix} -1 & 1 \\ 1 & -1 \end{bmatrix} \mathbf{y}$

C **37.** $\mathbf{y}' = \begin{bmatrix} -1 & -3 \\ 2 & 6 \end{bmatrix} \mathbf{y}$

C **38.** $\mathbf{y}' = \begin{bmatrix} 1 & -3 \\ -1 & 3 \end{bmatrix} \mathbf{y}$

C **39.** $\mathbf{y}' = \begin{bmatrix} 1 & -2 \\ -1 & 2 \end{bmatrix} \mathbf{y}$

C **40.** $\mathbf{y}' = \begin{bmatrix} -4 & -4 \\ 1 & 1 \end{bmatrix} \mathbf{y}$

C **41.** $\mathbf{y}' = \begin{bmatrix} 3 & -1 \\ -3 & 1 \end{bmatrix} \mathbf{y}$

L **42.** Let $P = P(t)$ and $Q = Q(t)$ be the populations of two species at time t, and assume that each population would grow exponentially if the other didn't exist; that is, in the absence of competition we would have

$$P' = aP \quad \text{and} \quad Q' = bQ, \tag{A}$$

where a and b are positive constants. One way to model the effect of competition is to assume that the growth rate of each population is reduced by an amount proportional to the other population, so (A) is replaced by

$$P' = \cdot aP - \alpha Q$$
$$Q' = -\beta P + bQ,$$

where α and β are positive constants. (Since negative population doesn't make sense, this system holds only while P and Q are both positive.) Now suppose that $P(0) = P_0 > 0$ and $Q(0) = Q_0 > 0$.

    (a) For several choices of a, b, α, and β, verify experimentally (by graphing trajectories of (A) in the PQ-plane) that there is a constant $\rho > 0$ (depending upon a, b, α, and β) with the following properties:

        (i) If $Q_0 > \rho P_0$ then P decreases monotonically to zero in finite time, during which Q remains positive.

        (ii) If $Q_0 < \rho P_0$ then Q decreases monotonically to zero in finite time, during which P remains positive.

    (b) Conclude from **(a)** that exactly one of the species becomes extinct in finite time if $Q_0 \neq \rho P_0$. Determine experimentally what happens if $Q_0 = \rho P_0$.

    (c) Confirm your experimental results and determine γ by expressing the eigenvalues and associated eigenvectors of

$$A = \begin{bmatrix} a & -\alpha \\ -\beta & b \end{bmatrix}$$

in terms of a, b, α, and β, and applying the geometric arguments developed at the end of this section.

10.5   Constant Coefficient Homogeneous Systems II

We saw in Section 10.4 that if an $n \times n$ constant matrix A has n real eigenvalues $\lambda_1, \lambda_2, \ldots, \lambda_n$ (which need not be distinct) with associated linearly independent eigenvectors $\mathbf{x}_1, \mathbf{x}_2, \ldots, \mathbf{x}_n$, then the general solution of $\mathbf{y}' = A\mathbf{y}$ is

$$\mathbf{y} = c_1 \mathbf{x}_1 e^{\lambda_1 t} + c_2 \mathbf{x}_2 e^{\lambda_2 t} + \cdots + c_n \mathbf{x}_n e^{\lambda_n t}.$$

In this section we consider the case where A has n real eigenvalues, but does not have n linearly independent eigenvectors. It is shown in linear algebra that this occurs if and only if A has at least one eigenvalue of multiplicity $r > 1$ such that the associated eigenspace has dimension less than r. In this case A is said to be **defective.** Since it is beyond the scope of this book to give a complete analysis of systems with defective coefficient matrices, we will restrict our attention to some commonly occurring special cases.

EXAMPLE 1   Show that the system

$$\mathbf{y}' = \begin{bmatrix} 11 & -25 \\ 4 & -9 \end{bmatrix} \mathbf{y} \tag{1}$$

does not have a fundamental set of solutions of the form $\{\mathbf{x}_1 e^{\lambda_1 t}, \mathbf{x}_2 e^{\lambda_2 t}\}$ where λ_1 and λ_2 are eigenvalues of the coefficient matrix A of (1) and $\{\mathbf{x}_1, \mathbf{x}_2\}$ are associated linearly independent eigenvectors.

Solution   The characteristic polynomial of A is

$$\begin{vmatrix} 11 - \lambda & -25 \\ 4 & -9 - \lambda \end{vmatrix} = (\lambda - 11)(\lambda + 9) + 100$$

$$= \lambda^2 - 2\lambda + 1 = (\lambda - 1)^2.$$

Hence, $\lambda = 1$ is the only eigenvalue of A. The augmented matrix of the system $(A - I)\mathbf{x} = \mathbf{0}$ is

$$\begin{bmatrix} 10 & -25 & \vdots & 0 \\ 4 & -10 & \vdots & 0 \end{bmatrix},$$

which is row equivalent to

$$\begin{bmatrix} 1 & -\frac{5}{2} & \vdots & 0 \\ 0 & 0 & \vdots & 0 \end{bmatrix}.$$

Hence, $x_1 = 5x_2/2$ where x_2 is arbitrary. Therefore all eigenvectors of A are scalar multiples of $\mathbf{x}_1 = \begin{bmatrix} 5 \\ 2 \end{bmatrix}$, so A does not have a set of two linearly independent eigenvectors. ∎

From Example 1 we know that all scalar multiples of $\mathbf{y}_1 = \begin{bmatrix} 5 \\ 2 \end{bmatrix} e^t$ are solutions of (1); however, to find the general solution we must find a second solution $\mathbf{y}_2$ such that $\{\mathbf{y}_1, \mathbf{y}_2\}$ is linearly independent. Based on your recollection of the procedure for solving a constant coefficient scalar equation

$$ay'' + by' + cy = 0$$

in the case where the characteristic polynomial has a repeated root, you might expect to obtain a second solution of (1) by multiplying the first solution by t. However, this yields $\mathbf{y}_2 = \begin{bmatrix} 5 \\ 2 \end{bmatrix} te^t$, which doesn't work, since

$$\mathbf{y}_2' = \begin{bmatrix} 5 \\ 2 \end{bmatrix}(te^t + e^t), \qquad \text{while} \qquad \begin{bmatrix} 11 & -25 \\ 4 & -9 \end{bmatrix} \mathbf{y}_2 = \begin{bmatrix} 5 \\ 2 \end{bmatrix} te^t.$$

The following theorem shows what to do in this situation.

THEOREM 10.5.1

Suppose that the $n \times n$ matrix A has an eigenvalue λ_1 of multiplicity ≥ 2 and the associated eigenspace has dimension 1; that is, all λ_1-eigenvectors of A are scalar multiples of an eigenvector $\mathbf{x}$. Then there are infinitely many vectors $\mathbf{u}$ such that

$$(A - \lambda_1 I)\mathbf{u} = \mathbf{x}. \tag{2}$$

Moreover, if $\mathbf{u}$ is any such vector then

$$\mathbf{y}_1 = \mathbf{x}e^{\lambda_1 t} \quad and \quad \mathbf{y}_2 = \mathbf{u}e^{\lambda_1 t} + \mathbf{x}te^{\lambda_1 t} \tag{3}$$

are linearly independent solutions of $\mathbf{y}' = A\mathbf{y}$.

A complete proof of this theorem is beyond the scope of this book. The difficulty is in proving that there is a vector $\mathbf{u}$ satisfying (2), since $\det(A - \lambda_1 I) = 0$. We will take this without proof and verify the other assertions of the theorem.

We already know that $\mathbf{y}_1$ in (3) is a solution of $\mathbf{y}' = A\mathbf{y}$. To see that $\mathbf{y}_2$ is also a solution, we compute

$$\mathbf{y}_2' - A\mathbf{y}_2 = \lambda_1 \mathbf{u}e^{\lambda_1 t} + \mathbf{x}e^{\lambda_1 t} + \lambda_1 \mathbf{x}te^{\lambda_1 t} - A\mathbf{u}e^{\lambda_1 t} - A\mathbf{x}te^{\lambda_1 t}$$
$$= (\lambda_1 \mathbf{u} + \mathbf{x} - A\mathbf{u})e^{\lambda_1 t} + (\lambda_1 \mathbf{x} - A\mathbf{x})te^{\lambda_1 t}.$$

Since $A\mathbf{x} = \lambda_1 \mathbf{x}$ this can be written as

$$\mathbf{y}_2' - A\mathbf{y}_2 = -((A - \lambda_1 I)\mathbf{u} - \mathbf{x})e^{\lambda_1 t},$$

and now (2) implies that $\mathbf{y}_2' = A\mathbf{y}_2$.

To see that $\mathbf{y}_1$ and $\mathbf{y}_2$ are linearly independent, suppose that c_1 and c_2 are constants such that

$$c_1 \mathbf{y}_1 + c_2 \mathbf{y}_2 = c_1 \mathbf{x}e^{\lambda_1 t} + c_2(\mathbf{u}e^{\lambda_1 t} + \mathbf{x}te^{\lambda_1 t}) = \mathbf{0}. \tag{4}$$

We must show that $c_1 = c_2 = 0$. Multiplying (4) by $e^{-\lambda_1 t}$ shows that

$$c_1 \mathbf{x} + c_2(\mathbf{u} + \mathbf{x}t) = \mathbf{0}. \tag{5}$$

By differentiating this with respect to t we see that $c_2 \mathbf{x} = \mathbf{0}$, which implies $c_2 = 0$, because $\mathbf{x} \neq \mathbf{0}$. Substituting $c_2 = 0$ into (5) yields $c_1 \mathbf{x} = \mathbf{0}$, which implies that $c_1 = 0$, again because $\mathbf{x} \neq \mathbf{0}$.

EXAMPLE 2 Use Theorem 10.5.1 to find the general solution of the system

$$\mathbf{y}' = \begin{bmatrix} 11 & -25 \\ 4 & -9 \end{bmatrix} \mathbf{y} \tag{6}$$

considered in Example 1.

Solution In Example 1 we saw that $\lambda_1 = 1$ is an eigenvalue of multiplicity 2 of the coefficient matrix A in (6), and that all of the eigenvectors of A are multiples of

$$\mathbf{x} = \begin{bmatrix} 5 \\ 2 \end{bmatrix}.$$

Therefore

$$\mathbf{y}_1 = \begin{bmatrix} 5 \\ 2 \end{bmatrix} e^t$$

is a solution of (6). From Theorem 10.5.1, a second solution is given by $\mathbf{y}_2 = \mathbf{u}e^t + \mathbf{x}te^t$, where $(A - I)\mathbf{u} = \mathbf{x}$. The augmented matrix of this system is

$$\left[\begin{array}{cc:c} 10 & -25 & 5 \\ 4 & -10 & 2 \end{array}\right],$$

which is row equivalent to

$$\left[\begin{array}{cc:c} 1 & -\frac{5}{2} & \frac{1}{2} \\ 0 & 0 & 0 \end{array}\right].$$

Therefore the components of $\mathbf{u}$ must satisfy

$$u_1 - \frac{5}{2}u_2 = \frac{1}{2},$$

where u_2 is arbitrary. We choose $u_2 = 0$, so that $u_1 = 1/2$ and

$$\mathbf{u} = \left[\begin{array}{c} \frac{1}{2} \\ 0 \end{array}\right].$$

Thus,

$$\mathbf{y}_2 = \left[\begin{array}{c} 1 \\ 0 \end{array}\right]\frac{e^t}{2} + \left[\begin{array}{c} 5 \\ 2 \end{array}\right]te^t.$$

Since $\mathbf{y}_1$ and $\mathbf{y}_2$ are linearly independent by Theorem 10.5.1, they form a fundamental set of solutions of (6). Therefore the general solution of (6) is

$$\mathbf{y} = c_1\left[\begin{array}{c} 5 \\ 2 \end{array}\right]e^t + c_2\left(\left[\begin{array}{c} 1 \\ 0 \end{array}\right]\frac{e^t}{2} + \left[\begin{array}{c} 5 \\ 2 \end{array}\right]te^t\right). \qquad \blacksquare$$

Note that choosing the arbitrary constant u_2 to be nonzero is equivalent to adding a scalar multiple of $\mathbf{y}_1$ to the second solution $\mathbf{y}_2$ (Exercise 33).

EXAMPLE 3 Find the general solution of

$$\mathbf{y}' = \left[\begin{array}{ccc} 3 & 4 & -10 \\ 2 & 1 & -2 \\ 2 & 2 & -5 \end{array}\right]\mathbf{y}. \qquad (7)$$

Solution The characteristic polynomial of the coefficient matrix in (7) is

$$\left|\begin{array}{ccc} 3 - \lambda & 4 & -10 \\ 2 & 1 - \lambda & -2 \\ 2 & 2 & -5 - \lambda \end{array}\right| = -(\lambda - 1)(\lambda + 1)^2.$$

Hence, the eigenvalues are $\lambda_1 = 1$ with multiplicity 1 and $\lambda_2 = -1$ with multiplicity 2.

Eigenvectors associated with $\lambda_1 = 1$ must satisfy $(A - I)\mathbf{x} = \mathbf{0}$. The augmented matrix of this system is

$$\left[\begin{array}{ccc:c} 2 & 4 & -10 & 0 \\ 2 & 0 & -2 & 0 \\ 2 & 2 & -6 & 0 \end{array}\right],$$

which is row equivalent to

$$\left[\begin{array}{ccc:c} 1 & 0 & -1 & 0 \\ 0 & 1 & -2 & 0 \\ 0 & 0 & 0 & 0 \end{array}\right].$$

Hence, $x_1 = x_3$ and $x_2 = 2x_3$, where x_3 is arbitrary. Choosing $x_3 = 1$ yields the eigenvector

$$\mathbf{x}_1 = \begin{bmatrix} 1 \\ 2 \\ 1 \end{bmatrix}.$$

Therefore

$$\mathbf{y}_1 = \begin{bmatrix} 1 \\ 2 \\ 1 \end{bmatrix} e^t$$

is a solution of (7).

Eigenvectors associated with $\lambda_2 = -1$ satisfy $(A + I)\mathbf{x} = \mathbf{0}$. The augmented matrix of this system is

$$\left[\begin{array}{ccc:c} 4 & 4 & -10 & 0 \\ 2 & 2 & -2 & 0 \\ 2 & 2 & -4 & 0 \end{array} \right],$$

which is row equivalent to

$$\left[\begin{array}{ccc:c} 1 & 1 & 0 & 0 \\ 0 & 0 & 1 & 0 \\ 0 & 0 & 0 & 0 \end{array} \right].$$

Hence, $x_3 = 0$ and $x_1 = -x_2$, where x_2 is arbitrary. Choosing $x_2 = 1$ yields the eigenvector

$$\mathbf{x}_2 = \begin{bmatrix} -1 \\ 1 \\ 0 \end{bmatrix},$$

so

$$\mathbf{y}_2 = \begin{bmatrix} -1 \\ 1 \\ 0 \end{bmatrix} e^{-t}$$

is a solution of (7).

Since all the eigenvectors of A associated with $\lambda_2 = -1$ are multiples of $\mathbf{x}_2$, we must now use Theorem 10.5.1 to find a third solution of (7) in the form

$$\mathbf{y}_3 = \mathbf{u}e^{-t} + \begin{bmatrix} -1 \\ 1 \\ 0 \end{bmatrix} te^{-t}, \tag{8}$$

where $\mathbf{u}$ is a solution of $(A + I)\mathbf{u} = \mathbf{x}_2$. The augmented matrix of this system is

$$\left[\begin{array}{ccc:c} 4 & 4 & -10 & -1 \\ 2 & 2 & -2 & 1 \\ 2 & 2 & -4 & 0 \end{array} \right],$$

which is row equivalent to

$$\begin{bmatrix} 1 & 1 & 0 & \vdots & 1 \\ 0 & 0 & 1 & \vdots & \frac{1}{2} \\ 0 & 0 & 0 & \vdots & 0 \end{bmatrix}.$$

Hence, $u_3 = 1/2$ and $u_1 = 1 - u_2$, where u_2 is arbitrary. Choosing $u_2 = 0$ yields

$$\mathbf{u} = \begin{bmatrix} 1 \\ 0 \\ \frac{1}{2} \end{bmatrix},$$

and substituting this into (8) yields the solution

$$\mathbf{y}_3 = \begin{bmatrix} 2 \\ 0 \\ 1 \end{bmatrix} \frac{e^{-t}}{2} + \begin{bmatrix} -1 \\ 1 \\ 0 \end{bmatrix} te^{-t}$$

of (7).

Since the Wronskian of $\{\mathbf{y}_1, \mathbf{y}_2, \mathbf{y}_3\}$ at $t = 0$ is

$$\begin{vmatrix} 1 & -1 & 1 \\ 2 & 1 & 0 \\ 1 & 0 & \frac{1}{2} \end{vmatrix} = \frac{1}{2},$$

it follows that $\{\mathbf{y}_1, \mathbf{y}_2, \mathbf{y}_3\}$ is a fundamental set of solutions of (7). Therefore the general solution of (7) is

$$\mathbf{y} = c_1 \begin{bmatrix} 1 \\ 2 \\ 1 \end{bmatrix} e^t + c_2 \begin{bmatrix} -1 \\ 1 \\ 0 \end{bmatrix} e^{-t} + c_3 \left(\begin{bmatrix} 2 \\ 0 \\ 1 \end{bmatrix} \frac{e^{-t}}{2} + \begin{bmatrix} -1 \\ 1 \\ 0 \end{bmatrix} te^{-t} \right). \quad \blacksquare$$

THEOREM 10.5.2

Suppose that the $n \times n$ matrix A has an eigenvalue λ_1 of multiplicity ≥ 3 and the associated eigenspace is one-dimensional; that is, all eigenvectors associated with λ_1 are scalar multiples of the eigenvector $\mathbf{x}$. Then there are infinitely many vectors $\mathbf{u}$ such that

$$(A - \lambda_1 I)\mathbf{u} = \mathbf{x}, \tag{9}$$

and if $\mathbf{u}$ is any such vector then there are infinitely many vectors $\mathbf{v}$ such that

$$(A - \lambda_1 I)\mathbf{v} = \mathbf{u}. \tag{10}$$

If $\mathbf{u}$ satisfies (9) and $\mathbf{v}$ satisfies (10) then

$$\mathbf{y}_1 = \mathbf{x}e^{\lambda_1 t},$$
$$\mathbf{y}_2 = \mathbf{u}e^{\lambda_1 t} + \mathbf{x}te^{\lambda_1 t},$$

and

$$\mathbf{y}_3 = \mathbf{v}e^{\lambda_1 t} + \mathbf{u}te^{\lambda_1 t} + \mathbf{x}\frac{t^2 e^{\lambda_1 t}}{2}$$

are linearly independent solutions of $\mathbf{y}' = A\mathbf{y}$.

Again, it is beyond the scope of this book to prove that there are vectors **u** and **v** that satisfy (9) and (10). Theorem 10.5.1 implies that $\mathbf{y}_1$ and $\mathbf{y}_2$ are solutions of $\mathbf{y}' = A\mathbf{y}$. We leave the rest of the proof to you (Exercise 34).

EXAMPLE 4 Use Theorem 10.5.2 to find the general solution of

$$\mathbf{y}' = \begin{bmatrix} 1 & 1 & 1 \\ 1 & 3 & -1 \\ 0 & 2 & 2 \end{bmatrix} \mathbf{y}. \tag{11}$$

Solution The characteristic polynomial of the coefficient matrix A in (11) is

$$\begin{vmatrix} 1-\lambda & 1 & 1 \\ 1 & 3-\lambda & -1 \\ 0 & 2 & 2-\lambda \end{vmatrix} = -(\lambda - 2)^3.$$

Hence, $\lambda_1 = 2$ is an eigenvalue of multiplicity 3. The associated eigenvectors satisfy $(A - 2I)\mathbf{x} = \mathbf{0}$. The augmented matrix of this system is

$$\begin{bmatrix} -1 & 1 & 1 & \vdots & 0 \\ 1 & 1 & -1 & \vdots & 0 \\ 0 & 2 & 0 & \vdots & 0 \end{bmatrix},$$

which is row equivalent to

$$\begin{bmatrix} 1 & 0 & -1 & \vdots & 0 \\ 0 & 1 & 0 & \vdots & 0 \\ 0 & 0 & 0 & \vdots & 0 \end{bmatrix}.$$

Hence $x_1 = x_3$ and $x_2 = 0$, so the eigenvectors are all scalar multiples of

$$\mathbf{x}_1 = \begin{bmatrix} 1 \\ 0 \\ 1 \end{bmatrix}.$$

Therefore

$$\mathbf{y}_1 = \begin{bmatrix} 1 \\ 0 \\ 1 \end{bmatrix} e^{2t}$$

is a solution of (11).

We now find a second solution of (11) in the form

$$\mathbf{y}_2 = \mathbf{u}e^{2t} + \begin{bmatrix} 1 \\ 0 \\ 1 \end{bmatrix} te^{2t}$$

where **u** satisfies $(A - 2I)\mathbf{u} = \mathbf{x}_1$. The augmented matrix of this system is

$$\begin{bmatrix} -1 & 1 & 1 & \vdots & 1 \\ 1 & 1 & -1 & \vdots & 0 \\ 0 & 2 & 0 & \vdots & 1 \end{bmatrix},$$

which is row equivalent to

$$\begin{bmatrix} 1 & 0 & -1 & \vdots & -\frac{1}{2} \\ 0 & 1 & 0 & \vdots & \frac{1}{2} \\ 0 & 0 & 0 & \vdots & 0 \end{bmatrix}.$$

Letting $u_3 = 0$, we have $u_1 = -1/2$ and $u_2 = 1/2$; hence

$$\mathbf{u} = \frac{1}{2} \begin{bmatrix} -1 \\ 1 \\ 0 \end{bmatrix}$$

and

$$\mathbf{y}_2 = \begin{bmatrix} -1 \\ 1 \\ 0 \end{bmatrix} \frac{e^{2t}}{2} + \begin{bmatrix} 1 \\ 0 \\ 1 \end{bmatrix} te^{2t}$$

is a solution of (11).

We now find a third solution of (11) in the form

$$\mathbf{y}_3 = \mathbf{v}e^{2t} + \begin{bmatrix} -1 \\ 1 \\ 0 \end{bmatrix} \frac{te^{2t}}{2} + \begin{bmatrix} 1 \\ 0 \\ 1 \end{bmatrix} \frac{t^2 e^{2t}}{2}$$

where $\mathbf{v}$ satisfies $(A - 2I)\mathbf{v} = \mathbf{u}$. The augmented matrix of this system is

$$\begin{bmatrix} -1 & 1 & 1 & \vdots & -\frac{1}{2} \\ 1 & 1 & -1 & \vdots & \frac{1}{2} \\ 0 & 2 & 0 & \vdots & 0 \end{bmatrix}.$$

which is row equivalent to

$$\begin{bmatrix} 1 & 0 & -1 & \vdots & \frac{1}{2} \\ 0 & 1 & 0 & \vdots & 0 \\ 0 & 0 & 0 & \vdots & 0 \end{bmatrix},$$

Letting $v_3 = 0$, we have $v_1 = 1/2$ and $v_2 = 0$; hence

$$\mathbf{v} = \frac{1}{2} \begin{bmatrix} 1 \\ 0 \\ 0 \end{bmatrix}.$$

Therefore

$$\mathbf{y}_3 = \begin{bmatrix} 1 \\ 0 \\ 0 \end{bmatrix} \frac{e^{2t}}{2} + \begin{bmatrix} -1 \\ 1 \\ 0 \end{bmatrix} \frac{te^{2t}}{2} + \begin{bmatrix} 1 \\ 0 \\ 1 \end{bmatrix} \frac{t^2 e^{2t}}{2}$$

is a solution of (11). Since $\mathbf{y}_1, \mathbf{y}_2,$ and $\mathbf{y}_3$ are linearly independent by Theorem 10.5.2, they form a fundamental set of solutions of (11). Therefore the general solution of (11) is

$$\mathbf{y} = c_1 \begin{bmatrix} 1 \\ 0 \\ 1 \end{bmatrix} e^{2t} + c_2 \left(\begin{bmatrix} -1 \\ 1 \\ 0 \end{bmatrix} \frac{e^{2t}}{2} + \begin{bmatrix} 1 \\ 0 \\ 1 \end{bmatrix} t e^{2t} \right)$$

$$+ c_3 \left(\begin{bmatrix} 1 \\ 0 \\ 0 \end{bmatrix} \frac{e^{2t}}{2} + \begin{bmatrix} -1 \\ 1 \\ 0 \end{bmatrix} \frac{t e^{2t}}{2} + \begin{bmatrix} 1 \\ 0 \\ 1 \end{bmatrix} \frac{t^2 e^{2t}}{2} \right). \qquad \blacksquare$$

THEOREM 10.5.3

Suppose that the $n \times n$ matrix A has an eigenvalue λ_1 of multiplicity ≥ 3 and the associated eigenspace is two-dimensional; that is, all eigenvectors of A associated with λ_1 are linear combinations of two linearly independent eigenvectors $\mathbf{x}_1$ and $\mathbf{x}_2$. Then there are constants α and β (not both zero) such that if

$$\mathbf{x}_3 = \alpha \mathbf{x}_1 + \beta \mathbf{x}_2 \qquad (12)$$

then there are infinitely many vectors $\mathbf{u}$ such that

$$(A - \lambda_1 I)\mathbf{u} = \mathbf{x}_3. \qquad (13)$$

If $\mathbf{u}$ satisfies (13) then

$$\mathbf{y}_1 = \mathbf{x}_1 e^{\lambda_1 t},$$
$$\mathbf{y}_2 = \mathbf{x}_2 e^{\lambda_1 t},$$

and

$$\mathbf{y}_3 = \mathbf{u} e^{\lambda_1 t} + \mathbf{x}_3 t e^{\lambda_1 t}, \qquad (14)$$

are linearly independent solutions of $\mathbf{y}' = A\mathbf{y}$.

We omit the proof of this theorem.

EXAMPLE 5 Use Theorem 10.5.3 to find the general solution of

$$\mathbf{y}' = \begin{bmatrix} 0 & 0 & 1 \\ -1 & 1 & 1 \\ -1 & 0 & 2 \end{bmatrix} \mathbf{y}. \qquad (15)$$

Solution The characteristic polynomial of the coefficient matrix A in (15) is

$$\begin{vmatrix} -\lambda & 0 & 1 \\ -1 & 1-\lambda & 1 \\ -1 & 0 & 2-\lambda \end{vmatrix} = -(\lambda - 1)^3.$$

Hence, $\lambda_1 = 1$ is an eigenvalue of multiplicity 3. The associated eigenvectors satisfy $(A - I)\mathbf{x} = \mathbf{0}$. The augmented matrix of this system is

$$\begin{bmatrix} -1 & 0 & 1 & \vdots & 0 \\ -1 & 0 & 1 & \vdots & 0 \\ -1 & 0 & 1 & \vdots & 0 \end{bmatrix},$$

which is row equivalent to

$$\begin{bmatrix} 1 & 0 & -1 & \vdots & 0 \\ 0 & 0 & 0 & \vdots & 0 \\ 0 & 0 & 0 & \vdots & 0 \end{bmatrix}.$$

Hence, $x_1 = x_3$ and x_2 is arbitrary, so the eigenvectors are of the form

$$\mathbf{x}_1 = \begin{bmatrix} x_3 \\ x_2 \\ x_3 \end{bmatrix} = x_3 \begin{bmatrix} 1 \\ 0 \\ 1 \end{bmatrix} + x_2 \begin{bmatrix} 0 \\ 1 \\ 0 \end{bmatrix}.$$

Therefore the vectors

$$\mathbf{x}_1 = \begin{bmatrix} 1 \\ 0 \\ 1 \end{bmatrix} \quad \text{and} \quad \mathbf{x}_2 = \begin{bmatrix} 0 \\ 1 \\ 0 \end{bmatrix} \tag{16}$$

form a basis for the eigenspace, and

$$\mathbf{y}_1 = \begin{bmatrix} 1 \\ 0 \\ 1 \end{bmatrix} e^t \quad \text{and} \quad \mathbf{y}_2 = \begin{bmatrix} 0 \\ 1 \\ 0 \end{bmatrix} e^t$$

are linearly independent solutions of (15).

To find a third linearly independent solution of (15) we must find constants α and β (not both zero) such that the system

$$(A - I)\mathbf{u} = \alpha\mathbf{x}_1 + \beta\mathbf{x}_2 \tag{17}$$

has a solution $\mathbf{u}$. The augmented matrix of this system is

$$\begin{bmatrix} -1 & 0 & 1 & \vdots & \alpha \\ -1 & 0 & 1 & \vdots & \beta \\ -1 & 0 & 1 & \vdots & \alpha \end{bmatrix},$$

which is row equivalent to

$$\begin{bmatrix} 1 & 0 & -1 & \vdots & -\alpha \\ 0 & 0 & 0 & \vdots & \beta - \alpha \\ 0 & 0 & 0 & \vdots & 0 \end{bmatrix}. \tag{18}$$

Therefore (17) has a solution if and only if $\beta = \alpha$, where α is arbitrary. If we choose $\alpha = \beta = 1$ then (12) and (16) yield

$$\mathbf{x}_3 = \mathbf{x}_1 + \mathbf{x}_2 = \begin{bmatrix} 1 \\ 0 \\ 1 \end{bmatrix} + \begin{bmatrix} 0 \\ 1 \\ 0 \end{bmatrix} = \begin{bmatrix} 1 \\ 1 \\ 1 \end{bmatrix},$$

and the augmented matrix (18) becomes

$$\begin{bmatrix} 1 & 0 & -1 & \vdots & -1 \\ 0 & 0 & 0 & \vdots & 0 \\ 0 & 0 & 0 & \vdots & 0 \end{bmatrix}.$$

This implies that $u_1 = -1 + u_3$, while u_2 and u_3 are arbitrary. Choosing $u_2 = u_3 = 0$ yields

$$\mathbf{u} = \begin{bmatrix} -1 \\ 0 \\ 0 \end{bmatrix}.$$

Therefore (14) implies that

$$\mathbf{y}_3 = \mathbf{u}e^t + \mathbf{x}_3 te^t = \begin{bmatrix} -1 \\ 0 \\ 0 \end{bmatrix} e^t + \begin{bmatrix} 1 \\ 1 \\ 1 \end{bmatrix} te^t$$

is a solution of (15). Since $\mathbf{y}_1, \mathbf{y}_2$, and $\mathbf{y}_3$ are linearly independent by Theorem 10.5.3, they form a fundamental set of solutions for (15). Therefore the general solution of (15) is

$$\mathbf{y} = c_1 \begin{bmatrix} 1 \\ 0 \\ 1 \end{bmatrix} e^t + c_2 \begin{bmatrix} 0 \\ 1 \\ 0 \end{bmatrix} e^t + c_3 \left(\begin{bmatrix} -1 \\ 0 \\ 0 \end{bmatrix} e^t + \begin{bmatrix} 1 \\ 1 \\ 1 \end{bmatrix} te^t \right). \qquad \blacksquare$$

GEOMETRIC PROPERTIES OF SOLUTIONS WHEN $n = 2$

We will now consider the geometric properties of solutions of a 2×2 constant coefficient system

$$\begin{bmatrix} y_1' \\ y_2' \end{bmatrix} = \begin{bmatrix} a_{11} & a_{12} \\ a_{21} & a_{22} \end{bmatrix} \begin{bmatrix} y_1 \\ y_2 \end{bmatrix} \qquad (19)$$

under the assumptions of this section; that is, when the matrix

$$A = \begin{bmatrix} a_{11} & a_{12} \\ a_{21} & a_{22} \end{bmatrix}$$

has a repeated eigenvalue λ_1 and the associated eigenspace is one-dimensional. In this case we know from Theorem 10.5.1 that the general solution of (19) is

$$\mathbf{y} = c_1 \mathbf{x} e^{\lambda_1 t} + c_2 (\mathbf{u} e^{\lambda_1 t} + \mathbf{x} t e^{\lambda_1 t}), \qquad (20)$$

where $\mathbf{x}$ is an eigenvector of A and $\mathbf{u}$ is any one of the infinitely many solutions of

$$(A - \lambda_1 I)\mathbf{u} = \mathbf{x}. \qquad (21)$$

We assume that $\lambda_1 \neq 0$.

Let L denote the line through the origin parallel to $\mathbf{x}$. By a **half-line** of L we mean either of the rays obtained by removing the origin from L. If $c_2 = 0$ then (20) is a parametric equation of the half-line of L in the direction of $\mathbf{x}$ if $c_1 > 0$, or of the half-line of L in the direction of $-\mathbf{x}$ if $c_1 < 0$. The origin is the trajectory of the trivial solution $\mathbf{y} \equiv \mathbf{0}$.

Henceforth we assume that $c_2 \neq 0$. In this case the trajectory of (20) can't intersect L, since every point of L is on a trajectory obtained by setting $c_2 = 0$. Therefore, the trajectory of (20) must lie entirely in one of the open half-planes bounded by L, but not containing any point on L. Since the initial point $(y_1(0), y_2(0))$ defined by $\mathbf{y}(0) = c_1\mathbf{x}_1 + c_2\mathbf{u}$ is on the trajectory, we can determine which half-plane contains the trajectory from the sign of c_2, as shown in Figure 1. For convenience we'll call the half-plane in which $c_2 > 0$ the **positive half-plane.** Similarly, the half-plane in which $c_2 < 0$ is the ***negative half-plane.*** You should convince yourself (Exercise 35) that even though there are infinitely many vectors **u** that satisfy (21), they all define the same positive and negative half-planes. In the figures simply regard **u** as an arrow pointing to the positive half-plane, since we haven't attempted to give **u** its proper length or direction in comparison with **x**. For our purposes here only the relative orientation of **x** and **u** is important; that is, whether the positive half-plane is to the right of an observer facing the direction of **x** (as in Figures 2 and 5), or to the left of the observer (as in Figures 3 and 4).

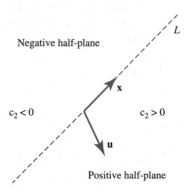

Figure I Positive and negative half-planes

Multiplying (20) by $e^{-\lambda_1 t}$ yields

$$e^{-\lambda_1 t}\mathbf{y}(t) = c_1\mathbf{x} + c_2\mathbf{u} + c_2 t\mathbf{x}.$$

Since the last term on the right is dominant when $|t|$ is large, this provides the following information on the direction of $\mathbf{y}(t)$:

1. Along trajectories in the positive half-plane ($c_2 > 0$) the direction of $\mathbf{y}(t)$ approaches the direction of **x** as $t \to \infty$ and the direction of $-\mathbf{x}$ as $t \to -\infty$.
2. Along trajectories in the negative half-plane ($c_2 < 0$) the direction of $\mathbf{y}(t)$ approaches the direction of $-\mathbf{x}$ as $t \to \infty$ and the direction of **x** as $t \to -\infty$.

Since

$$\lim_{t \to \infty} \|\mathbf{y}(t)\| = \infty \quad \text{and} \quad \lim_{t \to -\infty} \mathbf{y}(t) = 0 \quad \text{if } \lambda_1 > 0,$$

or

$$\lim_{t \to -\infty} \|\mathbf{y}(t)\| = \infty \quad \text{and} \quad \lim_{t \to \infty} \mathbf{y}(t) = 0 \quad \text{if } \lambda_1 < 0,$$

it follows that there are four possible patterns for the trajectories of (19), depending upon the signs of c_2 and λ_1. Figures 2–5 illustrate these patterns, and reveal the following principle:

If λ_1 and c_2 have the same sign then the direction of the trajectory approaches the direction of $-\mathbf{x}$ as $\|\mathbf{y}\| \to 0$ and the direction of $\mathbf{x}$ as $\|\mathbf{y}\| \to \infty$. If λ_1 and c_2 have opposite signs then the direction of the trajectory approaches the direction of $\mathbf{x}$ as $\|\mathbf{y}\| \to 0$ and the direction of $-\mathbf{x}$ as $\|\mathbf{y}\| \to \infty$.

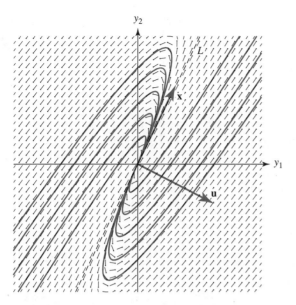

Figure 2 Positive eigenvalue; motion away from the origin

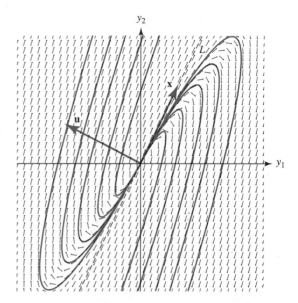

Figure 3 Positive eigenvalue; motion away from the origin

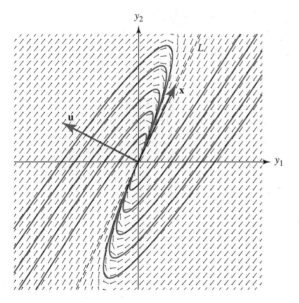

Figure 4 Negative eigenvalue; motion toward the origin

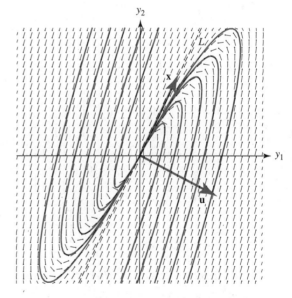

Figure 5 Negative eigenvalue; motion toward the origin

10.5 EXERCISES

In Exercises 1–12 find the general solution.

1. $\mathbf{y}' = \begin{bmatrix} 3 & 4 \\ -1 & 7 \end{bmatrix} \mathbf{y}$

2. $\mathbf{y}' = \begin{bmatrix} 0 & -1 \\ 1 & -2 \end{bmatrix} \mathbf{y}$

3. $\mathbf{y}' = \begin{bmatrix} -7 & 4 \\ -1 & -11 \end{bmatrix} \mathbf{y}$

4. $\mathbf{y}' = \begin{bmatrix} 3 & 1 \\ -1 & 1 \end{bmatrix} \mathbf{y}$

5. $\mathbf{y}' = \begin{bmatrix} 4 & 12 \\ -3 & -8 \end{bmatrix} \mathbf{y}$

6. $\mathbf{y}' = \begin{bmatrix} -10 & 9 \\ -4 & 2 \end{bmatrix} \mathbf{y}$

7. $\mathbf{y}' = \begin{bmatrix} -13 & 16 \\ -9 & 11 \end{bmatrix} \mathbf{y}$

8. $\mathbf{y}' = \begin{bmatrix} 0 & 2 & 1 \\ -4 & 6 & 1 \\ 0 & 4 & 2 \end{bmatrix} \mathbf{y}$

9. $\mathbf{y}' = \dfrac{1}{3} \begin{bmatrix} 1 & 1 & -3 \\ -4 & -4 & 3 \\ -2 & 1 & 0 \end{bmatrix} \mathbf{y}$

10. $\mathbf{y}' = \begin{bmatrix} -1 & 1 & -1 \\ -2 & 0 & 2 \\ -1 & 3 & -1 \end{bmatrix} \mathbf{y}$

11. $\mathbf{y}' = \begin{bmatrix} 4 & -2 & -2 \\ -2 & 3 & -1 \\ 2 & -1 & 3 \end{bmatrix} \mathbf{y}$

12. $\mathbf{y}' = \begin{bmatrix} 6 & -5 & 3 \\ 2 & -1 & 3 \\ 2 & 1 & 1 \end{bmatrix} \mathbf{y}$

In Exercises 13–23 solve the initial value problem.

13. $\mathbf{y}' = \begin{bmatrix} -11 & 8 \\ -2 & -3 \end{bmatrix} \mathbf{y}, \quad \mathbf{y}(0) = \begin{bmatrix} 6 \\ 2 \end{bmatrix}$

14. $\mathbf{y}' = \begin{bmatrix} 15 & -9 \\ 16 & -9 \end{bmatrix} \mathbf{y}, \quad \mathbf{y}(0) = \begin{bmatrix} 5 \\ 8 \end{bmatrix}$

15. $\mathbf{y}' = \begin{bmatrix} -3 & -4 \\ 1 & -7 \end{bmatrix} \mathbf{y}, \quad \mathbf{y}(0) = \begin{bmatrix} 2 \\ 3 \end{bmatrix}$

16. $\mathbf{y}' = \begin{bmatrix} -7 & 24 \\ -6 & 17 \end{bmatrix} \mathbf{y}, \quad \mathbf{y}(0) = \begin{bmatrix} 3 \\ 1 \end{bmatrix}$

17. $\mathbf{y}' = \begin{bmatrix} -7 & 3 \\ -3 & -1 \end{bmatrix} \mathbf{y}, \quad \mathbf{y}(0) = \begin{bmatrix} 0 \\ 2 \end{bmatrix}$

18. $y' = \begin{bmatrix} -1 & 1 & 0 \\ 1 & -1 & -2 \\ -1 & -1 & -1 \end{bmatrix} \mathbf{y}, \quad \mathbf{y}(0) = \begin{bmatrix} 6 \\ 5 \\ -7 \end{bmatrix}$

19. $\mathbf{y}' = \begin{bmatrix} -2 & 2 & 1 \\ -2 & 2 & 1 \\ -3 & 3 & 2 \end{bmatrix} \mathbf{y}, \quad \mathbf{y}(0) = \begin{bmatrix} -6 \\ -2 \\ 0 \end{bmatrix}$

20. $\mathbf{y}' = \begin{bmatrix} -7 & -4 & 4 \\ -1 & 0 & 1 \\ -9 & -5 & 6 \end{bmatrix} \mathbf{y}, \quad \mathbf{y}(0) = \begin{bmatrix} -6 \\ 9 \\ -1 \end{bmatrix}$

21. $\mathbf{y}' = \begin{bmatrix} -1 & -4 & -1 \\ 3 & 6 & 1 \\ -3 & -2 & 3 \end{bmatrix} \mathbf{y}, \quad \mathbf{y}(0) = \begin{bmatrix} -2 \\ 1 \\ 3 \end{bmatrix}$

22. $\mathbf{y}' = \begin{bmatrix} 4 & -8 & -4 \\ -3 & -1 & -3 \\ 1 & -1 & 9 \end{bmatrix} \mathbf{y}, \quad \mathbf{y}(0) = \begin{bmatrix} -4 \\ 1 \\ -3 \end{bmatrix}$

23. $\mathbf{y}' = \begin{bmatrix} -5 & -1 & 11 \\ -7 & 1 & 13 \\ -4 & 0 & 8 \end{bmatrix} \mathbf{y}, \quad \mathbf{y}(0) = \begin{bmatrix} 0 \\ 2 \\ 2 \end{bmatrix}$

The coefficient matrices in Exercises 24–32 have eigenvalues of multiplicity 3. Find the general solution.

24. $\mathbf{y}' = \begin{bmatrix} 5 & -1 & 1 \\ -1 & 9 & -3 \\ -2 & 2 & 4 \end{bmatrix} \mathbf{y}$

25. $\mathbf{y}' = \begin{bmatrix} 1 & 10 & -12 \\ 2 & 2 & 3 \\ 2 & -1 & 6 \end{bmatrix} \mathbf{y}$

26. $\mathbf{y}' = \begin{bmatrix} -6 & -4 & -4 \\ 2 & -1 & 1 \\ 2 & 3 & 1 \end{bmatrix} \mathbf{y}$

27. $\mathbf{y}' = \begin{bmatrix} 0 & 2 & -2 \\ -1 & 5 & -3 \\ 1 & 1 & 1 \end{bmatrix} \mathbf{y}$

28. $\mathbf{y}' = \begin{bmatrix} -2 & -12 & 10 \\ 2 & -24 & 11 \\ 2 & -24 & 8 \end{bmatrix} \mathbf{y}$

29. $\mathbf{y}' = \begin{bmatrix} -1 & -12 & 8 \\ 1 & -9 & 4 \\ 1 & -6 & 1 \end{bmatrix} \mathbf{y}$

30. $\mathbf{y}' = \begin{bmatrix} -4 & 0 & -1 \\ -1 & -3 & -1 \\ 1 & 0 & -2 \end{bmatrix} \mathbf{y}$

31. $\mathbf{y}' = \begin{bmatrix} -3 & -3 & 4 \\ 4 & 5 & -8 \\ 2 & 3 & -5 \end{bmatrix} \mathbf{y}$

32. $\mathbf{y}' = \begin{bmatrix} -3 & -1 & 0 \\ 1 & -1 & 0 \\ -1 & -1 & -2 \end{bmatrix} \mathbf{y}$

33. Under the assumptions of Theorem 10.5.1, suppose that $\mathbf{u}$ and $\hat{\mathbf{u}}$ are vectors such that

$$(A - \lambda_1 I)\mathbf{u} = \mathbf{x} \quad \text{and} \quad (A - \lambda_1 I)\hat{\mathbf{u}} = \mathbf{x},$$

and let

$$\mathbf{y}_2 = \mathbf{u}e^{\lambda_1 t} + \mathbf{x}te^{\lambda_1 t} \quad \text{and} \quad \hat{\mathbf{y}}_2 = \hat{\mathbf{u}}e^{\lambda_1 t} + \mathbf{x}te^{\lambda_1 t}.$$

Show that $\mathbf{y}_2 - \hat{\mathbf{y}}_2$ is a scalar multiple of $\mathbf{y}_1 = \mathbf{x}e^{\lambda_1 t}$.

34. Under the assumptions of Theorem 10.5.2, let

$$\mathbf{y}_1 = \mathbf{x}e^{\lambda_1 t},$$
$$\mathbf{y}_2 = \mathbf{u}e^{\lambda_1 t} + \mathbf{x}te^{\lambda_1 t},$$

and

$$\mathbf{y}_3 = \mathbf{v}e^{\lambda_1 t} + \mathbf{u}te^{\lambda_1 t} + \mathbf{x}\frac{t^2 e^{\lambda_1 t}}{2}.$$

Complete the proof of Theorem 10.5.2 by showing that $\mathbf{y}_3$ is a solution of $\mathbf{y}' = A\mathbf{y}$ and that $\{\mathbf{y}_1, \mathbf{y}_2, \mathbf{y}_3\}$ is linearly independent.

35. Suppose that the matrix

$$A = \begin{bmatrix} a_{11} & a_{12} \\ a_{21} & a_{22} \end{bmatrix}$$

has a repeated eigenvalue λ_1 and the associated eigenspace is one-dimensional. Let $\mathbf{x}$ be an eigenvector of A. Show that if $(A - \lambda_1 I)\mathbf{u}_1 = \mathbf{x}$ and $(A - \lambda_1 I)\mathbf{u}_2 = \mathbf{x}$ then $\mathbf{u}_2 - \mathbf{u}_1$ is parallel to $\mathbf{x}$. Conclude from this that all vectors $\mathbf{u}$ such that $(A - \lambda_1 I)\mathbf{u} = \mathbf{x}$ define the same positive and negative half-planes with respect to the line L through the origin parallel to $\mathbf{x}$.

In Exercises 36–45 graph trajectories of the given system.

C **36.** $\mathbf{y}' = \begin{bmatrix} -3 & -1 \\ 4 & 1 \end{bmatrix} \mathbf{y}$

C **37.** $\mathbf{y}' = \begin{bmatrix} 2 & -1 \\ 1 & 0 \end{bmatrix} \mathbf{y}$

C **38.** $\mathbf{y}' = \begin{bmatrix} -1 & -3 \\ 3 & 5 \end{bmatrix} \mathbf{y}$

C **39.** $\mathbf{y}' = \begin{bmatrix} -5 & 3 \\ -3 & 1 \end{bmatrix} \mathbf{y}$

C **40.** $\mathbf{y}' = \begin{bmatrix} -2 & -3 \\ 3 & 4 \end{bmatrix} \mathbf{y}$

C **41.** $\mathbf{y}' = \begin{bmatrix} -4 & -3 \\ 3 & 2 \end{bmatrix} \mathbf{y}$

C **42.** $\mathbf{y}' = \begin{bmatrix} 0 & -1 \\ 1 & -2 \end{bmatrix} \mathbf{y}$

C **43.** $\mathbf{y}' = \begin{bmatrix} 0 & 1 \\ -1 & 2 \end{bmatrix} \mathbf{y}$

C **44.** $\mathbf{y}' = \begin{bmatrix} -2 & 1 \\ -1 & 0 \end{bmatrix} \mathbf{y}$

C **45.** $\mathbf{y}' = \begin{bmatrix} 0 & -4 \\ 1 & -4 \end{bmatrix} \mathbf{y}$

10.6   Constant Coefficient Homogeneous Systems III

We now consider the system $\mathbf{y}' = A\mathbf{y}$ where A has a complex eigenvalue $\lambda = \alpha + i\beta$ with $\beta \neq 0$. We continue to assume that A has real entries, so the characteristic polynomial of A has real coefficients. This implies that $\overline{\lambda} = \alpha - i\beta$ is also an eigenvalue of A.

An eigenvector $\mathbf{x}$ of A associated with $\lambda = \alpha + i\beta$ will have complex entries, so we will write

$$\mathbf{x} = \mathbf{u} + i\mathbf{v}$$

where $\mathbf{u}$ and $\mathbf{v}$ have real entries; that is, $\mathbf{u}$ and $\mathbf{v}$ are the real and imaginary parts of $\mathbf{x}$. Since $A\mathbf{x} = \lambda\mathbf{x}$,

$$A(\mathbf{u} + i\mathbf{v}) = (\alpha + i\beta)(\mathbf{u} + i\mathbf{v}). \tag{1}$$

Taking complex conjugates here and recalling that A has real entries yields

$$A(\mathbf{u} - i\mathbf{v}) = (\alpha - i\beta)(\mathbf{u} - i\mathbf{v}),$$

which shows that $\mathbf{x} = \mathbf{u} - i\mathbf{v}$ is an eigenvector associated with $\overline{\lambda} = \alpha - i\beta$. The complex conjugate eigenvalues λ and $\overline{\lambda}$ can be separately associated with linearly independent solutions $\mathbf{y}' = A\mathbf{y}$; however, we won't pursue this approach, since solutions obtained in this way turn out to be complex-valued. Instead, we will obtain solutions of $\mathbf{y}' = A\mathbf{y}$ in the form

$$\mathbf{y} = f_1\mathbf{u} + f_2\mathbf{v} \tag{2}$$

where f_1 and f_2 are real-valued scalar functions. The following theorem shows how to do this.

THEOREM 10.6.1

Let A be an $n \times n$ matrix with real entries. Let $\lambda = \alpha + i\beta$ ($\beta \neq 0$) be a complex eigenvalue of A and let $\mathbf{x} = \mathbf{u} + i\mathbf{v}$ be an associated eigenvector, where $\mathbf{u}$ and $\mathbf{v}$ have real components. Then $\mathbf{u}$ and $\mathbf{v}$ are both nonzero and

$$\mathbf{y}_1 = e^{\alpha t}(\mathbf{u} \cos \beta t - \mathbf{v} \sin \beta t) \quad \text{and} \quad \mathbf{y}_2 = e^{\alpha t}(\mathbf{u} \sin \beta t + \mathbf{v} \cos \beta t),$$

which are the real and imaginary parts of

$$e^{\alpha t}(\cos \beta t + i \sin \beta t)(\mathbf{u} + i\mathbf{v}), \tag{3}$$

are linearly independent solutions of $\mathbf{y}' = A\mathbf{y}$.

PROOF. A function of the form (2) is a solution of $\mathbf{y}' = A\mathbf{y}$ if and only if

$$f_1'\mathbf{u} + f_2'\mathbf{v} = f_1 A\mathbf{u} + f_2 A\mathbf{v}. \tag{4}$$

Carrying out the multiplication indicated on the right side of (1) and collecting the real and imaginary parts of the result yields

$$A(\mathbf{u} + i\mathbf{v}) = (\alpha\mathbf{u} - \beta\mathbf{v}) + i(\alpha\mathbf{v} + \beta\mathbf{u}).$$

Equating real and imaginary parts on the two sides of this equation yields

$$A\mathbf{u} = \alpha\mathbf{u} - \beta\mathbf{v}$$
$$A\mathbf{v} = \alpha\mathbf{v} + \beta\mathbf{u}.$$

We leave it to you (Exercise 25) to show from this that $\mathbf{u}$ and $\mathbf{v}$ are both nonzero. Substituting from these equations into (4) yields

$$f_1'\mathbf{u} + f_2'\mathbf{v} = f_1(\alpha\mathbf{u} - \beta\mathbf{v}) + f_2(\alpha\mathbf{v} + \beta\mathbf{u})$$
$$= (\alpha f_1 + \beta f_2)\mathbf{u} + (-\beta f_1 + \alpha f_2)\mathbf{v}.$$

This will hold if

$$\begin{array}{lll} f_1' = & \alpha f_1 + \beta f_2 & \\ f_2' = & -\beta f_1 + \alpha f_2, & \end{array} \quad \text{or, equivalently,} \quad \begin{array}{l} f_1' - \alpha f_1 = \quad \beta f_2 \\ f_2' - \alpha f_2 = -\beta f_1. \end{array}$$

If we let $f_1 = g_1 e^{\alpha t}$ and $f_2 = g_2 e^{\alpha t}$ where g_1 and g_2 are to be determined, then the last two equations become

$$\begin{array}{l} g_1' = \quad \beta g_2 \\ g_2' = -\beta g_1, \end{array}$$

which implies that

$$g_1'' = \beta g_2' = -\beta^2 g_1,$$

so

$$g_1'' + \beta^2 g_1 = 0.$$

The general solution of this equation is

$$g_1 = c_1 \cos \beta t + c_2 \sin \beta t.$$

Moreover, since $g_2 = g_1'/\beta$, we have

$$g_2 = -c_1 \sin \beta t + c_2 \cos \beta t.$$

Multiplying g_1 and g_2 by $e^{\alpha t}$ shows that

$$f_1 = e^{\alpha t}(\quad c_1 \cos \beta t + c_2 \sin \beta t),$$
$$f_2 = e^{\alpha t}(-c_1 \sin \beta t + c_2 \cos \beta t).$$

Substituting these into (2) shows that

$$\mathbf{y} = e^{\alpha t}[(c_1 \cos \beta t + c_2 \sin \beta t)\mathbf{u} + (-c_1 \sin \beta t + c_2 \cos \beta t)\mathbf{v}]$$
$$= c_1 e^{\alpha t}(\mathbf{u} \cos \beta t - \mathbf{v} \sin \beta t) + c_2 e^{\alpha t}(\mathbf{u} \sin \beta t + \mathbf{v} \cos \beta t) \tag{5}$$

is a solution of $\mathbf{y}' = A\mathbf{y}$ for any choice of the constants c_1 and c_2. In particular, by first taking $c_1 = 1$ and $c_2 = 0$ and then taking $c_1 = 0$ and $c_2 = 1$, we see that $\mathbf{y}_1$ and $\mathbf{y}_2$ are solutions of $\mathbf{y}' = A\mathbf{y}$. We leave it to you to verify that they are, respectively, the real and imaginary parts of (3) (Exercise 26), and that they are linearly independent (Exercise 27). ☐

EXAMPLE 1   Find the general solution of

$$\mathbf{y}' = \begin{bmatrix} 4 & -5 \\ 5 & -2 \end{bmatrix} \mathbf{y}. \tag{6}$$

Solution The characteristic polynomial of the coefficient matrix A in (6) is

$$\begin{vmatrix} 4 - \lambda & -5 \\ 5 & -2 - \lambda \end{vmatrix} = (\lambda - 1)^2 + 16.$$

Hence, $\lambda = 1 + 4i$ is an eigenvalue of A. The associated eigenvectors satisfy $(A - (1 + 4i)I)\mathbf{x} = \mathbf{0}$. The augmented matrix of this system is

$$\begin{bmatrix} 3 - 4i & -5 & \vdots & 0 \\ 5 & -3 - 4i & \vdots & 0 \end{bmatrix},$$

which is row equivalent to

$$\begin{bmatrix} 1 & -\frac{3 + 4i}{5} & \vdots & 0 \\ 0 & 0 & \vdots & 0 \end{bmatrix}.$$

Therefore $x_1 = (3 + 4i)x_2/5$. Taking $x_2 = 5$ yields $x_1 = 3 + 4i$, so

$$\mathbf{x} = \begin{bmatrix} 3 + 4i \\ 5 \end{bmatrix}$$

is an eigenvector. The real and imaginary parts of

$$e^t (\cos 4t + i \sin 4t) \begin{bmatrix} 3 + 4i \\ 5 \end{bmatrix}$$

are

$$\mathbf{y}_1 = e^t \begin{bmatrix} 3 \cos 4t - 4 \sin 4t \\ 5 \cos 4t \end{bmatrix} \quad \text{and} \quad \mathbf{y}_2 = e^t \begin{bmatrix} 3 \sin 4t + 4 \cos 4t \\ 5 \sin 4t \end{bmatrix},$$

which are linearly independent solutions of (6). The general solution of (6) is

$$\mathbf{y} = c_1 e^t \begin{bmatrix} 3 \cos 4t - 4 \sin 4t \\ 5 \cos 4t \end{bmatrix} + c_2 e^t \begin{bmatrix} 3 \sin 4t + 4 \cos 4t \\ 5 \sin 4t \end{bmatrix}.$$

EXAMPLE 2 Find the general solution of

$$\mathbf{y}' = \begin{bmatrix} -14 & 39 \\ -6 & 16 \end{bmatrix} \mathbf{y}. \tag{7}$$

Solution The characteristic polynomial of the coefficient matrix A in (7) is

$$\begin{vmatrix} -14 - \lambda & 39 \\ -6 & 16 - \lambda \end{vmatrix} = (\lambda - 1)^2 + 9.$$

Hence, $\lambda = 1 + 3i$ is an eigenvalue of A. The associated eigenvectors satisfy $(A - (1 + 3i)I)\mathbf{x} = \mathbf{0}$. The augmented matrix of this system is

$$\begin{bmatrix} -15 - 3i & 39 & \vdots & 0 \\ -6 & 15 - 3i & \vdots & 0 \end{bmatrix},$$

which is row equivalent to

$$\begin{bmatrix} 1 & \frac{-5 + i}{2} & \vdots & 0 \\ 0 & 0 & \vdots & 0 \end{bmatrix}.$$

Therefore $x_1 = (5 - i)/2$. Taking $x_2 = 2$ yields $x_1 = 5 - i$, so

$$\mathbf{x} = \begin{bmatrix} 5 - i \\ 2 \end{bmatrix}$$

is an eigenvector. The real and imaginary parts of

$$e^t(\cos 3t + i \sin 3t) \begin{bmatrix} 5 - i \\ 2 \end{bmatrix}$$

are

$$\mathbf{y}_1 = e^t \begin{bmatrix} \sin 3t + 5 \cos 3t \\ 2 \cos 3t \end{bmatrix} \quad \text{and} \quad \mathbf{y}_2 = e^t \begin{bmatrix} -\cos 3t + 5 \sin 3t \\ 2 \sin 3t \end{bmatrix},$$

which are linearly independent solutions of (7). The general solution of (7) is

$$\mathbf{y} = c_1 e^t \begin{bmatrix} \sin 3t + 5 \cos 3t \\ 2 \cos 3t \end{bmatrix} + c_2 e^t \begin{bmatrix} -\cos 3t + 5 \sin 3t \\ 2 \sin 3t \end{bmatrix}.$$

EXAMPLE 3 Find the general solution of

$$\mathbf{y}' = \begin{bmatrix} -5 & 5 & 4 \\ -8 & 7 & 6 \\ 1 & 0 & 0 \end{bmatrix} \mathbf{y}. \tag{8}$$

Solution The characteristic polynomial of the coefficient matrix A in (8) is

$$\begin{vmatrix} -5 - \lambda & 5 & 4 \\ -8 & 7 - \lambda & 6 \\ 1 & 0 & -\lambda \end{vmatrix} = -(\lambda - 2)(\lambda^2 + 1).$$

Hence, the eigenvalues of A are $\lambda_1 = 2, \lambda_2 = i$, and $\lambda_3 = -i$. The augmented matrix of $(A - 2I)\mathbf{x} = \mathbf{0}$ is

$$\begin{bmatrix} -7 & 5 & 4 & \vdots & 0 \\ -8 & 5 & 6 & \vdots & 0 \\ 1 & 0 & -2 & \vdots & 0 \end{bmatrix},$$

which is row equivalent to

$$\begin{bmatrix} 1 & 0 & -2 & \vdots & 0 \\ 0 & 1 & -2 & \vdots & 0 \\ 0 & 0 & 0 & \vdots & 0 \end{bmatrix}.$$

Therefore $x_1 = x_2 = 2x_3$. Taking $x_3 = 1$ yields

$$\mathbf{x}_1 = \begin{bmatrix} 2 \\ 2 \\ 1 \end{bmatrix},$$

so

$$\mathbf{y}_1 = \begin{bmatrix} 2 \\ 2 \\ 1 \end{bmatrix} e^{2t}$$

is a solution of (8).

The augmented matrix of $(A - iI)\mathbf{x} = \mathbf{0}$ is

$$\begin{bmatrix} -5 - i & 5 & 4 & \vdots & 0 \\ -8 & 7 - i & 6 & \vdots & 0 \\ 1 & 0 & -i & \vdots & 0 \end{bmatrix},$$

which is row equivalent to

$$\begin{bmatrix} 1 & 0 & -i & \vdots & 0 \\ 0 & 1 & 1 - i & \vdots & 0 \\ 0 & 0 & 0 & \vdots & 0 \end{bmatrix}.$$

Therefore $x_1 = ix_3$ and $x_2 = -(1 - i)x_3$. Taking $x_3 = 1$ yields the eigenvector

$$\mathbf{x}_2 = \begin{bmatrix} i \\ -1 + i \\ 1 \end{bmatrix}.$$

The real and imaginary parts of

$$(\cos t + i \sin t) \begin{bmatrix} i \\ -1 + i \\ 1 \end{bmatrix}$$

are

$$\mathbf{y}_2 = \begin{bmatrix} -\sin t \\ -\cos t - \sin t \\ \cos t \end{bmatrix} \quad \text{and} \quad \mathbf{y}_3 = \begin{bmatrix} \cos t \\ \cos t - \sin t \\ \sin t \end{bmatrix},$$

which are solutions of (8). Since the Wronskian of $\{\mathbf{y}_1, \mathbf{y}_2, \mathbf{y}_3\}$ at $t = 0$ is

$$\begin{vmatrix} 2 & 0 & 1 \\ 2 & -1 & 1 \\ 1 & 1 & 0 \end{vmatrix} = 1,$$

it follows that $\{\mathbf{y}_1, \mathbf{y}_2, \mathbf{y}_3\}$ is a fundamental set of solutions of (8). The general solution of (8) is

$$\mathbf{y} = c_1 \begin{bmatrix} 2 \\ 2 \\ 1 \end{bmatrix} e^{2t} + c_2 \begin{bmatrix} -\sin t \\ -\cos t - \sin t \\ \cos t \end{bmatrix} + c_3 \begin{bmatrix} \cos t \\ \cos t - \sin t \\ \sin t \end{bmatrix}.$$

EXAMPLE 4   Find the general solution of

$$\mathbf{y}' = \begin{bmatrix} 1 & -1 & -2 \\ 1 & 3 & 2 \\ 1 & -1 & 2 \end{bmatrix} \mathbf{y}. \tag{9}$$

Solution   The characteristic polynomial of the coefficient matrix A in (9) is

$$\begin{vmatrix} 1 - \lambda & -1 & -2 \\ 1 & 3 - \lambda & 2 \\ 1 & -1 & 2 - \lambda \end{vmatrix} = -(\lambda - 2)((\lambda - 2)^2 + 4).$$

Hence, the eigenvalues of A are $\lambda_1 = 2$, $\lambda_2 = 2 + 2i$, and $\lambda_3 = 2 - 2i$. The augmented matrix of $(A - 2I)\mathbf{x} = \mathbf{0}$ is

$$\left[\begin{array}{ccc:c} -1 & -1 & -2 & 0 \\ 1 & 1 & 2 & 0 \\ 1 & -1 & 0 & 0 \end{array}\right],$$

which is row equivalent to

$$\left[\begin{array}{ccc:c} 1 & 0 & 1 & 0 \\ 0 & 1 & 1 & 0 \\ 0 & 0 & 0 & 0 \end{array}\right].$$

Therefore $x_1 = x_2 = -x_3$. Taking $x_3 = 1$ yields

$$\mathbf{x}_1 = \begin{bmatrix} -1 \\ -1 \\ 1 \end{bmatrix},$$

so

$$\mathbf{y}_1 = \begin{bmatrix} -1 \\ -1 \\ 1 \end{bmatrix} e^{2t}$$

is a solution of (9).

The augmented matrix of $(A - (2 + 2i)I)\mathbf{x} = \mathbf{0}$ is

$$\left[\begin{array}{ccc:c} -1 - 2i & -1 & -2 & 0 \\ 1 & 1 - 2i & 2 & 0 \\ 1 & -1 & -2i & 0 \end{array}\right],$$

which is row equivalent to

$$\left[\begin{array}{ccc:c} 1 & 0 & -i & 0 \\ 0 & 1 & i & 0 \\ 0 & 0 & 0 & 0 \end{array}\right].$$

Therefore $x_1 = ix_3$ and $x_2 = -ix_3$. Taking $x_3 = 1$ yields the eigenvector

$$\mathbf{x}_2 = \begin{bmatrix} i \\ -i \\ 1 \end{bmatrix}.$$

The real and imaginary parts of

$$e^{2t}(\cos 2t + i \sin 2t) \begin{bmatrix} i \\ -i \\ 1 \end{bmatrix}$$

are

$$\mathbf{y}_2 = e^{2t} \begin{bmatrix} -\sin 2t \\ \sin 2t \\ \cos 2t \end{bmatrix} \quad \text{and} \quad \mathbf{y}_2 = e^{2t} \begin{bmatrix} \cos 2t \\ -\cos 2t \\ \sin 2t \end{bmatrix},$$

which are solutions of (9). Since the Wronskian of $\{\mathbf{y}_1, \mathbf{y}_2, \mathbf{y}_3\}$ at $t = 0$ is

$$\begin{vmatrix} -1 & 0 & 1 \\ -1 & 0 & -1 \\ 1 & 1 & 0 \end{vmatrix} = -2,$$

it follows that $\{\mathbf{y}_1, \mathbf{y}_2, \mathbf{y}_3\}$ is a fundamental set of solutions of (9). The general solution of (9) is

$$\mathbf{y} = c_1 \begin{bmatrix} -1 \\ -1 \\ 1 \end{bmatrix} e^{2t} + c_2 e^{2t} \begin{bmatrix} -\sin 2t \\ \sin 2t \\ \cos 2t \end{bmatrix} + c_3 e^{2t} \begin{bmatrix} \cos 2t \\ -\cos 2t \\ \sin 2t \end{bmatrix}. \qquad \blacksquare$$

GEOMETRIC PROPERTIES OF SOLUTIONS WHEN $n = 2$

We will now consider the geometric properties of solutions of a 2×2 constant coefficient system

$$\begin{bmatrix} y_1' \\ y_2' \end{bmatrix} = \begin{bmatrix} a_{11} & a_{12} \\ a_{21} & a_{22} \end{bmatrix} \begin{bmatrix} y_1 \\ y_2 \end{bmatrix} \tag{10}$$

under the assumptions of this section; that is, when the matrix

$$A = \begin{bmatrix} a_{11} & a_{12} \\ a_{21} & a_{22} \end{bmatrix}$$

has a complex eigenvalue $\lambda = \alpha + i\beta$ ($\beta \neq 0$) and $\mathbf{x} = \mathbf{u} + i\mathbf{v}$ is an associated eigenvector, where $\mathbf{u}$ and $\mathbf{v}$ have real components. To describe the trajectories accurately it is necessary to introduce a new rectangular coordinate system in the $y_1 y_2$-plane. This raises a point that hasn't come up before: It is always possible to choose $\mathbf{x}$ so that $(\mathbf{u}, \mathbf{v}) = 0$. A special effort is required to do this, since not every eigenvector has this property. However, if we know an eigenvector that doesn't, we can multiply it by a suitable complex constant to obtain one that does. To see this, note that if $\mathbf{x}$ is a λ-eigenvector of A and k is an arbitrary real number, then

$$\mathbf{x}_1 = (1 + ik)\mathbf{x} = (1 + ik)(\mathbf{u} + i\mathbf{v}) = (\mathbf{u} - k\mathbf{v}) + i(\mathbf{v} + k\mathbf{u})$$

is also a λ-eigenvector of A, since

$$A\mathbf{x}_1 = A((1 + ik)\mathbf{x}) = (1 + ik)A\mathbf{x} = (1 + ik)\lambda\mathbf{x} = \lambda((1 + ik)\mathbf{x}) = \lambda\mathbf{x}_1.$$

The real and imaginary parts of $\mathbf{x}_1$ are

$$\mathbf{u}_1 = \mathbf{u} - k\mathbf{v} \quad \text{and} \quad \mathbf{v}_1 = \mathbf{v} + k\mathbf{u}, \tag{11}$$

so

$$(\mathbf{u}_1, \mathbf{v}_1) = (\mathbf{u} - k\mathbf{v}, \mathbf{v} + k\mathbf{u}) = -[(\mathbf{u}, \mathbf{v})k^2 + (\|\mathbf{v}\|^2 - \|\mathbf{u}\|^2)k - (\mathbf{u}, \mathbf{v})].$$

Therefore $(\mathbf{u}_1, \mathbf{v}_1) = 0$ if

$$(\mathbf{u}, \mathbf{v})k^2 + (\|\mathbf{v}\|^2 - \|\mathbf{u}\|^2)k - (\mathbf{u}, \mathbf{v}) = 0. \tag{12}$$

If $(\mathbf{u}, \mathbf{v}) \neq 0$ we can use the quadratic formula to find two real values of k such that $(\mathbf{u}_1, \mathbf{v}_1) = 0$ (Exercise 28).

EXAMPLE 5 In Example 1 we found the eigenvector

$$\mathbf{x} = \begin{bmatrix} 3 + 4i \\ 5 \end{bmatrix} = \begin{bmatrix} 3 \\ 5 \end{bmatrix} + i \begin{bmatrix} 4 \\ 0 \end{bmatrix}$$

for the matrix of the system (6). Here $\mathbf{u} = \begin{bmatrix} 3 \\ 5 \end{bmatrix}$ and $\mathbf{v} = \begin{bmatrix} 4 \\ 0 \end{bmatrix}$ are not orthogonal, since $(\mathbf{u}, \mathbf{v}) = 12$. Since $\|\mathbf{v}\|^2 - \|\mathbf{u}\|^2 = -18$, (12) is equivalent to

$$2k^2 - 3k - 2 = 0.$$

The zeros of this equation are $k_1 = 2$ and $k_2 = -1/2$. Letting $k = 2$ in (11) yields

$$\mathbf{u}_1 = \mathbf{u} - 2\mathbf{v} = \begin{bmatrix} -5 \\ 5 \end{bmatrix} \quad \text{and} \quad \mathbf{v}_1 = \mathbf{v} + 2\mathbf{u} = \begin{bmatrix} 10 \\ 10 \end{bmatrix},$$

and $(\mathbf{u}_1, \mathbf{v}_1) = 0$. Letting $k = -1/2$ in (11) yields

$$\mathbf{u}_1 = \mathbf{u} + \frac{\mathbf{v}}{2} = \begin{bmatrix} 5 \\ 5 \end{bmatrix} \quad \text{and} \quad \mathbf{v}_1 = \mathbf{v} - \frac{\mathbf{u}}{2} = \frac{1}{2}\begin{bmatrix} 5 \\ -5 \end{bmatrix},$$

and again $(\mathbf{u}_1, \mathbf{v}_1) = 0$. ■

(Usually the numbers don't work out as nicely as in this example. You'll need a calculator or computer to do Exercises 29–40.)

Henceforth we'll assume that $(\mathbf{u}, \mathbf{v}) = 0$. Let $\mathbf{U}$ and $\mathbf{V}$ be unit vectors in the directions of $\mathbf{u}$ and $\mathbf{v}$, respectively; that is, $\mathbf{U} = \mathbf{u}/\|\mathbf{u}\|$ and $\mathbf{V} = \mathbf{v}/\|\mathbf{v}\|$. The new rectangular coordinate system will have the same origin as the y_1y_2-system. The coordinates of a point in this system will be denoted by (z_1, z_2), where z_1 and z_2 are the displacements in the directions of $\mathbf{U}$ and $\mathbf{V}$, respectively.

From (5), the solutions of (10) are given by

$$\mathbf{y} = e^{\alpha t}[(c_1 \cos \beta t + c_2 \sin \beta t)\mathbf{u} + (-c_1 \sin \beta t + c_2 \cos \beta t)\mathbf{v}]. \tag{13}$$

For convenience let's call the curve traversed by $e^{-\alpha t}\mathbf{y}(t)$ a **shadow trajectory** of (10). Multiplying (13) by $e^{-\alpha t}$ yields

$$e^{-\alpha t}\mathbf{y}(t) = z_1(t)\mathbf{U} + z_2(t)\mathbf{V},$$

where

$$z_1(t) = \|\mathbf{u}\|(c_1 \cos \beta t + c_2 \sin \beta t)$$
$$z_2(t) = \|\mathbf{v}\|(-c_1 \sin \beta t + c_2 \cos \beta t).$$

Therefore

$$\frac{(z_1(t))^2}{\|\mathbf{u}\|^2} + \frac{(z_2(t))^2}{\|\mathbf{v}\|^2} = c_1^2 + c_2^2$$

(verify!), which means that the shadow trajectories of (10) are ellipses centered at the origin, with axes of symmetry parallel to $\mathbf{U}$ and $\mathbf{V}$. Since

$$z_1' = \frac{\beta\|\mathbf{u}\|}{\|\mathbf{v}\|} z_2 \quad \text{and} \quad z_2' = -\frac{\beta\|\mathbf{v}\|}{\|\mathbf{u}\|} z_1,$$

the vector from the origin to a point on the shadow ellipse rotates in the same direction that $\mathbf{V}$ would have to be rotated by $\pi/2$ radians to bring it into coincidence with $\mathbf{U}$ (Figures 1 and 2).

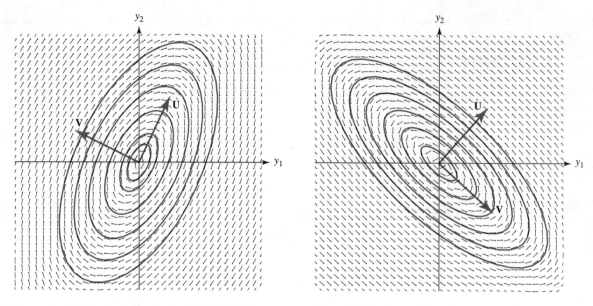

Figure 1 Shadow trajectories traversed clockwise

Figure 2 Shadow trajectories traversed counter-clockwise

If $\alpha = 0$ then any trajectory of (10) is a shadow trajectory of (10); therefore, if λ is purely imaginary then the trajectories of (10) are ellipses traversed periodically as indicated in Figures 1 and 2.

If $\alpha > 0$ then

$$\lim_{t \to \infty} \|\mathbf{y}(t)\| = \infty \qquad \text{and} \qquad \lim_{t \to -\infty} \mathbf{y}(t) = 0,$$

so the trajectory spirals away from the origin as t varies from $-\infty$ to ∞. The direction of the spiral depends upon the relative orientation of $\mathbf{U}$ and $\mathbf{V}$, as shown in Figures 3 and 4.

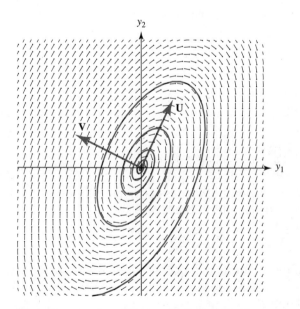

Figure 3 $\alpha > 0$; trajectory spiraling outward

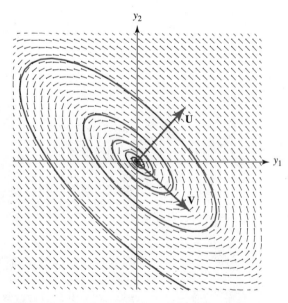

Figure 4 $\alpha > 0$; trajectory spiraling outward

If $\alpha < 0$ then

$$\lim_{t \to -\infty} \|\mathbf{y}(t)\| = \infty \qquad \text{and} \qquad \lim_{t \to \infty} \mathbf{y}(t) = 0,$$

so the trajectory spirals toward the origin as t varies from $-\infty$ to ∞. Again, the direction of the spiral depends upon the relative orientation of $\mathbf{U}$ and $\mathbf{V}$, as shown in Figures 5 and 6.

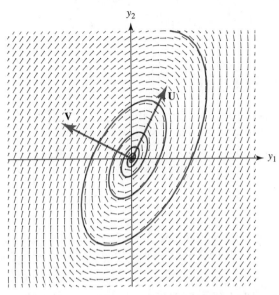

Figure 5 $\alpha < 0$; trajectory spiraling inward

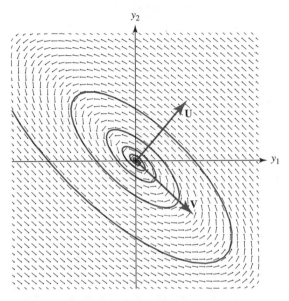

Figure 6 $\alpha < 0$; trajectory spiraling inward

10.6 EXERCISES

In Exercises 1–16 find the general solution.

1. $\mathbf{y}' = \begin{bmatrix} -1 & 2 \\ -5 & 5 \end{bmatrix} \mathbf{y}$

2. $\mathbf{y}' = \begin{bmatrix} -11 & 4 \\ -26 & 9 \end{bmatrix} \mathbf{y}$

3. $\mathbf{y}' = \begin{bmatrix} 1 & 2 \\ -4 & 5 \end{bmatrix} \mathbf{y}$

4. $\mathbf{y}' = \begin{bmatrix} 5 & -6 \\ 3 & -1 \end{bmatrix} \mathbf{y}$

5. $\mathbf{y}' = \begin{bmatrix} 3 & -3 & 1 \\ 0 & 2 & 2 \\ 5 & 1 & 1 \end{bmatrix} \mathbf{y}$

6. $\mathbf{y}' = \begin{bmatrix} -3 & 3 & 1 \\ 1 & -5 & -3 \\ -3 & 7 & 3 \end{bmatrix} \mathbf{y}$

7. $\mathbf{y}' = \begin{bmatrix} 2 & 1 & -1 \\ 0 & 1 & 1 \\ 1 & 0 & 1 \end{bmatrix} \mathbf{y}$

8. $\mathbf{y}' = \begin{bmatrix} -3 & 1 & -3 \\ 4 & -1 & 2 \\ 4 & -2 & 3 \end{bmatrix} \mathbf{y}$

9. $\mathbf{y}' = \begin{bmatrix} 5 & -4 \\ 10 & 1 \end{bmatrix} \mathbf{y}$

10. $\mathbf{y}' = \frac{1}{3} \begin{bmatrix} 7 & -5 \\ 2 & 5 \end{bmatrix} \mathbf{y}$

11. $\mathbf{y}' = \begin{bmatrix} 3 & 2 \\ -5 & 1 \end{bmatrix} \mathbf{y}$

12. $\mathbf{y}' = \begin{bmatrix} 34 & 52 \\ -20 & -30 \end{bmatrix} \mathbf{y}$

13. $\mathbf{y}' = \begin{bmatrix} 1 & 1 & 2 \\ 1 & 0 & -1 \\ -1 & -2 & -1 \end{bmatrix} \mathbf{y}$

14. $\mathbf{y}' = \begin{bmatrix} 3 & -4 & -2 \\ -5 & 7 & -8 \\ -10 & 13 & -8 \end{bmatrix} \mathbf{y}$

15. $\mathbf{y}' = \begin{bmatrix} 6 & 0 & -3 \\ -3 & 3 & 3 \\ 1 & -2 & 6 \end{bmatrix} \mathbf{y}$

16. $\mathbf{y}' = \begin{bmatrix} 1 & 2 & -2 \\ 0 & 2 & -1 \\ 1 & 0 & 0 \end{bmatrix} \mathbf{y}$

In Exercises 17–24 solve the initial value problem.

17. $\mathbf{y}' = \begin{bmatrix} 4 & -6 \\ 3 & -2 \end{bmatrix} \mathbf{y}, \quad \mathbf{y}(0) = \begin{bmatrix} 5 \\ 2 \end{bmatrix}$

18. $\mathbf{y}' = \begin{bmatrix} 7 & 15 \\ -3 & 1 \end{bmatrix} \mathbf{y}, \quad \mathbf{y}(0) = \begin{bmatrix} 5 \\ 1 \end{bmatrix}$

19. $\mathbf{y}' = \begin{bmatrix} 7 & -15 \\ 3 & -5 \end{bmatrix} \mathbf{y}, \quad \mathbf{y}(0) = \begin{bmatrix} 17 \\ 7 \end{bmatrix}$

20. $\mathbf{y}' = \dfrac{1}{6} \begin{bmatrix} 4 & -2 \\ 5 & 2 \end{bmatrix} \mathbf{y}, \quad \mathbf{y}(0) = \begin{bmatrix} 1 \\ -1 \end{bmatrix}$

21. $\mathbf{y}' = \begin{bmatrix} 5 & 2 & -1 \\ -3 & 2 & 2 \\ 1 & 3 & 2 \end{bmatrix} \mathbf{y}, \quad \mathbf{y}(0) = \begin{bmatrix} 4 \\ 0 \\ 6 \end{bmatrix}$

22. $\mathbf{y}' = \begin{bmatrix} 4 & 4 & 0 \\ 8 & 10 & -20 \\ 2 & 3 & -2 \end{bmatrix} \mathbf{y}, \quad \mathbf{y}(0) = \begin{bmatrix} 8 \\ 6 \\ 5 \end{bmatrix}$

23. $\mathbf{y}' = \begin{bmatrix} 1 & 15 & -15 \\ -6 & 18 & -22 \\ -3 & 11 & -15 \end{bmatrix} \mathbf{y}, \quad \mathbf{y}(0) = \begin{bmatrix} 15 \\ 17 \\ 10 \end{bmatrix}$

24. $\mathbf{y}' = \begin{bmatrix} 4 & -4 & 4 \\ -10 & 3 & 15 \\ 2 & -3 & 1 \end{bmatrix} \mathbf{y}, \quad \mathbf{y}(0) = \begin{bmatrix} 16 \\ 14 \\ 6 \end{bmatrix}$

25. Suppose that an $n \times n$ matrix A with real entries has a complex eigenvalue $\lambda = \alpha + i\beta$ ($\beta \neq 0$) with associated eigenvector $\mathbf{x} = \mathbf{u} + i\mathbf{v}$, where $\mathbf{u}$ and $\mathbf{v}$ have real components. Show that $\mathbf{u}$ and $\mathbf{v}$ are both nonzero.

26. Verify that

$$\mathbf{y}_1 = e^{\alpha t}(\mathbf{u} \cos \beta t - \mathbf{v} \sin \beta t) \quad \text{and} \quad \mathbf{y}_2 = e^{\alpha t}(\mathbf{u} \sin \beta t + \mathbf{v} \cos \beta t),$$

are the real and imaginary parts of

$$e^{\alpha t}(\cos \beta t + i \sin \beta t)(\mathbf{u} + i\mathbf{v}).$$

27. Show that if the vectors $\mathbf{u}$ and $\mathbf{v}$ are not both $\mathbf{0}$ and $\beta \neq 0$ then the vector functions

$$\mathbf{y}_1 = e^{\alpha t}(\mathbf{u} \cos \beta t - \mathbf{v} \sin \beta t) \quad \text{and} \quad \mathbf{y}_2 = e^{\alpha t}(\mathbf{u} \sin \beta t + \mathbf{v} \cos \beta t)$$

are linearly independent on every interval.

Hint: *There are two cases to consider:* (i) $\{\mathbf{u}, \mathbf{v}\}$ *linearly independent, and* (ii) $\{\mathbf{u}, \mathbf{v}\}$ *linearly dependent. In either case, exploit the fact that* $\{\cos \beta t, \sin \beta t\}$ *is linearly independent on every interval.*

28. Suppose that $\mathbf{u} = \begin{bmatrix} u_1 \\ u_2 \end{bmatrix}$ and $\mathbf{v} = \begin{bmatrix} v_1 \\ v_2 \end{bmatrix}$ are not orthogonal; that is, $(\mathbf{u}, \mathbf{v}) \neq 0$.

(a) Show that the quadratic equation

$$(\mathbf{u}, \mathbf{v})k^2 + (\|\mathbf{v}\|^2 - \|\mathbf{u}\|^2)k - (\mathbf{u}, \mathbf{v}) = 0$$

has a positive root k_1 and a negative root $k_2 = -1/k_1$.

(b) Let $\mathbf{u}_1^{(1)} = \mathbf{u} - k_1\mathbf{v}, \mathbf{v}_1^{(1)} = \mathbf{v} + k_1\mathbf{u}, \mathbf{u}_1^{(2)} = \mathbf{u} - k_2\mathbf{v}$, and $\mathbf{v}_1^{(2)} = \mathbf{v} + k_2\mathbf{u}$, so that $(\mathbf{u}_1^{(1)}, \mathbf{v}_1^{(1)}) = (\mathbf{u}_1^{(2)}, \mathbf{v}_1^{(2)}) = 0$, from the discussion given above. Show that

$$\mathbf{u}_1^{(2)} = \frac{\mathbf{v}_1^{(1)}}{k_1} \quad \text{and} \quad \mathbf{v}_1^{(2)} = -\frac{\mathbf{u}_1^{(1)}}{k_1}.$$

(c) Let $\mathbf{U}_1, \mathbf{V}_1, \mathbf{U}_2$, and $\mathbf{V}_2$ be unit vectors in the directions of $\mathbf{u}_1^{(1)}, \mathbf{v}_1^{(1)}, \mathbf{u}_1^{(2)}$, and $\mathbf{v}_1^{(2)}$, respectively. Conclude from **(a)** that $\mathbf{U}_2 = \mathbf{V}_1$ and $\mathbf{V}_2 = -\mathbf{U}_1$, and that therefore the counterclockwise angles from $\mathbf{U}_1$ to $\mathbf{V}_1$ and from $\mathbf{U}_2$ to $\mathbf{V}_2$ are both $\pi/2$ or both $-\pi/2$.

*In Exercises 29–32 find vectors **U** and **V** parallel to the axes of symmetry of the trajectories, and plot some typical trajectories.*

C **29.** $\mathbf{y}' = \begin{bmatrix} 3 & -5 \\ 5 & -3 \end{bmatrix} \mathbf{y}$

C **30.** $\mathbf{y}' = \begin{bmatrix} -15 & 10 \\ -25 & 15 \end{bmatrix} \mathbf{y}$

C **31.** $\mathbf{y}' = \begin{bmatrix} -4 & 8 \\ -4 & 4 \end{bmatrix} \mathbf{y}$

C **32.** $\mathbf{y}' = \begin{bmatrix} -3 & -15 \\ 3 & 3 \end{bmatrix} \mathbf{y}$

*In Exercises 33–40 find vectors **U** and **V** parallel to the axes of symmetry of the shadow trajectories, and plot a typical trajectory.*

C **33.** $\mathbf{y}' = \begin{bmatrix} -5 & 6 \\ -12 & 7 \end{bmatrix} \mathbf{y}$

C **34.** $\mathbf{y}' = \begin{bmatrix} 5 & -12 \\ 6 & -7 \end{bmatrix} \mathbf{y}$

C **35.** $\mathbf{y}' = \begin{bmatrix} 4 & -5 \\ 9 & -2 \end{bmatrix} \mathbf{y}$

C **36.** $\mathbf{y}' = \begin{bmatrix} -4 & 9 \\ -5 & 2 \end{bmatrix} \mathbf{y}$

C **37.** $\mathbf{y}' = \begin{bmatrix} -1 & 10 \\ -10 & -1 \end{bmatrix} \mathbf{y}$

C **38.** $\mathbf{y}' = \begin{bmatrix} -1 & -5 \\ 20 & -1 \end{bmatrix} \mathbf{y}$

C **39.** $\mathbf{y}' = \begin{bmatrix} -7 & 10 \\ -10 & 9 \end{bmatrix} \mathbf{y}$

C **40.** $\mathbf{y}' = \begin{bmatrix} -7 & 6 \\ -12 & 5 \end{bmatrix} \mathbf{y}$

10.7 Variation of Parameters for Nonhomogeneous Linear Systems

We now consider the nonhomogeneous linear system

$$\mathbf{y}' = A(t)\mathbf{y} + \mathbf{f}(t),$$

where A is an $n \times n$ matrix function and $\mathbf{f}$ is an n-vector forcing function. Associated with this system is the ***complementary system*** $\mathbf{y}' = A(t)\mathbf{y}$.

The following theorem is analogous to Theorems 5.3.2. and 9.1.5. It shows how to find the general solution of $\mathbf{y}' = A(t)\mathbf{y} + \mathbf{f}(t)$ if we know a particular solution of $\mathbf{y}' = A(t)\mathbf{y} + \mathbf{f}(t)$ and a fundamental set of solutions of the complementary system. We leave the proof as an exercise (Exercise 21).

THEOREM 10.7.1

Suppose that the $n \times n$ matrix function A and the n-vector function $\mathbf{f}$ are continuous on (a,b). Let $\mathbf{y}_p$ be a particular solution of $\mathbf{y}' = A(t)\mathbf{y} + \mathbf{f}(t)$ on (a,b), and let $\{\mathbf{y}_1, \mathbf{y}_2, \ldots, \mathbf{y}_n\}$ be a fundamental set of solutions of the complementary equation $\mathbf{y}' = A(t)\mathbf{y}$ on (a,b). Then $\mathbf{y}$ is a solution of $\mathbf{y}' = A(t)y + \mathbf{f}(t)$ on (a,b) if and only if

$$\mathbf{y} = \mathbf{y}_p + c_1\mathbf{y}_1 + c_2\mathbf{y}_2 + \cdots + c_n\mathbf{y}_n,$$

where $c_1, c_2, \ldots, c_n$ are constants.

FINDING A PARTICULAR SOLUTION OF A NONHOMOGENEOUS SYSTEM

We now discuss an extension of the method of variation of parameters to linear nonhomogeneous systems. This method will produce a particular solution of a nonhomogenous system $\mathbf{y}' = A(t)\mathbf{y} + \mathbf{f}(t)$ provided that we know a fundamental matrix for the complementary system. To derive the method, suppose that Y is a fundamental matrix for the complementary system; that is,

$$Y = \begin{bmatrix} y_{11} & y_{12} & \cdots & y_{1n} \\ y_{21} & y_{22} & \cdots & y_{2n} \\ \vdots & \vdots & \ddots & \vdots \\ y_{n1} & y_{n2} & \cdots & y_{nn} \end{bmatrix},$$

where

$$\mathbf{y}_1 = \begin{bmatrix} y_{11} \\ y_{21} \\ \vdots \\ y_{n1} \end{bmatrix}, \qquad \mathbf{y}_2 = \begin{bmatrix} y_{12} \\ y_{22} \\ \vdots \\ y_{n2} \end{bmatrix}, \ldots, \qquad \mathbf{y}_n = \begin{bmatrix} y_{1n} \\ y_{2n} \\ \vdots \\ y_{nn} \end{bmatrix}$$

is a fundamental set of solutions of the complementary system. In Section 10.3 we saw that $Y' = A(t)Y$. We seek a particular solution of

$$\mathbf{y}' = A(t)\mathbf{y} + \mathbf{f}(t) \tag{1}$$

of the form

$$\mathbf{y}_p = Y\mathbf{u} \tag{2}$$

where $\mathbf{u}$ is to be determined. Differentiating (2) yields

$$\begin{aligned} \mathbf{y}_p' &= Y'\mathbf{u} + Y\mathbf{u}' \\ &= AY\mathbf{u} + Y\mathbf{u}' \quad \text{(since } Y' = AY\text{)} \\ &= A\mathbf{y}_p + Y\mathbf{u}' \quad \text{(since } Y\mathbf{u} = \mathbf{y}_p\text{)}. \end{aligned}$$

Comparing this with (1) shows that $\mathbf{y}_p = Y\mathbf{u}$ is a solution of (1) if and only if

$$Y\mathbf{u}' = \mathbf{f}.$$

Thus, we can find a particular solution $\mathbf{y}_p$ by solving this equation for $\mathbf{u}'$, integrating to obtain $\mathbf{u}$, and computing $Y\mathbf{u}$. We can take all constants of integration to be zero, since any particular solution will suffice.

Exercise 22 sketches a proof that this method is analogous to the method of variation of parameters discussed in Sections 5.7 and 9.4 for scalar linear equations.

Example 1   (a) Find a particular solution of the system

$$\mathbf{y}' = \begin{bmatrix} 1 & 2 \\ 2 & 1 \end{bmatrix} \mathbf{y} + \begin{bmatrix} 2e^{4t} \\ e^{4t} \end{bmatrix}, \tag{3}$$

which we considered in Example 1 of Section 10.2.
(b) Find the general solution of (3).

Solution   (a) The complementary system is

$$\mathbf{y}' = \begin{bmatrix} 1 & 2 \\ 2 & 1 \end{bmatrix} \mathbf{y}. \tag{4}$$

The characteristic polynomial of the coefficient matrix is

$$\begin{vmatrix} 1 - \lambda & 2 \\ 2 & 1 - \lambda \end{vmatrix} = (\lambda + 1)(\lambda - 3).$$

Using the method of Section 10.4, we find that

$$\mathbf{y}_1 = \begin{bmatrix} e^{3t} \\ e^{3t} \end{bmatrix} \quad \text{and} \quad \mathbf{y}_2 = \begin{bmatrix} e^{-t} \\ -e^{-t} \end{bmatrix}$$

are linearly independent solutions of (4). Therefore

$$Y = \begin{bmatrix} e^{3t} & e^{-t} \\ e^{3t} & -e^{-t} \end{bmatrix}$$

is a fundamental matrix for (4). We seek a particular solution $\mathbf{y}_p = Y\mathbf{u}$ of (3), where $Y\mathbf{u}' = \mathbf{f}$; that is,

$$\begin{bmatrix} e^{3t} & e^{-t} \\ e^{3t} & -e^{-t} \end{bmatrix} \begin{bmatrix} u_1' \\ u_2' \end{bmatrix} = \begin{bmatrix} 2e^{4t} \\ e^{4t} \end{bmatrix}$$

The determinant of Y is the Wronskian

$$\begin{vmatrix} e^{3t} & e^{-t} \\ e^{3t} & -e^{-t} \end{vmatrix} = -2e^{2t}.$$

By Cramer's rule,

$$u_1' = -\frac{1}{2e^{2t}} \begin{vmatrix} 2e^{4t} & e^{-t} \\ e^{4t} & -e^{-t} \end{vmatrix} = \frac{3e^{3t}}{2e^{2t}} = \frac{3}{2}e^{t},$$

$$u_2' = -\frac{1}{2e^{2t}} \begin{vmatrix} e^{3t} & 2e^{4t} \\ e^{3t} & e^{4t} \end{vmatrix} = \frac{e^{7t}}{2e^{2t}} = \frac{1}{2}e^{5t}.$$

Therefore

$$\mathbf{u}' = \frac{1}{2} \begin{bmatrix} 3e^{t} \\ e^{5t} \end{bmatrix}.$$

Integrating and taking the constants of integration to be zero yields

$$\mathbf{u} = \frac{1}{10} \begin{bmatrix} 15e^{t} \\ e^{5t} \end{bmatrix},$$

so

$$\mathbf{y}_p = Y\mathbf{u} = \frac{1}{10} \begin{bmatrix} e^{3t} & e^{-t} \\ e^{3t} & -e^{-t} \end{bmatrix} \begin{bmatrix} 15e^{t} \\ e^{5t} \end{bmatrix} = \frac{1}{5} \begin{bmatrix} 8e^{4t} \\ 7e^{4t} \end{bmatrix}$$

is a particular solution of (3).

Solution   **(b)** From Theorem 10.7.1 the general solution of (3) is

$$\mathbf{y} = \mathbf{y}_p + c_1\mathbf{y}_1 + c_2\mathbf{y}_2 = \frac{1}{5} \begin{bmatrix} 8e^{4t} \\ 7e^{4t} \end{bmatrix} + c_1 \begin{bmatrix} e^{3t} \\ e^{3t} \end{bmatrix} + c_2 \begin{bmatrix} e^{-t} \\ -e^{-t} \end{bmatrix}, \tag{5}$$

which can also be written as

$$\mathbf{y} = \frac{1}{5} \begin{bmatrix} 8e^{4t} \\ 7e^{4t} \end{bmatrix} + \begin{bmatrix} e^{3t} & e^{-t} \\ e^{3t} & -e^{-t} \end{bmatrix} \mathbf{c},$$

where $\mathbf{c}$ is an arbitrary constant vector.

Writing (5) in terms of coordinates yields

$$y_1 = \frac{8}{5}e^{4t} + c_1 e^{3t} + c_2 e^{-t},$$

$$y_2 = \frac{7}{5}e^{4t} + c_1 e^{3t} - c_2 e^{-t},$$

so our result is consistent with Example 1 of Section 10.2.   ■

If A is not a constant matrix then it is usually difficult to find a fundamental set of solutions for the system $\mathbf{y}' = A(t)\mathbf{y}$. It is beyond the scope of this text to discuss methods for doing this. Therefore, in the following examples and in the exercises involving systems with variable coefficient matrices we will provide fundamental matrices for the complementary systems without explaining how they were obtained.

EXAMPLE 2   Find a particular solution of

$$\mathbf{y}' = \begin{bmatrix} 2 & 2e^{-2t} \\ 2e^{2t} & 4 \end{bmatrix} \mathbf{y} + \begin{bmatrix} 1 \\ 1 \end{bmatrix}, \tag{6}$$

given that

$$Y = \begin{bmatrix} e^{4t} & -1 \\ e^{6t} & e^{2t} \end{bmatrix}$$

is a fundamental matrix for the complementary system.

Solution   We seek a particular solution $\mathbf{y}_p = Y\mathbf{u}$ of (6) where $Y\mathbf{u}' = \mathbf{f}$; that is,

$$\begin{bmatrix} e^{4t} & -1 \\ e^{6t} & e^{2t} \end{bmatrix} \begin{bmatrix} u_1' \\ u_2' \end{bmatrix} = \begin{bmatrix} 1 \\ 1 \end{bmatrix}.$$

The determinant of Y is the Wronskian

$$\begin{vmatrix} e^{4t} & -1 \\ e^{6t} & e^{2t} \end{vmatrix} = 2e^{6t}.$$

By Cramer's rule,

$$u_1' = \frac{1}{2e^{6t}} \begin{vmatrix} 1 & -1 \\ 1 & e^{2t} \end{vmatrix} = \frac{e^{2t} + 1}{2e^{6t}} = \frac{e^{-4t} + e^{-6t}}{2},$$

$$u_2' = \frac{1}{2e^{6t}} \begin{vmatrix} e^{4t} & 1 \\ e^{6t} & 1 \end{vmatrix} = \frac{e^{4t} - e^{6t}}{2e^{6t}} = \frac{e^{-2t} - 1}{2}.$$

Therefore

$$\mathbf{u}' = \frac{1}{2} \begin{bmatrix} e^{-4t} + e^{-6t} \\ e^{-2t} - 1 \end{bmatrix}.$$

Integrating and taking the constants of integration to be zero yields

$$\mathbf{u} = -\frac{1}{24} \begin{bmatrix} 3e^{-4t} + 2e^{-6t} \\ 6e^{-2t} + 12t \end{bmatrix},$$

so

$$\mathbf{y}_p = Y\mathbf{u} = -\frac{1}{24} \begin{bmatrix} e^{4t} & -1 \\ e^{6t} & e^{2t} \end{bmatrix} \begin{bmatrix} 3e^{-4t} + 2e^{-6t} \\ 6e^{-2t} + 12t \end{bmatrix} = \frac{1}{24} \begin{bmatrix} 4e^{-2t} + 12t - 3 \\ -3e^{2t}(4t + 1) - 8 \end{bmatrix}$$

is a particular solution of (6).

EXAMPLE 3   Find a particular solution of

$$\mathbf{y}' = -\frac{2}{t^2}\begin{bmatrix} t & -3t^2 \\ 1 & -2t \end{bmatrix}\mathbf{y} + t^2\begin{bmatrix} 1 \\ 1 \end{bmatrix}, \tag{7}$$

given that

$$Y = \begin{bmatrix} 2t & 3t^2 \\ 1 & 2t \end{bmatrix}$$

is a fundamental matrix for the complementary system on $(-\infty, 0)$ and $(0, \infty)$.

Solution   We seek a particular solution $\mathbf{y}_p = Y\mathbf{u}$ of (7) where $Y\mathbf{u}' = \mathbf{f}$; that is,

$$\begin{bmatrix} 2t & 3t^2 \\ 1 & 2t \end{bmatrix}\begin{bmatrix} u_1' \\ u_2' \end{bmatrix} = \begin{bmatrix} t^2 \\ t^2 \end{bmatrix}.$$

The determinant of Y is the Wronskian

$$\begin{vmatrix} 2t & 3t^2 \\ 1 & 2t \end{vmatrix} = t^2.$$

By Cramer's rule,

$$u_1' = \frac{1}{t^2}\begin{vmatrix} t^2 & 3t^2 \\ t^2 & 2t \end{vmatrix} = \frac{2t^3 - 3t^4}{t^2} = 2t - 3t^2,$$

$$u_2' = \frac{1}{t^2}\begin{vmatrix} 2t & t^2 \\ 1 & t^2 \end{vmatrix} = \frac{2t^3 - t^2}{t^2} = 2t - 1.$$

Therefore

$$\mathbf{u}' = \begin{bmatrix} 2t - 3t^2 \\ 2t - 1 \end{bmatrix}.$$

Integrating and taking the constants of integration to be zero yields

$$\mathbf{u} = \begin{bmatrix} t^2 - t^3 \\ t^2 - t \end{bmatrix},$$

so

$$\mathbf{y}_p = Y\mathbf{u} = \begin{bmatrix} 2t & 3t^2 \\ 1 & 2t \end{bmatrix}\begin{bmatrix} t^2 - t^3 \\ t^2 - t \end{bmatrix} = \begin{bmatrix} t^3(t - 1) \\ t^2(t - 1) \end{bmatrix}$$

is a particular solution of (7).

EXAMPLE 4   **(a)** Find a particular solution of

$$\mathbf{y}' = \begin{bmatrix} 2 & -1 & -1 \\ 1 & 0 & -1 \\ 1 & -1 & 0 \end{bmatrix}\mathbf{y} + \begin{bmatrix} e^t \\ 0 \\ e^{-t} \end{bmatrix}. \tag{8}$$

(b) Find the general solution of (8).

Solution   **(a)** The complementary system for (8) is

$$\mathbf{y}' = \begin{bmatrix} 2 & -1 & -1 \\ 1 & 0 & -1 \\ 1 & -1 & 0 \end{bmatrix}\mathbf{y}. \tag{9}$$

The characteristic polynomial of the coefficient matrix is

$$\begin{vmatrix} 2 - \lambda & -1 & -1 \\ 1 & -\lambda & -1 \\ 1 & -1 & -\lambda \end{vmatrix} = -\lambda(\lambda - 1)^2.$$

By using the method of Section 10.4 we find that

$$\mathbf{y}_1 = \begin{bmatrix} 1 \\ 1 \\ 1 \end{bmatrix}, \qquad \mathbf{y}_2 = \begin{bmatrix} e^t \\ e^t \\ 0 \end{bmatrix}, \qquad \text{and} \qquad \mathbf{y}_3 = \begin{bmatrix} e^t \\ 0 \\ e^t \end{bmatrix}$$

are linearly independent solutions of (9). Therefore

$$Y = \begin{bmatrix} 1 & e^t & e^t \\ 1 & e^t & 0 \\ 1 & 0 & e^t \end{bmatrix}$$

is a fundamental matrix for (9). We seek a particular solution $\mathbf{y}_p = Y\mathbf{u}$ of (8), where $Y\mathbf{u}' = \mathbf{f}$; that is,

$$\begin{bmatrix} 1 & e^t & e^t \\ 1 & e^t & 0 \\ 1 & 0 & e^t \end{bmatrix} \begin{bmatrix} u_1' \\ u_2' \\ u_3' \end{bmatrix} = \begin{bmatrix} e^t \\ 0 \\ e^{-t} \end{bmatrix}.$$

The determinant of Y is the Wronskian

$$\begin{vmatrix} 1 & e^t & e^t \\ 1 & e^t & 0 \\ 1 & 0 & e^t \end{vmatrix} = -e^{2t}.$$

By Cramer's rule,

$$u_1' = -\frac{1}{e^{2t}} \begin{vmatrix} e^t & e^t & e^t \\ 0 & e^t & 0 \\ e^{-t} & 0 & e^t \end{vmatrix} = -\frac{e^{3t} - e^t}{e^{2t}} = e^{-t} - e^t,$$

$$u_2' = -\frac{1}{e^{2t}} \begin{vmatrix} 1 & e^t & e^t \\ 1 & 0 & 0 \\ 1 & e^{-t} & e^t \end{vmatrix} = -\frac{1 - e^{2t}}{e^{2t}} = 1 - e^{-2t}$$

$$u_3' = -\frac{1}{e^{2t}} \begin{vmatrix} 1 & e^t & e^t \\ 1 & e^t & 0 \\ 1 & 0 & e^{-t} \end{vmatrix} = \frac{e^{2t}}{e^{2t}} = 1.$$

Therefore

$$\mathbf{u}' = \begin{bmatrix} e^{-t} - e^t \\ 1 - e^{-2t} \\ 1 \end{bmatrix}.$$

Integrating and taking the constants of integration to be zero yields

$$\mathbf{u} = \begin{bmatrix} -e^t - e^{-t} \\ \dfrac{e^{-2t}}{2} + t \\ t \end{bmatrix},$$

so

$$\mathbf{y}_p = Y\mathbf{u} = \begin{bmatrix} 1 & e^t & e^t \\ 1 & e^t & 0 \\ 1 & 0 & e^t \end{bmatrix} \begin{bmatrix} -e^t - e^{-t} \\ \dfrac{e^{-2t}}{2} + t \\ t \end{bmatrix} = \begin{bmatrix} e^t(2t-1) - \dfrac{e^{-t}}{2} \\ e^t(t-1) - \dfrac{e^{-t}}{2} \\ e^t(t-1) - e^{-t} \end{bmatrix}$$

is a particular solution of (8).

Solution   **(b)** From Theorem 10.7.1 the general solution of (8) is

$$\mathbf{y} = \mathbf{y}_p + c_1\mathbf{y}_1 + c_2\mathbf{y}_2 + c_3\mathbf{y}_3$$

$$= \begin{bmatrix} e^t(2t-1) - \dfrac{e^{-t}}{2} \\ e^t(t-1) - \dfrac{e^{-t}}{2} \\ e^t(t-1) - e^{-t} \end{bmatrix} + c_1 \begin{bmatrix} 1 \\ 1 \\ 1 \end{bmatrix} + c_2 \begin{bmatrix} e^t \\ e^t \\ 0 \end{bmatrix} + c_3 \begin{bmatrix} e^t \\ 0 \\ e^t \end{bmatrix},$$

which can also be written as

$$\mathbf{y} = \mathbf{y}_p + Y\mathbf{c} = \begin{bmatrix} e^t(2t-1) - \dfrac{e^{-t}}{2} \\ e^t(t-1) - \dfrac{e^{-t}}{2} \\ e^t(t-1) - e^{-t} \end{bmatrix} + \begin{bmatrix} 1 & e^t & e^t \\ 1 & e^t & 0 \\ 1 & 0 & e^t \end{bmatrix} \mathbf{c}$$

where **c** is an arbitrary constant vector.

EXAMPLE 5   Find a particular solution of

$$\mathbf{y}' = \frac{1}{2} \begin{bmatrix} 3 & e^{-t} & -e^{2t} \\ 0 & 6 & 0 \\ -e^{-2t} & e^{-3t} & -1 \end{bmatrix} \mathbf{y} + \begin{bmatrix} 1 \\ e^t \\ e^{-t} \end{bmatrix}, \tag{10}$$

given that

$$Y = \begin{bmatrix} e^t & 0 & e^{2t} \\ 0 & e^{3t} & e^{3t} \\ e^{-t} & 1 & 0 \end{bmatrix}$$

is a fundamental matrix for the complementary system.

Solution   We seek a particular solution of (10) in the form $\mathbf{y}_p = Y\mathbf{u}$, where $Y\mathbf{u}' = \mathbf{f}$; that is,

$$\begin{bmatrix} e^t & 0 & e^{2t} \\ 0 & e^{3t} & e^{3t} \\ e^{-t} & 1 & 0 \end{bmatrix} \begin{bmatrix} u_1' \\ u_2' \\ u_3' \end{bmatrix} = \begin{bmatrix} 1 \\ e^t \\ e^{-t} \end{bmatrix}.$$

The determinant of Y is the Wronskian

$$\begin{vmatrix} e^t & 0 & e^{2t} \\ 0 & e^{3t} & e^{3t} \\ e^{-t} & 1 & 0 \end{vmatrix} = -2e^{4t}.$$

By Cramer's rule,

$$u_1' = -\frac{1}{2e^{4t}} \begin{vmatrix} 1 & 0 & e^{2t} \\ e^t & e^{3t} & e^{3t} \\ e^{-t} & 1 & 0 \end{vmatrix} = \frac{e^{4t}}{2e^{4t}} = \frac{1}{2},$$

$$u_2' = -\frac{1}{2e^{4t}} \begin{vmatrix} e^t & 1 & e^{2t} \\ 0 & e^t & e^{3t} \\ e^{-t} & e^{-t} & 0 \end{vmatrix} = \frac{e^{3t}}{2e^{4t}} = \frac{1}{2}e^{-t},$$

$$u_3' = -\frac{1}{2e^{4t}} \begin{vmatrix} e^t & 0 & 1 \\ 0 & e^{3t} & e^t \\ e^{-t} & 1 & e^{-t} \end{vmatrix} = -\frac{e^{3t} - 2e^{2t}}{2e^{4t}} = \frac{2e^{-2t} - e^{-t}}{2}.$$

Therefore

$$\mathbf{u}' = \frac{1}{2}\begin{bmatrix} 1 \\ e^{-t} \\ 2e^{-2t} - e^{-t} \end{bmatrix}.$$

Integrating and taking the constants of integration to be zero yields

$$\mathbf{u} = \frac{1}{2}\begin{bmatrix} t \\ -e^{-t} \\ e^{-t} - e^{-2t} \end{bmatrix},$$

so

$$\mathbf{y}_p = Y\mathbf{u} = \frac{1}{2}\begin{bmatrix} e^t & 0 & e^{2t} \\ 0 & e^{3t} & e^{3t} \\ e^{-t} & 1 & 0 \end{bmatrix}\begin{bmatrix} t \\ -e^{-t} \\ e^{-t} - e^{-2t} \end{bmatrix} = \frac{1}{2}\begin{bmatrix} e^t(t+1) - 1 \\ -e^t \\ e^{-t}(t-1) \end{bmatrix}$$

is a particular solution of (10).   ■

10.7 EXERCISES

In Exercises 1–10 find a particular solution.

1. $\mathbf{y}' = \begin{bmatrix} -1 & -4 \\ -1 & -1 \end{bmatrix}\mathbf{y} + \begin{bmatrix} 21e^{4t} \\ 8e^{-3t} \end{bmatrix}$

2. $\mathbf{y}' = \frac{1}{5}\begin{bmatrix} -4 & 3 \\ -2 & -11 \end{bmatrix}\mathbf{y} + \begin{bmatrix} 50e^{3t} \\ 10e^{-3t} \end{bmatrix}$

3. $\mathbf{y}' = \begin{bmatrix} 1 & 2 \\ 2 & 1 \end{bmatrix}\mathbf{y} + \begin{bmatrix} 1 \\ t \end{bmatrix}$

4. $\mathbf{y}' = \begin{bmatrix} -4 & -3 \\ 6 & 5 \end{bmatrix}\mathbf{y} + \begin{bmatrix} 2 \\ -2e^t \end{bmatrix}$

5. $\mathbf{y}' = \begin{bmatrix} -6 & -3 \\ 1 & -2 \end{bmatrix}\mathbf{y} + \begin{bmatrix} 4e^{-3t} \\ 4e^{-5t} \end{bmatrix}$

6. $\mathbf{y}' = \begin{bmatrix} 0 & 1 \\ -1 & 0 \end{bmatrix}\mathbf{y} + \begin{bmatrix} 1 \\ t \end{bmatrix}$

7. $\mathbf{y}' = \begin{bmatrix} 3 & 1 & -1 \\ 3 & 5 & 1 \\ -6 & 2 & 4 \end{bmatrix}\mathbf{y} + \begin{bmatrix} 3 \\ 6 \\ 3 \end{bmatrix}$

8. $\mathbf{y}' = \begin{bmatrix} 3 & -1 & -1 \\ -2 & 3 & 2 \\ 4 & -1 & -2 \end{bmatrix}\mathbf{y} + \begin{bmatrix} 1 \\ e^t \\ e^t \end{bmatrix}$

9. $\mathbf{y}' = \begin{bmatrix} -3 & 2 & 2 \\ 2 & -3 & 2 \\ 2 & 2 & -3 \end{bmatrix}\mathbf{y} + \begin{bmatrix} e^t \\ e^{-5t} \\ e^t \end{bmatrix}$

10. $\mathbf{y}' = \frac{1}{3}\begin{bmatrix} 1 & 1 & -3 \\ -4 & -4 & 3 \\ -2 & 1 & 0 \end{bmatrix}\mathbf{y} + \begin{bmatrix} e^t \\ e^t \\ e^t \end{bmatrix}$

In Exercises 11–20 find a particular solution, given that Y is a fundamental matrix for the complementary system.

11. $y' = \dfrac{1}{t}\begin{bmatrix} 1 & t \\ -t & 1 \end{bmatrix} y + t\begin{bmatrix} \cos t \\ \sin t \end{bmatrix}; \quad Y = t\begin{bmatrix} \cos t & \sin t \\ -\sin t & \cos t \end{bmatrix}$

12. $y' = \dfrac{1}{t}\begin{bmatrix} 1 & t \\ t & 1 \end{bmatrix} y + \begin{bmatrix} t \\ t^2 \end{bmatrix}; \quad Y = t\begin{bmatrix} e^t & e^{-t} \\ e^t & -e^{-t} \end{bmatrix}$

13. $y' = \dfrac{1}{t^2 - 1}\begin{bmatrix} t & -1 \\ -1 & t \end{bmatrix} y + t\begin{bmatrix} 1 \\ -1 \end{bmatrix}; \quad Y = \begin{bmatrix} t & 1 \\ 1 & t \end{bmatrix}$

14. $y' = \dfrac{1}{3}\begin{bmatrix} 1 & -2e^{-t} \\ 2e^t & -1 \end{bmatrix} y + \begin{bmatrix} e^{2t} \\ e^{-2t} \end{bmatrix}; \quad Y = \begin{bmatrix} 2 & e^{-t} \\ e^t & 2 \end{bmatrix}$

15. $y' = \dfrac{1}{2t^4}\begin{bmatrix} 3t^3 & t^6 \\ 1 & -3t^3 \end{bmatrix} y + \dfrac{1}{t}\begin{bmatrix} t^2 \\ 1 \end{bmatrix}; \quad Y = \dfrac{1}{t^2}\begin{bmatrix} t^3 & t^4 \\ -1 & t \end{bmatrix}$

16. $y' = \begin{bmatrix} \dfrac{1}{t-1} & -\dfrac{e^{-t}}{t-1} \\ \dfrac{e^t}{t+1} & \dfrac{1}{t+1} \end{bmatrix} y + \begin{bmatrix} t^2 - 1 \\ t^2 - 1 \end{bmatrix}; \quad Y = \begin{bmatrix} t & e^{-t} \\ e^t & t \end{bmatrix}$

17. $y' = \dfrac{1}{t}\begin{bmatrix} 1 & 1 & 0 \\ 0 & 2 & 1 \\ -2 & 2 & 2 \end{bmatrix} y + \begin{bmatrix} 1 \\ 2 \\ 1 \end{bmatrix}; \quad Y = \begin{bmatrix} t^2 & t^3 & 1 \\ t^2 & 2t^3 & -1 \\ 0 & 2t^3 & 2 \end{bmatrix}$

18. $y' = \begin{bmatrix} 3 & e^t & e^{2t} \\ e^{-t} & 2 & e^t \\ e^{-2t} & e^{-t} & 1 \end{bmatrix} y + \begin{bmatrix} e^{3t} \\ 0 \\ 0 \end{bmatrix}; \quad Y = \begin{bmatrix} e^{5t} & e^{2t} & 0 \\ e^{4t} & 0 & e^t \\ e^{3t} & -1 & -1 \end{bmatrix}$

19. $y' = \dfrac{1}{t}\begin{bmatrix} 1 & t & 0 \\ 0 & 1 & t \\ 0 & -t & 1 \end{bmatrix} y + \begin{bmatrix} t \\ t \\ t \end{bmatrix}; \quad Y = t\begin{bmatrix} 1 & \cos t & \sin t \\ 0 & -\sin t & \cos t \\ 0 & -\cos t & -\sin t \end{bmatrix}$

20. $y' = -\dfrac{1}{t}\begin{bmatrix} e^{-t} & -t & 1-e^{-t} \\ e^{-t} & 1 & -t-e^{-t} \\ e^{-t} & -t & 1-e^{-t} \end{bmatrix} y + \dfrac{1}{t}\begin{bmatrix} e^t \\ 0 \\ e^t \end{bmatrix}; \quad Y = \dfrac{1}{t}\begin{bmatrix} e^t & e^{-t} & t \\ e^t & -e^{-t} & e^{-t} \\ e^t & e^{-t} & 0 \end{bmatrix}$

21. Prove Theorem 10.7.1.

22. (a) Convert the scalar equation

$$P_0(t)y^{(n)} + P_1(t)y^{(n-1)} + \cdots + P_n(t)y = F(t) \tag{A}$$

into an equivalent $n \times n$ system

$$y' = A(t)y + f(t). \tag{B}$$

(b) Suppose that (A) is normal on an interval (a, b) and $\{y_1, y_2, \ldots, y_n\}$ is a fundamental set of solutions of

$$P_0(t)y^{(n)} + P_1(t)y^{(n-1)} + \cdots + P_n(t)y = 0 \tag{C}$$

on (a,b). Find a corresponding fundamental matrix Y for

$$\mathbf{y}' = A(t)\mathbf{y} \tag{D}$$

on (a,b) such that

$$y = c_1y_1 + c_2y_2 + \cdots + c_ny_n$$

is a solution of (C) if and only if $\mathbf{y} = Y\mathbf{c}$ with

$$\mathbf{c} = \begin{bmatrix} c_1 \\ c_2 \\ \vdots \\ c_n \end{bmatrix}$$

is a solution of (D).

(c) Let $y_p = u_1y_1 + u_1y_2 + \cdots + u_ny_n$ be a particular solution of (A), obtained by the method of variation of parameters for scalar equations as given in Section 9.4, and define

$$\mathbf{u} = \begin{bmatrix} u_1 \\ u_2 \\ \vdots \\ u_n \end{bmatrix}.$$

Show that $\mathbf{y}_p = Y\mathbf{u}$ is a solution of (B).

(d) Let $\mathbf{y}_p = Y\mathbf{u}$ be a particular solution of (B), obtained by the method of variation of parameters for systems as given in this section. Show that $y_p = u_1y_1 + u_1y_2 + \cdots + u_ny_n$ is a solution of (A).

23. Suppose that the $n \times n$ matrix function A and the n-vector function $\mathbf{f}$ are continuous on (a,b). Let t_0 be in (a,b), let $\mathbf{k}$ be an arbitrary constant vector, and let Y be a fundamental matrix for the homogeneous system $\mathbf{y}' = A(t)\mathbf{y}$. Use variation of parameters to show that the solution of the initial value problem

$$\mathbf{y}' = A(t)\mathbf{y} + \mathbf{f}(t), \qquad \mathbf{y}(t_0) = \mathbf{k}$$

is

$$\mathbf{y}(t) = Y(t)\left(Y^{-1}(t_0)\mathbf{k} + \int_{t_0}^{t} Y^{-1}(s)\mathbf{f}(s)\, ds \right).$$

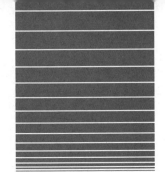

Answers to Selected Exercises

Section 1.2 Answers (page 14)

1. (a) 3 (b) 2 (c) 1 (d) 2

3. (a) $y = -\dfrac{x^2}{2} + c$ (b) $y = x\cos x - \sin x + c$ (c) $y = \dfrac{x^2}{2}\ln x - \dfrac{x^2}{4} + c$

(d) $y = -x\cos x + 2\sin x + c_1 + c_2 x$ (e) $y = (2x - 4)e^x + c_1 + c_2 x$ (f) $y = \dfrac{x^3}{3} - \sin x + e^x + c_1 + c_2 x$

(g) $y = \sin x + c_1 + c_2 x + c_3 x^2$ (h) $y = -\dfrac{x^5}{60} + e^x + c_1 + c_2 x + c_3 x^2$ (i) $y = \dfrac{7}{64}e^{4x} + c_1 + c_2 x + c_3 x^2$

4. (a) $y = -(x-1)e^x$ (b) $y = 1 - \dfrac{1}{2}\cos x^2$ (c) $y = 3 - \ln(\sqrt{2}\cos x)$ (d) $y = -\dfrac{47}{15} - \dfrac{37}{5}(x-2) + \dfrac{x^6}{30}$

(e) $y = \dfrac{1}{4}xe^{2x} - \dfrac{1}{4}e^{2x} + \dfrac{5}{4}x + \dfrac{29}{4}$ (f) $y = x\sin x + 2\cos x - 3x - 1$

(g) $y = (x^2 - 6x + 12)e^x + \dfrac{x^2}{2} - 8x - 11$ (h) $y = \dfrac{x^3}{3} + \dfrac{\cos 2x}{8} + \dfrac{7}{4}x^2 - 6x + \dfrac{7}{8}$

(i) $y = \dfrac{x^4}{12} + \dfrac{x^3}{6} + \dfrac{1}{2}(x-2)^2 - \dfrac{26}{3}(x-2) - \dfrac{5}{3}$

7. (a) 576 ft (b) 10 s **8.** (b) $y \equiv 0$ **10.** (a) $(-2c - 2, \infty)$ (b) $(-\infty, \infty)$

Section 2.1 Answers (page 37)

1. $y = e^{-ax}$ **2.** $y = ce^{-x^3}$ **3.** $y = ce^{-(\ln x)^2/2}$ **4.** $y = \dfrac{c}{x^3}$ **5.** $y = ce^{1/x}$ **6.** $y = \dfrac{e^{-(x-1)}}{x}$ **7.** $y = \dfrac{e}{x\ln x}$

8. $y = \dfrac{\pi}{x\sin x}$ **9.** $y = 2(1 + x^2)$ **10.** $y = 3x^{-k}$ **11.** $y = c(\cos kx)^{1/k}$ **12.** $y = \dfrac{1}{3} + ce^{-3x}$

13. $y = \dfrac{2}{x} + \dfrac{c}{x}e^x$ **14.** $y = e^{-x^2}\left(\dfrac{x^2}{2} + c\right)$ **15.** $y = -\dfrac{e^{-x} + c}{1 + x^2}$ **16.** $y = \dfrac{7\ln|x|}{x} + \dfrac{3}{2}x + \dfrac{c}{x}$

17. $y = (x-1)^{-4}(\ln|x-1| - \cos x + c)$ **18.** $y = e^{-x^2}\left(\dfrac{x^3}{4} + \dfrac{c}{x}\right)$ **19.** $y = \dfrac{2\ln|x|}{x^2} + \dfrac{1}{2} + \dfrac{c}{x^2}$

20. $y = (x + c)\cos x$ **21.** $y = \dfrac{c - \cos x}{(1 + x)^2}$ **22.** $y = -\dfrac{1}{2}\dfrac{(x-2)^3}{x-1} + c\dfrac{(x-2)^5}{x-1}$ **23.** $y = (x + c)e^{-\sin^2 x}$

24. $y = \dfrac{e^x}{x^2} - \dfrac{e^x}{x^3} + \dfrac{c}{x^3}$ **25.** $y = \dfrac{1}{10}(e^{3x} - e^{-7x})$ **26.** $y = \dfrac{2x + 1}{(1 + x^2)^2}$ **27.** $y = \dfrac{1}{x^3}\ln\left(\dfrac{1 + x^2}{2}\right)$

28. $y = \dfrac{1}{2}(\sin x + \csc x)$ **29.** $y = \dfrac{2\ln|x|}{x} + \dfrac{x}{2} - \dfrac{1}{2x}$ **30.** $y = (x-1)^{-3}[\ln(1 - x) - \cos x]$

31. $y = 2x^2 + \dfrac{1}{x^2}$; $(0, \infty)$ **32.** $y = x^2(1 - \ln x)$ **33.** $y = \dfrac{1}{2} + \dfrac{5}{2}e^{-x^2}$ **34.** $y = \dfrac{\ln|x - 1| + \tan x + 1}{(x - 1)^3}$

35. $y = \dfrac{\ln|x| + x^2 + 1}{(x+2)^4}$ **36.** $y = (x^2-1)\left(\dfrac{1}{2}\ln|x^2-1| - 4\right)$ **37.** $y = -(x^2-5)(7 + \ln|x^2-5|)$

38. $y = e^{-x^2}\left(3 + \displaystyle\int_0^x t^2 e^{t^2}\, dt\right)$ **39.** $y = \dfrac{1}{x}\left(2 + \displaystyle\int_1^x \dfrac{\sin t}{t}\, dt\right)$ **40.** $y = e^{-x}\displaystyle\int_1^x \dfrac{\tan t}{t}\, dt$

41. $y = \dfrac{1}{1+x^2}\left(1 + \displaystyle\int_0^x \dfrac{e^t}{1+t^2}\, dt\right)$ **42.** $y = \dfrac{1}{x}\left(2e^{-(x-1)} + e^{-x}\displaystyle\int_1^x e^t e^{t^2}\, dt\right)$

43. $G = \dfrac{r}{\lambda} + \left(G_0 - \dfrac{r}{\lambda}\right)e^{-\lambda t}$; $\lim_{t\to\infty} G(t) = \dfrac{r}{\lambda}$ **45. (a)** $y = y_0 e^{-a(x-x_0)} + e^{-ax}\displaystyle\int_{x_0}^x e^{at} f(t)\, dt$

48. (a) $y = \tan^{-1}\left(\dfrac{1}{3} + ce^{3x}\right)$ **(b)** $y = \pm\left[\ln\left(\dfrac{1}{x} + \dfrac{c}{x^2}\right)\right]^{1/2}$ **(c)** $y = \exp\left(x^2 + \dfrac{c}{x^2}\right)$ **(d)** $y = -1 + \dfrac{x}{c + 3\ln|x|}$

Section 2.2 Answers   (page 47)

1. $y = 2 \pm \sqrt{2(x^3 + x^2 + x + c)}$ **2.** $\ln(|\sin y|) = \cos x + c$; $y \equiv k\pi$ $(k = \text{integer})$ **3.** $y = \dfrac{c}{x-c}$; $y \equiv -1$

4. $\dfrac{(\ln y)^2}{2} = -\dfrac{x^3}{3} + c$ **5.** $y^3 + 3\sin y + \ln|y| + \ln(1+x^2) + \tan^{-1} x = c$; $y \equiv 0$

6. $y = \pm\left(1 + \left(\dfrac{x}{1+cx}\right)^2\right)^{1/2}$; $y \equiv \pm 1$ **7.** $y = \tan\left(\dfrac{x^3}{3} + c\right)$ **8.** $y = \dfrac{c}{\sqrt{1+x^2}}$

9. $y = \dfrac{2 - ce^{(x-1)^2/2}}{1 - ce^{(x-1)^2/2}}$; $y \equiv 1$ **10.** $y = 1 + (3x^2 + 9x + c)^{1/3}$ **11.** $y = 2 + \sqrt{\dfrac{2}{3}x^3 + 3x^2 + 4x - \dfrac{11}{3}}$

12. $y = \dfrac{e^{-(x^2-4)/2}}{2 - e^{-(x^2-4)/2}}$ **13.** $y^3 + 2y^2 + x^2 + \sin x = 3$ **14.** $(y+1)(y-1)^{-3}(y-2)^2 = -256(x+1)^{-6}$

15. $y = -1 + 3e^{-x^2}$ **16.** $y = \dfrac{1}{\sqrt{2e^{-2x^2} - 1}}$ **17.** $y \equiv -1$; $(-\infty, \infty)$ **18.** $y = \dfrac{4 - e^{-x^2}}{2 - e^{-x^2}}$; $(-\infty, \infty)$

19. $y = \dfrac{-1 + \sqrt{4x^2 - 15}}{2}$; $\left(\dfrac{\sqrt{15}}{2}, \infty\right)$ **20.** $y = \dfrac{2}{1 + e^{-2x}}$; $(-\infty, \infty)$ **21.** $y = -\sqrt{25 - x^2}$; $(-5, 5)$

22. $y \equiv 2$; $(-\infty, \infty)$ **23.** $y = 3\left(\dfrac{x+1}{2x-4}\right)^{1/3}$; $(-\infty, 2)$ **24.** $y = \dfrac{x+c}{1-cx}$

25. $y = -x\cos c + \sqrt{1-x^2}\sin c$; $y \equiv 1$, $y \equiv -1$ **26.** $y = -x + 3\pi/2$

28. $P = \dfrac{P_0}{\alpha P_0 + (1 - \alpha P_0)e^{-at}}$; $\lim_{t\to\infty} P(t) = 1/\alpha$ **29.** $I = \dfrac{SI_0}{I_0 + (S - I_0)e^{-rSt}}$

30. If $q = rS$ then $I = I_0/(1 + rI_0 t)$ and $\lim_{t\to\infty} I(t) = 0$. If $q \ne Rs$ then $I = \alpha I_0/(I_0 + (\alpha - I_0)e^{-rat})$. If $q < rs$ then $\lim_{t\to\infty} I(t) = \alpha = S - q/r$; if $q > rS$ then $\lim_{t\to\infty} I(t) = 0$.

34. $f = ap$, where $a = $ constant   **35.** $y = e^{-x}(-1 \pm \sqrt{2x^2 + c})$ **36.** $y = x^2(-1 + \sqrt{x^2 + c})$

37. $y = e^x(-1 + (3xe^x + c)^{1/3})$ **38.** $y = e^{2x}(1 \pm \sqrt{c - x^2})$

39. (a) $y_1 = 1/x$; $g(x) = h(x)$ **(b)** $y_1 = x$; $g(x) = h(x)/x^2$ **(c)** $y_1 = e^{-x}$; $g(x) = e^x h(x)$

   (d) $y_1 = x^{-r}$; $g(x) = x^{r-1} h(x)$ **(e)** $y_1 = 1/v(x)$; $g(x) = v(x)h(x)$

Section 2.3 Answers   (page 55)

1. (a), (b) $x_0 \ne k\pi$ $(k = \text{integer})$ **2. (a), (b)** $(x_0, y_0) \ne (0, 0)$

3. (a), (b) $x_0 y_0 \ne (2k+1)\dfrac{\pi}{2}$ $(k = \text{integer})$ **4. (a), (b)** $x_0 y_0 > 0$ and $x_0 y_0 \ne 1$

5. (a) all (x_0, y_0) **(b)** (x_0, y_0) with $y_0 \ne 0$ **6. (a), (b)** all (x_0, y_0)

7. (a), (b) all (x_0, y_0) **8. (a), (b)** (x_0, y_0) such that $x_0 \ne 4y_0$

9. (a) all (x_0, y_0) **(b)** all $(x_0, y_0) \neq (0,0)$ **10. (a)** all (x_0, y_0) **(b)** all (x_0, y_0) with $y_0 \neq \pm 1$

11. (a), **(b)** all (x_0, y_0) **12. (a)**, **(b)** all (x_0, y_0) such that $x_0 + y_0 > 0$

13. (a), **(b)** all (x_0, y_0) with $x_0 \neq 1$, $y_0 \neq (2k+1)\dfrac{\pi}{2}$ (k = integer)

16. $y = \left(\dfrac{3}{5}x + 1\right)^{5/3}$, $-\infty < x < \infty$ is a solution. Also,

$$y = \begin{cases} 0, & -\infty < x \leq -\frac{5}{3}, \\ (\frac{3}{5}x + 1)^{5/3}, & -\frac{5}{3} < x < \infty \end{cases}$$

is a solution. For every $a \geq 5/3$, the following function is also a solution:

$$y = \begin{cases} (\frac{3}{5}(x + a))^{5/3}, & -\infty < x < -a, \\ 0, & -a \leq x \leq -\frac{5}{3}, \\ (\frac{3}{5}x + 1)^{5/3}, & -\frac{5}{3} < x < \infty. \end{cases}$$

17. (a) all (x_0, y_0) **(b)** all (x_0, y_0) with $y_0 \neq 1$

18. $y_1 \equiv 1$; $y_2 = 1 + |x|^3$; $y_3 = 1 - |x|^3$; $y_4 = 1 + x^3$; $y_5 = 1 - x^3$;   **19.** $y = 1 + (x^2 + 4)^{3/2}$, $-\infty < x < \infty$

$$y_6 = \begin{cases} 1 + x^3, & x \geq 0, \\ 1, & x < 0 \end{cases} \; ; \quad y_7 = \begin{cases} 1 - x^3, & x \geq 0, \\ 1, & x < 0 \end{cases} \; ;$$

$$y_8 = \begin{cases} 1, & x \geq 0, \\ 1 + x^3, & x < 0 \end{cases} \; ; \quad y_9 = \begin{cases} 1, & x \geq 0, \\ 1 - x^3, & x < 0 \end{cases}$$

20. (a) The solution is unique on $(0, \infty)$. It is given by

$$y = \begin{cases} 1, & 0 < x \leq \sqrt{5}, \\ 1 - (x^2 - 5)^{3/2}, & \sqrt{5} < x < \infty \end{cases}$$

(b)

$$y = \begin{cases} 1, & -\infty < x \leq \sqrt{5}, \\ 1 - (x^2 - 5)^{3/2}, & \sqrt{5} < x < \infty \end{cases}$$

is a solution of (A) on $(-\infty, \infty)$. If $\alpha \geq 0$, then

$$y = \begin{cases} 1 + (x^2 - \alpha^2)^{3/2}, & -\infty < x < -\alpha, \\ 1, & -\alpha \leq x \leq \sqrt{5}, \\ 1 - (x^2 - 5)^{3/2}, & \sqrt{5} < x < \infty, \end{cases}$$

and

$$y = \begin{cases} 1 - (x^2 - \alpha^2)^{3/2}, & -\infty < x < -\alpha, \\ 1, & -\alpha \leq x \leq \sqrt{5}, \\ 1 - (x^2 - 5)^{3/2}, & \sqrt{5} < x < \infty, \end{cases}$$

are also solutions of (A) on $(-\infty, \infty)$.

Section 2.4 Answers   (page 63)

1. $y = \dfrac{1}{1 - ce^x}$ **2.** $y = x^{2/7}(c - \ln |x|)^{1/7}$ **3.** $y = e^{2/x}(c - 1/x)^2$ **4.** $y = \pm\dfrac{\sqrt{2x + c}}{1 + x^2}$

5. $y = \pm(1 - x^2 + ce^{-x^2})^{-1/2}$ **6.** $y = \left[\dfrac{x}{3(1 - x) + ce^{-x}}\right]^{1/3}$ **7.** $y = \dfrac{2\sqrt{2}}{\sqrt{1 - 4x}}$

8. $y = \left[1 - \dfrac{3}{2}e^{-(x^2 - 1)/4}\right]^{-2}$ **9.** $y = \dfrac{1}{x(11 - 3x)^{1/3}}$ **10.** $y = (2e^x - 1)^2$ **11.** $y = (2e^{12x} - 1 - 12x)^{1/3}$

12. $y = \left[\dfrac{5x}{2(1 + 4x^5)}\right]^{1/2}$ **13.** $y = (4e^{x/2} - x - 2)^2$

14. $P = \dfrac{P_0 e^{at}}{1 + aP_0 \int_0^t \alpha(\tau)e^{a\tau}\,d\tau};$ $\lim_{t\to\infty} P(t) = \begin{cases} \infty & \text{if } L = 0, \\ 0 & \text{if } L = \infty, \\ 1/aL & \text{if } 0 < L < \infty. \end{cases}$

15. $y = x(\ln|x| + c)$ **16.** $y = \dfrac{cx^2}{1 - cx};$ $y = -x$ **17.** $y = \pm x(4\ln|x| + c)^{1/4}$ **18.** $y = x\sin^{-1}(\ln|x| + c)$

19. $y = x\tan(\ln|x| + c)$ **20.** $y = \pm x\sqrt{cx^2 - 1}$ **21.** $y = \pm x\ln(\ln|x| + c)$ **22.** $y = -\dfrac{2x}{2\ln|x| + 1}$

23. $y = x(3\ln x + 27)^{1/3}$ **24.** $y = \dfrac{1}{x}\left(\dfrac{9 - x^4}{2}\right)^{1/2}$ **25.** $y = -x$ **26.** $y = -\dfrac{x(4x - 3)}{(2x - 3)}$

27. $y = x\sqrt{4x^6 - 1}$ **28.** $\tan^{-1}\dfrac{y}{x} - \dfrac{1}{2}\ln(x^2 + y^2) = c$ **29.** $(x + y)\ln|x| + y(1 - \ln|y|) + cx = 0$

30. $(y + x)^3 = 3x^3(\ln|x| + c)$ **31.** $(y + x) = c(y - x)^3;$ $y = x;$ $y = -x$

32. $y^2(y - 3x) = c;$ $y \equiv 0;$ $y = 3x$ **33.** $(x - y)^3(x + y) = cy^2x^4;$ $y = 0;$ $y = x;$ $y = -x$

34. $\dfrac{y}{x} + \dfrac{y^3}{x^3} = \ln|x| + c$ **40.** Choose X_0 and Y_0 so that $\begin{aligned} aX_0 + bY_0 &= \alpha \\ cX_0 + dY_0 &= \beta. \end{aligned}$

41. $(y + 2x + 1)^4(2y - 6x - 3) = c;$ $y = 3x + 3/2;$ $y = -2x - 1$

42. $(y + x - 1)(y - x - 5)^3 = c;$ $y = x + 5;$ $y = -x + 1$

43. $\ln|y - x - 6| - \dfrac{2(x + 2)}{y - x - 6} = c;$ $y = x + 6$ **44.** $y_1 = x^{1/3};$ $y = x^{1/3}(\ln|x| + c)^{1/3}$

45. $y_1 = x^3;$ $y = \pm x^3\sqrt{cx^6 - 1}$ **46.** $y_1 = x^2;$ $y = \dfrac{x^2(1 + cx^4)}{1 - cx^4};$ $y = -x^2$

47. $y_1 = e^x;$ $y = -\dfrac{e^x(1 - 2ce^x)}{1 - ce^x};$ $y = -2e^x$ **48.** $y_1 = \tan x;$ $y = \tan x\tan(\ln|\tan x| + c)$

49. $y_1 = \ln x;$ $y = \dfrac{2\ln x\,(1 + c(\ln x)^4)}{1 - c(\ln x)^4};$ $y = -2\ln x$ **50.** $y_1 = x^{1/2};$ $y = x^{1/2}(-2 \pm \sqrt{\ln|x| + c})$

51. $y_1 = e^{x^2};$ $y = e^{x^2}(-1 \pm \sqrt{2x^2 + c})$ **52.** $y = \dfrac{-3 + \sqrt{1 + 60x}}{2x}$ **53.** $y = \dfrac{-5 + \sqrt{1 + 48x}}{2x^2}$

56. $y = 1 + \dfrac{1}{x + 1 + ce^x}$ **57.** $y = e^x - \dfrac{1}{1 + ce^{-x}}$ **58.** $y = 1 - \dfrac{1}{x(1 - cx)}$ **59.** $y = x - \dfrac{2x}{x^2 + c}$

Section 2.5 Answers   (page 73)

1. $2x^3y^2 = c$ **2.** $3y\sin x + 2x^2e^x + 3y = c$ **3.** Not exact   **4.** $x^2 - 2xy^2 + 4y^3 = c$ **5.** $x + y = c$

6. Not exact   **7.** $2y^2\cos x + 3xy^3 - x^2 = c$ **8.** Not exact   **9.** $x^3 + x^2y + 4xy^2 + 9y^2 = c$

10. Not exact   **11.** $\ln|xy| + x^2 + y^2 = c$ **12.** Not exact   **13.** $x^2 + y^2 = c$ **14.** $x^2y^2e^x + 2y + 3x^2 = c$

15. $x^3e^{x^2+y} - 4y^2 + 2x^2 = c$ **16.** $x^4e^{xy} + 3xy = c$ **17.** $x^3\cos xy + 4y^2 + 2x^2 = c$

18. $y = \dfrac{x + \sqrt{2x^2 + 3x - 1}}{x^2}$ **19.** $y = \sin x - \sqrt{1 - \dfrac{\tan x}{2}}$ **20.** $y = \left(\dfrac{e^x - 1}{e^x + 1}\right)^{1/3}$ **21.** $y = 1 + 2\tan x$

22. $y = \dfrac{x^2 - x + 6}{(x + 2)(x - 3)}$ **23.** $\dfrac{7x^2}{2} + 4xy + \dfrac{3y^2}{2} = c$ **24.** $(x^4y^2 + 1)e^x + y^2 = c$

29. **(a)** $M(x, y) = 2xy + f(x)$ **(b)** $M(x, y) = 2(\sin x + x\cos x)(y\sin y + \cos y) + f(x)$

   (c) $M(x, y) = ye^x - e^y\cos x + f(x)$

30. (a) $N(x,y) = \dfrac{x^4 y}{2} + x^2 + 6xy + g(y)$ **(b)** $N(x,y) = \dfrac{x}{y} + 2y \sin x + g(y)$

(c) $N(x,y) = x(\sin y + y \cos y) + g(y)$

33. $B = C$ **34.** $B = 2D,\quad E = 2C$ **37. (a)** $2x^2 + x^4 y^4 + y^2 = c$ **(b)** $x^3 + 3xy^2 = c$ **(c)** $x^3 + y^2 + 2xy = c$

38. $y = -1 - \dfrac{1}{x^2}$ **39.** $y = x^3\left(\dfrac{-3(x^2+1) + \sqrt{9x^4 + 34x^2 + 21}}{2}\right)$ **40.** $y = -e^{-x^2}\left(\dfrac{2x + \sqrt{9 - 5x^2}}{3}\right)$

44. (a) $G(x,y) = 2xy + c$ **(b)** $G(x,y) = e^x \sin y + c$ **(c)** $G(x,y) = 3x^2 y - y^3 + c$

(d) $G(x,y) = -\sin x \sinh y + c$ **(e)** $G(x,y) = \cos x \sinh y + c$

Section 2.6 Answers   (page 84)

3. $\mu(x) = 1/x^2;\quad y = cx$ and $\mu(y) = 1/y^2;\quad x = cy$ **4.** $\mu(x) = x^{-3/2};\quad x^{3/2} y = c$ **5.** $\mu(y) = 1/y^3;\quad y^3 e^{2x} = c$

6. $\mu(x) = e^{5x/2};\quad e^{5x/2}(xy + 1) = c$ **7.** $\mu(x) = e^x;\quad e^x(xy + y + x) = c$

8. $\mu(x) = x;\quad x^2 y^2 (9x + 4y) = c$ **9.** $\mu(y) = y^2;\quad y^3(3x^2 y + 2x + 1) = c$

10. $\mu(y) = ye^y;\quad e^y(xy^3 + 1) = c$ **11.** $\mu(y) = y^2;\quad y^3(3x^4 + 8x^3 y + y) = c$

12. $\mu(x) = xe^x;\quad x^2 y(x + 1)e^x = c$ **13.** $\mu(x) = (x^3 - 1)^{-4/3};\quad xy(x^3 - 1)^{-1/3} = c$ and $x \equiv 1$

14. $\mu(y) = e^y;\quad e^y(\sin x \cos y + y - 1) = c$ **15.** $\mu(y) = e^{-y^2};\quad xye^{-y^2}(x + y) = c$

16. $\dfrac{xy}{\sin y} = c$ and $y \equiv k\pi$ $(k = \text{integer})$ **17.** $\mu(x,y) = x^4 y^3;\quad x^5 y^4 \ln x = c$

18. $\mu(x,y) = 1/xy;\quad |x|^\alpha |y|^\beta e^{\gamma x} e^{\delta y} = c$ and $x \equiv 0,\quad y \equiv 0$

19. $\mu(x,y) = x^{-2} y^{-3};\quad 3x^2 y^2 + y = 1 + cxy^2$ and $x \equiv 0,\quad y \equiv 0$

20. $\mu(x,y) = x^{-2} y^{-1};\quad -\dfrac{2}{x} + y^3 + 3\ln|y| = c$ and $x \equiv 0,\quad y \equiv 0$ **21.** $\mu(x,y) = e^{ax} e^{by};\quad e^{ax} e^{by} \cos xy = c$

22. $\mu(x,y) = x^{-4} y^{-3}$ (and others); $xy = c$ **23.** $\mu(x,y) = xe^y;\quad x^2 y e^y \sin x = c$

24. $\mu(x) = 1/x^2;\quad \dfrac{x^3 y^3}{3} - \dfrac{y}{x} = c$ **25.** $\mu(x) = x + 1;\quad y(x + 1)^2(x + y) = c$

26. $\mu(x,y) = x^2 y^2;\quad x^3 y^3(3x + 2y^2) = c$ **27.** $\mu(x,y) = x^{-2} y^{-2};\quad 3x^2 y = cxy + 2$ and $x \equiv 0,\quad y \equiv 0$

Section 3.1 Answers   (page 100)

1. $y_1 = 1.450000000,\quad y_2 = 2.085625000,\quad y_3 = 3.079099746$

2. $y_1 = 1.200000000,\quad y_2 = 1.440415946,\quad y_3 = 1.729880994$

3. $y_1 = 1.900000000,\quad y_2 = 1.781375000,\quad y_3 = 1.646612970$

4. $y_1 = 2.962500000,\quad y_2 = 2.922635828,\quad y_3 = 2.880205639$

5. $y_1 = 2.513274123,\quad y_2 = 1.814517822,\quad y_3 = 1.216364496$

6.

x	$h = 0.1$	$h = 0.05$	$h = 0.025$	**Exact**
1.0	48.298147362	51.492825643	53.076673685	54.647937102

7.

x	$h = 0.1$	$h = 0.05$	$h = 0.025$	**Exact**
2.0	1.390242009	1.370996758	1.361921132	1.353193719

8.

x	$h = 0.05$	$h = 0.025$	$h = 0.0125$	**Exact**
1.50	7.886170437	8.852463793	9.548039907	10.500000000

9.

x	$h = 0.1$	$h = 0.05$	$h = 0.025$	$h = 0.1$	$h = 0.05$	$h = 0.025$
3.0	1.469458241	1.462514486	1.459217010	0.3210	0.1537	0.0753
	Approximate Solutions			Residuals		

10.

x	$h = 0.1$	$h = 0.05$	$h = 0.025$	$h = 0.1$	$h = 0.05$	$h = 0.025$
2.0	0.473456737	0.483227470	0.487986391	-0.3129	-0.1563	-0.0781
	Approximate Solutions			Residuals		

11.

x	$h = 0.1$	$h = 0.05$	$h = 0.025$	"Exact"
1.0	0.691066797	0.676269516	0.668327471	0.659957689

12.

x	$h = 0.1$	$h = 0.05$	$h = 0.025$	"Exact"
2.0	-0.772381768	-0.761510960	-0.756179726	-0.750912371

13. Applying variation of parameters to the given initial value problem yields $y = ue^{-3x}$, where (A) $u' = 7$, $u(0) = 6$. Since $u'' = 0$, Euler's method yields the exact solution of (A). Therefore the Euler semilinear method produces the exact solution of the given problem.

Euler's method			
x	$h = 0.1$	$h = 0.05$	$h = 0.025$	Exact
1.0	0.538871178	0.593002325	0.620131525	0.647231889

Euler semilinear method			
x	$h = 0.1$	$h = 0.05$	$h = 0.025$	Exact
1.0	0.647231889	0.647231889	0.647231889	0.647231889

14.

Euler's method			
x	$h = 0.1$	$h = 0.05$	$h = 0.025$	"Exact"
3.0	12.804226135	13.912944662	14.559623055	15.282004826

Euler semilinear method			
x	$h = 0.1$	$h = 0.05$	$h = 0.025$	"Exact"
3.0	15.354122287	15.317257705	15.299429421	15.282004826

15.

Euler's method			
x	$h = 0.2$	$h = 0.1$	$h = 0.05$	"Exact"
2.0	0.867565004	0.885719263	0.895024772	0.904276722

Euler semilinear method			
x	$h = 0.2$	$h = 0.1$	$h = 0.05$	"Exact"
2.0	0.569670789	0.720861858	0.808438261	0.904276722

16.

	Euler's method			
x	$h = 0.2$	$h = 0.1$	$h = 0.05$	"Exact"
3.0	0.922094379	0.945604800	0.956752868	0.967523153

	Euler semilinear method			
x	$h = 0.2$	$h = 0.1$	$h = 0.05$	"Exact"
3.0	0.993954754	0.980751307	0.974140320	0.967523153

17.

	Euler's method			
x	$h = 0.05$	$h = 0.025$	$h = 0.0125$	"Exact"
1.50	0.319892131	0.330797109	0.337020123	0.343780513

	Euler semilinear method			
x	$h = 0.05$	$h = 0.025$	$h = 0.0125$	"Exact"
1.50	0.305596953	0.323340268	0.333204519	0.343780513

18.

	Euler's method			
x	$h = 0.2$	$h = 0.1$	$h = 0.05$	"Exact"
2.0	0.754572560	0.743869878	0.738303914	0.732638628

	Euler semilinear method			
x	$h = 0.2$	$h = 0.1$	$h = 0.05$	"Exact"
2.0	0.722610454	0.727742966	0.730220211	0.732638628

19.

	Euler's method			
x	$h = 0.05$	$h = 0.025$	$h = 0.0125$	"Exact"
1.50	2.175959970	2.210259554	2.227207500	2.244023982

	Euler semilinear method			
x	$h = 0.05$	$h = 0.025$	$h = 0.0125$	"Exact"
1.50	2.117953342	2.179844585	2.211647904	2.244023982

20.

	Euler's method			
x	$h = 0.1$	$h = 0.05$	$h = 0.025$	"Exact"
1.0	0.032105117	0.043997045	0.050159310	0.056415515

	Euler semilinear method			
x	$h = 0.1$	$h = 0.05$	$h = 0.025$	"Exact"
1.0	0.056020154	0.056243980	0.056336491	0.056415515

21.

Euler's method				
x	$h = 0.1$	$h = 0.05$	$h = 0.025$	"Exact"
1.0	28.987816656	38.426957516	45.367269688	54.729594761

Euler semilinear method				
x	$h = 0.1$	$h = 0.05$	$h = 0.025$	"Exact"
1.0	54.709134946	54.724150485	54.728228015	54.729594761

22.

Euler's method				
x	$h = 0.1$	$h = 0.05$	$h = 0.025$	"Exact"
3.0	1.361427907	1.361320824	1.361332589	1.361383810

Euler semilinear method				
x	$h = 0.1$	$h = 0.05$	$h = 0.025$	"Exact"
3.0	1.291345518	1.326535737	1.344004102	1.361383810

Section 3.2 Answers   (page 110)

1. $y_1 = 1.542812500$, $y_2 = 2.421622101$, $y_3 = 4.208020541$

2. $y_1 = 1.220207973$, $y_2 = 1.489578775$, $y_3 = 1.819337186$

3. $y_1 = 1.890687500$, $y_2 = 1.763784003$, $y_3 = 1.622698378$

4. $y_1 = 3.075082231$, $y_2 = 3.150338981$, $y_3 = 3.225786092$

5. $y_1 = 2.478055238$, $y_2 = 1.844042564$, $y_3 = 1.313882333$

6.

x	$h = 0.1$	$h = 0.05$	$h = 0.025$	Exact
1.0	56.134480009	55.003390448	54.734674836	54.647937102

7.

x	$h = 0.1$	$h = 0.05$	$h = 0.025$	Exact
2.0	1.353501839	1.353288493	1.353219485	1.353193719

8.

x	$h = 0.05$	$h = 0.025$	$h = 0.0125$	Exact
1.50	10.141969585	10.396770409	10.472502111	10.500000000

9.

x	$h = 0.1$	$h = 0.05$	$h = 0.025$	$h = 0.1$	$h = 0.05$	$h = 0.025$
3.0	1.455674816	1.455935127	1.456001289	-0.00818	-0.00207	-0.000518
	Approximate Solutions			Residuals		

10.

x	$h = 0.1$	$h = 0.05$	$h = 0.025$	$h = 0.1$	$h = 0.05$	$h = 0.025$
2.0	0.492862999	0.492709931	0.492674855	0.00335	0.000777	0.000187
	Approximate Solutions			Residuals		

11.

x	$h = 0.1$	$h = 0.05$	$h = 0.025$	"Exact"
1.0	0.660268159	0.660028505	0.659974464	0.659957689

12.

x	$h = 0.1$	$h = 0.05$	$h = 0.025$	"Exact"
2.0	-0.749751364	-0.750637632	-0.750845571	-0.750912371

13. Applying variation of parameters to the given initial value problem yields $y = ue^{-3x}$, where (A) $u' = 1 - 2x$, $u(0) = 2$. Since $u''' = 0$, the improved Euler method yields the exact solution of (A). Therefore the improved Euler semilinear method produces the exact solution of the given problem.

	Improved Euler method			
x	$h = 0.1$	$h = 0.05$	$h = 0.025$	Exact
1.0	0.105660401	0.100924399	0.099893685	0.099574137

	Improved Euler semilinear method			
x	$h = 0.1$	$h = 0.05$	$h = 0.025$	Exact
1.0	0.099574137	0.099574137	0.099574137	0.099574137

14.

	Improved Euler method			
x	$h = 0.1$	$h = 0.05$	$h = 0.025$	"Exact"
3.0	15.107600968	15.234856000	15.269755072	15.282004826

	Improved Euler semilinear method			
x	$h = 0.1$	$h = 0.05$	$h = 0.025$	"Exact"
3.0	15.285231726	15.282812424	15.282206780	15.282004826

15.

	Improved Euler method			
x	$h = 0.2$	$h = 0.1$	$h = 0.05$	"Exact"
2.0	0.924335375	0.907866081	0.905058201	0.904276722

	Improved Euler semilinear method			
x	$h = 0.2$	$h = 0.1$	$h = 0.05$	"Exact"
2.0	0.969670789	0.920861858	0.908438261	0.904276722

16.

	Improved Euler method			
x	$h = 0.2$	$h = 0.1$	$h = 0.05$	"Exact"
3.0	0.967473721	0.967510790	0.967520062	0.967523153

	Improved Euler semilinear method			
x	$h = 0.2$	$h = 0.1$	$h = 0.05$	"Exact"
3.0	0.967473721	0.967510790	0.967520062	0.967523153

17.

Improved Euler method			
x	$h = 0.05$	$h = 0.025$	$h = 0.0125$	"Exact"
1.50	0.349176060	0.345171664	0.344131282	0.343780513

Improved Euler semilinear method			
x	$h = 0.05$	$h = 0.025$	$h = 0.0125$	"Exact"
1.50	0.349350206	0.345216894	0.344142832	0.343780513

18.

Improved Euler method			
x	$h = 0.2$	$h = 0.1$	$h = 0.05$	"Exact"
2.0	0.732679223	0.732721613	0.732667905	0.732638628

Improved Euler semilinear method			
x	$h = 0.2$	$h = 0.1$	$h = 0.05$	"Exact"
2.0	0.732166678	0.732521078	0.732609267	0.732638628

19.

Improved Euler method			
x	$h = 0.05$	$h = 0.025$	$h = 0.0125$	"Exact"
1.50	2.247880315	2.244975181	2.244260143	2.244023982

Improved Euler semilinear method			
x	$h = 0.05$	$h = 0.025$	$h = 0.0125$	"Exact"
1.50	2.248603585	2.245169707	2.244310465	2.244023982

20.

Improved Euler method			
x	$h = 0.1$	$h = 0.05$	$h = 0.025$	"Exact"
1.0	0.059071894	0.056999028	0.056553023	0.056415515

Improved Euler semilinear method			
x	$h = 0.1$	$h = 0.05$	$h = 0.025$	"Exact"
1.0	0.056295914	0.056385765	0.056408124	0.056415515

21.

Improved Euler method			
x	$h = 0.1$	$h = 0.05$	$h = 0.025$	"Exact"
1.0	50.534556346	53.483947013	54.391544440	54.729594761

Improved Euler semilinear method			
x	$h = 0.1$	$h = 0.05$	$h = 0.025$	"Exact"
1.0	54.709041434	54.724083572	54.728191366	54.729594761

22.

	Improved Euler method			
x	$h = 0.1$	$h = 0.05$	$h = 0.025$	"Exact"
3.0	1.361395309	1.361379259	1.361382239	1.361383810

	Improved Euler semilinear method			
x	$h = 0.1$	$h = 0.05$	$h = 0.025$	"Exact"
3.0	1.375699933	1.364730937	1.362193997	1.361383810

23.

x	$h = 0.1$	$h = 0.05$	$h = 0.025$	Exact
2.0	1.349489056	1.352345900	1.352990822	1.353193719

24.

x	$h = 0.1$	$h = 0.05$	$h = 0.025$	Exact
2.0	1.350890736	1.352667599	1.353067951	1.353193719

25.

x	$h = 0.05$	$h = 0.025$	$h = 0.0125$	Exact
1.50	10.133021311	10.391655098	10.470731411	10.500000000

26.

x	$h = 0.05$	$h = 0.025$	$h = 0.0125$	Exact
1.50	10.136329642	10.393419681	10.470731411	10.500000000

27.

x	$h = 0.1$	$h = 0.05$	$h = 0.025$	"Exact"
1.0	0.660846835	0.660189749	0.660016904	0.659957689

28.

x	$h = 0.1$	$h = 0.05$	$h = 0.025$	"Exact"
1.0	0.660658411	0.660136630	0.660002840	0.659957689

29.

x	$h = 0.1$	$h = 0.05$	$h = 0.025$	"Exact"
2.0	-0.750626284	-0.750844513	-0.750895864	-0.751331499

30.

x	$h = 0.1$	$h = 0.05$	$h = 0.025$	"Exact"
2.0	-0.750335016	-0.750775571	-0.750879100	-0.751331499

Section 3.3 Answers   (page 118)

1. $y_1 = 1.550598190$, $y_2 = 2.469649729$ **2.** $y_1 = 1.221551366$, $y_2 = 1.492920208$

3. $y_1 = 1.890339767$, $y_2 = 1.763094323$ **4.** $y_1 = 3.075081399$, $y_2 = 3.150337608$

5. $y_1 = 2.475605264$, $y_2 = 1.825992433$

6.

x	$h = 0.1$	$h = 0.05$	$h = 0.025$	Exact
1.0	54.654509699	54.648344019	54.647962328	54.647937102

7.

x	$h = 0.1$	$h = 0.05$	$h = 0.025$	Exact
2.0	1.353191745	1.353193606	1.353193712	1.353193719

8.

x	$h = 0.05$	$h = 0.025$	$h = 0.0125$	Exact
1.50	10.498658198	10.499906266	10.499993820	10.500000000

9.

x	$h = 0.1$	$h = 0.05$	$h = 0.025$	$h = 0.1$	$h = 0.05$	$h = 0.025$
3.0	1.456023907	1.456023403	1.456023379	0.0000124	0.000000611	0.0000000333
	Approximate Solutions			Residuals		

10.

x	$h = 0.1$	$h = 0.05$	$h = 0.025$	$h = 0.1$	$h = 0.05$	$h = 0.025$
2.0	0.492663789	0.492663738	0.492663736	0.000000902	0.0000000508	0.00000000302
	Approximate Solutions			Residuals		

11.

x	$h = 0.1$	$h = 0.05$	$h = 0.025$	"Exact"
1.0	0.659957046	0.659957646	0.659957686	0.659957689

12.

x	$h = 0.1$	$h = 0.05$	$h = 0.025$	"Exact"
2.0	-0.750911103	-0.750912294	-0.750912367	-0.750912371

13. Applying variation of parameters to the given initial value problem yields $y = ue^{-3x}$, where (A) $u' = 1 - 4x + 3x^2 - 4x^3$, $u(0) = -3$. Since $u^{(5)} = 0$, the Runge–Kutta method yields the exact solution of (A). Therefore, the Runge–Kutta semilinear method produces the exact solution of the given problem.

Runge–Kutta method			
x	$h = 0.1$	$h = 0.05$	$h = 0.025$	Exact
1.0	-0.199187198	-0.199150401	-0.199148398	-0.199148273

Runge–Kutta semilinear method			
x	$h = 0.1$	$h = 0.05$	$h = 0.025$	Exact
1.0	-0.199148273	-0.199148273	-0.199148273	-0.199148273

14.

Runge–Kutta method			
x	$h = 0.1$	$h = 0.05$	$h = 0.025$	"Exact"
3.0	15.281660036	15.281981407	15.282003300	15.282004826

Runge–Kutta semilinear method			
x	$h = 0.1$	$h = 0.05$	$h = 0.025$	"Exact"
3.0	15.282005990	15.282004899	15.282004831	15.282004826

15.

Runge–Kutta method			
x	$h = 0.2$	$h = 0.1$	$h = 0.05$	"Exact"
2.0	0.904678156	0.904295772	0.904277759	0.904276722

Runge–Kutta semilinear method			
x	h = 0.2	h = 0.1	h = 0.05	"Exact"
2.0	0.904592215	0.904297062	0.904278004	0.904276722

16.

Runge–Kutta method			
x	h = 0.2	h = 0.1	h = 0.05	"Exact"
3.0	0.967523147	0.967523152	0.967523153	0.967523153

Runge–Kutta semilinear method			
x	h = 0.2	h = 0.1	h = 0.05	"Exact"
3.0	0.967523147	0.967523152	0.967523153	0.967523153

17.

Runge–Kutta method			
x	h = 0.05	h = 0.025	h = 0.0125	"Exact"
1.50	0.343839158	0.343784814	0.343780796	0.343780513

Runge–Kutta semilinear method			
x	h = 0.05	h = 0.025	h = 0.0125	"Exact"
1.50	0.343839124	0.343784811	0.343780796	0.343780513

18.

Runge–Kutta method			
x	h = 0.2	h = 0.1	h = 0.05	"Exact"
2.0	0.732633229	0.732638318	0.732638609	0.732638628

Runge–Kutta semilinear method			
x	h = 0.2	h = 0.1	h = 0.05	"Exact"
2.0	0.732639212	0.732638663	0.732638630	0.732638628

19.

Runge–Kutta method			
x	h = 0.05	h = 0.025	h = 0.0125	"Exact"
1.50	2.244025683	2.244024088	2.244023989	2.244023982

Runge–Kutta semilinear method			
x	h = 0.05	h = 0.025	h = 0.0125	"Exact"
1.50	2.244025081	2.244024051	2.244023987	2.244023982

20.

Runge–Kutta method			
x	h = 0.1	h = 0.05	h = 0.025	"Exact"
1.0	0.056426886	0.056416137	0.056415552	0.056415515

Runge–Kutta semilinear method			
x	$h = 0.1$	$h = 0.05$	$h = 0.025$	"Exact"
1.0	0.056415185	0.056415495	0.056415514	0.056415515

21.

Runge–Kutta method			
x	$h = 0.1$	$h = 0.05$	$h = 0.025$	"Exact"
1.0	54.695901186	54.727111858	54.729426250	54.729594761

Runge–Kutta semilinear method			
x	$h = 0.1$	$h = 0.05$	$h = 0.025$	"Exact"
1.0	54.729099966	54.729561720	54.729592658	54.729594761

22.

Runge–Kutta method			
x	$h = 0.1$	$h = 0.05$	$h = 0.025$	"Exact"
3.0	1.361384082	1.361383812	1.361383809	1.361383810

Runge–Kutta semilinear method			
x	$h = 0.1$	$h = 0.05$	$h = 0.025$	"Exact"
3.0	1.361456502	1.361388196	1.361384079	1.361383810

24.

x	$h = 0.1$	$h = 0.05$	$h = 0.025$	Exact
1.00	0.142854841	0.142857001	0.142857134	0.142857143

25.

x	$h = 0.1$	$h = 0.05$	$h = 0.025$	"Exact"
0.00	3.612827656	3.612969291	3.612978701	3.612979347

26.

x	$h = 0.1$	$h = 0.05$	$h = 0.025$	Exact
0.50	-8.954103230	-8.954063245	-8.954060698	-8.954060528
1.50	4.142171279	4.142170553	4.142170508	4.142170505

27.

x	$h = 0.1$	$h = 0.05$	$h = 0.025$	Exact
1.0	-9.999949359	-9.999996584	-9.999999779	-10.000000000
3.0	16.666666988	16.666666687	16.666666668	16.666666667

Section 4.1 Answers   (page 131)

1. $Q = 20e^{-(t \ln 2)/3200}$ g   **2.** $\dfrac{2 \ln 10}{\ln 2}$ days   **3.** $\tau = 10 \dfrac{\ln 2}{\ln 4/3}$ minutes   **4.** $\tau \dfrac{\ln p_0/p_1}{\ln 2}$ **5.** $\dfrac{t_p}{t_q} = \dfrac{\ln p}{\ln q}$

6. $k = \dfrac{1}{t_2 - t_1} \ln \dfrac{Q_1}{Q_2}$ **7.** 20 g   **8.** $\dfrac{50 \ln 2}{3}$ yrs   **9.** $\dfrac{25}{2} \ln 2\%$ **10.** **(a)** $20 \ln 3$ yr **(b)** $Q_0 = 100000e^{-.5}$

11. **(a)** $Q(t) = 5000 - 4750e^{-t/10}$ **(b)** 5000 lb   **12.** $\dfrac{1}{25}$ yr   **13.** $V = V_0 e^{t \ln 10/2}$;   4 hours

14. $\dfrac{1500 \ln 4/3}{\ln 2}$ yr; $2^{-4/3}Q_0$ **15.** $W(t) = 20 - 19e^{-t/20}$; $\lim_{t \to \infty} W(t) = 20$ ounces

16. $S(t) = 10(1 + e^{-t/10})$; $\lim_{t \to \infty} S(t) = 10$ g   **17.** 10 gal   **18.** $V(t) = 15000 + 10000e^{t/20}$

19. $W(t) = 4 \times 10^6 (t + 1)^2$ dollars t years from now   **20.** $p = \dfrac{100}{25 - 24e^{-t/2}}$

21. **(a)** $P(t) = 1000e^{.06t} + 50 \dfrac{e^{.06t} - 1}{e^{.06/52} - 1}$ **(b)** 5.64×10^{-4}

22. **(a)** $P' = rP - 12M$ **(b)** $P = \dfrac{12M}{r}(1 - e^{rt}) + P_0 e^{rt}$ **(c)** $M \approx \dfrac{rP_0}{12(1 - e^{-rN})}$

   (d) For **(i)** approximate $M = \$402.25$, exact $M = \$402.80$; for **(ii)** approximate $M = \$1206.05$, exact $M = \$1206.93$.

23. **(a)** $T(\alpha) = -\dfrac{1}{r} \ln(1 - (1 - e^{-rN})/\alpha)$ years; $S(\alpha) = \dfrac{P_0}{(1 - e^{-rN})}[rN + \alpha \ln(1 - (1 - e^{-rN})/\alpha)]$

   (b) $T(1.05) = 13.69$ yr, $S(1.05) = \$3579.94$; $T(1.10) = 12.61$ yr, $S(1.10) = \$6476.63$; $T(1.15) = 11.70$ yr,

   $S(1.15) = \$8874.98$ **24.** $P_0 = \begin{cases} \dfrac{S_0(1 - e^{(a-r)T})}{r - a} & \text{if } a \neq r, \\ S_0 T & \text{if } a = r. \end{cases}$

Section 4.2 Answers   (page 139)

1. $\approx 15.15°F$ **2.** $T = -10 + 110e^{-t \ln 11/9}$ **3.** $\approx 24.33°F$

4. **(a)** $91.30°F$ **(b)** 8.99 minutes after being placed outside **(c)** never   **5.** **(a)** 12:11:32 **(b)** 12:47:33

6. $(85/3)°C$ **7.** $32°F$ **8.** $Q(t) = 40(1 - e^{-3t/40})$ **9.** $Q(t) = 30 - 20e^{-t/10}$ **10.** $K(t) = .3 - .2e^{-t/20}$

11. $Q(50) = 47.5$ (lb)   **12.** 50 gal   **13.** min $q_2 = q_1/\bar{c}$ **14.** $Q = t + 300 - \dfrac{234 \times 10^5}{(t + 300)^2}$, $0 \leq t \leq 300$

15. **(a)** $Q' + \dfrac{2}{25}Q = 6 - 2e^{-t/25}$ **(b)** $Q = 75 - 50e^{-t/25} - 25e^{-2t/25}$ **(c)** 75

16. **(a)** $T = T_m + (T_0 - T_m)e^{-kt} + \dfrac{k(S_0 - T_m)}{k - k_m}(e^{-k_m t} - e^{-kt})$ **(b)** $T = T_m + k(S_0 - T_m)te^{-kt} + (T_0 - T_m)e^{-kt}$

   (c) $\lim_{t \to \infty} T(t) = \lim_{t \to \infty} S(t) = T_m$

17. **(a)** $T' = -k\left(1 + \dfrac{a}{a_m}\right)T + k\left(T_{m0} + \dfrac{a}{a_m}T_0\right)$ **(b)** $T = \dfrac{aT_0 + a_m T_{m0}}{a + a_m} + \dfrac{a_m(T_0 - T_{m0})}{a + a_m}e^{-k(1 + a/a_m)t}$,

   $T_m = \dfrac{aT_0 + a_m T_{m0}}{a + a_m} + \dfrac{a(T_{m0} - T_0)}{a + a_m}e^{-k(1 + a/a_m)t}$ **(c)** $\lim_{t \to \infty} T(t) = \lim_{t \to \infty} T_m(t) = \dfrac{aT_0 + a_m T_{m0}}{a + a_m}$

18. $V = \dfrac{a}{b}\dfrac{V_0}{V_0 - (V_0 - a/b)e^{-at}}$; $\lim_{t \to \infty} V(t) = a/b$ **19.** $c_1 = c(1 - e^{-rt/W})$, $c_2 = c\left(1 - e^{-rt/W} - \dfrac{r}{W}te^{-rt/W}\right)$

20. **(a)** $c_n = c\left(1 - e^{-rt/W}\sum_{j=0}^{n-1}\dfrac{1}{j!}\left(\dfrac{rt}{W}\right)^j\right)$ **(b)** c **(c)** 0

21. Let $c_\infty = (c_1 W_1 + c_2 W_2)/(W_1 + W_2)$, $\alpha = (c_2 W_2^2 - c_1 W_1^2)/(W_1 + W_2)$,   and $\beta = (W_1 + W_2)/(W_1 W_2)$. Then:

   (a) $k_1(t) = c_\infty + \dfrac{\alpha}{W_1}e^{-r\beta t}$, $k_2(t) = c_\infty - \dfrac{\alpha}{W_2}e^{-r\beta t}$ **(b)** $\lim_{t \to \infty} k_1(t) = \lim_{t \to \infty} k_2(t) = c_\infty$

Section 4.3 Answers   (page 151)

1. $v = -\dfrac{384}{5}(1 - e^{-5t/12});\quad -\dfrac{384}{5}$ ft/s   **2.** $k = 12;\quad v = -16(1 - e^{-2t})$ **3.** $v = 25(1 - e^{-t});\quad 25$ ft/s

4. $v = 20 - 27e^{-t/40}$ **5.** ≈ 17.10 ft   **6.** $v = -\dfrac{40(13 + 3e^{-4t/5})}{13 - 3e^{-4t/5}};\quad -40$ ft/s   **7.** $v = -128(1 - e^{-t/4})$

9. $T = \dfrac{m}{k}\ln\!\left(1 + \dfrac{v_0 k}{mg}\right);\quad y_m = y_0 + \dfrac{m}{k}\left[v_0 - \dfrac{mg}{k}\ln\!\left(1 + \dfrac{v_0 k}{mg}\right)\right]$ **10.** $v = -\dfrac{64(1 - e^{-t})}{1 + e^{-t}};\quad -64$ ft/s

11. $v = \alpha\,\dfrac{v_0(1 + e^{-\beta t}) - \alpha(1 - e^{-\beta t})}{\alpha(1 + e^{-\beta t}) - v_0(1 - e^{-\beta t})};\quad -\alpha$, where $\alpha = \sqrt{\dfrac{mg}{k}}$ and $\beta = 2\sqrt{\dfrac{kg}{m}}$.

12. $T = \sqrt{\dfrac{m}{kg}}\tan^{-1}\!\left(v_0\sqrt{\dfrac{k}{mg}}\right);\quad v = -\sqrt{\dfrac{mg}{k}}\,\dfrac{1 - e^{-2\sqrt{gk/m}\,(t-T)}}{1 + e^{-2\sqrt{gk/m}\,(t-T)}}$ **13.** $ms' = mg - \dfrac{as}{s + 1};\quad a_0 = mg$

14. (a) $ms' = mg - f(s)$ **15. (a)** $v' = -9.8 + v^4/81$ **(b)** $v_T \approx -5.308$ m/s

16. (a) $v' = -32 + 8\sqrt{|v|};\quad v_T = -16$ ft/s **(b)** From Exercise 14(c), v_T is the negative number such that $-32 + 8\sqrt{|v_T|} = 0;\quad$ thus, $v_T = -16$ ft/s.

17. ≈ 6.76 miles/s   **18.** ≈ 1.47 miles/s   **20.** $\alpha = \dfrac{gR^2}{(y_m + R)^2}$

Section 4.4 Answers   (page 165)

1. $\bar{y} = 0$ is a stable equilibrium; trajectories are $v^2 + y^4/4 = c$

2. $\bar{y} = 0$ is an unstable equilibrium; trajectories are $v^2 + 2y^3/3 = c$

3. $\bar{y} = 0$ is a stable equilibrium; trajectories are $v^2 + 2|y|^3/3 = c$

4. $\bar{y} = 0$ is a stable equilibrium; trajectories are $v^2 - e^{-y}(y + 1) = c$

5. equilibria: 0 (stable) and $-2, 2$ (unstable); trajectories: $2v^2 - y^4 + 8y^2 = c$; separatrix: $2v^2 - y^4 + 8y^2 = 16$

6. equilibria: 0 (unstable) and $-2, 2$ (stable); trajectories: $2v^2 + y^4 - 8y^2 = c$; separatrix: $2v^2 + y^4 - 8y^2 = 0$

7. equilibria: $0, -2, 2$ (stable), $-1, 1$ (unstable); trajectories: $6v^2 + y^2(2y^4 - 15y^2 + 24) = c$; separatrix: $6v^2 + y^2(2y^4 - 15y^2 + 24) = 11$

8. equilibria: $0, 2$ (stable) and $-2, 1$ (unstable); trajectories: $30v^2 + y^2(12y^3 - 15y^2 - 80y + 120) = c$; separatrices: $30v^2 + y^2(12y^3 - 15y^2 - 80y + 120) = 496$ and $30v^2 + y^2(12y^3 - 15y^2 - 80y + 120) = 37$

9. No equilibria if $a < 0$; 0 is unstable if $a = 0$; $\sqrt{a}$ is stable and $-\sqrt{a}$ is unstable if $a > 0$.

10. 0 is a stable equilibrium if $a \le 0$; $-\sqrt{a}$ and $\sqrt{a}$ are stable and 0 is unstable if $a > 0$.

11. 0 is unstable if $a \le 0$; $-\sqrt{a}$ and $\sqrt{a}$ are unstable and 0 is stable if $a > 0$.

12. 0 is stable if $a \le 0$; 0 is stable and $-\sqrt{a}$ and $\sqrt{a}$ are unstable if $a > 0$.

22. An equilibrium solution $\bar{y}$ of $y'' + p(y) = 0$ is *unstable* if there is an $\epsilon > 0$ such that for every $\delta > 0$ there is a solution of (A) with $\sqrt{(y(0) - \bar{y})^2 + v^2(0)} < \delta$ but $\sqrt{(y(t) - \bar{y})^2 + v^2(t)} \ge \epsilon$ for some $t > 0$.

Section 4.5 Answers   (page 180)

1. $y' = -\dfrac{2xy}{x^2 + 3y^2}$ **2.** $y' = -\dfrac{y^2}{(xy - 1)}$ **3.** $y' = -\dfrac{y(x^2 + y^2 - 2x^2\ln|xy|)}{x(x^2 + y^2 - 2y^2\ln|xy|)}$ **4.** $xy' - y = -\dfrac{x^{1/2}}{2}$

5. $y' + 2xy = 4xe^{x^2}$ **6.** $xy' + y = 4x^3$ **7.** $y' - y = \cos x - \sin x$ **8.** $(1 + x^2)y' - 2xy = (1 - x)^2 e^x$

10. $y'g - yg' = f'g - fg'$.   **11.** $(x - x_0)y' = y - y_0$ **12.** $y'(y^2 - x^2 + 1) + 2xy = 0$

13. $2x(y - 1)y' - y^2 + x^2 + 2y = 0$

14. (a) $y = -81 + 18x$, $(9,81)$; $y = -1 + 2x$, $(1,1)$ **(b)** $y = -121 + 22x$, $(11,121)$; $y = -1 + 2x$, $(1,1)$

(c) $y = -100 - 20x$, $(-10,100)$; $y = -4 - 4x$, $(-2,4)$

(d) $y = -25 - 10x$, $(-5,25)$; $y = -1 - 2x$, $(-1,1)$

15. (e) $y = \dfrac{5 + 3x}{4}$, $(-3/5, 4/5)$; $y = -\dfrac{5 - 4x}{3}$, $(4/5, -3/5)$

17. (a) $y = -\dfrac{1}{2}(1 + x)$, $(1,-1)$; $y = \dfrac{5}{2} + \dfrac{x}{10}$, $(25,5)$ **(b)** $y = \dfrac{1}{4}(4 + x)$, $(4,2)$; $y = -\dfrac{1}{4}(4 + x)$, $(4,-2)$

(c) $y = \dfrac{1}{2}(1 + x)$, $(1,1)$; $y = \dfrac{7}{2} + \dfrac{x}{14}$, $(49,7)$ **(d)** $y = -\dfrac{1}{2}(1 + x)$, $(1,-1)$; $y = -\dfrac{5}{2} - \dfrac{x}{10}$, $(25,-5)$

18. $y = 2x^2$ **19.** $y = \dfrac{cx}{\sqrt{|x^2 - 1|}}$ **20.** $y = y_1 + c(x - x_1)$ **21.** $y = -\dfrac{x^3}{2} - \dfrac{x}{2}$ **22.** $y = -x \ln|x| + cx$

23. $y = \sqrt{2x + 4}$ **24.** $y = \sqrt{x^2 - 3}$ **25.** $y = kx^2$ **26.** $(y - x)^3(y + x) = k$ **27.** $y^2 = -x + k$

28. $y^2 = -\dfrac{1}{2}\ln(1 + 2x^2) + k$ **29.** $y^2 = -2x - \ln(x - 1)^2 + k$ **30.** $y = 1 + \sqrt{\dfrac{9 - x^2}{2}}$; those with $c > 0$

33. $\tan^{-1}\dfrac{y}{x} - \dfrac{1}{2}\ln(x^2 + y^2) = k$ **34.** $\dfrac{1}{2}\ln(x^2 + y^2) + \tan\alpha\,\tan^{-1}\dfrac{y}{x} = k$

Section 5.1 Answers   (page 194)

1. (c) $y = -2e^{2x} + e^{5x}$ **(d)** $y = (5k_0 - k_1)\dfrac{e^{2x}}{3} + (k_1 - 2k_0)\dfrac{e^{5x}}{3}$

2. (c) $y = e^x(3\cos x - 5\sin x)$ **(d)** $y = e^x(k_0\cos x + (k_1 - k_0)\sin x)$

3. (c) $y = e^x(7 - 3x)$ **(d)** $y = e^x(k_0 + (k_1 - k_0)x)$ **4. (a)** $y = \dfrac{c_1}{x - 1} + \dfrac{c_2}{x + 1}$ **(b)** $y = \dfrac{2}{x - 1} - \dfrac{3}{x + 1}$; $(-1,1)$

5. (a) e^x **(b)** $e^{2x}\cos x$ **(c)** $x^2 + 2x - 2$ **(d)** $-\dfrac{5}{6}x^{-5/6}$ **(e)** $-\dfrac{1}{x^2}$ **(f)** $(x\ln|x|)^2$ **(g)** $\dfrac{e^{2x}}{2\sqrt{x}}$

6. 0   **7.** $W(x) = (1 - x^2)^{-1}$ **8.** $W(x) = \dfrac{1}{x}$ **10.** $y_2 = e^{-x}$ **11.** $y_2 = xe^{3x}$ **12.** $y_2 = xe^{ax}$ **13.** $y_2 = \dfrac{1}{x}$

14. $y_2 = x\ln x$ **15.** $y_2 = x^a\ln x$ **16.** $y_2 = x^{1/2}e^{-2x}$ **17.** $y_2 = x$ **18.** $y_2 = x\sin x$ **19.** $y_2 = x^{1/2}\cos x$

20. $y_2 = xe^{-x}$ **21.** $y_2 = \dfrac{1}{x^2 - 4}$ **22.** $y_2 = e^{2x}$ **23.** $y_2 = x^2$

35. (a) $y'' - 2y' + 5y = 0$ **(b)** $(2x - 1)y'' - 4xy' + 4y = 0$ **(c)** $x^2y'' - xy' + y = 0$ **(d)** $x^2y'' + xy' + y = 0$

(e) $y'' - y = 0$ **(f)** $xy'' - y' = 0$

37. (c) $y = k_0y_1 + k_1y_2$ **38.** $y_1 = 1$, $y_2 = x - x_0$; $y = k_0 + k_1(x - x_0)$

39. $y_1 = \cosh(x - x_0)$, $y_2 = \sinh(x - x_0)$; $y = k_0\cosh(x - x_0) + k_1\sinh(x - x_0)$

40. $y_1 = \cos\omega(x - x_0)$, $y_2 = \dfrac{1}{\omega}\sin\omega(x - x_0)$, $y = k_0\cos\omega(x - x_0) + \dfrac{k_1}{\omega}\sin\omega(x - x_0)$

41. $y_1 = \dfrac{1}{1 - x^2}$, $y_2 = \dfrac{x}{1 - x^2}$; $y = \dfrac{k_0 + k_1x}{1 - x^2}$

42. (c) $k_0 = k_1 = 0$; $y = \begin{cases} c_1x^2 + c_2x^3, & x \geq 0, \\ c_1x^2 + c_3x^3, & x < 0 \end{cases}$ **(d)** $(0,\infty)$ if $x_0 > 0$, $(-\infty,0)$ if $x_0 < 0$

43. (c) $k_0 = 0$, k_1 arbitrary; $y = k_1x + c_2x^2$

44. (c) $k_0 = k_1 = 0$; $y = \begin{cases} a_1x^3 + a_2x^4, & x \geq 0, \\ b_1x^3 + b_2x^4, & x < 0 \end{cases}$ **(d)** $(0,\infty)$ if $x_0 > 0$, $(-\infty,0)$ if $x_0 < 0$

Section 5.2 Answers   (page 206)

1. $y = c_1 e^{-6x} + c_2 e^x$ **2.** $y = e^{2x}(c_1 \cos x + c_2 \sin x)$ **3.** $y = c_1 e^{-7x} + c_2 e^{-x}$ **4.** $y = e^{2x}(c_1 + c_2 x)$

5. $y = e^{-x}(c_1 \cos 3x + c_2 \sin 3x)$ **6.** $y = e^{-3x}(c_1 \cos x + c_2 \sin x)$ **7.** $y = e^{4x}(c_1 + c_2 x)$ **8.** $y = c_1 + c_2 e^{-x}$

9. $y = e^x(c_1 \cos \sqrt{2}x + c_2 \sin \sqrt{2}x)$ **10.** $y = e^{-3x}(c_1 \cos 2x + c_2 \sin 2x)$ **11.** $y = e^{-x/2}\left(c_1 \cos \dfrac{3x}{2} + c_2 \sin \dfrac{3x}{2}\right)$

12. $y = c_1 e^{-x/5} + c_2 e^{x/2}$ **13.** $y = e^{-7x}(2 \cos x - 3 \sin x)$ **14.** $y = 4e^{x/2} + 6e^{-x/3}$ **15.** $y = 3e^{x/3} - 4e^{-x/2}$

16. $y = \dfrac{e^{-x/2}}{3} + \dfrac{3e^{3x/2}}{4}$ **17.** $y = e^{3x/2}(3 - 2x)$ **18.** $y = 3e^{-4x} - 4e^{-3x}$ **19.** $y = 2xe^{3x}$

20. $y = e^{x/6}(3 + 2x)$ **21.** $y = e^{-2x}\left(3 \cos \sqrt{6}x + \dfrac{2\sqrt{6}}{3} \sin \sqrt{6}x\right)$ **23.** $y = 2e^{-(x-1)} - 3e^{-2(x-1)}$

24. $y = \dfrac{1}{3}e^{-(x-2)} - \dfrac{2}{3}e^{7(x-2)}$ **25.** $y = e^{7(x-1)}(2 - 3(x - 1))$ **26.** $y = e^{-(x-2)/3}(2 - 4(x - 2))$

27. $y = 2 \cos \dfrac{2}{3}\left(x - \dfrac{\pi}{4}\right) - 3 \sin \dfrac{2}{3}\left(x - \dfrac{\pi}{4}\right)$ **28.** $y = 2 \cos \sqrt{3}\left(x - \dfrac{\pi}{3}\right) - \dfrac{1}{\sqrt{3}} \sin \sqrt{3}\left(x - \dfrac{\pi}{3}\right)$

30. $y = \dfrac{k_0}{r_2 - r_1}(r_2 e^{r_1(x-x_0)} - r_1 e^{r_2(x-x_0)}) + \dfrac{k_1}{r_2 - r_1}(e^{r_2(x-x_0)} - e^{r_1(x-x_0)})$ **31.** $y = e^{r_1(x-x_0)}[k_0 + (k_1 - r_1 k_0)(x - x_0)]$

32. $y = e^{\lambda(x-x_0)}\left[k_0 \cos \omega(x - x_0) + \left(\dfrac{k_1 - \lambda k_0}{\omega}\right) \sin \omega(x - x_0)\right]$

Section 5.3 Answers   (page 215)

1. $y_p = -1 + 2x + 3x^2$; $y = -1 + 2x + 3x^2 + c_1 e^{-6x} + c_2 e^x$

2. $y_p = 1 + x$; $y = 1 + x + e^{2x}(c_1 \cos x + c_2 \sin x)$ **3.** $y_p = -x + x^3$; $y = -x + x^3 + c_1 e^{-7x} + c_2 e^{-x}$

4. $y_p = 1 - x^2$; $y = 1 - x^2 + e^{2x}(c_1 + c_2 x)$

5. $y_p = 2x + x^3$; $y = 2x + x^3 + e^{-x}(c_1 \cos 3x + c_2 \sin 3x)$; $y = 2x + x^3 + e^{-x}(2 \cos 3x + 3 \sin 3x)$

6. $y_p = 1 + 2x$; $y = 1 + 2x + e^{-3x}(c_1 \cos x + c_2 \sin x)$; $y = 1 + 2x + e^{-3x}(\cos x - \sin x)$

8. $y_p = \dfrac{2}{x}$ **9.** $y_p = 4x^{1/2}$ **10.** $y_p = \dfrac{x^3}{2}$ **11.** $y_p = \dfrac{1}{x^3}$ **12.** $y_p = 9x^{1/3}$ **13.** $y_p = \dfrac{2x^4}{13}$

16. $y_p = \dfrac{e^{3x}}{3}$; $y = \dfrac{e^{3x}}{3} + c_1 e^{-6x} + c_2 e^x$ **17.** $y_p = e^{2x}$; $y = e^{2x}(1 + c_1 \cos x + c_2 \sin x)$

18. $y_p = -2e^{-2x}$; $y = -2e^{-2x} + c_1 e^{-7x} + c_2 e^{-x}$; $y = -2e^{-2x} - e^{-7x} + e^{-x}$

19. $y_p = e^x$; $y = e^x + e^{2x}(c_1 + c_2 x)$; $y = e^x + e^{2x}(1 - 3x)$

20. $y_p = \dfrac{4}{45}e^{x/2}$; $y = \dfrac{4}{45}e^{x/2} + e^{-x}(c_1 \cos 3x + c_2 \sin 3x)$ **21.** $y_p = e^{-3x}$; $y = e^{-3x}(1 + c_1 \cos x + c_2 \sin x)$

24. $y_p = \cos x - \sin x$; $y = \cos x - \sin x + e^{4x}(c_1 + c_2 x)$

25. $y_p = \cos 2x - 2 \sin 2x$; $y = \cos 2x - 2 \sin 2x + c_1 + c_2 e^{-x}$

26. $y_p = \cos 3x$; $y = \cos 3x + e^x(c_1 \cos \sqrt{2}x + c_2 \sin \sqrt{2}x)$

27. $y_p = \cos x + \sin x$; $y = \cos x + \sin x + e^{-3x}(c_1 \cos 2x + c_2 \sin 2x)$

28. $y_p = -2 \cos 2x + \sin 2x$; $y = -2 \cos 2x + \sin 2x + c_1 e^{-4x} + c_2 e^{-3x}$; $y = -2 \cos 2x + \sin 2x + 2e^{-4x} - 3e^{-3x}$

29. $y_p = \cos 3x - \sin 3x$; $y = \cos 3x - \sin 3x + e^{3x}(c_1 + c_2 x)$; $y = \cos 3x - \sin 3x + e^{3x}(1 + 2x)$

30. $y = \dfrac{1}{\omega_0^2 - \omega^2}(M \cos \omega x + N \sin \omega x) + c_1 \cos \omega_0 x + c_2 \sin \omega_0 x$

33. $y_p = -1 + 2x + 3x^2 + \dfrac{e^{3x}}{3}$; $y = -1 + 2x + 3x^2 + \dfrac{e^{3x}}{3} + c_1 e^{-6x} + c_2 e^x$

34. $y_p = 1 + x + e^{2x}$; $y = 1 + x + e^{2x}(1 + c_1 \cos x + c_2 \sin x)$

35. $y_p = -x + x^3 - 2e^{-2x}$; $y = -x + x^3 - 2e^{-2x} + c_1 e^{-7x} + c_2 e^{-x}$

36. $y_p = 1 - x^2 + e^x$; $y = 1 - x^2 + e^x + e^{2x}(c_1 + c_2 x)$

37. $y_p = 2x + x^3 + \dfrac{4}{45} e^{x/2}$; $y = 2x + x^3 + \dfrac{4}{45} e^{x/2} + e^{-x}(c_1 \cos 3x + c_2 \sin 3x)$

38. $y_p = 1 + 2x + e^{-3x}$; $y = 1 + 2x + e^{-3x}(1 + c_1 \cos x + c_2 \sin x)$

Section 5.4 Answers   (page 222)

1. $y_p = e^{3x}\left(-\dfrac{1}{4} + \dfrac{x}{2}\right)$ **2.** $y_p = e^{-3x}\left(1 - \dfrac{x}{4}\right)$ **3.** $y_p = e^x\left(2 - \dfrac{3x}{4}\right)$ **4.** $y_p = e^{2x}(1 - 3x + x^2)$

5. $y_p = e^{-x}(1 + x^2)$ **6.** $y_p = e^x(-2 + x + 2x^2)$ **7.** $y_p = xe^{-x}\left(\dfrac{1}{6} + \dfrac{x}{2}\right)$ **8.** $y_p = xe^x(1 + 2x)$

9. $y_p = xe^{3x}\left(-1 + \dfrac{x}{2}\right)$ **10.** $y_p = xe^{2x}(-2 + x)$ **11.** $y_p = x^2 e^{-x}\left(1 + \dfrac{x}{2}\right)$ **12.** $y_p = x^2 e^x\left(\dfrac{1}{2} - x\right)$

13. $y_p = \dfrac{x^2 e^{2x}}{2}(1 - x + x^2)$ **14.** $y_p = \dfrac{x^2 e^{-x/3}}{27}(3 - 2x + x^2)$ **15.** $y = \dfrac{e^{3x}}{4}(-1 + 2x) + c_1 e^x + c_2 e^{2x}$

16. $y = e^x(1 - 2x) + c_1 e^{2x} + c_2 e^{4x}$ **17.** $y = \dfrac{e^{2x}}{5}(1 - x) + e^{-3x}(c_1 + c_2 x)$ **18.** $y = xe^x(1 - 2x) + c_1 e^x + c_2 e^{-3x}$

19. $y = e^x[x^2(1 - 2x) + c_1 + c_2 x]$ **20.** $y = -e^{2x}(1 + x) + 2e^{-x} - e^{5x}$ **21.** $y = xe^{2x} + 3e^x - e^{-4x}$

22. $y = e^{-x}(2 + x - 2x^2) - e^{-3x}$ **23.** $y = e^{-2x}(3 - x) - 2e^{5x}$ **24.** $y_p = -\dfrac{e^x}{3}(1 - x) + e^{-x}(3 + 2x)$

25. $y_p = e^x(3 + 7x) + xe^{3x}$ **26.** $y_p = x^3 e^{4x} + 1 + 2x + x^2$ **27.** $y_p = xe^{2x}(1 - 2x) + xe^x$

28. $y_p = e^x(1 + x) + x^2 e^{-x}$ **29.** $y_p = x^2 e^{-x} + e^{3x}(1 - x^2)$ **31.** $y_p = 2e^{2x}$ **32.** $y_p = 5xe^{4x}$

33. $y_p = x^2 e^{4x}$ **34.** $y_p = -\dfrac{e^{3x}}{4}(1 + 2x - 2x^2)$ **35.** $y_p = xe^{3x}(4 - x + 2x^2)$

36. $y_p = x^2 e^{-x/2}(-1 + 2x + 3x^2)$

37. (a) $y = e^{-x}\left(\dfrac{4}{3} x^{3/2} + c_1 x + c_2\right)$ **(b)** $y = e^{-3x}\left[\dfrac{x^2}{4}(2 \ln x - 3) + c_1 x + c_2\right]$

   (c) $y = e^{2x}[(x + 1) \ln |x + 1| + c_1 x + c_2]$ **(d)** $y = e^{-x/2}\left(x \ln |x| + \dfrac{x^3}{6} + c_1 x + c_2\right)$

39. (a) $e^x(3 + x) + c$ **(b)** $-e^{-x}(1 + x)^2 + c$ **(c)** $-\dfrac{e^{-2x}}{8}(3 + 6x + 6x^2 + 4x^3) + c$ **(d)** $e^x(1 + x^2) + c$

   (e) $e^{3x}(-6 + 4x + 9x^2) + c$ **(f)** $-e^{-x}(1 - 2x^3 + 3x^4) + c$

40. $\dfrac{(-1)^k k!\, e^{\alpha x}}{\alpha^{k+1}} \sum\limits_{r=0}^{k} \dfrac{(-\alpha x)^r}{r!} + c$

Section 5.5 Answers   (page 232)

1. $y_p = \cos x + 2 \sin x$ **2.** $y_p = \cos x + (2 - 2x) \sin x$ **3.** $y_p = e^x(-2 \cos x + 3 \sin x)$

4. $y_p = \dfrac{e^{2x}}{2}(\cos 2x - \sin 2x)$ **5.** $y_p = -e^x(x \cos x - \sin x)$ **6.** $y_p = e^{-2x}(1 - 2x)(\cos 3x - \sin 3x)$

7. $y_p = x(\cos 2x - 3 \sin 2x)$ **8.** $y_p = -x[(2 - x) \cos x + (3 - 2x) \sin x]$ **9.** $y_p = x\left[x \cos\left(\dfrac{x}{2}\right) - 3 \sin\left(\dfrac{x}{2}\right)\right]$

10. $y_p = xe^{-x}(3 \cos x + 4 \sin x)$ **11.** $y_p = xe^x[(-1 + x) \cos 2x + (1 + x) \sin 2x]$

12. $y_p = -(14 - 10x) \cos x - (2 + 8x - 4x^2) \sin x$ **13.** $y_p = (1 + 2x + x^2) \cos x + (1 + 3x^2) \sin x$

14. $y_p = \dfrac{x^2}{2}(\cos 2x - \sin 2x)$ **15.** $y_p = e^x(x^2 \cos x + 2 \sin x)$ **16.** $y_p = e^x(1 - x^2)(\cos x + \sin x)$

17. $y_p = e^x(x^2 - x^3)(\cos x + \sin x)$ **18.** $y_p = e^{-x}[(1 + 2x) \cos x - (1 - 3x) \sin x]$

19. $y_p = x(2 \cos 3x - \sin 3x)$ **20.** $y_p = -x^3 \cos x + (x + 2x^2) \sin x$

21. $y_p = -e^{-x}[(x + x^2) \cos x - (1 + 2x) \sin x]$ **22.** $y = e^x(2 \cos x + 3 \sin x) + 3e^x - e^{6x}$

23. $y = e^x[(1 + 2x) \cos x + (1 - 3x) \sin x]$ **24.** $y = e^x(\cos x - 2 \sin x) + e^{-3x}(\cos x + \sin x)$

25. $y = e^{3x}[(2 + 2x) \cos x - (1 + 3x) \sin x]$ **26.** $y = e^{3x}[(2 + 3x) \cos x + (4 - x) \sin x] + 3e^x - 5e^{2x}$

27. $y_p = xe^{3x} - \dfrac{e^x}{5}(\cos x - 2 \sin x)$ **28.** $y_p = x(\cos x + 2 \sin x) - \dfrac{e^x}{2}(1 - x) + \dfrac{e^{-x}}{2}$

29. $y_p = -\dfrac{xe^x}{2}(2 + x) + 2xe^{2x} + \dfrac{1}{10}(3 \cos x + \sin x)$

30. $y_p = xe^x(\cos x + x \sin x) + \dfrac{e^{-x}}{25}(4 + 5x) + 1 + x + \dfrac{x^2}{2}$

31. $y_p = \dfrac{x^2 e^{2x}}{6}(3 + x) - e^{2x}(\cos x - \sin x) + 3e^{3x} + \dfrac{1}{4}(2 + x)$

32. $y = (1 - 2x + 3x^2)e^{2x} + 4 \cos x + 3 \sin x$ **33.** $y = xe^{-2x} \cos x + 3 \cos 2x$

34. $y = -\dfrac{3}{8} \cos 2x + \dfrac{1}{4} \sin 2x + e^{-x} - \dfrac{13}{8}e^{-2x} - \dfrac{3}{4}xe^{-2x}$

35. $y = e^x(1 + 3x) + (2 + 3x) \cos 2x - (1 - 2x) \sin 2x$

40. (a) $2x \cos x - (2 - x^2) \sin x + c$ **(b)** $-\dfrac{e^x}{2}[(1 - x^2) \cos x - (1 - x)^2 \sin x] + c$

(c) $-\dfrac{e^{-x}}{25}[(4 + 10x) \cos 2x - (3 - 5x) \sin 2x] + c$ **(d)** $-\dfrac{e^{-x}}{2}[(1 + x)^2 \cos x - (1 - x^2) \sin x] + c$

(e) $-\dfrac{e^x}{2}[x(3 - 3x + x^2) \cos x - (3 - 3x + x^3) \sin x] + c$ **(f)** $-e^x[(1 - 2x) \cos x + (1 + x) \sin x] + c$

(g) $e^{-x}[x \cos x + x(1 + x) \sin x] + c$

Section 5.6 Answers   (page 239)

1. $y = 1 - 2x + c_1 e^{-x} + c_2 x e^x$; $\{e^{-x}, xe^x\}$ **2.** $y = \dfrac{4}{3x^2} + c_1 x + \dfrac{c_2}{x}$; $\{x, 1/x\}$

3. $y = \dfrac{x(\ln|x|)^2}{2} + c_1 x + c_2 x \ln|x|$; $\{x, x \ln|x|\}$ **4.** $y = (e^{2x} + e^x) \ln(1 + e^{-x}) + c_1 e^{2x} + c_2 e^x$; $\{e^{2x}, e^x\}$

5. $y = e^x\left(\dfrac{4}{5}x^{7/2} + c_1 + c_2 x\right)$; $\{e^x, xe^x\}$ **6.** $y = e^x(2x^{3/2} + x^{1/2} \ln x + c_1 x^{1/2} + c_2 x^{-1/2})$; $\{x^{1/2}e^x, x^{-1/2}e^{-x}\}$

7. $y = e^x(x \sin x + \cos x \ln|\cos x| + c_1 \cos x + c_2 \sin x)$; $\{e^x \cos x, e^x \sin x\}$

8. $y = e^{-x^2}(2e^{-2x} + c_1 + c_2 x)$; $\{e^{-x^2}, xe^{-x^2}\}$ **9.** $y = 2x + 1 + c_1 x^2 + \dfrac{c_2}{x^2}$; $\{x^2, 1/x^2\}$

10. $y = \dfrac{xe^{2x}}{9} + xe^{-x}(c_1 + c_2 x)$; $\{xe^{-x}, x^2 e^{-x}\}$ **11.** $y = xe^x\left(\dfrac{x}{3} + c_1 + \dfrac{c_2}{x^2}\right)$; $\{xe^x, e^x/x\}$

12. $y = -\dfrac{(2x - 1)^2 e^x}{8} + c_1 e^x + c_2 x e^{-x}$; $\{e^x, xe^{-x}\}$ **13.** $y = x^4 + c_1 x^2 + c_2 x^2 \ln|x|$; $\{x^2, x^2 \ln|x|\}$

14. $y = e^{-x}(x^{3/2} + c_1 + c_2 x^{1/2})$; $\{e^{-x}, x^{1/2}e^{-x}\}$ **15.** $y = e^x(x + c_1 + c_2 x^2)$; $\{e^x, x^2 e^x\}$

16. $y = x^{1/2}\left(\dfrac{e^{2x}}{2} + c_1 + c_2 e^x\right)$; $\{x^{1/2}, x^{1/2}e^x\}$ **17.** $y = -2x^2 \ln x + c_1 x^2 + c_2 x^4$; $\{x^2, x^4\}$ **18.** $\{e^x, e^x/x\}$

19. $\{x^2, x^3\}$ **20.** $\{\ln|x|, x\ln|x|\}$ **21.** $\{\sin\sqrt{x}, \cos\sqrt{x}\}$ **22.** $\{e^x, x^3e^x\}$ **23.** $\{x^a, x^a\ln x\}$

24. $\{x\sin x, x\cos x\}$ **25.** $\{e^{2x}, x^2e^{2x}\}$ **26.** $\{x^{1/2}, x^{1/2}\cos x\}$ **27.** $\{x^{1/2}e^{2x}, x^{1/2}e^{-2x}\}$ **28.** $\{1/x, e^{2x}\}$

29. $\{e^x, x^2\}$ **30.** $\{e^{2x}, x^2e^{2x}\}$ **31.** $y = x^4 + 6x^2 - 8x^2\ln|x|$ **32.** $y = 2e^{2x} - xe^{-x}$

33. $y = \dfrac{(x+1)}{4}\left[-e^x(3-2x) + 7e^{-x}\right]$ **34.** $y = \dfrac{x^2}{4} + x$ **35.** $y = \dfrac{(x+2)^2}{6(x-2)} + \dfrac{2x}{x^2-4}$

38. (a) $y = \dfrac{-kc_1\sin kx + kc_2\cos kx}{c_1\cos kx + c_2\sin kx}$ **(b)** $y = \dfrac{c_1 + 2c_2e^x}{c_1 + c_2e^x}$ **(c)** $y = \dfrac{-6c_1 + c_2e^{7x}}{c_1 + c_2e^{7x}}$ **(d)** $y = -\dfrac{7c_1 + c_2e^{6x}}{c_1 + c_2e^{6x}}$

(e) $y = -\dfrac{(7c_1 - c_2)\cos x + (c_1 + 7c_2)\sin x}{c_1\cos x + c_2\sin x}$ **(f)** $y = \dfrac{-2c_1 + 3c_2e^{5x/6}}{6(c_1 + c_2e^{5x/6})}$ **(g)** $y = \dfrac{c_1 + c_2(x+6)}{6(c_1 + c_2x)}$

39. (a) $y = \dfrac{c_1 + c_2e^x(1+x)}{x(c_1 + c_2e^x)}$ **(b)** $y = \dfrac{-2c_1x + c_2(1-2x^2)}{c_1 + c_2x}$ **(c)** $y = \dfrac{-c_1 + c_2e^{2x}(x+1)}{c_1 + c_2xe^{2x}}$

(d) $y = \dfrac{2c_1 + c_2e^{-3x}(1-x)}{c_1 + c_2xe^{-3x}}$ **(e)** $y = \dfrac{(2c_2x - c_1)\cos x - (2c_1x + c_2)\sin x}{2x(c_1\cos x + c_2\sin x)}$ **(f)** $y = \dfrac{c_1 + 7c_2x^6}{x(c_1 + c_2x^6)}$

Section 5.7 Answers   (page 249)

1. $y_p = \dfrac{-\cos 3x\,\ln|\sec 3x + \tan 3x|}{9}$ **2.** $y_p = -\dfrac{\sin 2x\,\ln|\cos 2x|}{4} + \dfrac{x\cos 2x}{2}$

3. $y_p = 4e^x(1 + e^x)\ln(1 + e^{-x})$ **4.** $y_p = 3e^x(\cos x\,\ln|\cos x| + x\sin x)$ **5.** $y_p = \dfrac{8}{5}x^{7/2}e^x$

6. $y_p = e^x\ln(1 - e^{-2x}) - e^{-x}\ln(e^{2x} - 1)$ **7.** $y_p = \dfrac{2(x^2-3)}{3}$ **8.** $y_p = \dfrac{e^{2x}}{x}$ **9.** $y_p = x^{1/2}e^x\ln x$

10. $y_p = e^{-x(x+2)}$ **11.** $y_p = -4x^{5/2}$ **12.** $y_p = -2x^2\sin x - 2x\cos x$

13. $y_p = -\dfrac{xe^{-x}(x+1)}{2}$ **14.** $y_p = -\dfrac{\sqrt{x}\cos\sqrt{x}}{2}$ **15.** $y_p = \dfrac{3x^4e^x}{2}$ **16.** $y_p = x^{a+1}$

17. $y_p = \dfrac{x^2\sin x}{2}$ **18.** $y_p = -2x^2$ **19.** $y_p = -e^{-x}\sin x$ **20.** $y_p = -\dfrac{\sqrt{x}}{2}$ **21.** $y_p = \dfrac{x^{3/2}}{4}$

22. $y_p = -3x^2$ **23.** $y_p = \dfrac{x^3e^x}{2}$ **24.** $y_p = -\dfrac{4x^{3/2}}{15}$ **25.** $y_p = x^3e^x$ **26.** $y_p = xe^x$

27. $y_p = x^2$ **28.** $y_p = xe^x(x-2)$ **29.** $y_p = \sqrt{x}e^x(x-1)/4$

30. $y = \dfrac{e^{2x}(3x^2 - 2x + 6)}{6} + \dfrac{xe^{-x}}{3}$ **31.** $y = (x-1)^2\ln(1-x) + 2x^2 - 5x + 3$

32. $y = (x^2 - 1)e^x - 5(x-1)$ **33.** $y = \dfrac{x(x^2+6)}{3(x^2-1)}$ **34.** $y = -\dfrac{x^2}{2} + x + \dfrac{1}{2x^2}$

35. $y = \dfrac{x^2(4x+9)}{6(x+1)}$

38. (a) $y = k_0\cosh x + k_1\sinh x + \displaystyle\int_0^x \sinh(x-t)f(t)\,dt$

(b) $y' = k_0\sinh x + k_1\cosh x + \displaystyle\int_0^x \cosh(x-t)f(t)\,dt$

39. (a) $y(x) = k_0\cos x + k_1\sin x + \displaystyle\int_0^x \sin(x-t)f(t)\,dt$

(b) $y'(x) = -k_0\sin x + k_1\cos x + \displaystyle\int_0^x \cos(x-t)f(t)\,dt$

Section 6.1 Answers (page 262)

1. $y = 3 \cos 4\sqrt{6}t - \dfrac{1}{2\sqrt{6}} \sin 4\sqrt{6}t$ ft **2.** $y = -\dfrac{1}{4} \cos 8\sqrt{5}t - \dfrac{1}{4\sqrt{5}} \sin 8\sqrt{5}t$ ft

3. $y = 1.5 \cos 14\sqrt{10}t$ cm

4. $y = \dfrac{1}{4} \cos 8t - \dfrac{1}{16} \sin 8t$ ft; $R = \dfrac{\sqrt{17}}{16}$ ft; $\omega_0 = 8$ rad/s; $T = \pi/4$ s; $\phi \approx -.245$ rad $\approx -14.04°$

5. $y = 10 \cos 14t + \dfrac{25}{14} \sin 14t$ cm; $R = \dfrac{5}{14}\sqrt{809}$ cm; $\omega_0 = 14$ rad/s; $T = \pi/7$ s; $\phi \approx .177$ rad $\approx 10.12°$

6. $y = -\dfrac{1}{4} \cos \sqrt{70}t + \dfrac{2}{\sqrt{70}} \sin \sqrt{70}t$ m; $R = \dfrac{1}{4}\sqrt{\dfrac{67}{35}}$ m; $\omega_0 = \sqrt{70}$ rad/s;

$T = 2\pi/\sqrt{70}$ s; $\phi \approx 2.38$ rad $\approx 136.28°$

7. $y = \dfrac{2}{3} \cos 16t - \dfrac{1}{4} \sin 16t$ ft **8.** $y = \dfrac{1}{2} \cos 8t - \dfrac{3}{8} \sin 8t$ ft

9. .72 m **10.** $y = \dfrac{1}{3} \sin t + \dfrac{1}{2} \cos 2t + \dfrac{5}{6} \sin 2t$ ft **11.** $y = \dfrac{16}{5}\left(4 \sin \dfrac{t}{4} - \sin t\right)$

12. $y = -\dfrac{1}{16} \sin 8t + \dfrac{1}{3} \cos 4\sqrt{2}t - \dfrac{1}{8\sqrt{2}} \sin 4\sqrt{2}t$ **13.** $y = -t \cos 8t - \dfrac{1}{6} \cos 8t + \dfrac{1}{8} \sin 8t$ ft

14. $T = 4\sqrt{2}$ s **15.** $\omega = 8$ rad/s; $y = -\dfrac{t}{16}(-\cos 8t + 2 \sin 8t) + \dfrac{1}{128} \sin 8t$ ft

16. $\omega = 4\sqrt{6}$ rad/s; $y = -\dfrac{t}{\sqrt{6}}\left[\dfrac{8}{3} \cos 4\sqrt{6}t + 4 \sin 4\sqrt{6}t\right] + \dfrac{1}{9} \sin 4\sqrt{6}t$ ft

17. $y = \dfrac{t}{2} \cos 2t - \dfrac{t}{4} \sin 2t + 3 \cos 2t + 2 \sin 2t$ m

18. $y = y_0 \cos \omega_0 t + \dfrac{v_0}{\omega_0} \sin \omega_0 t$; $R = \dfrac{1}{\omega_0}\sqrt{(\omega_0 y_0)^2 + (v_0)^2}$; $\cos \phi = \dfrac{y_0 \omega_0}{\sqrt{(\omega_0 y_0)^2 + (v_0)^2}}$; $\sin \phi = \dfrac{v_0}{\sqrt{(\omega_0 y_0)^2 + (v_0)^2}}$

19. The object with the longer period weighs 4 times as much as the other.

20. $T_2 = \sqrt{2}T_1$, where T_1 is the period of the smaller object.

21. $k_1 = 9k_2$, where k_1 is the spring constant of the system with the shorter period.

Section 6.2 Answers (page 271)

1. $y = \dfrac{e^{-2t}}{2}(3 \cos 2t - \sin 2t)$ ft; $\sqrt{\dfrac{5}{2}}e^{-2t}$ ft **2.** $y = -e^{-t}\left(3 \cos 3t + \dfrac{1}{3} \sin 3t\right)$ ft; $\dfrac{\sqrt{82}}{3}e^{-t}$ ft

3. $y = e^{-16t}\left(\dfrac{1}{4} + 10t\right)$ ft **4.** $y = -\dfrac{e^{-3t}}{4}(5 \cos t + 63 \sin t)$ ft **5.** $0 \le c < 8$ lb-s/ft

6. $y = \dfrac{1}{2}e^{-3t}\left(\cos \sqrt{91}t + \dfrac{11}{\sqrt{91}} \sin \sqrt{91}t\right)$ ft **7.** $y = -\dfrac{e^{-4t}}{3}(2 + 8t)$ ft

8. $y = e^{-10t}\left(9 \cos 4\sqrt{6}t + \dfrac{45}{2\sqrt{6}} \sin 4\sqrt{6}t\right)$ cm **9.** $y = e^{-3t/2}\left(\dfrac{3}{2} \cos \dfrac{\sqrt{41}}{2}t + \dfrac{9}{2\sqrt{41}} \sin \dfrac{\sqrt{41}}{2}t\right)$ ft

10. $y = e^{-3t/2}\left(\dfrac{1}{2} \cos \dfrac{\sqrt{119}}{2}t - \dfrac{9}{2\sqrt{119}} \sin \dfrac{\sqrt{119}}{2}t\right)$ ft **11.** $y = e^{-8t}\left(\dfrac{1}{4} \cos 8\sqrt{2}t - \dfrac{1}{4\sqrt{2}} \sin 8\sqrt{2}t\right)$ ft

12. $y = e^{-t}\left(-\dfrac{1}{3} \cos 3\sqrt{11}t + \dfrac{14}{9\sqrt{11}} \sin 3\sqrt{11}t\right)$ ft **13.** $y_p = \dfrac{22}{61} \cos 2t + \dfrac{2}{61} \sin 2t$ ft **14.** $y = -\dfrac{2}{3}(e^{-8t} - 2e^{-4t})$

15. $y = e^{-2t}\left(\dfrac{1}{10} \cos 4t - \dfrac{1}{5} \sin 4t\right)$ m **16.** $y = e^{-3t}(10 \cos t - 70 \sin t)$ cm **17.** $y_p = -\dfrac{2}{15} \cos 3t + \dfrac{1}{15} \sin 3t$ ft

18. $y_p = \dfrac{11}{100}\cos 4t + \dfrac{27}{100}\sin 4t$ cm   **19.** $y_p = \dfrac{42}{73}\cos t + \dfrac{39}{73}\sin t$ ft   **20.** $y = -\dfrac{1}{2}\cos 2t + \dfrac{1}{4}\sin 2t$ m

21. $y_p = \dfrac{1}{c\omega_0}(-\beta\cos\omega_0 t + \alpha\sin\omega_0 t)$ **24.** $y = e^{-ct/2m}\left(y_0\cos\omega_1 t + \dfrac{1}{\omega_1}\left(v_0 + \dfrac{cy_0}{2m}\right)\sin\omega_1 t\right)$

25. $y = \dfrac{r_2 y_0 - v_0}{r_2 - r_1}e^{r_1 t} + \dfrac{v_0 - r_1 y_0}{r_2 - r_1}e^{r_2 t}$ **26.** $y = e^{r_1 t}(y_0 + (v_0 - r_1 y_0)t)$

Section 6.3 Answers   (page 278)

1. $I = e^{-15t}\left(2\cos 5\sqrt{15}t - \dfrac{6}{\sqrt{31}}\sin 5\sqrt{31}t\right)$ **2.** $I = e^{-20t}(2\cos 40t - 101\sin 40t)$ **3.** $I = -\dfrac{200}{3}e^{-10t}\sin 30t$

4. $I = -10e^{-30t}(\cos 40t + 18\sin 40t)$ **5.** $I = -e^{-40t}(2\cos 30t - 86\sin 30t)$ **6.** $I_p = -\dfrac{1}{3}(\cos 10t + 2\sin 10t)$

7. $I_p = \dfrac{20}{37}(\cos 25t - 6\sin 25t)$ **8.** $I_p = \dfrac{8}{13}(3\cos 50t - \sin 50t)$ **9.** $I_p = \dfrac{20}{123}(17\sin 100t - 11\cos 100t)$

10. $I_p = -\dfrac{45}{52}(\cos 30t + 8\sin 30t)$ **12.** $\omega_0 = 1/\sqrt{LC}$;   maximum amplitude $= \sqrt{U^2 + V^2}/R$

Section 6.4 Answers   (page 285)

1. If $e = 1$ then $Y^2 = \rho(\rho - 2X)$;   if $e \ne 1$ then $\left(X + \dfrac{e\rho}{1 - e^2}\right)^2 + \dfrac{Y^2}{1 - e^2} = \dfrac{\rho^2}{1 - e^2}$;   if $e < 1$ let $X_0 = -\dfrac{e\rho}{1 - e^2}$,

$a = \dfrac{\rho}{1 - e^2}$, $b = \dfrac{\rho}{\sqrt{1 - e^2}}$

2. Let $h = r_0^2\theta_0'$; then $\rho = \dfrac{h^2}{k}$, $e = \left[\left(\dfrac{\rho}{r_0} - 1\right)^2 + \left(\dfrac{\rho r_0'}{h}\right)^2\right]^{1/2}$. If $e = 0$ then θ_0 is undefined, but also irrelevant; if $e \ne 0$

then $\phi = \theta_0 - \alpha$, where $-\pi \le \alpha < \pi$, $\cos\alpha = \dfrac{1}{e}\left(\dfrac{\rho}{r_0} - 1\right)$ and $\sin\alpha = \dfrac{\rho r_0'}{eh}$

3. (a) $e = \dfrac{\gamma_2 - \gamma_1}{\gamma_1 + \gamma_2}$ **(b)** $r_0 = R\gamma_1$, $r_0' = 0$, θ_0 arbitrary, $\theta_0' = \left[\dfrac{2g\gamma_2}{R\gamma_1^3(\gamma_1 + \gamma_2)}\right]^{1/2}$

4. $f(r) = -mh^2\left(\dfrac{6c}{r^4} + \dfrac{1}{r^3}\right)$ **5.** $f(r) = -\dfrac{mh^2(\gamma^2 + 1)}{r^3}$

6. (a) $\dfrac{d^2u}{d\theta^2} + \left(1 - \dfrac{k}{h^2}\right)u = 0$, $u(\theta_0) = \dfrac{1}{r_0}$, $\dfrac{du(\theta_0)}{d\theta} = -\dfrac{r_0'}{h}$.

(b) with $\gamma = \left|1 - \dfrac{k}{h^2}\right|^{1/2}$:   **(i)** $r = r_0\left(\cosh\gamma(\theta - \theta_0) - \dfrac{r_0 r_0'}{\gamma h}\sinh\gamma(\theta - \theta_0)\right)^{-1}$;   **(ii)** $r = r_0\left(1 - \dfrac{r_0 r_0'}{h}(\theta - \theta_0)\right)^{-1}$;

(iii) $r = r_0\left(\cos\gamma(\theta - \theta_0) - \dfrac{r_0 r_0'}{\gamma h}\sin\gamma(\theta - \theta_0)\right)^{-1}$

Section 7.1 Answers   (page 298)

1. (a) $R = 2$; $I = (-1, 3)$ **(b)** $R = 1/2$; $I = (3/2, 5/2)$ **(c)** $R = 0$ **(d)** $R = 16$; $I = (-14, 18)$

(e) $R = \infty$; $I = (-\infty, \infty)$ **(f)** $R = 4/3$; $I = (-25/3, -17/3)$

3. (a) $R = 1$; $I = (0, 2)$ **(b)** $R = \sqrt{2}$; $I = (-2 - \sqrt{2}, -2 + \sqrt{2})$ **(c)** $R = \infty$; $I = (-\infty, \infty)$ **(d)** $R = 0$

(e) $R = \sqrt{3}$; $I = (-\sqrt{3}, \sqrt{3})$ **(f)** $R = 1$; $I = (0, 2)$

5. (a) $R = 3$; $I = (0, 6)$ **(b)** $R = 1$; $I = (-1, 1)$ **(c)** $R = 1/\sqrt{3}$; $I = (3 - 1/\sqrt{3}, 3 + 1/\sqrt{3})$

(d) $R = \infty$; $I = (-\infty, \infty)$ **(e)** $R = 0$ **(f)** $R = 2$; $I = (-1, 3)$

11. $b_n = 2(n + 2)(n + 1)a_{n+2} + (n + 1)na_{n+1} + (n + 3)a_n$

12. $b_0 = 2a_2 - 2a_0$
$b_n = (n + 2)(n + 1)a_{n+2} + [3n(n - 1) - 2]a_n + 3(n - 1)a_{n-1}, \quad n \geq 1$

13. $b_n = (n + 2)(n + 1)a_{n+2} + 2(n + 1)a_{n+1} + (2n^2 - 5n + 4)a_n$

14. $b_n = (n + 2)(n + 1)a_{n+2} + 2(n + 1)a_{n+1} + (n^2 - 2n + 3)a_n$

15. $b_n = (n + 2)(n + 1)a_{n+2} + (3n^2 - 5n + 4)a_n$

16. $b_0 = -2a_2 + 2a_1 + a_0$
$b_n = -(n + 2)(n + 1)a_{n+2} + (n + 1)(n + 2)a_{n+1} + (2n + 1)a_n + a_{n-1}, \quad n \geq 2$

17. $b_0 = 8a_2 + 4a_1 - 6a_0$
$b_n = 4(n + 2)(n + 1)a_{n+2} + 4(n + 1)^2a_{n+1} + (n^2 + n - 6)a_n - 3a_{n-1}, \quad n \geq 1$

21. $b_0 = (r + 1)(r + 2)a_0$
$b_n = (n + r + 1)(n + r + 2)a_n - (n + r - 2)^2a_{n-1}, \quad n \geq 1$

22. $b_0 = (r - 2)(r + 2)a_0$
$b_n = (n + r - 2)(n + r + 2)a_n + (n + r + 2)(n + r - 3)a_{n-1}, \quad n \geq 1$

23. $b_0 = (r - 1)^2a_0$
$b_1 = r^2a_1 + (r + 2)(r + 3)a_0$
$b_n = (n + r - 1)^2a_n + (n + r + 1)(n + r + 2)a_{n-1} + (n + r - 1)a_{n-2}, \quad n \geq 2$

24. $b_0 = r(r + 1)a_0$
$b_1 = (r + 1)(r + 2)a_1 + 3(r + 1)(r + 2)a_0$
$b_n = (n + r)(n + r + 1)a_n + 3(n + r)(n + r + 1)a_{n-1} + (n + r)a_{n-2}, \quad n \geq 2$

25. $b_0 = (r + 2)(r + 1)a_0$
$b_1 = (r + 3)(r + 2)a_1$
$b_n = (n + r + 2)(n + r + 1)a_n + 2(n + r - 1)(n + r - 3)a_{n-2}, \quad n \geq 2$

26. $b_0 = 2(r + 1)(r + 3)a_0$
$b_1 = 2(r + 2)(r + 4)a_1$
$b_n = 2(n + r + 1)(n + r + 3)a_n + (n + r - 3)(n + r)a_{n-2}, \quad n \geq 2$

Section 7.2 Answers   (page 312)

1. $y = a_0 \sum_{m=0}^{\infty} (-1)^m(2m + 1)x^{2m} + a_1 \sum_{m=0}^{\infty} (-1)^m(m + 1)x^{2m+1}$ **2.** $y = a_0 \sum_{m=0}^{\infty} (-1)^{m+1}\dfrac{x^{2m}}{2m - 1} + a_1x$

3. $y = a_0(1 - 10x^2 + 5x^4) + a_1\left(x - 2x^3 + \dfrac{1}{5}x^5\right)$

4. $y = a_0 \sum_{m=0}^{\infty} (m + 1)(2m + 1)x^{2m} + \dfrac{a_1}{3} \sum_{m=0}^{\infty} (m + 1)(2m + 3)x^{2m+1}$

5. $y = a_0 \sum_{m=0}^{\infty}(-1)^m\left[\prod_{j=0}^{m-1}\dfrac{4j + 1}{2j + 1}\right]x^{2m} + a_1 \sum_{m=0}^{\infty}(-1)^m\left[\prod_{j=0}^{m-1}(4j + 3)\right]\dfrac{x^{2m+1}}{2^mm!}$

6. $y = a_0 \sum_{m=0}^{\infty}(-1)^m\left[\prod_{j=0}^{m-1}\dfrac{(4j + 1)^2}{2j + 1}\right]\dfrac{x^{2m}}{8^mm!} + a_1 \sum_{m=0}^{\infty}(-1)^m\left[\prod_{j=0}^{m-1}\dfrac{(4j + 3)^2}{2j + 3}\right]\dfrac{x^{2m+1}}{8^mm!}$

7. $y = a_0 \sum_{m=0}^{\infty}\dfrac{2^mm!}{\prod_{j=0}^{m-1}(2j + 1)}x^{2m} + a_1 \sum_{m=0}^{\infty}\dfrac{\prod_{j=0}^{m-1}(2j + 3)}{2^mm!}x^{2m+1}$

8. $y = a_0\left(1 - 14x^2 + \dfrac{35}{3}x^4\right) + a_1\left(x - 3x^3 + \dfrac{3}{5}x^5 + \dfrac{1}{35}x^7\right)$

9. (a) $y = a_0 \sum_{m=0}^{\infty} (-1)^m\dfrac{x^{2m}}{\prod_{j=0}^{m-1}(2j + 1)} + a_1 \sum_{m=0}^{\infty} (-1)^m\dfrac{x^{2m+1}}{2^mm!}$

10. (a) $y = a_0 \sum_{m=0}^{\infty} (-1)^m \left[\prod_{j=0}^{m-1} \frac{4j+3}{2j+1} \right] \frac{x^{2m}}{2^m m!} + a_1 \sum_{m=0}^{\infty} (-1)^m \left[\prod_{j=0}^{m-1} \frac{4j+5}{2j+3} \right] \frac{x^{2m+1}}{2^m m!}$

11. $y = 2 - x - x^2 + \frac{1}{3} x^3 + \frac{5}{12} x^4 - \frac{1}{6} x^5 - \frac{17}{72} x^6 + \frac{13}{126} x^7 + \cdots$

12. $y = 1 - x + 3x^2 - \frac{5}{2} x^3 + 5x^4 - \frac{21}{8} x^5 + 3x^6 - \frac{11}{16} x^7 + \cdots$

13. $y = 2 - x - 2x^2 + \frac{1}{3} x^3 + 3x^4 - \frac{5}{6} x^5 - \frac{49}{5} x^6 + \frac{45}{14} x^7 + \cdots$

16. $y = a_0 \sum_{m=0}^{\infty} \frac{(x-3)^{2m}}{(2m)!} + a_1 \sum_{m=0}^{\infty} \frac{(x-3)^{2m+1}}{(2m+1)!}$ **17.** $y = a_0 \sum_{m=0}^{\infty} \frac{(x-3)^{2m}}{2^m m!} + a_1 \sum_{m=0}^{\infty} \frac{(x-3)^{2m+1}}{\prod_{j=0}^{m-1}(2j+3)}$

18. $y = a_0 \sum_{m=0}^{\infty} \left[\prod_{j=0}^{m-1} (2j+3) \right] \frac{(x-1)^{2m}}{m!} + a_1 \sum_{m=0}^{\infty} \frac{4^m (m+1)!}{\prod_{j=0}^{m-1}(2j+3)} (x-1)^{2m+1}$

19. $y = a_0 \left(1 - 6(x-2)^2 + \frac{4}{3}(x-2)^4 + \frac{8}{135}(x-2)^6 \right) + a_1 \left((x-2) - \frac{10}{9}(x-2)^3 \right)$

20. $y = a_0 \sum_{m=0}^{\infty} (-1)^m \left[\prod_{j=0}^{m-1} (2j+1) \right] \frac{3^m}{4^m m!} (x+1)^{2m} + a_1 \sum_{m=0}^{\infty} (-1)^m \frac{3^m m!}{\prod_{j=0}^{m-1}(2j+3)} (x+1)^{2m+1}$

21. $y = -1 + 2x + \frac{3}{8} x^2 - \frac{1}{3} x^3 - \frac{3}{128} x^4 - \frac{1}{1024} x^6 + \cdots$

22. $y = -2 + 3(x-3) + 3(x-3)^2 - 2(x-3)^3 - \frac{5}{4}(x-3)^4 + \frac{3}{5}(x-3)^5 + \frac{7}{24}(x-3)^6 - \frac{4}{35}(x-3)^7 + \cdots$

23. $y = -1 + (x-1) + 3(x-1)^2 - \frac{5}{2}(x-1)^3 - \frac{27}{4}(x-1)^4 + \frac{21}{4}(x-1)^5 + \frac{27}{2}(x-1)^6 - \frac{81}{8}(x-1)^7 + \cdots$

24. $y = 4 - 6(x-3) - 2(x-3)^2 + (x-3)^3 + \frac{3}{2}(x-3)^4 - \frac{5}{4}(x-3)^5 - \frac{49}{20}(x-3)^6 + \frac{135}{56}(x-3)^7 + \cdots$

25. $y = 3 - 4(x-4) + 15(x-4)^2 - 4(x-4)^3 + \frac{15}{4}(x-4)^4 - \frac{1}{5}(x-4)^5$

26. $y = 3 - 3(x+1) - 30(x+1)^2 + \frac{20}{3}(x+1)^3 + 20(x+1)^4 - \frac{4}{3}(x+1)^5 - \frac{8}{9}(x+1)^6$

27. (a) $y = a_0 \sum_{m=0}^{\infty} (-1)^m x^{2m} + a_1 \sum_{m=0}^{\infty} (-1)^m x^{2m+1}$ **(b)** $y = \frac{a_0 + a_1 x}{1 + x^2}$

33. $y = a_0 \sum_{m=0}^{\infty} \frac{x^{3m}}{3^m m! \, \prod_{j=0}^{m-1}(3j+2)} + a_1 \sum_{m=0}^{\infty} \frac{x^{3m+1}}{3^m m! \, \prod_{j=0}^{m-1}(3j+4)}$

34. $y = a_0 \sum_{m=0}^{\infty} \left(\frac{2}{3} \right)^m \left[\prod_{j=0}^{m-1} (3j+2) \right] \frac{x^{3m}}{m!} + a_1 \sum_{m=0}^{\infty} \frac{6^m m!}{\prod_{j=0}^{m-1}(3j+4)} x^{3m+1}$

35. $y = a_0 \sum_{m=0}^{\infty} (-1)^m \frac{3^m m!}{\prod_{j=0}^{m-1}(3j+2)} x^{3m} + a_1 \sum_{m=0}^{\infty} (-1)^m \left[\prod_{j=0}^{m-1} (3j+4) \right] \frac{x^{3m+1}}{3^m m!}$

36. $y = a_0(1 - 4x^3 + 4x^6) + a_1 \sum_{m=0}^{\infty} 2^m \left[\prod_{j=0}^{m-1} \frac{3j-5}{3j+4} \right] x^{3m+1}$

37. $y = a_0 \left(1 + \frac{21}{2} x^3 + \frac{42}{5} x^6 + \frac{7}{20} x^9 \right) + a_1 \left(x + 4x^4 + \frac{10}{7} x^7 \right)$

39. $y = a_0 \sum_{m=0}^{\infty} (-2)^m \left[\prod_{j=0}^{m-1} \frac{5j+1}{5j+4} \right] x^{5m} + a_1 \sum_{m=0}^{\infty} \left(-\frac{2}{5} \right)^m \left[\prod_{j=0}^{m-1} (5j+2) \right] \frac{x^{5m+1}}{m!}$

40. $y = a_0 \sum_{m=0}^{\infty} (-1)^m \frac{x^{4m}}{4^m m! \, \prod_{j=0}^{m-1}(4j+3)} + a_1 \sum_{m=0}^{\infty} (-1)^m \frac{x^{4m+1}}{4^m m! \, \prod_{j=0}^{m-1}(4j+5)}$

41. $y = a_0 \sum_{m=0}^{\infty} (-1)^m \dfrac{x^{7m}}{\prod_{j=0}^{m-1} (7j+6)} + a_1 \sum_{m=0}^{\infty} (-1)^m \dfrac{x^{7m+1}}{7^m m!}$ **42.** $y = a_0\left(1 - \dfrac{9}{7}x^8\right) + a_1\left(x - \dfrac{7}{9}x^9\right)$

43. $y = a_0 \sum_{m=0}^{\infty} x^{6m} + a_1 \sum_{m=0}^{\infty} x^{6m+1}$ **44.** $y = a_0 \sum_{m=0}^{\infty} (-1)^m \dfrac{x^{6m}}{\prod_{j=0}^{m-1} (6j+5)} + a_1 \sum_{m=0}^{\infty} (-1)^m \dfrac{x^{6m+1}}{6^m m!}$

Section 7.3 Answers   (page 319)

1. $y = 2 - 3x - 2x^2 + \dfrac{7}{2}x^3 - \dfrac{55}{12}x^4 + \dfrac{59}{8}x^5 - \dfrac{83}{6}x^6 + \dfrac{9547}{336}x^7 + \cdots$

2. $y = -1 + 2x - 4x^3 + 4x^4 + 4x^5 - 12x^6 + 4x^7 + \cdots$

3. $y = 1 + x^2 - \dfrac{2}{3}x^3 + \dfrac{11}{6}x^4 - \dfrac{9}{5}x^5 + \dfrac{329}{90}x^6 - \dfrac{1301}{315}x^7 + \cdots$

4. $y = x - x^2 - \dfrac{7}{2}x^3 + \dfrac{15}{2}x^4 + \dfrac{45}{8}x^5 - \dfrac{261}{8}x^6 + \dfrac{207}{16}x^7 + \cdots$

5. $y = 4 + 3x - \dfrac{15}{4}x^2 + \dfrac{1}{4}x^3 + \dfrac{11}{16}x^4 - \dfrac{5}{16}x^5 + \dfrac{1}{20}x^6 + \dfrac{1}{120}x^7 + \cdots$

6. $y = 7 + 3x - \dfrac{16}{3}x^2 + \dfrac{13}{3}x^3 - \dfrac{23}{9}x^4 + \dfrac{10}{9}x^5 - \dfrac{7}{27}x^6 - \dfrac{1}{9}x^7 + \cdots$

7. $y = 2 + 5x - \dfrac{7}{4}x^2 - \dfrac{3}{16}x^3 + \dfrac{37}{192}x^4 - \dfrac{7}{192}x^5 - \dfrac{1}{1920}x^6 + \dfrac{19}{11520}x^7 + \cdots$

8. $y = 1 - (x-1) + \dfrac{4}{3}(x-1)^3 - \dfrac{4}{3}(x-1)^4 - \dfrac{4}{5}(x-1)^5 + \dfrac{136}{45}(x-1)^6 - \dfrac{104}{63}(x-1)^7 + \cdots$

9. $y = 1 - (x+1) + 4(x+1)^2 - \dfrac{13}{3}(x+1)^3 + \dfrac{77}{6}(x+1)^4 - \dfrac{278}{15}(x+1)^5 + \dfrac{1942}{45}(x+1)^6 - \dfrac{23332}{315}(x+1)^7 + \cdots$

10. $y = 2 - (x-1) - \dfrac{1}{2}(x-1)^2 + \dfrac{5}{3}(x-1)^3 - \dfrac{19}{12}(x-1)^4 + \dfrac{7}{30}(x-1)^5 + \dfrac{59}{45}(x-1)^6 - \dfrac{1091}{630}(x-1)^7 + \cdots$

11. $y = -2 + 3(x+1) - \dfrac{1}{2}(x+1)^2 - \dfrac{2}{3}(x+1)^3 + \dfrac{5}{8}(x+1)^4 - \dfrac{11}{30}(x+1)^5 + \dfrac{29}{144}(x+1)^6 - \dfrac{101}{840}(x+1)^7 + \cdots$

12. $y = 1 - 2(x-1) - 3(x-1)^2 + 8(x-1)^3 - 4(x-1)^4 - \dfrac{42}{5}(x-1)^5 + 19(x-1)^6 - \dfrac{604}{35}(x-1)^7 + \cdots$

19. $y = 2 - 7x - 4x^2 - \dfrac{17}{6}x^3 - \dfrac{3}{4}x^4 - \dfrac{9}{40}x^5 + \cdots$

20. $y = 1 - 2(x-1) + \dfrac{1}{2}(x-1)^2 - \dfrac{1}{6}(x-1)^3 + \dfrac{5}{36}(x-1)^4 - \dfrac{73}{1080}(x-1)^5 + \cdots$

21. $y = 2 - (x+2) - \dfrac{7}{2}(x+2)^2 + \dfrac{4}{3}(x+2)^3 - \dfrac{1}{24}(x+2)^4 + \dfrac{1}{60}(x+2)^5 + \cdots$

22. $y = 2 - 2(x+3) - (x+3)^2 + (x+3)^3 - \dfrac{11}{12}(x+3)^4 + \dfrac{67}{60}(x+3)^5 + \cdots$

23. $y = -1 + 2x + \dfrac{1}{3}x^3 - \dfrac{5}{12}x^4 + \dfrac{2}{5}x^5 + \cdots$

24. $y = 2 - 3(x+1) + \dfrac{7}{2}(x+1)^2 - 5(x+1)^3 + \dfrac{197}{24}(x+1)^4 - \dfrac{287}{20}(x+1)^5 + \cdots$

25. $y = -2 + 3(x+2) - \dfrac{9}{2}(x+2)^2 + \dfrac{11}{6}(x+2)^3 + \dfrac{5}{24}(x+2)^4 + \dfrac{7}{20}(x+2)^5 + \cdots$

26. $y = 2 - 4(x - 2) - \dfrac{1}{2}(x - 2)^2 + \dfrac{2}{9}(x - 2)^3 + \dfrac{49}{432}(x - 2)^4 + \dfrac{23}{1080}(x - 2)^5 + \cdots$

27. $y = 1 + 2(x + 4) - \dfrac{1}{6}(x + 4)^2 - \dfrac{10}{27}(x + 4)^3 + \dfrac{19}{648}(x + 4)^4 + \dfrac{13}{324}(x + 4)^5 + \cdots$

28. $y = -1 + 2(x + 1) - \dfrac{1}{4}(x + 1)^2 + \dfrac{1}{2}(x + 1)^3 - \dfrac{65}{96}(x + 1)^4 + \dfrac{67}{80}(x + 1)^5 + \cdots$

31. (a) $y = \dfrac{c_1}{1 + x} + \dfrac{c_2}{1 + 2x}$ **(b)** $y = \dfrac{c_1}{1 - 2x} + \dfrac{c_2}{1 - 3x}$ **(c)** $y = \dfrac{c_1}{1 - 2x} + \dfrac{c_2 x}{(1 - 2x)^2}$ **(d)** $y = \dfrac{c_1}{2 + x} + \dfrac{c_2 x}{(2 + x)^2}$

   (e) $y = \dfrac{c_1}{2 + x} + \dfrac{c_2}{2 + 3x}$

32. $y = 1 - 2x - \dfrac{3}{2}x^2 + \dfrac{5}{3}x^3 + \dfrac{17}{24}x^4 - \dfrac{11}{20}x^5 + \cdots$ **33.** $y = 1 - 2x - \dfrac{5}{2}x^2 + \dfrac{2}{3}x^3 - \dfrac{3}{8}x^4 + \dfrac{1}{3}x^5 + \cdots$

34. $y = 6 - 2x + 9x^2 + \dfrac{2}{3}x^3 - \dfrac{23}{4}x^4 - \dfrac{3}{10}x^5 + \cdots$ **35.** $y = 2 - 5x + 2x^2 - \dfrac{10}{3}x^3 + \dfrac{3}{2}x^4 - \dfrac{25}{12}x^5 + \cdots$

36. $y = 3 + 6x - 3x^2 + x^3 - 2x^4 - \dfrac{17}{20}x^5 + \cdots$ **37.** $y = 3 - 2x - 3x^2 + \dfrac{3}{2}x^3 + \dfrac{3}{2}x^4 - \dfrac{49}{80}x^5 + \cdots$

38. $y = -2 + 3x + \dfrac{4}{3}x^2 - x^3 - \dfrac{19}{54}x^4 + \dfrac{13}{60}x^5 + \cdots$ **39.** $y_1 = \displaystyle\sum_{m=0}^{\infty} \dfrac{(-1)^m x^{2m}}{m!} = e^{-x^2}, \quad y_2 = \displaystyle\sum_{m=0}^{\infty} \dfrac{(-1)^m x^{2m+1}}{m!} = xe^{-x^2}$

40. $y = -2 + 3x + x^2 - \dfrac{1}{6}x^3 - \dfrac{3}{4}x^4 + \dfrac{31}{120}x^5 + \cdots$ **41.** $y = 2 + 3x - \dfrac{7}{2}x^2 - \dfrac{5}{6}x^3 + \dfrac{41}{24}x^4 + \dfrac{41}{120}x^5 + \cdots$

42. $y = -3 + 5x - 5x^2 + \dfrac{23}{6}x^3 - \dfrac{23}{12}x^4 + \dfrac{11}{30}x^5 + \cdots$

43. $y = -2 + 3(x - 1) + \dfrac{3}{2}(x - 1)^2 - \dfrac{17}{12}(x - 1)^3 - \dfrac{1}{12}(x - 1)^4 + \dfrac{1}{8}(x - 1)^5 + \cdots$

44. $y = 2 - 3(x + 2) + \dfrac{1}{2}(x + 2)^2 - \dfrac{1}{3}(x + 2)^3 + \dfrac{31}{24}(x + 2)^4 - \dfrac{53}{120}(x + 2)^5 + \cdots$

45. $y = 1 - 2x + \dfrac{3}{2}x^2 - \dfrac{11}{6}x^3 + \dfrac{15}{8}x^4 - \dfrac{71}{60}x^5 + \cdots$

46. $y = 2 - (x + 2) - \dfrac{7}{2}(x + 2)^2 - \dfrac{43}{6}(x + 2)^3 + \dfrac{203}{24}(x + 2)^4 - \dfrac{167}{30}(x + 2)^5 + \cdots$

47. $y = 2 - x - x^2 + \dfrac{7}{6}x^3 - x^4 + \dfrac{89}{120}x^5 + \cdots$ **48.** $y = 1 + \dfrac{3}{2}(x - 1)^2 + \dfrac{1}{6}(x - 1)^3 - \dfrac{1}{8}(x - 1)^5 + \cdots$

49. $y = 1 - 2(x - 3) + \dfrac{1}{2}(x - 3)^2 - \dfrac{1}{6}(x - 3)^3 + \dfrac{1}{4}(x - 3)^4 - \dfrac{1}{6}(x - 3)^5 + \cdots$

Section 7.4 Answers (page 327)

1. $y = c_1 x^{-4} + c_2 x^{-2}$ **2.** $y = c_1 x + c_2 x^7$ **3.** $y = x(c_1 + c_2 \ln x)$ **4.** $y = x^{-2}(c_1 + c_2 \ln x)$

5. $y = c_1 \cos(\ln x) + c_2 \sin(\ln x)$ **6.** $y = x^2[c_1 \cos(3 \ln x) + c_2 \sin(3 \ln x)]$ **7.** $y = c_1 x + \dfrac{c_2}{x^3}$

8. $y = c_1 x^{2/3} + c_2 x^{3/4}$ **9.** $y = x^{-1/2}(c_1 + c_2 \ln x)$ **10.** $y = c_1 x + c_2 x^{1/3}$ **11.** $y = c_1 x^2 + c_2 x^{1/2}$

12. $y = \dfrac{1}{x}[c_1 \cos(2 \ln x) + c_2 \sin(2 \ln x)]$ **13.** $y = x^{-1/3}(c_1 + c_2 \ln x)$ **14.** $y = x[c_1 \cos(3 \ln x) + c_2 \sin(3 \ln x)]$

15. $y = c_1 x^3 + \dfrac{c_2}{x^2}$ **16.** $y = \dfrac{c_1}{x} + c_2 x^{1/2}$ **17.** $y = x^2(c_1 + c_2 \ln x)$

18. $y = \dfrac{1}{x^2}\left[c_1 \cos\left(\dfrac{1}{\sqrt{2}} \ln x\right) + c_2 \sin\left(\dfrac{1}{\sqrt{2}} \ln x\right)\right]$

Section 7.5 Answers (page 339)

1. $y_1 = x^{1/2}\left(1 - \dfrac{1}{5}x - \dfrac{2}{35}x^2 + \dfrac{31}{315}x^3 + \cdots\right)$; $\quad y_2 = x^{-1}\left(1 + x + \dfrac{1}{2}x^2 - \dfrac{1}{6}x^3 + \cdots\right)$

2. $y_1 = x^{1/3}\left(1 - \dfrac{2}{3}x + \dfrac{8}{9}x^2 - \dfrac{40}{81}x^3 + \cdots\right)$; $\quad y_2 = 1 - x + \dfrac{6}{5}x^2 - \dfrac{4}{5}x^3 + \cdots$

3. $y_1 = x^{1/3}\left(1 - \dfrac{4}{7}x - \dfrac{7}{45}x^2 + \dfrac{970}{2457}x^3 + \cdots\right)$; $\quad y_2 = x^{-1}\left(1 - x^2 + \dfrac{2}{3}x^3 + \cdots\right)$

4. $y_1 = x^{1/4}\left(1 - \dfrac{1}{2}x - \dfrac{19}{104}x^2 + \dfrac{1571}{10608}x^3 + \cdots\right)$; $\quad y_2 = x^{-1}\left(1 + 2x - \dfrac{11}{6}x^2 - \dfrac{1}{7}x^3 + \cdots\right)$

5. $y_1 = x^{1/3}\left(1 - x + \dfrac{28}{31}x^2 - \dfrac{1111}{1333}x^3 + \cdots\right)$; $\quad y_2 = x^{-1/4}\left(1 - x + \dfrac{7}{8}x^2 - \dfrac{19}{24}x^3 + \cdots\right)$;

6. $y_1 = x^{1/5}\left(1 - \dfrac{6}{25}x - \dfrac{1217}{625}x^2 + \dfrac{41972}{46875}x^3 + \cdots\right)$; $\quad y_2 = x - \dfrac{1}{4}x^2 - \dfrac{35}{18}x^3 + \dfrac{11}{12}x^4 + \cdots$

7. $y_1 = x^{3/2}\left(1 - x + \dfrac{11}{26}x^2 - \dfrac{109}{1326}x^3 + \cdots\right)$; $\quad y_2 = x^{1/4}\left(1 + 4x - \dfrac{131}{24}x^2 + \dfrac{39}{14}x^3 + \cdots\right)$

8. $y_1 = x^{1/3}\left(1 - \dfrac{1}{3}x + \dfrac{2}{15}x^2 - \dfrac{5}{63}x^3 + \cdots\right)$; $\quad y_2 = x^{-1/6}\left(1 - \dfrac{1}{12}x^2 + \dfrac{1}{18}x^3 + \cdots\right)$

9. $y_1 = 1 - \dfrac{1}{14}x^2 + \dfrac{1}{105}x^3 + \cdots$; $\quad y_2 = x^{-1/3}\left(1 - \dfrac{1}{18}x - \dfrac{71}{405}x^2 + \dfrac{719}{34992}x^3 + \cdots\right)$

10. $y_1 = x^{1/5}\left(1 + \dfrac{3}{17}x - \dfrac{7}{153}x^2 - \dfrac{547}{5661}x^3 + \cdots\right)$; $\quad y_2 = x^{-1/2}\left(1 + x + \dfrac{14}{13}x^2 - \dfrac{556}{897}x^3 + \cdots\right)$

14. $y_1 = x^{1/2}\displaystyle\sum_{n=0}^{\infty} \dfrac{(-2)^n}{\prod_{j=1}^{n}(2j+3)}x^n$; $\quad y_2 = x^{-1}\displaystyle\sum_{n=0}^{\infty}\dfrac{(-1)^n}{n!}x^n$ $\qquad$ **15.** $y_1 = x^{1/3}\displaystyle\sum_{n=0}^{\infty}\dfrac{(-1)^n\prod_{j=1}^{n}(3j+1)}{9^n n!}x^n$; $\quad y_2 = x^{-1}$

16. $y_1 = x^{1/2}\displaystyle\sum_{n=0}^{\infty}\dfrac{(-1)^n}{2^n n!}x^n$; $\quad y_2 = \dfrac{1}{x^2}\displaystyle\sum_{n=0}^{\infty}\dfrac{(-1)^n}{\prod_{j=1}^{n}(2j-5)}x^n$

17. $y_1 = x\displaystyle\sum_{n=0}^{\infty}\dfrac{(-1)^n}{\prod_{j=1}^{n}(3j+4)}x^n$; $\quad y_2 = x^{-1/3}\displaystyle\sum_{n=0}^{\infty}\dfrac{(-1)^n}{3^n n!}$

18. $y_1 = x\displaystyle\sum_{n=0}^{\infty}\dfrac{2^n}{n!\,\prod_{j=1}^{n}(2j+1)}x^n$; $\quad y_2 = x^{1/2}\displaystyle\sum_{n=0}^{\infty}\dfrac{2^n}{n!\,\prod_{j=1}^{n}(2j-1)}x^n$

19. $y_1 = x^{1/3}\displaystyle\sum_{n=0}^{\infty}\dfrac{1}{n!\,\prod_{j=1}^{n}(3j+2)}x^n$; $\quad y_2 = x^{-1/3}\displaystyle\sum_{n=0}^{\infty}\dfrac{1}{n!\,\prod_{j=1}^{n}(3j-2)}x^n$

20. $y_1 = x\left(1 + \dfrac{2}{7}x + \dfrac{1}{70}x^2\right)$; $\quad y_2 = x^{-1/3}\displaystyle\sum_{n=0}^{\infty}\dfrac{(-1)^n}{3^n n!}\left(\prod_{j=1}^{n}\dfrac{3j-13}{3j-4}\right)x^n$

21. $y_1 = x^{1/2}\displaystyle\sum_{n=0}^{\infty}(-1)^n\left(\prod_{j=1}^{n}\dfrac{2j+1}{6j+1}\right)x^n$; $\quad y_2 = x^{1/3}\displaystyle\sum_{n=0}^{\infty}\dfrac{(-1)^n}{9^n n!}\left(\prod_{j=1}^{n}(3j+1)\right)x^n$

22. $y_1 = x\displaystyle\sum_{n=0}^{\infty}\dfrac{(-1)^n(n+2)!}{2\,\prod_{j=1}^{n}(4j+3)}x^n$; $\quad y_2 = x^{1/4}\displaystyle\sum_{n=0}^{\infty}\dfrac{(-1)^n}{16^n n!}\left(\prod_{j=1}^{n}(4j+5)\right)x^n$

23. $y_1 = x^{-1/2}\displaystyle\sum_{n=0}^{\infty}\dfrac{(-1)^n}{n!\,\prod_{j=1}^{n}(2j+1)}x^n$; $\quad y_2 = x^{-1}\displaystyle\sum_{n=0}^{\infty}\dfrac{(-1)^n}{n!\,\prod_{j=1}^{n}(2j-1)}x^n$

24. $y_1 = x^{1/3}\displaystyle\sum_{n=0}^{\infty}\dfrac{(-1)^n}{n!}\left(\dfrac{2}{9}\right)^n\left(\prod_{j=1}^{n}(6j+5)\right)x^n$; $\quad y_2 = x^{-1}\displaystyle\sum_{n=0}^{\infty}(-1)^n 2^n\left(\prod_{j=1}^{n}\dfrac{2j-1}{3j-4}\right)x^n$

25. $y_1 = 4x^{1/3}\displaystyle\sum_{n=0}^{\infty}\dfrac{1}{6^n n!(3n+4)}x^n$; $\quad y_2 = x^{-1}$

28. $y_1 = x^{1/2}\left(1 - \dfrac{9}{40}x + \dfrac{5}{128}x^2 - \dfrac{245}{39936}x^3 + \cdots\right);$ $\quad y_2 = x^{1/4}\left(1 - \dfrac{25}{96}x + \dfrac{675}{14336}x^2 - \dfrac{38025}{5046272}x^3 + \cdots\right)$

29. $y_1 = x^{1/3}\left(1 + \dfrac{32}{117}x - \dfrac{28}{1053}x^2 + \dfrac{4480}{540189}x^3 + \cdots\right);$ $\quad y_2 = x^{-3}\left(1 + \dfrac{32}{7}x + \dfrac{48}{7}x^2\right)$

30. $y_1 = x^{1/2}\left(1 - \dfrac{5}{8}x + \dfrac{55}{96}x^2 - \dfrac{935}{1536}x^3 + \cdots\right);$ $\quad y_2 = x^{-1/2}\left(1 + \dfrac{1}{4}x + \dfrac{5}{32}x^2 - \dfrac{55}{384}x^3 + \cdots\right)$

31. $y_1 = x^{1/2}\left(1 - \dfrac{3}{4}x + \dfrac{5}{96}x^2 + \dfrac{5}{4224}x^3 + \cdots\right);$ $\quad y_2 = x^{-2}(1 + 8x + 60x^2 - 160x^3 + \cdots)$

32. $y_1 = x^{-1/3}\left(1 - \dfrac{10}{63}x + \dfrac{200}{7371}x^2 - \dfrac{17600}{3781323}x^3 + \cdots\right);$ $\quad y_2 = x^{-1/2}\left(1 - \dfrac{3}{20}x + \dfrac{9}{352}x^2 - \dfrac{105}{23936}x^3 + \cdots\right)$

33. $y_1 = x^{1/2}\displaystyle\sum_{m=0}^{\infty}\dfrac{(-1)^m}{8^m m!}\left(\prod_{j=1}^{m}\dfrac{4j-3}{8j+1}\right)x^{2m};$ $\quad y_2 = x^{1/4}\displaystyle\sum_{m=0}^{\infty}\dfrac{(-1)^m}{16^m m!}\left(\prod_{j=1}^{m}\dfrac{8j-7}{8j-1}\right)x^{2m}$

34. $y_1 = x^{1/2}\displaystyle\sum_{m=0}^{\infty}\left(\prod_{j=1}^{m}\dfrac{8j-3}{8j+1}\right)x^{2m};$ $\quad y_2 = x^{1/4}\displaystyle\sum_{m=0}^{\infty}\dfrac{1}{2^m m!}\left(\prod_{j=1}^{m}(2j-1)\right)x^{2m}$

35. $y_1 = x^4\displaystyle\sum_{m=0}^{\infty}(-1)^m(m+1)x^{2m};$ $\quad y_2 = -x\displaystyle\sum_{m=0}^{\infty}(-1)^m(2m-1)x^{2m}$

36. $y_1 = x^{1/3}\displaystyle\sum_{m=0}^{\infty}\dfrac{(-1)^m}{18^m m!}\left(\prod_{j=1}^{m}(6j-17)\right)x^{2m};$ $\quad y_2 = 1 + \dfrac{4}{5}x^2 + \dfrac{8}{55}x^4$

37. $y_1 = x^{1/4}\displaystyle\sum_{m=0}^{\infty}\left(\prod_{j=1}^{m}\dfrac{8j+1}{8j+5}\right)x^{2m};$ $\quad y_2 = x^{-1}\displaystyle\sum_{m=0}^{\infty}\dfrac{\prod_{j=1}^{m}(2j-1)}{2^m m!}x^{2m}$

38. $y_1 = x^{1/2}\displaystyle\sum_{m=0}^{\infty}\dfrac{1}{8^m m!}\left(\prod_{j=1}^{m}(4j-1)\right)x^{2m};$ $\quad y_2 = x^{1/3}\displaystyle\sum_{m=0}^{\infty}2^m\left(\prod_{j=1}^{m}\dfrac{3j-1}{12j-1}\right)x^{2m}$

39. $y_1 = x^{7/2}\displaystyle\sum_{m=0}^{\infty}(-1)^m\dfrac{\prod_{j=1}^{m}(4j+5)}{8^m m!}x^{2m};$ $\quad y_2 = x^{1/2}\displaystyle\sum_{m=0}^{\infty}\dfrac{(-1)^m}{4^m}\left(\prod_{j=1}^{m}\dfrac{4j-1}{2j-3}\right)x^{2m}$

40. $y_1 = x^{1/2}\displaystyle\sum_{m=0}^{\infty}\dfrac{(-1)^m}{4^m}\left(\prod_{j=1}^{m}\dfrac{4j-1}{2j+1}\right)x^{2m};$ $\quad y_2 = x^{-1/2}\displaystyle\sum_{m=0}^{\infty}\dfrac{(-1)^m}{8^m m!}\left(\prod_{j=1}^{m}(4j-3)\right)x^{2m}$

41. $y_1 = x^{1/2}\displaystyle\sum_{m=0}^{\infty}\dfrac{(-1)^m}{m!}\left(\prod_{j=1}^{m}(2j+1)\right)x^{2m};$ $\quad y_2 = \dfrac{1}{x^2}\displaystyle\sum_{m=0}^{\infty}(-2)^m\left(\prod_{j=1}^{m}\dfrac{4j-3}{4j-5}\right)x^{2m}$

42. $y_1 = x^{1/3}\displaystyle\sum_{m=0}^{\infty}(-1)^m\left(\prod_{j=1}^{m}\dfrac{3j-4}{3j+2}\right)x^{2m};$ $\quad y_2 = x^{-1}(1 + x^2)$

43. $y_1 = \displaystyle\sum_{m=0}^{\infty}(-1)^m\dfrac{2^m(m+1)!}{\prod_{j=1}^{m}(2j+3)}x^{2m};$ $\quad y_2 = \dfrac{1}{x^3}\displaystyle\sum_{m=0}^{\infty}(-1)^m\dfrac{\prod_{j=1}^{m}(2j-1)}{2^m m!}x^{2m}$

44. $y_1 = x^{1/2}\displaystyle\sum_{m=0}^{\infty}\dfrac{(-1)^m}{8^m m!}\left(\prod_{j=1}^{m}\dfrac{(4j-3)^2}{4j+3}\right)x^{2m};$ $\quad y_2 = x^{-1}\displaystyle\sum_{m=0}^{\infty}\dfrac{(-1)^m}{2^m m!}\left(\prod_{j=1}^{m}\dfrac{(2j-3)^2}{4j-3}\right)x^{2m}$

45. $y_1 = x\displaystyle\sum_{m=0}^{\infty}(-2)^m\left(\prod_{j=1}^{m}\dfrac{2j+1}{4j+5}\right)x^{2m};$ $\quad y_2 = x^{-3/2}\displaystyle\sum_{m=0}^{\infty}\dfrac{(-1)^m}{4^m m!}\left(\prod_{j=1}^{m}(4j-3)\right)x^{2m}$

46. $y_1 = x^{1/3}\displaystyle\sum_{m=0}^{\infty}\dfrac{(-1)^m}{2^m\prod_{j=1}^{m}(3j+1)}x^{2m};$ $\quad y_2 = x^{-1/3}\displaystyle\sum_{m=0}^{\infty}\dfrac{(-1)^m}{6^m m!}x^{2m}$

47. $y_1 = x^{1/2}\left(1 - \dfrac{6}{13}x^2 + \dfrac{36}{325}x^4 - \dfrac{216}{12025}x^6 + \cdots\right);$ $\quad y_2 = x^{1/3}\left(1 - \dfrac{1}{2}x^2 + \dfrac{1}{8}x^4 - \dfrac{1}{48}x^6 + \cdots\right)$

48. $y_1 = x^{1/4}\left(1 - \dfrac{13}{64}x^2 + \dfrac{273}{8192}x^4 - \dfrac{2639}{524288}x^6 + \cdots\right);$ $\quad y_2 = x^{-1}\left(1 - \dfrac{1}{3}x^2 + \dfrac{2}{33}x^4 - \dfrac{2}{209}x^6 + \cdots\right)$

49. $y_1 = x^{1/3}\left(1 - \dfrac{3}{4}x^2 + \dfrac{9}{14}x^4 - \dfrac{81}{140}x^6 + \cdots\right);\quad y_2 = x^{-1/3}\left(1 - \dfrac{2}{3}x^2 + \dfrac{5}{9}x^4 - \dfrac{40}{81}x^6 + \cdots\right)$

50. $y_1 = x^{1/2}\left(1 - \dfrac{3}{2}x^2 + \dfrac{15}{8}x^4 - \dfrac{35}{16}x^6 + \cdots\right);\quad y_2 = x^{-1/2}\left(1 - 2x^2 + \dfrac{8}{3}x^4 - \dfrac{16}{5}x^6 + \cdots\right)$

51. $y_1 = x^{1/4}\left(1 - x^2 + \dfrac{3}{2}x^4 - \dfrac{5}{2}x^6 + \cdots\right);\quad y_2 = x^{-1/2}\left(1 - \dfrac{2}{5}x^2 + \dfrac{36}{65}x^4 - \dfrac{408}{455}x^6 + \cdots\right)$

53. (a) $y_1 = x^\nu \displaystyle\sum_{m=0}^\infty \dfrac{(-1)^m}{4^m m!\, \prod_{j=1}^m (j+\nu)} x^{2m};\quad y_2 = x^{-\nu} \displaystyle\sum_{m=0}^\infty \dfrac{(-1)^m}{4^m m!\, \prod_{j=1}^m (j-\nu)} x^{2m};$ **(b)** $\dfrac{\sin x}{\sqrt{x}};\ \dfrac{\cos x}{\sqrt{x}}$

61. $y_1 = \dfrac{x^{1/2}}{1+x};\quad y_2 = \dfrac{x}{1+x}$ **62.** $y_1 = \dfrac{x^{1/3}}{1+2x^2};\quad y_2 = \dfrac{x^{1/2}}{1+2x^2}$

63. $y_1 = \dfrac{x^{1/4}}{1-3x};\quad y_2 = \dfrac{x^2}{1-3x}$ **64.** $y_1 = \dfrac{x^{1/3}}{5+x};\quad y_2 = \dfrac{x^{-1/3}}{5+x}$

65. $y_1 = \dfrac{x^{1/4}}{2-x^2};\quad y_2 = \dfrac{x^{-1/2}}{2-x^2}$ **66.** $y_1 = \dfrac{x^{1/2}}{1+3x+x^2};\quad y_2 = \dfrac{x^{3/2}}{1+3x+x^2}$

67. $y_1 = \dfrac{x}{(1+x)^2};\quad y_2 = \dfrac{x^{1/3}}{(1+x)^2}$ **68.** $y_1 = \dfrac{x}{3+2x+x^2};\quad y_2 = \dfrac{x^{1/4}}{3+2x+x^2}$

Section 7.6 Answers (page 354)

1. $y_1 = x\left(1 - x + \dfrac{3}{4}x^2 - \dfrac{13}{36}x^3 + \cdots\right);\quad y_2 = y_1 \ln x + x^2\left(1 - x + \dfrac{65}{108}x^2 + \cdots\right)$

2. $y_1 = x^{-1}\left(1 - 2x + \dfrac{9}{2}x^2 - \dfrac{20}{3}x^3 + \cdots\right);\quad y_2 = y_1 \ln x + 1 - \dfrac{15}{4}x + \dfrac{133}{18}x^2 + \cdots$

3. $y_1 = 1 + x - x^2 + \dfrac{1}{3}x^3 + \cdots;\quad y_2 = y_1 \ln x - x\left(3 - \dfrac{1}{2}x - \dfrac{31}{18}x^2 + \cdots\right)$

4. $y_1 = x^{1/2}\left(1 - 2x + \dfrac{5}{2}x^2 - 2x^3 + \cdots\right);\quad y_2 = y_1 \ln x + x^{3/2}\left(1 - \dfrac{9}{4}x + \dfrac{17}{6}x^2 + \cdots\right)$

5. $y_1 = x\left(1 - 4x + \dfrac{19}{2}x^2 - \dfrac{49}{3}x^3 + \cdots\right);\quad y_2 = y_1 \ln x + x^2\left(3 - \dfrac{43}{4}x + \dfrac{208}{9}x^2 + \cdots\right)$

6. $y_1 = x^{-1/3}\left(1 - x + \dfrac{5}{6}x^2 - \dfrac{1}{2}x^3 + \cdots\right);\quad y_2 = y_1 \ln x + x^{2/3}\left(1 - \dfrac{11}{12}x + \dfrac{25}{36}x^2 + \cdots\right)$

7. $y_1 = 1 - 2x + \dfrac{7}{4}x^2 - \dfrac{7}{9}x^3 + \cdots;\quad y_2 = y_1 \ln x + x\left(3 - \dfrac{15}{4}x + \dfrac{239}{108}x^2 + \cdots\right)$

8. $y_1 = x^{-2}\left(1 - 2x + \dfrac{5}{2}x^2 - 3x^3 + \cdots\right);\quad y_2 = y_1 \ln x + \dfrac{3}{4} - \dfrac{13}{6}x + \cdots$

9. $y_1 = x^{-1/2}\left(1 - x + \dfrac{1}{4}x^2 + \dfrac{1}{18}x^3 + \cdots\right);\quad y_2 = y_1 \ln x + x^{1/2}\left(\dfrac{3}{2} - \dfrac{13}{16}x + \dfrac{1}{54}x^2 + \cdots\right)$

10. $y_1 = x^{-1/4}\left(1 - \dfrac{1}{4}x - \dfrac{7}{32}x^2 + \dfrac{23}{384}x^3 + \cdots\right);\quad y_2 = y_1 \ln x + x^{3/4}\left(\dfrac{1}{4} + \dfrac{5}{64}x - \dfrac{157}{2304}x^2 + \cdots\right)$

11. $y_1 = x^{-1/3}\left(1 - x + \dfrac{7}{6}x^2 - \dfrac{23}{18}x^3 + \cdots\right);\quad y_2 = y_1 \ln x - x^{5/3}\left(\dfrac{1}{12} - \dfrac{13}{108}x + \cdots\right)$

12. $y_1 = x^{1/2} \displaystyle\sum_{n=0}^\infty \dfrac{(-1)^n}{(n!)^2} x^n;\quad y_2 = y_1 \ln x - 2x^{1/2} \displaystyle\sum_{n=1}^\infty \dfrac{(-1)^n}{(n!)^2}\left(\sum_{j=1}^n \dfrac{1}{j}\right)x^n;$

13. $y_1 = x^{1/6} \displaystyle\sum_{n=0}^\infty \left(\dfrac{2}{3}\right)^n \dfrac{\prod_{j=1}^n (3j+1)}{n!} x^n;\quad y_2 = y_1 \ln x - x^{1/6} \displaystyle\sum_{n=1}^\infty \left(\dfrac{2}{3}\right)^n \dfrac{\prod_{j=1}^n (3j+1)}{n!}\left(\sum_{j=1}^n \dfrac{1}{j(3j+1)}\right)x^n$

14. $y_1 = x^2 \sum_{n=0}^{\infty} (-1)^n (n+1)^2 x^n$; $y_2 = y_1 \ln x - 2x^2 \sum_{n=1}^{\infty} (-1)^n n(n+1)x^n$

15. $y_1 = x^3 \sum_{n=0}^{\infty} 2^n (n+1)x^n$; $y_2 = y_1 \ln x - x^3 \sum_{n=1}^{\infty} 2^n n x^n$

16. $y_1 = x^{1/5} \sum_{n=0}^{\infty} \frac{(-1)^n \prod_{j=1}^{n} (5j+1)}{125^n (n!)^2} x^n$; $y_2 = y_1 \ln x - x^{1/5} \sum_{n=1}^{\infty} \frac{(-1)^n \prod_{j=1}^{n} (5j+1)}{125^n (n!)^2} \left(\sum_{j=1}^{n} \frac{5j+2}{j(5j+1)} \right) x^n$

17. $y_1 = x^{1/2} \sum_{n=0}^{\infty} \frac{(-1)^n \prod_{j=1}^{n} (2j-3)}{4^n n!} x^n$; $y_2 = y_1 \ln x + 3x^{1/2} \sum_{n=1}^{\infty} \frac{(-1)^n \prod_{j=1}^{n} (2j-3)}{4^n n!} \left(\sum_{j=1}^{n} \frac{1}{j(2j-3)} \right) x^n$

18. $y_1 = x^{1/3} \sum_{n=0}^{\infty} \frac{(-1)^n \prod_{j=1}^{n} (6j-7)^2}{81^n (n!)^2} x^n$;

$y_2 = y_1 \ln x + 14x^{1/3} \sum_{n=1}^{\infty} \frac{(-1)^n \prod_{j=1}^{n} (6j-7)^2}{81^n (n!)^2} \left(\sum_{j=1}^{n} \frac{1}{j(6j-7)} \right) x^n$

19. $y_1 = x^2 \sum_{n=0}^{\infty} \frac{(-1)^n \prod_{j=1}^{n} (2j+5)}{(n!)^2} x^n$; $y_2 = y_1 \ln x - 2x^2 \sum_{n=1}^{\infty} \frac{(-1)^n \prod_{j=1}^{n} (2j+5)}{(n!)^2} \left(\sum_{j=1}^{n} \frac{(j+5)}{j(2j+5)} \right) x^n$

20. $y_1 = \frac{1}{x} \sum_{n=0}^{\infty} \frac{2^n \prod_{j=1}^{n} (2j-1)}{n!} x^n$; $y_2 = y_1 \ln x + \frac{1}{x} \sum_{n=1}^{\infty} \frac{2^n \prod_{j=1}^{n} (2j-1)}{n!} \left(\sum_{j=1}^{n} \frac{1}{j(2j-1)} \right) x^n$

21. $y_1 = \frac{1}{x} \sum_{n=0}^{\infty} \frac{(-1)^n \prod_{j=1}^{n} (2j-5)}{n!} x^n$; $y_2 = y_1 \ln x + \frac{5}{x} \sum_{n=1}^{\infty} \frac{(-1)^n \prod_{j=1}^{n} (2j-5)}{n!} \left(\sum_{j=1}^{n} \frac{1}{j(2j-5)} \right) x^n$

22. $y_1 = x^2 \sum_{n=0}^{\infty} \frac{(-1)^n \prod_{j=1}^{n} (2j+3)}{2^n n!} x^n$; $y_2 = y_1 \ln x - 3x^2 \sum_{n=0}^{\infty} \frac{(-1)^n \prod_{j=1}^{n} (2j+3)}{2^n n!} \left(\sum_{j=1}^{n} \frac{1}{j(2j+3)} \right) x^n$;

23. $y_1 = x^{-2} \left(1 + 3x + \frac{3}{2}x^2 - \frac{1}{2}x^3 + \cdots \right)$; $y_2 = y_1 \ln x - 5x^{-1} \left(1 + \frac{5}{4}x - \frac{1}{4}x^2 + \cdots \right)$

24. $y_1 = x^3 (1 + 20x + 180x^2 + 1120x^3 + \cdots)$; $y_2 = y_1 \ln x - x^4 \left(26 + 324x + \frac{6968}{3}x^2 + \cdots \right)$

25. $y_1 = x \left(1 - 5x + \frac{85}{4}x^2 - \frac{3145}{36}x^3 + \cdots \right)$; $y_2 = y_1 \ln x + x^2 \left(2 - \frac{39}{4}x + \frac{4499}{108}x^2 + \cdots \right)$

26. $y_1 = 1 - x + \frac{3}{4}x^2 - \frac{7}{12}x^3 + \cdots$; $y_2 = y_1 \ln x + x \left(1 - \frac{3}{4}x + \frac{5}{9}x^2 + \cdots \right)$

27. $y_1 = x^{-3} (1 + 16x + 36x^2 + 16x^3 + \cdots)$; $y_2 = y_1 \ln x - x^{-2} \left(40 + 150x + \frac{280}{3}x^2 + \cdots \right)$

28. $y_1 = x \sum_{m=0}^{\infty} \frac{(-1)^m}{2^m m!} x^{2m}$; $y_2 = y_1 \ln x - \frac{x}{2} \sum_{m=1}^{\infty} \frac{(-1)^m}{2^m m!} \left(\sum_{j=1}^{m} \frac{1}{j} \right) x^{2m}$

29. $y_1 = x^2 \sum_{m=0}^{\infty} (-1)^m (m+1)x^{2m}$; $y_2 = y_1 \ln x - \frac{x^2}{2} \sum_{m=1}^{\infty} (-1)^m m x^{2m}$

30. $y_1 = x^{1/2} \sum_{m=0}^{\infty} \frac{(-1)^m}{4^m m!} x^{2m}$; $y_2 = y_1 \ln x - \frac{x^{1/2}}{2} \sum_{m=1}^{\infty} \frac{(-1)^m}{4^m m!} \left(\sum_{j=1}^{m} \frac{1}{j} \right) x^{2m}$

31. $y_1 = x \sum_{m=0}^{\infty} \frac{(-1)^m \prod_{j=1}^{m} (2j-1)}{2^m m!} x^{2m}$; $y_2 = y_1 \ln x + \frac{x}{2} \sum_{m=1}^{\infty} \frac{(-1)^m \prod_{j=1}^{m} (2j-1)}{2^m m!} \left(\sum_{j=1}^{m} \frac{1}{j(2j-1)} \right) x^{2m}$

32. $y_1 = x^{1/2} \sum_{m=0}^{\infty} \frac{(-1)^m \prod_{j=1}^{m} (4j-1)}{8^m m!} x^{2m}$; $y_2 = y_1 \ln x + \frac{x^{1/2}}{2} \sum_{m=1}^{\infty} \frac{(-1)^m \prod_{j=1}^{m} (4j-1)}{8^m m!} \left(\sum_{j=1}^{m} \frac{1}{j(4j-1)} \right) x^{2m}$

33. $y_1 = x \sum_{m=0}^{\infty} \frac{(-1)^m \prod_{j=1}^{m} (2j+1)}{2^m m!} x^{2m}$; $y_2 = y_1 \ln x - \frac{x}{2} \sum_{m=1}^{\infty} \frac{(-1)^m \prod_{j=1}^{m} (2j+1)}{2^m m!} \left(\sum_{j=1}^{m} \frac{1}{j(2j+1)} \right) x^{2m}$

34. $y_1 = x^{-1/4} \sum_{m=0}^{\infty} \frac{(-1)^m \prod_{j=1}^{m} (8j - 13)}{(32)^m m!} x^{2m}$; $y_2 = y_1 \ln x + \frac{13}{2} x^{-1/4} \sum_{m=1}^{\infty} \frac{(-1)^m \prod_{j=1}^{m} (8j - 13)}{(32)^m m!} \left(\sum_{j=1}^{m} \frac{1}{j(8j - 13)} \right) x^{2m}$

35. $y_1 = x^{1/3} \sum_{m=0}^{\infty} \frac{(-1)^m \prod_{j=1}^{m} (3j - 1)}{9^m m!} x^{2m}$; $y_2 = y_1 \ln x + \frac{x^{1/3}}{2} \sum_{m=1}^{\infty} \frac{(-1)^m \prod_{j=1}^{m} (3j - 1)}{9^m m!} \left(\sum_{j=1}^{m} \frac{1}{j(3j - 1)} \right) x^{2m}$

36. $y_1 = x^{1/2} \sum_{m=0}^{\infty} \frac{(-1)^m \prod_{j=1}^{m} (4j - 3)(4j - 1)}{4^m (m!)^2} x^{2m}$;

$y_2 = y_1 \ln x + x^{1/2} \sum_{m=1}^{\infty} \frac{(-1)^m \prod_{j=1}^{m} (4j - 3)(4j - 1)}{4^m (m!)^2} \left(\sum_{j=1}^{m} \frac{8j - 3}{j(4j - 3)(4j - 1)} \right) x^{2m}$

37. $y_1 = x^{5/3} \sum_{m=0}^{\infty} \frac{(-1)^m}{3^m m!} x^{2m}$; $y_2 = y_1 \ln x - \frac{x^{5/3}}{2} \sum_{m=1}^{\infty} \frac{(-1)^m}{3^m m!} \left(\sum_{j=1}^{m} \frac{1}{j} \right) x^{2m}$

38. $y_1 = \frac{1}{x} \sum_{m=0}^{\infty} \frac{(-1)^m \prod_{j=1}^{m} (4j - 7)}{2^m m!} x^{2m}$; $y_2 = y_1 \ln x + \frac{7}{2x} \sum_{m=1}^{\infty} \frac{(-1)^m \prod_{j=1}^{m} (4j - 7)}{2^m m!} \left(\sum_{j=1}^{m} \frac{1}{j(4j - 7)} \right) x^{2m}$

39. $y_1 = x^{-1} \left(1 - \frac{3}{2} x^2 + \frac{15}{8} x^4 - \frac{35}{16} x^6 + \cdots \right)$; $y_2 = y_1 \ln x + x \left(\frac{1}{4} - \frac{13}{32} x^2 + \frac{101}{192} x^4 + \cdots \right)$

40. $y_1 = x \left(1 - \frac{1}{2} x^2 + \frac{1}{8} x^4 - \frac{1}{48} x^6 + \cdots \right)$; $y_2 = y_1 \ln x + x^3 \left(\frac{1}{4} - \frac{3}{32} x^2 + \frac{11}{576} x^4 + \cdots \right)$

41. $y_1 = x^{-2} \left(1 - \frac{3}{4} x^2 - \frac{9}{64} x^4 - \frac{25}{256} x^6 + \cdots \right)$; $y_2 = y_1 \ln x + \frac{1}{2} - \frac{21}{128} x^2 - \frac{215}{1536} x^4 + \cdots$

42. $y_1 = x^{-3} \left(1 - \frac{17}{8} x^2 + \frac{85}{256} x^4 - \frac{85}{18432} x^6 + \cdots \right)$; $y_2 = y_1 \ln x + x^{-1} \left(\frac{25}{8} - \frac{471}{512} x^2 + \frac{1583}{110592} x^4 + \cdots \right)$

43. $y_1 = x^{-1} \left(1 - \frac{3}{4} x^2 + \frac{45}{64} x^4 - \frac{175}{256} x^6 + \cdots \right)$; $y_2 = y_1 \ln x - x \left(\frac{1}{4} - \frac{33}{128} x^2 + \frac{395}{1536} x^4 + \cdots \right)$

44. $y_1 = \frac{1}{x}$; $y_2 = y_1 \ln x - 6 + 6x - \frac{8}{3} x^2$ **45.** $y_1 = 1 - x$; $y_2 = y_1 \ln x + 4x$

46. $y_1 = \frac{(x - 1)^2}{x}$; $y_2 = y_1 \ln x + 3 - 3x + 2 \sum_{n=2}^{\infty} \frac{1}{n(n^2 - 1)} x^n$

47. $y_1 = x^{1/2}(x + 1)^2$; $y_2 = y_1 \ln x - x^{3/2} \left(3 + 3x + 2 \sum_{n=2}^{\infty} \frac{(-1)^n}{n(n^2 - 1)} x^n \right)$

48. $y_1 = x^2(1 - x)^3$; $y_2 = y_1 \ln x + x^3 \left(4 - 7x + \frac{11}{3} x^2 - 6 \sum_{n=3}^{\infty} \frac{1}{n(n - 2)(n^2 - 1)} x^n \right)$

49. $y_1 = x - 4x^3 + x^5$; $y_2 = y_1 \ln x + 6x^3 - 3x^5$

50. $y_1 = x^{1/3} \left(1 - \frac{1}{6} x^2 \right)$; $y_2 = y_1 \ln x + x^{7/3} \left(\frac{1}{4} - \frac{1}{12} \sum_{m=1}^{\infty} \frac{1}{6^m m(m + 1)(m + 1)!} x^{2m} \right)$

51. $y_1 = (1 + x^2)^2$; $y_2 = y_1 \ln x - \frac{3}{2} x^2 - \frac{3}{2} x^4 + \sum_{m=3}^{\infty} \frac{(-1)^m}{m(m - 1)(m - 2)} x^{2m}$

52. $y_1 = x^{-1/2} \left(1 - \frac{1}{2} x^2 + \frac{1}{32} x^4 \right)$; $y_2 = y_1 \ln x + x^{3/2} \left(\frac{5}{8} - \frac{9}{128} x^2 + \sum_{m=2}^{\infty} \frac{1}{4^{m+1}(m - 1)m(m + 1)(m + 1)!} x^{2m} \right)$

56. $y_1 = \sum_{m=0}^{\infty} \frac{(-1)^m}{4^m (m!)^2} x^{2m}$; $y_2 = y_1 \ln x - \sum_{m=1}^{\infty} \frac{(-1)^m}{4^m (m!)^2} \left(\sum_{j=1}^{m} \frac{1}{j} \right) x^{2m}$

58. $y_1 = \frac{x^{1/2}}{1 + x}$; $y_2 = \frac{x^{1/2} \ln x}{1 + x}$ **59.** $y_1 = \frac{x^{1/3}}{3 + x}$; $y_2 = \frac{x^{1/3} \ln x}{3 + x}$ **60.** $y_1 = \frac{x}{2 - x^2}$; $y_2 = \frac{x \ln x}{2 - x^2}$

61. $y_1 = \frac{x^{1/4}}{1 + x^2}$; $y_2 = \frac{x^{1/4} \ln x}{1 + x^2}$ **62.** $y_1 = \frac{x}{4 + 3x}$; $y_2 = \frac{x \ln x}{4 + 3x}$ **63.** $y_1 = \frac{x^{1/2}}{1 + 3x + x^2}$; $y_2 = \frac{x^{1/2} \ln x}{1 + 3x + x^2}$

64. $y_1 = \frac{x}{(1 - x)^2}$; $y_2 = \frac{x \ln x}{(1 - x)^2}$ **65.** $y_1 = \frac{x^{1/3}}{1 + x + x^2}$; $y_2 = \frac{x^{1/3} \ln x}{1 + x + x^2}$

Section 7.7 Answers (page 367)

1. $y_1 = 2x^3 \displaystyle\sum_{n=0}^{\infty} \frac{(-4)^n}{n!(n+2)!} x^n$; $y_2 = x + 4x^2 - 8\left(y_1 \ln x - 4 \displaystyle\sum_{n=1}^{\infty} \frac{(-4)^n}{n!(n+2)!} \left(\displaystyle\sum_{j=1}^{n} \frac{j+1}{j(j+2)} \right) x^n \right)$

2. $y_1 = x \displaystyle\sum_{n=0}^{\infty} \frac{(-1)^n}{n!(n+1)!} x^n$; $y_2 = 1 - y_1 \ln x + x \displaystyle\sum_{n=1}^{\infty} \frac{(-1)^n}{n!(n+1)!} \left(\displaystyle\sum_{j=1}^{n} \frac{2j+1}{j(j+1)} \right) x^n$

3. $y_1 = x^{1/2}$; $y_2 = x^{-1/2} + y_1 \ln x + x^{1/2} \displaystyle\sum_{n=1}^{\infty} \frac{(-1)^n}{n} x^n$

4. $y_1 = x \displaystyle\sum_{n=0}^{\infty} \frac{(-1)^n}{n!} x^n = xe^{-x}$; $y_2 = 1 - y_1 \ln x + x \displaystyle\sum_{n=1}^{\infty} \frac{(-1)^n}{n!} \left(\displaystyle\sum_{j=1}^{\infty} \frac{1}{j} \right) x^n$

5. $y_1 = x^{1/2} \displaystyle\sum_{n=0}^{\infty} \left(-\frac{3}{4}\right)^n \frac{\prod_{j=1}^{n}(2j+1)}{n!} x^n$;

$y_2 = x^{-1/2} - \frac{3}{4}\left(y_1 \ln x - x^{1/2} \displaystyle\sum_{n=1}^{\infty} \left(-\frac{3}{4}\right)^n \frac{\prod_{j=1}^{n}(2j+1)}{n!} \left(\displaystyle\sum_{j=1}^{n} \frac{1}{j(2j+1)} \right) x^n \right)$

6. $y_1 = x \displaystyle\sum_{n=0}^{\infty} \frac{(-1)^n}{n!} x^n = xe^{-x}$; $y_2 = x^{-2}\left(1 + \frac{1}{2}x + \frac{1}{2}x^2 \right) - \frac{1}{2}\left(y_1 \ln x - x \displaystyle\sum_{n=1}^{\infty} \frac{(-1)^n}{n!} \left(\displaystyle\sum_{j=1}^{n} \frac{1}{j} \right) x^n \right)$

7. $y_1 = 6x^{3/2} \displaystyle\sum_{n=0}^{\infty} \frac{(-1)^n}{4^n n!(n+3)!} x^n$;

$y_2 = x^{-3/2}\left(1 + \frac{1}{8}x + \frac{1}{64}x^2 \right) - \frac{1}{768}\left(y_1 \ln x - 6x^{3/2} \displaystyle\sum_{n=1}^{\infty} \frac{(-1)^n}{4^n n!(n+3)!} \left(\displaystyle\sum_{j=1}^{n} \frac{2j+3}{j(j+3)} \right) x^n \right)$

8. $y_1 = \dfrac{120}{x^2} \displaystyle\sum_{n=0}^{\infty} \frac{(-1)^n}{n!(n+5)!} x^n$;

$y_2 = x^{-7}\left(1 + \frac{1}{4}x + \frac{1}{24}x^2 + \frac{1}{144}x^3 + \frac{1}{576}x^4 \right) - \frac{1}{2880}\left(y_1 \ln x - \frac{120}{x^2} \displaystyle\sum_{n=1}^{\infty} \frac{(-1)^n}{n!(n+5)!} \left(\displaystyle\sum_{j=1}^{n} \frac{2j+5}{j(j+5)} \right) x^n \right)$

9. $y_1 = \dfrac{x^{1/2}}{6} \displaystyle\sum_{n=0}^{\infty} (-1)^n (n+1)(n+2)(n+3)x^n$;

$y_2 = x^{-5/2}\left(1 + \frac{1}{2}x + x^2 \right) - 3y_1 \ln x + \frac{3}{2}x^{1/2} \displaystyle\sum_{n=1}^{\infty} (-1)^n (n+1)(n+2)(n+3) \left(\displaystyle\sum_{j=1}^{n} \frac{1}{j(j+3)} \right) x^n$

10. $y_1 = x^4\left(1 - \frac{2}{5}x \right)$; $y_2 = 1 + 10x + 50x^2 + 200x^3 - 300\left(y_1 \ln x + \frac{27}{25}x^5 - \frac{1}{30}x^6 \right)$

11. $y_1 = x^3$; $y_2 = x^{-3}\left(1 - \frac{6}{5}x + \frac{3}{4}x^2 - \frac{1}{3}x^3 + \frac{1}{8}x^4 - \frac{1}{20}x^5 \right) - \frac{1}{120}\left(y_1 \ln x + x^3 \displaystyle\sum_{n=1}^{\infty} \frac{(-1)^n 6!}{n(n+6)!} x^n \right)$

12. $y_1 = x^2 \displaystyle\sum_{n=0}^{\infty} \frac{1}{n!} \left(\displaystyle\prod_{j=1}^{n} \frac{2j+3}{j+4} \right) x^n$;

$y_2 = x^{-2}\left(1 + x + \frac{1}{4}x^2 - \frac{1}{12}x^3 \right) - \frac{1}{16}y_1 \ln x + \frac{x^2}{8} \displaystyle\sum_{n=1}^{\infty} \frac{1}{n!} \left(\displaystyle\prod_{j=1}^{n} \frac{2j+3}{j+4} \right) \left(\displaystyle\sum_{j=1}^{n} \frac{j^2+3j+6}{j(j+4)(2j+3)} \right) x^n$

13. $y_1 = x^5 \displaystyle\sum_{n=0}^{\infty} (-1)^n (n+1)(n+2)x^n$; $y_2 = 1 - \dfrac{x}{2} + \dfrac{x^2}{6}$

14. $y_1 = \dfrac{1}{x} \displaystyle\sum_{n=0}^{\infty} \frac{(-1)^n}{n!} \left(\displaystyle\prod_{j=1}^{n} \frac{(j+3)(2j-3)}{j+6} \right) x^n$; $y_2 = x^{-7}\left(1 + \frac{26}{5}x + \frac{143}{20}x^2 \right)$

15. $y_1 = 24x^{7/2} \displaystyle\sum_{n=0}^{\infty} \frac{(-1)^n}{2^n(n+4)!} x^n$; $y_2 = x^{-1/2}\left(1 - \frac{1}{2}x + \frac{1}{8}x^2 - \frac{1}{48}x^3 \right)$

16. $y_1 = x^{10/3} \displaystyle\sum_{n=0}^{\infty} \frac{(-1)^n(n+1)}{9^n} \left(\displaystyle\prod_{j=1}^{n} \frac{3j+7}{j+4} \right) x^n$; $y_2 = x^{-2/3}\left(1 + \frac{4}{27}x - \frac{1}{243}x^2 \right)$

17. $y_1 = x^3 \sum\limits_{n=0}^{7} (-1)^n (n+1) \left(\prod\limits_{j=1}^{n} \dfrac{j-8}{j+6} \right) x^n$; $\quad y_2 = \dfrac{1}{x^3} \sum\limits_{n=0}^{4} \dfrac{(-1)^n}{n!} \left(\prod\limits_{j=1}^{n} \dfrac{(j-14)(j-5)}{j-6} \right) x^n$

18. $y_1 = x^3 \sum\limits_{n=0}^{\infty} \dfrac{(-1)^n}{n!} \left(\prod\limits_{j=1}^{n} \dfrac{(j+3)^2}{j+5} \right) x^n$; $\quad y_2 = x^{-2} \left(1 + \dfrac{1}{4} x \right)$

19. $y_1 = x^6 \sum\limits_{n=0}^{4} (-1)^n 2^n \left(\prod\limits_{j=1}^{n} \dfrac{j-5}{j+5} \right) x^n$; $\quad y_2 = x \sum\limits_{n=0}^{9} \dfrac{(-1)^n 2^n \, \Pi_{j=1}^{n} (j-10)}{n!} x^n$

20. $y_1 = x^6 \left(1 + \dfrac{2}{3} x + \dfrac{1}{7} x^2 \right)$; $\quad y_2 = x \left(1 + \dfrac{21}{4} x + \dfrac{21}{2} x^2 + \dfrac{35}{4} x^3 \right)$

21. $y_1 = x^{7/2} \sum\limits_{n=0}^{\infty} (-1)^n (n+1) x^n$; $\quad y_2 = x^{-7/2} \sum\limits_{n=0}^{5} (-1)^n \left(\prod\limits_{j=1}^{n} \dfrac{j-6}{j-10} \right) x^n$

22. $y_1 = x^{10} \sum\limits_{n=0}^{\infty} (-1)^n 2^n (n+1)(n+2)(n+3) x^n$; $\quad y_2 = \sum\limits_{n=0}^{6} (-1)^n 2^n \left(\prod\limits_{j=1}^{n} \dfrac{j-7}{j-10} \right) x^n$

23. $y_1 = x^6 \sum\limits_{m=0}^{\infty} \dfrac{(-1)^m \, \Pi_{j=1}^{m} (2j+5)}{2^m m!} x^{2m}$;

$y_2 = x^2 \left(1 + \dfrac{3}{2} x^2 \right) - \dfrac{15}{2} y_1 \ln x + \dfrac{75}{2} x^6 \sum\limits_{m=1}^{\infty} \dfrac{(-1)^m \, \Pi_{j=1}^{m} (2j+5)}{2^{m+1} m!} \left(\sum\limits_{j=1}^{m} \dfrac{1}{j(2j+5)} \right) x^{2m}$

24. $y_1 = x^6 \sum\limits_{m=0}^{\infty} \dfrac{(-1)^m}{2^m m!} x^{2m} = x^6 e^{-x^2/2}$;

$y_2 = x^2 \left(1 + \dfrac{1}{2} x^2 \right) - \dfrac{1}{2} y_1 \ln x + \dfrac{x^6}{4} \sum\limits_{m=1}^{\infty} \dfrac{(-1)^m}{2^m m!} \left(\sum\limits_{j=1}^{m} \dfrac{1}{j} \right) x^{2m}$

25. $y_1 = 6 x^6 \sum\limits_{m=0}^{\infty} \dfrac{(-1)^m}{4^m m! (m+3)!} x^{2m}$;

$y_2 = 1 + \dfrac{1}{8} x^2 + \dfrac{1}{64} x^4 - \dfrac{1}{384} \left(y_1 \ln x - 3 x^6 \sum\limits_{m=1}^{\infty} \dfrac{(-1)^m}{4^m m! (m+3)!} \left(\sum\limits_{j=1}^{m} \dfrac{2j+3}{j(j+3)} \right) x^{2m} \right)$

26. $y_1 = \dfrac{x}{2} \sum\limits_{m=0}^{\infty} \dfrac{(-1)^m (m+2)}{m!} x^{2m}$;

$y_2 = x^{-1} - 4 y_1 \ln x + x \sum\limits_{m=1}^{\infty} \dfrac{(-1)^m (m+2)}{m!} \left(\sum\limits_{j=1}^{m} \dfrac{j^2 + 4j + 2}{j(j+1)(j+2)} \right) x^{2m}$

27. $y_1 = 2 x^3 \sum\limits_{m=0}^{\infty} \dfrac{(-1)^m}{4^m m! (m+2)!} x^{2m}$;

$y_2 = x^{-1} \left(1 + \dfrac{1}{4} x^2 \right) - \dfrac{1}{16} \left(y_1 \ln x - 2 x^3 \sum\limits_{m=1}^{\infty} \dfrac{(-1)^m}{4^m m! (m+2)!} \left(\sum\limits_{j=1}^{m} \dfrac{j+1}{j(j+2)} \right) x^{2m} \right)$

28. $y_1 = x^{-1/2} \sum\limits_{m=0}^{\infty} \dfrac{(-1)^m \, \Pi_{j=1}^{m} (2j-1)}{8^m m! (m+1)!} x^{2m}$;

$y_2 = x^{-5/2} + \dfrac{1}{4} y_1 \ln x - x^{-1/2} \sum\limits_{m=1}^{\infty} \dfrac{(-1)^m \, \Pi_{j=1}^{m} (2j-1)}{8^{m+1} m! (m+1)!} \left(\sum\limits_{j=1}^{m} \dfrac{2j^2 - 2j - 1}{j(j+1)(2j-1)} \right) x^{2m}$

29. $y_1 = x \sum\limits_{m=0}^{\infty} \dfrac{(-1)^m}{2^m m!} x^{2m} = x e^{-x^2/2}$; $\quad y_2 = x^{-1} - y_1 \ln x + \dfrac{x}{2} \sum\limits_{m=1}^{\infty} \dfrac{(-1)^m}{2^m m!} \left(\sum\limits_{j=1}^{m} \dfrac{1}{j} \right) x^{2m}$

30. $y_1 = x^2 \sum\limits_{m=0}^{\infty} \dfrac{1}{m!} x^{2m} = x^2 e^{x^2}$; $\quad y_2 = x^{-2} (1 - x^2) - 2 y_1 \ln x + x^2 \sum\limits_{m=1}^{\infty} \dfrac{1}{m!} \left(\sum\limits_{j=1}^{m} \dfrac{1}{j} \right) x^{2m}$

31. $y_1 = 6 x^{5/2} \sum\limits_{m=0}^{\infty} \dfrac{(-1)^m}{16^m m! (m+3)!} x^{2m}$;

$$y_2 = x^{-7/2}\left(1 + \frac{1}{32}x^2 + \frac{1}{1024}x^4\right) - \frac{1}{24576}\left(y_1 \ln x - 3x^{5/2} \sum_{m=1}^{\infty} \frac{(-1)^m}{16^m m!(m+3)!}\left(\sum_{j=1}^{m} \frac{2j+3}{j(j+3)}\right)x^{2m}\right)$$

32. $y_1 = 2x^{13/3} \sum_{m=0}^{\infty} \frac{\prod_{j=1}^{m}(3j+1)}{9^m m!(m+2)!}x^{2m};$

$\quad\quad y_2 = x^{1/3}\left(1 + \frac{2}{9}x^2\right) + \frac{2}{81}\left(y_1 \ln x - x^{13/3}\sum_{m=0}^{\infty} \frac{\prod_{j=1}^{m}(3j+1)}{9^m m!(m+2)!}\left(\sum_{j=1}^{m} \frac{3j^2+2j+2}{j(j+2)(3j+1)}\right)x^{2m}\right)$

33. $y_1 = x^2; \quad y_2 = x^{-2}(1 + 2x^2) - 2\left(y_1 \ln x + x^2 \sum_{m=1}^{\infty} \frac{1}{m(m+2)!}x^{2m}\right)$

34. $y_1 = x^2\left(1 - \frac{1}{2}x^2\right); \quad y_2 = x^{-2}\left(1 + \frac{9}{2}x^2\right) - \frac{27}{2}\left(y_1 \ln x + \frac{7}{12}x^4 - x^2 \sum_{m=2}^{\infty} \frac{(\frac{3}{2})^m}{m(m-1)(m+2)!}x^{2m}\right)$

35. $y_1 = \sum_{m=0}^{\infty} (-1)^m (m+1)x^{2m}; \quad y_2 = x^{-4}$

36. $y_1 = x^{5/2} \sum_{m=0}^{\infty} \frac{(-1)^m}{(m+1)(m+2)(m+3)}x^{2m}; \quad y_2 = x^{-7/2}(1 + x^2)^2$

37. $y_1 = x^7 \sum_{m=0}^{\infty} (-1)^m (m+5)x^{2m}; \quad y_2 = x^{-1}(1 - 2x^2 + 3x^4 - 4x^6)$

38. $y_1 = x^3 \sum_{m=0}^{\infty} (-1)^m \frac{m+1}{2^m}\left(\prod_{j=1}^{m} \frac{2j+1}{j+5}\right)x^{2m}; \quad y_2 = x^{-7}\left(1 + \frac{21}{8}x^2 + \frac{35}{16}x^4 + \frac{35}{64}x^6\right)$

39. $y_1 = x^4 \sum_{m=0}^{\infty} (-1)^m \frac{\prod_{j=1}^{m}(4j+5)}{2^m(m+2)!}x^{2m}; \quad y_2 = 1 - \frac{1}{2}x^2$

40. $y_1 = x^{3/2} \sum_{m=0}^{\infty} \frac{(-1)^m \prod_{j=1}^{m}(2j-1)}{2^{m-1}(m+2)!}x^{2m}; \quad y_2 = x^{-5/2}\left(1 + \frac{3}{2}x^2\right)$

42. $y_1 = x^\nu \sum_{m=0}^{\infty} \frac{(-1)^m}{4^m m! \prod_{j=1}^{m}(j+\nu)}x^{2m};$

$\quad\quad y_2 = x^{-\nu} \sum_{m=0}^{\nu-1} \frac{(-1)^m}{4^m m! \prod_{j=1}^{m}(j-\nu)}x^{2m} - \frac{2}{4^\nu \nu!(\nu-1)!}\left(y_1 \ln x - \frac{x^\nu}{2} \sum_{m=1}^{\infty} \frac{(-1)^m}{4^m m! \prod_{j=1}^{m}(j+\nu)}\left(\sum_{j=1}^{m} \frac{2j+\nu}{j(j+\nu)}\right)x^{2m}\right)$

Section 8.1 Answers   (page 381)

1. (a) $\dfrac{1}{s^2}$ **(b)** $\dfrac{1}{(s+1)^2}$ **(c)** $\dfrac{b}{s^2 - b^2}$ **(d)** $\dfrac{-2s+5}{(s-1)(s-2)}$ **(e)** $\dfrac{2}{s^3}$

2. (a) $\dfrac{s^2+2}{[(s-1)^2+1][(s+1)^2+1]}$ **(b)** $\dfrac{2}{s(s^2+4)}$ **(c)** $\dfrac{s^2+8}{s(s^2+16)}$ **(d)** $\dfrac{s^2-2}{s(s^2-4)}$ **(e)** $\dfrac{4s}{(s^2-4)^2}$ **(f)** $\dfrac{1}{s^2+4}$

$\quad$ **(g)** $\dfrac{1}{\sqrt{2}}\dfrac{s+1}{s^2+1}$ **(h)** $\dfrac{5s}{(s^2+4)(s^2+9)}$ **(i)** $\dfrac{s^3+2s^2+4s+32}{(s^2+4)(s^2+16)}$

4. (a) $f(3-) = -1, \quad f(3) = f(3+) = 1$ **(b)** $f(1-) = 3, \quad f(1) = 4, \quad f(1+) = 1$

$\quad$ **(c)** $f\left(\dfrac{\pi}{2}-\right) = 1, \quad f\left(\dfrac{\pi}{2}\right) = f\left(\dfrac{\pi}{2}+\right) = 2, \quad f(\pi-) = 0, \quad f(\pi) = f(\pi+) = -1$

$\quad$ **(d)** $f(1-) = 1, \quad f(1) = 2, \quad f(1+) = 1, \quad f(2-) = 0, \quad f(2) = 3, \quad f(2+) = 6$

5. (a) $\dfrac{1 - e^{-(s+1)}}{s+1} + \dfrac{e^{-(s+2)}}{s+2}$ **(b)** $\dfrac{1}{s} + e^{-4s}\left(\dfrac{1}{s^2} + \dfrac{3}{s}\right)$ **(c)** $\dfrac{1 - e^{-s}}{s^2}$ **(d)** $\dfrac{1 - e^{-(s-1)}}{(s-1)^2}$

7. $\mathcal{L}(te^{\lambda t} \cos \omega t) = \dfrac{(s-\lambda)^2 - \omega^2}{((s-\lambda)^2 + \omega^2)^2}, \quad \mathcal{L}(te^{\lambda t} \sin \omega t) = \dfrac{2\omega(s-\lambda)}{((s-\lambda)^2 + \omega^2)^2}$

15. (a) $\tan^{-1}\dfrac{\omega}{s}$, $s>0$ **(b)** $\dfrac{1}{2}\ln\dfrac{s^2}{s^2+\omega^2}$, $s>0$ **(c)** $\ln\dfrac{s-b}{s-a}$, $s>\max(a,b)$ **(d)** $\dfrac{1}{2}\ln\dfrac{s^2}{s^2-1}$, $s>1$

(e) $\dfrac{1}{4}\ln\dfrac{s^2}{s^2-4}$, $s>2$

18. (a) $\dfrac{1}{s^2}\tanh\dfrac{s}{2}$ **(b)** $\dfrac{1}{s}\tanh\dfrac{s}{4}$ **(c)** $\dfrac{1}{s^2+1}\coth\dfrac{\pi s}{2}$ **(d)** $\dfrac{1}{(s^2+1)(1-e^{-\pi s})}$

Section 8.2 Answers   (page 391)

1. (a) $\dfrac{1}{2}t^3e^{7t}$ **(b)** $2e^{2t}\cos 3t$ **(c)** $\dfrac{1}{4}e^{-2t}\sin 4t$ **(d)** $\dfrac{2}{3}\sin 3t$ **(e)** $t\cos t$ **(f)** $\dfrac{1}{2}e^{2t}\sinh 2t$ **(g)** $\dfrac{2}{3}e^{2t}t\sin 9t$

(h) $\dfrac{2}{3}e^{3t}\sinh 3t$ **(i)** $e^{2t}t\cos t$

2. (a) $t^2e^{7t}+\dfrac{17}{6}t^3e^{7t}$ **(b)** $e^{2t}\left(\dfrac{1}{6}t^3+\dfrac{1}{6}t^4+\dfrac{1}{40}t^5\right)$ **(c)** $e^{-3t}\left(\cos 3t+\dfrac{2}{3}\sin 3t\right)$ **(d)** $2\cos 3t+\dfrac{1}{3}\sin 3t$

(e) $e^{-t}(1-t)$ **(f)** $\cosh 3t+\dfrac{1}{3}\sinh 3t$ **(g)** $e^{-t}\left(1-t-t^2-\dfrac{1}{6}t^3\right)$ **(h)** $e^t\left(2\cos 2t+\dfrac{5}{2}\sin 2t\right)$ **(i)** $1-\cos t$

(j) $3\cosh t+4\sinh t$ **(k)** $3e^t+4\cos 3t+\dfrac{1}{3}\sin 3t$ **(l)** $3te^{-2t}-2\cos 2t-3\sin 2t$

3. (a) $\dfrac{1}{4}e^{2t}-\dfrac{1}{4}e^{-2t}-e^{-t}$ **(b)** $\dfrac{1}{5}e^{-4t}-\dfrac{41}{5}e^t+5e^{3t}$ **(c)** $-\dfrac{1}{2}e^{2t}-\dfrac{13}{10}e^{-2t}-\dfrac{1}{5}e^{3t}$ **(d)** $-\dfrac{2}{5}e^{-4t}-\dfrac{3}{5}e^t$

(e) $\dfrac{3}{20}e^{2t}-\dfrac{37}{12}e^{-2t}+\dfrac{1}{3}e^t+\dfrac{8}{5}e^{-3t}$ **(f)** $\dfrac{39}{10}e^t+\dfrac{3}{14}e^{3t}+\dfrac{23}{105}e^{-4t}-\dfrac{7}{3}e^{2t}$

4. (a) $\dfrac{4}{5}e^{-2t}-\dfrac{1}{2}e^{-t}-\dfrac{3}{10}\cos t+\dfrac{11}{10}\sin t$ **(b)** $\dfrac{2}{5}\sin t+\dfrac{6}{5}\cos t+\dfrac{7}{5}e^{-t}\sin t-\dfrac{6}{5}e^{-t}\cos t$

(c) $\dfrac{8}{13}e^{2t}-\dfrac{8}{13}e^{-t}\cos 2t+\dfrac{15}{26}e^{-t}\sin 2t$ **(d)** $\dfrac{1}{2}te^t+\dfrac{3}{8}e^t+e^{-2t}-\dfrac{11}{8}e^{-3t}$ **(e)** $\dfrac{2}{3}te^t+\dfrac{1}{9}e^t+te^{-2t}-\dfrac{1}{9}e^{-2t}$

(f) $-e^t+\dfrac{5}{2}te^t+\cos t-\dfrac{3}{2}\sin t$

5. (a) $\dfrac{3}{5}\cos 2t+\dfrac{1}{5}\sin 2t-\dfrac{3}{5}\cos 3t-\dfrac{2}{15}\sin 3t$ **(b)** $-\dfrac{4}{15}\cos t+\dfrac{1}{15}\sin t+\dfrac{4}{15}\cos 4t-\dfrac{1}{60}\sin 4t$

(c) $\dfrac{5}{3}\cos t+\sin t-\dfrac{5}{3}\cos 2t-\dfrac{1}{2}\sin 2t$ **(d)** $-\dfrac{1}{3}\cos\dfrac{t}{2}+\dfrac{2}{3}\sin\dfrac{t}{2}+\dfrac{1}{3}\cos t-\dfrac{1}{3}\sin t$

(e) $\dfrac{1}{15}\cos\dfrac{t}{4}-\dfrac{8}{15}\sin\dfrac{t}{4}-\dfrac{1}{15}\cos 4t+\dfrac{1}{30}\sin 4t$ **(f)** $\dfrac{2}{5}\cos\dfrac{t}{3}-\dfrac{3}{5}\sin\dfrac{t}{3}-\dfrac{2}{5}\cos\dfrac{t}{2}+\dfrac{2}{5}\sin\dfrac{t}{2}$

6. (a) $e^t(\cos 2t+\sin 2t)-e^{-t}\left(\cos 3t+\dfrac{4}{3}\sin 3t\right)$ **(b)** $e^{3t}\left(-\cos 2t+\dfrac{3}{2}\sin 2t\right)+e^{-t}\left(\cos 2t+\dfrac{1}{2}\sin 2t\right)$

(c) $e^{-2t}\left(\dfrac{1}{8}\cos t+\dfrac{1}{4}\sin t\right)-e^{2t}\left(\dfrac{1}{8}\cos 3t-\dfrac{1}{12}\sin 3t\right)$ **(d)** $e^{2t}\left(\cos t+\dfrac{1}{2}\sin t\right)-e^{3t}\left(\cos 2t-\dfrac{1}{4}\sin 2t\right)$

(e) $e^t\left(\dfrac{1}{5}\cos t+\dfrac{2}{5}\sin t\right)-e^{-t}\left(\dfrac{1}{5}\cos 2t+\dfrac{2}{5}\sin 2t\right)$ **(f)** $e^{t/2}\left(-\cos t+\dfrac{9}{8}\sin t\right)+e^{-t/2}\left(\cos t-\dfrac{1}{8}\sin t\right)$

7. (a) $1-\cos t$ **(b)** $\dfrac{e^t}{16}(1-\cos 4t)$ **(c)** $\dfrac{4}{9}e^{2t}+\dfrac{5}{9}e^{-t}\sin 3t-\dfrac{4}{9}e^{-t}\cos 3t$ **(d)** $3e^{t/2}-\dfrac{7}{2}e^t\sin 2t-3e^t\cos 2t$

(e) $\dfrac{1}{4}e^{3t}-\dfrac{1}{4}e^{-t}\cos 2t$ **(f)** $\dfrac{1}{9}e^{2t}-\dfrac{1}{9}e^{-t}\cos 3t+\dfrac{5}{9}e^{-t}\sin 3t$

8. (a) $-\dfrac{3}{10}\sin t + \dfrac{2}{5}\cos t - \dfrac{3}{4}e^t + \dfrac{7}{20}e^{3t}$ **(b)** $-\dfrac{3}{5}e^{-t}\sin t + \dfrac{1}{5}e^{-t}\cos t - \dfrac{1}{2}e^{-t} + \dfrac{3}{10}e^t$

(c) $-\dfrac{1}{10}e^t\sin t - \dfrac{7}{10}e^t\cos t + \dfrac{1}{5}e^{-t} + \dfrac{1}{2}e^{2t}$ **(d)** $-\dfrac{1}{2}e^t + \dfrac{7}{10}e^{-t} - \dfrac{1}{5}\cos 2t + \dfrac{3}{5}\sin 2t$

(e) $\dfrac{3}{10} + \dfrac{1}{10}e^{2t} + \dfrac{1}{10}e^t\sin 2t - \dfrac{2}{5}e^t\cos 2t$ **(f)** $-\dfrac{4}{9}e^{2t}\cos 3t + \dfrac{1}{3}e^{2t}\sin 3t - \dfrac{5}{9}e^{2t} + e^t$

9. $\dfrac{1}{a}e^{bt/a}f\left(\dfrac{t}{a}\right)$

Section 8.3 Answers   (page 398)

1. $y = \dfrac{1}{6}e^t - \dfrac{9}{2}e^{-t} + \dfrac{16}{3}e^{-2t}$ **2.** $y = -\dfrac{1}{3} + \dfrac{8}{15}e^{3t} + \dfrac{4}{5}e^{-2t}$ **3.** $y = -\dfrac{23}{15}e^{-2t} + \dfrac{1}{3}e^t + \dfrac{1}{5}e^{3t}$

4. $y = -\dfrac{1}{4}e^{2t} + \dfrac{17}{20}e^{-2t} + \dfrac{2}{5}e^{3t}$ **5.** $y = \dfrac{11}{15}e^{-2t} + \dfrac{1}{6}e^t + \dfrac{1}{10}e^{3t}$ **6.** $y = e^t + 2e^{-2t} - 2e^{-t}$

7. $y = \dfrac{5}{3}\sin t - \dfrac{1}{3}\sin 2t$ **8.** $y = 4e^t - 4e^{2t} + e^{3t}$ **9.** $y = -\dfrac{7}{2}e^{2t} + \dfrac{13}{3}e^t + \dfrac{1}{6}e^{4t}$ **10.** $y = \dfrac{5}{2}e^t - 4e^{2t} + \dfrac{1}{2}e^{3t}$

11. $y = \dfrac{1}{3}e^t - 2e^{-t} + \dfrac{5}{3}e^{-2t}$ **12.** $y = 2 - e^{-2t} + e^t$ **13.** $y = 1 - \cos 2t + \dfrac{1}{2}\sin 2t$

14. $y = -\dfrac{1}{3} + \dfrac{8}{15}e^{3t} + \dfrac{4}{5}e^{-2t}$ **15.** $y = \dfrac{1}{6}e^t - \dfrac{2}{3}e^{-2t} + \dfrac{1}{2}e^{-t}$ **16.** $y = -1 + e^t + e^{-t}$

17. $y = \cos 2t - \sin 2t + \sin t$ **18.** $y = \dfrac{7}{3} - \dfrac{7}{2}e^{-t} + \dfrac{1}{6}e^{3t}$ **19.** $y = 1 + \cos t$ **20.** $y = t + \sin t$

21. $y = t - 6\sin t + \cos t + \sin 2t$ **22.** $y = e^{-t} + 4e^{-2t} - 4e^{-3t}$ **23.** $y = -3\cos t - 2\sin t + e^{-t}(2 + 5t)$

24. $y = -\sin t - 2\cos t + 3e^{3t} + e^{-t}$ **25.** $y = (3t + 4)\sin t - (2t + 6)\cos t$

26. $y = -(2t + 2)\cos 2t + \sin 2t + 3\cos t$ **27.** $y = e^t(\cos t - 3\sin t) + e^{3t}$

28. $y = -1 + t + e^{-t}(3\cos t - 5\sin t)$ **29.** $y = 4\cos t - 3\sin t - e^t(3\cos t - 8\sin t)$

30. $y = e^{-t} - 2e^t + e^{-2t}\left(\cos 3t - \dfrac{11}{3}\sin 3t\right)$ **31.** $y = e^{-t}(\sin t - \cos t) + e^{-2t}(\cos t + 4\sin t)$

32. $y = \dfrac{1}{5}e^{2t} - \dfrac{4}{3}e^t + \dfrac{32}{15}e^{-t/2}$ **33.** $y = \dfrac{1}{7}e^{2t} - \dfrac{2}{5}e^{t/2} + \dfrac{9}{35}e^{-t/3}$ **34.** $y = e^{-t/2}(5\cos(t/2) - \sin(t/2)) + 2t - 4$

35. $y = \dfrac{1}{17}(12\cos t + 20\sin t - 3e^{t/2}(4\cos t + \sin t))$ **36.** $y = \dfrac{e^{-t/2}}{10}(5t + 26) - \dfrac{1}{5}(3\cos t + \sin t)$

37. $y = \dfrac{1}{100}(3e^{3t} - e^{t/3}(3 + 310t))$

Section 8.4 Answers   (page 406)

1. $1 + u(t - 4)(t - 1)$; $\dfrac{1}{s} + e^{-4s}\left(\dfrac{1}{s^2} + \dfrac{3}{s}\right)$ **2.** $t + u(t - 1)(1 - t)$; $\dfrac{1 - e^{-s}}{s^2}$

3. $2t - 1 - u(t - 2)(t - 1)$; $\left(\dfrac{2}{s^2} - \dfrac{1}{s}\right) - e^{-2s}\left(\dfrac{1}{s^2} + \dfrac{1}{s}\right)$ **4.** $1 + u(t - 1)(t + 1)$; $\dfrac{1}{s} + e^{-s}\left(\dfrac{1}{s^2} + \dfrac{2}{s}\right)$

5. $t - 1 + u(t - 2)(5 - t)$; $\dfrac{1}{s^2} - \dfrac{1}{s} - e^{-2s}\left(\dfrac{1}{s^2} - \dfrac{3}{s}\right)$ **6.** $t^2(1 - u(t - 1))$; $\dfrac{2}{s^3} - e^{-s}\left(\dfrac{2}{s^3} + \dfrac{2}{s^2} + \dfrac{1}{s}\right)$

7. $u(t - 2)(t^2 + 3t)$; $e^{-2s}\left(\dfrac{2}{s^3} + \dfrac{7}{s^2} + \dfrac{10}{s}\right)$ **8.** $t^2 + 2 + u(t - 1)(t - t^2 - 2)$; $\dfrac{2}{s^3} + \dfrac{2}{s} - e^{-s}\left(\dfrac{2}{s^3} + \dfrac{1}{s^2} + \dfrac{2}{s}\right)$

9. $te^t + u(t - 1)(e^t - te^t)$; $\dfrac{1 - e^{-(s-1)}}{(s - 1)^2}$

10. $e^{-t} + u(t - 1)(e^{-2t} - e^{-t})$; $\dfrac{1 - e^{-(s+1)}}{s + 1} + \dfrac{e^{-(s+2)}}{s + 2}$

11. $-t + 2u(t - 2)(t - 2) - u(t - 3)(t - 5)$; $-\dfrac{1}{s^2} + \dfrac{2e^{-2s}}{s^2} + e^{-3s}\left(\dfrac{2}{s} - \dfrac{1}{s^2}\right)$

12. $[u(t - 1) - u(t - 2)]t$; $e^{-s}\left(\dfrac{1}{s^2} + \dfrac{1}{s}\right) - e^{-2s}\left(\dfrac{1}{s^2} + \dfrac{2}{s}\right)$

13. $t + u(t - 1)(t^2 - t) - u(t - 2)t^2$; $\dfrac{1}{s^2} + e^{-s}\left(\dfrac{2}{s^3} + \dfrac{1}{s^2}\right) - e^{-2s}\left(\dfrac{2}{s^3} + \dfrac{4}{s^2} + \dfrac{4}{s}\right)$

14. $t + u(t - 1)(2 - 2t) + u(t - 2)(4 + t)$; $\dfrac{1}{s^2} - 2\dfrac{e^{-s}}{s^2} + e^{-2s}\left(\dfrac{1}{s^2} + \dfrac{6}{s}\right)$

15. $\sin t + u(t - \pi/2) \sin t + u(t - \pi)(\cos t - 2 \sin t)$; $\dfrac{1 + e^{-\pi s/2}s - e^{-\pi s}(s - 2)}{s^2 + 1}$

16. $2 - 2u(t - 1)t + u(t - 3)(5t - 2)$; $\dfrac{2}{s} - e^{-s}\left(\dfrac{2}{s^2} + \dfrac{2}{s}\right) + e^{-3s}\left(\dfrac{5}{s^2} + \dfrac{13}{s}\right)$

17. $3 + u(t - 2)(3t - 1) + u(t - 4)(t - 2)$; $\dfrac{3}{s} + e^{-2s}\left(\dfrac{3}{s^2} + \dfrac{5}{s}\right) + e^{-4s}\left(\dfrac{1}{s^2} + \dfrac{2}{s}\right)$

18. $(t + 1)^2 + u(t - 1)(2t + 3)$; $\dfrac{2}{s^3} + \dfrac{2}{s^2} + \dfrac{1}{s} + e^{-s}\left(\dfrac{2}{s^2} + \dfrac{5}{s}\right)$

19. $u(t - 2)e^{2(t-2)} = \begin{cases} 0, & 0 \le t < 2, \\ e^{2(t-2)}, & t \ge 2. \end{cases}$

20. $u(t - 1)(1 - e^{-(t-1)}) = \begin{cases} 0, & 0 \le t < 1, \\ 1 - e^{-(t-1)}, & t \ge 1. \end{cases}$

21. $u(t - 1)\dfrac{(t - 1)^2}{2} + u(t - 2)(t - 2) = \begin{cases} 0, & 0 \le t < 1, \\ \dfrac{(t - 1)^2}{2}, & 1 \le t < 2, \\ \dfrac{t^2 - 3}{2} & t \ge 2. \end{cases}$

22. $2 + t + u(t - 1)(4 - t) + u(t - 3)(t - 2) = \begin{cases} 2 + t, & 0 \le t < 1, \\ 6, & 1 \le t < 3, \\ t + 4, & t \ge 3. \end{cases}$

23. $5 - t + u(t - 3)(7t - 15) + \dfrac{3}{2}u(t - 6)(t - 6)^2 = \begin{cases} 5 - t, & 0 \le t < 3, \\ 6t - 10, & 3 \le t < 6, \\ 44 - 12t + \frac{3}{2}t^2, & t \ge 6. \end{cases}$

24. $u(t - \pi)e^{-2(t-\pi)}(2 \cos t - 5 \sin t) = \begin{cases} 0, & 0 \le t < \pi, \\ e^{-2(t-\pi)}(2 \cos t - 5 \sin t), & t \ge \pi. \end{cases}$

25. $1 - \cos t + u(t - \pi/2)(3 \sin t + \cos t) = \begin{cases} 1 - \cos t, & 0 \le t < \dfrac{\pi}{2}, \\ 1 + 3 \sin t, & t \ge \dfrac{\pi}{2}. \end{cases}$

26. $u(t - 2)(4e^{-(t-2)} - 4e^{2(t-2)} + 2e^{(t-2)}) = \begin{cases} 0, & 0 \le t < 2, \\ 4e^{-(t-2)} - 4e^{2(t-2)} + 2e^{(t-2)}, & t \ge 2. \end{cases}$

27. $1 + t + u(t - 1)(2t + 1) + u(t - 3)(3t - 5) = \begin{cases} t + 1, & 0 \le t < 1, \\ 3t + 2, & 1 \le t < 3, \\ 6t - 3, & t \ge 3, \end{cases}$

28. $1 - t^2 + u(t - 2)\left(-\dfrac{t^2}{2} + 2t + 1\right) + u(t - 4)(t - 4) = \begin{cases} 1 - t^2, & 0 \le t < 2, \\ -\dfrac{3t^2}{2} + 2t + 2, & 2 \le t < 4, \\ -\dfrac{3t^2}{2} + 3t - 2, & t \ge 4. \end{cases}$ **29.** $\dfrac{e^{-\tau s}}{s}$

30. For each t only finitely many terms are nonzero.   **33.** $1 + \displaystyle\sum_{m=1}^{\infty} u(t - m);\quad \dfrac{1}{s(1 - e^{-s})}$

34. $1 + 2\displaystyle\sum_{m=1}^{\infty} (-1)^m u(t - m);\quad \dfrac{1 - e^{-s}}{s(1 + e^{-s})}$ **35.** $1 + \displaystyle\sum_{m=1}^{\infty} (2m + 1)u(t - m);\quad \dfrac{1 + e^{-s}}{s(1 - e^{-s})^2}$

36. $\displaystyle\sum_{m=1}^{\infty} (-1)^m(2m - 1)u(t - m);\quad \dfrac{1 - e^{-s}}{s(1 + e^{s})^2}$

Section 8.5 Answers   (page 415)

1. $y = 3(1 - \cos t) - 3u(t - \pi)(1 + \cos t)$

2. $y = 3 - 2\cos t + 2u(t - 4)(t - 4 - \sin(t - 4))$ **3.** $y = -\dfrac{15}{2} + \dfrac{3}{2}e^{2t} - 2t + \dfrac{u(t - 1)}{2}\left(e^{2(t-1)} - 2t + 1\right)$

4. $y = \dfrac{1}{2}e^t + \dfrac{13}{6}e^{-t} + \dfrac{1}{3}e^{2t} + u(t - 2)\left(-1 + \dfrac{1}{2}e^{t-2} + \dfrac{1}{2}e^{-(t-2)} + \dfrac{1}{2}e^{t+2} - \dfrac{1}{6}e^{-(t-6)} - \dfrac{1}{3}e^{2t}\right)$

5. $y = -7e^t + 4e^{2t} + u(t - 1)\left(\dfrac{1}{2} - e^{t-1} + \dfrac{1}{2}e^{2(t-1)}\right) - 2u(t - 2)\left(\dfrac{1}{2} - e^{t-2} + \dfrac{1}{2}e^{2(t-2)}\right)$

6. $y = \dfrac{1}{3}\sin 2t - 3\cos 2t + \dfrac{1}{3}\sin t - 2u(t - \pi)\left(\dfrac{1}{3}\sin t + \dfrac{1}{6}\sin 2t\right) + u(t - 2\pi)\left(\dfrac{1}{3}\sin t - \dfrac{1}{6}\sin 2t\right)$

7. $y = \dfrac{1}{4} - \dfrac{31}{12}e^{4t} + \dfrac{16}{3}e^t + u(t - 1)\left(\dfrac{2}{3}e^{t-1} - \dfrac{1}{6}e^{4(t-1)} - \dfrac{1}{2}\right) + u(t - 2)\left(\dfrac{1}{4} + \dfrac{1}{12}e^{4(t-2)} - \dfrac{1}{3}e^{t-2}\right)$

8. $y = \dfrac{1}{8}(\cos t - \cos 3t) - \dfrac{1}{8}u\left(t - \dfrac{3\pi}{2}\right)\left(\sin t - \cos t + \sin 3t - \dfrac{1}{3}\cos 3t\right)$

9. $y = \dfrac{t}{4} - \dfrac{1}{8}\sin 2t + \dfrac{1}{8}u\left(t - \dfrac{\pi}{2}\right)(\pi\cos 2t - \sin 2t + 2\pi - 2t)$

10. $y = t - \sin t - 2u(t - \pi)(t + \sin t + \pi\cos t)$ **11.** $y = u(t - 2)\left(t - \dfrac{1}{2} + \dfrac{e^{2(t-2)}}{2} - 2e^{t-2}\right)$

12. $y = t + \sin t + \cos t - u(t - 2\pi)(3t - 3\sin t - 6\pi\cos t)$

13. $y = \dfrac{1}{2} + \dfrac{1}{2}e^{-2t} - e^{-t} + u(t - 2)(2e^{-(t-2)} - e^{-2(t-2)} - 1)$

14. $y = -\dfrac{1}{3} - \dfrac{1}{6}e^{3t} + \dfrac{1}{2}e^t + u(t - 1)\left(\dfrac{2}{3} + \dfrac{1}{3}e^{3(t-1)} - e^{t-1}\right)$ **15.** $y = \dfrac{1}{4}(e^t + e^{-t}(11 + 6t)) + u(t - 1)(te^{-(t-1)} - 1)$

16. $y = e^t - e^{-t} - 2te^{-t} - u(t - 1)(e^t - e^{-(t-2)} - 2(t - 1)e^{-(t-2)})$

17. $y = te^{-t} + e^{-2t} + u(t - 1)(e^{-t}(2 - t) - e^{-(2t-1)})$ **18.** $y = \dfrac{t^2 e^{2t}}{2} - te^{2t} - u(t - 2)(t - 2)^2 e^{2t}$

19. $y = \dfrac{t^4}{12} + 1 - \dfrac{1}{12}u(t - 1)(t^4 + 2t^3 - 10t + 7) + \dfrac{1}{6}u(t - 2)(2t^3 + 3t^2 - 36t + 44)$

20. $y = \dfrac{1}{2}e^{-t}(3\cos t + \sin t) + \dfrac{1}{2} - u(t - 2\pi)\left(e^{-(t-2\pi)}\left((\pi - 1)\cos t + \dfrac{2\pi - 1}{2}\sin t\right) + 1 - \dfrac{t}{2}\right)$
$\qquad - \dfrac{1}{2}u(t - 3\pi)(e^{-(t-3\pi)}(3\pi\cos t + (3\pi + 1)\sin t) + t)$

21. $y = \dfrac{t^2}{2} + \sum_{m=1}^{\infty} u(t - m)\dfrac{(t - m)^2}{2}$

22. (a) $y = \begin{cases} 2m + 1 - \cos t, & 2m\pi \le t < (2m + 1)\pi \quad (m = 0, 1, \ldots) \\ 2m, & (2m - 1)\pi \le t < 2m\pi \quad (m = 1, 2, \ldots) \end{cases}$

(b) $y = (m + 1)(t - \sin t - m\pi \cos t), \quad 2m\pi \le t < (2m + 2)\pi \quad (m = 0, 1, \ldots)$

(c) $y = (-1)^m - (2m + 1) \cos t, \quad m\pi \le t < (m + 1)\pi \quad (m = 0, 1, \ldots)$

(d) $y = \dfrac{e^{m+1} - 1}{2(e - 1)}(e^{t-m} + e^{-t}) - m - 1, \quad m \le t < m + 1 \quad (m = 0, 1, \ldots)$

(e) $y = \left(m + 1 - \left(\dfrac{e^{2(m+1)\pi} - 1}{e^{2\pi} - 1}\right)e^{-t}\right) \sin t, \quad 2m\pi \le t < 2(m + 1)\pi \quad (m = 0, 1, \ldots)$

(f) $y = \dfrac{m + 1}{2} - e^{t-m}\dfrac{e^{m+1} - 1}{e - 1} + \dfrac{1}{2}e^{2(t-m)}\dfrac{e^{2m+2} - 1}{e^2 - 1}, \quad m \le t < m + 1 \quad (m = 0, 1, \ldots)$

Section 8.6 Answers   (page 426)

1. (a) $\dfrac{1}{2}\displaystyle\int_0^t \tau \sin 2(t - \tau)\, d\tau$ **(b)** $\displaystyle\int_0^t e^{-2\tau} \cos 3(t - \tau)\, d\tau$

(c) $\dfrac{1}{2}\displaystyle\int_0^t \sin 2\tau \cos 3(t - \tau)\, d\tau$ or $\dfrac{1}{3}\displaystyle\int_0^t \sin 3\tau \cos 2(t - \tau)\, d\tau$ **(d)** $\displaystyle\int_0^t \cos \tau \sin(t - \tau)\, d\tau$ **(e)** $\displaystyle\int_0^t e^{a\tau}\, d\tau$

(f) $e^{-t}\displaystyle\int_0^t \sin(t - \tau)\, d\tau$ **(g)** $e^{-2t}\displaystyle\int_0^t \tau e^{\tau} \sin(t - \tau)\, d\tau$ **(h)** $\dfrac{e^{-2t}}{2}\displaystyle\int_0^t \tau^2(t - \tau)e^{3\tau}\, d\tau$ **(i)** $\displaystyle\int_0^t (t - \tau)e^{\tau} \cos \tau\, d\tau$

(j) $\displaystyle\int_0^t e^{-3\tau} \cos \tau \cos 2(t - \tau)\, d\tau$ **(k)** $\dfrac{1}{4!\, 5!}\displaystyle\int_0^t \tau^4(t - \tau)^5 e^{3\tau}\, d\tau$ **(l)** $\dfrac{1}{4}\displaystyle\int_0^t \tau^2 e^{\tau} \sin 2(t - \tau)\, d\tau$

(m) $\dfrac{1}{2}\displaystyle\int_0^t \tau(t - \tau)^2 e^{2(t-\tau)}\, d\tau$ **(n)** $\dfrac{1}{5!\, 6!}\displaystyle\int_0^t (t - \tau)^5 e^{2(t-\tau)}\tau^6\, d\tau$

2. (a) $\dfrac{as}{(s^2 + a^2)(s^2 + b^2)}$ **(b)** $\dfrac{a}{(s - 1)(s^2 + a^2)}$ **(c)** $\dfrac{as}{(s^2 - a^2)^2}$ **(d)** $\dfrac{2\omega s(s^2 - \omega^2)}{(s^2 + \omega^2)^4}$ **(e)** $\dfrac{(s - 1)\omega}{((s - 1)^2 + \omega^2)^2}$

(f) $\dfrac{2}{(s - 2)^3(s - 1)^2}$ **(g)** $\dfrac{s + 1}{(s + 2)^2[(s + 1)^2 + \omega^2]}$ **(h)** $\dfrac{1}{(s - 3)((s - 1)^2 - 1)}$ **(i)** $\dfrac{2}{(s - 2)^2(s^2 + 4)}$

(j) $\dfrac{6}{s^4(s - 1)}$ **(k)** $\dfrac{3 \cdot 6!}{s^7[(s + 1)^2 + 9]}$ **(l)** $\dfrac{12}{s^7}$ **(m)** $\dfrac{2 \cdot 7!}{s^8[(s + 1)^2 + 4]}$ **(n)** $\dfrac{48}{s^5(s^2 + 4)}$

3. (a) $y = \dfrac{2}{\sqrt{5}}\displaystyle\int_0^t f(t - \tau)e^{-3\tau/2} \sinh \dfrac{\sqrt{5}\tau}{2}\, d\tau$ **(b)** $y = \dfrac{1}{2}\displaystyle\int_0^t f(t - \tau) \sin 2\tau\, d\tau$ **(c)** $y = \displaystyle\int_0^t \tau e^{-\tau} f(t - \tau)\, d\tau$

(d) $y(t) = -\dfrac{1}{k} \sin kt + \cos kt + \dfrac{1}{k}\displaystyle\int_0^t f(t - \tau) \sin k\tau\, d\tau$ **(e)** $y = -2te^{-3t} + \displaystyle\int_0^t \tau e^{-3\tau} f(t - \tau)\, d\tau$

(f) $y = \dfrac{3}{2} \sinh 2t + \dfrac{1}{2}\displaystyle\int_0^t f(t - \tau) \sinh 2\tau\, d\tau$ **(g)** $y = e^{3t} + \displaystyle\int_0^t (e^{3\tau} - e^{2\tau})f(t - \tau)\, d\tau$

(h) $y = \dfrac{k_1}{\omega} \sin \omega t + k_0 \cos \omega t + \dfrac{1}{\omega}\displaystyle\int_0^t f(t - \tau) \sin \omega\tau\, d\tau$

4. (a) $y = \sin t$ **(b)** $y = te^{-t}$ **(c)** $y = 1 + 2te^t$ **(d)** $y = t + \dfrac{t^2}{2}$ **(e)** $y = 4 + \dfrac{5}{2}t^2 + \dfrac{1}{24}t^4$ **(f)** $y = 1 - t$

5. (a) $\dfrac{7!\, 8!}{16!}t^{16}$ **(b)** $\dfrac{13!\, 7!}{21!}t^{21}$ **(c)** $\dfrac{6!\, 7!}{14!}t^{14}$ **(d)** $\dfrac{1}{2}(e^{-t} + \sin t - \cos t)$ **(e)** $\dfrac{1}{3}(\cos t - \cos 2t)$

Section 8.7 Answers   (page 436)

1. $y = \dfrac{1}{2}e^{2t} - 4e^{-t} + \dfrac{11}{2}e^{-2t} + 2u(t-1)(e^{-(t-1)} - e^{-2(t-1)})$ **2.** $y = 2e^{-2t} + 5e^{-t} + \dfrac{5}{3}u(t-1)(e^{(t-1)} - e^{-2(t-1)})$

3. $y = \dfrac{1}{6}e^{2t} - \dfrac{2}{3}e^{-t} - \dfrac{1}{2}e^{-2t} + \dfrac{5}{2}u(t-1)\sinh 2(t-1)$ **4.** $y = \dfrac{1}{8}(8\cos t - 5\sin t - \sin 3t) - 2u(t - \pi/2)\cos t$

5. $y = 1 - \cos 2t + \dfrac{1}{2}\sin 2t + \dfrac{1}{2}u(t - 3\pi)\sin 2t$ **6.** $y = 4e^t + 3e^{-t} - 8 + 2u(t-2)\sinh(t-2)$

7. $y = \dfrac{1}{2}e^t - \dfrac{7}{2}e^{-t} + 2 + 3u(t-6)(1 - e^{-(t-6)})$ **8.** $y = e^{2t} + 7\cos 2t - \sin 2t - \dfrac{1}{2}u(t - \pi/2)\sin 2t$

9. $y = \dfrac{1}{2}(1 + e^{-2t}) + u(t-1)(e^{-(t-1)} - e^{-2(t-1)})$ **10.** $y = \dfrac{1}{4}e^t + \dfrac{1}{4}e^{-t}(2t - 5) + 2u(t-2)(t-2)e^{-(t-2)}$

11. $y = \dfrac{1}{6}(2\sin t + 5\sin 2t) - \dfrac{1}{2}u(t - \pi/2)\sin 2t$

12. $y = e^{-t}(\sin t - \cos t) - e^{-(t-\pi)}u(t - \pi)\sin t - 3u(t - 2\pi)e^{-(t-2\pi)}\sin t$

13. $y = e^{-2t}\left(\cos 3t + \dfrac{4}{3}\sin 3t\right) - \dfrac{1}{3}u(t - \pi/6)e^{-2(t-\pi/6)}\cos 3t - \dfrac{2}{3}u(t - \pi/3)e^{-2(t-\pi/3)}\sin 3t$

14. $y = \dfrac{7}{10}e^{2t} - \dfrac{6}{5}e^{-t/2} - \dfrac{1}{2} + \dfrac{1}{5}u(t-2)(e^{2(t-2)} - e^{-(t-2)/2})$

15. $y = \dfrac{1}{17}(12\cos t + 20\sin t) + \dfrac{1}{34}e^{t/2}(10\cos t - 11\sin t) - u(t - \pi/2)e^{(2t-\pi)/4}\cos t + u(t - \pi)e^{(t-\pi)/2}\sin t$

16. $y = \dfrac{1}{3}(\cos t - \cos 2t - 3\sin t) - 2u(t - \pi/2)\cos t + 3u(t - \pi)\sin t$

17. $y = e^t - e^{-t}(1 + 2t) - 5u(t-1)\sinh(t-1) + 3u(t-2)\sinh(t-2)$

18. $y = \dfrac{1}{4}(e^t - e^{-t}(1 + 6t)) - u(t-1)(t-1)e^{-(t-1)} + 2u(t-2)(t-2)e^{-(t-2)}$

19. $y = \dfrac{5}{3}\sin t - \dfrac{1}{3}\sin 2t + \dfrac{1}{3}u(t - \pi)(\sin 2t + 2\sin t) + u(t - 2\pi)\sin t$

20. $y = \dfrac{3}{4}\cos 2t - \dfrac{1}{2}\sin 2t + \dfrac{1}{4} + \dfrac{1}{4}u(t - \pi/2)(1 + \cos 2t) + \dfrac{1}{2}u(t - \pi)\sin 2t + \dfrac{3}{2}u(t - 3\pi/2)\sin 2t$

21. $y = \cos t - \sin t$ **22.** $y = \dfrac{1}{4}(8e^{2t} - 12e^{-2t})$ **23.** $y = 5(e^{-2t} - e^{-t})$ **24.** $y = e^{-2t}(1 + 6t)$

25. $y = \dfrac{1}{4}e^{-t/2}(4 - 19t)$ **29.** $(-1)^k m\omega_1 Re^{-c\tau/2m}\delta(t - \tau)$ if $\omega_1\tau - \phi = (2k + 1)\pi/2$ $(k = \text{integer})$

30. (a) $y = \dfrac{e^{m+1} - 1}{2(e-1)}(e^{t-m} - e^{-t}),$ $m \leq t < m + 1$ $(m = 0, 1, \ldots)$

   (b) $y = (m + 1)\sin t,$ $2m\pi \leq t < 2(m + 1)\pi$ $(m = 0, 1, \ldots)$

   (c) $y = e^{2(t-m)}\dfrac{e^{2m+2} - 1}{e^2 - 1} - e^{(t-m)}\dfrac{e^{m+1} - 1}{e - 1},$ $m \leq t < m + 1$ $(m = 0, 1, \ldots)$

   (d) $y = \begin{cases} 0, & 2m\pi \leq t < (2m + 1)\pi, \\ -\sin t, & (2m + 1)\pi \leq t < (2m + 2)\pi, \end{cases}$ $(m = 0, 1, \ldots)$

Section 9.1 Answers   (page 445)

2. $y = 2x^2 - 3x^3 + \dfrac{1}{x}$ **3.** $y = 2e^x + 3e^{-x} - e^{2x} + e^{-3x}$ **4.** $y_i = \dfrac{(x - x_0)^{i-1}}{(i - 1)!},$ $1 \leq i \leq n$

5. (b) $y_1 = -\dfrac{1}{2}x^3 + x^2 + \dfrac{1}{2x},$ $y_2 = \dfrac{1}{3}x^2 - \dfrac{1}{3x},$ $y_3 = \dfrac{1}{4}x^3 - \dfrac{1}{3}x^2 + \dfrac{1}{12x}$ **(c)** $y = k_0 y_1 + k_1 y_2 + k_2 y_3$

7. $2e^{-x^2}$ **8.** $\sqrt{2}K\cos x$ **9. (a)** $W(x) = 2e^{3x}$ **(d)** $y = e^x(c_1 + c_2 x + c_3 x^2)$

10. (a) 2   **(b)** $-e^{3x}$ **(c)** 4   **(d)** $4/x^2$ **(e)** 1   **(f)** $2x$ **(g)** $2/x^2$ **(h)** $e^x(x^2 - 2x + 2)$ **(i)** $-240/x^5$ **(j)** $6e^{2x}(2x - 1)$
(k) $-128x$

24. (a) $y''' = 0$ **(b)** $xy''' - y'' - xy' + y = 0$ **(c)** $(2x - 3)y''' - 2y'' - (2x - 5)y' = 0$
(d) $(x^2 - 2x + 2)y''' - x^2y'' + 2xy' - 2y = 0$ **(e)** $x^3y''' + x^2y'' - 2xy' + 2y = 0$
(f) $(3x - 1)y''' - (12x - 1)y'' + 9(x + 1)y' - 9y = 0$ **(g)** $x^4y^{(4)} + 5x^3y''' - 3x^2y'' - 6xy' + 6y = 0$
(h) $x^4y^{(4)} + 3x^2y''' - x^2y'' + 2xy' - 2y = 0$ **(i)** $(2x - 1)y^{(4)} - 4xy''' + (5 - 2x)y'' + 4xy' - 4y = 0$
(j) $xy^{(4)} - y''' - 4xy'' + 4y' = 0$

Section 9.2 Answers   (page 456)

1. $y = e^x(c_1 + c_2x + c_3x^2)$ **2.** $y = c_1e^x + c_2e^{-x} + c_3\cos 3x + c_4\sin 3x.$ **3.** $y = c_1e^x + c_2\cos 4x + c_3\sin 4x$
4. $y = c_1e^x + c_2e^{-x} + c_3e^{-3x/2}$ **5.** $y = c_1e^{-x} + e^{-2x}(c_1\cos x + c_2\sin x)$ **6.** $y = c_1e^x + e^{x/2}(c_2 + c_3x)$
7. $y = e^{-x/3}(c_1 + c_2x + c_3x^2)$ **8.** $y = c_1 + c_2x + c_3\cos x + c_4\sin x$
9. $y = c_1e^{2x} + c_2e^{-2x} + c_3\cos 2x + c_4\sin 2x$ **10.** $y = (c_1 + c_2x)\cos\sqrt{6}x + (c_3 + c_4x)\sin\sqrt{6}x$
11. $y = e^{3x/2}(c_1 + c_2x) + e^{-3x/2}(c_3 + c_4x)$ **12.** $y = c_1e^{-x/2} + c_2e^{-x/3} + c_3\cos x + c_4\sin x$
13. $y = c_1e^x + c_2e^{-2x} + c_3e^{-x/2} + c_4e^{-3x/2}$ **14.** $y = e^x(c_1 + c_2x + c_3\cos x + c_4\sin x)$
15. $y = \cos 2x - 2\sin 2x + e^{2x}$ **16.** $y = 2e^x + 3e^{-x} - 5e^{-3x}$ **17.** $y = 2e^x + 3xe^x - 4e^{-x}$
18. $y = 2e^{-x}\cos x - 3e^{-x}\sin x + 4e^{2x}$ **19.** $y = \dfrac{9}{5}e^{-5x/3} + e^x(1 + 2x)$ **20.** $y = e^{2x}(1 - 3x + 2x^2)$
21. $y = e^{3x}(2 - x) + 4e^{-x/2}$ **22.** $y = e^{x/2}(1 - 2x) + 3e^{-x/2}$ **23.** $y = \cos 2x - 2\sin 2x + e^{2x}$
24. $y = -4e^x + e^{2x} - e^{4x} + 2e^{-x}$ **25.** $y = 2e^x - e^{-x}$ **26.** $y = e^{2x} + e^{-2x} + e^{-x}(3\cos x + \sin x)$
27. $y = 2e^{-x/2} + \cos 2x - \sin 2x$
28. (a) $\{e^x, xe^x, e^{2x}\}$; 1   **(b)** $\{\cos 2x, \sin 2x, e^{3x}\}$; 26   **(c)** $\{e^{-x}\cos x, e^{-x}\sin x, e^x\}$; 5   **(d)** $\{1, x, x^2, e^x\}$; 2
(e) $\{e^x, e^{-x}, \cos x, \sin x\}$; -8 **(f)** $\{\cos x, \sin x, e^x\cos x, e^x\sin x\}$; 5
29. $\{e^{-3x}\cos 2x, e^{-3x}\sin 2x, e^{2x}, xe^{2x}, 1, x, x^2\}$ **30.** $\{e^x, xe^x, e^{x/2}, xe^{x/2}, x^2e^{x/2}, \cos x, \sin x\}$
31. $\{\cos 3x, x\cos 3x, x^2\cos 3x, \sin 3x, x\sin 3x, x^2\sin 3x, 1, x\}$ **32.** $\{e^{2x}, xe^{2x}, x^2e^{2x}, e^{-x}, xe^{-x}, 1\}$
33. $\{\cos x, \sin x, \cos 3x, x\cos 3x, \sin 3x, x\sin 3x, e^{2x}\}$ **34.** $\{e^{2x}, xe^{2x}, e^{-2x}, xe^{-2x}, \cos 2x, x\cos 2x, \sin 2x, x\sin 2x\}$
35. $\{e^{-x/2}\cos 2x, xe^{-x/2}\cos 2x, x^2e^{-x/2}\cos 2x, e^{-x/2}\sin 2x, xe^{-x/2}\sin 2x, x^2e^{-x/2}\sin 2x\}$
36. $\{1, x, x^2, e^{2x}, xe^{2x}, \cos 2x, x\cos 2x, \sin 2x, x\sin 2x\}$
37. $\{\cos(x/2), x\cos(x/2), \sin(x/2), x\sin(x/2), \cos(2x/3), x\cos(2x/3),$
    $x^2\cos(2x/3), \sin(2x/3), x\sin(2x/3), x^2\sin(2x/3)\}$
38. $\{e^{-x}, e^{3x}, e^x\cos 2x, e^x\sin 2x\}$ **39. (b)** $e^{(a_1 + a_2 + \cdots + a_n)x}\displaystyle\prod_{1 \le i < j \le n}(a_j - a_i)$

43. (a) $\left\{e^x, e^{-x/2}\cos\left(\dfrac{\sqrt{3}}{2}x\right), e^{-x/2}\sin\left(\dfrac{\sqrt{3}}{2}x\right)\right\}$ **(b)** $\left\{e^{-x}, e^{x/2}\cos\left(\dfrac{\sqrt{3}}{2}x\right), e^{x/2}\sin\left(\dfrac{\sqrt{3}}{2}x\right)\right\}$
(c) $\{e^{2x}\cos 2x, e^{2x}\sin 2x, e^{-2x}\cos 2x, e^{-2x}\sin 2x\}$
(d) $\left\{e^x, e^{-x}, e^{x/2}\cos\left(\dfrac{\sqrt{3}}{2}x\right), e^{x/2}\sin\left(\dfrac{\sqrt{3}}{2}x\right), e^{-x/2}\cos\left(\dfrac{\sqrt{3}}{2}x\right), e^{-x/2}\sin\left(\dfrac{\sqrt{3}}{2}x\right)\right\}$
(e) $\{\cos 2x, \sin 2x, e^{-\sqrt{3}x}\cos x, e^{-\sqrt{3}x}\sin x, e^{\sqrt{3}x}\cos x, e^{\sqrt{3}x}\sin x\}$
(f) $\left\{1, e^{2x}, e^{3x/2}\cos\left(\dfrac{\sqrt{3}}{2}x\right), e^{3x/2}\sin\left(\dfrac{\sqrt{3}}{2}x\right), e^{x/2}\cos\left(\dfrac{\sqrt{3}}{2}x\right), e^{x/2}\sin\left(\dfrac{\sqrt{3}}{2}x\right)\right\}$
(g) $\left\{e^{-x}, e^{x/2}\cos\left(\dfrac{\sqrt{3}}{2}x\right), e^{x/2}\sin\left(\dfrac{\sqrt{3}}{2}x\right), e^{-x/2}\cos\left(\dfrac{\sqrt{3}}{2}x\right), e^{-x/2}\sin\left(\dfrac{\sqrt{3}}{2}x\right)\right\}$
45. $y = c_1x^{r_1} + c_2x^{r_2} + c_3x^{r_3}$ $(r_1, r_2, r_3 \text{ distinct})$; $y = c_1x^{r_1} + (c_2 + c_3\ln x)x^{r_2}$ $(r_1, r_2 \text{ distinct})$;
    $y = [c_1 + c_2\ln x + c_3(\ln x)^2]x^{r_1}$; $y = c_1x^{r_1} + x^\lambda[c_2\cos(\omega\ln x) + c_3\sin(\omega\ln x)]$.

Section 9.3 Answers   (page 465)

1. $y_p = e^{-x}(2 + x - x^2)$ **2.** $y_p = -\dfrac{e^{-3x}}{4}(3 - x + x^2)$ **3.** $y_p = e^x(1 + x - x^2)$ **4.** $y_p = e^{-2x}(1 - 5x + x^2)$

5. $y_p = -\dfrac{xe^x}{2}(1 - x + x^2 - x^3)$ **6.** $y_p = x^2e^x(1 + x)$ **7.** $y_p = \dfrac{xe^{-2x}}{2}(2 + x)$ **8.** $y_p = \dfrac{x^2e^x}{2}(2 + x)$

9. $y_p = \dfrac{x^2e^{2x}}{2}(1 + 2x)$ **10.** $y_p = x^2e^{3x}(2 + x - x^2)$ **11.** $y_p = x^2e^{4x}(2 + x)$ **12.** $y_p = \dfrac{x^3e^{x/2}}{48}(1 + x)$

13. $y_p = e^{-x}(1 - 2x + x^2)$ **14.** $y_p = e^{2x}(1 - x)$ **15.** $y_p = e^{-2x}(1 + x + x^2 - x^3)$ **16.** $y_p = \dfrac{e^x}{3}(1 - x)$

17. $y_p = e^x(1 + x)^2$ **18.** $y_p = xe^x(1 + x^3)$ **19.** $y_p = xe^x(2 + x)$ **20.** $y_p = \dfrac{xe^{2x}}{6}(1 - x^2)$

21. $y_p = 4xe^{-x/2}(1 + x)$ **22.** $y_p = \dfrac{xe^x}{6}(1 + x^2)$ **23.** $y_p = \dfrac{x^2e^{2x}}{6}(1 + x + x^2)$ **24.** $y_p = \dfrac{x^2e^{2x}}{6}(3 + x + x^2)$

25. $y_p = \dfrac{x^3e^x}{48}(2 + x)$ **26.** $y_p = \dfrac{x^3e^x}{6}(1 + x)$ **27.** $y_p = -\dfrac{x^3e^{-x}}{6}(1 - x + x^2)$ **28.** $y_p = \dfrac{x^3e^{2x}}{12}(2 + x - x^2)$

29. $y_p = e^{-x}[(1 + x)\cos x + (2 - x)\sin x]$ **30.** $y_p = e^{-x}[(1 - x)\cos 2x + (1 + x)\sin 2x]$

31. $y_p = e^{2x}[(1 + x - x^2)\cos x + (1 + 2x)\sin x]$ **32.** $y_p = \dfrac{e^x}{2}[(1 + x)\cos 2x + (1 - x + x^2)\sin 2x]$

33. $y_p = \dfrac{x}{13}(8\cos 2x + 14\sin 2x)$ **34.** $y_p = xe^x[(1 + x)\cos x + (3 + x)\sin x]$

35. $y_p = \dfrac{xe^{2x}}{2}[(3 - x)\cos 2x + \sin 2x]$ **36.** $y_p = -\dfrac{xe^{3x}}{12}(x\cos 3x + \sin 3x)$ **37.** $y_p = -\dfrac{e^x}{10}(\cos x + 7\sin x)$

38. $y_p = \dfrac{e^x}{12}(\cos 2x - \sin 2x)$ **39.** $y_p = xe^{2x}\cos 2x$ **40.** $y_p = -\dfrac{e^{-x}}{2}[(1 + x)\cos x + (2 - x)\sin x]$

41. $y_p = \dfrac{xe^{-x}}{10}(\cos x + 2\sin x)$ **42.** $y_p = \dfrac{xe^x}{40}(3\cos 2x - \sin 2x)$

43. $y_p = \dfrac{xe^{-2x}}{8}[(1 - x)\cos 3x + (1 + x)\sin 3x]$ **44.** $y_p = -\dfrac{xe^x}{4}(1 + x)\sin 2x$

45. $y_p = \dfrac{x^2e^{-x}}{4}(\cos x - 2\sin x)$ **46.** $y_p = -\dfrac{x^2e^{2x}}{32}(\cos 2x - \sin 2x)$ **47.** $y_p = \dfrac{x^2e^{2x}}{8}(1 + x)\sin x$

48. $y_p = 2x^2e^x + xe^{2x} - \cos x$ **49.** $y_p = e^{2x} + xe^x + 2x\cos x$ **50.** $y_p = 2x + x^2 + 2xe^x - 3xe^{-x} + 4e^{3x}$

51. $y_p = xe^x(\cos 2x - 2\sin 2x) + 2xe^{2x} + 1$ **52.** $y_p = 2x^3e^{-2x} - \cos 2x + \sin 2x$

53. $y_p = 2x^2(1 + x)e^{-x} + x\cos x - 2\sin x$ **54.** $y_p = 2xe^x + xe^{-x} + \cos x$ **55.** $y_p = \dfrac{xe^x}{6}(\cos x + \sin 2x)$

56. $y_p = \dfrac{x^2}{54}[(2 + 2x)e^x + 3e^{-2x}]$ **57.** $y_p = \dfrac{x}{8}\sinh x \sin x$ **58.** $y_p = x^3(1 + x)e^{-x} + xe^{-2x}$

59. $y_p = xe^x(2x^2 + \cos x + \sin x)$ **60.** $y = e^{2x}(1 + x) + c_1e^{-x} + e^x(c_2 + c_3x)$

61. $y = e^{3x}\left(1 - x - \dfrac{x^2}{2}\right) + c_1e^x + e^{-x}(c_2\cos x + c_3\sin x)$ **62.** $y = xe^{2x}(1 + x)^2 + c_1e^x + c_2e^{2x} + c_3e^{3x}$

63. $y = x^2e^{-x}(1 - x)^2 + c_1 + e^{-x}(c_2 + c_3x)$ **64.** $y = \dfrac{x^3e^x}{24}(4 + x) + e^x(c_1 + c_2x + c_3x^2)$

65. $y = \dfrac{x^2e^{-x}}{16}(1 + 2x - x^2) + e^x(c_1 + c_2x) + e^{-x}(c_3 + c_4x)$

66. $y = e^{-2x}\left[\left(1 + \dfrac{x}{2}\right)\cos x + \left(\dfrac{3}{2} - 2x\right)\sin x\right] + c_1e^x + c_2e^{-x} + c_3e^{-2x}$

67. $y = -xe^x \sin 2x + c_1 + c_2 e^x + e^x(c_3 \cos x + c_4 \sin x)$

68. $y = -\dfrac{x^2 e^x}{16}(1 + x) \cos 2x + e^x[(c_1 + c_2 x) \cos 2x + (c_3 + c_4 x) \sin 2x]$ **69.** $y = (x^2 + 2)e^x - e^{-2x} + e^{3x}$

70. $y = e^{-x}(1 + x + x^2) + (1 - x)e^x$ **71.** $y = \left(\dfrac{x^2}{12} + 16\right)xe^{-x/2} - e^x$

72. $y = (2 - x)(x^2 + 1)e^{-x} + \cos x - \sin x$ **73.** $y = (2 - x) \cos x - (1 - 7x) \sin x + e^{-2x}$

74. $y = 2 + e^x[(1 + x) \cos x - \sin x - 1]$

Section 9.4 Answers   (page 473)

1. $y_p = 2x^3$ **2.** $y_p = \dfrac{8}{105}x^{7/2}e^{-x^2}$ **3.** $y_p = x \ln|x|$ **4.** $y_p = -\dfrac{2(x^2 + 2)}{x}$ **5.** $y_p = -\dfrac{xe^{-3x}}{64}$

6. $y_p = -\dfrac{2x^2}{3}$ **7.** $y_p = -\dfrac{e^{-x}(x + 1)}{x}$ **8.** $y_p = 2x^2 \ln|x|$ **9.** $y_p = x^2 + 1$ **10.** $y_p = \dfrac{2x^2 + 6}{3}$

11. $y_p = \dfrac{x^2 \ln|x|}{3}$ **12.** $y_p = -x^2 - 2$ **13.** $y_p = \dfrac{1}{4}x^3 \ln|x| - \dfrac{25}{48}x^3$ **14.** $y_p = \dfrac{x^{5/2}}{4}$ **15.** $y_p = \dfrac{x(12 - x^2)}{6}$

16. $y_p = \dfrac{x^4 \ln|x|}{6}$ **17.** $y_p = \dfrac{x^3 e^x}{2}$ **18.** $y_p = x^2 \ln|x|$ **19.** $y_p = \dfrac{xe^x}{2}$ **20.** $y_p = \dfrac{3xe^x}{2}$ **21.** $y_p = -x^3$

22. $y = -x(\ln x)^2 + 3x + x^3 - 2x \ln x$ **23.** $y = \dfrac{x^3}{2}(\ln x)^2 + x^2 - x^3 + 2x^3 \ln x$

24. $y = -\dfrac{1}{2}(3x + 1)xe^x - 3e^x - e^{2x} + 4xe^{-x}$ **25.** $y = \dfrac{3}{2}x^4(\ln x)^2 + 3x - x^4 + 2x^4 \ln x$

26. $y = -\dfrac{x^4 + 12}{6} + 3x - x^2 + 2e^x$ **27.** $y = \left(\dfrac{x^2}{3} - \dfrac{x}{2}\right) \ln|x| + 4x - 2x^2$

28. $y = -\dfrac{xe^x(1 + 3x)}{2} + \dfrac{x + 1}{2} - \dfrac{e^x}{4} + \dfrac{e^{3x}}{2}$ **29.** $y = -8x + 2x^2 - 2x^3 + 2e^x - e^{-x}$ **30.** $y = 3x^2 \ln x - 7x^2$

31. $y = \dfrac{3(4x^2 + 9)}{2} + \dfrac{x}{2} - \dfrac{e^x}{2} + \dfrac{e^{-x}}{2} + \dfrac{e^{2x}}{4}$ **32.** $y_p = x \ln x + x - \sqrt{x} + \dfrac{1}{x} + \dfrac{1}{\sqrt{x}}$

33. $y_p = x^3 \ln|x| + x - 2x^3 + \dfrac{1}{x} - \dfrac{1}{x^2}$ **35.** $y_p = \displaystyle\int_{x_0}^x \dfrac{e^{(x-t)} - 3e^{-(x-t)} + 2e^{-2(x-t)}}{6} F(t)\, dt$

36. $y_p = \displaystyle\int_{x_0}^x \dfrac{(x - t)^2(2x + t)}{6xt^3} F(t)\, dt$ **37.** $y_p = \displaystyle\int_{x_0}^x \dfrac{xe^{(x-t)} - x^2 + x(t - 1)}{t^4} F(t)\, dt$

38. $y_p = \displaystyle\int_{x_0}^x \dfrac{x^2 - t(t - 2) - 2te^{(x-t)}}{2x(t - 1)^2} F(t)\, dt$ **39.** $y_p = \displaystyle\int_{x_0}^x \dfrac{e^{2(x-t)} - 2e^{(x-t)} + 2e^{-(x-t)} - e^{-2(x-t)}}{12} F(t)\, dt$

40. $y_p = \displaystyle\int_{x_0}^x \dfrac{(x - t)^3}{6x} F(t)\, dt$ **41.** $y_p = \displaystyle\int_{x_0}^x \dfrac{(x + t)(x - t)^3}{12x^2 t^3} F(t)\, dt$

42. $y_p = \displaystyle\int_{x_0}^x \dfrac{e^{2(x-t)}(1 + 2t) + e^{-2(x-t)}(1 - 2t) - 4x^2 + 4t^2 - 2}{32t^2} F(t)\, dt$

Section 10.1 Answers   (page 484)

1. $Q_1' = 2 - \dfrac{1}{10}Q_1 + \dfrac{1}{25}Q_2$ **2.** $Q_1' = 12 - \dfrac{5}{100 + 2t}Q_1 + \dfrac{1}{100 + 3t}Q_2$

$\quad Q_2' = 6 + \dfrac{3}{50}Q_1 - \dfrac{1}{20}Q_2$ $\quad Q_2' = 5 + \dfrac{1}{50 + t}Q_1 - \dfrac{4}{100 + 3t}Q_2$

3. $m_1 y_1'' = -(c_1 + c_2)y_1' + c_2 y_2' - (k_1 + k_2)y_1 + k_2 y_2 + F_1$

$m_2 y_2'' = (c_2 - c_3)y_1' - (c_2 + c_3)y_2' + c_3 y_3' + (k_2 - k_3)y_1$

$\qquad - (k_2 + k_3)y_2 + k_3 y_3 + F_2$

$m_3 y_3'' = c_3 y_1' + c_3 y_2' - c_3 y_3' + k_3 y_1 + k_3 y_2 - k_3 y_3 + F_3$

4. $x'' = -\dfrac{\alpha}{m}x' - \dfrac{gR^2 x}{(x^2 + y^2 + z^2)^{3/2}}$

$y'' = -\dfrac{\alpha}{m}y' - \dfrac{gR^2 y}{(x^2 + y^2 + z^2)^{3/2}}$

$z'' = -\dfrac{\alpha}{m}z' - \dfrac{gR^2 z}{(x^2 + y^2 + z^2)^{3/2}}.$

5. (a) $x_1' = x_2$

$\quad x_2' = x_3$

$\quad x_3' = f(t, x_1, y_1, y_2)$

$\quad y_1' = y_2$

$\quad y_2' = g(t, y_1, y_2)$

(b) $u_1' = f(t, u_1, v_1, v_2, w_2)$

$\quad v_1' = v_2$

$\quad v_2' = g(t, u_1, v_1, v_2, w_1)$

$\quad w_1' = w_2$

$\quad w_2' = h(t, u_1, v_1, v_2, w_1, w_2)$

(c) $y_1' = y_2$

$\quad y_2' = y_3$

$\quad y_3' = f(t, y_1, y_2, y_3)$

(d) $y_1' = y_2$

$\quad y_2' = y_3$

$\quad y_3' = y_4$

$\quad y_4' = f(t, y_1)$

(e) $x_1' = x_2$

$\quad x_2' = f(t, x_1, y_1)$

$\quad y_1' = y_2$

$\quad y_2' = g(t, x_1, y_1)$

6. $x' = x_1$

$\quad y' = y_1$

$\quad z' = z_1$

$x_1' = -\dfrac{gR^2 x}{(x^2 + y^2 + z^2)^{3/2}}$

$y_1' = -\dfrac{gR^2 y}{(x^2 + y^2 + z^2)^{3/2}}$

$z_1' = -\dfrac{gR^2 z}{(x^2 + y^2 + z^2)^{3/2}}$

Section 10.2 Answers   (page 487)

1. (a) $\mathbf{y}' = \begin{bmatrix} 2 & 4 \\ 4 & 2 \end{bmatrix}\mathbf{y}$ **(b)** $\mathbf{y}' = \begin{bmatrix} -2 & -2 \\ -5 & 1 \end{bmatrix}\mathbf{y}$ **(c)** $\mathbf{y}' = \begin{bmatrix} -4 & -10 \\ 3 & 7 \end{bmatrix}\mathbf{y}$ **(d)** $\mathbf{y}' = \begin{bmatrix} 2 & 1 \\ 1 & 2 \end{bmatrix}\mathbf{y}$

2. (a) $\mathbf{y}' = \begin{bmatrix} -1 & 2 & 3 \\ 0 & 1 & 6 \\ 0 & 0 & -2 \end{bmatrix}\mathbf{y}$ **(b)** $\mathbf{y}' = \begin{bmatrix} 0 & 2 & 2 \\ 2 & 0 & 2 \\ 2 & 2 & 0 \end{bmatrix}\mathbf{y}$ **(c)** $\mathbf{y}' = \begin{bmatrix} -1 & 2 & 2 \\ 2 & -1 & 2 \\ 2 & 2 & -1 \end{bmatrix}\mathbf{y}$ **(d)** $\mathbf{y}' = \begin{bmatrix} 3 & -1 & -1 \\ -2 & 3 & 2 \\ 4 & -1 & -2 \end{bmatrix}\mathbf{y}$

3. (a) $\mathbf{y}' = \begin{bmatrix} 1 & 1 \\ -2 & 4 \end{bmatrix}\mathbf{y}, \quad \mathbf{y}(0) = \begin{bmatrix} 1 \\ 0 \end{bmatrix}$ **(b)** $\mathbf{y}' = \begin{bmatrix} 5 & 3 \\ -1 & 1 \end{bmatrix}\mathbf{y}, \quad \mathbf{y}(0) = \begin{bmatrix} 9 \\ -5 \end{bmatrix}$

4. (a) $\mathbf{y}' = \begin{bmatrix} 6 & 4 & 4 \\ -7 & -2 & -1 \\ 7 & 4 & 3 \end{bmatrix}\mathbf{y}, \quad \mathbf{y}(0) = \begin{bmatrix} 3 \\ -6 \\ 4 \end{bmatrix}$ **(b)** $\mathbf{y}' = \begin{bmatrix} 8 & 7 & 7 \\ -5 & -6 & -9 \\ 5 & 7 & 10 \end{bmatrix}\mathbf{y}, \quad \mathbf{y}(0) = \begin{bmatrix} 2 \\ -4 \\ 3 \end{bmatrix}$

5. (a) $\mathbf{y}' = \begin{bmatrix} -3 & 2 \\ -5 & 3 \end{bmatrix} + \begin{bmatrix} 3 - 2t \\ 6 - 3t \end{bmatrix}$ **(b)** $\mathbf{y}' = \begin{bmatrix} 3 & 1 \\ -1 & 1 \end{bmatrix}\mathbf{y} + \begin{bmatrix} -5e^t \\ e^t \end{bmatrix}$

10. (a) $\dfrac{d}{dt}Y^2 = Y'Y + YY'$ **(b)** $\dfrac{d}{dt}Y^n = Y'Y^{n-1} + YY'Y^{n-2} + Y^2Y'Y^{n-3} + \cdots + Y^{n-1}Y'$

$$= \sum_{r=0}^{n-1} Y^r Y' Y^{n-r-1}$$

13. $B = (P' + PA)P^{-1}.$

Section 10.3 Answers   (page 495)

2. $\mathbf{y}' = \begin{bmatrix} 0 & 1 \\ -\dfrac{P_2(x)}{P_0(x)} & -\dfrac{P_1(x)}{P_0(x)} \end{bmatrix}\mathbf{y}$ **3.** $\mathbf{y}' = \begin{bmatrix} 0 & 1 & \cdots & 0 \\ \vdots & \vdots & \ddots & \vdots \\ 0 & 0 & \cdots & 1 \\ -\dfrac{P_n(x)}{P_0(x)} & -\dfrac{P_{n-1}(x)}{P_0(x)} & \cdots & -\dfrac{P_1(x)}{P_0(x)} \end{bmatrix}\mathbf{y}$

7. (b) $\mathbf{y} = \begin{bmatrix} 3e^{6t} - 6e^{-2t} \\ 3e^{6t} + 6e^{-2t} \end{bmatrix}$ **(c)** $\mathbf{y} = \dfrac{1}{2}\begin{bmatrix} e^{6t} + e^{-2t} & e^{6t} - e^{-2t} \\ e^{6t} - e^{-2t} & e^{6t} + e^{-2t} \end{bmatrix}\mathbf{k}$

8. (b) $\mathbf{y} = \begin{bmatrix} 6e^{-4t} + 4e^{3t} \\ 6e^{-4t} - 10e^{3t} \end{bmatrix}$ **(c)** $\mathbf{y} = \dfrac{1}{7}\begin{bmatrix} 5e^{-4t} + 2e^{3t} & 2e^{-4t} - 2e^{3t} \\ 5e^{-4t} - 5e^{3t} & 2e^{-4t} + 5e^{3t} \end{bmatrix}\mathbf{k}$

9. (b) $\mathbf{y} = \begin{bmatrix} -15e^{2t} - 4e^{t} \\ 9e^{2t} + 2e^{t} \end{bmatrix}$ **(c)** $\mathbf{y} = \begin{bmatrix} -5e^{2t} + 6e^{t} & -10e^{2t} + 10e^{t} \\ 3e^{2t} - 3e^{t} & 6e^{2t} - 5e^{t} \end{bmatrix}\mathbf{k}$

10. (b) $\mathbf{y} = \begin{bmatrix} 5e^{3t} - 3e^{t} \\ 5e^{3t} + 3e^{t} \end{bmatrix}$ **(c)** $\mathbf{y} = \dfrac{1}{2}\begin{bmatrix} e^{3t} + e^{t} & e^{3t} - e^{t} \\ e^{3t} - e^{t} & e^{3t} + e^{t} \end{bmatrix}\mathbf{k}$

11. (b) $\mathbf{y} = \begin{bmatrix} e^{2t} - 2e^{3t} + 3e^{-t} \\ 2e^{3t} - 9e^{-t} \\ e^{2t} - 2e^{3t} + 21e^{-t} \end{bmatrix}$ **(c)** $\mathbf{y} = \dfrac{1}{6}\begin{bmatrix} 4e^{2t} + 3e^{3t} - e^{-t} & 6e^{2t} - 6e^{3t} & 2e^{2t} - 3e^{3t} + e^{-t} \\ -3e^{3t} + 3e^{-t} & 6e^{3t} & 3e^{3t} - 3e^{-t} \\ 4e^{2t} + 3e^{3t} - 7e^{-t} & 6e^{2t} - 6e^{3t} & 2e^{2t} - 3e^{3t} + 7e^{-t} \end{bmatrix}\mathbf{k}$

12. (b) $\mathbf{y} = \dfrac{1}{3}\begin{bmatrix} -e^{-2t} + e^{4t} \\ -10e^{-2t} + e^{4t} \\ 11e^{-2t} + e^{4t} \end{bmatrix}$ **(c)** $\mathbf{y} = \dfrac{1}{3}\begin{bmatrix} 2e^{-2t} + e^{4t} & -e^{-2t} + e^{4t} & -e^{-2t} + e^{4t} \\ -e^{-2t} + e^{4t} & 2e^{-2t} + e^{4t} & -e^{-2t} + e^{4t} \\ -e^{-2t} + e^{4t} & -e^{-2t} + e^{4t} & 2e^{-2t} + e^{4t} \end{bmatrix}\mathbf{k}$

13. (b) $\mathbf{y} = \begin{bmatrix} 3e^{t} + 3e^{-t} - e^{-2t} \\ 3e^{t} + 2e^{-2t} \\ -e^{-2t} \end{bmatrix}$ **(c)** $\mathbf{y} = \begin{bmatrix} e^{-t} & e^{t} - e^{-t} & 2e^{t} - 3e^{-t} + e^{-2t} \\ 0 & e^{t} & 2e^{t} - 2e^{-2t} \\ 0 & 0 & e^{-2t} \end{bmatrix}\mathbf{k}$

14. YZ^{-1} and ZY^{-1}

Section 10.4 Answers   (page 509)

1. $\mathbf{y} = c_1\begin{bmatrix} 1 \\ 1 \end{bmatrix}e^{3t} + c_2\begin{bmatrix} 1 \\ -1 \end{bmatrix}e^{-t}$ **2.** $\mathbf{y} = c_1\begin{bmatrix} 1 \\ 1 \end{bmatrix}e^{-t/2} + c_2\begin{bmatrix} -1 \\ 1 \end{bmatrix}e^{-2t}$ **3.** $\mathbf{y} = c_1\begin{bmatrix} -3 \\ 1 \end{bmatrix}e^{-t} + c_2\begin{bmatrix} -1 \\ 2 \end{bmatrix}e^{-2t}$

4. $\mathbf{y} = c_1\begin{bmatrix} 2 \\ 1 \end{bmatrix}e^{-3t} + c_2\begin{bmatrix} -2 \\ 1 \end{bmatrix}e^{t}$ **5.** $\mathbf{y} = c_1\begin{bmatrix} 1 \\ 1 \end{bmatrix}e^{-2t} + c_1\begin{bmatrix} -4 \\ 1 \end{bmatrix}e^{3t}$ **6.** $\mathbf{y} = c_1\begin{bmatrix} 3 \\ 2 \end{bmatrix}e^{2t} + c_2\begin{bmatrix} 1 \\ 1 \end{bmatrix}e^{t}$

7. $\mathbf{y} = c_1\begin{bmatrix} -3 \\ 1 \end{bmatrix}e^{-5t} + c_2\begin{bmatrix} -1 \\ 1 \end{bmatrix}e^{-3t}$ **8.** $\mathbf{y} = c_1\begin{bmatrix} 1 \\ 2 \\ 1 \end{bmatrix}e^{-3t} + c_2\begin{bmatrix} -1 \\ -4 \\ 1 \end{bmatrix}e^{-t} + c_3\begin{bmatrix} -1 \\ -1 \\ 1 \end{bmatrix}e^{2t}$

9. $\mathbf{y} = c_1\begin{bmatrix} 2 \\ 1 \\ 2 \end{bmatrix}e^{-16t} + c_2\begin{bmatrix} -1 \\ 2 \\ 0 \end{bmatrix}e^{2t} + c_3\begin{bmatrix} -1 \\ 0 \\ 1 \end{bmatrix}e^{2t}$ **10.** $\mathbf{y} = c_1\begin{bmatrix} -2 \\ -4 \\ 3 \end{bmatrix}e^{t} + c_2\begin{bmatrix} -1 \\ 1 \\ 0 \end{bmatrix}e^{-2t} + c_3\begin{bmatrix} -7 \\ -5 \\ 4 \end{bmatrix}e^{2t}$

11. $\mathbf{y} = c_1\begin{bmatrix} -1 \\ -1 \\ 1 \end{bmatrix}e^{-2t} + c_2\begin{bmatrix} -1 \\ -2 \\ 1 \end{bmatrix}e^{-3t} + c_3\begin{bmatrix} -2 \\ -6 \\ 3 \end{bmatrix}e^{-5t}$ **12.** $\mathbf{y} = c_1\begin{bmatrix} 11 \\ 7 \\ 1 \end{bmatrix}e^{3t} + c_2\begin{bmatrix} 1 \\ 2 \\ 1 \end{bmatrix}e^{-2t} + c_3\begin{bmatrix} 1 \\ 1 \\ 1 \end{bmatrix}e^{-t}$

13. $\mathbf{y} = c_1\begin{bmatrix} 4 \\ -1 \\ 1 \end{bmatrix}e^{-4t} + c_2\begin{bmatrix} -1 \\ -1 \\ 1 \end{bmatrix}e^{6t} + c_3\begin{bmatrix} -1 \\ 0 \\ 1 \end{bmatrix}e^{4t}$ **14.** $\mathbf{y} = c_1\begin{bmatrix} 1 \\ 1 \\ 5 \end{bmatrix}e^{-5t} + c_2\begin{bmatrix} -1 \\ 0 \\ 1 \end{bmatrix}e^{5t} + c_3\begin{bmatrix} 1 \\ 1 \\ 0 \end{bmatrix}e^{5t}$

15. $\mathbf{y} = c_1\begin{bmatrix} 1 \\ -1 \\ 2 \end{bmatrix} + c_2\begin{bmatrix} -1 \\ 0 \\ 3 \end{bmatrix}e^{6t} + c_3\begin{bmatrix} 1 \\ 3 \\ 0 \end{bmatrix}e^{6t}$ **16.** $\mathbf{y} = -\begin{bmatrix} 2 \\ 6 \end{bmatrix}e^{5t} + \begin{bmatrix} 4 \\ 2 \end{bmatrix}e^{-5t}$ **17.** $\mathbf{y} = \begin{bmatrix} 2 \\ -4 \end{bmatrix}e^{t/2} + \begin{bmatrix} -2 \\ 1 \end{bmatrix}e^{t}$

18. $\mathbf{y} = \begin{bmatrix} 7 \\ 7 \end{bmatrix}e^{9t} - \begin{bmatrix} 2 \\ 4 \end{bmatrix}e^{-3t}$ **19.** $\mathbf{y} = \begin{bmatrix} 3 \\ 9 \end{bmatrix}e^{5t} - \begin{bmatrix} 4 \\ 2 \end{bmatrix}e^{-5t}$ **20.** $\mathbf{y} = \begin{bmatrix} 5 \\ 5 \\ 0 \end{bmatrix}e^{t/2} + \begin{bmatrix} 0 \\ 0 \\ 1 \end{bmatrix}e^{t/2} + \begin{bmatrix} -1 \\ 2 \\ 0 \end{bmatrix}e^{-t/2}$

21. $\mathbf{y} = \begin{bmatrix} 3 \\ 3 \\ 3 \end{bmatrix} e^t + \begin{bmatrix} -2 \\ -2 \\ 2 \end{bmatrix} e^{-t}$ **22.** $\mathbf{y} = \begin{bmatrix} 2 \\ -2 \\ 2 \end{bmatrix} e^t - \begin{bmatrix} 3 \\ 0 \\ 3 \end{bmatrix} e^{-2t} + \begin{bmatrix} 1 \\ 1 \\ 0 \end{bmatrix} e^{3t}$ **23.** $\mathbf{y} = -\begin{bmatrix} 1 \\ 2 \\ 1 \end{bmatrix} e^t + \begin{bmatrix} 4 \\ 2 \\ 4 \end{bmatrix} e^{-t} + \begin{bmatrix} 1 \\ 1 \\ 0 \end{bmatrix} e^{2t}$

24. $\mathbf{y} = \begin{bmatrix} -2 \\ -2 \\ 2 \end{bmatrix} e^{2t} - \begin{bmatrix} 0 \\ 3 \\ 0 \end{bmatrix} e^{-2t} + \begin{bmatrix} 4 \\ 12 \\ 4 \end{bmatrix} e^{4t}$ **25.** $\mathbf{y} = \begin{bmatrix} -1 \\ -1 \\ 1 \end{bmatrix} e^{-6t} + \begin{bmatrix} 2 \\ -2 \\ 2 \end{bmatrix} e^{2t} + \begin{bmatrix} 7 \\ -7 \\ -7 \end{bmatrix} e^{4t}$

26. $\mathbf{y} = \begin{bmatrix} 1 \\ 4 \\ 4 \end{bmatrix} e^{-t} + \begin{bmatrix} 6 \\ 6 \\ -2 \end{bmatrix} e^{2t}$ **27.** $\mathbf{y} = \begin{bmatrix} 4 \\ -2 \\ 2 \end{bmatrix} + \begin{bmatrix} 3 \\ -9 \\ 6 \end{bmatrix} e^{4t} + \begin{bmatrix} -1 \\ 1 \\ -1 \end{bmatrix} e^{2t}$

29. Half-lines of L_1: $y_2 = y_1$ and L_2: $y_2 = -y_1$ are trajectories; other trajectories are asymptotically tangent to L_1 as $t \to -\infty$ and asymptotically tangent to L_2 as $t \to \infty$.

30. Half-lines of L_1: $y_2 = -2y_1$ and L_2: $y_2 = -y_1/3$ are trajectories; other trajectories are asymptotically parallel to L_1 as $t \to -\infty$ and asymptotically tangent to L_2 as $t \to \infty$.

31. Half-lines of L_1: $y_2 = y_1/3$ and L_2: $y_2 = -y_1$ are trajectories; other trajectories are asymptotically tangent to L_1 as $t \to -\infty$ and asymptotically parallel to L_2 as $t \to \infty$.

32. Half-lines of L_1: $y_2 = y_1/2$ and L_2: $y_2 = -y_1$ are trajectories; other trajectories are asymptotically tangent to L_1 as $t \to -\infty$ and asymptotically tangent to L_2 as $t \to \infty$.

33. Half-lines of L_1: $y_2 = -y_1/4$ and L_2: $y_2 = -y_1$ are trajectories; other trajectories are asymptotically tangent to L_1 as $t \to -\infty$ and asymptotically parallel to L_2 as $t \to \infty$.

34. Half-lines of L_1: $y_2 = -y_1$ and L_2: $y_2 = 3y_1$ are trajectories; other trajectories are asymptotically parallel to L_1 as $t \to -\infty$ and asymptotically tangent to L_2 as $t \to \infty$.

36. Points on L_2: $y_2 = y_1$ are trajectories of constant solutions. The trajectories of nonconstant solutions are half-lines on either side of L_1, parallel to $\begin{bmatrix} 1 \\ -1 \end{bmatrix}$, traversed toward L_1.

37. Points on L_1: $y_2 = -y_1/3$ are trajectories of constant solutions. The trajectories of nonconstant solutions are half-lines on either side of L_1, parallel to $\begin{bmatrix} -1 \\ 2 \end{bmatrix}$, traversed away from L_1.

38. Points on L_1: $y_2 = y_1/3$ are trajectories of constant solutions. The trajectories of nonconstant solutions are half-lines on either side of L_1, parallel to $\begin{bmatrix} 1 \\ -1 \end{bmatrix}$, traversed away from L_1.

39. Points on L_1: $y_2 = y_1/2$ are trajectories of constant solutions. The trajectories of nonconstant solutions are half-lines on either side of L_1, parallel to $\begin{bmatrix} 1 \\ -1 \end{bmatrix}$, traversed away from L_1.

40. Points on L_2: $y_2 = -y_1$ are trajectories of constant solutions. The trajectories of nonconstant solutions are half-lines on either side of L_2, parallel to $\begin{bmatrix} -4 \\ 1 \end{bmatrix}$, traversed toward L_1.

41. Points on L_1: $y_2 = 3y_1$ are trajectories of constant solutions. The trajectories of nonconstant solutions are half-lines on either side of L_1, parallel to $\begin{bmatrix} 1 \\ -1 \end{bmatrix}$, traversed away from L_1.

Section 10.5 Answers   (page 523)

1. $\mathbf{y} = c_1 \begin{bmatrix} 2 \\ 1 \end{bmatrix} e^{5t} + c_2 \left(\begin{bmatrix} -1 \\ 0 \end{bmatrix} e^{5t} + \begin{bmatrix} 2 \\ 1 \end{bmatrix} te^{5t} \right)$ **2.** $\mathbf{y} = c_1 \begin{bmatrix} 1 \\ 1 \end{bmatrix} e^{-t} + c_2 \left(\begin{bmatrix} 1 \\ 0 \end{bmatrix} e^{-t} + \begin{bmatrix} 1 \\ 1 \end{bmatrix} te^{-t} \right)$

3. $\mathbf{y} = c_1 \begin{bmatrix} -2 \\ 1 \end{bmatrix} e^{-9t} + c_2 \left(\begin{bmatrix} -1 \\ 0 \end{bmatrix} e^{-9t} + \begin{bmatrix} -2 \\ 1 \end{bmatrix} te^{-9t} \right)$ **4.** $\mathbf{y} = c_1 \begin{bmatrix} -1 \\ 1 \end{bmatrix} e^{2t} + c_2 \left(\begin{bmatrix} -1 \\ 0 \end{bmatrix} e^{2t} + \begin{bmatrix} -1 \\ 1 \end{bmatrix} te^{2t} \right)$

5. $y = c_1 \begin{bmatrix} -2 \\ 1 \end{bmatrix} e^{-2t} + c_2 \left(\begin{bmatrix} -1 \\ 0 \end{bmatrix} \frac{e^{-2t}}{3} + \begin{bmatrix} -2 \\ 1 \end{bmatrix} te^{-2t} \right)$ **6.** $y = c_1 \begin{bmatrix} 3 \\ 2 \end{bmatrix} e^{-4t} + c_2 \left(\begin{bmatrix} -1 \\ 0 \end{bmatrix} \frac{e^{-4t}}{2} + \begin{bmatrix} 3 \\ 2 \end{bmatrix} te^{-4t} \right)$

7. $y = c_1 \begin{bmatrix} 4 \\ 3 \end{bmatrix} e^{-t} + c_2 \left(\begin{bmatrix} -1 \\ 0 \end{bmatrix} \frac{e^{-t}}{3} + \begin{bmatrix} 4 \\ 3 \end{bmatrix} te^{-t} \right)$ **8.** $y = c_1 \begin{bmatrix} -1 \\ -1 \\ 2 \end{bmatrix} + c_2 \begin{bmatrix} 1 \\ 1 \\ 2 \end{bmatrix} e^{4t} + c_3 \left(\begin{bmatrix} 0 \\ 1 \\ 0 \end{bmatrix} \frac{e^{4t}}{2} + \begin{bmatrix} 1 \\ 1 \\ 2 \end{bmatrix} te^{4t} \right)$

9. $y = c_1 \begin{bmatrix} -1 \\ 1 \\ 1 \end{bmatrix} e^t + c_2 \begin{bmatrix} 1 \\ -1 \\ 1 \end{bmatrix} e^{-t} + c_3 \left(\begin{bmatrix} 0 \\ 3 \\ 0 \end{bmatrix} e^{-t} + \begin{bmatrix} 1 \\ -1 \\ 1 \end{bmatrix} te^{-t} \right)$

10. $y = c_1 \begin{bmatrix} 0 \\ 1 \\ 1 \end{bmatrix} e^{2t} + c_2 \begin{bmatrix} 1 \\ 0 \\ 1 \end{bmatrix} e^{-2t} + c_3 \left(\begin{bmatrix} 1 \\ 1 \\ 0 \end{bmatrix} \frac{e^{-2t}}{2} + \begin{bmatrix} 1 \\ 0 \\ 1 \end{bmatrix} te^{-2t} \right)$

11. $y = c_1 \begin{bmatrix} -2 \\ -3 \\ 1 \end{bmatrix} e^{2t} + c_2 \begin{bmatrix} 0 \\ -1 \\ 1 \end{bmatrix} e^{4t} + c_3 \left(\begin{bmatrix} 1 \\ 0 \\ 0 \end{bmatrix} \frac{e^{4t}}{2} + \begin{bmatrix} 0 \\ -1 \\ 1 \end{bmatrix} te^{4t} \right)$

12. $y = c_1 \begin{bmatrix} -1 \\ -1 \\ 1 \end{bmatrix} e^{-2t} + c_2 \begin{bmatrix} 1 \\ 1 \\ 1 \end{bmatrix} e^{4t} + c_3 \left(\begin{bmatrix} 1 \\ 0 \\ 0 \end{bmatrix} \frac{e^{4t}}{2} + \begin{bmatrix} 1 \\ 1 \\ 1 \end{bmatrix} te^{4t} \right)$

13. $y = \begin{bmatrix} 6 \\ 2 \end{bmatrix} e^{-7t} - \begin{bmatrix} 8 \\ 4 \end{bmatrix} te^{-7t}$ **14.** $y = \begin{bmatrix} 5 \\ 8 \end{bmatrix} e^{3t} - \begin{bmatrix} 12 \\ 16 \end{bmatrix} te^{3t}$ **15.** $y = \begin{bmatrix} 2 \\ 3 \end{bmatrix} e^{-5t} - \begin{bmatrix} 8 \\ 4 \end{bmatrix} te^{-5t}$

16. $y = \begin{bmatrix} 3 \\ 1 \end{bmatrix} e^{5t} - \begin{bmatrix} 12 \\ 6 \end{bmatrix} te^{5t}$ **17.** $y = \begin{bmatrix} 0 \\ 2 \end{bmatrix} e^{-4t} + \begin{bmatrix} 6 \\ 6 \end{bmatrix} te^{-4t}$ **18.** $y = \begin{bmatrix} 4 \\ 8 \\ -6 \end{bmatrix} e^t + \begin{bmatrix} 2 \\ -3 \\ -1 \end{bmatrix} e^{-2t} + \begin{bmatrix} -1 \\ 1 \\ 0 \end{bmatrix} te^{-2t}$

19. $y = \begin{bmatrix} 3 \\ 3 \\ 6 \end{bmatrix} e^{2t} - \begin{bmatrix} 9 \\ 5 \\ 6 \end{bmatrix} + \begin{bmatrix} 2 \\ 2 \\ 0 \end{bmatrix} t$ **20.** $y = - \begin{bmatrix} 2 \\ 0 \\ 2 \end{bmatrix} e^{-3t} + \begin{bmatrix} -4 \\ 9 \\ 1 \end{bmatrix} e^t - \begin{bmatrix} 0 \\ 4 \\ 4 \end{bmatrix} te^t$

21. $y = \begin{bmatrix} -2 \\ 2 \\ 2 \end{bmatrix} e^{4t} + \begin{bmatrix} 0 \\ -1 \\ 1 \end{bmatrix} e^{2t} + \begin{bmatrix} 3 \\ -3 \\ 3 \end{bmatrix} te^{2t}$ **22.** $y = - \begin{bmatrix} 1 \\ 1 \\ 0 \end{bmatrix} e^{-4t} + \begin{bmatrix} -3 \\ 2 \\ -3 \end{bmatrix} e^{8t} + \begin{bmatrix} 8 \\ 0 \\ -8 \end{bmatrix} te^{8t}$

23. $y = \begin{bmatrix} 3 \\ 6 \\ 3 \end{bmatrix} e^{4t} - \begin{bmatrix} 3 \\ 4 \\ 1 \end{bmatrix} + \begin{bmatrix} 8 \\ 4 \\ 4 \end{bmatrix} t$

24. $y = c_1 \begin{bmatrix} 0 \\ 1 \\ 1 \end{bmatrix} e^{6t} + c_2 \left(\begin{bmatrix} -1 \\ 1 \\ 0 \end{bmatrix} \frac{e^{6t}}{4} + \begin{bmatrix} 0 \\ 1 \\ 1 \end{bmatrix} te^{6t} \right) + c_3 \left(\begin{bmatrix} 1 \\ 1 \\ 0 \end{bmatrix} \frac{e^{6t}}{8} + \begin{bmatrix} -1 \\ 1 \\ 0 \end{bmatrix} \frac{te^{6t}}{4} + \begin{bmatrix} 0 \\ 1 \\ 1 \end{bmatrix} \frac{t^2 e^{6t}}{2} \right)$

25. $y = c_1 \begin{bmatrix} -1 \\ 1 \\ 1 \end{bmatrix} e^{3t} + c_2 \left(\begin{bmatrix} 1 \\ 0 \\ 0 \end{bmatrix} \frac{e^{3t}}{2} + \begin{bmatrix} -1 \\ 1 \\ 1 \end{bmatrix} te^{3t} \right) + c_3 \left(\begin{bmatrix} 1 \\ 2 \\ 0 \end{bmatrix} \frac{e^{3t}}{36} + \begin{bmatrix} 1 \\ 0 \\ 0 \end{bmatrix} \frac{te^{3t}}{2} + \begin{bmatrix} -1 \\ 1 \\ 1 \end{bmatrix} \frac{t^2 e^{3t}}{2} \right)$

26. $y = c_1 \begin{bmatrix} 0 \\ -1 \\ 1 \end{bmatrix} e^{-2t} + c_2 \left(\begin{bmatrix} -1 \\ 1 \\ 0 \end{bmatrix} e^{-2t} + \begin{bmatrix} 0 \\ -1 \\ 1 \end{bmatrix} te^{-2t} \right) + c_3 \left(\begin{bmatrix} 3 \\ -2 \\ 0 \end{bmatrix} \frac{e^{-2t}}{4} + \begin{bmatrix} -1 \\ 1 \\ 0 \end{bmatrix} te^{-2t} + \begin{bmatrix} 0 \\ -1 \\ 1 \end{bmatrix} \frac{t^2 e^{-2t}}{2} \right)$

27. $y = c_1 \begin{bmatrix} 0 \\ 1 \\ 1 \end{bmatrix} e^{2t} + c_2 \left(\begin{bmatrix} 1 \\ 1 \\ 0 \end{bmatrix} \frac{e^{2t}}{2} + \begin{bmatrix} 0 \\ 1 \\ 1 \end{bmatrix} te^{2t} \right) + c_3 \left(\begin{bmatrix} -1 \\ 1 \\ 0 \end{bmatrix} \frac{e^{2t}}{8} + \begin{bmatrix} 1 \\ 1 \\ 0 \end{bmatrix} \frac{te^{2t}}{2} + \begin{bmatrix} 0 \\ 1 \\ 1 \end{bmatrix} \frac{t^2 e^{2t}}{2} \right)$

28. $\mathbf{y} = c_1 \begin{bmatrix} -2 \\ 1 \\ 2 \end{bmatrix} e^{-6t} + c_2 \left(-\begin{bmatrix} 6 \\ 1 \\ 0 \end{bmatrix} \dfrac{e^{-6t}}{6} + \begin{bmatrix} -2 \\ 1 \\ 2 \end{bmatrix} te^{-6t} \right) + c_3 \left(-\begin{bmatrix} 12 \\ 1 \\ 0 \end{bmatrix} \dfrac{e^{-6t}}{36} - \begin{bmatrix} 6 \\ 1 \\ 0 \end{bmatrix} \dfrac{te^{-6t}}{6} + \begin{bmatrix} -2 \\ 1 \\ 2 \end{bmatrix} \dfrac{t^2 e^{-6t}}{2} \right)$

29. $\mathbf{y} = c_1 \begin{bmatrix} -4 \\ 0 \\ 1 \end{bmatrix} e^{-3t} + c_2 \begin{bmatrix} 6 \\ 1 \\ 0 \end{bmatrix} e^{-3t} + c_3 \left(\begin{bmatrix} 1 \\ 0 \\ 0 \end{bmatrix} e^{-3t} + \begin{bmatrix} 2 \\ 1 \\ 1 \end{bmatrix} te^{-3t} \right)$

30. $\mathbf{y} = c_1 \begin{bmatrix} -1 \\ 0 \\ 1 \end{bmatrix} e^{-3t} + c_2 \begin{bmatrix} 0 \\ 1 \\ 0 \end{bmatrix} e^{-3t} + c_3 \left(\begin{bmatrix} 1 \\ 0 \\ 0 \end{bmatrix} e^{-3t} + \begin{bmatrix} -1 \\ -1 \\ 1 \end{bmatrix} te^{-3t} \right)$

31. $\mathbf{y} = c_1 \begin{bmatrix} 2 \\ 0 \\ 1 \end{bmatrix} e^{-t} + c_2 \begin{bmatrix} -3 \\ 2 \\ 0 \end{bmatrix} e^{-t} + c_3 \left(\begin{bmatrix} 1 \\ 0 \\ 0 \end{bmatrix} \dfrac{e^{-t}}{2} + \begin{bmatrix} -1 \\ 2 \\ 1 \end{bmatrix} te^{-t} \right)$

32. $\mathbf{y} = c_1 \begin{bmatrix} -1 \\ 1 \\ 0 \end{bmatrix} e^{-2t} + c_2 \begin{bmatrix} 0 \\ 0 \\ 1 \end{bmatrix} e^{-2t} + c_3 \left(\begin{bmatrix} -1 \\ 0 \\ 0 \end{bmatrix} e^{-2t} + \begin{bmatrix} 1 \\ -1 \\ 1 \end{bmatrix} te^{-2t} \right)$

Section 10.6 Answers   (page 535)

1. $\mathbf{y} = c_1 e^{2t} \begin{bmatrix} 3\cos t + \sin t \\ 5\cos t \end{bmatrix} + c_2 e^{2t} \begin{bmatrix} 3\sin t - \cos t \\ 5\sin t \end{bmatrix}$

2. $\mathbf{y} = c_1 e^{-t} \begin{bmatrix} 5\cos 2t + \sin 2t \\ 13\cos 2t \end{bmatrix} + c_2 e^{-t} \begin{bmatrix} 5\sin 2t - \cos 2t \\ 13\sin 2t \end{bmatrix}$

3. $\mathbf{y} = c_1 e^{3t} \begin{bmatrix} \cos 2t + \sin 2t \\ 2\cos 2t \end{bmatrix} + c_2 e^{3t} \begin{bmatrix} \sin 2t - \cos 2t \\ 2\sin 2t \end{bmatrix}$

4. $\mathbf{y} = c_1 e^{2t} \begin{bmatrix} \cos 3t - \sin 3t \\ \cos 3t \end{bmatrix} + c_2 e^{2t} \begin{bmatrix} \sin 3t + \cos 3t \\ \sin 3t \end{bmatrix}$

5. $\mathbf{y} = c_1 \begin{bmatrix} -1 \\ -1 \\ 2 \end{bmatrix} e^{-2t} + c_2 e^{4t} \begin{bmatrix} \cos 2t - \sin 2t \\ \cos 2t + \sin 2t \\ 2\cos 2t \end{bmatrix} + c_3 e^{4t} \begin{bmatrix} \sin 2t + \cos 2t \\ \sin 2t - \cos 2t \\ 2\sin 2t \end{bmatrix}$

6. $\mathbf{y} = c_1 \begin{bmatrix} -1 \\ -1 \\ 1 \end{bmatrix} e^{-t} + c_2 e^{-2t} \begin{bmatrix} \cos 2t - \sin 2t \\ -\cos 2t - \sin 2t \\ 2\cos 2t \end{bmatrix} + c_3 e^{-2t} \begin{bmatrix} \sin 2t + \cos 2t \\ -\sin 2t + \cos 2t \\ 2\sin 2t \end{bmatrix}$

7. $\mathbf{y} = c_1 \begin{bmatrix} 1 \\ 1 \\ 1 \end{bmatrix} e^{2t} + c_2 e^t \begin{bmatrix} -\sin t \\ \sin t \\ \cos t \end{bmatrix} + c_3 e^t \begin{bmatrix} \cos t \\ -\cos t \\ \sin t \end{bmatrix}$

8. $\mathbf{y} = c_1 \begin{bmatrix} -1 \\ 1 \\ 1 \end{bmatrix} e^t + c_2 e^{-t} \begin{bmatrix} -\sin 2t - \cos 2t \\ 2\cos 2t \\ 2\cos 2t \end{bmatrix} + c_3 e^{-t} \begin{bmatrix} \cos 2t - \sin 2t \\ 2\sin 2t \\ 2\sin 2t \end{bmatrix}$

9. $\mathbf{y} = c_1 e^{3t} \begin{bmatrix} \cos 6t - 3\sin 6t \\ 5\cos 6t \end{bmatrix} + c_2 e^{3t} \begin{bmatrix} \sin 6t + 3\cos 6t \\ 5\sin 6t \end{bmatrix}$

10. $\mathbf{y} = c_1 e^{2t} \begin{bmatrix} \cos t - 3\sin t \\ 2\cos t \end{bmatrix} + c_2 e^{2t} \begin{bmatrix} \sin t + 3\cos t \\ 2\sin t \end{bmatrix}$

11. $\mathbf{y} = c_1 e^{2t} \begin{bmatrix} 3\sin 3t - \cos 3t \\ 5\cos 3t \end{bmatrix} + c_2 e^{2t} \begin{bmatrix} -3\cos 3t - \sin 3t \\ 5\sin 3t \end{bmatrix}$

12. $\mathbf{y} = c_1 e^{2t} \begin{bmatrix} \sin 4t - 8\cos 4t \\ 5\cos 4t \end{bmatrix} + c_2 e^{2t} \begin{bmatrix} -\cos 4t - 8\sin 4t \\ 5\sin 4t \end{bmatrix}$

13. $\mathbf{y} = c_1 \begin{bmatrix} -1 \\ 1 \\ 1 \end{bmatrix} e^{-2t} + c_2 e^t \begin{bmatrix} \sin t \\ -\cos t \\ \cos t \end{bmatrix} + c_3 e^t \begin{bmatrix} -\cos t \\ -\sin t \\ \sin t \end{bmatrix}$

14. $\mathbf{y} = c_1 \begin{bmatrix} 2 \\ 2 \\ 1 \end{bmatrix} e^{-2t} + c_2 e^{2t} \begin{bmatrix} -\cos 3t - \sin 3t \\ -\sin 3t \\ \cos 3t \end{bmatrix} + c_3 e^{2t} \begin{bmatrix} -\sin 3t + \cos 3t \\ \cos 3t \\ \sin 3t \end{bmatrix}$

15. $\mathbf{y} = c_1 \begin{bmatrix} 1 \\ 2 \\ 1 \end{bmatrix} e^{3t} + c_2 e^{6t} \begin{bmatrix} -\sin 3t \\ \sin 3t \\ \cos 3t \end{bmatrix} + c_3 e^{6t} \begin{bmatrix} \cos 3t \\ -\cos 3t \\ \sin 3t \end{bmatrix}$

16. $\mathbf{y} = c_1 \begin{bmatrix} 1 \\ 1 \\ 1 \end{bmatrix} e^t + c_2 e^t \begin{bmatrix} 2\cos t - 2\sin t \\ \cos t - \sin t \\ 2\cos t \end{bmatrix} + c_3 e^t \begin{bmatrix} 2\sin t + 2\cos t \\ \cos t + \sin t \\ 2\sin t \end{bmatrix}$

17. $\mathbf{y} = e^t \begin{bmatrix} 5\cos 3t + \sin 3t \\ 2\cos 3t + 3\sin 3t \end{bmatrix}$ **18.** $\mathbf{y} = e^{4t} \begin{bmatrix} 5\cos 6t + 5\sin 6t \\ \cos 6t - 3\sin 6t \end{bmatrix}$

19. $\mathbf{y} = e^t \begin{bmatrix} 17\cos 3t - \sin 3t \\ 7\cos 3t + 3\sin 3t \end{bmatrix}$ **20.** $\mathbf{y} = e^{t/2} \begin{bmatrix} \cos(t/2) + \sin(t/2) \\ -\cos(t/2) + 2\sin(t/2) \end{bmatrix}$

21. $\mathbf{y} = \begin{bmatrix} 1 \\ -1 \\ 2 \end{bmatrix} e^t + e^{4t} \begin{bmatrix} 3\cos t + \sin t \\ \cos t - 3\sin t \\ 4\cos t - 2\sin t \end{bmatrix}$ **22.** $\mathbf{y} = \begin{bmatrix} 4 \\ 4 \\ 2 \end{bmatrix} e^{8t} + e^{2t} \begin{bmatrix} 4\cos 2t + 8\sin 2t \\ -6\sin 2t + 2\cos 2t \\ 3\cos 2t + \sin 2t \end{bmatrix}$

23. $\mathbf{y} = \begin{bmatrix} 0 \\ 3 \\ 3 \end{bmatrix} e^{-4t} + e^{4t} \begin{bmatrix} 15\cos 6t + 10\sin 6t \\ 14\cos 6t - 8\sin 6t \\ 7\cos 6t - 4\sin 6t \end{bmatrix}$ **24.** $\mathbf{y} = \begin{bmatrix} 6 \\ -3 \\ 3 \end{bmatrix} e^{8t} + \begin{bmatrix} 10\cos 4t - 4\sin 4t \\ 17\cos 4t - \sin 4t \\ 3\cos 4t - 7\sin 4t \end{bmatrix}$

29. $\mathbf{U} = \dfrac{1}{\sqrt{2}} \begin{bmatrix} -1 \\ 1 \end{bmatrix}, \quad \mathbf{V} = \dfrac{1}{\sqrt{2}} \begin{bmatrix} 1 \\ 1 \end{bmatrix}$ **30.** $\mathbf{U} \approx \begin{bmatrix} .5257 \\ .8507 \end{bmatrix}, \quad \mathbf{V} \approx \begin{bmatrix} -.8507 \\ .5257 \end{bmatrix}$ **31.** $\mathbf{U} \approx \begin{bmatrix} .8507 \\ .5257 \end{bmatrix}, \quad \mathbf{V} \approx \begin{bmatrix} -.5257 \\ .8507 \end{bmatrix}$

32. $\mathbf{U} \approx \begin{bmatrix} -.9732 \\ .2298 \end{bmatrix}, \quad \mathbf{V} \approx \begin{bmatrix} .2298 \\ .9732 \end{bmatrix}$ **33.** $\mathbf{U} \approx \begin{bmatrix} .5257 \\ .8507 \end{bmatrix}, \quad \mathbf{V} \approx \begin{bmatrix} -.8507 \\ .5257 \end{bmatrix}$ **34.** $\mathbf{U} \approx \begin{bmatrix} -.5257 \\ .8507 \end{bmatrix}, \quad \mathbf{V} \approx \begin{bmatrix} .8507 \\ .5257 \end{bmatrix}$

35. $\mathbf{U} \approx \begin{bmatrix} -.8817 \\ .4719 \end{bmatrix}, \quad \mathbf{V} \approx \begin{bmatrix} .4719 \\ .8817 \end{bmatrix}$ **36.** $\mathbf{U} \approx \begin{bmatrix} .8817 \\ .4719 \end{bmatrix}, \quad \mathbf{V} \approx \begin{bmatrix} -.4719 \\ .8817 \end{bmatrix}$ **37.** $\mathbf{U} = \begin{bmatrix} 0 \\ 1 \end{bmatrix}, \quad \mathbf{V} = \begin{bmatrix} -1 \\ 0 \end{bmatrix}$

38. $\mathbf{U} = \begin{bmatrix} 0 \\ 1 \end{bmatrix}, \quad \mathbf{V} = \begin{bmatrix} 1 \\ 0 \end{bmatrix}$ **39.** $\mathbf{U} = \dfrac{1}{\sqrt{2}} \begin{bmatrix} 1 \\ 1 \end{bmatrix}, \quad \mathbf{V} = \dfrac{1}{\sqrt{2}} \begin{bmatrix} -1 \\ 1 \end{bmatrix}$ **40.** $\mathbf{U} \approx \begin{bmatrix} .5257 \\ .8507 \end{bmatrix}, \quad \mathbf{V} \approx \begin{bmatrix} -.8507 \\ .5257 \end{bmatrix}$

Section 10.7 Answers (page 544)

1. $\mathbf{y}_p = \begin{bmatrix} 5e^{4t} + e^{-3t}(2 + 8t) \\ -e^{4t} - e^{-3t}(1 - 4t) \end{bmatrix}$ **2.** $\mathbf{y}_p = \begin{bmatrix} 13e^{3t} + 3e^{-3t} \\ -e^{3t} - 11e^{-3t} \end{bmatrix}$ **3.** $\mathbf{y}_p = \dfrac{1}{9} \begin{bmatrix} 7 - 6t \\ -11 + 3t \end{bmatrix}$ **4.** $\mathbf{y}_p = \begin{bmatrix} 5 - 3e^t \\ -6 + 5e^t \end{bmatrix}$

5. $\mathbf{y}_p = \begin{bmatrix} e^{-5t}(3 + 6t) + e^{-3t}(3 - 2t) \\ -e^{-5t}(3 + 2t) - e^{-3t}(1 - 2t) \end{bmatrix}$ **6.** $\mathbf{y}_p = \begin{bmatrix} t \\ 0 \end{bmatrix}$ **7.** $\mathbf{y}_p = -\dfrac{1}{6} \begin{bmatrix} 2 - 6t \\ 7 + 6t \\ 1 - 12t \end{bmatrix}$ **8.** $\mathbf{y}_p = -\dfrac{1}{6} \begin{bmatrix} 3e^t + 4 \\ 6e^t - 4 \\ 10 \end{bmatrix}$

9. $\mathbf{y}_p = \dfrac{1}{18} \begin{bmatrix} e^t(1 + 12t) - e^{-5t}(1 + 6t) \\ -2e^t(1 - 6t) - e^{-5t}(1 - 12t) \\ e^t(1 + 12t) - e^{-5t}(1 + 6t) \end{bmatrix}$ **10.** $\mathbf{y}_p = \dfrac{1}{3} \begin{bmatrix} 2e^t \\ e^t \\ 2e^t \end{bmatrix}$ **11.** $\mathbf{y}_p = \begin{bmatrix} t\sin t \\ 0 \end{bmatrix}$ **12.** $\mathbf{y}_p = -\begin{bmatrix} t^2 \\ 2t \end{bmatrix}$

13. $\mathbf{y}_p = (t - 1)(\ln|t - 1| + t) \begin{bmatrix} 1 \\ -1 \end{bmatrix}$ **14.** $\mathbf{y}_p = \dfrac{1}{9} \begin{bmatrix} 5e^{2t} - e^{-3t} \\ e^{3t} - 5e^{-2t} \end{bmatrix}$ **15.** $\mathbf{y}_p = \dfrac{1}{4t} \begin{bmatrix} 2t^3 \ln|t| + t^3(t + 2) \\ 2\ln|t| + 3t - 2 \end{bmatrix}$

16. $\mathbf{y}_p = \dfrac{1}{2} \begin{bmatrix} te^{-t}(t+2) + t^3 - 2 \\ te^t(t-2) + t^3 + 2 \end{bmatrix}$ **17.** $\mathbf{y}_p = -\begin{bmatrix} t \\ t \\ t \end{bmatrix}$ **18.** $\mathbf{y}_p = \dfrac{1}{4} \begin{bmatrix} -3e^t \\ 1 \\ e^{-t} \end{bmatrix}$ **19.** $\mathbf{y}_p = \begin{bmatrix} 2t^2 + t \\ t \\ -t \end{bmatrix}$

20. $\mathbf{y}_p = \dfrac{e^t}{4t} \begin{bmatrix} 2t + 1 \\ 2t - 1 \\ 2t + 1 \end{bmatrix}$

22. (a) $\mathbf{y}' = \begin{bmatrix} 0 & 1 & \cdots & 0 \\ 0 & 0 & \cdots & 0 \\ \vdots & \vdots & \ddots & \vdots \\ 0 & 0 & \cdots & 1 \\ -P_n(t)/P_0(t) & -P_{n-1}/P_0(t) & \cdots & -P_1(t)/P_0(t) \end{bmatrix} \mathbf{y} + \begin{bmatrix} 0 \\ 0 \\ \vdots \\ F(t)/P_0(t) \end{bmatrix}$

(b) $Y = \begin{bmatrix} y_1 & y_2 & \cdots & y_n \\ y_1' & y_2' & \cdots & y_n' \\ \vdots & \vdots & \ddots & \vdots \\ y_1^{(n-1)} & y_2^{(n-1)} & \cdots & y_n^{(n-1)} \end{bmatrix}$

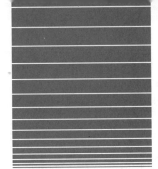

Index